Physiology in Health and Disease

Published on behalf of the American Physiological
Society by Springer

Physiology in Health and Disease

This book series is published on behalf of the American Physiological Society (APS) by Springer. Access to APS books published with Springer is free to APS members.

APS publishes three book series in partnership with Springer: *Physiology in Health and Disease* (formerly *Clinical Physiology*), *Methods in Physiology*, and *Perspectives in Physiology* (formerly *People and Ideas*), as well as general titles.

More information about this series at http://www.springer.com/series/11780

Kirk L. Hamilton • Daniel C. Devor

Editors

Basic Epithelial Ion Transport Principles and Function

Ion Channels and Transporters of Epithelia
in Health and Disease - Vol. 1

Second Edition

 Springer

american
physiological
society

Editors
Kirk L. Hamilton
Department of Physiology, School of
Biomedical Sciences
University of Otago
Dunedin, Otago, New Zealand

Daniel C. Devor
Department of Cell Biology
University of Pittsburgh
Pittsburgh, PA, USA

ISSN 2625-252X   ISSN 2625-2538   (electronic)
Physiology in Health and Disease
ISBN 978-3-030-52782-2   ISBN 978-3-030-52780-8   (eBook)
https://doi.org/10.1007/978-3-030-52780-8

This Springer imprint is published by the registered company Springer Nature Switzerland AG.
The registered company address is: Gewerbestrasse 11, 6330 Cham, Switzerland

We dedicate this second edition to our families ... Judy, Nathan, and Emma for KLH, and Cathy, Caitlin, Emily, and Daniel for DCD.

Preface to Second Edition—Volume 1

Our ultimate goal for the first edition of *Ion Channels and Transporters of Epithelia in Health and Disease* was to provide a comprehensive and authoritative volume that encapsulated the most recent research findings in basic molecular physiology of epithelial ion channels and transporters of molecular diseases from the laboratory bench top to the bedside. Additionally, we envisioned that the book would be very exciting and useful to a range of readers from undergraduate and postgraduate students, to postdoctoral fellows, to research and clinical scientists providing a wealth of up-to-date research information in the field of epithelial ion channels and transporters in health and disease. We firmly believe that the first edition fulfilled a niche that was crucially required. We have been informed that the first edition of the book has proven to be the best performing APS/Springer book based on downloaded chapters, to date. This is a direct testament to the world-class scientists and clinicians who contributed excellent chapters to that edition. Of course, there were many epithelial ion channels and transporters which were not included in the first edition, but certainly warranted inclusion.

With our second Edition, we have superseded our original expectations by increasing the number of chapters from 29 in the first edition to a 3-volume second edition including 54 chapters, resulting in 25 new chapters. All of the original chapters have been expanded. Again, we were very fortunate to recruit "key" outstanding scientists and clinicians who contributed excellent chapters, some of whom were unable to commit to the first edition. In the end, the second edition has a total of 128 authors from 13 countries across four continents and both hemispheres. We truly believe that this book series represents a worldwide collaboration of outstanding international scientists and clinicians.

Volume 1: Basic Epithelial Ion Transport Principles and Function

This is the first of three volumes highlighting the importance of epithelial ion channels and transporters in the basic physiology and pathophysiology of human diseases. This volume consists of 13 chapters (5 new chapters), including chapters focused on techniques used to study epithelial transport physiology, principles of epithelial transport function, the recent developments in the mathematical modeling of epithelia, the establishment and maintenance of epithelial polarity, protein sorting to specific membranes of epithelial cells, membrane protein folding, structure and endoplasmic reticulum-associated degradation, the fundamentals of transepithelial ion transport of chloride, sodium, and potassium, epithelial volume regulation, the fundamentals of bicarbonate secretion, and the role of non-coding RNA-dependent regulation of transport proteins. These chapters will set the "epithelial physiological" groundwork of the molecular participants, key concepts, and epithelial cell models that play critical roles in transepithelial ion transport function detailed throughout volumes 2 and 3 of this new edition.

It is our intent that the second edition continues to be the comprehensive and authoritative work that captures the recent research on the basic molecular physiology of epithelial ion channels and transporters of molecular diseases. We hope this new edition will be the "go-to" compendium that provides significant detailed research results about specific epithelial ion channels and transporters and how these proteins play roles in molecular disease in epithelial tissues.

As stated in the preface of the first edition, the massive undertaking of a book of this enormity would certainly be an "Everest" of work. We want to sincerely thank all of our authors, and their families, who have spared time from their very busy work and non-work schedules to provide exciting and dynamic chapters, which provide depth of knowledge, informative description, and coverage of the basic physiology and pathophysiology of the topic of their individual chapters.

We want to, again, thank Dr. Dee Silverthorn who planted the "initial seed" that developed into the first edition, which stemmed from a Featured Topic session entitled "Ion Channels in Health and Disease" held during the Experimental Biology meetings in Boston in April 2013 (chaired by KLH). Then, based on the performance of that edition, Dee "twisted" our arms, with love, to attempt a second edition in 2017. We, once again, want to extend our huge thanks, gratitude, and appreciation to the members of the American Physiology Society Book Committee for their continued faith in us to pursue such a monumental second edition.

As with the first edition, this 3-volume second edition would not have been possible without the excellent partnership between the American Physiological Society and Springer Nature and the publishing team in Heidelberg, Germany. Many thanks to Markus Spaeth, Associate Editor (Life Science and Books), and Dr. Andrea Schlitzberger, Project Coordinator (Book Production Germany and Asia), who guided us on our second book publication journey never dreaming that this edition would be a 3-volume book bonanza.

We extend special thanks to Anand Venkatachalam (Project Coordinator, Books, Chennai, India) at SPi Global who answered unending questions during the production process. We thank his production team who assisted us through the many stages of the publication of the second edition. We also thank Nancey Biswas (Project Management, SPi Content Solution, Puducherry, India), Nedounsejiane Narmadha (Production General, SPi Technologies, Puducherry, India), and Mahalakshmi Rajendran (Project Manager, SPi Technologies, Chennai, India) at Spi Global for their assistance for overseeing the production of the chapters during the final print and online file stages of the second edition.

We want to thank our mentors Douglas C. Eaton and the late Dale J. Benos for KLH; Michael E. Duffey and Raymond A. Frizzell for DCD; and our colleagues who guided us over the years to be able to undertake this book project.

Finally, and most importantly, we want to thank our families: Judy, Nathan, and Emma for KLH, and Cathy, Caitlin, Emily, and Daniel for DCD for all your love and support during this 8-year journey.

We dedicate this second edition to our families.

Dunedin, New Zealand   Kirk L. Hamilton
Pittsburgh, PA   Daniel C. Devor
July 2020

Preface

Ion channels and transporters play critical roles both in the homeostasis of normal function of the human body and during the disease process. Indeed, as of 2005, 16% of all Food and Drug Administration-approved drugs targeted ion channel and transporters, highlighting their importance in the disease process. Further, the Human Genome Project provided a wealth of genetic information that has since been utilized, and will again in the future, to describe the molecular pathophysiology of many human diseases. Over the recent years, our understanding of the pathophysiology of many diseases has been realized. The next great "step" is a combined scientific effort in basic, clinical, and pharmaceutical sciences to advance treatments of molecular diseases.

A number of unique ion channels and transporters are located within epithelial tissues of various organs including the kidney, intestine, pancreas, and respiratory tract, and all play crucial roles in various transport processes responsible for maintaining homeostasis. Ultimately, understanding the fundamentals of ion channels and transporters, in terms of function, modeling, regulation, molecular biology, trafficking, structure, and pharmacology, will shed light on the importance of ion channels and transporters in the basic physiology and pathophysiology of human diseases.

This book contains chapters written by notable world-leading scientists and clinicians in their respective research fields. The book consists of four sections. The first section of the book is entitled **Basic Epithelial Ion Transport Principles and Function** (Chapters 1–8) and spans the broad fundamentals of chloride, sodium, potassium, and bicarbonate transepithelial ion transport, the most recent developments in cell volume regulation, the mathematical modeling of these processes, the mechanisms by which these membrane proteins are correctly sorted to the apical and basolateral membranes, and protein folding of ion channels and transporters. The chapters in Section 1 provide the foundation of the molecular "participants" and epithelial cell models that play key roles in transepithelial ion transport function of epithelia detailed throughout the rest of this volume.

The second section is entitled **Epithelial Ion Channels and Transporters** and contains seventeen chapters (9–25) in which authors have concentrated their discussion on a particular ion channel or transporter ranging from chloride channels to the Na^+/K^+-ATPase, for example. Generally, the authors have initially provided a broad perspective of the physiology/biology of a particular ion channel or transporter in epithelial tissues, followed by a focused in-depth discussion of the latest physiology, cell biology, and molecular biology of the ion channel/transporter and then finish their discussion on aspects of pathophysiology and disease.

It will be appreciated following the discussion of the various ion channels and transporters that many of these transport proteins are potential pharmacological targets for possible treatment of disease. Therefore, the third section is entitled **Pharmacology of Potassium Channels** that consists of two chapters (26 and 27) that provide the latest developments on the pharmacology of calcium-activated potassium channels and small-molecule pharmacology of inward rectified potassium channels. It should be noted, however, that pharmacological information about various ion channels and transporters is also provided in some of the chapters found within Section 2 of this volume.

Finally, the last section in the book is entitled **Diseases in Epithelia** and consists of two chapters (28 and 29). These chapters are designed to bridge the basic cellular models and epithelial transport functions discussed throughout this volume with a compelling clinical perspective: from bench to bedside. In these chapters, Dr. Whitcomb discusses the role of ion channels and transporters in pancreatic disease, while Dr. Ameen and her colleagues similarly provide insights into the secretory diarrheas.

Our utmost goal, with this book, was to provide a comprehensive and authoritative volume that encapsulates the most recent research findings in the basic physiology of ion channels and transporters of molecular diseases from the laboratory bench top to the bedside. Additionally, we hope that the book will be very exciting and useful to a range of readers from students to research scientists providing a wealth of up-to-date research information in the field of epithelial ion channels and transporters in health and disease.

The undertaking of a book of this scale would always be a "mountain" of work. We want to give our heartfelt thanks to all of our authors who have taken time from their very busy work and non-work schedules to provide excellent chapters, which provided depth of knowledge, informative description, and coverage of the basic physiology and pathophysiology of the topic of their particular chapters.

We want to thank Dr. Dee Silverthorn who planted the "seed" that developed into this volume, which stemmed from a Featured Topic session entitled "Ion Channels in Health and Disease" held during the Experimental Biology meetings in Boston in April 2013 (chaired by KLH). We thank the members of the American Physiology Society (APS) Book Committee who had faith in us to pursue such an exciting book.

As with any book, this volume would not have been possible without the excellent partnership between the APS and Springer-Verlag and the publishing team at Heidelberg, Germany (Britta Mueller, Springer Editor, and Jutta Lindenborn, Project Coordinator). We wish to thank Portia Wong, our Developmental Editor at

Springer+Business Media (San Mateo, CA), and her team who assisted with the early stages of the publishing process that greatly added to this contribution. Finally, special thanks to Shanthi Ramamoorthy (Production Editor, Books) and Ramya Prakash (Project Manager) of Publishing—Springer, SPi Content Solutions—SPi Global and their production team who assisted us through the final stages of the publication of our book.

Finally, we want to thank our mentors Douglas C. Eaton and the late Dale J. Benos for KLH; Michael E. Duffey and Raymond A. Frizzell for DCD; and our colleagues who guided us over the years to be able to undertake this volume.

Dunedin, New Zealand   Kirk L. Hamilton
Pittsburgh, PA   Daniel C. Devor
June 2015

Contents

About the Editors

Kirk L. Hamilton was born in Baltimore, Maryland, in 1953. He gained his under-graduate (biology/chemistry) and M.Sc. (ecology) degrees from the University of Texas at Arlington. He obtained his Ph.D. at Utah State University under the tutelage of Dr. James A. Gessaman, where he studied incubation physiology of Barn owls. His first postdoctoral position was at the University of Texas Medical Branch in Galveston, Texas, under the mentorship of Dr. Douglas C. Eaton where he studied epithelial ion transport, specifically the epithelial sodium channel (ENaC). He then moved to the Department of Physiology at the University of Alabama, Birmingham, for additional postdoctoral training under the supervision of the late Dr. Dale J. Benos where he further studied ENaC, and non-specific cation channels. He took his first academic post in the Department of Biology at Xavier University of Louisiana in New Orleans (1990–1994). He then joined the Department of Physiology at the University of Otago in 1994, and he is currently an Associate Professor. He has focused his research on the molecular physiology and trafficking of potassium channels (specifically KCa3.1). He has published more than 60 papers and book chapters. His research work has been funded by the NIH, American Heart Association, Cystic Fibrosis Foundation, and Lottery Health Board New Zealand. Dr. Devor and he have been collaborators since 1999. When he not working, he enjoys playing guitar (blues and jazz) and volleyball. Kirk is married to Judith Rodda, a recent Ph.D. graduate in spatial ecology. They have two children, Nathan (b. 1995) and Emma (b. 1998).

Daniel C. Devor was born in Vandercook Lake, Michigan, in 1961. His education took him through the Southampton College of Long Island University, where he studied marine biology, before entering SUNY Buffalo for his Ph.D., under the guidance of Dr. Michael E. Duffey. During this time, he studied the role of basolateral potassium channels in regulating transepithelial ion transport. He subsequently did his postdoctoral work at the University of Alabama, Birmingham, under the mentorship of Dr. Raymond A. Frizzell, where he studied both apical CFTR and basolateral KCa3.1

in intestinal and airway epithelia. He joined the University of Pittsburgh faculty in 1995 where he is currently a Professor of Cell Biology. During this time, he has continued to study the regulation, gating, and trafficking of KCa3.1 as well as the related family member, KCa2.3, publishing more than 50 papers on these topics. These studies have been funded by the NIH, Cystic Fibrosis Foundation, American Heart Association, and pharmaceutical industry. When not in the lab, he enjoys photography and growing exotic plants. Dan is married to Catherine Seluga, an elementary school teacher. They have three children, Caitlin (b. 1990), Emily (b. 1993), and Daniel (b. 1997).

Chapter 1
Techniques of Epithelial Transport Physiology

Kirk L. Hamilton

Abstract Epithelial tissues play many roles in maintaining homeostasis of the human body. These tissues separate the body from the external environment (e.g., skin which protects the body), and of course, epithelial tissues separate body compartments, line the surfaces of organs, and line the inner surfaces of many hollow organs. Epithelial cells are polarized as there are specific transport proteins (ion channels and ion transporters) residing in the apical and basolateral membranes of the epithelial cells. Different epithelial cells perform specific functions in the regulation of absorption and secretion of ions, solutes, nutrients, and water. Understanding how these tissues (cells) function has been challenging and a number of techniques have been developed and/or adapted to study the functions of epithelial tissues and cells. Our ability to understand the physiology and the disease pathophysiology of epithelial tissues and cells is really reduced down to determining the fundamental characteristics and basic biology/physiology of the specific ion channels and ion transporters participating in overall epithelial transport physiology. This chapter provides a historical overview of various experimental techniques which have been instrumental and are still employed to discover intriguing aspects of epithelial ion transport physiology.

Keywords Epithelial transport · Ussing chamber · Radioisotopic studies · Short-circuit current · Micropuncture · Isolated perfused-tubule · Evert-sac preparation · Brush border vesicles · Site-directed metagenesis · PCR · Fluctuation analyses

K. L. Hamilton (✉)
Department of Physiology, School of Biomedical Sciences, University of Otago, Dunedin, New Zealand
e-mail: kirk.hamilton@otago.ac.nz

© The American Physiological Society 2020
K. L. Hamilton, D. C. Devor (eds.), *Basic Epithelial Ion Transport Principles and Function*, Physiology in Health and Disease,
https://doi.org/10.1007/978-3-030-52780-8_1

1.1 Introduction

Epithelial tissues are vitally important for proper homeostasis of the body. Our understanding of epithelial tissues can be pared down to the proper function of individual epithelial cells. These cells are polarized and maintain specific ion channels and transporters on the apical and basolateral membranes of the cell which are required for transepithelial ion, solutes, nutrients, and water transport (see Chaps. 2 and 3 of this volume for discussions about epithelial cell structure, functions, and the establishment of epithelial polarity). Currently, over 70% of the Federal Drug Administration-approved drugs are targeted to membrane proteins including ion channels, ion transporters, and receptors (Mushegian 2018). Therefore, determining the basic biology/physiology of epithelial ion channels and transporters is crucial in making the leap from basic epithelial function to understanding the pathophysiology of diseases of epithelial tissues.

As science progresses, scientists' ingenuity results in the development of experimental methods, techniques, and equipment that furthers scientific creativity. The great Danish physiologist, August Krogh, who received the Nobel Prize in Physiology or Medicine in 1920 for 'his discovery of the capillary motor regulating mechanism', stated that 'While it is undoubtedly true that the chief tool and weapon in research is thought and ideas and that a large amount of experimental work in biology is more or less wasted for lack of thought, it is less true that progress depends to a very large extent upon methods and that new methods may open up new and fruitful fields' (Krogh 1937a). He wrote those words in the context of the use of isotopes as indicators in biological research. Still, those words are appropriate even now when discussing the various experimental techniques that many scientists use today in their research. Of course, there are many techniques that are the backbone of molecular benchtop science. However, when envisioning studying biology/physiology at the animal, organ, tissue or cell level, additional technical approaches are required. With respect to epithelial tissues, one conducts experiments at the tissue and/or cell level to describe and confirm physiological phenomena. This chapter describes the historical concepts and technical aspects of commonly used methods for studying transport physiology of epithelial tissues and cells.

1.2 The Road to Epithelial Transport: How Did Epithelial Ion Transport Begin?

1.2.1 Early Research in Epithelial Transport

The age of epithelial transport physiology has had a long journey that began in the early nineteenth century. As early as 1826, Dutrochet (as cited in Reid 1890) reported that the cecum (containing milk) of the fowl, when placed in water, gained weight over a 36-h period suggesting the movement of water by osmosis.

Subsequently, Matteucci and Lima (1845), using frog skin, determined that the speed of 'osmotic transference of fluid' varied depending upon the orientation of the skin with respect to the bathing solutions, and that this difference was observed while the skin was freshly removed from the animal (Reid 1890). In their paper, Matteucci and Lima confirmed those same findings in the gastric mucous membrane of the lamb, cat, and dog. Therefore, even in the first quarter of the nineteenth century, scientists were intrigued with understanding the transport of water across epithelial tissues.

Stanley Schultz (1989) in his poignant review reminded the epithelial transport community of the work by E. Waymouth Reid. Schultz stressed that the contributions of Reid had not been fully appreciated during most of the twentieth century, although Reid actively investigated the field of epithelial transport physiology in the 1890s and early 1900s. His studies of frog skin osmosis (1890) and absorption without osmosis in the rabbit intestine (1892b) and the frog skin (1892a) confirmed the results of Matteucci and Lima (1845). From his own experimental results, Reid described four notable observations that have withstood the test of time for nearly 130 years. These observations are:

1. '…The normal direction of easier osmotic transference of fluid through the living skin of the frog is in the direction of outer to inner.'
2. '…The transference of fluid through the skin in the above direction is intimately associated with the physiological conditions of its tissues; conditions or agents tending to depress vitality diminish the transfer in the normal direction, while stimulants give rise to augmentation.'
3. '…The cause of the easier transference of fluid from the outer towards the inner surface, is probably to be found in the existence of an absorptive force dependent on protoplasmic activity, and comparable to the secretive force of the gland cell.'
4. Finally, '…In consequences of the absorptive force acting from without inwards, an alteration of the relations of the surfaces of the skin to the two fluids used in an osmosis experiments modifies the rapidity of the transfer of fluid from one to the other side of the membrane, according as the force exerted by the living tissues is with or against the osmotic stream.' (Quoted from Reid 1890.)

Additionally, Reid followed his earlier work with studies of fluid transport in various epithelia including the cat and rabbit ileum (inverted gut prep), gastric mucosa of the toad, and frog skin (Reid 1901). Indeed, Schultz (1989) revered the work of Reid by stating that '…Reid had…for the first time unambiguously demonstrated and recognized "active transport"'. As with the famous quote by Sir Isaac Newton '…If I had seen further than others, it is by standing on the shoulders of giants'; this is true for scientists today whom have learned and built upon the tremendous work of Dutrochet, Matteucci, Lima, and Reid.

It was, however, Krogh and de Veresy who were early innovators in the era of membrane physiology because of their radioactive isotopic studies; followed shortly by Hans Ussing and his famous 'chamber' and the short-circuit current technique (Schultz 1989).

1.3   Radioisotopes and Radioisotopic Tracers Studies

1.3.1   Early Pioneer Researchers in Radioactivity

When one ponders about radiation or radioactivity several names quickly spring to mind including Wilhem Röntgen, Henri Becquerel, and Marie and Pierre Curie. To begin a discussion of radioisotopes, we must go to the beginnings of the work by Röntgen who is credited with the discovery of 'X-rays' in 1895 (Röntgen 1895; Posner 1970). Röntgen was experimenting with different types of vacuum tubes including cathode-ray tubes when he discovered a glow on a black cardboard screen coated with fluorescent material in the distance away from the cathode ray (Posner 1970). Röntgen surmised that the resulting fluorescence could only have originated from one end of the cathode ray tube (Lentle and Aldrich 1997). The contribution of Röntgen's 'invisible ray' to the medical field has been incredible. Röntgen was awarded the 1st Nobel Prize in Physics in 1901 'in recognition of the extraordinary services he has rendered by the discovery of the remarkable rays subsequently named after him'.

In 1896, Becquerel had been wondering if phosphorescent material could emit similar 'invisible rays' (Becquerel 1903). He believed that uranium salts might be the suitable 'matter' for his investigations. Within a short period, after some key experiments, Becquerel soon realized that uranium salts were 'the source' of the emitted rays causing this phenomenon; essentially identifying 'radioactivity'. Becquerel published six scientific papers about his research on radioactivity in 1896 which are described by Myers (1976).

Marie Curie was intrigued with Becquerel's work on uranium and believed that other elements must emit these 'new Becquerel rays' (Curie 1905; Hevesy 1961). She began her quest for other elements exhibiting 'rays', and in 1898, she identified that thorium emitted rays. Then, she focused on minerals containing uranium including pitchblende (now known as uraninite) and torbernite (also known as chalcolite), and discovered that the intensity of radiation emitted by these minerals was much greater than uranium (Hevesy 1961). Also in 1898, Marie and Pierre Curie reported the discovery of the radioactive element polonium. Later that same year, the Curies reported the discovery of radium (Curie 1905; Hevesy 1961). Marie Curie continued her work on polonium and radium and was determined to isolate these elements. By 1910, she successfully isolated radium metal from pitchblende, but was never able to isolate polonium.

For their collective work on radioactivity, Becquerel shared the Nobel Prize in Physics in 1903 with Marie and Pierre Curie. Becquerel was awarded the Prize 'in recognition of the extraordinary services he has rendered by his discovery of spontaneous radioactivity'; while the Curies' received the Prize 'in recognition of the extraordinary services they have rendered by their joint researches on the radiation phenomena discovered by Professor Henri Becquerel'. Based on her work and the isolation of radium and attempts to isolate polonium, Marie was awarded the Nobel Prize in Chemistry in 1911 'in recognition of her services to the advancement of chemistry by the discovery of the elements radium and polonium, by the isolation of radium and the study of the nature and compounds of this remarkable element'. Interested readers are referred to Curie (1911) for her Nobel lecture.

How did radioactive elements aid our understanding of epithelial ion transport function?

1.3.2   Taking Radioisotopes Tracers into Chemistry and Biology: George de Hevesy

The application of radiotracers to physical chemistry and biology started shortly after the early pioneering work of the scientists mentioned in the previous section. George de Hevesy (referenced Hevesy in scientific publications) is credited as the first scientist to use isotopes as a tracer in biological studies in 1923 (Hevesy 1923; Levi 1985). However, earlier work commenced when Hevesy visited with Ernest Rutherford (Nobel Prize in Chemistry in 1908 'for his investigation into the disintegration of the elements, and the chemistry of radioactive substances') in 1911.

During 1912–1915, Hevesy began experimenting with mixtures of lead in the form of nitrate in water and adding a negligible amount of radium D and determined the radioactivity with an electroscope. An electroscope is an instrument which detects electrical charges (electrical potential) of an object, especially as an indication of the ionization of air by radioactivity. Hevesy noted that radioactivity was detected in the origin fraction of the lead nitrate. Later, Hevesy stated that that initial finding was the first hint of the potential power of what would become radio-labelled tracer studies (de Hevesy 1944). The next step was to explore this methodology in a biological application context.

Hevesy did not settle for working only in physical chemistry and, therefore, branched out into biological systems. In 1923, Hevesy published the first paper using a radioactive indicator in biological research (Hevesy 1923). In that paper, he reported the absorption and translocation of lead in horse-bean (*Vicia fiba*) plants. He used the element thorium B as the radioactive indicator of lead, and clearly demonstrated that lead was taken up by the roots and distributed throughout the plant. Hevesy also determined that the ordinary lead could displace the thorium B suggesting that the 'radioactive lead' was not incorporated within the bean plant. Krogh's (1937a) thoughts about results of de Hevesy's were that '. . .lead atoms are in reality never fixed anywhere, but are always on the move up and down the plant to and from the single cells, to and from the organic lead compounds which are continually formed and reformed.'

Hevesy continued his research on isotope tracer studies with deuterium in humans (Hevesy and Hofer 1934) and in frogs (Hevesy et al. 1935), and with phosphorus in rabbits (Hahn et al. 1937). Indeed, Krogh (1937a) applauded Hevesy and wrote that Hevesy '. . .was the first to see the possibilities offered in biology by recognizable isotopes and made the classical and fundamental experiments with radioactive lead'. Additional information and the use of radioactive tracers in experimental studies investigating ion transport in epithelia is found in Sect. 1.4.1.

Hevesy was the recipient the 1943 Nobel Prize in Chemistry 'for his work on the use of isotopes in the study of chemical processes'. For further information about Hevesy's scientific contributions, readers are directed to de Hevesy (1944), Cockcroft (1967), Levi (1985), and Niese (2006).

For those interested in learning more about early days of radioisotopes and radiotracers, and the use of these molecules in application to biological studies, readers are directed to the following excellent reviews by Krogh (1937a, b), Hevesy (1940, 1962), de Hevesy (1944), Popják (1948), Lyon (1949), Burris (1950), Cockcroft (1967), Posner (1970), Nitske (1971), Ussing (1980), Lentle and Aldrich (1997), Niese (2006), and Creager (2013).

1.4   The Epithelial Cell Begins to Open Up: Hans H. Ussing, the 'Black Box', the Ussing Chamber, and I_{sc}

It is interesting to observe how the scientific endeavors of one individual can affect/ change the course of an entire research field. Once such person was Hans Henriksen Ussing (1911–2000) whose pioneering work in ion transport physiology, and his development of innovative research tools and methods paved the way for epithelial transport physiologists/scientists to enter the physiological 'Black Box' of the epithelial cell (Larsen 2002, 2009). The monumental work of Ussing, beginning in the mid-1930s, which includes the Ussing chamber and the short-circuit current (I_{sc}) technique (discussed below), ion flux experiments, and ion exchange experiments, that undoubtedly, ushered in the 'age of epithelial ion transport physiology' as we know it (Ussing 1949a; Ussing and Zerahn 1951). Schultz (1989) claimed that it was Ussing, his work, and his conceptual cell model had set '. . .the birth of the modern era of epithelial transport physiology with the introduction of the Koefoed-Johnsen-Ussing (KJU) double membrane model' (Koefoed-Johnsen and Ussing 1958). The use of the Ussing chamber technique is still thriving today in countless numbers of epithelial physiology laboratories in both the academic and pharmaceutical environments throughout the world (Lenneräs 2007).

Others have stated the 'impact' of Ussing's work on science, and, in particular, epithelial ion transport physiology. Schultz (1998) wrote that Ussing's work of 1951 was '. . .his crowning triumph with Zerahn. . .[Ussing and Zerahn 1951]. . .the demonstration of the equivalence between the short-circuit current and active Na^+ transport across isolated frog skin. Na^+ could be transported across viable frog skin from the outer bathing solution to an identical inner bathing solution in the absence of external driving forces.' Schultz (1989) further declared that 'Ussing's demonstration of active transport was immediately embraced, and a new paradigm was born.' Lindemann (2001) exemplified Ussing's contribution to science simply with four words '...founder of epithelial transport...' Similarly, Palmer and Andersen (2008) paid homage to Ussing by writing 'In a sense the field of epithelial polarity began in 1958 with the Koefoed-Johnsen and Ussing paper.' (Koefoed-Johnsen and Ussing 1958). Clarke (2009), while describing the application of the Ussing chamber technique in intestinal tissues, acknowledged that 'As our understanding of the molecular interactions of transporters is refined, the methodology of the Ussing chamber will continue to provide a "gold standard" in the application of this knowledge to the physiological complexities of healthy and diseased intestinal mucosa'.

Finally, Jerrold Turner, who was the recipient of the 2015 'Hans H. Ussing Distinguished Lectureship' of the Epithelial Transport Group of the American Physiological Society, summed up his thoughts about Ussing's contribution to the field of epithelial transport physiology as follows, 'The great strides forward in our understanding of epithelial transport in the relatively short time since Ussing created his chamber are, nevertheless, remarkable. And, as is often the case, this knowledge has led to even more exciting frontiers for exploration.' (Hermann and Turner 2016). One could list more quotes from other eminent scientists who have highlighted Ussing's life and work with equally superlative remarks. How did Ussing change the epithelial transport world?

1.4.1 Ussing's Early Years in Preparation for His Chamber

Ussing's journey to epithelial transport physiology and the Ussing chamber was not a straight-forward pathway. He studied biology and geography at the University of Copenhagen (UCph) and graduated with a Master's in 1934. He also studied zoology, biochemistry, physiology, and physical chemistry (Larsen 2009). All these subjects would serve Ussing well over his 65+ year research career. During the summer of 1933, Ussing worked as a marine biologist and a hydrographer with the *Lauge Koch's* 3-year expedition to East Greenland, and he collected and analyzed zooplankton samples (Ussing 1980). He received additional samples for the remainder of the expedition which provided him with his research material culminating with his D. Phil. in 1938 from the UCph (Ussing 1980; Larsen 2002, 2009). One of the examiners of his doctorate thesis was August Krogh who thought very highly of Ussing (Larsen 2002). Krogh would play a major role and influence in Ussing's research future.

In 1935, Ussing joined Krogh at the Zoophysiological Laboratory at UCph at an auspicious time when Krogh had begun working with Niels Bohr (awarded the Nobel Prize in Physics in 1922 'for his services in the investigation of the structure of atoms and of the radiation emanating from them'), and Hevesy discussing radioisotopic studies. Earlier, Krogh had visited Harold C. Urey who had recently discovered 'heavy hydrogen' and followed that with heavy water, deuterium, D_2O (Urey 1925, 1933; Washburn and Urey 1932; Larsen 2009). Urey received the Nobel Prize for Chemistry in 1934 'for the discovery of heavy hydrogen' (Urey 1935). Krogh had asked Urey for some heavy water for proposed experiments to determine the permeability of living membranes using D_2O as a tracer for normal water (Ussing 1980; Larsen 2002, 2009). It was Hevesy who suggested the idea of using radioisotopes as tracers to examine biological processes (Hevesy 1923; Ussing 1980; Larsen 2002). The team was ready! Krogh, along with Hevesy and Hofer used D_2O to examine the water permeability of the frog skin and demonstrated that water movement was passive, although there was an inconsistency that normal water had a higher permeability than D_2O (Hevesy et al. 1935; Ussing 1980; Larsen 2002). For further details regarding the early historical aspects of radio-isotopic studies by the Copenhagen group, the reader is directed to Larsen (2002) and (2009).

During that period, Ussing noted that D_2O disappeared from the water and was incorporated into tissues, possibly into proteins (Ussing 1980). At that time, Ussing was a young member of the laboratory and was quite interested in biochemistry, therefore, he decided to pursue the use of D_2O in examining incorporation into amino acids and proteins (Ussing 1980). He quickly succeeded in developing new methods and protocols for measuring the incorporation of D_2O into amino acids and reported that D_2O-labelled amino acids were introduced into certain body proteins very quickly, both in mice and rats (Ussing 1938a; Larsen 2009). Larsen (2002) claimed that Ussing had '...provided the first evidence that body proteins are constantly synthesized and degraded in such a way that amino acids taken up via food are incorporated into new protein molecules, while at the same time others are catabolized.' Further, Larsen continued by stating that Ussing '...demonstrated how tracer technology provides fundamentally new opportunities for exploring the dynamic state of living cells.' Ussing continued his pursuit of protein biochemistry as evidenced by his series of seminal publications during the late 1930s and the 1940s (Ussing 1938a, b, 1941, 1943a, 1945a, b, 1946).

In the mid-1940s, Krogh, as a prominent Danish citizen, was advised to flee to Sweden due to the German occupation of Denmark, so he asked Ussing to oversee the radiotracer program, which Ussing accepted, although reluctantly (Larsen 2009; Larsen, Pers. Comm.). Krogh suggested that Ussing should focus on the transport of K^+ in the frog muscle. However, Ussing chose to examine the Na^+ transport instead (Ussing 1980; Larsen 2002). Ussing quickly published studies on the exchange of radioisotopic tracers in frog muscle tissue (Ussing 1947; Levi and Ussing 1948).

However, what about Ussing and epithelial transport? Throughout Ussing's time with Krogh, there had been work performed with frogs by the group (Hevesy et al. 1935; Krogh 1937b), and even Ussing had conducted experiments with amphibians (axolotl, *Ambystoma mexicanum*; Barker Jørgensen et al. 1946) and red blood cells (Ussing 1943b). Ussing realized that with the success of axolotls in which he demonstrated uptake of radioactive Na^+ (that is, $^{24}Na^+$) through the skin of animals in salt balance (Barker Jørgensen et al. 1946; Larsen 2002), the skin of amphibians would be the preparation to continue his studies. Ussing (1980) stated that 'I decided to begin with the isolated frog skin, mostly because I found it easier to skin a frog than an axolotl but also because Krogh had demonstrated that frogs take up both sodium and chloride from very dilute solution.' The epithelial transport world changed, forever, and Larsen (2009) reminded us that the '...frog skin became the preparation of choice of Ussing [and many others], which continued to challenge him for more than 50 years.' Ussing's nearly 20 years of experience with radioisotope tracer, ion exchange, and ion flux experiments, undoubtedly, gave him an excellent background in transport physiology before focusing his efforts on epithelial transport physiology.

1.4.2   The Ussing Chamber and the Short-Circuit Technique

In 1948, Ussing received a Rockefeller Scholarship and journeyed to the Donner Laboratory at the University of California at Berkeley. Shortly after arriving in Berkeley, he was introduced to Lund and Stapp's (1947) book entitled *Bioelectric Fields and Growth*. Ussing read the book and commented that someone '...had attempted to draw the electric current from frog skins via reversible lead-lead chloride electrodes.' (Ussing 1980). Ussing further stated 'When I recalculated the currents drawn from frog skins in terms of sodium fluxes they turned out to be roughly the same order of magnitude as isotope fluxes I had measured.... a plan took shape: If one could "short-circuit" the skin via suitable electrodes so that the potential drop across was reduced to zero and if the bathing solutions were identical then only actively transported ions could contribute to the current passing the skin; the flux ratio for passive ions would become one.' So, the idea of the I_{sc} technique was already germinating. This was the basic electrical 'workings' of the short-circuit current in which Ussing would use an external current to drive the transepithelial current generated by the frog skin back to zero, thus, determining the current of the actively transported ions.

Once deciding upon the frog skin, Ussing began applying the radiotracer technique to the frog skin and examined the ion transport across the flux of $^{24}Na^+$ and $^{38}Cl^-$ across the skin (Ussing 1949a). For that project, he developed an apparatus (Fig. 1.1) which was comprised of separate compartments in which the skin was isolated between two bathing solutions on either side of the frog skin. From that study, he demonstrated that the net Na^+ flux was from outside to inside and that was higher than the outflow of Na^+; he reported the pH dependence for Na^+ flux, and that the Cl^- influx was lower than Na^+ at times, but there was a large parallelism between the potential difference across the skin and the influx of Na^+ and Cl^- (Ussing 1949a). With this chamber, Ussing quickly reported the resting potential and ion movements of the frog skin (Levi and Ussing 1949), and using I^{131}, they demonstrated that inward movement of I was less than the outward movement and that this transport was not active (Ussing 1949b). However, after those studies, Ussing was eager to return to his earlier ideas of examining the 'electrical properties' of the frog skin. Although the chamber and technique used in those recent experiments were just a foreshadow of what was in store for the epithelial transport world through the rest of the twentieth century and still a prominent technique in the twenty-first century..... here comes the famous paper of 1951!

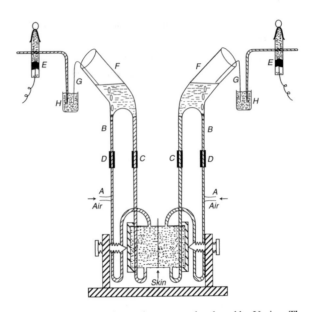

Fig. 1.1 This is a diagram of the experimental apparatus developed by Ussing. The solutions in the two chambers thus formed on either side of the skin are circulated by blowing air or any gas mixture wanted through the side tube. A. The solutions will then ascend to the funnels F through the tubes B and return to the chambers through the tubes C. From the funnels F samples can be drawn. If the total contents of for instance the inside circulation system is to be removed, an arterial clamp is placed on the rubber tube D. Then the pressure air cannot escape that way and will force the solution in the chamber up into the funnel. From here the solution is removed with a pipette or with the suction pump. The funnels are further used for making contacts between the solutions and two calomel electrodes E. The contacts are made through the capillary tubes G. One end dips in the funnel and the other end in a small tube H filled with saturated KCl solution. Before a measurement of the potential difference is made, H is lowered and the solution is allowed for a moment to flow through G before H is elevated again just so much that a little KCl solution penetrates into the end of the capillary [from Ussing (1949a), with permission from John Wiley & Sons, Inc., UK]

Prior to Ussing's work on the frog skin, others had laid down a foundation of information about the frog skin. Francis and Pumphrey (1933) wrote that the concept of a skin potential had been reported in the 1850s. Larsen communicated to this author that Dubois Raymond (1848) suggested the phenomena of a skin potential (Larsen, Pers. Comm.). Lund (1926) had demonstrated a cyanide concentration-dependent reversible decrease of the electrical polarity ('difference in electrical potential' as stated by Lund) of the isolated frog skin. Additionally, Francis (1933) and Francis and Pumphrey (1933), using calomel electrodes, examined the electrical properties and the resting skin potential of the isolated frog skin. Francis (1933) determined that the potential increases as the temperature was raised to 20°C and fell above that temperature. Furthermore, they demonstrated that oxygen deprivation rapidly reduced the potential and that removing Ca^{2+} and K^+ from the solutions resulted in a similar effect (Francis and Pumphrey 1933). Others have used frog skin to conduct experiments examining the resting potential and currents of the skin and the effects of changing ion concentrations across the skin (Stapp 1941; Meyer and Bernfeld 1946). Ussing was aware of these studies. However, and more importantly, Ussing made the 'major link' between electrical current and the active movement of specific ions across the frog skin. These ideas culminated in the famous research papers of 1951 by Ussing and Zerahn, and in 1958 by Koefoed-Johnsen and Ussing.

Before the 1951 paper, Ussing had reported that the influx of Na^+ transport across the frog skin was higher than the outflux of Na^+ using two symmetrical skin preparations for radioactive influx and efflux studies, respectively, and both monitored by the $^{24}Na^+$ isotope (Levi and Ussing, 1949). With the introduction of his modified chamber and rig from that used in 1949, Ussing and Zerahn (1951) were now able to implement the short-circuit current technique in parallel with previous used radiotracer methods (Fig. 1.2). This chamber provided the ideal apparatus for the cells of an epithelial tissue to be studied in a precisely defined way. Again, the isolated frog skin epithelium was mounted between two fluid-filled chambers with essentially the same design as still used today. Hence, the introduction of the Ussing chamber opened a vast new opportunity to study ion transport of epithelial tissues. Ussing enlisted the help of Karl Zerahn to aid in the wiring of the circuit to be able to short-circuit the tissue and Ussing made the glass chambers (Ussing 1980) (Fig. 1.2). They conducted the first investigation of ion transport physiology (Na^+, in this case) of an epithelium in which radioisotopes (Na^{24}) could be used while simultaneously monitoring the 'active Na^+ current' generated by the frog skin.

As one reads the 1951 paper, one realizes there are five central advances made by Ussing and Zerahn. (1) They applied the short-circuiting (I_{sc}) technique to an epithelial tissue preparation; (2) They described an apparatus (the Ussing chamber with an associated DC-current generator, i.e., a battery) that permitted them to measure, simultaneously, the electrical current and the net Na^+ flux through the frog skin; (3) They bathed the frog skin with identical solutions (NaCl Ringer's solution) on either sides of the epithelium, thus preventing transepithelial passive net-ion flows. Thus, under those experimental conditions, ions that move by active transport would continue, and the generated short-circuit current would result from a net transport of those ions; (4) They stated that one could calculate the electromotive force for Na^+ and also the resistance to the Na^+ current from the efflux of Na^+ and the I_{sc} current. Lastly, Ussing and Zerahn reported that the net active transport of Na^+ resulted in the I_{sc} current generated by the frog skin, and this was based on the ^{24}Na experiments compared with the I_{sc} current measurements. Therefore, Na^+ influx across the entire frog skin epithelium (from pond water to the blood) dominated the Na^+ efflux with little Na^+ being transported from the blood to the pond water. The cellular model of Na^+ absorption is reviewed in other chapters of this volume and Volume 3 (Chaps. 8 and 9 of this volume and Chap. 18 of Volume 3) of this series, thus, the 1958 paper will not be discussed here. Interested readers are directed to Palmer and Andersen (2008) who celebrated and highlighted the 50th anniversary of the Koefoed-Johnsen and Ussing 1958 paper.

Today, the Ussing chamber method has been applied to virtually every epithelium in the animal body, including the reproductive tract, exocrine/endocrine ducts,

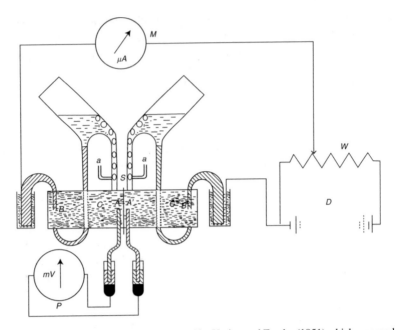

Fig. 1.2 Diagram of the chamber apparatus used by Ussing and Zerahn (1951) which was modified from chamber in Ussing's (1949a) paper. The apparatus used for determining Na flux and short-circuit current is shown diagrammatically in Fig. 1.2. The skin S, is placed as a membrane separating the Ringer solutions in the celluloid tubes C. Two narrow agar-Ringer bridges A and A4′ open on either side in the immediate vicinity of the skin. The outer ends of A and-A′ make contact with saturated KC1-calomel electrodes. The potential difference between the latter is read on the tube Potentiometer, P. Another pair of agar-Ringer bridges, B and B′ open at either end of the celluloid tubes, as far as possible from the skin. The outer ends of these bridges dip into beakers with saturated KC1 saturated with AgC1. Spirals of stout silver wire, immersed in these beakers, are used as electrodes through which an outer E. M. F. can be applied. The voltage is supplied from the dry battery D, and by aid of the potential divider, W, the voltage is adjusted so that the potential difference across the skin, as read on the potentiometer P, is maintained equal to zero. It is obvious that this is equal to a total short-circuiting of the skin potential. The current passing the skin at zero potential difference or any other potential difference desired is read on the microamperemeter M. Usually readings every five minutes suffice for a sufficiently accurate graphical integration, over the period between two Na^{24} samples, of the total amount of electricity crossing the skin. The potential was, however, under continuous observation during the experiments, so that also unexpected changes in current strength could be recorded, should they occur [from Ussing and Zerahn (1951), with permission from John Wiley & Sons, Inc., UK]

intestine, respiratory airway, eye, and choroid plexus. Furthermore, the method has been extensively used for studies of cultured epithelial cells (primary cells and stably transfected cell lines) where tight junction integrity maintains apical and basolateral membrane polarity.

As noted above, and will be presented further in this chapter, the discoveries of x-rays, the elements radium and polonium, heavy water, the development of research tools (site-directed mutagenesis and PCR) and techniques (e.g., patch-clamp technique, Sakmann and Neher, Nobel Prize for Physiology or Medicine, 1991 for their discoveries concerning the function of single ion channels in cells) which have ushered in new 'eras' of research have garnered Nobel Prizes. Considering the major impact of Ussing's work in the field of epithelial transport physiology, was he ever nominated for a Nobel Prize? Indeed, Ussing was nominated at least once. Erik Larsen informed this author that a group from Copenhagen led by Ulrik Lassen and Ove Sten-Knudsen with supporting letters from the USA, UK, and Germany was submitted to Stockholm in the 1970s for consideration for a Nobel Prize. Unfortunately, even with an extraordinarily strong case for Ussing, the nomination was unsuccessful (Larsen, Pers. Comm.).

For more information regarding the discoveries of Hans Ussing, see the Hans Ussing Memorial issue of the *Journal of Membrane Biology*, Issue 3 Vol. 184, 2001 (https://link.springer.com/journal/232/184/3/page/1). The following are reviews that may be of use to interested readers; Ussing et al. (1974), Ussing (1980), Lindeman (2001), Larsen (2002), Palmer and Andersen (2008), Clarke (2009), Larsen (2009), Hamilton (2011), Hermann and Turner (2016), and Zajac and Dolowy (2017).

1.5   The Micropuncture Technique

Prior to the Ussing's work, the quest to acquire scientific information about epithelia transport function was certainly flourishing as noted in the earlier sections of this chapter. When trying to understand the function of a complex organ such as the kidney, one must step back and think about the basic structure of the organ. As we know, each kidney is composed of $\sim 1 \times 10^6$ nephrons, which are the function units of the kidney. In the first quarter of the twentieth century, Wearn and Richards revolutionized the study of the kidney by introducing the micropuncture technique in 1924. This technique truly opened the kidney for scientific exploration.

1.5.1   Historical Aspects of the Micropuncture Technique

Starting to unravel the function of an organ (e.g., the kidney) or even a small structure (e.g., the nephron) within an organ presented a major technical challenge for scientists. Hence, advancement of knowledge is driven by the development of new technology. A case in point is our early understanding of the function of the single nephron of the kidney. In their 1924 seminal paper, Wearn and Richards established the renal micropuncture technique that allowed them to puncture surface nephrons of the frog kidney with pipettes to determine the composition of the ultrafiltrate. Some 80 years later, Sands (2004) stated that 'The development of the micropuncture by Wearn and Richards in 1924 ranks as one of the greatest advances in renal physiology during the 20th century.' Lorenz (2012) and others (Sands 2004; Vallon 2009) remarked that the work by Wearn and Richards provided the first experimental evidence that the glomerular filtrate was protein free and that the nephron exhibited tubular reabsorption, although Wearn and Richards reminded the reader that Carl Ludwig first developed the concept that the process of absorption occurred in tubules of the kidney in 1844 (see Wearn and Richards 1924 for the Ludwig references). Additionally, Wearn and Richards provided the first experimental evidence that the concentrations of Cl^-, K^+, glucose, urea, and the pH of the glomerular filtrate were similar to that of plasma. Many other milestones in renal tubular epithelial physiology were discovered by using the micropuncture technique (Vallon 2009). These discoveries included providing evidence of tubule reabsorption by demonstrating that Na, Cl, and glucose were present in the proximal segments of the nephron but absent from urine in the bladder (Wearn and Richards 1924). Walker and co-workers developed techniques that enabled them to microperfuse surface nephrons of *Necturus* along with the frog and perfused proximal and distal tubules in situ (Richards and Walker 1936). It was Walker and Hudson (1936) who first reported that the proximal convoluted tubule was the site of glucose reabsorption.

Lorenz (2012) reminded us that it was nearly 20 years later that Walker and colleagues (Walker et al. 1941; Walker and Oliver 1941), modified the micropuncture technique to allow collection of filtrate samples from the mammalian kidney. Walker and co-workers (Walker et al. 1941), using guinea pig kidney, collected filtrate from glomeruli and verified that a 'near ideal' ultrafiltrate was produced by the glomeruli. Further, Walker and Oliver (1941) provided evidence that the proximal tubule reabsorbed two-thirds of the filtered fluid produced by the glomeruli. As one might imagine, many other seminal experimental findings of tubular function were published using the micropuncture technique. For example, Gottschalk and Mylle (1959) provided the initial experimental evidence for the counter-current hypothesis for concentrating urine. Brenner and colleagues developed modifications of the microperfusion technique that allowed the direct measurement of the capillary hydraulic and oncotic pressures. They were able to assess the pressure gradients, flows and fluid, and solute permeabilities in individual nephrons (Brenner et al. 1971, 1972; Robertson et al. 1972). These ground-breaking papers provided the experimental findings used to establish the model for glomerular filtration which is as accurate today as it was nearly 50 years ago. These are only a few examples of the major 'break-through' research studies that have provided experimental data that lead to a greater understanding of the renal tubular physiology, all thanks of the micropuncture technique.

Readers interested in further aspects of the history of the micropuncture technique and the seminal advances in renal tubular physiology are referred to the following articles by Wearn and Richards (1924), Gottschalk and Lassiter (1973), Lang et al. (1978), Windhager (1987), Lorenz (2012), Stockand et al. (2012). Additionally, in 1971, there was a symposium on the renal micropuncture technique held in the Department of Physiology at Yale University School of Medicine. Papers from that symposium were published in the *Yale Journal of Biology and Medicine* in 1972 (Volume 45(3–4);Jun-Aug 1972). These papers can be accessed through the following website: https://www.ncbi.nlm.nih.gov/pmc/issues/174113/.

1.5.2 The Micropuncture Technique

All the technical details of the micropuncture technique will not be explained here, but will direct interested readers to a number of extensive review articles. In simple terms, the micropuncture technique permits the investigation of transport function of specific segments of the surface nephrons in situ in the context of the kidney including the complexity of the physiological nature and regulation of function by intra-renal and extra-renal mechanisms (Stockand et al. 2012). Unfortunately, this technique can only be used to examine transport function of the S1 and S2 segments of the proximal tubule, and the early distal tubule, connecting tubules and few early convoluted ducts (Stockand et al. 2012). Nonetheless, transport function of other segments of a specific surface nephron can still be obtained by sampling the filtrate leaving the proximal tubule and that which enters early distal tubule. Additionally, one can determine the glomerular filtration rate of a single nephron by monitoring the segment between the glomerulus and the S2 segment of that nephron (Lorenz 2012).

Throughout the last 95+ years, the basic theory of the micropuncture technique has remained essentially the same. Indeed, Wearn and Richards (1924), using frog kidneys, used a very simplistic micropuncture rig as seen in Fig. 1.3a. A 'sharp pointed capillary pipette' was used for the collection of glomerular filtrate fluid, using a microscope, after which, the fluid was then analyzed for protein, sugar, chloride, urea, and potassium. As noted by Stockand et al. (2012), now, most micropuncture experiments are conducted on mice and rats and data are extrapolated to humans. Although today, the micropuncture rig is much more sophisticated as seen in Fig. 1.3b. The components of the rig and the technical requirements for various types of experiments executed lead to advanced types of data acquisition and analyses systems, multiple experimental data types obtained, microchip transducers, monitoring blood pressure, microperfusion pumps, micromanipulators, temperature controller, among other electrophysiological apparatuses (Lorenz 2012).

The key to this technique is the quality of the pipettes used. A series of pipettes with specific functions need to be fabricated, and these include marking pipettes which are used to inject dyes into the segment of interest to determine which segment of the nephron is being studied; sample collecting pipettes for obtaining filtrate; pressure pipettes; perfusion pipettes filled with the appropriate stained artificial tubular fluid depending upon which segment is being studied; wax block pipettes are used to 'block' flow of filtrate to measure the stop flow pressure of the proximal tubule, for example (Lorenz 2012). Additionally, to measure biopotential differences and ion selectivity of filtrate, standard glass microelectrodes are used to construct micropipettes for measuring potential, and borosilicate or alumina silicate glass capillaries are used to make ion-selective ion micropipettes, made with liquid-selective resins, to measure specific ion concentrations of Na^+, Ca^{2+}, Cl^-, K^+, and pH (Lorenz 2012; Stockand et al. 2012). There are specific requirements for animal anesthesia, surgical preparation, and kidney isolation depending upon the particular planned experiment. Figure 1.4 illustrates the preparation of a surface nephron for

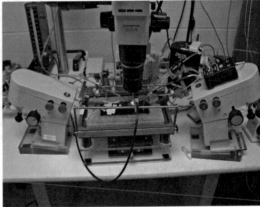

Fig. 1.3 (**a**) The apparatus used by Wearn and Richards (1924) for their initial micropuncture experiments with frog kidneys [from Wearn and Richards (1924)]. (**b**) Completed micropuncture setup for mouse, which is positioned on the heated table in a right lateral decubitus lie to expose the left flank, with head to the right and tail to the left. Left kidney is immobilized in a Lucite cup anchored from the rear, and is illuminated with a fiber optic light source from the front. The Leitz manipulator on the left is holding the marking pipet, and the Leitz manipulator on the right is holding the pressure pipet. The wax-block/paraffin press can be seen behind the pressure pipet and is attached to a Narishige hydraulic manipulator with controller at the far lower right (text and figure is from Lorenz 2012)

fluid collection from the proximal tubule and the distal tubule. Figure 1.4b is an example of micropuncture in situ experiment of a rat kidney.

Over the years, advances in micropuncture have occurred and certainly coupling microinjection of a predefined solution into the tubule with the micropuncture technique provided a new element for the investigator to address specific experimental questions of ion and solute transport in a given segment of the nephron (Stockand et al. 2012). Lorenz (2012) highlighted that gene delivery added a new dimension in which one can develop adenoviral vectors containing specific cDNA sequences and use the micropuncture technique to deliver these vectors into segments of the nephron to express proteins in the tubule epithelium. Ashworth and colleagues (Tanner et al. 2005) provided a novel approach and the first evidence demonstrating gene transfer and visualization of a protein product in the rat kidney. They successfully infused adenovirus containing GFP-cDNA constructs, (i.e., XAC (wt)GFP) into single rat proximal tubules by using micropuncture. Within 2 days, using two-photon microscopy, they identified XAC(wt)GFP expressing proximal tubule cells (Tanner et al. 2005). These investigators emphasized that their experimental approach allows the study of protein expression, cellular destination, and behavioral movements of fluorescently-tagged proteins in vivo within the kidney. These authors further suggested that their method will be advantageous in examining the in vivo behavior of a range of molecules in numerous pathophysiological conditions. Interested readers in the gene delivery and two-photon imaging techniques are directed to the following reviews: Ashworth et al. (2007), Peti-Peterdi et

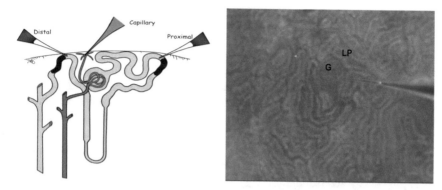

Fig. 1.4 (**a**) Pipet configuration for fluid collections from the proximal and distal tubules, and from the peritubular capillary (star vessel). Pipets for fluid collections are back-filled with stained oil, and a mobile oil block is inserted into the tubule before aspiration of tubular fluid begins in a timed collection. Fluid segments can be identified using a marking pipet to inject dyed fluid and allowing it to flow downstream. GFR can be measured by analyzing inulin concentration in fluid and plasma. The capillary pipette is filled with stained oil to prevent evaporation and to aid visualization, but a block is not inserted since the blood is not collected quantitatively (text and figure is from Lorenz 2012). (**b**) Surface of a Munich-Wistar rat kidney prepared for micropuncture measurement. The pipet, at right, has been inserted into a surface glomerulus (G), and dyed artificial tubular fluid has been injected and allowed to flow downstream; in this image, the dye fills part of the late proximal labyrinth, and the last surface loop of the proximal tubule is identified (LP) (text and figure is from Lorenz 2012)

al. (2015), Nakano and Nishiyama (2016), Sandoval and Molitoris (2017), and Matsushita et al. (2018).

1.5.3 Advantages and Limitations of the Micropuncture Technique

As with any technique, there are advantages and limitations and the micropuncture technique is no exception. As mentioned, the micropuncture technique has the advantage that enables scientists to examine the function of surface nephrons to understand renal transport processes of a specific segment of the nephron. This technique does not allow the study of deeper nephrons, so comparison of function from juxtamedullary nephrons and superficial nephrons must be considered (Jamison 1970). Of course, the micropuncture of the nephron is proceeded by animals being anesthetized and experiencing surgical procedures. Additionally, physical rupture of the nephron segment wall may alter the function of that epithelial cells in the part of the segment. Therefore, experimental results must be interpreted considering these factors (Lang et al. 1978).

1.5.4   The Future of the Micropuncture Technique

There is the worry that the younger generation of renal physiologists are not being trained in this especially important technique. Hopefully, with the popularity of genetic rat and mouse models, there will be a resurgence in interest in the micropuncture technique, and what experimental questions can be asked and answered by using this technique (Lorenz 2012). Therefore, there is the hope that more scientists will be learning the micropuncture technique over the next few years (Sands 2004; Vallon 2009). As Vallon (2009) stated that 'Application of micropuncture techniques in the intact kidney will remain a gold standard to answer questions related to renal physiology and pathophysiology'.

We apologize to those authors whom we have not referenced specific 'key' publications in this brief section about the micropuncture technique. For further historical perspectives and technical details of the micropuncture technique, the reader is directed to Wearn and Richards (1924) and the following excellent reviews by Lang et al. (1978), Vallon (2009), Lorenz (2012), and Stockand et al. (2012). Finally, interested readers are also directed to a dedicated volume of *Yale Journal of Biology and Medicine* (Volume 45, Issue (3–4), 1972) entitled 'Renal Micropuncture Techniques: A Symposium' (https://www.ncbi.nlm.nih.gov/pmc/issues/174113/).

1.6   Isolated Perfused Kidney Tubule: Maurice Burg

The micropuncture technique certainly allowed scientists to 'peek into the workings' of the nephron. Additionally, many fascinating discoveries of epithelial transport function of the nephron was and is still being made with the micropuncture technique. However, how do we move from examining the physiology function of the nephron to the functions of the epithelial cells of specific segments of the nephron? Again, development of technology was necessary to address experimental questions about the cells of nephron which could not be accomplished with micropuncture. Enter Maurice Burg who developed the isolated perfused tubule technique which was an epoch leap in the ability to precisely study the physiological functions of specific segments of the nephron in vitro which culminated in his seminal work of 1966 (Burg et al. 1966). Throughout the years, the physiological research on the isolated perfused tubule, of many species, has flourished, and the efforts of numerous scientists have increased our understanding of the physiological functions of the nephron, and thus, the function of the kidney. The significance and impact of Burg's work and 'first generation (and their scientific offspring)' of those who trained with him, and/or those who set-up isolated perfused tubule rigs based on Burg's work was reported by Hamilton and Moore (2016). Material from Hamilton and Moore (2016) formed the basis of this section. Other important reviews about the isolated perfused tubule preparation are provided later.

1.6.1 Maurice Burg's Scientific Training

Born in Boston, Burg traversed from good marks in high school to majoring in psychology for his AB (cum laude) from Harvard College (Burg 2018). This was followed by his entry into Harvard Medical School, and as a 4th-year medical student, he took an elective course in nephrology at a time when the technique of dialysis was in its infancy (Burg 2018). With hindsight, this was Burg's first step toward his illustrious career in nephrology. Following graduation from medical school, he started an internship at the Beth Israel Hospital in Boston. However, because of a bureaucratic glitch, he quickly entered a residency in nephrology and endocrine disorders at the Boston Veterans Administration (VA) Hospital (Burg 2018). At the VA, Burg worked with Maurice B. Strauss who was then the Chief of Medicine. It was Strauss who strongly recommended Burg to Robert Berliner and Jack Orloff of the Laboratory of Kidney and Electrolyte Metabolism (LKEM) at the National Heart Institute (NHI, now the National Heart, Lung and Blood Institute) at the National Institutes of Health where Burg worked from 1957 to 1959. Burg returned to the Boston VA for an additional year after which he returned to the NHI as an Investigator in the LKEM. As Burg (2018) states in his autobiography 'I found that I preferred research over clinical care, and I ended up remaining at NIH for 57 years, doing research in kidney physiology. Thus, happenstance, not fore-thought, was responsible for my career in science.'

1.6.2 Development of the Isolated Perfused Tubule Preparation

When Burg returned to the LKEM in 1960, he noted that the prominent method for studying kidney function was by clearance measurements and the micropuncture technique (Burg 1982). Although the micropuncture technique was a powerful renal tool, Burg was not able to use that technique as others at the LKEM dominated the micropuncture rigs, but Burg stated that '…monopoly on micropuncture turned out to be fortunate for me since it set me on the path to developing perfusion of isolate renal tubules…' and he further thought '…Reasoning that in vitro experiments can be more strictly controlled over in vivo ones, I began my search for an alternative to micropuncture….' (Burg 2018). He envisioned a different version of micropuncture that would allow scientists to investigate the functional properties of individual segments of the nephron (Burg 1982). Jared Grantham recalls that Burg became very enamored with the idea of perfusing renal tubules '…outside of their normal environment…' in the kidney (Grantham, Pers. Comm.). Burg's idea was based on two important experimental techniques that had opened a number of research avenues. The first was Ussing's marvelous experiments examining the ion transport properties of the frog skin epithelium with his chamber and the short-circuit tech-nique, described above (Ussing and Zerahn 1951; Koefoed-Johnsen and Ussing

1958). Second, Burg recognized the innovations made by Alan Hodgkin and Andrew Huxley and colleagues (1952a–d; Hodgkin et al. 1952) and their new methods used to examine the action potential of the giant squid axon. Their work along with John Eccles' work was awarded with the Nobel Prize in Physiology of Medicine in 1963 '...for their discoveries concerning the ionic mechanisms involved in expiration and inhibition in peripheral and central portions of the nerve cell membrane.' These technical advances developed to determine the properties of epithelia tissues and the axon lead Burg to write '...While these were single cells, not tubes of epithelia, and while they were much larger than kidney tubules the successes with the giant axons were well known by the late 1950s and suggested to me that in vitro perfusion might be an experimental approach to kidney tubules' (Burg 1982).

An extensive review of Burg's development of the isolated perfusion technique will not be attempted to be provided here (Burg et al. 1966), although interested readers of the specific details are directed to reviews written by Burg (Burg 1972, 1982; Burg and Knepper 1986; and Burg et al. 1997).

Here, the major developments of the isolated perfused tubule apparatus will be described. Burg had three major hurdles to overcome to establish the perfused tubule preparation. They were: (1) learning how to isolate specific segments of the nephron, (2) how to mount the isolated tubule within a perfusion system, and (3) how to collect fluid from the distal end of the perfused tubule. Burg began to isolate segments of the rabbit nephron with collagenase which was commonly used to isolate cells at that time. The collagenase approach was useful for attaining a tubule suspension, however, that approach was not effective for isolating single tubules, because the tubules were too fragile (Burg et al. 1966; Burg 1982). However, in 1963, and fortunate for Burg, Ivar Sperber, a renal anatomist, visited the LKEM and Burg talked to him about his tubule work. After which, Sperber taught Burg how to hand-microdissect individual nephrons with forceps rather than micromanipulators (Burg 1982). This good fortune for Burg was followed by the arrival of his first two fellows, Maurice Abramow and Jared J. Grantham, who joined the lab in 1963 and 1964, respectively. Grantham made a quick impact by dissecting rabbit tubules without collagenase and using Sylgard to electrically seal perfused tubules (Grantham, Pers. Comm.) Next, it was important to determine how to mount the isolated tubule in a chamber between holding pipettes to allow perfusion of the lumen of the tubule. Burg was able to construct glass micropipettes in an attempt to hold the isolated tubule on one end with one pipette while applying suction to the opposite end of the tubule with another pipette (Burg 1982), although he quickly realized that that approach was not suitable for securing the tubule properly. After time, Burg thought that concentric pipettes (Fig. 1.5) might be an easier way to apply suction all around the tubule, therefore, the inner pipette was used to cannulate the tubule while the outer pipette was used for suction (Burg 1982, ISN 2005). The next hurdle was finding an apparatus that would be able to move the concentric pipette system to execute axial positioning of the micropipettes with respect to each other. For this, Burg visited with Peter Davies (at Johns Hopkins University) who had already developed an apparatus that held and moved micropipettes precisely as Burg

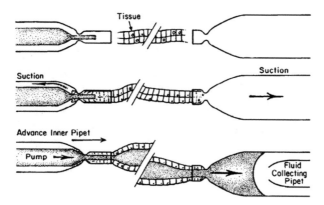

Fig. 1.5 Procedure for tubule perfusion. Concentric pipets are on the left. The end of the tubule is drawn by suction into the tip of the outer pipet, which supports it and seals the inner pipet within the tubulde lumen. Collecting pipets are on the right. The tubule is drawn into the tip of the outer collecting pipet by suction and remains lodged there when the suction is stopped. The inner pipet is introudced periodically to remove aliquots of fluid (from Burg et al. 1966)

had envisioned. Unfortunately, to conduct an experiment, Davies' delicate apparatus had to be taken apart and re-assembled for every experiment (Burg 1982). Therefore, when Burg returned to NIH, he met with Kenneth Bolen, head of the precision machining unit, and Jim White, and they designed a more robust pipette-holding apparatus that is essentially the same still used today (Fig. 1.6). Finally, Grantham solved the final hurdle by designing and developing the pipette used to collect fluid from the distal end of the perfused tubule. Burg gave the highest claim to Grantham by stating '...If anyone is responsible for the progress in that field, it's Jared...he... had one pipette that he sucked the tubule into, had oil in the pipette, and introduced a collection pipette to collect the fluid that had accumulated' (ISN 2005).

Over the subsequent years, scientists have developed modifications to the perfusion system or provided technology that enhanced the ability to analyze tubular

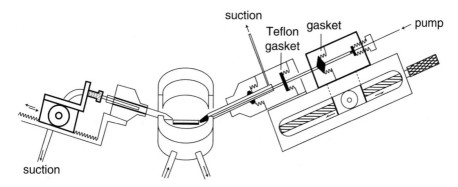

Fig. 1.6 Apparatus for perfusing isolated tubule fragments (from Burg et al. 1966)

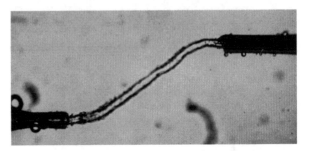

Fig. 1.7 Perfusion of a single collecting tubule. The direction of perfusion is from left to right (from Burg et al. 1966)

samples. Mark A. Knepper (a research fellow with Burg) reminded this author of the contributions of Gerald Vurek and Robert Bowman of the LKEM in the overall success of the analytical potential of the samples obtained via the isolated perfused tubule preparation. Knepper wrote that they '...produced multiple devices that allowed us to make measurements in the 5–100 nanoliter samples collected from the perfused tubules. The earliest ones that may be familiar to readers are the helium glow photometer (Bowman) and the picapnotherm (Vurek). Bowman and Vurek could be considered major players in the success of the isolated perfused tubule.' (Knepper, Pers. Comm.).

Of course, all the work by Burg, Grantham, Abramow, and Orloff resulted in the seminal 1966 paper (Burg et al. 1966). A perfused collecting tubule from the original paper is shown in Fig. 1.7). Figure 1.8 shows the isolated perfused tubule rig that Knepper inherited from Burg (Knepper, Pers. Comm.).

Readers are directed to Burg's autobiography (Burg 2018) and Burg and colleagues (Burg et al. 1997) in which the original Burg et al. (1966) paper was reprinted with commentaries by Burg and the late James Schafer, who was a visitor to the Burg Laboratory before setting up his own isolated perfused tubule rigs (Hamilton and Moore 2016). Grantham reflected on his research time with Burg and his own experiences and contributions to the development of the isolated perfused tubule technique in his autobiography (Grantham 2014) and in Grantham et al. (1978). Finally, interested readers are directed to a dedicated volume of *Kidney International* (Volume 22 Issue 5 Nov 1982) devoted to the advances of the perfused tubule preparation (https://www.sciencedirect.com/journal/kidney-international/vol/22/issue/5).

1.6.3 Advances in Renal Physiology with the Isolated Perfused Kidney Tubule

As noted above, Sands (2004) exclaimed that the micropuncture technique was '...one of the greatest advances in renal physiology during the 20th century...'. Sands' comments continued with '...greatest advances...along with the development of the

Fig. 1.8 A photo of an isolated perfused tubule rig. Perfusion is from right to left in this rig. This is the rig that Dr. Moe Burg gave to Dr. Mark A. Knepper for his original urea studies (photo courtesy of Dr. Mark A. Knepper, circ. 1980)

isolated perfused tubule by Burg and colleagues in 1966.' When Burg began work on the nephron, one would consider that there were 6–7 'functionally-different' segments of the nephron. The nephron is now considered to have 14 segments. Indeed, Knepper and colleagues (Lee et al. 2015) conducted a comprehensive profile gene expression analysis of each of the 14 renal tubule segments of the rat nephron. The Knepper study resulted in transcriptomic profiling of the different segments of the nephron resulting in an online database with the data deposited in the Gene Expression Omnibus. Knepper stated that this database provides '…the resources for future studies of renal systems biology, physiology and development.' So, in little more than 50 years, scientists have progressed from initial experiments with the isolated perfused tubule to now identifying distinct patterns of distribution of region-specific transcription factors, G protein-coupled receptors, and metabolic enzymes along the renal tubule, for example (Lee et al. 2015).

Many paragraphs would be required to discuss the advancements in our understanding the function of the nephron and the kidney, itself, discovered with the use of the isolated perfused tubule preparation. The issue of *Kidney International* that was devoted to the isolated perfused tubule preparation mentioned above was published only 16 years after the original Burg et al. (1966) paper. Harry R. Jacobson and Juha P. Kokko (also a research fellow of Burg's) provided the opening remarks for that issue. Jacobson and Kokko (1982) provided a list of the 'new developments' made with perfused tubule studies which are presented here: '(1) identification of active chloride transport in the thick ascending limb of Henle, (2) demonstration of intrinsic transport difference among diverse segments of the proximal tubule [(NB. which now applies to many segments of the nephron)], (3) description of the site and mechanism of action of various hormones and second messengers on salt and water transport [NB. aldosterone and arginine vasopressin, for examples], (4) proposal of a new model of the countercurrent multiplication system without active transport within the inner medulla, (5) detailed description of divalent anion and cation transport in the various nephron segments, (6) significant advancement in understanding the mechanism of H^+/HCO_3^- transport across segments of proximal and distal nephron, (7) furthering our understanding of organic anions and cation transport, (8) specific localization of the site of action of various diuretics, and (9) in vitro evaluation of nephronal adaptive response to in vivo manipulation such as uremia and changes in acid-base balance.' In addition to these discoveries, Grantham et al. (1974) reported that the net secretion of para-aminohippuric acid was enough to overcome the reabsorption of NaCl in the S2 and S3 segments of proximal tubule causing sustained net secretion of fluid into the tubule, a finding that had never been previously envisioned for the mammalian nephron. This finding led to the discovery of how fluid accumulates within the cysts of patients with polycystic kidney disease in response to cAMP-dependent chloride secretion. This is by no means an exhaustive list of the accomplishments, as nearly 40 years of research with perfused tubule technique has occurred since the Jacobson and Kokko's list was compiled in 1982. Some of the key papers that contributed to the list above were expanded upon by James A. Schafer in his commentary that accompanied the re-publication of the 1966 paper in 1997 (Burg et al. 1997). Additional advances contributed by the isolated perfused tubule to our understanding of the function of the nephron can be found in Hamilton and Moore (2016).

Interested readers in further information about the isolated perfused tubule and about Burg are directed to the American Physiological Society's web page 'Living History of Physiology' (http://www.the-aps.org/mm/Membership/Living-History/Burg) and the ISN Video Legacy Project (cybernephrology.ualberta.ca/ISN/VLP/Trans/burg.htm, ISN 2005) for interviews with Burg. Additionally, an appreciation of Dr. Burg's research impact, research collaborations (first and second generation of scientists), migrational visualization of first generation scientists who worked directly with Burg at LKEM, quantification of metrics indices of research collaborations, outputs and citation analyses, and the global radiation of the isolated perfused tubule technique are provided by Hamilton and Moore (2016).

1.7 In Vivo and In Vitro Intestinal Techniques and the Everted Intestinal Sac Technique: Gerald Wiseman

The anatomy of the human intestine has been of interest for hundreds of years. Indeed, Leonardo da Vinci studied the anatomy of the human body and animals. Most people will recognize two particular de Vinci paintings; one is the famous *Mona Lisa* and the other is the *Virtuvian Man* (the male figure with two superimposed figures within a circle, c 1490). Indeed, de Vinci produced a large number of drawings of the human abdomen and intestine (Keele and Roberts 1983). Not surprisingly, our understanding of the physiology of the intestine developed after the anatomical descriptions of the human body. The experimental progress has been fueled by a series of technical developments. There have been major advances in in vivo and in vitro techniques developed to study the physiological function of the intestine, and these will be briefly discussed here. However, simple sometimes can be easier and more informative than complicated technology. The everted intestinal sac preparation developed by Gerald Wiseman is a case in point (Wilson and Wiseman 1954), although a brief overview of the in vivo techniques and a more in-depth examination of in vitro techniques are discussed here.

1.7.1 In Vivo Techniques

For intestinal tissues, as described in Sect. 1.2.1, Reid (1890) cited that Dutrochet (1826) first examined the function of the cecum of the fowl in the early 1820s. Wilson (1962) reported that Thiry (1864) described the first surgical preparation of the blind loop intestinal fistula. Once the fistula was completed, Thiry was able to introduce test solutions and then after a time, samples were collected for analyses. Interested readers are directed to Anderson et al. (1962) for further discussions of the isolated intestinal loop preparation and Wilson (1962) for drawings from the Thiry's paper.

There was much progress made with in vivo methods to study the physiology of the intestine during the second half of the nineteenth century and during the early to mid-twentieth century. Wilson (1962) briefly described, with figures from the original publications, some of these methodological advances which included the intestinal cannulation technique with direct addition of fluid to the stomach (Cori 1925), intestinal intubation (Miller and Abbott 1934), tied-loop intestinal preparation (Verzár and McDougall 1936), and intestinal loop circulation (Sols and Polz 1947). These in vivo techniques have greatly enhanced the ability of scientists to explore and progress the field of physiology of the intestine.

1.7.2   In Vitro Techniques

The next advance in examining the physiology of intestinal physiology was in vitro preparations. Wilson (1962) claimed that the first important physiological information of the intestine was reported by Reid where he determined that fluid could be absorbed across the intestine 'against a hydrostatic pressure gradient'. As mentioned above, Reid (1892a) developed a modified version of an apparatus (Fig. 1.9) that was established by Matteucci and Lima (1845) to investigate fluid transport in the frog skin, and he also examined fluid movement in intestine of the cat and rabbit and the stomach of the toad (Reid 1892b, 1901). Reid (1892a) described that apparatus in considerable detail. In brief, Reid stated that the apparatus '…consists of two glass cylinders closed with baudruche [windbag], which has soaked for days in normal saline solution. The experimental membrane was held vertical between the layers of baudruche, the fluid pressure on the two sides being equal. Any transfer of fluid from one side of the membrane to the other was gauged by variation in the volume of fluid in the cylinders, measured by rise and fall of fluid in a tube thereto attached'. Reid (1901) did note that there were disadvantages of the apparatus such as: (1) the piece of tissue used was quite small and, therefore, observation tubes must be fine caliber, (2) one had to be aware of the temperature variations if using tissue from mammals

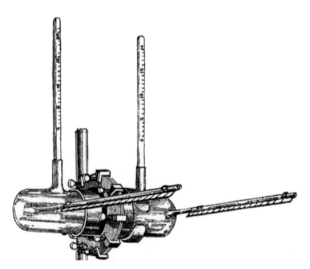

Fig. 1.9 A double layer of baudruche (stocked in saline to avoid imbibition error) forms a vertical membrane between the two halves of a horizon- tally placed cylindrical glass vessel, each half having a horizontal observation tube (figure). The epithelial membrane lies between the two layers of baudruche bathed with normal saline on both sides, and any transport of fluid, and the direction thereof in relation to the orientation of the cells of the membrane, should be evident, by diminution of volume of fluid in the vessel on one side, with concomitant increase of that on the other as gauged by the levels of the menisci in the observation tubes [figure from Reid (1901), with permission from John Wiley & Sons, Inc., UK]

as there was no way to control the temperature, and (3) that, with his apparatus, the preparation was not oxygenated.

Fisher and Parsons (1949) developed the first oxygenated in vitro circulation apparatus for studying transport function of intestinal tissues (Fig. 1.10). This procedure involved cannulation of an isolated segment of rat intestine which was then placed within the circulation apparatus. They stated that they were able to anesthetize the animal, prepare the isolated segment within the rat and have that isolated segment positioned within the circulation apparatus within 10 min. With the addition of the oxygenation, they suggested that the viability of the preparation was extended because the mucosal cells were not subjected to anoxic conditions. Fisher and Parsons (1949) also maintained from their histological studies that the epithelial preparation was viable for over an hour. Additionally, they stated evidence that '…the preparation contains actively metabolizing cells, capable of active translocation of solutes across the intestinal wall [membrane], and that there is no appreciable barrier to diffusion across the submucosa tissues of the intestinal wall'. Certainly, the

Fig. 1.10 Circulation unit of Fisher and Parsons the loop labeled 'l' = intestine tissue, the reader is referred to the original paper for definitions of the abbreviations [from Fisher and Parsons (1949), with permission from John Wiley & Sons, Inc., UK]

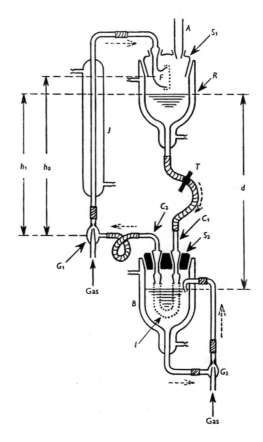

addition of oxygenation to the in vitro apparatus was a significant advance in studying the physiology of intestinal tissues. Others have made modifications to the circulation apparatus, as Wiseman (1953) and Darlington and Quastel (1953) modified the design which afforded a simpler use with having the advantage of yielding an increased area of intestine to the outer fluid and the overall ease of construction of the apparatus. Additionally, Wilson (1956) constructed a circulation apparatus where one could examine the effects of controlled hydrostatic pressure on fluid absorption of an everted intestine. All these technological advances provided the opportunity for more sophisticated experimentation.

1.7.3   The Historical Perspective of the Everted Sac Preparation: Gerald Wiseman

Gerald Wiseman spent his entire academic career (1948–1989) in the Department of Physiology (subsequently renamed Department of Biomedical Science) at the University of Sheffield (Christie and Tansey 2000; Angel 2001). Even while in medical school, Wiseman became enamored with the function of the intestine. At that time, he began questioning the then current mechanism of how nutrients, such as amino acids, were absorbed by the intestine and spent his research career pursuing this (Christie and Tansey 2000; Angel 2001). The dogma, at that time in the 1940s–1950s, was that amino acids were absorbed by diffusion within the intestine. Wiseman did not really agree with that idea. Although he was in the Department of Physiology, he spent much of his time with members of the Department of Biochemistry as those scientists were studying amino acid biochemistry. Wiseman (1953) had already modified the Fisher and Parsons circulation apparatus (1949) in which he suspended pieces of intestine to try to determine if the intestine could transfer (transport) amino acids against a concentration gradient (Fig. 1.11). As Roy Levin stated at the Welcome Witness Seminar on Intestinal Absorption in 1999 (Christie and Tansey 2000), that on one of Wiseman's ventures to Biochemistry, he visited Robert Davies who was studying acid secretion of the frog stomach. Levin (from Christie and Tansey 2000) described the encounter of the two scientists as follows, 'What attracted Gerald's attention was the fact that the manometers were being gassed with rubber tubes and as he [Davies] was doing manometry, he [Wiseman] thought "What on earth is he doing?"' So he popped into the laboratory and said to him, 'Why are you gassing it?' and Davies said 'Well, I have got stomachs inside'. Gerald said, 'Can you keep the stomachs alive?' and he [Davies] said, 'Oh yes! They go on for a long time'; and then he [Wiseman] thought 'Well that's a good idea, maybe I should, in fact take pieces of intestine and keep them alive outside of the body like Fisher and Parsons did'. Further Wiseman stated 'If I turn the intestine inside out I will have a small serosal fluid volume where I could pick up quite a large degree of change in the concentration [of a solute].' So, as Levin

Fig. 1.11 Circulation unit of Wiseman. The reader is referred to the original paper by Wiseman for definitions of the abbreviations [from Wiseman (1953), with permission from John Wiley & Sons, Inc., UK]

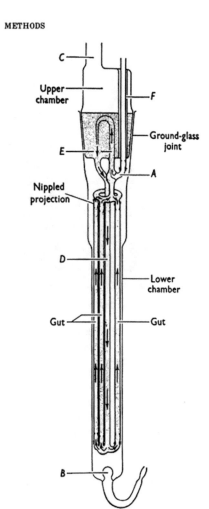

METHODS

stated, Wiseman 'had a method for estimating amino acids.' (Christie and Tansey 2000). The everted sac idea has now been seeded.

Next, Thomas Hastings Wilson had finished medical school and wanted to pursue a PhD and study absorption of purines by the intestine (Christie and Tansey 2000). He enquired to study with Hans Krebs (shared awardee of the Nobel Prize of Physiology or Medicine in 1953 'for his discovery of the citric acid cycle') who was the head of the Department of Biochemistry at Sheffield, at that time. Krebs invited Wilson over to Sheffield, and then informed him that he did not work on purines and directed Wilson to speak with Wisemen whose research interests were intestinal absorption of amino acids, peptides, and sugars (Christie and Tansey 2000). Similarly, Wilson was told by Wiseman that he did not study purines, but

Fig. 1.12 Dr. Gerald Wiseman (left) and Dr. Thomas Hastings Wilson (right) holding an enlarged photograph of a sac of everted intestine of the hamster. This photograph was taken in 1973 in Wilson's laboratory at Harvard Medical School in Boston, Massachusetts, USA. Photograph taken from Christie D A, Tansey E M. (eds) (2000) Intestinal absorption, p 24. London: Wellcome Trust, 81pp. Efforts to find the current copyright holder was unsuccessful

he was conducting experiments on the absorption of amino acids with an emerging technique (Christie and Tansey 2000). The rest is history as can be seen in Fig. 1.12 of Wiseman and Wilson with a photo of the everted sac preparation. There was considerable work conducted by them to convince themselves that the preparation was viable for the timeframe of the desired experiment. Indeed, Chris Cheeseman (Department of Physiology, University of Alberta) who was a PhD student at Sheffield and knew Wiseman considerable well, recalled that Wiseman was 'extremely thorough during all of the testing phases and development of the sac preparation' (Cheeseman, Pers. Comm.). Levin said that Wiseman and Wilson had tried several species including rats, mice, pigeons, and hamster. In fact, Levin stated '. . .in the end, because the sacs had to be shaken quite a lot to get good oxygenation, they had the problem of cell shredding, knocking off the mucosa. They determined that the best sacs and ones that got damaged the least, were those from the Golden hamster [*Mesocricetus auratus*].' (Christie and Tansey 2000). Additionally, Cheeseman noted that the Golden hamster was preferred over the rat because 'the use of anesthetics would unfavorably affect the tissue: therefore, Wiseman chose to use 'blunt trauma' to kill the animals and then quickly remove the intestine.' (Cheeseman, Pers. Comm.).

The everted sac preparation is quite simple to execute and is described in Wilson and Wiseman's classic paper (1954) by using the Golden hamster. Once the intestine was isolated, they used a stainless-steel rod to push the ileal end of the intestine into the lumen of the intestine until the intestine appeared in the duodenum. Next, the proximal end of the intestine was rolled on the rod. The everted intestine was placed in glucose saline, after which the intestine was separated into 2–3 cm lengths. A sac was prepared by tying a thread ligature on one end of the everted sac and filling the sac with a Ringer's solution with a 1 ml syringe with a blunt needle and tying the opposite end of the sac with a second thread ligature as the needle was being removed (Wilson and Wiseman 1954). Wilson and Wiseman noted that any issues with oxygenation was overcome as the delicate mucosal surface of the epithelium was expose to well-oxygenated bathing medium and the serosal surface of the sac was oxygenated by adding oxygen in the sac while filling the sac. Note the everted sac in Fig. 1.12. A detailed procedure of the everted sac preparation and all the associated equipment and supplies needed for conducting experiments can be found in Hamilton and Butt (2013).

1.7.4 Advantages and Disadvantages of the Everted Sac Preparation

Wilson and Wiseman (1954) provided some advantages of the everted sac preparation which included: (1) with the serosa surface of the epithelium facing the inside of the sac and the small volume, a rapid increase in concentration of the transported substance would occur, and (2) because small amounts of tissue are needed for an experiment, adjacent pieces of tissue can be studied simultaneously. Further, Sanford (1982) suggested other advantages which included: (1) that the fluid absorbed into the sac could be determined gravimetrically, (2) the volume bathing the mucosal surface of the epithelium can be quite large, therefore, the concentration of substances of the bath are not appreciably altered, and (3) the substances transported can be determined in both the 'gut wall' or the 'small volume' of the fluid bathing the serosa (fluid inside the sac), therefore, one can determine if substances are being transported against their concentration gradients (radio-isotopic studies). Sanford (1982) noted that a disadvantage with the everted sac was that the intestinal 'wall' is composed of many different cell types such as epithelial cells, muscle mucus-secreting cells, and connective tissue. Therefore, measured concentrations of solutions might not be just by the epithelial cells.

Interested readers are directed to Wilson and Wiseman (1954) and Wiseman (1961) for further information regarding the everted sac preparation. Alam et al. (2011) provided a highly informative review of the use of the everted sac preparation in pharmaceutical and pharmacokinetics research. Additionally, readers can explore Hamilton (2014) for further historical details of the everted sac preparation, the use of this preparation in basic science, and that publication was the basis of this section of this chapter. Lastly, Acra and Ghishan (1991) and Liu and Liu (2013) provide brief reviews of methods of investigation of intestinal transport that readers might find useful.

1.8 Brush Border Membrane Vesicles

1.8.1 Transition from the Epithelial Membrane to Membrane Vesicles

To understand the ion transport function of epithelial tissues, examining the function of the isolated tissues, such as Ussing and Wiseman did, is the beginning of the research journey to understand the physiological function of an individual membrane transport protein (e.g., K^+ channels, Na^+ channels, Na^+/HCO_3^- counter-transporter). However, to gain information about the physiological role(s) of individual membrane ion channels and transporters, scientists needed to gain access to ion transport proteins at the level of the plasma membrane. Thus, it was logical to pursue the isolation of membrane fractions and ultimately isolated membrane vesicles from the mucosal membrane containing the transport proteins of interest. As Acra and Ghishan (1991) stated 'Membrane vesicles offer a means of studying epithelial transport function without the interference of intracellular metabolism. This is done while manipulating with relative ease the composition of both the extra vesicular and intravesicular spaces.'

There might be discussions about which was the first research group to isolate brush border membrane vesicles (BBMVs), although Crane and Miller (Miller and Crane 1961) were the first to report transport function in isolated fragments of the brush border of the small intestine of the Golden hamster. They clearly demonstrated that the 'liberation' of monosaccharides by hydrolysis of disaccharides and sugar phosphate esters occurred within the brush border portion of enterocytes of the intestine. Wright and colleagues (Stevens et al. 1984) stated that 'A landmark in physiology was the introduction of brush-border membrane vesicles by Hopfer and co-workers in 1973.' Indeed, Hopfer et al. (1973) greatly altered a method for isolating brush border membranes by Forstner et al. (1968) by the fractionation of intact brush borders into membranes and cellular components. Hopfer et al. (1973) provided a detailed stepwise procedure for the preparation of highly purified brush border membranes from intestinal epithelial cells. Since Hopfer's paper, a number of protocols for isolating brush border membrane vesicles have been described, although at the time, the Ca^{2+} (Mg^{2+}) differential precipitation technique developed by Crane and colleagues (Schmitz et al., 1973) appeared to be the more popular method.

Acra and Ghishan (1991) summarized the Crane method, which is provided here. 'This method is based on the fact that Ca^{2+} and Mg^{2+} react with brush border membranes in a manner different than with other cellular membranes and organelles. The brush border membrane is probably protected from Ca^{2+} and Mg^{2+} by its glycocalyx while the basolateral plasma membrane fragments and intracellular organelles are cross linked by the divalent cations to form a coprecipitate. The methodology is straightforward and yields vesicles within 3 h of animal sacrifice. The intestine is inverted, and the mucosa is then scraped and homogenized in a blender. After adding Ca^{2+} or Mg^{2+} to the crude homogenate most of the nonbrushborder membranes precipitate and are spun down at low speeds. The brush border membrane vesicles are then collected from the supernatant by using higher centrifugal speeds. The vesicles are rehomogenized, Ca^{2+} (or Mg^{2+}) added once again and the differential centrifugation steps are repeated to insure greater purity of the vesicles.' Of course, there must be controls to determine the purity of the vesicles, providing evidence that the vesicles are actually of brush border origin. Acra and Ghishan (1991) further state '...The purity of the brush border membrane vesicles can be determined by measuring the extent of brush border enzyme marker enrichment (e.g., sucrose, aminopeptides, G-glutamly transpeptidase) as well impoverishment of markers of other cell components in final isolates when compared to crude homogenate.' The determination of purity of the vesicles is crucially important before conducting specific experiments.

As an example of the usefulness of the BBMV technique coupled with the use of radioisotopic studies, Hopfer et al. (1973) examined glucose and other sugar uptake into intestinal vesicles. They demonstrated that: (1) the specificity towards sugars tested with a rate of uptake of D-glucose > D-galactose >> L-glucose ~ D-mannitol; (2) Na^+ was the only cation that increased the rate of carrier-dependent sugar uptake; (3) phlorizin inhibited the glucose; (4) the uptake of sugars were reversible depending upon the experimental gradients; and (5) D-glucose and D-galactose competed for uptake of one another if present on the same side, but one sugar stimulated the uptake of the other if the sugars were on opposite sides of the membrane preparation. Hopfer et al. (1973) stated that their data highly suggested that '...These similarity to studies of intact cells strongly suggest that the specific intestinal glucose carrier system is retained intact by the isolate brush border membranes.' Additionally, the results provided by Hopfer et al. (1973) greatly strengthened the sodium gradient hypothesis proposed by Crane (1965).

In addition to radioisotopic studies, Murer and Kinne (1980) briefly reviewed other methods used to investigate the characteristics of transport proteins of isolated membrane vesicles. These techniques include: (1) rapid filtration technique (Kessler et al. 1978); (2) column chromatography (Gasko et al. 1976); (3) photometric techniques with fluorescent dyes (Beck and Sacktor 1978); and (4) ion selective electrodes can be used to examine transport of ions (Murer et al. 1976). Each of these techniques have unique merits that allow an investigator to explore specific questions of transport function of isolated membrane vesicles such as determining characteristics of specific transport proteins, the kinetic transport parameters of transporters, ion dependence, and blocker pharmacology (Acra and Ghishan 1991).

1.8.2 Advantages and Disadvantages to the Brush Border Membrane Vesicle Technique

As with any experimental preparation, there are advantages and disadvantages with the isolated BBMV preparation. As mentioned above, one must determine the purity and the orientation of the isolated membrane vesicle so that it possesses the same orientation (right side out) as the intact epithelium (Haase et al. 1978). The major advantage to using membrane vesicles is that the investigator can dictate the experimental conditions. Thus, one can examine the contribution of intracellular or extracellular factors of the transmembrane transport of a given ion or solute (Murer and Kinne 1980). Other advantages of the isolated brush border vesicle preparation include: (1) access to both the luminal and contraluminal sides of the plasma membrane to study separately; (2) the vesicles are absent of any metabolic enzymes; (3) the unstirred water layer is negligible compared to the intact tissue; and (4) the paracellular transport pathway cannot contribute to transport system being studies (Acra and Ghishan 1991).

Not surprisingly, there are distinct disadvantages with BBMV preparation which include: (1) contamination of vesicles with other membranes, therefore, vesicles need to be assessed for purity of brush border membrane origin; (2) the likelihood that initial velocities of rates of uptake deviate quickly from linearity because of the collapse of driving forces; and (3) vesicular membrane may become 'leaky' creating an 'artificial' efflux, therefore, the actual efflux rate might be over-estimated (Acra and Ghishan 1991).

In summary, the development of isolate BBMVs provided another methodology, which is still in use, that allows a unique way to examine characteristics and modulation of plasma membrane transport proteins. For further information about the BBMV technique, interested readers are directed to Forstner et al. (1968), Schmitz et al. (1973), Murer and Kinne (1980), Kinne and Kinne-Saffran (1981), Stevens et al. (1984), Murer and Gmaj (1986), and Acra and Ghishan (1991).

1.9 Additional Techniques Used in Epithelial Transport Physiology

The type of science that was conducted 50+ years ago using only a single research technique seemed more the norm then compared to today's research standard and expectations. Today, as an epithelial transport physiologist, one is not just limited to the techniques described earlier in this chapter. It seems the further various research fields evolve, scientists are asking even more advanced and challenging questions that require a broader palette of research techniques and tools to address specific questions. In this section, few techniques which are commonly used in the laboratory of an epithelial physiologist are highlighted.

1.9.1 Site-Directed Mutagenesis and Polymerase Chain Reaction

Since James D. Watson and Francis H. C. Crick published their two-page research article in *Nature* in 1953 (Watson and Crick 1953), science has never been the same. With their description of the molecular structure of deoxyribonucleic acid (DNA), countless numbers of research projects have explored the human genome. However, surprises in science and technology still remain to blossom following the innovative work of Watson and Crick. Here, the site-directed mutagenesis (SDM) and polymerase chain reaction (PCR) techniques which lead to the Nobel Prize in Chemistry of 1993 being awarded to Michael Smith ('for his fundamental contributions to the establishment of oligonucleotide-based, site-directed mutagenesis and its development for protein studies') and Kary B. Mullis ('for his invention of the polymerase chain reaction (PCR) method') are examined.

1.9.1.1 Oligonucleotide-Based Site-Directed Mutagenesis (SDM)

As seen for other techniques described in this chapter, SDM has revolutionized science and molecular biology and physiology, in particular. With SDM, scientists can and have explored the functions and dysfunctions of countless numbers of proteins, and this will, undoubtedly, continue for generations of scientists. Certainly, the pathophysiological basis of many diseases might not have been solved without experimental data produced by SDM.

Michael Smith credits that his illustrious career in nucleic acid research 'would not exist without the example and inspiration of Gobind Khorana. . .' as he stated during his acknowledgements of his Nobel Prize Lecture (Smith 1993). Indeed, in his lecture, Smith described his post-doctoral time, working with Khorana during the mid-1950s to the early 1960s. There, Smith was very productive as he worked on small molecule synthesis including nucleoside-5′ triphosphates (Smith and Khorana 1958), examined the first chemical synthesis approach of ribo-nucleotides (Smith and Khorana 1959), the synthesis and properties of ribonucleoside-3′,5′ cyclic phosphates (Smith et al. 1961), and the development of the methoxytrityl series of 5′-hydoxyl protecting groups (Smith et al. 1962; Smith 1993). These research experiences certainly well-equipped Smith to pursue what resulted, some 20 years later, as the oligonucleotide-directed site-directed mutagenesis technique (Hutchinson et al. 1978). Smith's early and continued admiration and respect for Khorana might have been a 'foreshadow' to the 1968 Nobel Prize Committee's announcement that Khorana would share the Nobel Prize in Physiology or Medicine of 1968 with Robert W. Holley and Marshall W. Nirenberg 'for their interpretation of the genetic code and its function in protein synthesis.'

When we think of the SDM technique, as science has arrived to the 2020s, SDM might appear as a 'simplistic' technique in the array of molecular techniques that are now available. However, the beauty of SDM now, as it was decades ago, is that a scientist can introduce a specific alteration (mutation) to the DNA sequence of a gene for a protein, which can cause a single change of an amino acid resulting in an omission or substitution of an amino acid, or an introduction of a stop codon; all of which can result in different effects on the downstream phenotype (function) of the protein of interest. Such changes to a DNA sequence might result in a mutation that causes a gain-of-function (such as hereditary xerocytosis, Andolfo et al. 2015; Glogowska et al. 2015; Rapetti-Mauss et al. 2015) or a loss(decrease)-of-function (cystic fibrosis, White et al. 1990). Mutations are generally classified as substitution, missense, nonsense, insertion, deletion, duplication, or frameshift mutations. SDM is not just limited to DNA but can be useful in examining biological activity and structure of RNA (Sanjuán 2010), and can be used in protein-engineering (Jimenez-Rosales and Flores-Merino 2018).

The approach for SDM is to engineer a short single-stranded oligonucleotide DNA primer (10–30 nucleotides in length, Smith 1993) that possesses the desired mutation and has a complementary sequence to the template DNA surrounding the desired mutation site of the gene of interest. The DNA primer can hybridize with the template DNA. After which, the single-stranded DNA is extended requiring a DNA polymerase that duplicates the remainder of the gene. A single point mutation can be a change of a single base pair of the nucleic acid sequence for the gene resulting in the substitution of an amino acid once the protein has been translated. Of course, it is paramount to ensure that the desired mutation has been successfully incorporated into the gene by having a sequence analysis of the DNA after the gene has been introduced into a cell via a vector. So, what inspired Smith to pursue his science that resulted with the SDM technique?

Smith (1993) documents his research journey from his beginnings in Khorana's lab to the SDM technique in substantial detail in his Nobel Lecture, which is a suggested reading for those interested. Only few highlights of his journey are mentioned here. Undoubtedly, Smith had a keen interest in small molecule bio-chemistry which he honed while in Khorana's lab that he carried during his time in science.

Smith suggested to Caroline Astell, a PhD student, to examine the specificity and stability of oligonucleotide duplexes of defined sequence using chemically engineered oligonucleotides. His idea was to determine whether '. . . a synthetically accessible oligonucleotide could be a tool for identifying and isolating a specific messenger RNA based on Watson-Crick hydrogen-bonding.' (Smith 1993). Smith and colleagues successfully established that deoxyribo-oligonucleotide-celluloses, using simple oligonucleotides, in the form of thermally-eluted columns, were able to determine the stability of duplexes with a number of complementary, or partially deoxyribo- and ribo-oligonucleotides (Astell and Smith 1971; 1972; Astell et al. 1973; Gilliam et al. 1975). Those experiments convinced Smith that '. . .it was possible to use a synthetic oligonucleotide to identify an RNA or a DNA containing the exact Watson-Crick complement of its sequence and to differentiate it from similar but not identical sequences.' (Smith 1993). Smith mentioned that '. . .at a later date, the data made me realize that it should be possible to use short synthetic deoxyribo-oligonucleotides as specific mutagens and to differentiate between a point mutant and wild-type DNA.' So, was this the initial idea of SDM in its infancy?

In late 1975, Smith visited Cambridge to spend a year with Fredrick Sanger (who was awarded the Nobel Prize in Chemistry for 1958 for 'his work on the structure for proteins, especially insulin') so that he could learn more about DNA sequencing (Smith 1993; Astell 2001). Smith joined Sanger and colleagues who were, at the time, pursuing the DNA sequence of the *Escherichia coli* phage ØX174. In fact, the entire sequence for ØX174 was determined by Sanger et al. (1977). Sanger shared the Nobel for Chemistry in 1980 with Paul Berg and Walter Gilbert. Berg's work was honored 'for his fundamental studies of biochemistry of nucleic acids, with particular regard to recombinant-DNA' while Sanger and Gilbert's work were honored 'for their contributions concerning the determination of base sequences in nucleic acids'.

With unique timing for Smith, Clyde Hutchinson III was also visiting Sanger. As Astell (2001) writes, 'Clyde had studied the molecular biology of ØX174 and developed a technique to recover mutant and wild-type genomes by transfecting fragments of about 50 nucleotides of phage DNA along with the phage genome into *Escherichia coli* cells. Recombination within the cells resulted in the recovery of mutant or wild-type phage- a type of marker rescue'. So, with Hutchinson's background with DNA phage ØX174 biology, Smith's background with shorter deoxyribooligonucleotides, and the work from his own laboratory by Gilliam (Gilliam et al. 1975), Smith believed that using short deoxyribooligonucleotides, it might be possible to create specific mutations within the ØX174 genome (Astell 2001). In fact, in his Nobel Lecture, Smith stated that based on the work with Gilliam et al. (1975) '...that oligonucleotide-directed mutagenesis should be possible' (Smith 1993). Indeed, a collaboration emerged between Smith and Hutchinson while at Sanger's lab, which resulted in their seminal paper entitled 'Mutagenesis at a specific position in a DNA sequence' with Hutchinson as the first author and Smith as the last author (Hutchinson et al. 1978). The paper was published in the *Journal of Biochemical Chemistry*. Ironically, Astell (2001) stated that the manuscript had been originally submitted to *Cell*, but the '...editors of the journal decided the paper was not of sufficient interest to warrant publication there!' Mistakes result in good fortune for others.

For further information about Smith and Hutchinson's research experiences during their time in Sanger's laboratory, please refer to Smith (1993) and Astell (2001). For additional papers from the Smith laboratory during the developmental stages and the refinement period of oligonucleotide-based, site-directed mutagenesis, see Gilliam et al. (1974); Gilliam and Smith (1974); Gilliam et al. (1975); Gilliam et al. (1979); Gilliam and Smith (1979a); Gilliam and Smith (1979b); and Zoller and Smith (1982, 1984).

Further information about oligonucleotide-based, site-directed mutagenesis technique, and the further development of in vitro mutagenesis, readers are encouraged to see Smith (1985); Trower (1996); Bachman (2013); and Reeves (2017).

1.9.1.2 Polymerase Chain Reaction

As stated, the advances in technology has ushered science to new heights even as research questions delve 'down deeper' to the molecular and sub-molecular levels. The concept of PCR is quite simple, which is to have a means to accurately and rapidly amplify a specific segment of DNA. With the PCR technique, one can amplify a significant amount of specific DNA in a matter of hours which generally required weeks/months of work with the traditional cloning techniques at that time (Arnheim and Erlich 1992). Ideally, this technique requires a DNA polymerase, initially the Klenow fragment of *Escherichia coli* (Saiki et al. 1985) and later with Taq polymerase from *Thermus aquaticus* a heat resistant DNA polymerase (Saiki et al. 1988), and two short oligonucleotide primers that are complementary to the $3'$ ends of each the antisense and sense strands of the DNA 'template', along with deoxyribonucleotide triphosphates (dNTPs). Generally, there are a series of thermal cycles in which the first step is the 'denaturation' to break the bonds between the complementary bases of the DNA template to yield two single-stranded DNA molecules. Next is the 'annealing' step where the two oligonucleotide primers anneal to their complementary sequences so that their $3'$ hydroxyl ends face the template DNA. The final step is 'extension/elongation' where the DNA polymerase along with free dNTPs are provided to the reaction mixture and the DNA polymerase synthesizes the 'new' DNA strands complementary to the DNA template, and this happens from the $5'$-to-$3'$ direction (Arnheim and Erlich 1992; Ishmael and Stellato 2008). This three-step cycle can be repeated many times, and with each cycle the original template strand of DNA and newly formed strands serve as template strands for the next elongation cycle for the chain reaction. Therefore, within about 30 cycles, one can generate 1×10^9 copies of the original double-stranded DNA template region. Not surprisingly, PCR has revolutionized molecular biology, and now, PCR is a routine technique in research and clinical laboratories throughout the world.

There has been considerable debate about the scientific history of the development of PCR technology (Rabinow 1996). Nonetheless, Mullis was the key person to appreciate the 'full potential' of the technique by his work in the early 1980s and he was awarded the Nobel Prize. Interestingly, less than 20 years after Watson and Crick's paper, scientists began to explore the possibilities of replicating short sequences of DNA with primers in vitro. Khorana was already contemplating the generation of short sequences of DNA. He and colleagues studied 'repair replication' of short synthetic DNA as catalyzed by DNA polymerases (Kleppe et al. 1971). Shortly thereafter, Khorana and Panet (Panet and Khorana 1974) conducted experiments to link deoxyribopolynucleotide templates to cellulose. They proposed that 'if the synthetic deoxyribopolynucleotide chains, which were to serve as the templates, could be linked in the proper direction to an insoluble support, then replication and transcription both could be initiated at the proper site by providing suitable deoxy- or ribo-oligonucleotides as primers. After completion of one round of polymerization, the DNA or the RNA product could be isolated simply by denaturing the duplex and separating the cellulose-bound template from the supernatant solution. The cycle of replication or transcription could, in principle, be repeated many times.' Khorana appeared to have been ahead of the 'Pre-PCR gang' of scientists pursuing the ability to amplify DNA.

For additional information about various aspects of the PCR technique, readers are directed to see Mullis et al. (1986), Mullis (1990), Mullis and Faloona (1987), Arnheim and Erlich (1992), Mullis et al. (1994), Rabinow (1996), Mullis (1998), and Ishmael and Stellato (2008).

1.9.2 Fluctuation (Noise Analysis) Analyses and Epithelia Ion Channels

Stevens (1977) suggested that the concept of 'fluctuation analysis' was a way 'to exploit the inherently probabilistic nature of the processes that underlie the membrane permeability changes in order to gain information about the molecular mechanisms responsible.' That was written 25 years after Hodgkin and Huxley published their seminal studies on the nerve impulse that was defined by the permeability of the ions of the nerve cell in context of the action potential of the giant squid axon in 1952 (Hodgkin and Huxley 1952a–d; Hodgkin et al. 1952). The idea of 'fluctuations' in excitability was first noted in nerve physiological data and the concept dates to the early 1930s (Blair and Erlanger 1932). The concept of 'fluctuations' progressed and was confirmed by Fatt and Katz (1952) and Hagiwara (1954) (Verveen and Derksen 1965). To use fluctuation (noise) analysis, it requires the basic understanding that the opening and closing of ion channels are stochastic (probabilistic) events that generate a 'non-stationary noise' in the recorded current. Noise analysis relates microscopic components of the total ionic current to microscopic parameters that include the single-channel current 'i', the number of functional channels in the membrane 'N', and the open probability that the channels are open under an experimental condition, 'P_{open}' (Heinemann and Conti 1992). Therefore, the total current (I) can be calculated as $I = NiP_O$. There are assumptions that are made when analyzing experimental data which include: (1) that there are independent channels (N) which are identical within the membrane; (2) the channels only have two conductance states—either 'open' or 'closed', (3) that the channel obeys binomial statistics; and (4) that the graded changes in current are because of graded changes in open probability (Jackson and Strange 1996).

Fluctuation analysis has flourished in the field of nerve physiology, and it is not surprising that this technique was quickly included as a research technique to analyze ion channel data in epithelial transport physiology. Lindemann (1980) emphasized that the spontaneous fluctuations of membrane current and membrane voltage can offer information about transport mechanisms, and indeed, 'mechanisms of control'. Certainly, the knowledge of ion transport protein kinetics and conformational configurations of proteins (ion channels) can provide in-depth understanding of overall epithelial function. The incorporation of noise analysis into the realm of analysis of single-channel data obtained with patch-clamp technique (recordings) broadens the information gained about the functioning of membrane ion channels, for instance. Lindemann (1980) related 'noise' and ion movement through a channel in the context that if there is deviation from the electrochemical equilibrium, there will be an increased microscopic current 'i' and that the 'shot-noise' will be more easily detected and analyzed. Additionally, he suggested '...that the valve-effect of spontaneously switching channels acts like a molecular amplifier, a rather fortuitous property which has, of course, promoted the applications of fluctuation analysis to ion transporting systems.' This seems appropriate for the channel that gates at a high frequency, but what about a channel that exhibits long open and closed times like the epithelial sodium channel? The introduction of a reversibly inhibiting channel blockers is a way to 'amplify' the 'noise' of a membrane transport protein (Lindemann 1980). A case in point was amiloride that Lindemann and van Driessche (1977) used to determine the 'Na turnover of individual transport sites' by an evaluation of current fluctuations of the frog skin. They demonstrated that the power density spectrum revealed that the transport of Na^+ was more than 10^6 ion/s; suggesting a 'pore' mechanism, of course, now known to be the epithelial Na^+ channel.

For additional information about the historical details of fluctuation analysis and the use of this technique when applied to epithelia, readers are directed to reviews by Lindemann (1980, 1984) and by van Driessche (van Driessche and Lindemann 1978; van Driessche and Gögelein 1980; van Driessche and Zeiske 1985).

1.10   Human Genome Project and the Physiology and Pathophysiology of Epithelia

The investigation of the function of epithelial tissues has occurred for at least 200 years. However, our understanding of the link between normal epithelial function and pathophysiology of diseases of epithelial tissues has really only blossomed in the past 30+ years as a result of the Human Genome Project (HGP) and the sequencing of the entire human genome.

During 1985–86, the forethought of a few scientists such as Charles DeLisi, David Smith, Robert Sinsheimer, and Mark Bitensky lead scientists into the 'Big Science' era. In 1985, Sinshiemer, the then Chancellor of University of California at Santa Cruz (UCSC), organized a workshop, held at the UCSC, in which he invited notable scientists to discuss the possibilities of sequencing the human genome (Sinsheimer 1989). Shortly thereafter, the 1986 Santa Fe Workshop was organized by Bitensky (at that time he was the head of life sciences at the Los Alamos National Laboratory in New Mexico) at the urging of DeLisi and Smith, both of the Department of Energy's Office of Health and Environmental Research, to discuss the concept of sequencing the human genome (DeLisi 1988, 2008). At the same time,

Renato Dulbecco (who shared the Nobel Laureate in Physiology or Medicine in 1975 with David Baltimore and Howard Temin 'for their discoveries concerning the interaction between tumour viruses and the genetic material of the cell') had also proposed the sequencing the human genome (Dulbecco 1986). Ultimately, in 1990, the HGP was launched and funded by the United States Department of Energy and the National Institutes of Health. The HGP (1990–2003) was a major international scientific effort with the implicit goal to determine the nucleotide sequence of all the base pairs that compose the human DNA. Additionally, the work of the members of the HGP successfully identified and mapped all the genes of the entire human genome (DeLisi 2001). Prior to the beginning of the HGP, Sinsheimer summarized his thoughts of sequencing the human genome by stating 'This achievement would be a landmark in human biology and the knowledge would be the basis of all human biology and medicine in the future' (Sinsheimer 1989).

With the information generated by the HGP, numerous ion channels and ion transporters have been described and pathophysiological roles have been explored. Subsequently, the physiological roles of many epithelial ion channels and transporters responsible for transepithelial ion/solute transport have been elucidated. Now there are a number of classification schemes of ion channels and ion transporters which categorized these transport proteins into specific 'families' based on protein structure and function. Below is a list of databases for ion channels and ion transporters.

http://www.guidetopharmacology.org/GRAC/FamilyDisplayForward?
    familyId=863
http://www.tcdb.org/
https://www.genenames.org/data/genegroup/#!/group/752
https://www.omim.org/
https://www.genenames.org/data/genegroup/#!/group/177
http://www.guidetopharmacology.org/GRAC/ReceptorFamiliesForward?type=IC

Since the initiation of the HGP, the pathophysiology of many diseases of epithelia caused by specific mutations of ion channels and ion transporters have been described. Specific information can be found in numerous gene disease databases, such as Online Mendelian Inheritance in Man (https://www.omim.org/) and The Human Gene Mutation Database (http://www.hgmd.cf.ac.uk/ac/index.php).

1.10.1 Impact of the HGP on Epithelial Diseases

One way to gauge the impact of the efforts of the HGP has had on science is to search PubMed for key words such as 'epithelial diseases'. Using those words, there have been 106,922 articles published as of 8/3/2020. In 1990, only 1,029 papers were published that year, although 7,668 articles were published during all of 2019. Space limits the number of epithelial diseases for which the pathophysiology has been described over the years. These include various renal tubulopathies, including

Bartter's syndrome (Kleta and Bockenhauer 2006; Welling and Ho 2009), Gitelman's syndrome (Graziana et al. 2010; Miller 2011), EAST syndrome (for infant Epilepsy, severe Ataxia, moderate Sensorineural deafness and renal salt wasting Tubulopathy) (Bockenhauer et al. 2009; Bandulik et al. 2011), SeSAME syndrome (for Seizures, Sensorineural deafness, Ataxia, Mental retardation and Electrolyte imbalance) (Scholl et al. 2009; Reichold et al. 2010; Williams et al. 2010), Liddle's disease (Shimkets et al. 1994; Bubien et al. 1996), nephrogenic diabetes insipidus (Bichet 2008; Loonen et al. 2008), and renal tubular acidosis syndrome (Warth et al. 2004). Other epithelial diseases include cystic fibrosis (Tsui 1992), Glucose-Galactose Malabsorption (Lam et al. 1999), and Dent's disease (Pook et al. 1993; Devuyst and Thakker 2010), for example. No doubt, without the huge efforts and progress made by all the international scientists and the funding agencies supporting the HGP, we would certainly not have described the pathophysiological basis of many diseases of epithelial tissues. These and more diseases of epithelial tissues will be examined in detail in many chapters of the three volumes of this book series.

1.11   Conclusions

It has been an exceptionally long journey for the field of epithelial transport physiology and this journey endures. Experimental methods and techniques continue to be developed as epithelial ion transport research evolves and scientists ask challenging and intriguing research questions. This chapter could have been greatly expanded to cover other techniques commonly used in epithelial ion transport physiology such as (1) patch-clamp technology, (2) organoid physiology (see Chapter 1, Volume 2), (3) ion-selective microelectrodes, (4) pHi-stat, (5) atomic force microscopy, and (6) oocyte two-electrode voltage clamp, for example. Some of these techniques may be briefly discussed in various chapters of this volume and/or the other two volumes of this book series. Nonetheless, it was the intent of this chapter to provide a rockbed of information of several research techniques which will complement the experimental evidence that is presented in numerous chapters of this book series.

Acknowledgements This chapter is dedicated to the memory of Jared J. Grantham, MD (1936–2017) for his contributions to the isolated perfused tubule preparation and to his lifelong research work on polycystic kidney disease. Thanks to Chris Cheeseman for his insightful memories of his interactions with Wiseman. Many thanks to Moe Burg for his comments and his friendship over the years. A special thanks to Mark Knepper for the photo of his isolated perfused tubule rig (Fig. 1.8) and the discussions of the isolated perfused tubule technique. Many thanks to Erik Larsen who provided impeccable historical insight and details on various aspects of Sect. 1.4 on Hans Ussing and other sections of this chapter. Thanks to Dee Silverthorn for her enlightening comments and editing of an earlier version of this chapter and the information about the transporter databases provided in Sect. 1.10. Additionally, I thank Dee for her 'kick starting' the now two editions of this APS/Springer-Verlag (now APS/Springer-Nature) book series back in 2013. Thanks to John

Hughes for generating the digital images for Figs. 1.9, 1.10, and 1.11. Thanks to the Department of Physiology and the School of Biomedical Sciences at the University of Otago for continued research support for many years.

References

Acra SA, Ghishan FK (1991) Methods of investigation intestinal transport. JPEN J Parenter Enteral Nutr 15:93S–98S

Alam MA, Al-Jenoobi FI, Al-mohizea AM (2011) Everted gut sac model as a tool in pharmaceutical research: limitations and applications. J Pharmac Pharmacol 64:326–336

Anderson RF, Diffenbaugh MD, Schmidtke WH (1962) Isolated intestinal loop. Arch Surg. 84:559–563

Andolfo I, Russo R, Manna F, Shmukler BF, Gambale A, Vitiello G, De Rosa G, Brugnara C, Alper SL, Snyder LM, Iolascon A (2015) Novel Gardos channel mutations linked to dehydrated hereditary stomatocytosis (xerocytosis). Am J Hematol 90:921–926

Angel A. (2001) Gerald Wiseman MB BS MD PhD—Obituary. Physiol Soc Newsl, Autumn, p 31

Arnheim N, Erlich H (1992) Polymerase chain reaction strategy. Annu Rev Biochem 61:131–156

Ashworth SL, Sandoval RM, Tanner GA, Molitoris BA (2007) Two-photon microscopy: visualization of kidney dynamics. Kidney Int 72:416–421

Astell CR (2001) Michael Smith 26 April 1932 – 4 October 2000. Biogr Mem Fellow R Soc 47:429–441

Astell C, Smith M (1971) Thermal elution of complementary sequences of nucleic acids from cellulose columns with covalently attached oligonucleotides of known length and sequence. J Biol Chem 246:1944–1946

Astell CR, Smith M (1972) Synthesis and properties of oligonucleotide-cellulose columns. Biochemistry 11:4114–4120

Astell CR, Doel MT, Jahnke PA, Smith M (1973) Further studies on the properties of oligonucleotide cellulose columns. Biochemistry 12:5068–5078

Bachman J (2013) Site-directed mutagenesis. Methods Enzymol 529:241–248

Bandulik S, Schmidt K, Bockenhauer D, Zbelik AA, Humberg E, Kleta R, Warth R, Reichold M (2011) The salt-wasting phenotype of EAST syndrome, a disease with multifaceted symptoms lined to the KCNJ10 K^+ channel. Pflügers Arch 461:423–435

Barker Jørgensen C, Levi H, Ussing HH (1946) On the influence of the neurohypophysial principles on the sodium metabolism in the axolotl (*Ambystoma mexicanum*). Acta Physiol Scand 12:350–371

Beck JC, Sacktor B (1978) The sodium electrochemical potential-mediated uphill transport of D-glucose in renal brush border membrane vesicles. J Biol Chem 253:5531–5535

Becquerel AH (1903) On radioactivity, a new property of matter. Nobel Prize in Physics 1903 Lecture. www.nobelprize.org/nobel_prizes/physics/laureates/1903/becquerel-lecture.html

Bichet DG (2008) Vasopressin receptor mutations in nephrogenic diabetes insipidus. Sem Nephrol 28:245–251

Blair EA, Erlanger J (1932) Responses of axons through their individual electrical responses. Proc Soc Exp Biol (NY) 29:926–927

Bockenhauer D, Feather S, Standescu HC, Bandulik S, Zdebik AA, Reichold M, Tobin J, Lieberer E, Sterner C, Landoure G, Arora R, Sirimanna T, Thompson D, Cross JH, van't Hoff W, Al Masri O, Tullus K, Yeung S, Anikster Y, Klootwijk E, Hubank M, Dillion MJ, Heitzmann D, Arcos-Burgos M, Knepper MA, Dobbie A, Gahl WA, Warth R, Sheridan E, Kleta R (2009) Epilepsy, ataxia, sensorineural deafness, tubulopathy and *KCNJ10* mutations. N Engl J Med 360:1960–1970

Brenner BM, Troy JL, Daugharty TM (1971) The dynamics of glomerular ultrafiltration in the rat. J Clin Invest 50:1776–1780

Brenner BM, Troy JL, Daugharty TM, Deen WM, Robertson CR (1972) Dynamics of glomerular ultrafiltration in the rat. II. Plasma-flow dependence of GFR. Am J Physiol 223:1184–1190

Bubien JK, Ismailov II, Berdiev BK, Cornwell T, Lifton RP, Fuller CM, Achard J-M, Benos DJ, Warnock DG (1996) Liddle's disease: abnormal regulation of amiloride-sensitive Na$^+$ channel by β-subunit mutation. Am J Physiol Cell Physiol 270:C208–C213

Burg MB (1972) Perfusion of isolated renal tubules. Yale J Biol Med 45:321–326

Burg MB (1982) Introduction: background and development of microperfusion technique. Kidney Int 22:414–424

Burg MB (2018) Autobiography of Maurice (Moe) B. Burg. CreateSpace Independent Publishing Platform, 124 pp. ISBN: 978-1982036317

Burg MB, Knepper MA (1986) Single tubule perfusion techniques. Kidney Int 30:166–170

Burg M, Grantham JJ, Abramow M, Orloff J (1966) Preparation and study of fragments of single rabbit nephrons. Am J Physiol 210:1293–1298

Burg M, Grantham JJ, Abramow M, Orloff J, Schafer JA (1997) Preparation and study of fragments of single rabbit nephrons. J Am Soc Neophrol 8:675–683 (reprint of the 1966 paper including commentaries by MB Burg and JA Schafer)

Burris RH (1950) Isotopes as tracers in plants. Botanical Rev 16:150–180

Christie DA, Tansey EM. (2000) Intestinal absorption. Wellcome Witness to Twentieth Century Medicine. Wellcome Trust, London, 81 pp

Clarke LL (2009) A guide to Ussing chamber studies of mouse intestine. Am J Physiol Gastrointest Liver Physiol 296:G1151–G1166

Cockcroft JD (1967) George De Hevesy, 1885-1966. Biogr Mems Fell R Soc 13:125–166

Cori CF (1925) The fate of sugar in the animal body. I. The rate of absorption of hexoses and pentoses from the intestinal tract. J Biol Chem 66:691–715

Crane RK (1965) Na$^+$-dependent transport in the intestine and other animal tissues. Fed Proc 24:1000–1006

Creager AN (2013) Life atomic: a history of radioisotopes in science and medicine. University of Chicago Press, Chicago, IL, 512 pp

Curie P (1905) Radioactive substances, especially radium. Nobel Prize in Physics 1903 Lecture presented in March 1905. www.nobelprize.org/nobel_prizes/physics/laureates/1903/pierre-curie-lecture.html

Curie M (1911) Radium and the new concepts in chemistry. Nobel Prize in Chemistry—1911 Lecture. www.nobelprize.org/nobel_prizes/chemistry/laureates/1911/marie-curie-lecture.html

Darlington WA, Quastel JH (1953) Absorption of sugars from isolated surviving intestine. Arch Biochem Biophys 43:194–207

de Hevesy G (1944) Some applications of isotopic indicators. Nobel Prize in Chemistry—1943 Lecture (awarded in 1944). www.nobelprize.org/nobel_prizes/chemistry/laureates/1943/hevesy-lecture.html

DeLisi C (1988) The human genome project. Am Sci 76:488–493

DeLisi C (2001) Genomes: 15 years—later a perspective by Charles DeLisi. Hum Genome News. 11:3–4

DeLisi C (2008) Sante Fe 1986: human genome baby-steps. Nature 455:876–877

Devuyst O, Thakker RV (2010) Dent's disease. Orphaner J Rare Dis 5:28. https://doi.org/10.1186/1750-1172-5-28

DuBois Raymond E (1848) Untersuchungen über tierisches elektrizität, Berlin

Dulbecco R (1986) Turning point in cancer research, sequencing the human genome. Science 231:1055–1056

Dutrochet H (1826) L'agent immédiat du mouvement vital dévoilé dans sa nature et dans son mode d'action, chez les vegétaux et chez les animaux. Paris (quoted by Reid EW, 1890)

Fatt P, Katz B (1952) Spontaneous subthreshold activity at motor nerve endings. J Physiol 117:109–128

Fisher RB, Parsons DS (1949) A preparation of surviving rat small intestine for the study of absorption. J Physiol 110:36–46

Forstner GG, Sabesin SM, Isselbacher KJ (1968) Rat intestinal microvillus membranes purification and biochemical characterization. Biochem J 106:381–390

Francis WL (1933) Output of electrical energy by frog-skin. Nature 131:805

Francis W, Pumphrey R (1933) The electrical properties of frog skin. Part I. Introductory. J Exp Biol 10:379–385

Gasko OD, Knowles AF, Shertzer HG, Suolinna EM, Racker E (1976) The use of ion-exchange resins for studying transport in biological systems. Ana Biochem 72:57–65

Gilliam S, Smith M (1974) enzymatic synthesis of deoxyribo-oligonucleotides of defined sequence. Properties of the enzyme. Nucleic Acids Res 1:1631–1648

Gilliam S, Smith M (1979a) Site-specific mutagenesis using synthetic oligodeoxyribonucleotide primers: I. Optimum conditions and minimum oligodeoxyribonucleotide length. Gene 8:81–97

Gilliam S, Smith M (1979b) Site-specific mutagenesis using synthetic oligodeoxyribonucleotide primers: II. In vitro selection of mutant DNA. Gene 8:99–106

Gilliam S, Waterman K, Doel M, Smith M (1974) Enzymatic synthesis of deoxyribo-oligonucleotides of defined sequence. Deoxyribo-oligonucleotide synthesis. Nucleic Acids Res 1:1649–1664

Gilliam S, Waterman K, Smith M (1975) The pair-pairing specificity of cellulose-pdT$_9$. Nucleic Acids Res 2:625–634

Gilliam S, Jahnke P, Astell C, Phillips S, Hutchinson CA III, Smith M (1979) Defined transversion mutations at a specific position in DNA using synthetic oligodeoxyribonucleotides as mutagens. Nucleic Acids Res 6(2):973–2985

Glogowska E, Lezon-Geyda K, Maksimova Y, Schulz VP, Gallagher PG (2015) Mutations in the Gardos channel (KCNN4) are associated with hereditary xerocytosis. Blood. 126:1281–1284

Gottschalk CW, Lassiter WE (1973) Micropuncture methodology. In: Orloff J, Berliner RW (eds) Handbook of physiology. Sect. 8: Renal physiology. American Physiological Society, Washington, DC, pp 129–143

Gottschalk CW, Mylle M (1959) Micropuncture study of the mammalian urinary concentrating mechanism: evidence for the countercurrent hypothesis. Am J Physiol 196:927–936

Grantham JJ (2014) Why I think about urine—and a treatment for polycystic kidney disease. Rockhill Books, Kansas City, MO, 280 pp. ISBN: 9781611691283

Grantham JJ, Qualizza PB, Irwin RL (1974) Net fluid secretion in proximal straight renal tubules in vitro: role of PAH. Am J Physiol 226:191–197

Grantham JJ, Irish JM III, Hall DA (1978) Studies of isolated renal tubules in vitro. Ann Rev Physiol 40:249–277

Graziana G, Fedell C, Moroni L, Cosmai L, Badlamenti S, Ponticelli C (2010) Gitelman syndrome: pathophysiological and clinical aspects. Q J Med 103:741–748

Haase W, Schäfer A, Murer H, Kinne R (1978) Studies on the orientation of brush-border membrane vesicles. Biochem J 172:57–62

Hagiwara S (1954) Analysis of interval fluctuation of the sensory nerve impulse. Jap J Physiol 4:234–240

Hahn LA, Hevesy GC, Lundsgaard EC (1937) The circulation of phosphorus in the body revealed by application of radioactive phosphorus as indicator. Biochem J 31:1705–1709

Hamilton KL (2011) Ussing's 'little chamber': 60 years+ old and counting. Front Physiol 2:6. https://doi.org/10.3389/fphys.2011.00006

Hamilton KL (2014) Even an old technique is suitable in the modern molecular world of Science: the everted sac preparation—turns 60 years old. Am J Physiol Cell Physiol 306:C715–C720

Hamilton KL, Butt AG (2013) Glucose transport into everted sacs of the small intestine of mice. Adv Physiol Educ 37:415–426

Hamilton KL, Moore AB (2016) 50 years of renal physiology from one man and the perfused tubule: Maurice B. Burg. Am J Physiol Renal Physiol 311:F291–F304

Heinemann SH, Conti F (1992) Nonstationary noise analysis and application to patch clamp recordings. Methods Enzymol 207:131–148

Hermann JR, Turner JR (2016) Beyond Ussing's chamber: contemporary thoughts on integration of transepithelial transport. Am J Physiol Cell Physiol 310:C424–C431

Hevesy G (1923) LIII. The absorption and translocation of lead by plants. A contribution to the application of the method of radioactive indicators in the investigation of the change of substance in plants. Biochem J 17:435–445

Hevesy G (1940) Application of radioactive indicators in biology. Ann Rev Biochem 9:641–662

Hevesy GC (1961) Marie Curie and her contemporaries: the Becquerel-Curie memorial lecture. J Nucl Med 25:116–131

Hevesy G (1962) Adventures in radioisotope research, vol I. Pergamon, New York

Hevesy G, Hofer E (1934) Elimination of water from the human body. Nature 135:879

Hevesy GV, Hofer E, Krogh A (1935) The permeability of the skin of frogs to water as determined by D_2O and H_2O. Skand Arch Physiol (now Acta Physiol Scand) 72: 199–214 Skandinavisches Archiv Für Physiologie

Hodgkin AL, Huxley AF (1952a) Currents carried by sodium and potassium ions through the membrane of the giant axon of *Loligo*. J Physiol 116:449–472

Hodgkin AL, Huxley AF (1952b) The components of the membrane conductance in the giant axon of *Loligo*. J Physiol 116:473–496

Hodgkin AL, Huxley AF (1952c) The dual effects of membrane potential on sodium conductance in the giant axon of *Loligo*. J Physiol 116:497–506

Hodgkin AL, Huxley AF (1952d) A quantitative description of membrane current and its application to conduction and excitation in nerve. J Physiol 117:500–544

Hodgkin AL, Huxley AF, Katz B (1952) Measurement of current-voltage relations in the membrane of the giant axon of *Liligo*. J Physiol 116:424–448

Hopfer U, Nelson K, Perrotto J, Isselbacher KJ (1973) Glucose transport in isolated brush border membrane from rat small intestine. J Biol Chem 248:25–32

Hutchinson CA III, Phillips S, Edge MH, Gilliam S, Jahnke P, Smith M (1978) Mutagenesis at a specific position in a DNA sequence. J Biol Chem 253:6551–6560

Ishmael FT, Stellato C (2008) Principles and applications of polymerase chain reaction: basic science for the practicing physician. Ann Allergy Asthma Immunol 101:437–443

ISN (2005) International Society of Nephrology Video Legacy Project. Dr. Maurice Burg interviewed by Dr. Mark Knepper. https://cybernephrology/ualberta.ca/ISN/LVP/Trans/burg.htm

Jackson PS, Strange K (1996) Single channel properties of a volume sensitive anion channel: lessons from noise analysis. Kidney Int 49:1695–1699

Jacobson HR, Kokko JP (1982) Initial remarks and acknowledgements: isolated perfused tubule symposium. Kidney Int 22:415–416

Jamison RL (1970) Micropuncture study of superficial and juxtamedullary nephrons in the rat. Am J Physiol 218:46–55

Jimenez-Rosales A, Flores-Merino MV (2018) Tailoring proteins to re-evolve nature: a short review. Mol Biotech 60:946–974

Keele KD, Roberts J (1983) Leonardo Di Vinci: anatomical drawings from the Royal Library Windsor Castle. The Metropolitan Museum of Art, New York, 167 pp. ISBN: 0-87099-362-3

Kessler M, Tannenbaum V, Tannenbaum C (1978) A simple apparatus for performing short time (1-2 seconds) uptake measurements in small volume; its application to D-glucose transport studies in brush border vesicles from rabbit jejunum and ileum. Biochim Biophys Acta 509:348–359

Kinne R, Kinne-Saffran E (1981) Membrane vesicles as tools to elucidate epithelial cell function. Eur J Cell Biol 25:346–352

Kleppe K, Ohtsuka E, Kleppe R, Molineux I, Khorana HG (1971) Studies of polynucleotides. XCVI. Repair replication of short synthetic DNA's as catalysed by DNA polymerases. J Mol Biol 56:341–361

Kleta R, Bockenhauer D (2006) Bartter syndromes and other salt-losing tubulopathies. Nephron Physiol 104:73–80

Koefoed-Johnsen V, Ussing HH (1958) The nature of the frog skin potential. Acta Physiol Scand 42:298–308

Krogh A (1937a) The use of isotopes as indicators in biological research. Science 85:187–191. https://doi.org/10.1126/science.85.2199.187

Krogh A (1937b) Osmotic regulation in the frog (*Rana esculenta*) by active absorption of chloride ion. Skand Arch Physiol (now Acta Physiol Scand) 76:60–74

Lam JT, Martin MG, Turk E, Hirayama BA, Bosshard NU, Steinmann B, Wright EM (1999) Missense mutations in SGLT1 cause glucose-galactose malabsorption by trafficking defects. Biochim Biophys Acta 1453:297–303

Lang F, Greger R, Lechene C, Knox FG (1978) Micropuncture techniques. In: Renal pharmacology, Martinez-Maldonado M.. Chap. 4, p. 75–103, New York: Plenum

Larsen EH (2002) Hans H. Ussing—scientific work: contemporary significance and perspectives. Biochem Biophys Acta 1566:2–15

Larsen EH (2009) Hans Henriksen Ussing. 30 December 1911 – 22 December 2000. Biog Mems Fell R Soc 55:305–300

Lee JW, Chou C-L, Knepper MA (2015) Deep sequencing in microdissected renal tubules identifies nephron segment-specific transcriptomes. J Am Soc Nephrol 26:2669–2677

Lennerås H (2007) Animal data: the contributions of the Ussing chamber and perfusion systems to predicting human oral drug delivery *in vivo*. Adv Drug Deliv Rev 59:1103–1120

Lentle B, Aldrich J (1997) Radiological sciences, past and present. Lancet 350:280–285

Levi H (1985) George de Hevesy—life and work: a biography. Rhodos, Copenhagen. George de Hevesy—Life and work. A biography by Hilde Levi. Rhodos, Copenhagen, 147 pp. ISBN: 87 7245 0541

Levi H, Ussing HH (1948) The exchange of sodium and chloride ions across the fibre membrane of the isolated frog sartorius. Acta Physiol Scand 16:232–249

Levi H, Ussing HH (1949) Resting potential and ion movements in the frog skin. Nature 164:928–930

Lindemann B (1980) The beginning of fluctuation analysis of epithelial transport. J Membr Biol 54:1–11

Lindemann B (1984) Fluctuation analysis of sodium channels in epithelia. Annu Rev Physiol 46:497–515

Lindemann B (2001) Hans Ussing, experiments and models. J Membr Biol 184:203–210

Lindemann B, van Driessche W (1977) Sodium-specific membrane channels of frog skin are pores: current fluctuations reveal high turnover. Science 195:292–294

Liu Z, Liu K (2013) The transporters of intestinal tract and techniques applied to evaluate interactions between drugs and transporters. Asian J Pharmac Sci 8:151–158

Loonen AJN, Knoers NVAM, Van Os CH, Deen PMT (2008) Aquaporin 2 mutations in nephrogenic diabetes insipidus. Sem Nephrol 28:252–265

Lorenz JN (2012) Micropuncture of the kidney: a primer on techniques. Comp Physiol 2:621–637

Lund EJ (1926) the electrical polarity of *Obelia* and frog's skin and its reversible inhibition by cyanide, ether, and chloroform. J Exptl Zool 44:383–396

Lund EJ, Stapp P (1947) Use of the iodine coulometer in the measurement of bioelectrical energy and the efficiency of the bioelectrical process. In: Lund EJ (ed) Bioelectric fields and growth. University of Texas Press, Austin, TX, pp 235–280

Lyon GM (1949) Radioisotopes in biology and medicine. Am Biology Teacher 11:70–75

Matsushita K, Golgotiu K, Orton DJ, Smith RD, Rodland JD, Piehowski PD, Hutchens MP (2018) Micropuncture of Bowman's space in mice facilitated by 2 photon microscopy. J Vis Exp 140: e58206. https://doi.org/10.3791/58206

Matteucci MM, Lima A (1845) Mémoire sur l'endosmose. Annales de chimie et de physique XIII:63–86

Meyer KH, Bernfeld P (1946) The potentiometric analysis of membrane structure and its application to living animal membranes. J Gen Physiol 29:353–378

Miller RT (2011) Genetic disorders of NaCl transport in the distal convoluted tubule. Nephron Physiol 118:15–21

Miller TG, Abbott WO (1934) Intestinal intubation: a practical technique. Am J Med Sci 187:595–598

Miller D, Crane RK (1961) The digestive function of the epithelium of the small intestine. Biochim Biophys Acta 11:293–298

Mullis KB (1990) The unusual origin of the polymerase chain reaction. Sci Am 262(56–61):64–65

Mullis KB (1998) Dancing in the mind field. Pantheon, New York, 222pp. ISBN: 978-0-679-44255-4

Mullis KB, Faloona FA (1987) Specific synthesis of DNA *in vitro* via a polymerase-catalyzed chain reaction. Meth Enzymol 155:335–350

Mullis K, Faloona F, Sccharf S, Saiki R, Horn G, Erlich H (1986) Specific enzymatic amplification of DNA in vitro: the polymerase chain reaction. Cold Spring Harb Symp Quant Biol 51:263–273

Mullis KB, Ferré F, Gibbs RA (1994) The polymerase chain reaction. Springer, New York. ISBN: 978-0-8176-3750-7. https://doi.org/10.1007/978-1-4612-0257-8

Murer H, Gmaj P (1986) Transport studies in plasma membrane vesicles isolated from renal cortex. Kidney Int 30:171–186

Murer H, Kinne R (1980) The use of isolated membrane vesicles to study epithelial transport processes. J Membr Biol 55:81–95

Murer H, Hopfer U, Kinne R (1976) Sodium/proton antiport in brush-border-membrane vesicles isolated from rat small intestine and kidney. Biochem J 154:597–604

Mushegian S (2018) Simulations might speed drug target discovery. Asbmbtoday 17:10. https://www.asbmb.org/asbmb-today/science/050118/jbc-simulations-might-speed-drug-target-discovery

Myers WG (1976) Becquerel's discovery of radioactivity in 1896. J Nucl Med 17:579–572

Nakano D, Nishiyama A (2016) Multiphoton imaging of kidney pathophysiology. J Pharmacol Sci 132:1–5. https://doi.org/10.1016/j.jphs.2016.08.001

Niese S (2006) George de Hevesy (1885-1966), Founder of radioanalytical chemistry. Czechoslovak J Phys 5(suppl D):D3–D11

Nitske RW (1971) The life of Wilhem Conrad Röntgen, Discoverer of the X-Ray. University of Arizona Press, Tucson, AZ, 355pp. ISBN: 10:0816502595

Palmer LG, Andersen OS (2008) The two-membrane model of epithelial transport: Koefoed-Johnsen and Ussing (1958). J Gen Physiol 132:607–617

Panet A, Khorana HG (1974) Studies on polynucleotides. The linkage of deoxyribopolynucleotide templates to cellulose and its use in their replication. J Biol Chem 249:5213–5221

Peti-Peterdi J, Kidokoro K, Riquier-Brison A (2015) Novel *in vivo* techniques to visualize kidney anatomy and function. Kidney Int 88:44–51

Pook MA, Wrong O, Wooding C, Norden AGW, Feest TG, Thakker R (1993) Dent's disease, a renal Fanconi syndrome with nephrocalcinosis and kidney stones, is associated with a microdeletion involving DXS255 and maps to Xp11.22. Hum Mol Genet 2:2129–2134

Popják G (1948) The use of isotopes in biology. Sci Progr 36:239–261

Posner R (1970) Reception of Röntgen's discovery in Britain and U.S.A. Brit Med J 4:357–360

Rabinow P (1996) Making PCR: a story of biotechnology. University of Chicago Press, Chicago, IL, 190pp. ISBN: 0-226-70146-8

Rapetti-Mauss R, Lacoste C, Picard V, Guitton C, Lombard E, Loosveld M, Nivaggioni V, Dasilva N, Salgado D, Desvignes JP, Béroud C, Viout P, Bernard M, Soriani O, Lacroze V, Feneant-Thibault M, Thuret I, Guizouam H, Badens C (2015) A mutation in the Gardos channel is associated with hereditary xerocytosis. Blood 126:1273–1280

Reeves A (ed) (2017) In vitro mutagenesis methods and protocols. Humana, 511pp. Hardcover book ISBN: 978-1-4939-6470-3, eBook ISBN: 978-1-4939-6472-7

Reichold M, Zdebik AA, Lieberer E, Rapedius M, Schmidt K, Bandulik S, Sterner C, Tegtmeier I, Penton D, Baukrowitz T, Hulton SA, Witzgall R, Ben-Zeev B, Howie AJ, Kleta R, Bockenhauer D, Warth R (2010) KCNJ10 gene mutations causing EAST syndrome (epilepsy, ataxia, sensorineural deafness, and tubulopathy) disrupt channel function. Proc Natl Acd Sci USA 107:14490–14495

Reid EW (1890) Osmosis experiments with living and dead membranes. J Physiol 11(4–5):312–400.11

Reid EW (1892a) Report on experiments upon "absorption without osmosis". Br Med J 13:1 (1624):323–326

Reid EW (1892b) Preliminary report on experiments upon intestinal absorption without osmosis. Br Med J 1(1639):1133–1134

Reid EW (1901) Transport of fluid by certain epithelia. J Physiol 26:436–444

Richards AN, Walker AM (1936) Methods of collecting fluid from known regions of the renal tubules of Amphibia and of perfusing the lumen of a single tubule. Am J Physiol 118:111–120

Robertson CR, Deen WM, Troy JL, Brenner BM (1972) Dynamics of glomerular ultrafiltration in the rat. 3. Hemodynamics and autoregulation. Am J Physiol 223:1191–1200

Röntgen W (1895) Ueber eine Art von Strahlen. 'Vorläufige Mitteilung'. In: Aus den Sitzungsberichten der Würzburger Physik.-medic. Gesellschaft Würzburg, pp 137–147

Saiki RK, Scharf S, Faloona F, Mullis KB, Horn GT, Erlich H, Arnheim N (1985) Enzymatic amplification of β-globin genomic sequences and restriction site analysis for diagnosis of sickle cell anemia. Science 230:1350–1354

Saiki RK, Gelfand DH, Stoffel S, Scharf SJ, Higuchi R, Horn GT, Mullis KB, Erlich H (1988) Primer-directed enzymatic amplification of DNA with a thermostable DNA polymerase. Science 239:487–491

Sandoval RM, Molitoris BA (2017) Intravital multiphoton microscopy as a tool for studying renal physiology and pathophysiology. Methods 128:30–32

Sands JM (2004) Micropuncture: unlocking the secrets of renal function. Am J Physiol Renal Physiol 287:F866–F867

Sanford PA (1982) Digestive system physiology, Physiological principles in medicine series. Edward Arnold, London

Sanger F, Air GM, Barrell BG, Brown NL, Coulson AR, Fiddes CA, Hutchinson CA, Slocombe PM, Smith M (1977) Nucleotide sequence of bacteriophage phi X174 DNA. Nature 265:687–695

Sanjuán R (2010) Mutational fitness effects in RNA and single-stranded DNA viruses: common patterns revealed by site-directed mutagenesis studies. Philos Trans R Soc B 365:1975–1982

Schmitz J, Preiser H, Maestracci D, Ghosh BK, Cerda JJ, Crane RK (1973) Purification of the human intestinal brush border membrane. Biochim Biophys Acta 323:98–112

Scholl UI, Choi M, Liu T, Ramaekers VT, Hausler MG, Grimmer J, Tobe SW, Farhi A, Nelson-Williams C, Lifton RP (2009) Seizures, sensorineural deafness, ataxia, mental retardation, and electrolyte imbalance (SeSAME syndrome) caused by mutations in KCNJ10. Proc Natl Acad Sci USA 106:5842–5847

Schultz SG (1998) A century of (epithelial) transport physiology: from vitalism to molecular cloning. Am J Physiol Cell Physiol 274:C13–C23

Shimkets RA, Warnock DG, Bositis CM, Nelson-Williams C, Hansson JH, Schambelan M, Gill JR Jr, Ulick S, Milora RV, Findling JW, Canessa CM, Rossier BC, Lifton RP (1994) Liddle's syndrome: heritable human hypertension caused by mutations in the β subunit of the epithelial sodium channel. Cell 79:407–414

Sinsheimer R (1989) The Santa Cruz workshop, May 1985. Genomics 5:954–956

Smith M (1985) In vitro mutagenesis. Ann Rev Genet 19:423–462

Smith M (1993) Synthetic DNA and biology. https://www.nobelprize.org/prizes/chemistry/1993/smith/lecture/

Smith M, Khorana HG (1958) Nucleoside polyphosphates. VI. An improved and general method for the synthesis of ribo- and deoxyribonuleoside 5′ triphosphates. J Am Chem Soc 80:1141–1145

Smith M, Khorana HG (1959) Specific synthesis of the C_5'-C_3' inter-ribonucleotide linkage: the synthesis of uridylyl-(5'→3')-uridine. J Am Chem Soc 81:2911–2912

Smith M, Drummond GI, Khorana HG (1961) Cyclic phosphates. IV. Ribonucleoside-3',5' cyclic phosphate. A general method of synthesis and some properties. J Am Chem Soc 83:698–706

Smith M, Rammler DH, Goldberg IH, Khorana HG (1962) Studies on polynucleotides. XIV. Specific synthesis of the C_3'-C_5' Inter-ribonucleotide linkage: syntheses of uridylyl-(3'→5')-uridine and uridylyl-(3'→5')-adenosine. J Am Chem Soc 84:430–440

Sols A, Polz F (1947) A new method for the study of intestinal absorption. Rev Espan Fisiol 3:207–211

Stapp P (1941) Efficiency of electrical energy production by surviving frog skin, measured by iodine coulometer. Exp Biol Med (Maywood) 46:382–384

Stevens CF (1977) Study of membrane permeability changes by fluctuation analysis. Nature 270:391–396

Stevens BR, Kaunitz JD, Wright EM (1984) Intestinal transport of amino acids and sugars: advances using membrane vesicles. Ann Rev Physiol 46:417–433

Stockand DJ, Vallon V, Oritz P (2012) In vivo and ex vivo analysis of tubule function. Comp Physiol 2:2495–2525

Tanner GA, Sandoval RM, Molitoris BA, Bamburg JR, Ashworth SL (2005) Micropuncture gene delivery and intravital two-photon visualization of protein expression in rat kidney. Am J Physiol Renal Physiol 289:F638–F643

Thiry L (1864) Über eine neue Methode, den Dünndarm zu isolieren Sitz Akad Wien, Math-Natur KI I 50:77–96

Trower MK (ed) (1996) In vitro mutagenesis protocols. Methods in molecular biology 57. Humana, Mahwah, NJ, 408 pp. ISBN: 10:0896033325, eBook ISBN: 978-1-59259-544-0

Tsui L-C (1992) Mutation and sequence variations detected in the cystic fibrosis transmembrane conductance regulator (CFTR) gene: a report from the cystic fibrosis genetic analysis consortium. Hum Mutat 1:197–203

Urey HC (1925) The structure of the hydrogen molecule ion. Proc Natl Acad Sci USA 11:618–621

Urey HC (1933) The separation and properties of the isotopes of hydrogen. Science 78:566–571

Urey HC (1935) Some thermodynamic properties of hydrogen and deuterium. Nobel Prize in Chemistry 1934 Lecture. www.nobelprize.org/prizes/chemistry/1934/urey/lecture

Ussing HH (1938a) Use of amino acids containing deuterium to follow protein production in the organism. Nature 142:399

Ussing HH (1938b) The exchange of H and D atoms between water and protein in vivo and in vitro. Skand Arch Physiol 78:225–241

Ussing HH (1941) The rate of protein renewal in mice and rats studied by means of heavy hydrogen. Acta Physiol Scand 2:209–221

Ussing HH (1943a) On the partition of certain amino acids between blood and tissues. Acta Physiol Scand 6:222–232

Ussing HH (1943b) The nature of the amino nitrogen of red corpuscles. Acta Physiol Scand 5:335–351

Ussing HH (1945a) The reabsorption of glycine and other amino acids in the kidney of man. Acta Physiol Scand 9:193–212

Ussing HH (1945b) The determination of chloroform in tissues and blood. Acta Physiol Scand 9:214–220

Ussing HH (1946) Amino acids and related compounds in the haemolymph of Oryctes nasicornis and Melolontha vulgaris. Acta Physiol Scand 11:61–84

Ussing HH (1947) Interpretation of the exchange of radio-sodium in the isolated muscle. Nature 160:262–263

Ussing HH (1949a) The active ion transport through the isolated frog skin in the light of tracer studies. Acta Physiol Scand 17:1–37

Ussing HH (1949b) The distinction by means of tracers between active transport and diffusion. Acta Physiol Scand 19:43–56

Ussing HH (1980) Life with tracers. Ann Rev Physiol 42:1–16

Ussing HH, Zerahn K (1951) Active transport of sodium as the source of electric current in the short- circuited isolated frog skin. Acta Physiol Scand 23:110–127

Ussing HH, Erlij D, Lassen H (1974) Transport pathways in biological membranes. Ann Rev Physiol 36:17–79

Vallon V (2009) Micropuncturing the nephron. Pflügers Arch 458:189–201

van Driessche W, Gögelein H (1980) Attenuation of current and voltage noise signals recorded from epithelia. J Theor Biol 86:629–648

van Driessche W, Lindemann B (1978) Low-noise amplification of voltage and current fluctuations arising in epithelia. Rev Sci Instrum 49:53–57

van Driessche W, Zeiske W (1985) Ionic channels in epithelial cell membranes. Physiol Rev 65:833–903

Verveen AA, Derksen HE (1965) Fluctuations in membrane potential of axons and the problems of coding. Kybernetik 4:152–160

Verzár F, McDougall EJ (1936) Absorption from the intestine. Longmans, Green, London

Walker AM, Hudson CL (1936) The reabsorption of glucose from the renal tubule in Amphibia and the action of phlorhizin upon it. Am J Physiol 118:130–143

Walker AM, Oliver J (1941) Methods for the collection of fluid from single glomeruli and tubules of the mammalian kidney. Am J Physiol 134:562–579

Walker AM, Bott PA, Oliver J, MacDowell MC (1941) The collection and analysis of fluid from single nephrons of the mammalian kidney. Am J Physiol 134:580–595

Warth R, Barriere H, Meneton P, Bloch M, Thomas J, Tauc M, Heitzmann D, Romeo E, Verrey F, Mengual R, Guy N, Vendahhou S, Lesage F, Pouljeol P, Barhanin J (2004) Proximal renal tubular acidosis in TASK2 K^+ channel-deficient mice reveals a mechanism for stabilizing bicarbonate transport. Proc Natl Acad Sci USA 101:8215–8220

Washburn EW, Urey HC (1932) Concentration of the H^2 isotope of hydrogen by fractional electrolysis of water. Proc Natl Acad Sci USA 18:496–498

Watson JD, Crick FHC (1953) Molecular structure of nucleic acids. A structure for deoxyribose nucleic acid. Nature 171:737–738

Wearn JT, Richards AN (1924) Observations on the composition of glomerular urine, with particular reference to the problem of reabsorption in the renal tubules. Am J Physiol 71:209–227

Welling PA, Ho K (2009) A comprehensive guide to the ROMK potassium channel: form and function in health and disease. Am J Physiol Renal Physiol 297:F849–F863

White MB, Amos J, Hsu JMC, Gerrard B, Finn P, Dean M (1990) A frame-shift mutation in the cystic fibrosis gene. Nature 344:665–667

Williams DM, Lopes CMB, Rosenhouse-Dantsker A, Connelly HL, Matavel A, O-Uchi J, McBeath E, Gray DA (2010) Molecular basis of decreased Kir4.1 function in SeSAME/EAST syndrome. J Am Soc Nephrol 21:2117–2129

Wilson TH (1956) A modified method for study of intestinal absorption in vitro. J Appl Physiol 9:137–149

Wilson TH (1962) Intestinal absorption. W.B. Saunders, Philadelphia. Chapter 2, pp 20–39

Wilson TH, Wiseman G (1954) The use of sacs of everted small intestine for the study of the transference of substances from the mucosal to the serosal surface. J Physiol 123:116–125

Windhager EE (1987) Micropuncture and microperfusion. In: Gottschalk CW, Berliner RW, Giebisch GH (eds) Renal physiology. People and ideas. American Physiological Society, Bethesda, pp 101–129

Wiseman G (1953) Absorption of amino-acids using an in vitro technique. J Physiol 120:63–72

Wiseman G (1961) Sac of everted intestine: technic for study of intestinal absorption in vitro. In Quastel JH (ed) Methods of medical research, vol 9. Year Book Medical, Chicago, IL, pp 287–292.

Zajac M, Dolowy K (2017) Measurement of ion fluxes across epithelia. Prog Biophys Mol Biol 127:1–11

Zoller MJ, Smith M (1982) Oligonucleotide-directed mutagenesis using M13-derived vectors: an efficient and general procedure for the production of point mutations in any fragment of DNA. Nucleic Acids Res 10:6487–6500
Zoller MJ, Smith M (1984) Oligonucleotide-directed mutagenesis: a simple method using two oligonucleotide primers and a single-stranded DNA template. DNA 3:479–488

Chapter 2
Principles of Epithelial Transport

Dee U. Silverthorn

Abstract This chapter provides a general overview of epithelium and transport in epithelial tissues. It is intended to be an introduction to those new in the field or an outline of what to teach to beginning students. Most of the general principles introduced here are described in detail in later chapters of this book. Epithelial transport is an ideal subject for teaching basic biological concepts (for example, electrical and chemical gradients) and for correcting common misconceptions that students develop, such as the idea that only the classic excitable tissues of nerve and muscle generate electrical signals. The chapter begins with an introduction to the structure and function of epithelial cells and tissues. It then considers the various mechanisms by which substances cross plasma membranes. The final sections examine various patterns of transport across epithelia and explain how to teach general principles of transport using compartmental models.

Keywords Epithelia · Transcellular compartments · Tight junctions · Membrane transport · Paracellular transport

2.1 Introduction

Animals are composed of four tissue types: nerve, muscle, connective tissue, and epithelium, each identifiable morphologically by cell shape, size, and the amount of extracellular matrix associated with the tissue. Epithelial tissues form the body's interface with the outside world, and they separate internal compartments. Specialized epithelial cells produce chemicals and solutions that range from tears and sweat to hormones and digestive enzymes. This chapter provides a broad overview of the properties of epithelial tissues. Many of the concepts introduced here are expanded upon in later chapters of this three-volume book series.

D. U. Silverthorn (✉)
Dell Medical School, University of Texas – Austin, Austin, TX, USA
e-mail: silverthorn@utexas.edu

© The American Physiological Society 2020
K. L. Hamilton, D. C. Devor (eds.), *Basic Epithelial Ion Transport Principles and Function*, Physiology in Health and Disease,
https://doi.org/10.1007/978-3-030-52780-8_2

53

This chapter is intended to be an introduction to epithelia for those new in the field or for physiology educators. Epithelial transport is an ideal subject for teaching basic biological concepts (for example, electrical and chemical gradients) and for correcting common misconceptions that students develop, such as the idea that only the classic excitable tissues of nerve and muscle generate electrical signals.

The chapter begins with an introduction to the structure and function of epithelial cells and tissues. It then considers the various mechanisms by which substances cross plasma membranes and describes how those mechanisms apply to transport across epithelia. The final section illustrates how to teach epithelial transport and help students develop problem-solving skills by asking them to apply the general principles of transport to specific scenarios.

2.2 What Are Epithelia?

Epithelia are tissues found in most multicellular animals (the Metazoa), where they form sheets or tubes that separate compartments. Homeostasis, the maintenance of a relatively stable internal environment, would not be possible without epithelia that shield the body's internal compartments from the outside world. The stability of internal compartments whose composition differs from the external environment requires a selectively permeable barrier, so all exchange between compartments takes place across an epithelium.

2.2.1 Epithelial Anatomy

Epithelial cells are distinguished by three characteristics: they are polarized into distinct regions, connected to each other with cadherin-based cell-cell junctions, and supported by an extracellular matrix layer called the basal lamina (Mescher 2018). Histologically, epithelial tissue is most easily recognized as sheets or tubes of connected cells with a basal lamina beneath one surface. Microscopic classification of epithelia depends on the number of cell layers, cell shape, and modifications of the cell membrane (Fawcett 1994). Nuclear shape and location provide useful information. At higher levels of resolution, the distribution of organelles within the cell also provides clues to the polarization of cell function.

Cell Layers and Shapes Epithelia with a single layer of cells are described as simple epithelia. Those with multiple cell layers and only one layer in contact with the basal lamina are called stratified. Pseudostratified epithelia appear to have multiple layers but on close inspection, all cells in the epithelium are in contact with the basal lamina.

Most epithelial cells come in one of three shapes: flattened or *squamous*, from the Latin word for scale; cuboidal; or columnar, with cells that are taller than they are wide. Simple squamous epithelia are ideal for rapid exchange across the cell. Cuboidal and columnar epithelia have increased volume for organelles such as mitochondria and therefore are more likely to be associated with transport and secretion. In epithelia specialized for transport, finger-like microvilli on the apical surface and convoluted folds on the basolateral side greatly expand the membrane surface area.

Epithelial cells are described as having apical and basolateral sides, distinguished both morphologically (apical microvilli or cilia, for example) and biochemically, by membrane protein distribution. The demarcation between apical and basolateral sides of the cell is an apical cell-cell junction visible with electron microscopy. Typically, the apical surface of the epithelium is exposed to either the external environment or to the lumen of some internal space, such as the vasculature or one of the transcellular spaces. (*Transcellular spaces* are body cavities or compartments that are sealed off from the external environment, such as the peritoneal cavity or pericardial sac.) The apical surface is called the luminal or mucosal side of the epithelium. The basolateral side usually faces the extracellular fluid and is also known as the serosal surface.

The Basal Lamina An acellular matrix called basal lamina is secreted from the basal side of epithelial cells to provide structural support for the tissue. The primary components of the basal lamina are type IV collagen, the glycoprotein laminin, and various proteoglycans and proteins (Pozzi et al. 2017; Randles et al. 2017). In some tissues, the basal lamina is sandwiched between the epithelial cells and a second layer of extracellular matrix, the reticular lamina, that is secreted by underlying connective tissue. In light microscope images, the two laminae are indistinguishable and are collectively referred to as the basement membrane (Mescher 2018). Normally blood vessels do not penetrate the basal lamina, which requires substances entering and leaving the epithelium to diffuse a short distance into the underlying tissue to reach the nearest capillaries.

Cell Junctions Epithelial cells are connected to each other by cell-cell junctions and to the underlying basal lamina by cell-matrix junctions (Fig. 2.1). In vertebrates, a series of cell-cell junctions occur near the apical surface and collectively are called the apical junctional complex (Zihni et al. 2016). A complex consists of two or three different types of cell-cell junctions, each with distinctive membrane proteins and links to the cytoskeleton (Fig. 2.2). Apical junctional complexes contribute to cell polarity by acting as a barrier to migration of membrane proteins between the apical and basolateral zones.

The junction nearest the apical surface is the tight junction (zonula occludins), followed closely by an adherens junction (zonula adherens) that forms a band or belt around the cell. Tight and adherens junctions connect to actin fibers of the cell cytoskeleton. Desmosomes form spot junctions rather than continuous belts around the cell, and they occur more basally along the lateral membrane, sometimes not even close to the adherens junction. Desmosomes attach to intermediate filaments of the cytoskeleton.

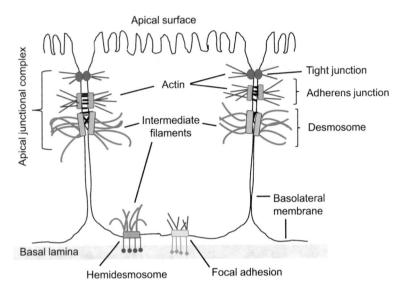

Fig. 2.1 Cell-cell and cell-matrix junctions in vertebrates

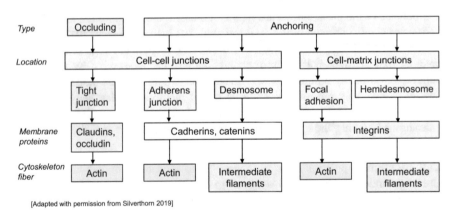

[Adapted with permission from Silverthorn 2019]

Fig. 2.2 Proteins of cell-cell and cell-matrix junctions in vertebrates

Adherens junctions and desmosomes are anchored by membrane-spanning cell adhesion glycoproteins known as cadherins, short for *calcium-dependent cell adhesion molecules* (Brasch et al. 2012). The integrity of cadherin-based cell junctions requires extracellular calcium, and if an epithelium is exposed to a calcium-chelating agent or a zero-calcium saline, the junctions dissociate and the barrier separating apical and basolateral compartments is lost (Gonzalez-Mariscal et al. 1990; Fawcett 1994). The prototype animal cadherin is E-cadherin, E for epithelial. The extracellular domains of E-cadherin interact with similar domains on adjacent cells to create the cell-cell junction. The cytoplasmic domain of E-cadherin links to actin filaments of the cytoskeleton by complexing with catenin proteins inside the cell (van Roy and Berx 2008).

Cell-matrix junctions bind the basal side of the epithelium to the underlying basal lamina. Hemidesmosomes link the matrix to intermediate filaments of the cytoskeleton like desmosomes do, but the hemidesmosome membrane-spanning protein is integrin $\alpha6\beta4$ instead of cadherin, and the intracellular linking protein is plectin rather than catenin (Borradori and Sonnenberg 1999; Walko et al. 2015). Focal adhesions are cell structures composed of integrins that connect to actin filaments of the cytoskeleton using talin and vinculin protein links. Focal adhesions can transiently link to microtubules and they play a role in cell migration (Wehrle-Haller 2012; Armitage and Plotnikov 2016).

Defects in or disruption of cell-cell junction proteins are implicated in a variety of disease states, ranging from inflammatory bowel disease (Landy et al. 2016) and digestive dysfunction (Zeisel et al. 2019) to male infertility (Zihni et al. 2016). Mutations in tight junction proteins of renal epithelia are often associated with electrolyte imbalances, such as hypomagnesaemia or hypercalciuria (Denker and Sabath 2011). Some bacteria target tight junction proteins as part of their pathogenic mechanism (Eichner et al. 2017). Of growing interest are the changes in epithelial junctions that occur in different cancers, especially as these changes are associated with conversion to metastatic states (Martin 2014; Salvador et al. 2016).

Polarity Epithelial cells are polarized, with membrane proteins, lipids, and intracellular organelles organized into distinct regions. Cell polarity is not unique to epithelia and is also seen in embryonic development, morphogenesis, and migrating cells. Many of the complex signaling mechanisms for developing and maintaining polarity are highly conserved throughout the animal kingdom (Rodriguez-Boulan and Macara 2014; Blasky et al. 2015; Sebé-Pedrós et al. 2017).

The formation and maintenance of epithelial cell polarity depend on three major polarity protein complexes, nicknamed Crumbs, Par, and Scribble, plus RhoGTPases (Hall 1998; Vacca et al. 2015; Campanale et al. 2017; Ebnet et al. 2018; Chang et al. 2019). These major players interact with a host of intracellular signal molecules that, along with cell adhesion proteins, create distinct cell domains. The Crumbs or Crb3 complex is associated with the apical domain of the epithelial cell. The Par complex proteins (Par3–aPKC–Par6) are found with the tight junctions and adherens junctions that separate the two poles of the cell, and the Scribble complex proteins (Scrib/Lgl/Dlg) localize to the basolateral membrane. The polarity complexes regulate the localization of RhoGTPases, which in turn regulate the organization of the actin cytoskeleton that forms the belt stabilizing the apical junctional complex, among other functions. Once polarity has been established, signaling between the domains keeps the polarity complexes in their proper place. For more on the development of cellular polarity, see Chap. 3 of this volume.

2.2.2   Evolution and Developmental Biology of Epithelia

Epithelia appeared early in Metazoan evolution. Jellyfish, sea anemones, and other members of the phylum Cnidaria have a single compartment body plan, with two epithelial cell layers, the epidermis and gastrodermis, separated by a gelatinous extracellular matrix (Technau and Steele 2011). As animal body plans developed in complexity and progressed from radial to bilateral symmetry, organisms acquired a third germ layer, the mesoderm, along with the internal compartment called the coelom, setting the stage for the development of complex organs (Tyler 2003).

Evolution and Epithelia   The evolution of multicellular animals from unicellular ancestors required epithelia, sheets of tissue, that could form a boundary between external environment and the animal's internal compartments. It was not a major evolutionary leap with all-new mechanisms—cell adhesion molecules that link cells together are older than Metazoans, with early members of cadherin, catenin, and integrin families found in the Protista, single-celled organisms that range from *Amoeba* to colony-forming organisms such as the choanoflagellates (Nichols et al. 2012; Murray and Zaidel-Bar 2014; Sebé-Pedrós et al. 2017; Pennisi 2018). The evolutionary transition from protists to the least complex metazoans, such as sponges (phylum Porifera), was marked by the appearance of a single surface layer of cells with the characteristics of a primitive epithelium, namely cell-cell junctions and cellular polarity. A basal lamina, the third defining characteristic of epithelial tissue, is absent in most sponges, and the tightness of sponge cell junctions has been questioned (Leys et al. 2009; Belahbib et al. 2018). However, there is one report of a freshwater sponge that in culture appears to form occluding junctions and can transport ions, resulting in a transepithelial potential difference (Adams et al. 2010). Functionally, a tight epithelium in freshwater sponges makes sense because of the osmotic challenge of life in a hypotonic environment.

True epithelial tissue isolating an internal compartment from the environment is first obvious in Cnidarians, the phylum that includes jellyfish, corals, and sea anemones. Cnidarians have a surface epithelium that specializes into two germ layers: ectoderm that forms the epidermis, and endoderm (the gastrodermis) modified for digestive function (Technau and Steele 2011). The internal compartment of Cnidaria is filled with jelly-like mesoglea and protein fibrils similar to those found in basal lamina (Tucker and Adams 2014).

All Metazoan phyla more complex than Porifera and Cnidaria, from worms and arthropods to the vertebrates, are collectively called the Bilateria because of their bilateral symmetry. Their body plan has mirror-image left and right sides, usually streamlined into a distinct head and tail. Embryologically the Bilateria are distinguished from the basal animal phyla by the development of mesoderm, a third germ layer that gives rise to complex internalized organs (Tyler 2003; Technau and Steele 2011). These animals all have classic epithelia with polarized cells connected by cell junctions and supported on a basal lamina (Rodriguez-Boulan and Macara 2014). The structural details of the junctions differ between vertebrates and invertebrates, but many of the proteins are similar (Müller and Bossinger 2003).

Epithelia and Embryonic Development Epithelium is the first identifiable tissue type to appear during embryonic development of animals. Most of what we know about the developmental biology of epithelia comes from comparative studies in animals ranging from sponges (Leys et al. 2009; Belahbib et al. 2018) and other invertebrates to vertebrates ranging from fish and amphibians to birds and mammals (Müller and Bossinger 2003; Kiecker et al. 2016). The primary function of embryonic epithelium is to compartmentalize the developing organism by separating inner cell populations and cavities from the external environment.

Polarized cells that display epithelial traits such as a basal lamina and cadherin-based adhesive junctions arise as early as the eight-cell stage in vertebrate embryonic development. It is at this stage that a separate internal compartment first becomes apparent. The primary epithelium of these early embryonic cells has many of the molecular hallmarks of mature epithelium, including cadherins and other proteins needed to form junctional complexes (Fleming and Johnson 1988; Shook and Keller 2003).

As development proceeds to the gastrulation stage, changes in gene expression cause some of the surface epithelium cells to lose their epithelial phenotype. Polarity disappears, the apical cell junctions release, basal lamina is disrupted, and cytoskeletal changes allow the altered cells to migrate into the internal compartment and transform into primary mesenchyme. This process is called *epithelial-mesenchymal transition*, or EMT. Mesenchymal cells no longer resemble epithelium and look more like fibroblasts: they are spindle-shaped, motile, and have enhanced ability to secrete components of extracellular matrix. These internalized mesenchymal cells of the gastrula then become the third germ layer, the mesoderm. Meanwhile, the remaining epithelium transforms into the germ layers of ectoderm and endoderm (Shook and Keller 2003; Kalluri and Weinberg 2009; Thiery et al. 2009; Chen et al. 2017; Kim et al. 2018).

The primary mesenchymal cells of mesoderm have multiple fates. They are the primary source of connective tissue in the developing body (Mescher 2018), and they migrate to join cells of the other germ layers to form tissues or organs. In many tissues, such as the developing kidney (Faa et al. 2012), peripheral nervous system, and skeletal tissues (Kalcheim 2016), some embryonic mesenchyme re-differentiates into secondary epithelium through *mesenchymal-epithelial transition* (MET), the reverse of EMT. Both EMT and MET are complex processes involving multiple genes, transcription factors, and chemical signaling pathways (Thiery et al. 2009; Kalluri and Weinberg 2009; Lamouille et al. 2014).

The developmental biology of epithelia is of considerable interest in translational research. Mesenchymal cells persist into adulthood and are considered adult stem cells for tissues of mesodermal origin (Ullah et al. 2015; Chen et al. 2017), although their efficacy as therapeutic agents in human clinical trials has yet to be established (Galipeau and Sensebe 2018). Post-embryonic epithelial-mesenchymal transition is associated with extracellular matrix production and inflammation during wound healing and with cell changes in cancer, although the process differs from embryonic EMT (Kalluri and Weinberg 2009; Kim et al. 2018). The probable role of EMT in cancer metastasis is currently a hot topic due to its potential as a target for future cancer therapies (Jiang et al. 2015; Brabletz et al. 2018; Li and Balazsi 2018; Aiello and Kang 2019).

2.2.3   Functional Classification of Epithelia

Beginning students of physiology, asked to name and give the functions of epithelia, are likely to describe the protective epithelium of skin, the ciliated epithelia of the airways, and the transporting epithelia of digestive tracts and excretory organs, such as vertebrate kidney and intestine, insect Malpighian tubules, or the gills of aquatic animals. Many physiology students are not aware that endocrine glands are specialized secretory epithelium or that the endothelium lining blood and lymph vessels is a type of epithelium. Even scientists sometimes forget that some highly specialized cells are epithelia, such as the choroid plexus (Vol. 2, Chap. 10), the retinal pigment epithelium (Vol. 2, Chap. 9), taste buds, and inner ear hair cells (Vol. 2, Chap. 8). Functionally, epithelia can be grouped into six broad categories: protective, exchange, transporting, secretory, ciliated, and specialized (Table 2.1).

Protective and Exchange Epithelia   The protective and exchange epithelia represent the two extremes of epithelial barrier function. Protective epithelia of skin and mucous membranes are organized in layers of cells to minimize exchange between the internal and external environments. The exchange epithelia, at the other morphological and functional extreme, are a single layer of thin, flattened cells, designed

Table 2.1   Functional classification of epithelia

Classification	Function	Examples
Protective	Restrict movement between compartments	Skin, mucous membranes, cornea
Exchange	Permit rapid exchange and bulk flow between compartments	Endothelium of blood and lymph vessels; type I alveolar cells
Transporting	Allow selective movement between compartments	Kidney, intestine, choroid plexus
Secretory	Manufacture and release chemicals • Into the blood (endocrine glands and cells) • Into the external environment (exocrine glands and cells)	• *Endocrine*: anterior pituitary, thyroid gland, enteroendocrine cells, endocrine pancreas • *Exocrine*: salivary and sweat glands, goblet cells, exocrine pancreas, type II alveolar cells
Ciliated	Multiple cilia move fluid over the surface	Airways and reproductive tract; ventricles of brain
Specialized	• Sensory functions • Contractile myoepithelium • Secrete dental tooth enamel	• Taste buds, hair cells, retinal pigment epithelium • Mammary, salivary, sweat, prostate, and lacrimal glands • Ameloblasts

Table 2.2 Transcellular compartments and their epithelia

Transcellular compartment	Exchange epithelium	Plasma ultrafiltrate name	Fluid composition relative to plasma
Cochlea of ear	Blood vessels and cells of the limbus spiralis (Firbas et al. 1981)	Perilymph	Lower protein and Ca^{2+}, higher Cl^- (Wangemann and Marcus 2017)
Joint capsule	Synovial membrane	Synovial fluid	Lower protein. Hyaluronate secreted by type B synoviocytes (Levick and McDonald 1995; Bullough 2010)
Pleural and pericardial spaces; peritoneum	Mesothelial lining of the compartment	• Pleural • Pericardial • Peritoneal	• Lower protein and Na^+, higher HCO_3^- (Zocchi 2002) • Slightly higher in K^+ (Gibson and Segal 1978). • Lower protein and Ca^{2+}, higher Cl^- (Kelton et al. 1978)

Transcellular compartment	Transporting epithelium	Fluid name	Fluid composition relative to plasma
Ventricles of brain and central canal of spinal cord	Choroid plexus	Cerebrospinal fluid	Higher Na^+, Cl^-, Mg^{2+}, and amino acids. Lower K^+ and Ca^{2+} (Sakkaa et al. 2011; Spector et al. 2015b)
Cochlea of ear	Stria vascularis	Endolymph	Lower Na^+ and Ca^{2+}. Higher K^+, Cl^-, and HCO_3^- (Wangemann and Marcus 2017)
Intraocular	Ciliary epithelium	Aqueous humor	Hyperosmotic. Very low in protein, low in glucose. High in ascorbate. (Bito 1982; Macknight et al. 2000; Goel et al. 2010)

for rapid exchange by diffusion or bulk flow of fluids between the compartments they separate. Exchange epithelia include the type I alveolar cells for gas exchange in the lungs, the endothelium lining blood and lymph vessels, and the mesothelium that lines body cavities such as the peritoneum. In the lung, the basal laminae of the alveolar and endothelial cell layers fuse to minimize the diffusion distance for gases (Mescher 2018).

Ultrafiltration of plasma through two layers of exchange epithelia, usually endothelium and mesothelium, is responsible for the production of fluid in several of the transcellular spaces of the body (Table 2.2). The ultrafiltrate found in these spaces is usually lower in protein than plasma due to the sieve-like structure of the epithelial membranes. Ion concentrations are similar to those in plasma except that Cl^- may be higher, replacing unfiltered anionic proteins, and Ca^{2+} lower because some Ca^{2+} remains bound to plasma proteins and does not filter.

Transporting Epithelia Transporting epithelia selectively control movement of material between the compartments they separate. The classic examples of transporting epithelia are the nephron and the mucosal lining of the gut. These tissues form a selective barrier between the internal and external environments.

Less well-known examples of transporting epithelia are responsible for secreting specialized fluids from the extracellular fluid (ECF) into transcellular compartments within the body (Table 2.2). Examples of transporting epithelia that supply transcellular compartments include the choroid plexus and the ciliary body of the eye. The choroid plexus of the central nervous system secretes cerebrospinal fluid by transporting sodium and other ions into the ventricles of the brain (Vol. 2, Chap. 10; Spector et al. 2015a; Praetorius and Damkier 2017). The ciliary epithelium secretes the aqueous humor of the eye. Specialized tissues in the cochlea of the ear secrete endolymph and perilymph (see Vol. 2, Chap. 8 on inner ear epithelia).

Secretory Epithelia Secretory epithelium, in which the cells manufacture and release a chemical product, occurs both as organized glands and as isolated cells. Secretion is classified as exocrine or endocrine, with exocrine secretions released to the external environment and endocrine secretions entering the blood. During embryonic development, exocrine glands arise by budding from surface epithelium but they remain connected to the epithelium by ducts that open to the external environment (Wang and Laurie 2004). Major exocrine glands include the exocrine pancreas and liver, salivary, mammary, prostate, and lacrimal glands, and smaller glands such as gastric glands and sweat glands in the skin. Single exocrine cells, particularly mucus-secreting goblet cells, are found mixed with other cell types in epithelia of the gut, airways, and other organs. For example, type II alveolar cells in the lung secrete surfactant and also absorb sodium and water to keep the alveolar air space free of fluid, while the thin, flat type I cells participate in gas exchange (Alcorn 2017; Downs 2017).

Endocrine glands are derived from primitive epithelium (see Sect. 2.2.2) but during development they separate from the parent tissue layer and migrate to different locations within the body (Kiecker et al. 2016). In some instances, particularly in the wall of the gut, isolated endocrine cells remain interspersed with other cells (Mescher 2018). Endocrine glands and cells secrete their products into the blood for transport to distant targets rather than into the external environment.

Ciliated Epithelia Ciliated epithelia have the cell surface that faces the lumen modified into motile cilia whose coordinated movements move fluid along the surface of the epithelium. Cilia facing the external environment include the ciliated epithelia of the upper respiratory system and male and female reproductive tracts. Ciliated ependymal cells line the ventricles of the brain and central canal of the spinal cord, where they help circulate cerebrospinal fluid secreted by the choroid plexus. Ependymal cells are considered modified epithelium because they lack the basal lamina typical of epithelia (Mescher 2018).

Specialized Epithelia Specialized epithelial cells include myoepithelium, neuroepithelial cells, and various unique cells, such as podocytes of the glomerulus (Greka and Mundel 2012; Reiser and Altintas 2016), umbrella cells of the bladder urothelium (Lewis 2000), and ameloblasts that lay down protein matrix as the foundation for dental enamel (He et al. 2010; Lacruz et al. 2017; see Vol. 2, Chap. 11). Myoepithelium is contractile and has many of the features and biochemical markers of smooth muscle. Myoepithelial cells are found in the mammary gland, where they aid ejection of milk, and in the salivary, sweat, lacrimal, and prostate glands (Balachander et al. 2015; Chitturi et al. 2015).

Neuroepithelial cells in adult animals are specialized for functions related to the special senses — taste receptor cells (Chaudhari and Roper 2010), hair cells of the inner ear (Goutman et al. 2015; Burns and Stone 2017), and the retinal pigment epithelium that underlies the photoreceptor layers of the retina (Strauss 2005; see Vol. 2, Chap. 9). Taste receptor cells and hair cells use ion channels or membrane receptors linked to second messenger systems to release neurotransmitter-like chemicals onto sensory neurons. The retinal pigment epithelium (RPE) has multiple functions, including a key role in the recycling of retinal after its activation by light. The RPE also transports ions, nutrients, and waste products between the blood and the intraocular fluid bathing the retina.

The various functional types of epithelial cells are often found mixed in different organs. For example, the airways combine ciliated epithelium and goblet cells with underlying exocrine glands. The pancreas is an example of an exocrine organ with isolated islands of endocrine cells. The mucosal lining of the small intestine includes absorptive enterocytes, isolated enteroendocrine cells, and exocrine cells such as mucus-secreting goblet cells and Paneth cells whose granules contains enzymes and other defensive peptides (Mescher 2018).

2.3   Epithelial Transport

Compartmentation, the division of an entity into compartments with or without visible walls, is a core concept in biological sciences. At its simplest level, the living organism is one compartment and its outside world a second compartment, with constant exchange of energy and matter between the two. The human body furthers subdivides into extracellular fluid—the body's "internal environment"—and the intracellular compartment. The extracellular fluid compartment has its own separation into interstitial fluid bathing the cells, and plasma, the liquid matrix of blood. The interstitial fluid also includes compartments filled with specialized *transcellular fluids:* cerebrospinal, pericardial, synovial, intraocular, pleural, and pericardial fluids, peritoneal fluid, and cochlear endolymph and perilymph.

Epithelia form the boundaries between all these compartments. Some, like the protective epithelium of the skin, are designed to minimize exchange. Others, like the capillary endothelium that separates plasma from interstitial fluid, place minimal restrictions on transfer between compartments. Most other epithelia fall between those two extremes, providing a selective barrier that helps control what passes between compartments.

Epithelia can change the composition of the substances they transport, and abnormal changes in the composition of fluid produced by an epithelium may reflect underlying disease processes. Urine, blood, and cerebrospinal fluid are routinely analyzed, as well as amniotic fluid in pregnant women. Less commonly, sweat, synovial fluid, tears, pleural and peritoneal fluid are examined (DiVenere et al. 2018). For example, the sweat chloride concentration test is the first step in diagnosis of cystic fibrosis (Farrell et al. 2017), and synovial fluid analysis is commonly used for diagnosis of joint paint (Seidman and Limaiem 2019).

2.3.1 Transcellular vs. Paracellular Transport

Transport across a sheet of epithelial cells can take one of two pathways: through the cells themselves (transcellular transport) or between the cells (paracellular transport) (Fig. 2.3). Tight junctions limit the paracellular movement of substances between cells. Some epithelia are truly tight and prevent most movement but others have varying degrees of "leakiness" (Frömter and Diamond 1972; Anderson and Van Itallie 2009). In certain tissues, specific ions and water are able to pass through the tight junctions in response to transepithelial electrical and osmotic gradients. For example, calcium reabsorption in the kidney depends largely on Ca^{2+} movement through tight junctions (Bleich et al. 2012).

Selective transport through tight junctions appears to be mostly closely associated with the junctional claudin proteins (Anderson and Van Itallie 2009; Tsukita et al. 2019). Polymers of claudins create linear tight junction "strands" with varying properties that depend on the particular claudin protein subtypes involved. Some combinations of claudins form tight barriers but other combinations create anion-selective or cation-selective channels that allow specific ions and water to pass between the apical and basolateral compartments.

What is the adaptive advantage of paracellular transport for an epithelium? In some tissues, such as the kidney, allowing paracellular movement of solutes and water appears to be a mechanism for minimizing energy expenditure (Pei et al. 2016; Yu 2017). In the case of calcium, keeping Ca^{2+} ions out of the cells avoids calcium signaling that can trigger intracellular processes ranging from exocytosis and con-tractility to enzyme regulation and gene expression (Carafoli and Krebs 2016).

Fig. 2.3 Pathways for transepithelial transport

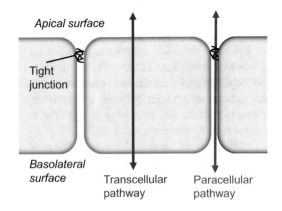

Substances taking the transcellular pathway through the epithelial cell must cross two plasma membranes: the apical and basolateral. Lipophilic molecules can simply pass through the phospholipid bilayer and do not require special mechanisms for transport, but water, ions, and lipophobic molecules require help to cross the phospholipid bilayer of the membrane. This assistance takes the form of either membrane protein transporters or vesicles. Vesicular transport includes various forms of endocytosis, exocytosis, and phagocytosis.

2.3.2   Energy for Membrane Transport

Transport of material across the cell membrane can be classified by energy source (active or passive) and by mechanism—whether transport takes place through the phospholipid bilayer or with the aid of membrane proteins or vesicles (Fig. 2.4).

Net movement of a substance across cell membranes requires a source of energy other than random molecular motion. If transport uses only the potential energy stored in a concentration or electrochemical gradient, and if net movement stops when the gradient is dispersed, we consider the process to be a passive one. Passive transport includes simple diffusion of lipophilic substances across the phospholipid bilayer, membrane carrier-mediated facilitated diffusion, and movement primarily of small ions and water through open membrane channels or pores.

Active transport requires the input, directly or indirectly, of additional energy from the high-phosphate bond of ATP. Active transport may be protein-mediated or may utilize membranous vesicles to transfer large molecules and particles between the intracellular and extracellular compartments. Direct or primary active transport uses enzyme-carriers (ATPases) that hydrolyze the high-energy phosphate bond of ATP and couple the energy released to transport of one or more substances against

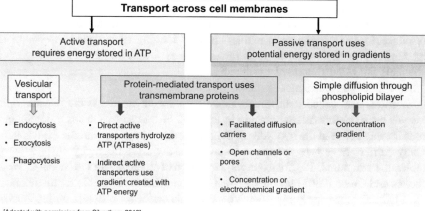

[Adapted with permission from Silverthorn 2019]

Fig. 2.4 Energy sources and mechanisms for membrane transport

their gradients. The iconic ATPase is the Na^+-K^+-ATPase that exchanges 3 Na^+ for 2 K^+, both moving against their electrochemical gradients (see Vol. 3, Chap. 1).

Indirect (secondary or tertiary) *active transport* uses the potential energy stored in a gradient created by ATP to couple one substance moving down its gradient to other substances moving uphill, against their gradient. An oft-used example of indirect active transport is the Na-dependent glucose transporter (SGLT) of intestine and kidney (see Vol. 1, Chap. 9 and Vol. 3, Chap. 6). This transporter uses the potential energy of the Na^+ gradient maintained by the Na^+-K^+-ATPase to concentrate glucose inside the cell. Because the energy from ATP hydrolysis is one step removed from the SGLT, this example is also called secondary active transport.

In tertiary active transport, the ATPase is two steps removed. This can be seen in the nephron, where the Na^+ gradient created using the Na^+-K^+-ATPase drives the concentration of α-ketoglutarate inside the cell using a Na^+-dependent dicarboxylate cotransporter, NaDC3 (Moe et al. 2016). The NaDC3 is a secondary active transporter. The tertiary active transporter is an organic anion transporter (OAT) that couples energy released by α-ketoglutarate leaving the cell down its gradient to movement of organic anions into the cell against their concentration gradient. For more on organic anion transport, see Vol. 3, Chap. 12.

2.3.3   Protein-Mediated Transport

Lipophobic molecules that are unable to pass through the phospholipid bilayer of the cell membrane must enter or leave cells using either protein-mediated transport or vesicular transport, as shown in Fig. 2.4. Integral membrane protein transporters can be classified into two groups: channels or carriers. Channels allow substrates to cross the membrane through an open, water-filled passageway, which in general limits substances that can move through the channels to smaller ions and a few small organic molecules (Stillwell 2016). Channels may spend most of their time in an open state (pores or leak channels) or they may have gates that open or close in response to an appropriate stimulus, such as binding of a ligand, a change in membrane potential, or mechanical forces such as stretch on the membrane. Gated channels are often described by the stimulus that controls the gate: ligand-gated, voltage-gated, or mechanically gated.

Carriers, also called transporters, move their substrates across the phospholipid bilayer by a conformational change in the protein that never creates a continuous passageway between the two sides of the membrane. The molecular mechanisms for many carriers have been described following x-ray crystallography elucidation of molecular structures. See, for example, Jorgensen et al. (2003) for the Na-K-ATPase, Stokes and Green (2003) for the Ca^{2+}-ATPase, and Bai et al. (2017) for members of the SLC solute carrier family. See Vol. 3 of this series for chapters on the current state of knowledge for many different carriers and channels.

2.3.4   Transporter Terminology

The conventions for naming membrane transporters are numerous and confusing, especially for novices. Early descriptions used the term *cotransport* for movement of two or more substrates in the same direction, and *counter-transport* if substrates moved in opposite directions (Byrne and Schultz 1988). More specific terms currently in use are *uniport* (one substrate being transported), *symport* (two or more substrates moving in the same direction), and *antiport* (two or more substrates moving in opposite directions across the membrane) (Mitchell 1967; Saier et al. 2016).

Classification schemes for transport proteins depend on the source. Although it is possible to find transport proteins called permeases in the literature, the Nomenclature Committee of the International Union of Biochemistry and Molecular Biology discourages this term and others that use the *-ase* suffix for what it calls "phenomenases," groups of proteins associated with physiological processes that are not identifiable chemical reactions (NC-IUBMB 2019). The transport proteins that extract energy from the high-energy phosphate bond of ATP correctly follow the IUBMB nomenclature rules and are called ATPases, where the *-ase* suffix is attached to the substrate for the chemical reaction. Scientists studying transport usually name transporters by the substrates they carry, such as the GLUT family of glucose and other hexose transporters.

Three different classification schemes for membrane transporters have been proposed by different organizations: the International Union of Basic and Clinical Pharmacology (Alexander et al. 2017), the International Union of Biochemistry and Molecular Biology (Saier et al. 2016), and HUGO, the Human Genome Organization (Yates et al. 2017). In 1987 the International Union of Basic and Clinical Pharmacology (IUPHAR) created a Committee on Receptor Nomenclature and Drug Classification. The committee worked with The British Pharmacological Society (BPS) to create a database of drug targets, the IUPHAR/BPS Guide to Pharmacology (www.guidetopharmacology.org). The IUPHAR/BPS Guide divides transporters into two broad groups: ion channels and transporters (Alexander et al. 2017). The ion channels are subdivided into three families: ligand-gated channels, voltage-gated channels, and "other," which includes aquaporins, mechanically gated channels such as piezo channels, and store-operated channels such as ORAI calcium channels (see Vol. 3, Chap. 26). The IUPHAR/BPS classification divides transporters into four families: the ATP-binding cassette or ABC family, two families of ATPases, and the solute carriers or SLC family.

In 2001, the International Union of Biochemistry and Molecular Biology (IUBMB) initiated the Transporter Classification Database (TCDB, www.tcdb.org) in an effort to standardize the classification of membrane transporters using both functional and phylogenic information (Saier et al. 2016). The TCBD recognizes nine classes of transporters with more than 1000 transporter families, many of which occur only in prokaryotes. The first three classes in the TCDB are most of interest in epithelial transport: Class 1 channels/pores, Class 2 electrochemical potential-driven transporters, and Class 3 primary active transporters. Class 2.A proteins are the "porters" or carriers that use potential energy stored in electrochemical gradients to move substrates. Class 3 primary active transporters use chemical, electrical, or light energy to power substrate translocation. The main source of chemical energy in eukaryotes is the high-energy phosphate bond of ATP. Electrical active transport uses energy from exothermic electron transfer between substrates, which occurs in eukaryotic organelles such as mitochondria and chloroplasts. Photosynthetic proteins and rhodopsin are examples of transporters that use light energy.

Finally, the Human Genome Organization (HUGO) Gene Nomenclature Committee, the HGNC, has taken responsibility for developing a standardized nomenclature of symbols and names for human genes, searchable at www.genenames.org (Yates et al. 2017). The HGNC solute carrier (SLC) group currently contains 417 genes, each with a HGNC number and approved symbol. Unfortunately, the HUGO nomenclature frequently does not match other commonly used names and symbols. For example, the Na-dependent glucose transporter found primarily on intestinal enterocytes is called SGLT1 in the biomedical literature and the IUPHAR/BPS Guide. Its HUGO approved gene name is solute carrier family 5 member 1, or SLC5A1, and its gene is given the number HGNC:11036. The protein is called 2.A.21.3.1 in the IUBMB Transporter Classification Database, where it is also listed under the aliases SL51, SGLT, SLC5A1 and SGLT1. The possibilities for confusion are endless, and people new to the field need to be alert to the alternate terminologies.

2.3.5   Conventions for Drawing Transport

The conventions for drawing epithelial transport are not standardized and therefore, like the nomenclature, can be confusing to those new to the field. Figure 2.5

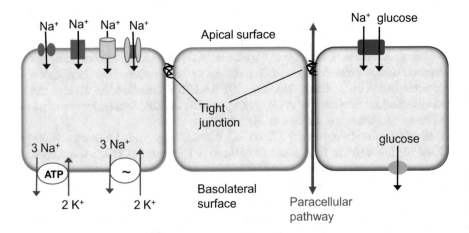

Fig. 2.5 Conventions for drawing transport

illustrates some of the most common ways to draw the different types of membrane proteins. Typical drawings will show two or more cells, connected by tight junctions, with the apical and basolateral sides labeled. The apical side may simply say "lumen" or "mucosal surface," while the basolateral side may be labeled "blood" or "extracellular fluid."

Channels, such as the pink Na^+ channels in the figure, have the most variable representation but usually have some indication of the open passageway between the cytoplasm and outside. Carriers are presented as solid shapes. Primary active transporters may have ATP or a tilde (~) inside the shape representing the protein. (A tilde indicates a high-energy phosphate bond, as in $ADP\sim P_i$ for ATP.)

Directional movement of substrates is indicated by arrows. Sometimes numbers indicate the stoichiometry of transport, as shown here for ATP, but many figures do not include the stoichiometry. Arrows for paracellular transport go between the cells.

2.3.6 Transepithelial Transport

Movement of material from one compartment across the epithelium to the other compartment is known as transepithelial transport. In most cases, transport of material from the extracellular fluid into a lumen is described as secretion, while movement from the lumen into the ECF is called *absorption*, or in the case of the kidney, *reabsorption* (because solutes in the lumen have just been filtered there from the ECF). The luminal compartment is most commonly the external environment but for some specialized epithelia, such as the choroid plexus, the lumen is a transcellular compartment (see Table 2.2).

The term *secretion* is another point of potential confusion for novices in the field. As just mentioned, secretion can refer to transepithelial movement of material from ECF into a lumen. However, it also describes the release of cell products by a cell. Endocrine and exocrine epithelial cells synthesize hormones or other molecules and secrete them from the cytoplasm into an adjacent compartment, either the lumen of a duct or the ECF (Fig. 2.6a). The secretory epithelial are represented by a two-compartment model, with the cell as one compartment and the ECF or lumen as the second compartment.

Most transporting and exchange epithelia are best represented with a three-compartment model, first proposed in 1962 (Curran and MacIntosh 1962). Usually the three compartments are the extracellular fluid compartment on the basolateral side of the epithelium and a luminal compartment on the apical side, with the epithelial cell layer sandwiched between (Fig. 2.6b). The classic apparatus for studying epithelial transport, Hans Ussing's "little chamber," is the literal representation of the theoretical three-compartment model (Li et al. 2004; Hamilton 2011). Transporting epithelia can be primarily absorptive, moving material from the external environment into the body (for example, the proximal tubule of the kidney); primarily secretory, like the choroid plexus; or mixed, like the collecting duct of the

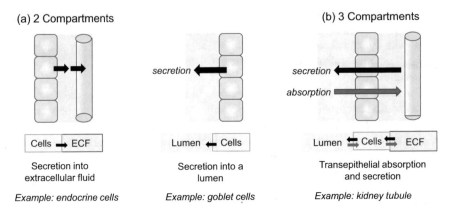

Fig. 2.6 Compartment models for epithelial transport. Key: ECF, extracellular fluid

nephron that uses regulated absorption and secretion to maintain ion and water homeostasis.

Cell polarity, with different populations of transport proteins on the two surfaces of the epithelial layer, is critical for directional transport across the epithelium. For example, absorptive epithelia such as intestine and nephron have the Na^+-K^+-ATPase localized on the basolateral membrane, while secretory epithelia such as the choroid plexus put the ATPase on the apical membrane (Steffensen et al. 2018). The two types of intercalated cells of the renal collecting duct direct H^+-ATPase and Cl^--HCO_3^- transporters to apical and basolateral membranes according to whether the cell is destined to compensate for acidosis (type A cells) or alkalosis (type B cells) (Giebisch et al. 2017). Type A cells put the H^+-ATPase on the apical membrane and the Cl^--HCO_3^- exchanger on the basolateral side, while type B cells reverse this orientation.

Cell polarity plays a role in vesicular transport as well. Some endosomes, the vesicles created by endocytosis, remain closely associated with either the apical or basolateral domain and have the ability to recycle the endosomal membrane and associated cargo back to the surface from which the vesicle was formed. A well-known example of such membrane recycling is the insertion and removal of aquaporin water channels in the renal collecting duct under the control of the hormone vasopressin (Brown 1989). In other tissues, endosomes formed on one side of the cell can send their cargo to the opposite side of the cell, a process known as transcytosis. Substances that cross epithelia using transcytosis include immunoglobulins, large proteins such as the hormone insulin, and some viruses and bacterial toxins (Garcia-Castillo et al. 2017).

2.3.7 Transepithelial Potential Differences

One of the earliest properties observed in epithelia as far back as the mid-1800s was the ability to create transepithelial electrical potential differences, that is, uneven distribution of charge between the two compartments separated by the epithelium. Hans Ussing and Karl Zerahn (1951) described how isolated living frog skin suspended between two compartments with identical Ringer's saline solutions developed a transepithelial potential difference. The basolateral (serosal) solution became as much as 100 mV positive, relative to the apical (external environment) solution, presumably due to the transport of Na^+ across the skin.

The role of transepithelial potentials in normal physiology and pathophysiology is mostly overlooked outside the epithelial transport community, despite the influence these electrical gradients have on fluid-electrolyte balance and in various secretory processes. Among epithelial researchers, manipulation of the transepithelial potential has proved to be a key element in developing our understanding of transport processes.

In that same 1951 paper, Ussing and Zerahn introduced the apparatus now known as the Ussing chamber, in which a flat sheet of transporting epithelium is mounted between chambers filled with apical and basolateral bathing solutions (Fig. 2.7). Their use of current-passing electrodes in addition to recording electrodes marked the beginning of a new era in the study of epithelial transport. By passing current that returned the transepithelial potential to zero—the short-circuit current—they were able to quantitatively estimate the magnitude of ion transport across the tissue. Additional details about the history and use of Ussing chambers in the study of transport can be found in Chap. 1 of this volume as well as in Cotton and Reuss

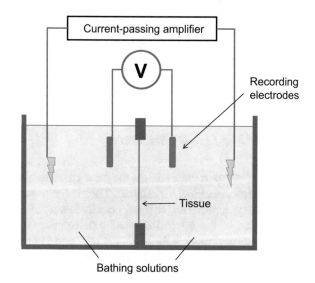

Fig. 2.7 Schematic of a Ussing chamber. A flat sheet of tissue is suspended between two chambers filled with bathing solutions whose composition can be manipulated. Recording electrodes and a voltmeter monitor the transepithelial potential difference. Current-passing electrodes can be used to short-circuit the epithelium

Current-passing amplifier

V

Recording electrodes

Tissue

Bathing solutions

Fig. 2.8 Electrical circuit model of the epithelium. The total resistance of the tissue (E) is determined by the transcellular resistance and the paracellular (junctional, R_J) resistance working in parallel. The transcellular resistance consists of the apical (R_A) and basolateral membrane resistances (R_B) in series

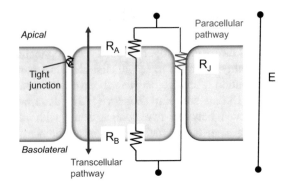

1996; Lewis 1996b; Larsen 2002; Li et al. 2004; Palmer and Andersen 2008; and Clarke 2009.

Transporting epithelia can be modeled with an electrical circuit model, illustrated in Fig. 2.8. The total transepithelial resistance to current flow (E) consists of two resistances in parallel: a transcellular resistance and a paracellular resistance. Transcellular resistance is determined by the resistances of the apical and basolateral membranes in series, where $R_{transcellular} = R_A + R_B$. The transcellular resistance is usually high but depends on the membrane transporters of the apical and basolateral membranes. The paracellular resistance (R_J) comes primarily from the resistance of the junctional complex and ranges from very low to very high, depending on the tissue.

Epithelia can be categorized as 'tight' or 'leaky' according to the magnitude of the transepithelial potential they generate when identical solutions are placed on both surfaces (Lewis 1996a). In most instances 'leaky' epithelia have low-resistance cell-cell junctions and 'tight' epithelia have high-resistance junctions. The potential difference created by tight epithelia requires electrogenic active transport of ions by the cells, coupled with tight junctions that prevent the paracellular movement of ions attracted by the electrical gradient. Significant active transport can also be found in leaky epithelia, but movement of ions through the paracellular junctions dissipates the transepithelial potential created by active transport (Frömter and Diamond 1972).

2.3.8 Flow Down Gradients

Epithelial transport depends on the flow of solutes and water down gradients—concentration, electrical, or osmotic. The gradients in turn are created or maintained by the energy input of active transport. Given a set of membrane transporters and the composition of the intracellular and extracellular compartments, it is possible to create functional models for epithelial transport in any epithelium by using combinations of paracellular and transcellular transport, and the concentration and electrical gradients.

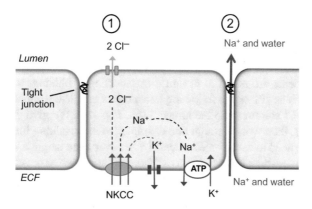

Fig. 2.9 Combined transcellular and paracellular secretion of fluid. In various tissues, such as the exocrine pancreas and salivary gland, cellular transport mechanisms move Cl⁻ from the extracellular fluid to the lumen of the gland (1). This creates a lumen-negative transepithelial potential difference that attracts Na⁺ through the cation-selective pores of the claudins that make up the tight junctions (2). Transfer of solute from the ECF to the lumen produces an osmotic gradient so that water follows Na⁺ through the paracellular pathway into the lumen. Key: NKCC, Na-K-2Cl cotransporter; ECF, extracellular fluid

An example of combined transcellular and paracellular ion transport is seen in Fig. 2.9, which illustrates fluid and electrolyte secretion by pancreatic exocrine cells and salivary glands (Lee et al. 2012). Transcellular transport mechanisms first move Cl^- from the extracellular fluid compartment to the lumen. The Na^+-K^+-2 Cl^- symporter NKCC concentrates Cl^- in the cell, while an apical Cl^- channel (CFTR in the pancreas, a Ca^{2+}-activated Cl^- channel in the salivary acinar cells) allows the Cl^- to exit into the lumen. The K^+ and Na^+ that enter the cell on the NKCC recycle back to the basolateral compartment via a K^+ channel and the Na^+-K^+-ATPase. Net movement of Cl⁻ from ECF to lumen creates a lumen-negative electrical gradient that attracts Na^+ through claudin pores of the tight junction. Translocation of NaCl from ECF to lumen causes an osmotic gradient that draws water through the paracellular pathway along with the Na^+ ions. Similar combinations of transcellular and paracellular transport create a pattern seen in many fluid-secreting epithelia.

2.4 Teaching Epithelial Transport

In this final section we change gears from what we know about epithelial transport to how we can teach it to students. Epithelial transport is a fundamental concept in animal biology but is seldom taught as a unified theme. Most students are exposed to transepithelial transport when they study the kidney and digestive tract but all too often, they come to think of absorption and secretion in the nephron, intestine, and exocrine pancreas as unique and unrelated processes. Learning the transport mechanisms in different tissues becomes an exercise in rote memorization of what transporters occur on which surface of the cell. How much simpler it would be to teach students to think about the electrochemical gradients and place the appropriate transporters on the correct surface using logic.

What follows is a description of this approach to teaching the principles of transport. Students are asked to design a transporting epithelium to accomplish a given task, such as "Place the following transporters on a leaky epithelium so that it secretes an NaCl solution into the lumen." (see Fig. 2.9). We give them a series of steps to follow, then work through one example together, talking them through the questions they need to ask. Some students have never been taught a systematic way to problem-solve—they use a random or 'shotgun' approach to develop their model rather than applying logic to narrow down the possibilities. We model the thought process with the example, then give them some 'unknown' epithelia to design. Usually we simplify the model and limit their possibilities by restricting the number of transporters they can use, but other times we give them multiple transporters so that they can figure out more than one possible arrangement to accomplish the task. The emphasis is on what works and what does not, not on what the actual mechanism is in the living tissue.

Here is one example and how we walk through it with the students.

The task:

1. Design an epithelium that absorbs NaCl from the lumen.
2. Assumptions for this epithelium: Assume normal intracellular and extra-cellular ion concentrations (i.e., K^+ high in intracellular fluid, Na^+ and Cl^- high in the ECF), and that the lumen contains the equivalent of ECF. Cell junctions are tight and do not allow paracellular transport of water and ions.
3. Available transporters: Na^+-K^+-2 Cl^- (NKCC) symporter, K^+ leak channel, Na^+ leak channel, Na^+-K^+-ATPase, Cl^- channel. You may not need to use all of them.

Instructions to students:

1. Draw the epithelium with three cells and label the compartments.
2. At one edge of the epithelium, note the task (absorb NaCl from the lumen). Use directional arrows.
3. Label the compartments with ion concentrations and note whether cell junctions are tight or leaky.
   (At this point, students should have a drawing that looks like Fig. 2.10.)
4. Look at the list of available transporters and describe which way they move ions relative to the gradients.

   • Na^+-K^+-2 Cl^- (NKCC) symporter: secondary active transporter that uses the energy of the Na^+ gradient to move K^+ and Cl^- against their gradients. Which way do the ions move? (*Answer*: all move into the cell)

(continued)

- K⁺ leak channel: which way does K⁺ move when the channel is open? (*Answer:* out of the cell)
- Na⁺ leak channel: which way does Na⁺ move when the channel is open? (*Answer:* into the cell)
- Na⁺-K⁺-ATPase: Which direction do the ions move? (*Answer:* Na⁺ out of the cell and K⁺ into the cell)
- Cl⁻ channel: which way does Cl⁻ move? (*Answer:* Cl⁻ movement follows Cl⁻ electrochemical gradients. When Cl⁻ has been concentrated in the cell by the NKCC, the Cl⁻ channel will allow it to leak back out.)

5. Arrange the transporters on the cell so that Na⁺ and Cl⁻ are moved from lumen to ECF without allowing K⁺ to build up within the cell.

The preferred student answer is shown in Fig. 2.11. An apical NKCC brings Cl⁻ and K⁺ into the cell using energy from the Na⁺ gradient. Concentrating Cl⁻ in the cell then allows it to exit through a basolateral Cl⁻ channel. The cytoplasmic Na⁺ is moved to the ECF using the Na⁺-K⁺-ATPase. Apical and basolateral K⁺ leak channels recycle K⁺, preventing a change in cell resting membrane potential. This cell model that students create in this task is very similar to transport mechanisms seen in cells of the thick ascending limb of the loop of Henle (Silverthorn 2019).

Talking with students as they work on the task is an excellent way to uncover student misconceptions about transport. They may want to place a Na⁺ channel on the basolateral membrane to have Na⁺ leave the cell, but the electrochemical gradient is in the wrong direction—Na⁺ will enter the cell. Sometimes students try to reverse the direction of ion exchange on the Na⁺-K⁺-ATPase, forgetting why this transporter

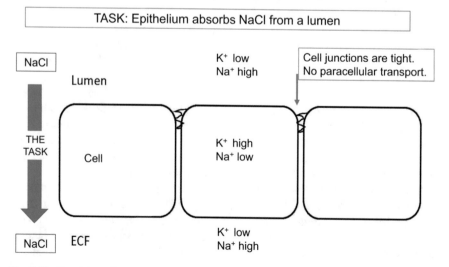

Fig. 2.10 Step 1 in a "design a cell" problem. ECF, extracellular fluid

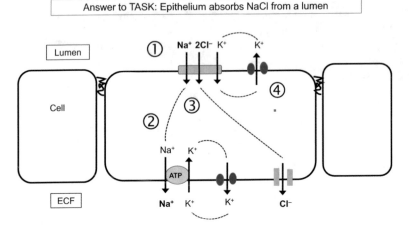

Fig. 2.11 Preferred student answer to "design a cell" task. Key: ECF, extracellular fluid

requires energy from ATP. One acceptable solution not shown is to place a Na^+ leak channel on the apical membrane, but we point out that although this brings Na^+ into the cell, it will also decrease the Na^+ gradient for the NKCC and possibly decrease Cl^- absorption into the cell.

We introduce this "design a cell" activity early in the curriculum as a discovery learning exercise (Svinicki 1998). During the cell physiology introduction to membrane transport (similar to Sects. 2.3.2 and 2.3.3), we ask students to apply what they have learned about active transport and facilitated diffusion to move Na^+ and glucose across an epithelium using only three transporters: GLUT, Na^+-K^+-ATPase, and SGLT. This exercise occurs before the students have been assigned the epithelial transport section to study, so they are independently "discovering" the process by thinking about it and applying what they have learned about transporters and gradients. This SGLT-mediated mechanism for glucose absorption appears again later in the course when students study the proximal tubule and the small intestine, but by that point it is familiar and less intimidating.

We also use the "design a cell" discovery method for selected other processes in the kidney and digestive tract. Processes that lend themselves particularly well to the strategy include aldosterone-mediated Na^+ reabsorption and K^+ secretion in renal principal cells, CFTR-mediated secretory diarrhea, and pancreatic bicarbonate secretion. But how do we know that students have really learned the transport principles and not simply memorized the cell models that appear in their reading?

The answer is to use transporting epithelia other than kidney and GI tract, most of which receive little attention in the typical physiology curriculum. How many students have been introduced to secretion of cerebrospinal fluid by the choroid plexus or milk production and secretion in the mammary gland? These oft-ignored tissues provide excellent examples for testing student understanding: present students with the transport goal of the tissue plus a set of transport proteins, and see if they can transfer their understanding of electrochemical and osmotic gradients to create a cell that carries out the desired process without disrupting the cell's membrane potential.

Mammary gland milk secretion is a particularly interesting example because it includes transcytosis of immunoglobulins produced by subepithelial immune cells as well as intracellular protein synthesis and packaging (Silverthorn 2003). Choroid plexus fluid secretion provides an opportunity for students to see that the Na^+-K^+-ATPase can occur on the apical membrane as well as the basolateral membrane. Once students learn to look at the gradients and think about the transporters, they have moved beyond simple memorization and are applying the basic principles of epithelial transport to solve problems.

Acknowledgements Thanks to Jan Melton Machart, Andrew Bergemann, and Steve Tyler for their comments and suggestions that improved this manuscript.

References

Adams EDM, Goss GG, Leys SP (2010) Freshwater sponges have functional, sealing epithelia with high transepithelial resistance and negative transepithelial potential. PLoS One 5:e15040. https://doi.org/10.1371/journal.pone.0015040

Aiello NM, Kang Y (2019) Context-dependent EMT programs in cancer metastasis. J Exp Med. Downloaded from jem.rupress.org. https://doi.org/10.1084/jem.20181827

Alcorn JL (2017) Pulmonary surfactant trafficking and homeostasis. In: Sidhaye VK, Koval M (eds) Lung epithelial biology in the pathogenesis of pulmonary disease. Academic Press, New York, pp 59–75. https://doi.org/10.1016/B978-0-12-803809-3.00004-X

Alexander SPH, Kelly E, Marrion NV, Peters JA, Faccenda E, Harding SD, Pawson AJ, Sharman JL, Southan C, Buneman OP, Cidlowski JA, Christopoulos A, Davenport AP, Fabbro D, Spedding M, Striessnig J, Davies JA, Collaborators CGTP (2017) The concise guide to pharmacology 2017/18. Br J Pharmacol 174(Suppl 1):S1–S446

Anderson JM, Van Itallie CM (2009) Physiology and function of the tight junction. Cold Spring Harb Perspect Biol 1:a002584. https://doi.org/10.1101/cshperspect.a002584

Armitage SK, Plotnikov SV (2016) Focal adhesions: bridging the gap. eLife 5:e19733. https://doi.org/10.7554/eLife.19733

Bai X, Moraes TF, Reithmeier RAF (2017) Structural biology of solute carrier (SLC) membrane transport proteins. Mol Membr Biol 34:1–32. https://doi.org/10.1080/09687688.2018.1448123

Balachander N, Masthan KMK, Aravindha Babu N, Anbazhagan V (2015) Myoepithelial cells in pathology. J Pharm Bioallied Sci 7(Suppl 1):S190–S193. https://doi.org/10.4103/0975-7406.155898

Belahbib H, Renard E, Santini S, Jourda C, Claverie J-M, Borchiellini C, LeBivic A (2018) New genomic data and analyses challenge the traditional vision of animal epithelium evolution. BMC Genomics 19:393. https://doi.org/10.1186/s12864-018-4715-9

Bito LZ (1982) Composition of intraocular fluids and the microenvironment of the retina. In: Lajtha A (ed) Handbook of neurochemistry, Neurochemical systems, vol 8, 2nd edn. Springer, Boston, pp 477–506. https://doi.org/10.1007/978-1-4684-7018-5

Blasky AJ, Mangan A, Prekeris R (2015) Polarized protein transport and lumen formation during epithelial tissue morphogenesis. Annu Rev Cell Dev Biol 31:575–591. https://doi.org/10.1146/annurev-cellbio-100814-125323

Bleich M, Shan Q, Himmerkus N (2012) Calcium regulation of tight junction permeability. Ann NY Acad Sci 1258:93–99. https://doi.org/10.1111/j.1749-6632.2012.06539.x

Borradori L, Sonnenberg A (1999) Structure and function of hemidesmosomes: more than simple adhesion complexes. J Invest Dermatol 112:411–418

Brabletz T, Kalluri R, Nieto MA, Weinberg RA (2018) EMT in cancer. Nat Rev Cancer 18:128–134. https://doi.org/10.1038/nrc.2017.118

Brasch J, Harrison OJ, Honig B, Shapiro L (2012) Thinking outside the cell: how cadherins drive adhesion. Trends Cell Biol. 22:299–310

Brown D (1989) Membrane recycling and epithelial cell function. Am J Physiol Renal Physiol 256: F1–F12. https://doi.org/10.1152/ajprenal.1989.256.1.F1

Bullough PG (2010) Orthopaedic pathology, 5th edn. Elsevier, New York

Burns JC, Stone JS (2017) Development and regeneration of vestibular hair cells in mammals. Sem Cell Dev Biol 65:96–105

Byrne JH, Schultz SG (1988) An introduction to membrane transport and bioelectricity. Raven, New York

Campanale JP, Sun TY, Montell DJ (2017) Development and dynamics of cell polarity at a glance. J Cell Sci 130:1201–1207. https://doi.org/10.1242/jcs.188599

Carafoli E, Krebs J (2016) Why calcium? How calcium became the best communicator. J Cell Biol 291:20849–20857. https://doi.org/10.1074/jbc.R116.735894

Chang B, Svoboda KKH, Liu X (2019) Cell polarization: from epithelial cells to odontoblasts. Eur J Cell Biol 98:1–11

Chaudhari N, Roper SD (2010) The cell biology of taste. J Cell Biol 190(3):285–296. www.jcb.org/cgi/doi/10.1083/jcb.201003144

Chen T, You Y, Jiang H, Wang ZZ (2017) Epithelial–mesenchymal transition (EMT): a biological process in the development, stem cell differentiation, and tumorigenesis. J Cell Physiol 232:3261–3272. https://doi.org/10.1002/jcp.25797

Chitturi RT, Veeravarmal V, Nirmal RM, Reddy BVR (2015) Myoepithelial cells (MEC) of the salivary glands in health and tumours. J Clin Diagn Res 9:ZE14–ZE18. https://doi.org/10.7860/JCDR/2015/11372.5707

Clarke LL (2009) A guide to Ussing chamber studies of mouse intestine. Am J Physiol Gastrointest Liver Physiol 296:G1151–G1166. https://doi.org/10.1152/ajpgi.90649.2008

Cotton CU, Reuss L (1996) Characterization of epithelial ion transport. In: Wills NK, Reuss L, Lewis SA (eds) Epithelial transport: a guide to methods and experimental analysis. Chapman and Hall, London, pp 70–92

Curran PF, MacIntosh JR (1962) A model system for biological water transport. Nature 193:347–348

Denker BM, Sabath E (2011) The biology of epithelial cell tight junctions in the kidney. J Am Soc Nephrol 22:622–625. https://doi.org/10.1681/ASN.2010090922

DiVenere M, Viglio S, Cagnone M, Bardoni A, Salvini R, Iadarola P (2018) Advances in the analysis of "less-conventional" human body fluids: an overview of the CE- and HPLC-MS applications in the years 2015–2017. Electrophoresis 39:160–178. https://doi.org/10.1002/elps.201700276

Downs C (2017) Ion transport and lung fluid balance. In: Sidhaye VK, Koval M (eds) Lung epithelial biology in the pathogenesis of pulmonary disease. Academic, New York, pp 21–31. https://doi.org/10.1016/B978-0-12-803809-3.00004-X

Ebnet K, Kummera D, Steinbacher T, Singh A, Nakayama M, Matis M (2018) Regulation of cell polarity by cell adhesion receptors. Sem Cell Dev Biol 81:2–12. https://doi.org/10.1016/j.semcdb.2017.07.032

Eichner M, Protze J, Piontek A, Krause G, Piontek J (2017) Targeting and alteration of tight junctions by bacteria and their virulence factors such as *Clostridium perfringens* enterotoxin. Pflugers Arch 469:77–90. https://doi.org/10.1007/s00424-016-1902-x

Faa G, Gerosa C, Fanni D, Monga G, Zaffanello M, Van Eyken P, Fanos V (2012) Morphogenesis and molecular mechanisms involved in human kidney development. J Cell Physiol 227:1257–1268. https://doi.org/10.1002/jcp.22985

Farrell PM, White TB, Ren CL, Hempstead SE, Accurso F, Derichs N, Howenstine M, McColley SA, Rock M, Rosenfeld M, Sermet-Gaudelus I, Southern KW, Marshall BC, Sosnay PR (2017) Diagnosis of cystic fibrosis: consensus guidelines from the Cystic Fibrosis Foundation. J Pediatr 181S:S4–S15. https://doi.org/10.1016/j.jpeds.2016.09.064

Fawcett DW (1994) Bloom and Fawcett: a textbook of histology, 12th edn. Chapman & Hall, New York

Firbas W, Gruber H, Wicke W (1981) The blood vessels of the limbus spiralis. Arch Otorhinolaryngol 232:131–137

Fleming TP, Johnson MH (1988) From egg to epithelium. Ann Rev Cell Biol 4:459–485

Frömter E, Diamond J (1972) Route of passive ion permeation in epithelia. Nat New Biol 235:9–13

Galipeau J, Sensebe L (2018) Mesenchymal stromal cells: clinical challenges and therapeutic opportunities. Cell Stem Cell 22:824–833. https://doi.org/10.1016/j.stem.2018.05.004

Garcia-Castillo MD, Chinnapen DJ-F, Lencer WI (2017) Membrane transport across polarized epithelia. Cold Spring Harb Perspect Biol 9:a027912. https://doi.org/10.1101/cshperspect.a027912

Gibson AT, Segal MB (1978) A study of the composition of pericardial fluid, with special reference to the probable mechanism of fluid formation. J Physiol 277:367–377

Giebisch G, Windhager EE, Aronson PS (2017) Transport of acids and bases. In: Boron WF, Boulpaep EL (eds) Medical physiology, 3rd edn. Elsevier, New York, pp 821–835

Goel M, Picciani RG, Lee RK, Bhattacharya SK (2010) Aqueous humor dynamics: a review. Open Ophthalmol J 4:52–59

Gonzalez-Mariscal L, Contreras RG, Bolívar JJ, Ponce A, Chávez DeRamirez B, Cereijido M (1990) Role of calcium in tight junction formation between epithelial cells. Am J Physiol 259:C978–C986. https://doi.org/10.1152/ajpcell.1990.259.6.C978

Goutman JD, Elgoyhen AB, Gómez-Casati ME (2015) Cochlear hair cells: the sound-sensing machines. FEBS Lett 589:3354–3361. https://doi.org/10.1016/j.febslet.2015.08.030

Greka A, Mundel P (2012) Cell biology and pathology of podocytes. Annu Rev Physiol 74:299–323. https://doi.org/10.1146/annurev-physiol-020911-153238

Hall A (1998) Rho GTPases and the actin cytoskeleton. Science 279:509–514. https://doi.org/10.1126/science.279.5350.509

Hamilton KL (2011) Ussing's "little chamber": 60 years+ old and counting. Front Physiol 2:6. https://doi.org/10.3389/fphys.2011.00006

He P, Zhang Y, Kim SO, Radlanski RJ, Butcher K, Schneider RA, DenBesten PK (2010) Ameloblast differentiation in the human developing tooth: effects of extracellular matrices. Matrix Biol 29:411–419. https://doi.org/10.1016/j.matbio.2010.03.001

Jiang WG, Sanders AJ, Katoh M, Ungefroren H, Gieseler F, Prince M, Thompson SK, Zollo M, Spano D, Dhawan P, Sliva D, Subbarayan PR, Sarkar M, Honoki K, Fujii H, Georgakilas AG, Amedei A, Niccolai E, Amin A, Ashraf SS, Ye L, Helferich WG, Yang X, Boosani CS, Guha G, Ciriolo MR, Aquilano K, Chen S, Azmi AS, Keith WN, Bilsland A, Bhakta D, Halicka D, Nowsheen S, Pantano F, Santini D (2015) Tissue invasion and *metastasis*: molecular, biological and clinical perspectives. Semin Cancer Biol 35:S244–S275. https://doi.org/10.1016/j.semcancer.2015.03.008

Jorgensen PL, Håkansson KO, Karlish SJD (2003) Structure and mechanism of Na,K-ATPase: functional sites and their interactions. Ann Rev Physiol 65:817–849. https://doi.org/10.1146/annurev.physiol.65.092101.142558

Kalcheim C (2016) Epithelial–mesenchymal transitions during neural crest and somite development. J Clin Med 5:1–15. https://doi.org/10.3390/jcm5010001

Kalluri R, Weinberg RA (2009) The basics of epithelial-mesenchymal transition. J Clin Invest 119:1420–1428. https://doi.org/10.1172/JCI39104

Kelton JG, Ulan R, Stiller C, Holmes E (1978) Comparison of chemical composition of peritoneal fluid and serum: a method for monitoring dialysis patients and a tool for assessing binding to serum proteins in vivo. Ann Intern Med. 89:67–70

Kiecker C, Bates T, Bell E (2016) Molecular specification of germ layers in vertebrate embryos. Cell Mol Life Sci 73:923–947. https://doi.org/10.1007/s00018-015-2092-y

Kim DH, Xing T, Yang Z, Dudek R, Lu Q, Chen Y-H (2018) Epithelial mesenchymal transition in embryonic development, tissue repair and cancer: a comprehensive overview. J Clin Med 7:1–25. https://doi.org/10.3390/jcm7010001

Lacruz RS, Habelitz S, Wright JT, Paine ML (2017) Dental enamel formation and implications for oral health and disease. Physiol Rev 97:939–993. https://doi.org/10.1152/physrev.00030.2016

Lamouille S, Xu J, Derynck R (2014) Molecular mechanisms of epithelial–mesenchymal transition. Nat Rev Mol Cell Biol 15:178–196

Landy J, Ronde E, English N, Clark SK, Hart AL, Knight SC, Ciclitira PJ, Al-Hassi HO (2016) Tight junctions in inflammatory bowel diseases and inflammatory bowel disease associated colorectal cancer. World J Gastroenterol 22:3117–3126. https://doi.org/10.3748/wjg.v22.i11. 3117

Larsen EH (2002) Hans H. Ussing—scientific work: contemporary significance and perspectives. Biochim Biophys Acta 1566:2–15

Lee MG, Ohana E, Park HW, Yang D, Muallem S (2012) Molecular mechanism of pancreatic and salivary gland fluid and HCO_3 secretion. Physiol Rev 92:39–74

Levick JR, McDonald JN (1995) Fluid movement across synovium in healthy joints: role of synovial fluid macromolecules. Ann Rheum Dis 54:417–423

Lewis SA (1996a) Epithelial structure and function. In: Wills NK, Reuss L, Lewis SA (eds) Epithelial transport: a guide to methods and experimental analysis. Chapman and Hall, London, pp 1–20

Lewis SA (1996b) Epithelial electrophysiology. In: Wills NK, Reuss L, Lewis SA (eds) Epithelial transport: a guide to methods and experimental analysis. Chapman and Hall, London, pp 93–117

Lewis SA (2000) Everything you wanted to know about the bladder epithelium but were afraid to ask. Am J Physiol Renal Physiol 278:F867–F874

Leys SP, Nichols SA, Adams EDM (2009) Epithelia and integration in sponges. Integr Comp Biol 49:167–177. https://doi.org/10.1093/icb/icp038

Li C, Balazsi G (2018) A landscape view on the interplay between EMT and cancer metastasis. NPJ Syst Biol Appl 4:34. https://doi.org/10.1038/s41540-018-0068-x

Li H, Sheppard DN, Hug MJ (2004) Transepithelial electrical measurements with the Ussing chamber. J Cystic Fibrosis 3:123–126. https://doi.org/10.1016/j.jcf.2004.05.026

Macknight ADC, McLaughlin CW, Peart D, Purves RD, Carré DA, Civan MM (2000) Formation of the aqueous humor. Clin Exp Pharm Physiol 27:100–106

Martin TA (2014) The role of tight junctions in cancer metastasis. Semin Cell Dev Biol 36:224–231. https://doi.org/10.1016/j.semcdb.2014.09.008

Mescher AL (2018) Junqueira's basic histology: text and atlas, 15th edn. McGraw-Hill, New York

Mitchell P (1967) Translocations through natural membranes. Adv Enzymol 29:33–87

Moe OW, Wright SH, Palacín M (2016) Renal handling of organic solutes. In: Skorecki K, Chertow GM, Marsden PA, Taal MW, Yu ASL (eds) Brenner and Rector's the kidney, 10th edn. Elsevier, New York

Müller H-AJ, Bossinger O (2003) Molecular networks controlling epithelial cell polarity in development. Mech Dev 120:1231–1256

Murray PS, Zaidel-Bar R (2014) Pre-metazoan origins and evolution of the cadherin adhesome. Biol Open 3:1183–1195. https://doi.org/10.1242/bio.20149761

NC-IUBMB Nomenclature Committee of the International Union of Biochemistry and Molecular Biology (2019) Classification and nomenclature of enzymes by the reactions they catalyse. https://www.qmul.ac.uk/sbcs/iubmb/enzyme/rules.html. Accessed 5 Jan 2019

Nichols SA, Roberts BW, Richter DJ, Fairclough SR, King N (2012) Origin of metazoan cadherin diversity and the antiquity of the classical cadherin/β-catenin complex. Proc Natl Acad Sci USA 109:13046–13051. https://doi.org/10.1073/pnas.1120685109

Palmer LG, Andersen OS (2008) The two-membrane model of epithelial transport: Koefoed-Johnsen and Ussing (1958). J Gen Physiol 132:607–612. www.jgp.org/cgi/doi/10.1085/jgp.200810149

Pei L, Solis G, Nguyen MTX, Kamat N, Magenheimer L, Zhuo M, Li J, Curry J, McDonough AA, Fields TA, Welch WJ, Yu ASL (2016) Paracellular epithelial sodium transport maximizes energy efficiency in the kidney. J Clin Invest 126:2509–2518. https://doi.org/10.1172/JCI83942

Pennisi E (2018) The power of many. Science 360(6396):1388–1391

Pozzi A, Yurchenco PD, Iozzo RV (2017) The nature and biology of basement membranes. Matrix Biol 57–58:1–11. https://doi.org/10.1016/j.matbio.2016.12.009

Praetorius J, Damkier HH (2017) Transport across the choroid plexus epithelium. Am J Physiol Cell Physiol 312:C673–C686. https://doi.org/10.1152/ajpcell.00041.2017

Randles MJ, Humphries MJ, Lennon R (2017) Proteomic definitions of basement membrane composition in health and disease. Matrix Biol 57–58:12–28. https://doi.org/10.1016/j.matbio.2016.08.006

Reiser J, Altintas MM (2016) Podocytes [version 1; peer review: 2 approved]. F1000Research 5 (F1000 Faculty Rev):114. https://doi.org/10.12688/f1000research.7255.1

Rodriguez-Boulan E, Macara IG (2014) Organization and execution of the epithelial polarity programme. Nat Rev Mol Cell Biol 15:225–242. https://doi.org/10.1038/nrm3775

Saier MH Jr, Reddy VS, Tsu BV, Ahmed MS, Li C, Moreno-Hagelsieb G (2016) The transporter classification database (TCDB): recent advances. Nucleic Acids Res 44(D1):D372–D379. https://doi.org/10.1093/nar/gkv1103

Sakkaa L, Coll G, Chazal J (2011) Anatomy and physiology of cerebrospinal fluid. Eur Ann Otorhinolaryngol Head Neck Dis 128:309–316. https://doi.org/10.1016/j.anorl.2011.03.002

Salvador E, Burek M, Förster CY (2016) Tight junctions and the tumor microenvironment. Curr Pathobiol Rep 4:135–145. https://doi.org/10.1007/s40139-016-0106-6

Sebé-Pedrós A, Degnan BM, Ruiz-Trillo I (2017) The origin of Metazoa: a unicellular perspective. Nat Rev Genetics 18:498–512

Seidman AJ, Limaiem F (2019) Synovial fluid analysis. In: StatPearls [Internet]. StatPearls, Treasure Island, FL. Available from: https://www.ncbi.nlm.nih.gov/books/NBK537114/. Accessed 28 Apr 2019

Shook D, Keller R (2003) Mechanisms, mechanics and function of epithelial–mesenchymal transitions in early development. Mech Dev 120:1351–1383. https://doi.org/10.1016/j.mod.2003.06.005

Silverthorn DU (2003) Milk secretion—a transport question. Life Science Teaching Resource Community. https://www.lifescitrc.org/resource.cfm?submissionID=125

Silverthorn DU (2019) Human physiology: an integrated approach, 8th edn. Pearson, New York

Spector R, Keep RF, Snodgrass SR, Smith QR, Johanson CE (2015a) A balanced view of choroid plexus structure and function: focus on adult humans. Exp Neurol 273:78–86

Spector R, Snodgrass RS, Johanson CE (2015b) A balanced view of the cerebrospinal fluid composition and functions: focus on adult humans. Exp Neurol 273:57–68. https://doi.org/10.1016/j.expneurol.2015.07.027

Steffensen AB, Oernbo EK, Stoica A, Gerkau NJ, Barbuskaite D, Tritsaris K, Rose CR, MacAulay N (2018) Cotransporter-mediated water transport underlying cerebrospinal fluid formation. Nat Commun 9:2167–2179. https://doi.org/10.1038/s41467-018-04677-9

Stillwell W (2016) An introduction to biological membranes: composition, structure and function, 2nd edn. Elsevier, New York

Stokes DL, Green NM (2003) Structure and function of the calcium pump. Ann Rev Biophys Biomol Struct 32:445–468. https://doi.org/10.1146/annurev.biophys.32.110601.142433

Strauss O (2005) The retinal pigment epithelium in visual function. Physiol Rev 85:845–881. https://doi.org/10.1152/physrev.00021.2004

Svinicki MD (1998) A theoretical foundation for discovery learning. Adv Physiol Educ 20:S4–S7

Technau U, Steele RE (2011) Evolutionary crossroads in developmental biology: Cnidaria. Development 138:1447–1458. https://doi.org/10.1242/dev.090472

Thiery JP, Acloque H, Huang RYJ, Nieto MA (2009) Epithelial-mesenchymal transitions in development and disease. Cell 139:871–890. https://doi.org/10.1016/j.cell.2009.11.007

Tsukita S, Tanaka H, Tamura A (2019) The claudins: from tight junctions to biological systems. Trends Biochem Sci 44:141–152. https://doi.org/10.1016/j.tibs.2018.09.008 141

Tucker RP, Adams JC (2014) Adhesion networks of Cnidarians: a postgenomic view. Intl Rev Cell Mol Biol 308:323–377. https://doi.org/10.1016/B978-0-12-800097-7.00008-7

Tyler S (2003) Epithelium—the primary building block for Metazoan complexity. Integr Comp Biol 43:55–63

Ullah I, Subbarao RB, Rho GJ (2015) Human mesenchymal stem cells—current trends and future prospective. Biosci Rep 35/art:e00191. https://doi.org/10.1042/BSR20150025

Ussing HH, Zerahn K (1951) Active transport of sodium as the source of electric current in the short-circuited isolated frog skin. Acta Physiol Scand 23:110–127. https://doi.org/10.1111/j.1748-1716.1951.tb00800.x

Vacca B, Barthélémy-Requin M, Burcklé C, Massey-Harroche D, Bivic AL (2015) The Crumbs3 complex. In: Ebnet K (ed) Cell polarity 1. Springer, Cham. https://doi.org/10.1007/978-3-319-14463-4_3

van Roy F, Berx G (2008) The cell-cell adhesion molecule E-cadherin. Cell Mol Life Sci 65:3756–3788. https://doi.org/10.1007/s00018-008-8281-1

Walko G, Castañón MJ, Wiche G (2015) Molecular architecture and function of the hemidesmosome. Cell Tissue Res 360:363–378. https://doi.org/10.1007/s00441-014-2061-z

Wang J, Laurie GW (2004) Organogenesis of the exocrine gland. Dev Biol 273:1–22

Wangemann P, Marcus DC (2017) Ion and fluid homeostasis in the cochlea. In: Manley GA, Gummer A, Popper A, Fay R (eds) Understanding the cochlea, Springer handbook of auditory research, vol 62. Springer International, Heidelberg, pp 253–286. https://doi.org/10.1007/978-3-319-52073-5_9

Wehrle-Haller B (2012) Structure and function of focal adhesions. Curr Opin Cell Biol 24:116–124. https://doi.org/10.1016/j.ceb.2011.11.001

Yates B, Braschi B, Gray K, Seal R, Tweedie S, Bruford E (2017) Genenames.org: the HGNC and VGNC resources in 2017. Nucleic Acids Res 45(D1):D619–D625

Yu ASL (2017) Paracellular transport as a strategy for energy conservation by multicellular organisms? Tissue Barrier 5:e1301852. https://doi.org/10.1080/21688370.2017.1301852

Zeisel MB, Dhawan P, Baumert TF (2019) Tight junction proteins in gastrointestinal and liver disease. Gut 68:547–561. https://doi.org/10.1136/gutjnl-2018-316906

Zihni C, Mills C, Matter K, Balda MS (2016) Tight junctions: from simple barriers to multifunctional molecular gates. Nat Rev Mol Cell Biol 17:564–580. https://doi.org/10.1038/nrm.2016.80

Zocchi L (2002) Physiology and pathophysiology of pleural fluid turnover. Eur Respir J 20:1545–1558. https://doi.org/10.1183/09031936.02.00062102

Chapter 3
Establishment and Maintenance of Epithelial Polarization

Andrew D. Bergemann

Abstract For the movement of solutes across epithelial layers to be effective, it must have the potential to be vectorial, which in turn means that epithelial layers must be polarized in the apical-basolateral axis. Of course, this is indeed the case—most ion channels and transporters localize specifically to either the apical or the basal membrane domains. The polarized distribution of membrane components is the result of a finely balanced developmental program, involving an antagonistic relationship between two apical protein complexes, PAR and CRUMBS, and a basolateral protein complex, SCRIBBLE. Their interactions specify the position of the adherens and tight junctions, and thereby the boundary between the apical and basolateral membrane domains. The three complexes also interact to shape the epithelial cell's microfilament and microtubule cytoskeletons, and in doing so control the polarized delivery of cargo to specific membrane domains. The effects of the complexes on the cytoskeleton are heavily dependent upon the activities of atypical protein kinase C (aPKC), and on small GTPases of the Rho family. In combination these proteins mediate many of the effects of the complexes upon the cytoskeleton, through phosphorylation of cytoskeletal components, or through activation of cytoskeletal remodeling proteins such as N-WASP. Emphasizing the importance of epithelial polarization in pathophysiology, the three components of the SCRIBBLE complex, the Scribble, Dlg, and Lgl proteins, are all known tumor suppressors. Their antagonism of oncogenesis in part reflects the importance of the basal and lateral membranes for the prevention of the epithelial-mesenchymal transition, a common feature of metastasis.

Keywords Polarity · Planar · Apical · Baso-lateral · Tight junction · Adherens junctions · Tumor suppressor

A. D. Bergemann (✉)
Department of Medical Education, Dell Medical School, University of Texas at Austin, Austin, TX, USA
e-mail: andrew.bergemann@austin.utexas.edu

© The American Physiological Society 2020
K. L. Hamilton, D. C. Devor (eds.), *Basic Epithelial Ion Transport Principles and Function*, Physiology in Health and Disease,
https://doi.org/10.1007/978-3-030-52780-8_3

3.1   Introduction

Epithelial layers play critical roles in every bodily process. Inherent to their usefulness is the capacity to create barriers, either between two bodies of fluids, or between
an organism and its external environment. Typically, these barriers separate a protein
and polysaccharide-rich fluid, termed extracellular matrix material, from a less-
viscous fluid. The less-viscous fluid is commonly contained in a hollow tube or
duct, the lumen of which is surrounded by the epithelial layer. The pole of the cell
facing the extracellular matrix is termed the basolateral compartment, while the pole
facing the lumen of the hollow tube or duct is termed the apical compartment. These
apical and basolateral compartments of each epithelial cell are biochemically distinct, the distinctions reflecting the specialized functions of each compartment.
Indeed, without the apical-basolateral polarization, epithelial cells could not perform
any of their extensive array of functions. For example, without specialization of the
apical and basolateral membranes, enterocytes of the gut could not vectorially
transport nutrients from the lumen of the gut to the lymph and the portal system
blood vessels. Even the simple act of creating a barrier would be compromised by a
loss of polarization, as the basement membrane, which is created by an interaction of
epithelial cells with underlying stromal cells, is essential for mechanical strength of
epithelial structures (Morrissey and Sherwood 2015).

The recognized major apical-basolateral polarity boundary of epithelial cells is a
junctional complex that includes the principle barrier to paracellular flow between
cells of the epithelial layer. In mammals, this complex is formed from apically-
located tight junctions and the adherens junctions. The adherens junctions, formed
principally from the intercellular interactions of cell adhesion molecules, is positioned immediately basal to the tight junctions (see Fig. 3.1). In *Drosophila (Drosophila melanogaster)*, the arrangement is slightly different, with adherens junctions
positioned apically to the septate junctions, the functional equivalent of the tight
junction (St Johnston and Ahringer 2010a). In both deuterostomes and protostomes,
the boundary is associated with an underlying cortical actomyosin band. The

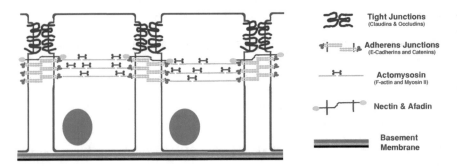

Fig. 3.1 Schematic view of a mammalian simple epithelial layer showing the relationship of the
tight junctions and adherens junctions. Schematic view of epithelial layer in the plane of the
epithelium

boundary effectively separates apical and basolateral membrane components, although the mechanism of separation is not clear (St Johnston and Ahringer 2010a). Thus, specific delivery of cargo to the membrane on one side of the boundary by the cell's transport vesicle apparatus can affect the biochemical composition of that membrane compartment, and thereby engender it with specific functions. This mechanism applies equally to conferring specific functions to either the apical or the basolateral compartments of the cell's membrane.

Despite the enormous importance of apical-basolateral polarization for epithelial function, the molecular programs controlling the formation of the polarization, and stabilizing the polarization through the life of the cell, have only been revealed over the last 30 years. In fact, the major regulators of the program were not initially revealed in studies of epithelia, but rather in studies of other examples of cellular polarization. In particular, it was the use of a powerful genetic model of cellular polarization, the polarization of the *Caenorhabditis elegans* (*C. elegans*) one-cell embryo, that revealed several master regulators of polarization, encoded by six Partitioning-defective genes (*Par1* to *Par6*) (Kemphues et al. 1988; Etemad-Moghadam et al. 1995; Watts et al. 2000). Since the initial discovery of the *Par* genes, it has been recognized that many of the key regulators have been conserved across protostomes and deuterostomes. Accordingly, studies of the mechanisms underlying polarization have extended into mammalian epithelial tissue culture experiments, mouse genetic experiments, and *Drosophila* genetics. The power of *Drosophila* genetics, in particular, has contributed to a rapid expansion of our knowledge of the molecular underpinnings of cellular polarization.

Because of numerous studies, three key protein complexes or "modules" have emerged as central to the establishment and maintenance of apical-basolateral polarity in diverse species: the PAR, CRUMBS, and SCRIBBLE complexes (Pieczynski and Margolis 2011). The PAR complex, which is essential for the establishment of the apical pole, is composed of Par3, Par6, and the protein kinase, aPKC (atypical protein kinase C). Par3 and Par6 are linker proteins holding the complex together and mediating interactions with lipids and other proteins. Particularly important are the interactions of the Par complex with Cdc42 (cell division control protein 42), a small, signal transducing GTPase (Nunes de Almeida et al. 2019), and some authors regard it as an integral component of the complex (Pichaud 2018). Much of the polarizing activity of the PAR complex is mediated through the actions of Cdc42 and aPKC on a large array of effector proteins, particularly proteins affecting remodeling of the actomyosin cytoskeleton. The CRUMBS complex, which works with the PAR complex to maintain the apical pole, is comprised of the proteins Crumbs, PALS1 (protein associated with lin-7, also called in various species, stardust, sdt, or MPP-5), and PATJ (Pals1-associated tight junction protein) (Ivanova et al. 2015). While Crumbs is a transmembrane protein, PALS1 and PATJ are cytoplasmic linker proteins (Sen et al. 2012). The last of these three complexes, the SCRIBBLE complex, is essential for maintenance of the basolateral pole and is comprised of the Dlg (discs large), Scribble, and Lgl (lethal giant larvae—the human equivalents are Llgl proteins) proteins, all three of which are large cytoplasmic scaffold proteins. As will be presented in this chapter, evidence is mounting for the importance of a fourth complex, a complex comprised of Cdc42, Par6 and aPKC (Nunes de Almeida et al. 2019).

The nomenclature of polarity proteins varies for human, mouse, and *C. elegans* and *D. melanogaster*. For ease of interpretation, in this chapter, wherever we are discussing a protein in a species-specific situation, we will use the species-specific name followed by the generic term. For example, for the *Drosophila* orthologue of Par3, called Baz or Bazooka, we will refer to as Baz/Par3 throughout the text.

3.2 Major Molecular Determinants of Polarity

Apical-basolateral polarization is a product of the interactions of two systems. One system, comprised of the PAR, CRUMBS and SCRIBBLE complexes, controls the behavior and localization of cytoskeletal elements, particularly the actomyosin circumferential belt associated with adherens junctions. The second system, a system of transport vesicles, targets the insertion of membrane proteins and phospholipids in a polarized fashion, to membrane domains either apical or basolateral relative to the adherens junctions. The interactions of these two systems are complex, although in at least some examples of polarized cells, genetic studies point to the PAR/CRUM BS/SCRIBBLE system, which directly regulates the cytoskeletal structure, as being the predominant system for the establishment of polarity (Nelson 2009). Further complexity is added by uncertainty as to the degree to which apical-basolateral polarization is cell autonomous. It is likely, therefore, that the actual initiating event for apical-basolateral polarization may vary significantly with cell type, development stage, physical location, environmental factors, species, and in the presence of a disease state, pathophysiology (St Johnston and Ahringer 2010a). Indeed, genetic studies have established that the relative importance of particular polarizing factors does vary between tissue types and developmental stages (Franz and Riechmann 2010). However, what is clear is that many of the decision points in apical-basolateral polarity are mutually re-enforcing, co-operating to cement the location of the apical-basolateral boundary. Due to the complexity of the apical-basolateral system, we will first consider the individual properties of the major players before assembling their interactions to describe the polarization process.

3.2.1 The aPKC Protein Kinases

The protein kinase C family is comprised of fifteen members in humans. The family includes well-known components of G-protein-coupled protein receptor (GPCR) and receptor tyrosine kinase (RTK) signal transduction pathways (Newton 2018). In these pathways, the protein kinase C activity is activated by a combination of diacylglycerol (DAG) and calcium ions, both of which are the products of the activation of enzymes of the phospholipase C family. The DAG is a direct product of the action of phospholipase C on the membrane phospholipid, phosphatidylinositol 3,4-bisphosphate (PIP2), while the calcium ions are released from intracellular stores (chiefly in the endoplasmic reticulum), in response to the presence of inositol-3-phosphate, another product of the phospholipase C reaction. However, the prominent PKC family members in epithelial polarization are aPKCs, or atypical PKCs (Hong 2018), which are distinctive in that they require neither DAG, nor calcium for activation. Two of the human PKCs, PKCiota and PKCzeta, fall into this category. The aPKCs are activated through protein-protein interactions. In terms of the apical-basolateral system, Par3, Par6, and Cdc42 are key regulators of aPKC activity. For the most part, the literature reflects a stimulatory role for Cdc42, an inhibitory role for Par3, and Par6 acting as an inhibitor or activator depending upon circumstances (Graybill et al. 2012; Rodriguez et al. 2017). Variability in the effect of these aPKC interactors reflects the fact that the details of apical-basolateral polarization vary depending on tissue type, developmental stage, organism and environmental cues.

A large proportion of apical-basolateral determinant proteins are aPKC substrates, as would be expected of a critical kinase in a complex, multi-faceted process. Among the relevant known substrates of aPKCs are Lgl/Llgl, Par1, Par2, Par3, Par6, and Crumbs (Hong 2018). Lgl/Llgl (and a number of other target proteins that are inhibited by aPKC-mediated phosphorylation) depends upon a highly basic region for functioning (see Fig. 3.2). The basic region promotes the electrostatic interaction of Lgl/Llgl with the negatively-charged inner leaflet of the membranes. A critical aPKC-mediated phosphorylation site on Lgl/Llgl is in the basic domain, well placed to interfere with the electrostatic attractions (Bailey and Prehoda 2015; Hong 2018).

For other proteins inhibited by aPKC, such as the basolateral-specific kinase Par1, the inhibition depends upon the phosphorylation of Par1 creating a binding site for the 14-3-3 protein Par5. Binding by 14-3-3 prevents the migration of active Par1 into the apical domain (Suzuki et al. 2004; Motegi et al. 2011; Jiang et al. 2015; Wu et al. 2016). Regulation of protein activity by phosphorylation-dependent sequestration by 14-3-3 proteins is a well-established mechanism of regulation for a wide range of proteins.

Phosphorylation of Baz/Par3 by aPKC is important for epithelial cell maturation as it inhibits the association of Baz/Par3 with apical complexes, facilitating the movement of Baz/Par3 to the adherens junction. However, at least in *Drosophila* follicular cells, there are questions as to the actual importance of aPKC-mediated phosphorylation of Baz/Par3. Cells expressing a kinase-dead mutant form of aPKC display normal distribution of Baz/Par3 (Kim et al. 2009).

3.2.2 The Small Signal-Transducing GTPases

Many stages of apical/basolateral polarization are under the control of signal transducing GTPases (often called small, monomeric G-proteins) of the RAS (rat sarcoma virus oncogene) superfamily (Colicelli 2004). Like other signal-transducing

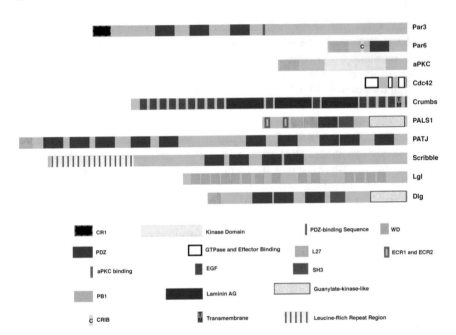

Fig. 3.2 Schematic of the domain components of the major determinants of apical-basolateral patterning. Note that typical structures have been selected for representation, as many of the represented proteins can vary with isoforms, splice variants, and species. Sizes are approximately to scale

GTPases, members of the RAS superfamily are only active when bound to GTP, not GDP. Like other GTPases, these small GTPases function through protein-protein interactions that activate effector proteins.

Activation happens through guanine-nucleotide exchange factors (GEFs), proteins which catalyze the exchange of guanine nucleotides on the small GTPases. As cytoplasmic GTP levels exceed GDP levels, free exchange favors the formation of the active form of the protein. The GTPases are negatively-self-regulating as they possess a GTPase activity that restores the protein to the inactive state. The GTPase activity of these proteins, therefore, is not involved in the activation of effector molecules, but is instead purely regulatory. The endogenous GTPase activity is typically weak in the absence of stimulatory cofactor proteins called GTPase activating proteins (GAPs) (Toma-Fukai and Shimizu 2019). Hence, through balancing the local activities of GEFs and GAPs, the cell can fine tune the level and localization of activity of RAS superfamily of GTPases (Goryachev and Leda 2019).

While the classic small GTPases are the products of the RAS proto-oncogenes, the RAS superfamily members centrally involved in epithelial cell polarization belong to the Rho (RAS homologue) and Rab (RAS genes from the rat brain) families. Rho family members, particularly Rac1 (RAS-related C3 botulinum toxin substrate 1), RhoA (RAS homologue 1), and Cdc42, are key controllers of the cytoskeleton, particularly the actomyosin skeleton (Quiros and Nusrat 2014). Rab family members on the other hand are central to the differential delivery of proteins and lipids to the apical or basolateral membranes, through their targeting of transport vesicles to membranes (Li and Marlin 2015). Members of both families function through attachment to the inner leaflet of the plasma membrane. However, these are large families of signaling proteins and their regulatory mechanisms vary significantly both within and between the families. In both families, members are negatively regulated in the GDP-bound form by GDI proteins (GDP disassociation inhibitors), that localize the GTPases away from the plasma membrane. The GTP-bound, membrane-bound forms are free of inhibitors and able to interact with a wide range of effectors.

Rho family members are well-known regulators of actin cytoskeleton dynamics. In total there are now twenty known Rho family members expressed in mammals (Aspenström 2019). RhoA is known for its role in the formation of focal adhesions and stress fibers, Cdc42 for its role in the formation of filipodia, and Rac1 for its role in the formation of lamellipodia (Ridley 1995). Clearly, these functions are not directly relevant to the role of these proteins in producing the polarization of stable epithelial sheets, where their roles pertain more to mediating phenomena such as positioning the adherens junction, and thereby the apical-basolateral boundary. Nonetheless, some of the biochemical activities used by these proteins to reshape the migrating cell are no doubt relevant to their roles in epithelial cell polarization. Rho family members are often activated in response to extracellular signals. For example, RhoA is regulated by phosphorylation of one of its GEFs, in response to activation of Eph (erythropoietin-producing hepatocellular) RTKs.

The Rho family members act in cell polarity chiefly through dictating the arrangement of the cell's actomyosin skeleton, contributing to the localization of the cellular junctions. Through the re-arrangement of the cytoskeleton, the Rho family members also impact the delivery of vesicle cargo to target membranes (Croisé et al. 2014). Their effect on the cytoskeleton is mediated through the localization or activation of a wide variety of effector proteins. The number of Rho family effector molecules is indeed large, and clearly an area of on-going study, with some effectors much better understood than others (Watson et al. 2017). Among the better understood families of effectors are the actin cytoskeleton re-modelling proteins of the WASP (Wiskott-Aldrich Syndrome Protein) and formin (limb deformity homology proteins) families, and protein kinases of the PAK (p21 activated kinases), MRCK (myotonic dystrophy kinase related Cdc42-binding kinases), PKC (protein kinase C), and ROCK (Rho-associated protein kinase) families. The WASP proteins are particularly important effectors for Cdc42, which activates WASP proteins through inducing dramatic changes in their conformation (Rohatgi et al. 2000). The conformational changes lead to the localization and activation of the ARP 2/3 (actin-related proteins 2 and 3) complex, an actin nucleator. Because the ARP2/3 complex associates with the lateral surface of existing microfilaments, its actions encourage the arrangement of actin into branched, dendritic structures (Goode and Eck 2007).

A particularly important effect of Rho family members on epithelial polarization is the RhoA-dependent formation of the peri-junctional actomyosin ring associated with the tight and adherens junctions (and thereby, the formation of the apical-basolateral border). The p114RhoGEF is localized to regions closely associated with tight junctions, causing the localized activation of RhoA. In columnar epithelial Caco-2 and HCE epithelial cells, depletion of p114RhoGEF resulted in an altered distribution of ZO-1 (zona occludins protein-1—ZO proteins are cytoplasmic proteins closely associated with the tight junctions), reduced peri-junctional actin microfilaments, and increased formation of stress fibers (Terry et al. 2011). Depletion of p114RhoGEF also resulted in a redistribution of activated RhoA from junctional complexes to more basal locations in the cell and also in the redistribution of myosin II away from points of cell contact (Terry et al. 2011). Consistent with a Rho-mediated activation of Myosin II, artificial overexpression of p114RhoGEF led to a contracted peri-junctional ring (Terry et al. 2011).

A similar story has emerged in *Drosophila*, where two apically activated Rho-GEFs, Wireless/p114RhoGEF and RhoGEF2, control the formation of actomyosin structures. While Wireless/p114RhoGEF is needed for proper formation of the actomyosin belt associated with adherens junctions, RhoGEF2 is needed for the actomyosin cytoskeleton that is more medially located in the apical domain (Garcia De Las Bayonas et al. 2019; Levayer et al. 2011), which is associated with the formation of structures such as microvilli and cilia.

The Rab family members are specific for particular transport vesicles, and are typically involved in directing those vesicles to fuse with very particular target membranes (Lamber et al. 2019; Bento et al. 2013). There are more than 60 Rab family members, and their association with the cytoplasmic side of vesicles provide specificity to the vesicles, particularly in terms of site of origin. Rab and Rab-effector proteins guide vesicles to their targets. Of particular importance in apical-basolateral polarization are Rab5, Rab7, Rab8 and Rab11. Rab5 and Rab7 are markers of early and late endosomes, respectively. Rab8 functions in the targeting of secretory vesicles from the trans Golgi network to the plasma membrane. Rab11 is a marker of recycling endosomes. Like most RAS superfamily members, Rab activation depends upon associated GEFs. However, some Rab-GEFs play additional roles in the physical targeting of Rabs to the correct organelle or membrane domain (Lamber et al. 2019).

Rab actions require the activities of effector proteins. Many Rab effector proteins are tethering proteins involved in direct physical interactions between Rab-marked vesicles and molecular markers on the target membranes. For example, effector EEA1 (early endosome antigen-1), which is recruited to vesicles by activated Rab5, identifies target membranes by binding to a phospholipid, phosphatidylinositol 3,4,5-triphosphate (PIP3). In an extended conformation, EEA1 can capture a membrane at a distance. Rab5 can then induce conformational changes in EEA1 that cause it to collapse, drawing the two membranes closer together (Murray et al. 2016).

After Rabs and Rab effector proteins initiate the first interactions between a vesicle and a target membrane, SNARE (SNAP receptor) proteins mediate vesicle fusion (Ungermann and Kümmel 2019). SNARES on a vesicle intertwine with SN ARES on the target membrane, and catalyze the membrane fusions. During docking, a Rab-GAP protein causes the hydrolysis of GTP, leading to the inactivation of the Rab and its disassociation from the vesicular membrane. While SNARES drive the membrane fusion reaction, Rab effector proteins provide much of the specificity that ensures the correct targeting of vesicles.

The importance of Rab GTPases to the polarization process is highlighted by the actions of Rab8 and Rab11 in positioning components of the PAR complex at the apical surface (Bryant et al. 2010; Colombié et al. 2017). This process can be studied in MDCK (Madin-Darbin Canine Kidney) epithelial cells plated in 3D culture, which undergo a reversal of polarity as aggregates of cells form lumens. Because polarity-reversal requires a transcellular transport of membrane components, the involvement of Rab11 is not surprising—Rab11 has been observed functioning in multiple examples of similar processes. Reduced expression of Rab11 and Rab8 in MDCK cells in 3D culture caused disorganized cell aggregates with multiple lumens (Bryant et al. 2010). Complementation experiments indicated that Rab8 probably acts downstream of Rab11 in the MDCK-3D system, most probably through a GEF-mediated recruitment of active Rab8 to sub-apical Rab11-positive vesicles. The sub-apical vesicles are positioned for fusion with the apical membrane and cargo deposition. In combination, Rab8 and Rab11 target Par3 to the apical membrane, and control the apical activation of Cdc42. The process of reverse polarization of the epithelial cells involves Par3 moving to the new apical membrane, and subsequently to junction sites as the lumen expands (Bryant et al. 2010). The self-re-enforcing nature of polarity determination is emphasized by the dependence of the actions of Rab11 upon Par3 function, indicating a circular inter-regulation between Par3 and the Rab protein (Bryant et al. 2010). A similar process of basal to apical transport of Baz/Par3 has also been observed as an early component of epithelial morphogenesis in *Drosophila* (McKinley and Harris 2012).

3.2.3   Par3 and Par6: Scaffold Proteins of the PAR Complex

Many of the components of the three master regulator complexes, PAR, CRUMBS, and SCRIBBLE, are scaffold proteins, proteins with many protein-protein interaction domains, that facilitate the juxtaposition of signaling components (see Fig. 3.2). Without doubt, the capability of these proteins to form functional conglomerations is central to the polarizing capabilities of the three complexes. The scaffold proteins of the complexes include Par3, Par6, Crumbs, PALS1, PATJ, Dlg, Scribble and Llgl. Crumbs is distinguished from the others in being a transmembrane protein.

The Par3 protein includes three PDZ (Psd95-Dlg/ZO1) domains, a phosphatidyl-inositol phosphate binding domain, an oligomerization domain (designated CR1), and an aPKC binding domain (Yu et al. 2014). PDZ domains are typically protein-binding domains that specialize in binding the carboxy-terminal tail of other proteins, frequently the carboxy-terminal tail of transmembrane proteins. However, certain PDZ domains can also bind lipids (Wu et al. 2007). The first PDZ domain of Par3 is required for the binding of Par3 to Par6. While the second PDZ domain of

Par3 binds a broad array of phosphoinositol phospholipids, including PIP2 and PIP3, the third PDZ domain binds the PIP3 phosphatase, PTEN (phosphatase and tensin homologue) phosphatase (Wu et al. 2007).

Par6 has one PDZ domain, one CRIB domain (Cdc42 and Rac-Interactive Binding motif domain), and one PB1 (Phox and Bem1) domain. The PDZ domain can bind to the C-terminal tail of the transmembrane Crumbs protein, a component of the complex interactions between the two apical complexes. Importantly, binding of activated (i.e., GTP-bound) Cdc42 to the CRIB domain of Par6 stabilizes the interaction of Par6 and Crumbs protein (Whitney et al. 2016). This provides the potential for a regulated physical interaction between the two major apical complexes. Par6 also binds aPKC through the Par6 PB1 domain binding the aPKC PB1 domain (Graybill et al. 2012).

3.2.4 Crumbs and the Scaffold Proteins of the CRUMBS Complex

The CRUMBS complex is composed of the transmembrane protein Crumbs, and two associated cytoplasmic scaffold proteins, PALS1/Sdt/Stardust and PATJ. Crumbs itself is important for the localization of the complex, and other proteins, to the regions of the adherens and tight junctions. PALS1/Sdt/Stardust binds Par6 through conserved amino-terminal regions of PALS1 called ECR1 and ECR2 (evolutionarily conserved regions 1 and 2) (see Wang et al. 2004; Gao and Macara 2004; Koch et al. 2016). PALS1 also has two L27 domains, the first of which is essential for the interaction with PATJ, binding to the L27 domain of PATJ. PALS1 has a single PDZ domain which can bind either the Crumbs protein or the Baz/Par3 protein (Sen et al. 2015). Additionally, PALS1 has GUK (guanylate kinase like) and SH3 (src-homology-3) domains that stabilize the interaction with the Crumbs protein (Li et al. 2014).

As mentioned previously, PATJ has a L27 domain that mediates binding to PALS1. Additionally, depending upon the species, PATJ has ten to thirteen PDZ domains, creating an enormous capacity for complex assembly. Included in its binding partners are the tight-junction associated scaffold protein ZO-3 and nectins (nectin cell adhesion molecules) (Adachi et al. 2009).

3.2.5 The Scaffold Proteins of the SCRIBBLE Complex

The major components of the Scribble complex, Scribble protein, Dlg, and Llgl/Lgl, are large scaffold proteins present throughout the animal kingdom. All three were originally discovered through genetic screens in *Drosophila* - *Drosophila* embryos with mutations in the genes encoding any of the three proteins display aberrant, excessive, cellular proliferation (Mechler et al. 1985; Woods and Bryant 1989; Bilder and Perrimon 2000).

Scribble protein includes a leucine-rich repeat domain and four PDZ domains (Caria et al. 2019). While most of Scribble's protein interactions are mediated by its PDZ domains, the leucine-rich repeats mediate the interaction with Lgl/Llgl (Stephens et al. 2018). Dlg includes one L27 domain, three PDZ domains, an SH3 domain, and a GUK domain. Lgl/Llgl interacts with other proteins through a WD40 repeat region, containing up to fourteen repeats, that potentially forms two seven-bladed WD40 propellers (Cao et al. 2015; Jossin et al. 2017; Stephens et al. 2018). The combined direct interactions of the three Scribble complex proteins comprises of an enormous number of binding partners, with several of significant import to apical/basolateral axis formation. Amongst those with significant roles in the establishment or maintenance of polarity are Rab5, beta-catenin, PTEN, ZO-2 (zona-occludens protein-2, an adaptor protein associated with tight junctions), E-cadherin, Par6, Myosin II, and the t-SNARE, syntaxin-4 (Stephens et al. 2018).

3.2.6   The Phosphatidyl-Phosphoinositols

Localization of the three master modules is partially dictated by the distribution of phosphatidyl phosphoinositols, particularly PIP2 and PIP3 (Martin-Belmonte et al. 2007; Ruch et al. 2017). PIP2 is enriched in the apical membrane, while PIP3 is enriched in the basolateral membrane. The two phospholipids are readily interchangeable, with PI3 kinase creating PIP3 from PIP2, and PTEN phosphatase restoring PIP3 back to PIP2. The polarized distribution of these lipids reflects the localized distributions of PI3 kinase and PTEN phosphatase activities. These localizations of PI3 kinase to the basolateral compartment and PTEN phosphatase to the apical compartment are directed by small GTPases of the Rho family. As phosphatidyl phosphoinositols play roles in the regulation of vesicular transport, this provides a second mechanism, in addition to cytoskeletal remodeling, by which Rho family GTPases may regulate transport (Croisé et al. 2014).

The importance of PIP2 and PIP3 in polarization also provides a mechanism for external signals to impact polarization. PI3 kinase activity is highly regulated by receptor tyrosine kinases (including receptors for many growth factors and insulin), by GPCRs, and by integrins (Guo and Giancotti 2004; Fritsch et al. 2013). The PIP2 at the apical domain can stabilize the location of the PAR module through a bridging molecule, annexin-2, linking the PIP2 to the Cdc-42 component of the PAR module.

3.3 Co-ordination of the Molecular Interactions That Prescribe the Apical-Basolateral Axis

As mentioned previously, the key processes in the establishment and maintenance of apical-basolateral polarity is a competition between two apical protein complexes, the PAR and CRUMBS complexes, and a basolateral complex, SCRIBBLE. While the apical complexes confer appropriate biochemical properties to the apical membrane and specify the location of the tight and adherens junctions, the basolateral complex prevents the invasion of apical proteins and structures into the basolateral domain. The apical-basolateral polarization therefore is the product of a finely-tuned balance between the three complexes.

This chapter is focused upon the segregation of the epithelial cell into apical and basolateral divisions, because of the relevance of this division to channel and transporter biology. However, in reality more sub-divisions of the epithelial cell membrane unquestionably exist, particularly in the region of the junctional complexes (St Johnston and Ahringer 2010a).

3.3.1 Initialization of Axis Formation

It is natural to ask whether one or other of the key modules has primacy, initiating the polarization process and dictating the cellular localization and behavior of the others. As will be outlined in this section, current evidence suggests the processes vary significantly according to species, developmental stage and context.

Central to the question of apical-basolateral axis morphogenesis is the positioning of the boundary between the two domains, corresponding to the positioning of the adherens junctions, and the associated tight or septate junctions. In *Drosophila*, part of the maturation process for a developing epithelium is the dislocation of Baz/Par3 from apical domains to associate with the intercellular junctions, specifically the adherens junctions (Harris and Peifer 2005; Walther et al. 2016). This dislocation process for Baz/Par3, which is dependent upon aPKC-mediated phosphorylation, is observed in several examples of epithelial morphogenesis in *Drosophila*, including the development of the follicular epithelium and of the photoreceptors (Walther and Pichaud 2010; Krahn et al. 2010; Morais-de-Sá et al. 2010). This localization of Baz/Par3, which mediates formation and stability of the adherens junctions, positions Baz/Par3 basal to both Crumbs and Par6, and presumably only allows interactions at a border region where their localizations intersect (Bryant et al. 2010; Walther et al. 2016).

In mammalian cells, the localization of Par3 to junctions during epithelial morphogenesis is critical for tight junction formation. At tight junctions, Par3 functions by activating PI3kinase, by spatially restricting PTEN, and by controlling the activity of Tiam1 (a Rac1 GEF protein), and Bcr1 (a Rac1 GTPase activating protein). These proteins function in the polarization of a range of cell types

(Narayanan et al. 2013). The Par3 regulation of Rac1, through Tiam1 and Bcr1, is essential for tight junction assembly (Chen and Macara 2005).

However, studies of the developing *Drosophila* photoreceptor have provided strong genetic evidence for an additional, sufficient pathway for adherens junctions positioning and maintenance. In addition to the pathway dependent upon Baz/Par3 (presumably acting after disassociation from the PAR complex), an alternative pathway dependent upon a Cdc42-dependent kinase (Mbt, mushroom bodies tiny, the *Drosophila* equivalent of Pak4) is also functional in specifying the adherens junctions localization (Walther et al. 2016). The Mbt/Pak4 functions to stabilize E-cadherin at the junctions through phosphorylation of beta-catenin (Selamat et al. 2015). Both Baz/Par3 and Mbt/Pak4 dependent localizations are dependent upon an apical localization of activated Cdc42.

A recently proposed model of polarization, based on experiments in the *Drosophila* photoreceptor, suggests a certain level of primacy for Cdc42, at least in this particular tissue. The proposed model begins with an apical localization of Cdc42 that is independent of Par6, Baz, aPKC, or Crumbs (Nunes de Almeida et al. 2019). The apically located Cdc42 is active and acts to recruit PAR complex (that is, Baz/Par3-Par6-aPKC). An exchange occurs in which Par6-aPKC is transferred from the Par complex to Cdc42 to form an alternative apical complex (Cdc42-Par6-aPKC). Subsequently, in tissue morphogenesis, Cdc42 facilitates the transfer of Par6-aPKC to the transmembrane molecule Crumbs, which in turn stabilizes Crumbs at the membrane. This model emphasizes the importance of Cdc42, both as an initiator and as a mediator of many subsequent steps (Nunes de Almeida et al. 2019). This model also invokes Par6-aPKC being shuttled sequentially between three complexes—the classic PAR module, a complex including Cdc42 and a complex involving the Crumbs protein (possibly the CRUMBS complex). The proposition of a Cdc42-Par6-aPKC complex is also supported by the observation of an equivalent complex in *C. elegans* zygotes, where the distribution of Par and PKC proteins reflect an anterior-posterior polarization (Rodriguez et al. 2017; Wang et al. 2017). The traditional view of discrete PAR and CRUMBS complexes may need to be revisited as the interactions of Crumbs protein, Par3, Par6, aPKC, and Cdc42 increasingly appear fluid and dynamic (Walther et al. 2016; Rodriguez et al. 2017; Pichaud et al. 2019).

How might Cdc42 be localized to the apical membrane, independent of PAR or CRUMBS complexes? One possibility is suggested by a different polarized system, the migrating astrocyte, in which Cdc42 is specifically delivered to the leading edge. Delivery of Cdc42 is mediated by endosomes, and is dependent upon Rab5 (Osmani et al. 2010; Baschieri et al. 2014). Alternatively, apical localization of appropriate GEFs could mediate an apical distribution of activated Cdc42 (Bryant et al. 2010; Zihni et al. 2014). Similarly, localization of Cdc42 GAP activity has been observed to localize activated Cdc42 through localized inactivation (Wells et al. 2006; Anderson et al. 2008; Elbediwy et al. 2012; Beatty et al. 2013; Klompstra et al. 2015). Finally, annexin-2 is thought to cross-link Cdc42 to PIP2, allowing the polarized distribution of phosphatidyl phosphoinositols to contribute to Cdc42 distribution (Pinal et al. 2006; Nunes de Almeida et al. 2019).

While the *Drosophila* photoreceptor system points to the early importance of Cdc42, two other systems emphasize the early importance of Par3 (or its *Drosophila* equivalent, Baz). MDCK cells initiating epithelialization reveal the importance of both Par3 and of intercellular interactions. Amongst the earliest events in these cells is the homophilic, intercellular interactions of nectin proteins. Subsequently, Par3 localizes to the nectin junctions through interactions of the first PDZ domain of Par3 with the carboxy-terminal tails of the nectins (Takai et al. 2008). The genesis of the primary epithelium of *Drosophila* also reflects an early importance for Baz/Par3. However, in this system polarization seems to be much more cell autonomous. The Baz/Par3 protein is delivered to the nascent apical surface by microtubules that are oriented by their attachment to the centrosome, which is situated above the nucleus (Harris and Peifer 2004; McGill et al. 2009). Baz/Par3 subsequently recruits cadherins to form adherens junctions.

The tissue-specific nature of epithelial polarization is also highlighted by the distribution of phosphatidyl phosphoinositols in *Drosophila* epithelial retinal cells. While in many epithelia, the localization of PIP2 to the apical membrane is critical for normal polarization, in these retinal cells, the apical membranes are enriched for PIP3 (Pinal et al. 2006).

In summary, therefore, current evidence suggests that the initiating events do indeed vary with environment, tissue-type, and developmental stage (Kim et al. 2009; Chen et al. 2018). Variation occurs as to the degree of cell autonomy, as to the requirement for either cell-cell or cell-matrix contacts, and as to the sequential ordering and hierarchy of polarizing factors. However, a consistent feature of most of the studied programs is the restricted localization of components of the PAR complex or its close associate Cdc42 early in polarity formation.

While PAR complex proteins appear to be common features of most, possibly even all examples of polarizing cells, the CRUMBS complex (Crumbs protein with PALS1 and PATJ), is more specifically important to epithelial polarization (St Johnston and Ahringer 2010a). In *Drosophila*, Crumbs protein is essential for the formation of the apical domain. Over-expression of Crumbs protein in *Drosophila* epithelial cells leads to an expansion of the apical domain (Wodarz et al. 1995), and its overexpression in MDCK cells disrupts the normal polarization process (Lemmers et al. 2004; Roh et al. 2003). Crumbs protein binds spectrin, connecting the CRUMBS complex to the actomyosin cytoskeleton (Izaddoost et al. 2002). PALS1 on the other hands binds to the tight junction-associated proteins ZO-3 and claudin. Combined with the localization of the CRUMBS complex to the lateral, not central, regions of the apical membrane, these protein associations suggest the complex may be responsible for positioning and stabilizing the tight and adherens junctions (Shin et al. 2005). Supporting this hypothesis, ectopic expression of Crumbs protein in mammalian cells is sufficient to cause the formation of tight junction-like structures in cells that do not normally have them (Fogg et al. 2005).

An important function of Crumbs protein is the exclusion of Baz/Par3 from binding to Par6, which results from competition for binding sites on Par6 with Crumbs protein (Morais-de-Sá et al. 2010). This allows for aPKC to phosphorylate Crumbs protein, which activates it, and also allows for the movement of Baz/Par3 to a location that is slightly more basal (as previously mentioned, a feature of matured, fully polarized epithelia) (St Johnston and Ahringer 2010a). The "less-apically" localized Par3 collaborates with Crumbs protein to stabilize the tight and adherens junctions, through Par3 mediated-localization of active Rac1.

3.3.2 SCRIBBLE Complex Function

SCRIBBLE complex components such as Lgl/Llgl, function in restricting apical factors such as the PAR complex to the apical domain, and antagonize PAR activity (Bilder and Perrimon 2000; Hutterer et al. 2004; Yamanaka et al. 2006). Lgl/Llgl and the basolateral-specific kinase, Par1, both contribute to the exclusion of Par3 and Par6 from the basolateral domain of epithelia and from the posterior domain of *Drosophila* oocytes (in many ways a molecular parallel to the basolateral domain of epithelia) (Bayraktar et al. 2006; Tian and Deng 2008; Doerflinger et al. 2003, 2006, 2010). Par1 phosphorylates Par3, a phosphorylation that excludes PAR complex from Par1 domains. Reciprocally, Lgl/Llgl is excluded from the apical or anterior domains by a mechanism that is probably dependent upon aPKC phosphorylation (Benton and St Johnston 2003; Plant et al. 2003; Suzuki et al. 2004; Hutterer et al. 2004; Tian and Deng 2008).

As would be expected, the SCRIBBLE complex components not only prevent basolateral spread of PAR and CRUMBS complex components, but are also responsible for localization of basolateral membrane components. In mammals, as is typical for many proteins, several equivalents to Scribble protein are present. In humans, for example, three Scribble proteins are present, human Scribble (Scrib), Erbin and Lano. Knockout studies in DLD1 cells (a human colorectal cancer cell line) indicate that the three proteins complement each other in apical-basolateral polarization (Choi et al. 2019). Basolateral to the adherens junctions in these cells are a second group of adherens junctions, the spot-like adherens junctions (SLAJ). Single knockouts of Scribble family members do not affect the formation of SLAJs, but combined Scrib/Erbin/Lano knockouts result in mis-localization of SLAJs. Consistent with *Drosophila* data, the triple knockout Scribble family members also affect the cellular distribution of PAR and CRUMBS components, allowing Par6, aPKC, and Pals1 to move basolaterally.

Genetic studies have revealed that SCRIBBLE-independent pathways in *Drosophila* embryos can drive formation of functioning basolateral domains, including a pathway dependent upon a Crumbs-binding protein, Yurt (Tanentzapf and Tepass 2003; Laprise et al. 2009; Perez-Vale and Peifer 2018). This pathway appears more important in the later stages of development.

3.3.3 The Microtubule Cytoskeleton Directs Cargo to Apical and Basolateral Membranes

Within columnar epithelial cells, the bulk of the microtubules run in an apical-basal orientation, with the positive end oriented towards the basal membrane (Müsch 2004). Most remaining microtubules form two web-like horizontal structures, one adjacent to the apical membrane, the other adjacent to the basal membrane. An important step therefore in the establishment of a differentiated epithelial cell is the redistribution of microtubules away from centrosomes, with most moving towards apically located non-centrosomal MTOCs (microtubule organizing centers).

The orientation and polarity of the apical-basolateral microtubules dictate the polarized delivery of cargo to the apical or basolateral membrane domains, which is essential for their biochemical and functional differentiation. Not surprisingly, therefore, the correct orientation and structure of the microtubules is controlled by the polarity complexes, in both non-epithelial and epithelial systems (Doerflinger et al. 2003, 2010). In *Drosophila* oogenesis, all normal microtubule polarization is lost in mid-development eggs with mutated forms of Par1 that cannot be phosphorylated by aPKC. Similarly, normal microtubule polarization is also lost in eggs with Baz/Par3 mutations that cannot be phosphorylated by Par1. Epistasis experiments indicated that Par1 acts downstream of Baz/Par3 to direct microtubule organization and polarity (Doerflinger et al. 2010). A critical component of establishing the polarized distribution of microtubules in *Drosophila* oocytes is the Par1-mediated polarized distribution of a spektraplakin, Short stop (Shot). Shot localizes cortically due to an F-actin-binding activity, and is excluded from regions where Par1 is active. Together with the microtubule binding protein Patronin, Shot forms non-centrosomal MTOCs, anchoring the minus ends of microtubules to the anterior cortex in *Drosophila* oocytes, and the apical cortex of epithelial cells (Nashchekin et al. 2016). A mammalian homologue of Patronin, CAMSAP3, is required for the apical-basal orientation of microtubules in epithelial cells, suggesting this model of microtubule polarization may be evolutionarily conserved (Toya et al. 2016). The kinase aPKC is also important for the formation of non-centrosomal MTOCs, as in early embryonic epithelia in *Drosophila*, aPKC is necessary to facilitate a movement of microtubules away from centrosomes to more cortical locations (Harris and Peifer 2007).

3.4 The Horizontal Axis of Polarization

Discussions around epithelial polarization have classically focused on the apical-basolateral axis. However, for many epithelial tissues, cellular polarization within the horizontal plane is also central to function. In most epithelia, the most obvious cytological indicator of horizontal plane polarization is the positioning of cilia or hairs on the apical membrane surface (Devenport 2014). This horizontal polarization, referred to as planar cell polarity (PCP), is central to the function of many epithelial tissues as it regulates the positioning, orientation, rotation, and co-ordination of the cilia, amongst other structures. As PCP is not as functionally important for ion channel and transporter function as apical-basolateral patterning, only a brief overview of the establishment and control of PCP will be provided here (the current state of PCP establishment and maintenance has been provided elsewhere—see Butler and Wallingford 2017; Blair and McNeill 2018; Humphries and Mlodzik 2018). The chief concern here will instead be on the interactions of PCP with the apical-basolateral system.

3.4.1 Establishment of Planar Cell Polarity in Epithelial Sheets

The core components regulating the establishment and maintenance of PCP are transmembrane molecules that are asymmetrically distributed in the horizontal plane, at the apical end of the cell (Devenport 2014). Marking one side of the apical membrane is the transmembrane protein Frizzled (Fz). Marking the opposite end of the apical membrane is the transmembrane protein Van Gough (Vang) (see Fig. 3.3). Both Fz and Vang interact with a third transmembrane protein, Flamingo (Fmi). While Fz/Fmi complexes are concentrated on one surface by interactions with the intracellular proteins Dishevelled (Dsh) and Diego (Dgo), Vang/Fmi complexes are concentrated on the opposite surface by interactions with the intracellular protein prickle (Pk). Although these interactions have largely been discerned through *Drosophila* genetics, particularly studies of the developing *Drosophila* wing and epidermis, clear support exists for homologous proteins establishing and maintaining horizontal polarization in many vertebrate systems (Theisen et al. 1994; Usui et al. 1999; Henderson et al. 2018).

For many purposes, functionality of a field of planar polarized cells requires that all the cells of the field be polarized in the same orientation. Therefore, intercellular interactions between epithelial cells are essential for the co-ordination of polarity

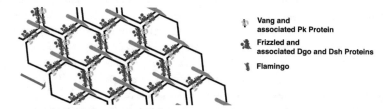

Fig. 3.3 Schematic of special relationships of key planar cell polarity (PCP) proteins to each other in a planar polarized epithelium. Schematic view looking down on the apical face of a planar polarized epithelium. Note the separation in the plane of the membrane protein Vang and associated cytoplasmic proteins, from the membrane protein Frizzled and associated proteins. Cilia are displayed in orange. Pk, Prickle; Dsh, Dishevelled; Dgo, Diego

across an epithelial field. These intercellular interactions occur at the intercellular junctions, and involve Fmi, Fz, and Vang. Fmi forms intercellular homophilic interactions that facilitate the transmission of the polarity information from one cell of a field to the immediately adjacent cells (Devenport and Fuchs 2008; Vladar et al. 2012; Devenport 2014). The nature of the signaling Fmi, and therefore the cellular asymmetry, is dependent upon the other proteins with which it is complexed.

External factors impinge upon, and where needed, alter the PCP of particular epithelia. These external factors include signaling from the Wnt family of secreted signaling molecules and anisotropic strain across the epithelial sheet (Butler and Wallingford 2017). A second PCP system, based on different molecules from the core PCP system, is now recognized as dominant in certain processes. This second system, the Fat, Dachsous, and Four-Jointed PCP (Ft-Ds-Fj) system, interacts with the original core system through currently unresolved mechanisms (Matakatsu and Blair 2004; Casal et al. 2006; Butler and Wallingford 2017).

3.4.2   Points of Intersection Between the Two Plains of Polarization

The PCP polarization is almost certainly downstream of the apical-basolateral (A-BL) polarization. This is borne out by the numerous examples of genetic lesions causing lost or inappropriate PCP patterning of bristles, hair or cilia, with the structures still forming at the apical end of the epithelial cell (Gubb and García-Bellido 1982). At the molecular level, there are also clear examples of components of the A-BL system affecting the PCP system (Courbard et al. 2009; Djiane et al. 2005). For example, a key mechanism for the regulation of PCP by Wnt5A requires the Par6 protein. Signaling induced by Wnt5A induces the formation of a complex including Par6 that directs the degradation of PK1 (a mouse homologue of Pk) (Narimatsu et al. 2009, Liu et al. 2014).

Curiously, genetic ablation of *Par3* in two mammal systems resulted in disruption of PCP, but no overt disruption of A-BL axis formation (Hikita et al. 2018; Malt et al. 2019). One of these two systems is murine endothelial vasculature. Endothelial cells normally align organelles such as the Golgi with the direction of flow, but failed to do so in a *Par3* conditional mutant in some regions of the vasculature. Similarly, conditional knockout of *Par3* disrupted the planar cell polarity of murine inner ear hair cells. Mutagenesis of *Scribl*, a murine homologue of the gene encoding the Scribble protein also interferes with planar cell polarity in the cochlea (Montcouquiol and Kelley 2003). This phenomenon of A-BL determinants assuming PCP functions in mammals is clearly an area for further investigation.

While the impact of A-BL components upon PCP is clear, the reverse is less clear. Nonetheless, the processes controlling the genesis of the two axes clearly converge upon Rho family GTPases as key components of both of their outputs (Strutt et al. 1997; Zihni et al. 2017). For both processes, rearrangements of the same cytoskeletal structures manifest their respective polarizations. Hence, each program must impact the other at a functionally meaningful level.

3.5 Outputs of Polarization

Polarization impacts every aspect of epithelial structure. This section will expand on the importance of polarization to four examples of epithelial cell function—tissue morphogenesis, mitotic spindle orientation, basement membrane formation, and solute transepithelial transport.

3.5.1 Tissue Morphogenesis

Polarization drives many of morphological specializations of epithelial tissues. For instance, polarization drives the formation of a medial, apical actomyosin belt which is necessary for the creation of microvilli in intestinal cells (Schneeberger et al. 2018). However, a more sophisticated level of tissue morphogenesis can occur through coordinated shape changes mediated by the polarization systems acting across an epithelial sheet. Indeed, co-ordination of the polarization systems across cells plays a major role in critical morphogenenic processes, including lumen formation and convergent extension (Navis and Nelson 2016; Huebner and Wallingford 2019). Convergent extension involves the extension of a field of epithelial cells in one plane, while it narrows in a second. While gross cell migration can be involved in convergent extension, the major component of convergent extension seems to be an intercalation of the epithelial cells. The intercalation results from small, polarized movements of cells relative to each other (Huebner and Wallingford 2019). The intercalation allows for tissue elongation, without new cell proliferation. Most famously, convergent extension controls change in the neuroepithelium of the vertebrate central nervous system (CNS), including changes that elongate the CNS and changes that facilitate neural tube closure (Juriloff and Harris 2012).

The necessary coordinated changes in cell shape to achieve convergent extension are largely under the control of components of the PCP pathway. Consistent with the importance of the pathway for convergent extension, one in five human newborns with a very severe neural tube defect (craniorachischisis) have deleterious mutations in components of the PCP pathway (Juriloff and Harris 2012). Similarly, neural tube defects are common features of mice with knock-outs of genes encoding components of the pathway, including mutations in *Scrib1*, the murine orthologue of Scribble. Once again, this would appear to reflect a vertebrate-specific role for Scribble protein in PCP, rather than its classical role in apical-basolateral axis formation (Juriloff and Harris 2018).

3.5.2 Mitotic Spindle Orientation

A second point of functional interaction between the two planes of polarization is cell division (Ségalen and Bellaïche 2009). Cells in simple epithelia mostly divide in the plane of the epithelial sheet (mitotic spindle aligns in the plane of the epithelial sheet, cytokinesis occurs in the plane of the apical-basolateral axis). However, cells in certain complex epithelia, such as the vertebrate embryonic epidermis, regularly divide in either a planar axis or a perpendicular axis. When the cells divide in a perpendicular axis, the mitotic spindle aligns with the apical-basolateral axis, and cytokinesis is in the same plane as the cell sheet. Planar divisions are typically symmetrical, and can be used for example for the expansion of an epithelial layer's stem cell population. Perpendicular divisions are typically asymmetrical, and often are used to generate cells committed to differentiation, such as the keratinocytes of the stratum spinosum. Cells committed to mitotic division within a complex epithelial layer, therefore, must make a choice between planar and perpendicular division.

Proper tissue formation clearly depends upon an appropriate relative number of planar and perpendicular divisions. These decisions act through changes in the location of spindle anchoring protein complexes, particularly mInsc/Insc-LGN/Pins-NuMA/Mud complex (Bergstralh et al. 2013; Williams et al. 2014; Saadaoui et al. 2014).

Apical localization of mInsc/Insc-LGN/Pins-NuMA/Mud favors perpendicular division. Apical polarity factors, such as Par3/Baz, recruit mInsc/Insc to the apical pole, which in turn recruits LGN/Pins and NuMA/Mud, forming the complex that anchors spindle astral microtubules to the cell cortex (Williams et al. 2014; Saadaoui et al. 2014, 2017). The actions of polarization factors may integrate with information from cellular architecture or tissue mechanical tension to dictate the orientation of cellular division (Dias Gomes et al. 2019; Box et al. 2019; Li et al. 2019; Finegan and Bergstralh 2019).

Lateral localization of LGN/Pins-NuMA/Mud favors planar cell division. In most tissues, this basal localization is regulated through the localization of the dynein-interacting protein LGN/Pins, which can be achieved by an inhibitory phosphorylation by the apical kinase, aPKC, (Hao et al. 2010) or through interactions with the basolateral determinant Dlg (Bergstralh et al. 2013).

Drosophila genetic evidence also points to a role for the planar cell polarity regulating spindle orientation (Bellaïche et al. 2001a, b, 2004; Ségalen et al. 2010). Similarly, mice defective in *Vangl2* (a mouse homologue of the PCP gene encoding Vang) display increased numbers of perpendicular divisions and as a result a hyper-thickened epidermis (Box et al. 2019). However, PCP pathway disruptions may interfere with the spindle orientation indirectly, through effects on tissue architecture (Box et al. 2019).

3.5.3 Formation of the Basement Lamina

Amongst the most important products delivered to the basolateral membrane are the components of the basement membrane, otherwise known as the basement lamina. Epithelial cell sheets attached basally to the basement membrane, and depend heavily on it for strength (Halfter et al. 2015). The predominant proteins of the basal lamina (the component of the basement membrane produced by the epithelial cells) are the Col IV, Laminin, Nidogen, and heparan sulfate proteoglycans. Advances have indicated that the basal membrane is more complicated than once thought, with a polarity of its own. While the epithelial-oriented face is rich in components that make contact with epithelial-presented proteins such as integrins, the stromal-oriented face is rich in components that interact with stroma (Halfter et al. 2015). It is interesting to question whether the integrin-basal lamina interactions represent a positively re-enforcing loop for basolateral membrane identity (O'Brien et al. 2001; Schneider et al. 2006). Integrin signaling has the potential to stimulate PI3kinase, leading to a localized enrichment of PIP3, a critical identifier of the basolateral membrane domain.

A fascinating aspect of basal lamina synthesis can be observed in *Drosophila* egg chamber morphogenesis—namely that endoplasmic reticulum and the Golgi apparatus are also polarized, in the sense that specific subdomains of each are responsible for the production and sorting of basal lamina proteins (Lerner et al. 2013). This localized translation depends upon the presence of a specialized discontinuous Golgi apparatus (Kondylis and Rabouille 2009). A similar specialized translation zone may act in the polarized distribution of at least some other proteins, as has been described for Wingless protein destined for apical secretion (Wilkie and Davis 2001; Herpers and Rabouille 2004; Bonifacino 2014). The basal lamina components are transported to the basal membrane and exported through a specialized Rab10-dependent vesicular transport designed to handle exceptionally large cargo proteins (Lerner et al. 2013; Devergne et al. 2014).

In many tissues, the presence of the basement membrane re-enforces the existing apical-basolateral axis decision, and may be important for its reinstatement following cell division. In these tissues the basal localization of laminin is probably decisive in determining polarization, as is the basal localization of the integrins that interact with it (Yu et al. 2005).

The movement of a cell towards cancer depends upon breaking interactions with the basal extracellular matrix (ECM), as well as disrupting other basal structures such as the desmosome. Hence, basal changes are important components of the epithelial-mesenchymal transition (EMT). Full or partial EMT contribute to most examples of metastasis (Heikenwalder and Lorentzen 2019). The importance of basal interactions in preventing the EMT process may contribute to the tumor suppressive activities of the three components of the basal SCRIBBLE complex. Each of Scribble, Lgl/Llgl, and Dlg were first identified through their inhibition of inappropriate cell proliferation in *Drosophila* embryos. The expression of human homologues of each of these proteins has been observed to be decreased in certain cancers (Lin et al. 2015). More recently, it has become recognized that Scribble and Dlg proteins are both important targets for HPV E6-mediated degradation, a phenomenon typically associate with proteins that are important tumor suppressors (Nakagawa and Huibregtse 2000, Matsumoto et al., 2006). Hence, the SCRIBBLE complex and its derived structures would appear to be particularly important in the suppression of cancer.

3.5.4 Transepithelial Solute Transport

As stated at the outset of this chapter, solute transport across an epithelial sheet must have the capacity to be vectorial to be effective in most of its functions. Vectorial transport depends upon apical-basolateral axis formation, and a resulting differential localization of pumps, transporters and channels to the apical and basolateral domains. This chapter will not review the specific pathways currently known for the localization of these proteins, which better belongs in chapters devoted to each. In broad overview these proteins are directed to the appropriate membrane subdomains through the interactions of Rab GTPases, myosins (including myosin II and myosin Vc) and other motor proteins, actin microfilaments and microtubules. The combination of these factors indicates several points at which the apical-basolateral complex may affect the transport of carriers, pumps, and channels to specific membrane subdomains. For example, the apical-basolateral axis predetermines the polarity of microtubules, the distribution of microfilaments, and the activity of Rho family effectors of myosin activity such as MLCK. Interested readers are directed to Chap. 5 (this volume) by Heike Fölsch for information about sorting of proteins to the apical or basolateral membranes of epithelial cells.

3.6 Conclusions

It is now increasingly clear that polarization is an aspect of most, possibly all, Eukaryotic cells, regardless of cell type or circumstance (Heikenwalder and Lorentzen 2019). Therefore, the early experiments in *C. elegans* and *D. melanogaster* should be viewed as visionary, and have appropriately inspired vibrant research communities. Enormous progress has resulted, but some areas clearly remain to be addressed. In particular, future comparisons of system specialities, such as comparing differences in the details of polarization between different epithelial tissues will deepen further the understanding of the polarization process.

Cells undergoing epithelial meseschymal transitions (EMT) have an apical-basolateral polarization, which morphs into an anterior-posterior patterning following the transition (Heikenwalder and Lorentzen 2019). Thus, the cellular manipulation of polarity is central to the process of becoming migratory, and therefore also central to the process of becoming metastatic. As such, proteins directing apical-basolateral patterning would seem to present potential targets for chemotherapeutic drug development. However, investigations into this field have been relatively limited. In part this reflects questions whether polarity proteins can be used to specifically target cancer cells, avoiding excessive adverse effects. In part, it may also reflect that metastasis in general has gained relatively little attention as a property of cancer to target, compared to other properties such as sustained proliferation (Steeg 2016). However, the potential large number of new targets should encourage pursuit of polarity factors as the basis of new treatments for cancer and other ailments, and some development has occurred for a small number of polarity factors including aPKC (Lin et al. 2018). Interest in polarity proteins as targets should also be spurred by the recent recognition of the role of polarity in genome stability (Dias Gomes et al. 2019).

Acknowledgements I would like to thank Dr. Dee Silverthorn and Dr. Kirk Hamilton for their helpful suggestions and editing of this chapter.

References

Adachi M, Hamazaki Y, Kobayashi Y, Itoh M, Tsukita S, Furuse M, Tsukita S (2009) Similar and distinct properties of MUPP1 and Patj, two homologous PDZ domain-containing tight-junction proteins. Mol Cell Biol 29:2372–2389. https://doi.org/10.1128/MCB.01505-08

Anderson DC, Gill JS, Cinalli RM, Nance J (2008) Polarization of the *C. elegans* embryo by RhoGAP-mediated exclusion of PAR-6 from cell contacts. Science 320:1771–1774. https://doi.org/10.1126/science.1156063

Aspenström P (2019) The Intrinsic GDP/GTP exchange activities of Cdc42 and Rac1 are critical determinants for their specific effects on mobilization of the actin filament system. Cells 8:759. https://doi.org/10.3390/cells8070759

Bailey MJ, Prehoda KE (2015) Establishment of Par-polarized cortical domains via phosphoregulated membrane motifs. Dev Cell 35:199–210. https://doi.org/10.1016/j.devcel.2015.09.016

Baschieri F, Confalonieri S, Bertalot G, Di Fiore PP, Dietmaier W, Leist M, Crespo P, Macara IG, Farhan H (2014) Spatial control of Cdc42 signalling by a GM130-RasGRF complex regulates polarity and tumorigenesis. Nat Commun 5:4839. https://doi.org/10.1038/ncomms5839

Bayraktar J, Zygmunt D, Carthew RW (2006) Par-1 kinase establishes cell polarity and functions in Notch signaling in the Drosophila embryo. J Cell Sci 119:711–721. https://doi.org/10.1242/jcs.02789

Beatty A, Morton DG, Kemphues K (2013) PAR-2, LGL-1 and the CDC-42 GAP CHIN-1 act in distinct pathways to maintain polarity in the *C. elegans* embryo. Development 140:2005–2014. https://doi.org/10.1242/dev.088310

Bellaïche Y, Gho M, Kaltschmidt JA, Brand AH, Schweisguth F (2001a) Frizzled regulates localization of cell-fate determinants and mitotic spindle rotation during asymmetric cell division. Nat Cell Biol 3:50–57. https://doi.org/10.1038/35050558

Bellaïche Y, Radovic A, Woods DF, Hough CD, Parmentier ML, O'Kane CJ, Bryant PJ, Schweisguth F (2001b) The partner of inscuteable/discs-large complex is required to establish planar polarity during asymmetric cell division in Drosophila. Cell 106:355–366. https://doi.org/10.1016/s0092-8674(01)00444-5

Bellaïche Y, Beaudoin-Massiani O, Stuttem I, Schweisguth F (2004) The planar cell polarity protein Strabismus promotes Pins anterior localization during asymmetric division of sensory organ precursor cells in Drosophila. Development 131:469–478. https://doi.org/10.1242/dev. 00928

Bento CF, Puri C, Moreau K, Rubinsztein DC (2013) The role of membrane-trafficking small GTPases in the regulation of autophagy. J Cell Sci 126:1059–1069. https://doi.org/10.1242/jcs. 123075

Benton R, St Johnston D (2003) Drosophila PAR-1 and 14-3-3 inhibit Bazooka/PAR-3 to establish complementary cortical domains in polarized cells. Cell 115:691–704. https://doi.org/10.1016/s0092-8674(03)00938-3

Bergstralh DT, Haack T, St Johnston D (2013) Epithelial polarity and spindle orientation: intersecting pathways. Philos Trans R Soc Lond B Biol Sci 368. https://doi.org/10.1098/rstb. 2013.0291

Bilder D, Perrimon N (2000) Localization of apical epithelial determinants by the basolateral PDZ protein Scribble. Nature 403:676–680. https://doi.org/10.1038/35001108

Blair S, McNeill H (2018) Big roles for Fat cadherins. Curr Opin Cell Biol 51:73–80. https://doi.org/10.1016/j.ceb.2017.11.006

Bonifacino JS (2014) Adaptor proteins involved in polarized sorting. J Cell Biol 204:7–17. https://doi.org/10.1083/jcb.201310021

Box K, Joyce BW, Devenport D (2019) Epithelial geometry regulates spindle orientation and progenitor fate during formation of the mammalian epidermis. Elife. https://doi.org/10.7554/eLife.47102

Bryant DM, Datta A, Rodríguez-Fraticelli AE, Peränen J, Martín-Belmonte F, Mostov KE (2010) A molecular network for de novo generation of the apical surface and lumen. Nat Cell Biol 12:1035–1045. https://doi.org/10.1038/ncb2106

Butler MT, Wallingford JB (2017) Planar cell polarity in development and disease. Nat Rev Mol Cell Biol 18:375–388. https://doi.org/10.1038/nrm.2017.11

Cao F, Miao Y, Xu K, Liu P (2015) Lethal (2) giant larvae: an indispensable regulator of cell polarity and cancer development. Int J Biol Sci 11:380–389. https://doi.org/10.7150/ijbs.11243

Caria S, Stewart BZ, Jin R, Smith BJ, Humbert PO, Kvansakul M (2019) Structural analysis of phosphorylation-associated interactions of human MCC with Scribble PDZ domains. FEBS J. https://doi.org/10.1111/febs.15002. epub ahead of print

Casal J, Lawrence PA, Struhl G (2006) Two separate molecular systems, Dachsous/Fat and Starry night/Frizzled, act independently to confer planar cell polarity. Development 133:4561–4572

Chen X, Macara IG (2005) Par-3 controls tight junction assembly through the Rac exchange factor Tiam1. Nat Cell Biol 7:262–269. https://doi.org/10.1038/ncb1226

Chen J, Sayadian AC, Lowe N, Lovegrove HE, St Johnston D (2018) An alternative mode of epithelial polarity in the Drosophila midgut. PLoS Biol 16:e3000041. https://doi.org/10.1371/journal.pbio.3000041

Choi J, Troyanovsky RB, Indra I, Mitchell BJ, Troyanovsky SM (2019) Scribble, Erbin, and Lano redundantly regulate epithelial polarity and apical adhesion complex. J Cell Biol 8:2277–2293. https://doi.org/10.1083/jcb.201804201

Colicelli J (2004) Human RAS superfamily proteins and related GTPases. Sci STKE RE13. https://doi.org/10.1126/stke.2502004re13

Colombié N, Choesmel-Cadamuro V, Series J, Emery G, Wang X, Ramel D (2017) Non-autonomous role of Cdc42 in cell-cell communication during collective migration. Dev Biol 423:12–18. https://doi.org/10.1016/j.ydbio.2017.01.018

Courbard JR, Djiane A, Wu J, Mlodzik M (2009) The apical/basal-polarity determinant Scribble cooperates with the PCP core factor Stbm/Vang and functions as one of its effectors. Dev Biol 333:67–77. https://doi.org/10.1016/j.ydbio.2009.06.024

Croisé P, Estay-Ahumada C, Gasman S, Ory S (2014) Rho GTPases, phosphoinositides, and actin: a tripartite framework for efficient vesicular trafficking. Small GTPases 5:e29469. https://doi.org/10.4161/sgtp.29469

Devenport D (2014) The cell biology of planar cell polarity. J Cell Biol 207:171–179. https://doi.org/10.1083/jcb.201408039

Devenport D, Fuchs E (2008) Planar polarization in embryonic epidermis orchestrates global asymmetric morphogenesis of hair follicles. Nat Cell Biol 10:1257–1268. https://doi.org/10.1038/ncb1784

Devergne O, Tsung K, Barcelo G, Schüpbach T (2014) Polarized deposition of basement membrane proteins depends on Phosphatidylinositol synthase and the levels of Phosphatidylinositol 4,5-bisphosphate. Proc Natl Acad Sci U S A 111:7689–7694. https://doi.org/10.1073/pnas.1407351111

Dias Gomes M, Letzian S, Saynisch M, Iden S (2019) Polarity signaling ensures epidermal homeostasis by coupling cellular mechanics and genomic integrity. Nat Commun 10:3362. https://doi.org/10.1038/s41467-019-11325-3

Djiane A, Yogev S, Mlodzik M (2005) The apical determinants aPKC and dPatj regulate Frizzled-dependent planar cell polarity in the Drosophila eye. Cell 121:621–631. https://doi.org/10.1016/j.cell.2005.03.014

Doerflinger H, Benton R, Shulman JM, St Johnston D (2003) The role of PAR-1 in regulating the polarised microtubule cytoskeleton in the Drosophila follicular epithelium. Development 130:3965–3975. https://doi.org/10.1242/dev.00616

Doerflinger H, Benton R, Torres IL, Zwart MF, St Johnston D (2006) Drosophila anterior-posterior polarity requires actin-dependent PAR-1 recruitment to the oocyte posterior. Curr Biol 16:1090–1095. https://doi.org/10.1016/j.cub.2006.04.001

Doerflinger H, Vogt N, Torres IL, Mirouse V, Koch I, Nüsslein-Volhard C, St Johnston D (2010) Bazooka is required for polarisation of the Drosophila anterior-posterior axis. Development 137:1765–1773. https://doi.org/10.1242/dev.045807

Elbediwy A, Zihni C, Terry SJ, Clark P, Matter K, Balda MS (2012) Epithelial junction formation requires confinement of Cdc42 activity by a novel SH3BP1 complex. J Cell Biol 198:677–693. https://doi.org/10.1083/jcb.201202094

Etemad-Moghadam B, Guo S, Kemphues KJ (1995) Asymmetrically distributed PAR-3 protein contributes to cell polarity and spindle alignment in early C. elegans embryos. Cell 83:743–752. https://doi.org/10.1016/0092-8674(95)90187

Finegan TM, Bergstralh DT (2019) Division orientation: disentangling shape and mechanical forces. Cell Cycle 18:1187–1198. https://doi.org/10.1080/15384101.2019.1617006

Fogg VC, Liu CJ, Margolis B (2005) Multiple regions of Crumbs3 are required for tight junction formation in MCF10A cells. J Cell Sci 118:2859–2869. https://doi.org/10.1242/jcs.02412

Franz A, Riechmann V (2010) Stepwise polarisation of the Drosophila follicular epithelium. Dev Biol 338:136–147. https://doi.org/10.1016/j.ydbio.2009.11.027

Fritsch R, de Krijger I, Fritsch K, George R, Reason B, Kumar MS, Diefenbacher M, Stamp G, Downward J (2013) RAS and RHO families of GTPases directly regulate distinct phosphoinositide 3-kinase isoforms. Cell 153:1050–1063. https://doi.org/10.1016/j.cell.2013.04.031

Gao L, Macara IG (2004) Isoforms of the polarity protein par6 have distinct functions. J Biol Chem 279:41557–41562. https://doi.org/10.1074/jbc.M403723200

Garcia De Las Bayonas A, Philippe JM, Lellouch AC, Lecuit T (2019) Distinct RhoGEFs activate apical and junctional contractility under control of G Proteins during epithelial morphogenesis. Curr Biol 29:3370–3385. https://doi.org/10.1016/j.cub.2019.08.017

Goode BL, Eck MJ (2007) Mechanism and function of formins in the control of actin assembly. Annu Rev Biochem 76:593–627. https://doi.org/10.1146/annurev.biochem.75.103004.142647

Goryachev AB, Leda M (2019) Autoactivation of small GTPases by the GEF-effector positive feedback modules. F1000Res pii: F1000 Faculty Rev-1676. https://doi.org/10.12688/f1000research.20003.1

Graybill C, Wee B, Atwood SX, Prehoda KE (2012) Partitioning-defective protein 6 (Par-6) activates atypical protein kinase C (aPKC) by pseudosubstrate displacement. J Biol Chem 287:21003–20111. https://doi.org/10.1074/jbc.M112.360495

Gubb D, García-Bellido A (1982) A genetic analysis of the determination of cuticular polarity during development in Drosophila melanogaster. J Embryol Exp Morphol 68:37–57

Guo W, Giancotti FG (2004) Integrin signalling during tumour progression. Nat Rev Mol Cell Biol 5:816–826. https://doi.org/10.1038/nrm1490

Halfter W, Oertle P, Monnier CA, Camenzind L, Reyes-Lua M, Hu H, Candiello J, Labilloy A, Balasubramani M, Henrich PB, Plodinec M (2015) New concepts in basement membrane biology. FEBS J 282:4466–4479. https://doi.org/10.1111/febs.13495

Hao Y, Du Q, Chen X, Zheng Z, Balsbaugh JL, Maitra S, Shabanowitz J, Hunt DF, Macara IG (2010) Par3 controls epithelial spindle orientation by aPKC-mediated phosphorylation of apical Pins. Curr Biol 20:1809–1818. https://doi.org/10.1016/j.cub.2010.09.032

Harris TJ, Peifer M (2004) Adherens junction-dependent and -independent steps in the establishment of epithelial cell polarity in Drosophila. J Cell Biol 167:135–147. https://doi.org/10.1083/jcb.200406024

Harris TJ, Peifer M (2005) The positioning and segregation of apical cues during epithelial polarity establishment in Drosophila. J Cell Biol 170:813–823

Harris TJ, Peifer M (2007) aPKC controls microtubule organization to balance adherens junction symmetry and planar polarity during development. Dev Cell 12:727–738. https://doi.org/10.1016/j.devcel.2007.02.011

Heikenwalder M, Lorentzen A (2019) The role of polarisation of circulating tumour cells in cancer metastasis. Cell Mol Life Sci 76:3765–3781. https://doi.org/10.1007/s00018-019-03169-3

Henderson DJ, Long DA, Dean CH (2018) Planar cell polarity in organ formation. Curr Opin Cell Biol 55:96–103. https://doi.org/10.1016/j.ceb.2018.06.011

Herpers B, Rabouille C (2004) mRNA localization and ER-based protein sorting mechanisms dictate the use of transitional endoplasmic reticulum-golgi units involved in gurken transport in Drosophila oocytes. Mol Biol Cell 15:5306–5317. https://doi.org/10.1091/mbc.e04-05-0398

Hikita T, Mirzapourshafiyi F, Barbacena P, Riddell M, Pasha A, Li M, Kawamura T, Brandes RP, Hirose T, Ohno S, Gerhardt H, Matsuda M, Franco CA, Nakayama M (2018) PAR-3 controls endothelial planar polarity and vascular inflammation under laminar flow. EMBO Rep 19: e45253. https://doi.org/10.15252/embr.201745253

Hong Y (2018) aPKC: the kinase that phosphorylates cell polarity. F1000Res 7 F1000 Faculty Rev-903. https://doi.org/10.12688/f1000research.14427.1

Huebner RJ, Wallingford JB (2019) Coming to consensus: a unifying model emerges for convergent extension. Dev Cell Vol 48:126. https://doi.org/10.1016/j.devcel.2018.12.006

Humphries AC, Mlodzik M (2018) From instruction to output: Wnt/PCP signaling in development and cancer. Curr Opin Cell Biol 51:110–116. https://doi.org/10.1016/j.ceb.2017.12.005

Hutterer A, Betschinger J, Petronczki M, Knoblich JA (2004) Sequential roles of Cdc42, Par-6, aPKC, and Lgl in the establishment of epithelial polarity during Drosophila embryogenesis. Dev Cell 6:845–854. https://doi.org/10.1016/j.devcel.2004.05.003

Ivanova ME, Fletcher GC, O'Reilly N, Purkiss AG, Thompson BJ, McDonald NQ (2015) Structures of the human Pals1 PDZ domain with and without ligand suggest gated access of Crb to the PDZ peptide-binding groove. Acta Crystallogr D Biol Crystallogr 71:555–564. https://doi.org/10.1107/S139900471402776X

Izaddoost S, Nam SC, Bhat MA, Bellen HJ, Choi KW (2002) Drosophila crumbs is a positional cue in photoreceptor adherens junctions and rhabdomeres. Nature 416:178–183. https://doi.org/10.1091/mbc.e03-04-0235

Jiang T, McKinley RF, McGill MA, Angers S, Harris TJ (2015) A Par-1-Par-3-centrosome cell polarity pathway and its tuning for isotropic cell adhesion. Curr Biol 25:2701–2708. https://doi.org/10.1016/j.cub.2015.08.063

Jossin Y, Lee M, Klezovitch O, Kon E, Cossard A, Lien WH, Fernandez TE, Cooper JA, Vasioukhin V (2017) Llgl1 connects cell polarity with cell-cell adhesion in embryonic neural stem cells. Dev Cell 41:481–495. https://doi.org/10.1016/j.devcel.2017.05.002

Juriloff DM, Harris MJ (2012) A consideration of the evidence that genetic defects in planar cell polarity contribute to the etiology of human neural tube defects. Birth Defects Res A Clin Mol Teratol 94:824–840. https://doi.org/10.1002/bdra.23079

Juriloff DM, Harris MJ (2018) Insights into the etiology of mammalian neural tube closure Defects from developmental, genetic and evolutionary studies. J Dev Biol 6:pii: E22. https://doi.org/10.3390/jdb6030022

Kemphues KJ, Priess JR, Morton DG, Cheng N (1988) Identification of genes required for cytoplasmic localization in early C. elegans embryos. Cell 52:311–320. https://doi.org/10.1016/s0092-8674(88)80024-2

Kim S, Gailite I, Moussian B, Luschnig S, Goette M, Fricke K, Honemann-Capito M, Grubmüller H, Wodarz A (2009) Kinase-activity-independent functions of atypical protein kinase C in Drosophila. J Cell Sci 122:3759–1771. https://doi.org/10.1242/jcs.052514

Klompstra D, Anderson DC, Yeh JY, Zilberman Y, Nance J (2015) An instructive role for C. elegans E-cadherin in translating cell contact cues into cortical polarity. Nat Cell Biol 17:726–735. https://doi.org/10.1038/ncb3168

Koch L, Feicht S, Sun R, Sen A, Krahn MP (2016) Domain-specific functions of Stardust in Drosophila embryonic development. R Soc Open Sci 3:160776. https://doi.org/10.1098/rsos.160776

Kondylis V, Rabouille C (2009) The Golgi apparatus: lessons from Drosophila. FEBS Lett 583:3827–3838. https://doi.org/10.1016/j.febslet.2009.09.048

Krahn MP, Bückers J, Kastrup L, Wodarz A (2010) Formation of a Bazooka-Stardust complex is essential for plasma membrane polarity in epithelia. J Cell Biol 190:751–760. https://doi.org/10.1083/jcb.201006029

Lamber EP, Siedenburg AC, Barr FA (2019) Rab regulation by GEFs and GAPs during membrane traffic. Curr Opin Cell Biol 59:34–39. https://doi.org/10.1016/j.ceb.2019.03.004

Laprise P, Lau KM, Harris KP, Silva-Gagliardi NF, Paul SM, Beronja S, Beitel GJ, McGlade CJ, Tepass U (2009) Yurt, Coracle, Neurexin IV and the Na$^+$,K$^+$-ATPase form a novel group of epithelial polarity proteins. Nature 459:1141–1145. https://doi.org/10.1038/nature08067

Lemmers C, Michel D, Lane-Guermonprez L, Delgrossi MH, Médina E, Arsanto JP, Le Bivic A (2004) CRB3 binds directly to Par6 and regulates the morphogenesis of the tight junctions in mammalian epithelial cells. Mol Biol Cell 15:1324–1333. https://doi.org/10.1091/mbc.e03-04-0235

Lerner DW, McCoy D, Isabella AJ, Mahowald AP, Gerlach GF, Chaudhry TA, Horne-Badovinac S (2013) A Rab10-dependent mechanism for polarized basement membrane secretion during organ morphogenesis. Dev Cell 24:159–168. https://doi.org/10.1016/j.devcel.2012.12.005

Levayer R, Pelissier-Monier A, Lecuit T (2011) Spatial regulation of Dia and Myosin-II by RhoGEF2 controls initiation of E-cadherin endocytosis during epithelial morphogenesis. Nat Cell Biol 13:529–540. https://doi.org/10.1038/ncb2224

Li G, Marlin MC (2015) Rab family of GTPases. Methods Mol Biol 1298:1–15. https://doi.org/10.1007/978-1-4939-2569-8_1

Li Y, Wei Z, Yan Y, Wan Q, Du Q, Zhang M (2014) Structure of Crumbs tail in complex with the PALS1 PDZ-SH3-GK tandem reveals a highly specific assembly mechanism for the apical Crumbs complex. Proc Natl Acad Sci U S A 111:17444–17449. https://doi.org/10.1073/pnas.1416515111

Li J, Cheng L, Jiang H (2019) Cell shape and intercellular adhesion regulate mitotic spindle orientation. Mol Biol Cell 30:2458–2468. https://doi.org/10.1091/mbc.E19-04-0227

Lin WH, Asmann YW, Anastasiadis PZ (2015) Expression of polarity genes in human cancer. Cancer Inform 14(Suppl 3):15–28. https://doi.org/10.4137/CIN.S18964

Lin CM, Titchenell PM, Keil JM, Garcia-Ocaña A, Bolinger MT, Abcouwer SF, Antonetti DA (2018) Inhibition of atypical protein kinase C reduces inflammation-induced retinal vascular permeability. Am J Pathol 188:2392–2405. https://doi.org/10.1016/j.ajpath.2018.06.020

Liu C, Lin C, Gao C, May-Simera H, Swaroop A, Li T (2014) Null and hypomorph Prickle1 alleles in mice phenocopy human Robinow syndrome and disrupt signaling downstream of Wnt5a. Biol Open 3:861–870. https://doi.org/10.1242/bio.20148375

Malt AL, Dailey Z, Holbrook-Rasmussen J, Zheng Y, Hogan A, Du Q, Lu X (2019) Par3 is essential for the establishment of planar cell polarity of inner ear hair cells. Proc Natl Acad Sci U S A 116:4999–5008. https://doi.org/10.1073/pnas.1816333116

Martin-Belmonte F, Gassama A, Datta A, Yu W, Rescher U, Gerke V, Mostov K (2007) PTEN-mediated apical segregation of phosphoinositides controls epithelial morphogenesis through Cdc42. Cell 128:383–397. https://doi.org/10.1016/j.cell.2006.11.051

Matakatsu H, Blair SS (2004) Interactions between Fat and Dachsous and the regulation of planar cell polarity in the Drosophila wing. Development 131:3785–3794. https://doi.org/10.1242/dev.01254

Matsumoto Y, Nakagawa S, Yano T, Takizawa S, Nagasaka K, Nakagawa K, Minaguchi T, Wada O, Ooishi H, Matsumoto K, Yasugi T, Kanda T, Huibregtse JM, Taketani Y (2006) Involvement of a cellular ubiquitin-protein ligase E6AP in the ubiquitin-mediated degradation of extensive substrates of high-risk human papillomavirus E6. J Med Virol 78:501–507. https://doi.org/10.1002/jmv.20568

McGill MA, McKinley RF, Harris TJ (2009) Independent cadherin-catenin and Bazooka clusters interact to assemble adherens junctions. J Cell Biol 185:787–796. https://doi.org/10.1083/jcb.200812146

McKinley RF, Harris TJ (2012) Displacement of basolateral Bazooka/PAR-3 by regulated transport and dispersion during epithelial polarization in Drosophila. Mol Biol Cell 23:4465–4471. https://doi.org/10.1091/mbc.E12-09-0655

Mechler BM, McGinnis W, Gehring WJ (1985) Molecular cloning of lethal(2)giant larvae, a recessive oncogene of Drosophila melanogaster. EMBO J 4:1551–1557

Montcouquiol M, Kelley MW (2003) Planar and vertical signals control cellular differentiation and patterning in the mammalian cochlea. J Neurosci 23:9469–9478. https://doi.org/10.1523/JNEUROSCI.23-28-09469.2003

Morais-de-Sá E, Mirouse V, St Johnston D (2010) aPKC phosphorylation of Bazooka defines the apical/lateral border in Drosophila epithelial cells. Cell 143:509–523. https://doi.org/10.1016/j.cell.2010.02.040

Morrissey MA, Sherwood DR (2015) An active role for basement membrane assembly and modification in tissue sculpting. J Cell Sci 128:1661–1668. https://doi.org/10.1242/jcs.168021

Motegi F, Zonies S, Hao Y, Cuenca AA, Griffin E, Seydoux G (2011) Microtubules induce self-organization of polarized PAR domains in Caenorhabditis elegans zygotes. Nat Cell Biol 13:1361–1367. https://doi.org/10.1038/ncb2354

Murray DH, Jahnel M, Lauer J, Avellaneda MJ, Brouilly N, Cezanne A, Morales-Navarrete H, Perini ED, Ferguson C, Lupas AN, Kalaidzidis Y, Parton RG, Grill SW, Zerial M (2016) An endosomal tether undergoes an entropic collapse to bring vesicles together. Nature 537:107–111. https://doi.org/10.1038/nature19326

Müsch A (2004) Microtubule organization and function in epithelial cells. Traffic 5:1–9

Nakagawa S, Huibregtse JM (2000) Human scribble (Vartul) is targeted for ubiquitin-mediated degradation by the high-risk papillomavirus E6 proteins and the E6AP ubiquitin-protein ligase. Mol Cell Biol 20:8244–8253. https://doi.org/10.1128/mcb.20.21.8244-8253.2000

Narayanan AS, Reyes SB, Um K, McCarty JH, Tolias KF (2013) The Rac-GAP Bcr is a novel regulator of the Par complex that controls cell polarity. Mol Biol Cell 24:3857–3868. https://doi.org/10.1091/mbc.E13-06-0333

Narimatsu M, Bose R, Pye M, Zhang L, Miller B, Ching P, Sakuma R, Luga V, Roncari L, Attisano L, Wrana JL (2009) Regulation of planar cell polarity by Smurf ubiquitin ligases. Cell 137:295–307. https://doi.org/10.1016/j.cell.2009.02.025

Nashchekin D, Fernandes AR, St Johnston D (2016) Patronin/Shot cortical foci assemble the noncentrosomal microtubule array that specifies the Drosophila anterior-posterior axis. Dev Cell 38:61–72. https://doi.org/10.1016/j.devcel.2016.06.010

Navis A, Nelson CM (2016) Pulling together: tissue-generated forces that drive lumen morphogenesis. Semin Cell Dev Biol 55:139–147. https://doi.org/10.1016/j.semcdb.2016.01.002

Nelson WJ (2009) Remodeling epithelial cell organization: transitions between front–rear and apical–basal polarity. Cold Spring Harb Perspect Biol 1:a000513. https://doi.org/10.1101/cshperspect.a000513

Newton AC (2018) Protein kinase C: perfectly balanced. Crit Rev Biochem Mol Biol 53:208–230. https://doi.org/10.1080/10409238.2018.1442408

Nunes de Almeida F, Walther RF, Pressé MT, Vlassaks E, Pichaud F (2019) Cdc42 defines apical identity and regulates epithelial morphogenesis by promoting apical recruitment of Par6-aPKC and Crumbs. Development 146:dev175497 https://doi.org/10.1242/dev.175497

O'Brien LE, Jou TS, Pollack AL, Zhang Q, Hansen SH, Yurchenco P, Mostov KE (2001) Rac1 orientates epithelial apical polarity through effects on basolateral laminin assembly. Nat Cell Biol 3:831–838. https://doi.org/10.1038/ncb0901-831

Osmani N, Peglion F, Chavrier P, Etienne-Manneville S (2010) Cdc42 localization and cell polarity depend on membrane traffic. J Cell Biol 191:1261–1269. https://doi.org/10.1083/jcb.201003091

Perez-Vale KZ, Peifer M (2018) Modulating apical-basal polarity by building and deconstructing a Yurt. J Cell Biol 217:3772–3773. https://doi.org/10.1083/jcb.201810059

Pichaud F (2018) PAR-complex and crumbs function during photoreceptor morphogenesis and retinal degeneration. Front Cell Neurosci 12:90. https://doi.org/10.3389/fncel.2018.00090

Pichaud F, Walther RF, Nunes de Almeida F (2019) Regulation of Cdc42 and its effectors in epithelial morphogenesis. J Cell Sci 132 pii: jcs217869. https://doi.org/10.1242/jcs.217869

Pieczynski J, Margolis B (2011) Protein complexes that control renal epithelial polarity. Am J Physiol Renal Physiol 300:F589–F601. https://doi.org/10.1152/ajprenal.00615.2010

Pinal N, Goberdhan DC, Collinson L, Fujita Y, Cox IM, Wilson C, Pichaud F (2006) Regulated and polarized PtdIns(3,4,5)P3 accumulation is essential for apical membrane morphogenesis in photoreceptor epithelial cells. Curr Biol 16:140–149

Plant PJ, Fawcett JP, Lin DC, Holdorf AD, Binns K, Kulkarni S, Pawson T (2003) A polarity complex of mPar-6 and atypical PKC binds, phosphorylates and regulates mammalian Lgl. Nat Cell Biol 5:301–308. https://doi.org/10.1038/ncb948

Quiros M, Nusrat A (2014) RhoGTPases, actomyosin signaling and regulation of the epithelial Apical Junctional Complex. Semin Cell Dev Biol 36:194–203. https://doi.org/10.1016/j.semcdb.2014.09.003

Ridley AJ (1995) Rho-related proteins: actin cytoskeleton and cell cycle. Curr Opin Genet Dev 5:24–30. https://doi.org/10.1016/s0959-437x(95)90049-7

Rodriguez J, Peglion F, Martin J, Hubatsch L, Reich J, Hirani N, Gubieda AG, Roffey J, Fernandes AR, St Johnston D, Ahringer J, Goehring NW (2017) aPKC cycles between functionally distinct PAR protein assemblies to drive cell polarity. Dev Cell 42:400–415. https://doi.org/10.1016/j.devcel.2017.07.007

Roh MH, Fan S, Liu C, Margolis B (2003) The Crumbs3-Pals1 complex participates in the establishment of polarity in mammalian epithelial cells. J Cell Sci 116:2895–2906. https://doi.org/10.1242/jcs.00500

Rohatgi R, Ho HY, Kirschner MW (2000) Mechanism of N-WASP activation by CDC42 and phosphatidylinositol 4, 5-bisphosphate. J Cell Biol 150:1299–1310. https://doi.org/10.1083/jcb.150.6.1299

Ruch TR, Bryant DM, Mostov KE, Engel JN (2017) Par3 integrates Tiam1 and phosphatidylinositol 3-kinase signaling to change apical membrane identity. Mol Biol Cell 28:252–260. https://doi.org/10.1091/mbc.E16-07-0541

Saadaoui M, Machicoane M, di Pietro F, Etoc F, Echard A, Morin X (2014) Dlg1 controls planar spindle orientation in the neuroepithelium through direct interaction with LGN. J Cell Biol 206:707–717. https://doi.org/10.1083/jcb.201405060

Saadaoui M, Konno D, Loulier K, Goiame R, Jadhav V, Mapelli M, Matsuzaki F, Morin X (2017) Loss of the canonical spindle orientation function in the Pins/LGN homolog AGS3. EMBO Rep 18:1509–1520. https://doi.org/10.15252/embr.201643048

Schneeberger K, Roth S, Nieuwenhuis EES, Middendorp S (2018) Intestinal epithelial cell polarity defects in disease: lessons from microvillus inclusion disease. Dis Model Mech 11 pii: dmm031088. https://doi.org/10.1242/dmm.031088

Schneider M, Khalil AA, Poulton J, Castillejo-Lopez C, Egger-Adam D, Wodarz A, Deng WM, Baumgartner S (2006) Perlecan and Dystroglycan act at the basal side of the *Drosophila* follicular epithelium to maintain epithelial organization. Development 133:3805–3815. https://doi.org/10.1242/dev.02549

Ségalen M, Bellaïche Y (2009) Cell division orientation and planar cell polarity pathways. Semin Cell Dev Biol 20:972–977. https://doi.org/10.1016/j.semcdb.2009.03.018

Ségalen M, Johnston CA, Martin CA, Dumortier JG, Prehoda KE, David NB, Doe CQ, Bellaïche Y (2010) The Fz-Dsh planar cell polarity pathway induces oriented cell division via Mud/NuMA in Drosophila and zebrafish. Dev Cell 19:740–752. https://doi.org/10.1016/j.devcel.2010.10.004

Selamat W, Tay PL, Baskaran Y, Manser E (2015) The Cdc42 effector kinase PAK4 localizes to cell-cell junctions and contributes to establishing cell polarity. PLoS One 10:e0129634. https://doi.org/10.1371/journal.pone.0129634

Sen A, Nagy-Zsvér-Vadas Z, Krahn MP (2012) Drosophila PATJ supports adherens junction stability by modulating Myosin light chain activity. J Cell Biol 199:685–698. https://doi.org/10.1083/jcb.201206064

Sen A, Sun R, Krahn MP (2015) Localization and function of Pals1-associated tight junction protein in Drosophila is regulated by two distinct apical complexes. J Biol Chem 290:13224–13233. https://doi.org/10.1074/jbc.M114.629014

Shin K, Straight S, Margolis B (2005) PATJ regulates tight junction formation and polarity in mammalian epithelial cells. J Cell Biol 168:705–711

Steeg PS (2016) Targeting metastasis. Nat Rev Cancer 16:201–218. https://doi.org/10.1038/nrc.2016.5

Stephens R, Lim K, Portela M, Kvansakul M, Humbert PO, Richardson HE (2018) The scribble cell polarity module in the regulation of cell signaling in tissue development and tumorigenesis. J Mol Biol 430:3585–3612. https://doi.org/10.1016/j.jmb.2018.01.011

St Johnston D, Ahringer J (2010a) Cell polarity in eggs and epithelia: parallels and diversity. Cell 141:757–774. https://doi.org/10.1016/j.cell.2010.05.011

St Johnston D, Ahringer J (2010b) Cell polarity in eggs and epithelia: parallels and diversity. Cell 141:757–774. https://doi.org/10.1016/j.cell.2010.05.011

Strutt DI, Weber U, Mlodzik M (1997) The role of RhoA in tissue polarity and Frizzled signalling. Nature 387:292–295. https://doi.org/10.1038/387292a0

Suzuki A, Hirata M, Kamimura K, Maniwa R, Yamanaka T, Mizuno K, Kishikawa M, Hirose H, Amano Y, Izumi N, Miwa Y, Ohno S (2004) aPKC acts upstream of PAR-1b in both the establishment and maintenance of mammalian epithelial polarity. Curr Biol 14:1425–1435. https://doi.org/10.1016/j.cub.2004.08.021

Takai Y, Ikeda W, Ogita H, Rikitake Y (2008) The immunoglobulin-like cell adhesion molecule nectin and its associated protein afadin. Annu Rev Cell Dev Biol 24:309–342. https://doi.org/10.1146/annurev.cellbio.24.110707.175339

Tanentzapf G, Tepass U (2003) Interactions between the crumbs, lethal giant larvae and bazooka pathways in epithelial polarization. Nat Cell Biol 5:46–52. https://doi.org/10.1038/ncb896

Terry SJ, Zihni C, Elbediwy A, Vitiello E, Leefa Chong San IV, Balda MS, Matter K (2011) Spatially restricted activation of RhoA signalling at epithelial junctions by p114RhoGEF drives junction formation and morphogenesis. Nat Cell Biol 13:159–166. https://doi.org/10.1038/ncb2156

Theisen H, Purcell J, Bennett M, Kansagara D, Syed A, Marsh JL (1994) Dishevelled is required during wingless signaling to establish both cell polarity and cell identity. Development 120:347–360

Tian AG, Deng WM (2008) Lgl and its phosphorylation by aPKC regulate oocyte polarity formation in Drosophila. Development 135:463–471. https://doi.org/10.1242/dev.016253

Toma-Fukai S, Shimizu T (2019) Structural insights into the regulation mechanism of small GTPases by GEFs. Molecules 24:3308. https://doi.org/10.3390/molecules24183308

Toya M, Kobayashi S, Kawasaki M, Shioi G, Kaneko M, Ishiuchi T, Misaki K, Meng W, Takeichi M (2016) CAMSAP3 orients the apical-to-basal polarity of microtubule arrays in epithelial cells. Proc Natl Acad Sci U S A 113:332–337. https://doi.org/10.1073/pnas.1520638113

Ungermann C, Kümmel D (2019) Structure of membrane tethers and their role in fusion. Traffic 20:479–490. https://doi.org/10.1111/tra.12655

Usui T, Shima Y, Shimada Y, Hirano S, Burgess RW, Schwarz TL, Takeichi M, Uemura T (1999) Flamingo, a seven-pass transmembrane cadherin, regulates planar cell polarity under the control of Frizzled. Cell 98:585–595. https://doi.org/10.1016/s0092-8674(00)80046-x

Vladar EK, Bayly RD, Sangoram AM, Scott MP, Axelrod JD (2012) Microtubules enable the planar cell polarity of airway cilia. Curr Biol 22:2203–2212. https://doi.org/10.1016/j.cub.2012.09.046

Walther RF, Pichaud F (2010) Crumbs/DaPKC-dependent apical exclusion of Bazooka promotes photoreceptor polarity remodeling. Curr Biol 20:1065–1074. https://doi.org/10.1016/j.cub.2010.04.049

Walther RF, Nunes de Almeida F, Vlassaks E, Burden JJ, Pichaud F (2016) Pak4 is required during epithelial polarity remodeling through regulating AJ stability and Bazooka Retention at the ZA. Cell Rep 15:45–53. https://doi.org/10.1016/j.celrep.2016.03.014

Wang Q, Hurd TW, Margolis B (2004) Tight junction protein Par6 interacts with an evolutionarily conserved region in the amino terminus of PALS1/stardust. J Biol Chem 279:30715–30721. https://doi.org/10.1074/jbc.M401930200

Wang SC, Low TYF, Nishimura Y, Gole L, Yu W, Motegi F (2017) Cortical forces and CDC-42 control clustering of PAR proteins for Caenorhabditis elegans embryonic polarization. Nat Cell Biol 19:988–995. https://doi.org/10.1038/ncb3577

Watson JR, Owen D, Mott HR (2017) Cdc42 in actin dynamics: an ordered pathway governed by complex equilibria and directional effector handover. Small GTPases 8:237–244. https://doi.org/10.1080/21541248.2016.1215657

Watts JL, Morton DG, Bestman J, Kemphues KJ (2000) The C. elegans par-4 gene encodes a putative serine-threonine kinase required for establishing embryonic asymmetry. Development 127:1467–1475

Wells CD, Fawcett JP, Traweger A, Yamanaka Y, Goudreault M, Elder K, Kulkarni S, Gish G, Virag C, Lim C, Colwill K, Starostine A, Metalnikov P, Pawson T (2006) A Rich1/Amot complex regulates the Cdc42 GTPase and apical-polarity proteins in epithelial cells. Cell 125:535–548. https://doi.org/10.1016/j.cell.2006.02.045

Whitney DS, Peterson FC, Kittell AW, Egner JM, Prehoda KE, Volkman BF (2016) Binding of crumbs to the Par-6 CRIB-PDZ module is regulated by Cdc42. Biochemistry 55:1455–1461. https://doi.org/10.1021/acs.biochem.5b01342

Wilkie GS, Davis I (2001) Drosophila wingless and pair-rule transcripts localize apically by dynein-mediated transport of RNA particles. Cell 105:209–219. https://doi.org/10.1016/s0092-8674(01)00312-9

Williams SE, Ratliff LA, Postiglione MP, Knoblich JA, Fuchs E (2014) Par3-mInsc and Gαi3 cooperate to promote oriented epidermal cell divisions through LGN. Nat Cell Biol 16:758–769. https://doi.org/10.1038/ncb3001

Wodarz A, Hinz U, Engelbert M, Knust E (1995) Expression of crumbs confers apical character on plasma membrane domains of ectodermal epithelia of Drosophila. Cell 14:67–76. https://doi.org/10.1016/0092-8674(95)90053-5

Woods DF, Bryant PJ (1989) Molecular cloning of the lethal(1)discs large-1 oncogene of Drosophila. Dev Biol 134:222–235. https://doi.org/10.1016/0012-1606(89)90092-4

Wu H, Feng W, Chen J, Chan LN, Huang S, Zhang M (2007) PDZ domains of Par-3 as potential phosphoinositide signaling integrators. Mol Cell 28:886–898. https://doi.org/10.1016/j.molcel.2007.10.028

Wu JC, Espiritu EB, Rose LS (2016) The 14-3-3 protein PAR-5 regulates the asymmetric localization of the LET-99 spindle positioning protein. Dev Biol 412:288–297. https://doi.org/10.1016/j.ydbio.2016.02.020

Yamanaka T, Horikoshi Y, Izumi N, Suzuki A, Mizuno K, Ohno S (2006) Lgl mediates apical domain disassembly by suppressing the PAR-3-aPKC-PAR-6 complex to orient apical membrane polarity. J Cell Sci 119:2107–2118. https://doi.org/10.1242/jcs.02938

Yu W, Datta A, Leroy P, O'Brien LE, Mak G, Jou TS, Matlin KS, Mostov KE, Zegers MM (2005) Beta1-integrin orients epithelial polarity via Rac1 and laminin. Mol Biol Cell 16:433–445. https://doi.org/10.1091/mbc.e04-05-0435

Yu CG, Tonikian R, Felsensteiner C, Jhingree JR, Desveaux D, Sidhu SS, Harris TJ (2014) Peptide binding properties of the three PDZ domains of Bazooka (Drosophila Par-3). PLoS One 9: e86412. https://doi.org/10.1371/journal.pone.0086412

Zihni C, Munro PM, Elbediwy A, Keep NH, Terry SJ, Harris J, Balda MS, Matter K (2014) Dbl3 drives Cdc42 signaling at the apical margin to regulate junction position and apical differentiation. J Cell Biol 204:111–127. https://doi.org/10.1083/jcb.201304064

Zihni C, Vlassaks E, Terry S, Carlton J, Leung TKC, Olson M, Pichaud F, Balda MS, Matter K (2017) An apical MRCK-driven morphogenetic pathway controls epithelial polarity. Nat Cell Biol 19:1049–1060. https://doi.org/10.1038/ncb3592

Chapter 4
Mathematical Modeling of Epithelial Ion Transport

David P. Nickerson, Leyla Noroozbabaee, Dewan M. Sarwar, Kirk L. Hamilton, and Peter J. Hunter

Abstract In this chapter, we provide a general introduction to mathematical modeling of epithelial ion transport. The basic mathematical concepts are introduced and the instantiation of these concepts into numerical simulation is demonstrated. Tools and technologies that aid scientists in the creation and use of epithelial ion transport models are also discussed.

Keywords Epithelial · Mathematical modeling · Ion · Transport physiology

4.1 Introduction

Epithelial cells are polarised cells in which different transport proteins reside within the apical and basolateral membranes (Fig. 4.1). It is this discrete localisation of transport proteins which allows the specific reabsorption of a solute species across the epithelial cell and, ultimately, its return to the blood. For instance, the reabsorption of sodium from the filtrate of the lumen of the proximal tubule in the renal nephron back to the blood requires the concerted action of both apical-specific and basolateral-specific transporters. The sodium-potassium pump (Na^+/K^+-ATPase) is a primary active transport protein located within the basolateral membrane of a cell and uses ATP to transport sodium against its electrochemical gradient, from the cell into the interstitium (from whence it can diffuse back into the blood; serosal solution in Fig. 4.1). The action of the Na^+/K^+-ATPase lowers the

D. P. Nickerson (✉) · L. Noroozbabaee · D. M. Sarwar · P. J. Hunter
Auckland Bioengineering Institute, University of Auckland, Auckland, New Zealand
e-mail: d.nickerson@auckland.ac.nz; l.noroozbabaee@auckland.ac.nz;
dsar941@aucklanduni.ac.nz; p.hunter@auckland.ac.nz

K. L. Hamilton
Department of Physiology, School of Biomedical Sciences, University of Otago, Dunedin, New Zealand
e-mail: kirk.hamilton@otago.ac.nz

© The American Physiological Society 2020
K. L. Hamilton, D. C. Devor (eds.), *Basic Epithelial Ion Transport Principles and Function*, Physiology in Health and Disease,
https://doi.org/10.1007/978-3-030-52780-8_4

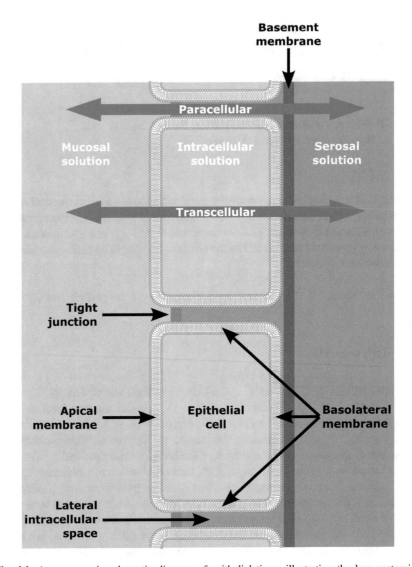

Fig. 4.1 A macroscopic schematic diagram of epithelial tissue illustrating the key anatomical features of the mathematical modeling framework discussed in this chapter. The black text labels indicate anatomical features and the white text indicates the functional compartments. The white text with solid arrows shows the two primary pathways for the transport of solutes between the mucosal and serosal solutions, either through the cell (transcellular) or via the tight junction (paracellular)

concentration of sodium inside the cell and thus establishes the electrochemical gradient favouring sodium entry from the lumen of the tubule (mucosal solution in Fig. 4.1) into the cell.

In this chapter, we present a general mathematical framework for modeling epithelial ion transport. The mathematical modeling framework is based on the principals of mass conservation and maintenance of electroneutrality. Building on the work of (Latta et al. 1984; Weinstein 1992; Weinstein et al. 2007), we discuss recent advances in model description technologies which enable arbitrary extension of the mathematical model.

Figure 4.1 presents a schematic diagram of a macroscopic view of epithelial tissue. While this image clearly shows individual cells and their membranes from which the epithelium is constructed, the mathematical framework presented here takes a Physiome-style multiscale view (de Bono and Hunter 2012; Hunter et al. 2010, 2013). In this approach, a *"cell"* represents the average behaviour of a region of tissue. Different modeling frameworks allow such cell models to be spatially distributed with varying properties to create tissue-scale models (Nickerson et al. 2011). Here, the cell-scale model is the upper level of the spatial hierarchy that we consider. The integration of subcelluar mechanisms is also discussed below.

Following the introduction of the mathematical modeling framework, we discuss (Cuellar et al. 2003) with the modeling framework, such simulation experiments build on previous work. We also refer the reader to some open-source software developed by the authors which provide an example implementation of the described methods. The software and example epithelial transport models are available online at: http://get.readthedocs.org/.

4.1.1 Model Exchange and Reproducible Science

The mathematical modeling framework presented here is able to represent models of epithelial transport of arbitrary complexity. For such models to be useful to the scientific community it is important that we are able to share, distribute, and reuse the models. Several community-driven standards for encoding mathematical models are widely accepted (e.g. Britten et al. 2013; Christie et al. 2009; Cuellar et al. 2003; Hucka et al. 2003; Nickerson et al. 2006). CellML (Cuellar et al. 2003) is a standard that is focused on encoding the mathematical relationships in a modular and reusable manner (Cooling et al. 2010, 2016; Nickerson and Buist 2008). The modular hierarchical nature of CellML makes it an ideal format for encoding epithelial transport models as the actual transporters can be archived and managed independently and then assembled together to fit the epithelial model requirements. Section 4.4 discusses some of the benefits of this approach to model description.

To be able to reproduce a previous modeling study, one needs to know not only the mathematical model, but also an unambiguous description of what tasks were performed with the model(s) to achieve the observed outcomes. In addition to the model encoding standards, guidelines have been established which define the minimal set of information required to enable a simulation experiment to be reproduced (Waltemath et al. 2011a). The simulation experiment description markup-language (SED-ML) is an instantiation of these guidelines into an XML-based format for

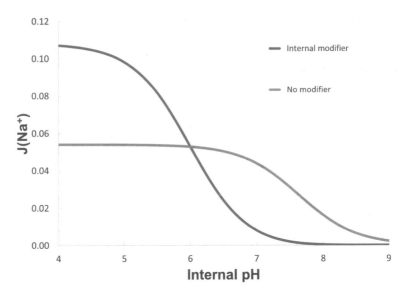

Fig. 4.2 Example reproducing Weinstein (1995, Fig. 5) using the model and simulation experiment available from: https://models.physiomeproject.org/e/5c3/

capturing descriptions of simulation experiments (Waltemath et al. 2011b). In developing the software tools discussed here, we endeavour to make use of standards such as CellML and SED-ML wherever possible to enable and support reproducible science.

An example of this is shown in Fig. 4.2 where we have reproduced Weinstein (1995, Fig. 5) using the CellML model and SED-ML simulation experiment description from https://models.physiomeproject.org/e/5c3/ with the tool OpenCOR (Garny and Hunter 2015 https://opencor.ws/).

4.2 Epithelial Cell Modeling

Many models can be found in the literature of mathematical models for specific types of epithelial cells (e.g. Hu et al. 2019; Layton and Layton 2019; Thomas 2009; Thomas and Dagher 1994; Weinstein 2010, 2011). As with similar modeling efforts in other areas of computational physiology and systems biology (e.g. Land et al. 2014; Nickerson and Hunter 2006; Swainston et al. 2013), these models have evolved to include a wide range of biophysical details and are capable of investigating novel transport mechanisms. The adoption of model encoding standards in this area lags that of some other fields (Nickerson and Hunter 2006), and therefore, scientists are required to either implement the models themselves or make use of provided computer codes which conflate the mathematical model and numerical simulation methods.

Fig. 4.3 A diagram of a general epithelial transport modeling framework, using the same anatomical schematic as presented in Fig. 4.1. See text for detailed description

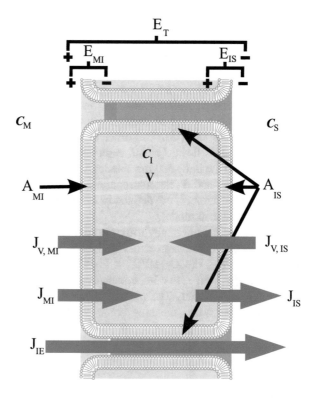

Here, we describe a general epithelial transport mathematical modeling framework derived from the core conservation laws upon which all the existing models are based. The framework is based on the derivation of Weinstein (1992); Weinstein et al. (2007) which is a comprehensive epithelial model with a wide range of biophysical details; however, the reproducibility of such models from the literature remains an issue. Investigating a number of related works (e.g. Weinstein 1986, 1992; Weinstein et al. 2007; Weinstein and Sontag 2009), one can often find ambiguities about the initial and boundary conditions and also the model's description of parameters and units. Regardless, the mathematical formulation of such models can be extracted and provides a generic tool for further exploration of epithelial transport.

Applying some assumptions, the Weinstein et al. (2007) model can be reduced to the Latta Na^+ (Latta et al. 1984) previously used to illustrate mathematical modeling of epithelial ion transport (Nickerson et al. 2016).

Understanding this general mathematical framework should help in the interpretation and reuse of the existing models and also encourage the adoption of a modular and declarative approach to create new models of epithelial transport. A graphical view of the mathematical modeling framework is presented in Fig. 4.3. The following sections define the various components in the modeling framework both in Wienstein et al. and Latta et al. system equations. The epithelial model (Weinstein

1992, 2010, 2011; Weinstein et al. 2007; Weinstein and Sontag 2009) consists of cellular and lateral intercellular compartments, between luminal and peritubular solutions. Figure 4.3 displays both configurations, in which cellular and lateral intercellular (LIS) compartments line the tubule lumen. Within each compartment, the concentration of species (i) is designated $C_\alpha(i)$, where α is lumen (M), interspace (E), cell (I), or peritubular solution (S) and the separating membranes the combination of characters such that luminal cell membrane (MI), tight junction (ME), or basal cell membrane (IS). The order of the two characters indicates the positive direction for mass flow. J and J_v represent the solute flow and water flux, respectively, through the indicated membrane; A the indicated membrane surface area; V is the cell volume; and E_T the transepithelial potential difference.

The specific symbols are defined in the following sections as they are introduced to the mathematical model.

Models can include many different solutes in the various compartments. In the current example (Weinstein et al. 2007) there are 15 model solutes, namely, Na^+, K^+, Cl^-, HCO_3^-, CO_2, H_2CO_3, HPO_4^{2-}, $H_2PO_4^-$, urea, NH_3, NH_4^+, H^+, HCO_2^-, H_2CO_2, and glucose, as well as two impermeant species within the cells, a nonreactive anion and a cytosolic buffer. The number of solutes included in a specific simulation experiment can vary, meaning some of the solutes are ignored under certain conditions.

4.2.1 State Equations

To start the state equations with Weinstein epithelial model, the model consists of two different compartments: cell and lateral intercellular space. To formulate the system equations of mass conservation within each compartment, the net generation of each species $S_\alpha(i)$ is defined as an intermediate variable Weinstein (1992) within compartment. The generation of multiple reacting solute is the sum of the net export of the flux plus the accumulation of that solute in each compartment.

$$S_I(i) = J_{IE}(i) + J_{IS}(i) - J_{MI}(i) + \frac{d}{dt}[V_I\, C_I(i)], \qquad (4.1)$$

$$S_E(i) = J_{ES}(i) - J_{ME}(i) - J_{IE}(i) + \frac{d}{dt}[V_E\, C_E(i)]. \qquad (4.2)$$

Here, $S_I(i)$ and $S_E(i)$ indicate the generation of the solutes in the cellular and interspace compartments. Within the epithelium, the flux of solute i across membrane $\alpha\beta$ is denoted as $J_{\alpha\beta}(i)$ (ml s^{-2} cm^{-2}) and V_α is the compartment volume $(cm^3\ cm^{-2})$. To consider the mass conservation of the volume, the following expressions are defined:

$$S_I (\nu) = J_{\nu IE} + J_{\nu IS} - J_{\nu MI} + \frac{d}{dt}[V_I], \tag{4.3}$$

$$S_E (\nu) = J_{\nu ES} - J_{\nu ME} - J_{\nu IE} + \frac{d}{dt}[V_E], \tag{4.4}$$

where $J_{\nu\alpha\beta}$ (mmol s^{-1} cm^{-1}) is denoted as the transmembrane volume flux. It is important to mention that for nonreacting solutes:

$$S_\alpha (i) = 0, \tag{4.5}$$

$$S_\alpha (\nu) = 0. \tag{4.6}$$

Applying the following assumptions, one can convert the current epithelial model (Weinstein 1992) to the reduced Latta Na$^+$ (Latta et al. 1984):

- Considering all the solutes as non-reacting solutes.
- Ignoring all the mass conservation equations related to the interspace compartment.

which leads us to a new set of equations as presented by Latta et al. (1984). The mass conservation then defines the changes rate of the concentration of the i'th species in the intracellular solution as the transport of solute i into and out of the cell through the apical and basolateral membrane, which is a direct transport of solutes through the membrane, and also a contribution from convective transport due to the flow of water through the membranes. This is modelled as Eq. (4.7).

$$Q_\nu = C_I(i)(J_{\nu MI} + J_{\nu IS}), \tag{4.7}$$

$$V\frac{dC_I(i)}{dt} = J_{MI}(i) - J_{IS}(i) - C_I(i)(J_{\nu MI} + J_{\nu IS}). \tag{4.8}$$

This equation holds for each solute being considered in a particular instantiation of the model. If required for a particular model, similar equations can be introduced for the solute concentrations in the mucosal and/or serosal solutions.

Consequently, the conservation of cellular water yields the equation below, with the rate of change of cell volume, V, defined as

$$\frac{dV}{dt} = J_{\nu MI} + J_{\nu IS}, \tag{4.9}$$

where each of the total membrane water fluxes, $J_{\nu\alpha\beta}$, is scaled by its respective membrane area to take into account the averaged behaviour of the representative membrane.

4.2.2 Buffer Pairs and pH Equilibrium

The Weinstein model defines different buffer pairs, the conservation of mass for buffer reactions takes the form:

$$S_\alpha\left(HPO_4^{2-}\right) = S_\alpha\left(HPO_4^-\right), \tag{4.10}$$

$$S_\alpha\left(HCO_2^-\right) = S_\alpha(H_2CO_2), \tag{4.11}$$

$$S_\alpha(NH_3) = S_\alpha\left(NH_4^+\right), \tag{4.12}$$

$$S_\alpha\left(Buf^-\right) + S_\alpha(HBuf) = 0. \tag{4.13}$$

All buffer species are assumed to be at chemical equilibrium. Within each compartment, there are four additional pH equilibrium relations, corresponding to the four buffer pairs. The algebraic relations of the model include the pH equilibria of four buffer pairs,

$$pH = pK + \log_{10} \frac{Base^-}{HBase}. \tag{4.14}$$

The collection of chosen buffer pairs and even the definition of mass conservation equation can be different over various studies depending on the specific focus of the study. As an example the mass conservation equation for $NH_3 : NH_4^+$ buffer pairs within a cell can be represented in the form of Eq. (4.12) (e.g. Weinstein 1992; Weinstein and Sontag 2009) or in the form

$$S_\alpha(NH_3) + S_\alpha\left(NH_4^+\right) = Q\left(NH_4^+\right), \tag{4.15}$$

where $Q\left(NH_4^+\right)$ is defined as an ammoniagenesis factor, see Weinstein et al. (2007). Defining a robust epithelial model which includes all buffer pairs is not straightforward due to a high level of inconsistencies the literature.

4.2.3 Electroneutrality Constraints

Equations (4.1–4.4) define a coupled system of differential equations ensuring that mass is conserved. However, when considering the movement of charged solutes, with valence z_i, this system is not sufficient to guarantee that the cell and interspace remain electrically neutral. Electroneutrality relation for the cell compartment is defined through the following definition:

$$\sum_i z_i C_I(i) + Z_{\text{Imp}} C_{\text{Imp}} - C_{\text{Buf}} = 0. \tag{4.16}$$

Here, C_{Imp} and C_{Buf} denote concentration of cell impermeant solute and cell unprotonated buffer, where z_i is the valence of species i. Electroneutrality relations for the interspace are defined as:

$$\sum_i z_i C_E(i) = 0, \tag{4.17}$$

and for all of the buffer reactions, there is conservation of protons as the following definition:

$$\sum_i z_i S_\alpha(i) = 0. \tag{4.18}$$

To ensure that the electroneutrality condition is not violated an additional constraint must be imposed to prevent net charge flux into or out of the cell.

The membrane charge fluxes can be represented as electrical currents using the following relationships:

$$I_{\text{In}} = I_{\text{MI}} + I_{\text{ME}} = F\left(\sum_i z_i J_{\text{MI}}(i) + \sum_i z_i J_{\text{ME}}(i)\right),$$

$$I_{\text{Out}} = I_{\text{IS}} + I_{\text{ES}} = F\left(\sum_i z_i J_{\text{IS}}(i) + \sum_i z_i J_{\text{ES}}(i)\right), \tag{4.19}$$

where F is Faraday's constant. Balancing the flow of charge into and out of the cell therefore results in the relationship

$$I_{\text{out}} = I_{\text{In}}, \tag{4.20}$$

which must hold true at all times. See Sect. 4.3 for a description of the numerical techniques that can be used to solve the membrane potential across each membrane which satisfies Eq. (4.20).

Equations (4.16–4.20) are a collection of the different electroneutrality constraints that are introduced in different studies (e.g. Weinstein 1992; Weinstein et al. 2007; Weinstein and Sontag 2009). However, it is important to note that not all these equations were utilised in all different studies, but they were chosen selectively based on the context of each study.

4.2.4 Model Specialisation

The basic principles of mass conservation, pH equilibrium of buffer species and maintenance of electroneutrality described above apply to epithelial transport in general. To instantiate the general model into a mathematical model for a specific epithelium, all that remains is to define the actual membrane solute and water fluxes of interest to create the specialised model.

In comparison to the model presented here (Weinstein 1992; Weinstein et al. 2007), Latta et al. (1984) present a minimal set of auxiliary flux equations that describe a specific model of the Koefoed-Johnsen and Ussing (1958) epithelial Na^+. As an example of the specialisation of the general model, we reproduce the Latta et al. (1984) Na^+ here and refer to the model as the Latta Na^+.

4.2.4.1 Water Fluxes

With respect to water flows, volume conservation equations for interspace and cell can be applied to compute the lateral interspace hydrostatic pressure, and cell volume. Across each cell membrane, the transmembrane volume fluxes are proportional to the hydrostatic, oncotic and osmotic driving forces

$$
\begin{aligned}
J^v_{\alpha\beta} = L_{p\alpha\beta}A_{\alpha\beta}\left(P_\alpha - P_\beta\right) - L_{p\alpha\beta}A_{\alpha\beta}\left(\pi_\alpha - \pi_\beta\right) \\
+ L_{p\alpha\beta}A_{\alpha\beta}\sum_i \sigma_{\alpha\beta}(i)\left(C_\beta(i) - C_\alpha(i)\right),
\end{aligned}
\tag{4.21}
$$

where P_α and π_α are the hydrostatic and oncotic pressure within compartment α, $L_{p\alpha\beta}$ is the membrane water permeability and $\sigma_{\alpha\beta}(i)$ is the reflection coefficient of membrane $\alpha\beta$ to solute i, and R and T are the gas constant and absolute temperature, respectively. The Latta Na^+ only osmotically induced water flux across the apical and basolateral membranes (Fig. 4.3) and assumes that the paracellular pathway is impermeable to water. It is also assumed that the cell membranes are unable to support hydrostatic gradients. Thus, the water fluxes are given by

$$
J^v_{\alpha\beta} = L_{p\alpha\beta}A_{\alpha\beta}\sum_i \sigma_{\alpha\beta}(i)\left(C_\beta(i) - C_\alpha(i)\right).
\tag{4.22}
$$

4.2.4.2 Convective Solute Fluxes

In the Weinstein et al. (2007) model, it is assumed that there are convective fluxes for all the intraepithelial solutes, having the water fluxes of Eq. (4.21), the convective flux can be represented as,

$$J^C_{\alpha\beta}(i) = J_{\nu\alpha\beta}\left(1 - \sigma_{\alpha\beta}\bar{C}_{\alpha\beta}(i)\right), \tag{4.23}$$

where

$$\bar{C}_{\alpha\beta}(i) = \frac{C_\alpha - C_\beta}{\log C_\alpha - \log C_\beta}. \tag{4.24}$$

Studying the reflection coefficient values $\sigma_{\alpha\beta}$ (defined as membrane properties), one can see that the reflection coefficient is mostly one in cell apical (MI), cell lateral (IE) and cell basal membrane. The most effective membrane to produce the convective fluxes is the interspace basement membrane (ES) with $\sigma_{\alpha\beta} = 0$, and then in the second place is the tight junction (ME).

4.2.4.3 Passive Solute Fluxes

In both the Weinstein (Weinstein and Sontag 2009) model and the Latta Na$^+$ model, passive solute fluxes across all membranes are assumed to occur by electrodiffusion and to conform to the Goldman-Hodgkin-Katz constant-field flux equation (Hodgkin and Katz 1949). Passive solute flux of the i'th species across the membrane is therefore given by

$$J^P_{\alpha\beta} = h_{\alpha\beta}(i)\zeta_{\alpha\beta}(i)\frac{C_\alpha(i) - C_\beta(i)\exp\left(-\zeta_{\alpha\beta}(i)\right)}{1 - \exp\left(-\zeta_{\alpha\beta}(i)\right)}, \tag{4.25}$$

where $h_{\alpha\beta}(i)$ (cm s^{-1}) is the membrane permeability which characterises the pathway. Weinstein et al. did not hold on to one solid definition for the permeability, in some cases permeability was multiplied by the area of that membrane $h_{\alpha\beta}(i)$ $A_{\alpha\beta}$ (10^{-5} cm^3 S^{-1} cm^{-2}) as an example see Weinstein et al. (2007).

For the uncoupled permeation of neutral solutes across membranes, a Ficklaw is utilised,

$$J^P_{\alpha\beta} = h_{\alpha\beta}(i)\left(C_\alpha(i) - C_\beta(i)\right). \tag{4.26}$$

Finally, under conditions where the junctional permeability to the solutes is finite (i.e. leaky or moderately tight epithelium), the Latta Na$^+$ the paracellular flux for the i'th species to also be passive and follows the above equation.

In the Latta Na$^+$, $h_{MS}(i)$ represents the solute permeability of the paracellular pathway determined by the combined properties of the tight junctions and lateral spaces. Comparing the permeability coefficients between Latta and Weinstein, we could not find any consistency between the same model variables. As an example, one can see that the permeability of Cl$^-$ the Weinstein model for IS membrane is zero, while there is a maximum permeability for all the solutes, which is 541×10^{-9} (cm S^{-1}) for the solute Cl$^-$.

4.2.5   Electrodiffusive Fluxes

Electrodiffusive fluxes in the Weinstein models include three different categories of transporters (simple cotransporter, simple exchanger and complex).

All of the coupled solute transporters in this model have been represented according to linear nonequilibrium thermodynamics, so that solute permeation rates are proportional to the electrochemical driving force of the aggregate species, with a single permeation coefficient.

Simple cotransporters consist of peritubular K^+-Cl^-, luminal Na^+-glucose and Na^+-$H_2PO_4^-$. As an example see Eq. (4.27),

$$
\begin{bmatrix} J_{IS}(K^+) \\ J_{IS}(Cl^-) \end{bmatrix} = L_{KCL} \begin{bmatrix} 1 & 1 \\ 1 & 1 \end{bmatrix} \begin{bmatrix} \bar{\mu}_{IS}(K^+) \\ \bar{\mu}_{IS}(Cl^-) \end{bmatrix}, \tag{4.27}
$$

where, $\bar{\mu}_{IS}(i)$ is the electrochemical potential difference of species i across the basal cell membrane.

$$
\bar{\mu}_{IS} = RT \ \ln \left(\frac{C_I(i)}{C_S(i)} \right) + z_i F \psi_{IS}, \tag{4.28}
$$

L_{KCl} is the rate coefficient identified as its permeability.

Simple exchangers such as luminal Na^+-NH_4^+ and Cl^--HCO_3^- are presented in this model as well, see Eq. (4.29).

$$
\begin{bmatrix} J_{MI}(Na^+) \\ J_{MI}(NH_4^+) \end{bmatrix} = L_{NaNH_4} \begin{bmatrix} 1 & -1 \\ -1 & 1 \end{bmatrix} \begin{bmatrix} \bar{\mu}_{MI}(Na^+) \\ \bar{\mu}_{MI}(NH_4^+) \end{bmatrix}. \tag{4.29}
$$

There are also two more complex transporters at the peritubular membrane: Na^+-HCO^-_3 and Na^+-$2HCO_3^-$/Cl^-, for more details see Weinstein (1992).

4.2.5.1   Active Solute Fluxes

The mathematical modeling framework we describe here is able to represent arbitrary contributions to transport across the specialised epithelium. In Sect. 4.4, we discuss the use of modeling standards to enable the assembly of specialised models from a library of available transporter models. The Latta Na^+ a sodium pump in the basolateral membrane defined by Lewis and Wills (1981) with the pump current given by

$$I^{\mathrm{A}} = \frac{I_{\max}}{[1 + (K_{\mathrm{Na}}/C_c(\mathrm{Na}))^{n_{\mathrm{Na}}}][1 + (K_{\mathrm{K}}/C_b(\mathrm{K}))^{n_{\mathrm{K}}}]}, \tag{4.30}$$

$I_{\max}$ is the maximum observable pump current, K_{Na} and K_{K} are the concentrations of intracellular Na$^+$ serosal K$^+$ which half the maximal pump current is produced. The corresponding basolateral membrane Na$^+$ and K$^+$ fluxes can then be expressed in terms of the pump current as

$$J^{\mathrm{A}}_{\mathrm{Na}} = n_{\mathrm{Na}} I^{\mathrm{A}}/F; \quad J^{\mathrm{A}}_{\mathrm{K}} = -n_{\mathrm{K}} I^{\mathrm{A}}/F. \tag{4.31}$$

4.2.5.2   Total Membrane Solute Fluxes

In summary, one can see that intraepithelial solute transport may be convective $J^{\mathrm{C}}_{\alpha\beta}$, passive $J^{\mathrm{P}}_{\alpha\beta}$, electrodiffusive $J^{\mathrm{E}}_{\alpha\beta}$, or metabolically driven $J^{\mathrm{A}}_{\alpha\beta}$. This is expressed in the Weinstein model by Eq. (4.32).

$$J_{\alpha\beta} = J^{\mathrm{C}}_{\alpha\beta} + J^{\mathrm{P}}_{\alpha\beta} + J^{\mathrm{E}}_{\alpha\beta} + J^{\mathrm{A}}_{\alpha\beta}. \tag{4.32}$$

The Latta Na$^+$ the solute flux of the i'th species across the apical membrane is entirely passive. Similarly, the paracellular flux for the i'th species is also assumed to be passive. This results in

$$J_{\mathrm{MI}} = J^{\mathrm{P}}_{\mathrm{MI}}, \tag{4.33}$$

$$J^{j}_{i} = J^{j}_{\mathrm{p},i}. \tag{4.34}$$

For the basolateral membrane, the passive flux is supplemented by the active sodium pump, giving the solute flux of the i'th species across the basolateral membrane as

$$J_{\mathrm{IS}} = J^{\mathrm{P}}_{\mathrm{IS}} + J^{\mathrm{A}}_{\mathrm{IS}}, \tag{4.35}$$

where $J^{\mathrm{A}}_{\mathrm{IS}}$ is given by Eq. (4.31) for Na$^+$ and K$^+$ and zero for all other species.

4.3   Computational Simulation

The equations of state described in Sect. 4.2.1 (Eqs. 4.9 and 4.8) can be numerically integrated as a function of time using any reasonable integration algorithm. We typically make use of the CVODE integrator from the SUNDIALS suite of solvers (Hindmarsh et al. 2005) or the Python SciPy package (Virtanen et al. 2019). The progression of the simulation broadly follows the following steps, as per the algorithm of Latta et al. (1984).

1. Initialise the model to the desired values at the initial time.
2. Compute the membrane potentials.
3. Compute the solute and water fluxes using the potentials just computed and the flux equations described in Sect. 4.2.4.
4. Integrate Eqs. (4.9) and (4.8) to the next desired time point.
5. Increment the simulation time to the next time point. If the end of the desired simulation interval is reached the computation ends; otherwise return to step 2 and continue the simulation.

We have implemented the algorithm described above as part of our freely available and open-source suite of tools available at: http://get.readthedocs.org/. In Fig. 4.4, we present results obtained using this implementation with the specific model as illustrated in Fig. 4.5.

4.4 Transporter Modeling

The mathematical modeling framework described in the preceding sections is well suited to the creation of models encoded in a hierarchical and modular manner. The general model defined by the state equations in Sect. 4.2.1 and the electroneutrality constraint defined in Sect. 4.2.3 provides the top-level model which varies only by the addition of different collections of solute species. The water and solute fluxes can then each be defined independently and integrated into the total flux equations, as shown in Sect. 4.3.

This concept of a hierarchical model description is shown graphically in Fig. 4.6. In Fig. 4.6, we take it one step further and illustrate how the epithelial cellular transport models discussed here can be further integrated into an example tubule model, such as the renal nephron, for example.

The modular and hierarchical nature of the CellML format (Cuellar et al. 2003) makes it an obvious candidate for encoding such multiscale epithelial cellular transport models. Indeed, the tools we develop do use CellML to encode the mathematical models (http://get.readthedocs.org/). As mentioned in Sect. 4.1.1, the use of such a standard format helps to ensure the models created using these tools are able to be widely disseminated and are able to be used by other scientists in their own work.

Using this approach, specialised models such as the described in Sect. 4.2.4 can be assembled by extracting the required constituent transporter models from a library or repository. The collection of transporter models can then be integrated into the cellular state equations (Eqs. 4.9 and 4.8) and simulations executed following the methods described in Sect. 4.3. Preliminary models are now freely available via the model library at http://get.readthedocs.org/ current work is focused on extending this library and enhancing the automated model assembly tools (Sarwar et al. 2019).

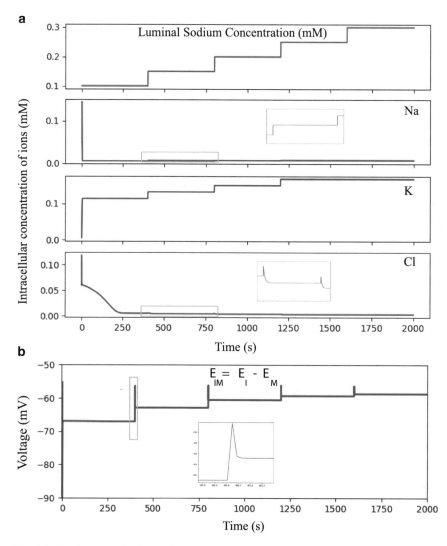

Fig. 4.4 Results from simulating the model illustrated in Fig. 4.5. The model is simulated to steady-state for step-wise increasing levels of luminal Na$^+$ (top panel). The top panels show the transient concentrations of the dynamically simulated intracellular concentrations (Na$^+$, K$^+$, and Cl$^-$; glucose results not shown). The lower panel shows the apical membrane potential, E_{IM}. In all plots, the inset shows an enlarged region following a step in the C_M(Na$^+$) to highlight the dynamic behaviour

Presenting such model libraries, modeling tools, and simulation experiments via interactive web-based user interfaces is also an ongoing project, building on the work of Nickerson et al. (2011).

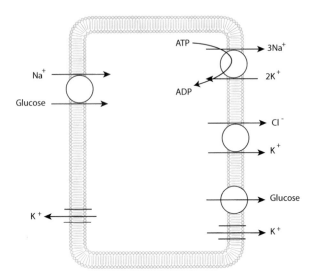

Fig. 4.5 An illustration of the specific model we use to demonstrate our implementation of the Weinstein model described above (Weinstein 1992; Weinstein et al. 2007)

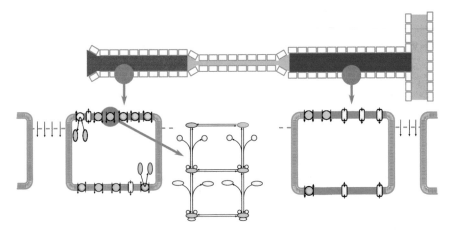

Fig. 4.6 A schematic diagram illustrating the spatial hierarchy and configuration of an example epithelial transport model. The wall of the tubule model displayed at the top of the figure consists of epithelial cells and the cells in different regions have different transport properties. The varying transport properties are due to the different transport mechanisms present in each of the cell variants—detail which can be captured in a hierarchical and modular description of the model

Acknowledgements DPN and LN are supported by a Aotearoa Foundation fellowship to DPN. DMS is supported by the Medical Technologies Centre of Research Excellence. KLH is supported by the Department of Physiology at the University of Otago.

References

Britten RD, Christie GR, Little C, Miller AK, Bradley C, Wu A, Yu T, Hunter P, Nielsen P (2013) FieldML, a proposed open standard for the physiome project for mathematical model representation. Med Biol Eng Comput 51:1191–1207. https://doi.org/10.1007/s11517-013-1097-7

Christie GR, Nielsen PMF, Blackett SA, Bradley CP, Hunter PJ (2009) FieldML: concepts and implementation. Philos Transact A Math Phys Eng Sci 367:1869–1884. https://doi.org/10.1098/rsta.2009.0025

Cooling MT, Rouilly V, Misirli G, Lawson J, Yu T, Hallinan J, Wipat A (2010) Standard virtual biological parts: a repository of modular modeling components for synthetic biology. Bioinformatics 26:925–931. https://doi.org/10.1093/bioinformatics/btq063

Cooling MT, Nickerson DP, Nielsen PMF, Hunter PJ (2016) Modular modelling with physiome standards: modular modelling with Physiome standards. J Physiol. https://doi.org/10.1113/JP272633

Cuellar AA, Lloyd CM, Nielsen PF, Bullivant DP, Nickerson DP, Hunter PJ (2003) An overview of CellML 1.1, a biological model description language. Simulation 79:740–747. https://doi.org/10.1177/0037549703040939

de Bono B, Hunter P (2012) Integrating knowledge representation and quantitative modelling in physiology. Biotechnol J 7:958–972. https://doi.org/10.1002/biot.201100304

Garny A, Hunter PJ (2015) OpenCOR: a modular and interoperable approach to computational biology. Comput Physiol Med 6:26. https://doi.org/10.3389/fphys.2015.00026

Hindmarsh AC, Brown PN, Grant KE, Lee SL, Serban R, Shumaker DE, Woodward CS (2005) SUNDIALS: suite of nonlinear and differential/algebraic equation solvers. ACM Trans Math Softw 31:363–396. https://doi.org/10.1145/1089014.1089020

Hodgkin AL, Katz B (1949) The effect of sodium ions on the electrical activity of the giant axon of the squid. J Physiol 108:37–77

Hu R, McDonough AA, Layton AT (2019) Functional implications of the differences in transporters' abundance along the rat nephron: modeling and analysis. Am J Physiol Renal Physiol 317:F1462–F1474. https://doi.org/10.1152/ajprenal.00352.2019

Hucka M, Finney A, Sauro HM, Bolouri H, Doyle JC, Kitano H, Arkin AP, Bornstein BJ, Bray D, Cornish-Bowden A, Cuellar AA, Dronov S, Gilles ED, Ginkel M, Gor V, Goryanin II, Hedley WJ, Hodgman TC, Hofmeyr J-H, Hunter PJ, Juty NS, Kasberger JL, Kremling A, Kummer U, Novère NL, Loew LM, Lucio D, Mendes P, Minch E, Mjolsness ED, Nakayama Y, Nelson MR, Nielsen PF, Sakurada T, Schaff JC, Shapiro BE, Shimizu TS, Spence HD, Stelling J, Takahashi K, Tomita M, Wagner J, Wang J (2003) The systems biology markup language (SBML): a medium for representation and exchange of biochemical network models. Bioinformatics 19:524–531. https://doi.org/10.1093/bioinformatics/btg015

Hunter P, Coveney PV, de Bono B, Diaz V, Fenner J, Frangi AF, Harris P, Hose R, Kohl P, Lawford P, McCormack K, Mendes M, Omholt S, Quarteroni A, Skår J, Tegner J, Randall Thomas S, Tollis I, Tsamardinos I, van Beek JHGM, Viceconti M (2010) A vision and strategy for the virtual physiological human in 2010 and beyond. Philos Transact A Math Phys Eng Sci 368:2595–2614. https://doi.org/10.1098/rsta.2010.0048

Hunter P, Chapman T, Coveney PV, de Bono B, Diaz V, Fenner J, Frangi AF, Harris P, Hose R, Kohl P, Lawford P, McCormack K, Mendes M, Omholt S, Quarteroni A, Shublaq N, Skår J, Stroetmann K, Tegner J, Thomas SR, Tollis I, Tsamardinos I, van Beek JHGM, Viceconti M (2013) A vision and strategy for the virtual physiological human: 2012 update. Interface Focus 3. https://doi.org/10.1098/rsfs.2013.0004

Koefoed-Johnsen V, Ussing HH (1958) The nature of the frog skin potential. Acta Physiol Scand 42:298–308. https://doi.org/10.1111/j.1748-1716.1958.tb01563.x

Land S, Niederer SA, Louch WE, Røe ÅT, Aronsen JM, Stuckey DJ, Sikkel MB, Tranter MH, Lyon AR, Harding SE, Smith NP (2014) Computational modeling of Takotsubo cardiomyopathy: effect of spatially varying β-adrenergic stimulation in the rat left ventricle. Am J Physiol Heart Circ Physiol 307:H1487–H1496. https://doi.org/10.1152/ajpheart.00443.2014

Latta R, Clausen C, Moore LC (1984) General method for the derivation and numerical solution of epithelial transport models. J Membr Biol 82:67–82. https://doi.org/10.1007/BF01870733

Layton AT, Layton HE (2019) A computational model of epithelial solute and water transport along a human nephron. PLoS Comput Biol 15:e1006108. https://doi.org/10.1371/journal.pcbi.1006108

Lewis SA, Wills NK (1981) Interaction between apical and basolateral membranes during sodium transport across tight epithelia. Soc Gen Physiol Ser 36:93–107

Nickerson D, Buist M (2008) Practical application of CellML 1.1: the integration of new mechanisms into a human ventricular myocyte model. Prog Biophys Mol Biol 98:38–51. https://doi.org/10.1016/j.pbiomolbio.2008.05.006

Nickerson DP, Hunter PJ (2006) The Noble cardiac ventricular electrophysiology models in CellML. Prog Biophys Mol Biol 90:346–359. https://doi.org/10.1016/j.pbiomolbio.2005.05.007

Nickerson D, Nash M, Nielsen P, Smith N, Hunter P (2006) Computational multiscale modeling in the IUPS Physiome Project: modeling cardiac electromechanics. IBM J Res Dev 50:617–630. https://doi.org/10.1147/rd.506.0617

Nickerson DP, Terkildsen JR, Hamilton KL, Hunter PJ (2011) A tool for multi-scale modelling of the renal nephron. Interface Focus 1:417–425. https://doi.org/10.1098/rsfs.2010.0032

Nickerson DP, Hamilton KL, Hunter PJ (2016) Mathematical modeling of epithelial ion transport. In: Ion channels and transporters of epithelia in health and disease. Springer, New York, pp 265–278

Sarwar DM, Kalbasi R, Gennari JH, Carlson BE, Neal ML, de Bono B, Atalag K, Hunter PJ, Nickerson DP (2019) Model annotation and discovery with the physiome model repository. BMC Bioinformatics 20:457. https://doi.org/10.1186/s12859-019-2987-y

Swainston N, Mendes P, Kell DB (2013) An analysis of a 'community-driven' reconstruction of the human metabolic network. Metabolomics 9:757–764. https://doi.org/10.1007/s11306-013-0564-3

Thomas SR (2009) Kidney modeling and systems physiology. Wiley Interdiscip Rev Syst Biol Med 1:172–190. https://doi.org/10.1002/wsbm.14

Thomas SR, Dagher G (1994) A kinetic model of rat proximal tubule transport—load-dependent bicarbonate reabsorption along the tubule. Bull Math Biol 56:431–458

Virtanen P, Gommers R, Oliphant TE, Haberland M, Reddy T, Cournapeau D, Burovski E, Peterson P, Weckesser W, Bright J, van der Walt SJ, Brett M, Wilson J, Millman KJ, Mayorov N, Nelson ARJ, Jones E, Kern R, Larson E, Carey CJ, Polat İ, Feng Y, Moore EW, VanderPlas J, Laxalde D, Perktold J, Cimrman R, Henriksen I, Quintero EA, Harris CR, Archibald AM, Ribeiro AH, Pedregosa F, van Mulbregt P, SciPy 1.0 Contributors (2019) SciPy 1.0—fundamental algorithms for scientific computing in python. arXiv:1907.10121. https://doi.org/10.1038/s41592-019-0686-2

Waltemath D, Adams R, Beard DA, Bergmann FT, Bhalla US, Britten R, Chelliah V, Cooling MT, Cooper J, Crampin EJ, Garny A, Hoops S, Hucka M, Hunter P, Klipp E, Laibe C, Miller AK, Moraru I, Nickerson D, Nielsen P, Nikolski M, Sahle S, Sauro HM, Schmidt H, Snoep JL, Tolle D, Wolkenhauer O, Le Novère N (2011a) Minimum information about a simulation experiment (MIASE). PLoS Comput Biol 7:e1001122. https://doi.org/10.1371/journal.pcbi.1001122

Waltemath D, Adams R, Bergmann FT, Hucka M, Kolpakov F, Miller AK, Moraru II, Nickerson D, Sahle S, Snoep JL, Novère NL (2011b) Reproducible computational biology experiments with SED-ML—the simulation experiment description markup language. BMC Syst Biol 5:198. https://doi.org/10.1186/1752-0509-5-198

Weinstein AM (1986) A mathematical model of the rat proximal tubule. Am J Physiol Ren Physiol 250:F860–F873

Weinstein AM (1992) Chloride transport in a mathematical model of the rat proximal tubule. Am J Physiol Ren Physiol 263:F784–F798. https://doi.org/10.1152/ajprenal.1992.263.5.F784

Weinstein AM (1995) A kinetically defined Na^+/H^+ antiporter within a mathematical model of the rat proximal tubule. J Gen Physiol 105:617–641. https://doi.org/10.1085/jgp.105.5.617

Weinstein AM (2010) A mathematical model of rat ascending Henle limb. I. Cotransporter function. Am J Physiol Ren Physiol 298:F512–F524. https://doi.org/10.1152/ajprenal.00230.2009

Weinstein AM (2011) Potassium deprivation: a systems approach. Am J Physiol Ren Physiol 301: F967–F968. https://doi.org/10.1152/ajprenal.00430.2011

Weinstein AM, Sontag ED (2009) Modeling proximal tubule cell homeostasis: tracking changes in luminal flow. Bull Math Biol 71:1285–1322. https://doi.org/10.1007/s11538-009-9402-1

Weinstein AM, Weinbaum S, Duan Y, Du Z, Yan Q, Wang T (2007) Flow-dependent transport in a mathematical model of rat proximal tubule. Am J Physiol Ren Physiol 292:F1164–F1181

Chapter 5
Molecular Mechanisms of Apical and Basolateral Sorting in Polarized Epithelial Cells

Ora A. Weisz and Heike Fölsch

Abstract The proper localization of ion channels and transporters to the apical or basolateral membrane domains is essential for correct ion transport through epithelia. Polarized sorting is achieved through elaborate intracellular sorting pathways encompassing the *trans*-Golgi network, early endosomes, and recycling endosomes that are fine-tuned for epithelial function and thus different from the secretory and endocytic systems in non-polarized cells. Polarized cells express epithelial-cell specific molecules, including sorting adaptors like the cytosolic clathrin adaptor AP-1B, which are required for efficient segregation and targeting of proteins to specific plasma membrane domains. In this book chapter, we will first discuss the various sorting stations in the cells followed by a review of apical and basolateral sorting signals and their interpretation at different compartments. Specific emphasis is placed on the trafficking of ion transporters.

Keywords Membrane traffic · Kidney · Biosynthetic · Endocytosis · Recycling · *trans*-Golgi network · Endosome · AP-1B · Apical · Basolateral

Abbreviations

AEE   Apical early endosome
ARE   Apical RE
BEE   Basolateral early endosome
CRE   Common RE
RE   Recycling endosome

O. A. Weisz
Renal-Electrolyte Division, Department of Medicine, University of Pittsburgh, Pittsburgh, PA, USA

H. Fölsch (✉)
Department of Cell and Developmental Biology, Northwestern University, Feinberg School of Medicine, Chicago, IL, USA
e-mail: h-folsch@northwestern.edu

© The American Physiological Society 2020
K. L. Hamilton, D. C. Devor (eds.), *Basic Epithelial Ion Transport Principles and Function*, Physiology in Health and Disease,
https://doi.org/10.1007/978-3-030-52780-8_5

TGN   *trans*-Golgi network
TJ   Tight junction

5.1   General Organization of Secretory and Endocytic Pathways

Typically, transmembrane proteins begin their journey to the cell surface in the endoplasmic reticulum (ER) during co-translational insertion into the ER membrane. In this compartment, N-glycans are added to consensus sequences within the protein ectodomains, and proteins are folded, and in many cases, assembled into oligomers. Properly folded proteins are transported via a coat protein complex II (COPII)-mediated pathway to the Golgi apparatus where they achieve their final glycosylation patterns. The most distal compartment of the Golgi complex, the *trans*-Golgi network (TGN), is the first major sorting station in the cells. From there, proteins may traffic to lysosomes, endosomes, and the cell surface. On their journey to the cell surface, they may traverse early or recycling endosomal populations. Often, the same sorting signals that are used during biosynthetic delivery are also used during endocytic recycling of transmembrane proteins (Matter et al. 1993; Potter et al. 2006b). In the case of polarized cells, an additional level of sorting exists to selectively direct proteins to and from the apical (luminal) and basolateral plasma membranes (reviewed in Ang and Fölsch 2012; Fölsch et al. 2009; Rodriguez-Boulan et al. 2005).

   Much of the work on polarized sorting in epithelial cells has been performed in the Madin-Darby canine kidney (MDCK) cell line, derived in the 1950s from a female cocker spaniel. The type II subclone of these cells is easily and rapidly cultured as well-differentiated monolayers when plated on permeable support membranes. Moreover, these cells express highly regular microvilli similar to that observed in the brush border of the proximal tubule and elaborate a primary cilium. In MDCK and other cells of kidney origin, most newly synthesized apical and basolateral proteins are targeted vectorially (albeit not necessarily directly) to the plasma membrane. The fidelity of delivery to a given domain is never 100%, and transcytotic retrieval pathways exist to reroute missorted proteins (Casanova et al. 1991). Other epithelial cell types make more use of this transcytotic pathway to deliver newly synthesized apical proteins. In hepatocytes, almost all transmembrane proteins are delivered first to the sinusoidal (= basolateral) membrane and bile canalicular (= apical) targeting follows after internalization (Bartles et al. 1987). An exception to this rule is the class of polytopic membrane proteins, including the bile acid transporters, which are targeted vectorially from the TGN to the apical surface of liver cells (Sai et al. 1999). Intestinal cells, including the polarized Caco-2 cell line, also use the transcytotic pathway to a significant extent to deliver apical proteins, but a sizable fraction of some proteins is also targeted via vectorial pathways (Le Bivic et al. 1990; Matter et al. 1990). In the remainder of this book chapter, we will focus our attention on kidney cells, and particularly on knowledge gained using MDCK cells.

After initial delivery to the apical or basolateral plasma membrane, transmembrane receptors may be internalized into early endosomes underlying the plasma membranes. Notably, epithelial cells maintain apical early endosomes (AEEs) and basolateral early endosomes (BEEs) that are biochemically distinct and physically segregated. Indeed, AEEs and BEEs did not fuse with one another in an in vitro fusion assay (Bomsel et al. 1990). Moreover, after treatment of MDCK cells with latrunculin B to disrupt the actin cytoskeleton, AEEs and BEEs were misplaced and intermixed in the cytoplasm, without undergoing fusion reactions (Sheff et al. 2002). Although it is currently unclear which proteins define AEEs and BEEs, their existence probably ensures correct delivery of internalized receptors back to their membrane of origin during rapid recycling from AEEs or BEEs.

In addition to rapid recycling, transmembrane receptors may be sorted further into the cells to reach lysosomes in case they are destined for degradation, or recycling endosomes (REs) for sorting back to either their membrane of origin or the opposing membrane during apical-to-basolateral or basolateral-to-apical transcytosis. Unlike in non-polarized cells, REs in polarized cells constitute a second major sorting station of equal importance to the TGN. Indeed, REs in polarized cells are organized into subdomains to accommodate the various sorting needs of cargo in polarized epithelial cells. Two views of REs in MDCK cells have emerged based on conflicting data. In one model, REs are considered to be a single population of interconnected endosomal membranes with different subdomains specialized for targeting to diverse compartments including domains that cluster cargo destined for the apical or basolateral domains (Fig. 5.1b, reviewed in Ang and Fölsch 2012). Notably, the basolateral targeting domain depends on and contains proteins that interact with the AP-1B clathrin adaptor (Fig. 5.1d, see also Sect. 5.3.4). The second model holds that two classes of REs exist: a transferrin receptor (TfnR)-positive common RE (CRE) that can receive internalized cargo from both apical and basolateral surface domains and a specialized apical RE (ARE), marked by Rab11a, that excludes basolaterally recycling cargo and communicates with the CRE and the apical domain (Fig. 5.1a, reviewed in Mostov et al. 2003). These models may not be mutually exclusive, and it is possible that the structure of the RE system in MDCK cells reflects their differentiated state and the need for communication between apical and basolateral surface domains. From here on, we will use the terms ARE and CRE or simply just REs interchangeably. Furthermore, if not noted otherwise, reviewed studies were performed in MDCK cells.

5.2 Sorting to the Apical Membrane

A wide variety of transporters, enzymes, and receptors are present at the apical surface of kidney cells. The steady-state distribution of these proteins can be modulated by changes in the fidelity of biosynthetic and postendocytic sorting as

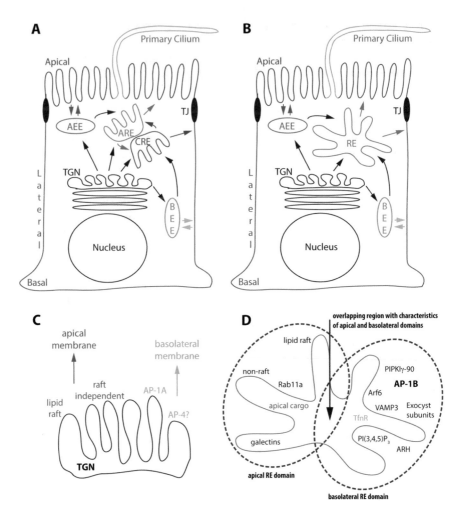

Fig. 5.1 This model figure depicts the intracellular sorting stations present in polarized epithelial cells. (**a**) Illustrates the model of two separated populations of REs, whereas (**b**) illustrates the model championing only one RE. (**c**) Depicts an expanded version of the TGN showing the different TGN domains responsible for apical and basolateral sorting. (**d**) Depicts an expanded version of the RE in (**b**) showing its subdomains. Note that the overlapping region between the apical and the basolateral sorting domain may contain any of the components listed for each domain including, but not necessarily limited to, Rab11a, Arf6, and exocyst subunits. See Sect. 5.3.4 for a detailed description of the basolateral RE domain

well as by alterations in the kinetics of delivery and removal from the surface. Dissecting the sorting signals on proteins that guide them to the apical or basolateral surface is further complicated by the observation that many proteins contain multiple signals of varying strengths. Thus, mutation or ablation of a given signal on a protein does not always lead to its nonpolarized distribution as one would expect, but rather can result in its profound retargeting to the opposing plasma membrane domain or to another compartment. These difficulties aside, we are slowly expanding our under-standing of the specific signals that direct proteins apically and how these signals are interpreted by the cellular machinery.

5.2.1 Apical Sorting Signals

The apical sorting information on proteins described to date constitutes a remarkably diverse collection of signals that includes peptide sequences, post-translational modifications, and structural information (reviewed in Weisz and Rodriguez-Boulan 2009).

5.2.2 GPI Anchors

The first known apical targeting signal to be identified was the presence of a glycosylphosphatidylinositol (GPI) lipid anchor on proteins expressed in MDCK cells (Lisanti et al. 1989). These anchors enable protein localization to glycolipid-enriched lipid microdomains (lipid rafts) and it was proposed that this in turn facilitates their segregation into apically-destined sorting platforms at the TGN and possibly at distal sites along the biosynthetic pathway as well (Simons and Ikonen 1997). However, subsequent studies have shown that this is an overly simplistic model. There is now considerable evidence that additional clustering of GPI-anchored proteins (GPI-APs) into high molecular weight oligomers is required for apical sorting (reviewed in Muniz and Zurzolo 2014). This clustering may be mediated by protein-protein, protein-lipid, and/or glycan-dependent interactions, and may be cell-type dependent (Imjeti et al. 2011; Paladino et al. 2004, 2008). Remodeling of GPI-AP acyl chains can also facilitate oligomer formation (Muniz and Zurzolo 2014).

5.2.3 Glycan-Dependent Sorting Signals

A second class of apical sorting signals is dependent on the addition or modification of N- or O-linked oligosaccharides (reviewed in Potter et al. 2006a). Dissection of these signals can be difficult as N-glycans are often essential for the proper folding and ER export of proteins. That said, specific N-glycan structures or linkages have directly been implicated in apical sorting of several proteins. For example, sialylation of N-glycans on the sialomucin endolyn has been demonstrated to be essential for its efficient apical sorting (Mo et al. 2012).

The paucity of reagents and approaches to perturb O-linked glycans has made the role of this modification more challenging to address. Specifically, it has been difficult to determine whether the oligosaccharides themselves, as opposed to the structural consequences of O-glycosylation at a specific region of a protein, is essential for polarized sorting (Yeaman et al. 1997; Youker et al. 2013). For example, the O-glycosylated imperfect tandem repeats of MUC1 contain transferable apical sorting information, but no specific sugar residues have yet been shown to be essential for sorting (Kinlough et al. 2011).

How glycans function to direct sorting remains unclear, although there is increasing evidence for a role of galectins in mediating the apical sorting of a subset of proteins (Delacour et al. 2009). Galectins 3 and 9 have been shown to function in apical sorting of cargos that do not partition into lipid rafts in MDCK cells (Delacour et al. 2006; Mo et al. 2012). Similar to the clustering mechanism proposed for lipid-raft associated proteins, galectin 3-mediated oligomerization of apical cargos into high molecular weight complexes has been demonstrated (Delacour et al. 2007). Interestingly, galectins 4 and 9 also bind to glycolipids in lipid rafts in intestinal and kidney cells, respectively, and may also facilitate the clustering of raft-associated proteins (Delacour et al. 2005; Mishra et al. 2010; Stechly et al. 2009).

5.2.4 Peptide-Based Sorting Signals

Peptide sequences present in either the luminal, transmembrane, or cytoplasmically exposed regions of many proteins have been identified as important for their apical localization. Indeed, peptide-based signals are the most common type of sorting information identified for polytopic proteins, including many ion transporters and G-protein coupled receptors. However, no unifying mechanism has emerged for how these sequences function to direct apical localization. Rather, these identified motifs have been variously implicated in polarized sorting, surface delivery, and cell surface retention. Sorting of proteins via selective retention at the plasma membrane will be discussed in more detail in Sect. 5.4.

The observation that some splice variants of ion transporters are targeted to opposing domains of epithelial cells has provided a unique opportunity to identify apical and basolateral sorting information encoded within these proteins. For example, the presence of a 45 amino acid insertion in the first cytosolic loop of the Ca^{2+}-ATPase, PMCA2 directs apical sorting whereas splice variants lacking this sequence are targeted basolaterally in MDCK cells (Chicka and Strehler 2003). While this example might imply that the determination of peptide-based apical sorting information is straightforward, this is far from the case. Indeed, there are only a few cases where linear amino acid sequences that confer transferable apical localization have been demonstrated, and even then, the identification of key residues required for sorting has been fraught. For example, the sole difference between the basolaterally-targeted human GLUT9a and the apically-targeted splice variant GLUT9b is the absence of 29 amino acids within the amino terminus of GLUT9b. The shorter amino terminus of GLUT9b could redirect the normally basolaterally-trafficked GLUT1 transporter to the apical surface, demonstrating that it contains transferable apical sorting information. However, no specific sequence unique to GLUT9b could be identified as essential for apical sorting (Bibee et al. 2013). One possible conclusion of these studies is that the amino terminal domain of GLUT9b interacts with or positions other regions of the transporter to create a conformation that somehow promotes apical sorting. These observations are consistent with a model in which apical sorting of some proteins is conferred by information dependent on a specific protein conformation as opposed to a specific amino acid sequence.

For other ion transporters, short stretches of amino acids have been identified most frequently within cytosolically oriented loops or tails, which direct their apical localization. A ten-amino acid sequence (PIKPVFKGFS) that includes a putative β-turn within the C-terminus of the human sodium-dependent vitamin C transporter hSVCT1 is required for apical delivery in MDCK cells (Subramanian et al. 2004). Interestingly, similar motifs have also been shown to be important for apical targeting of the neuronal sodium-dependent glutamate transporter EAAT3 (Cheng et al. 2002) and in the rat sodium-dependent bile acid transporter (Sun et al. 2003). In addition, short and apparently unrelated peptide sequences direct the apical localization of the potassium MaxiK channel (Kwon and Guggino 2004), the mouse sodium-phosphate co-transporter NaPi-IIc (WLHSL; Ito et al. 2010), and the cystic fibrosis transmembrane conductance regulator (CFTR; Milewski et al. 2000). It is not yet known how these non-terminal short sequences control apical polarity. One possibility is that they interact directly with apical targeting or retention machinery. In support of this, a five-amino acid (SQDAL) sequence within a cytoplasmic loop of the ATP-binding cassette transporter isoform C2 (ABCC2; aka MRP2) is required for apical sorting in HepG2 cells via an apparently saturable mechanism (Bandler et al. 2008; Emi et al. 2012).

Despite the sizable literature on polarized sorting, we still know little about the mechanisms by which apical sorting determinants on polytopic proteins, including ion transporters, are interpreted. The difficulty in identifying specific sequences or transferable motifs that confer apical sorting of polytopic proteins may in part be due to the intrinsic challenges of generating mutations that do not disrupt their folding or interactions with accessory subunits. Alternatively, the sequences or domains required for apical sorting may influence targeting via steric or conformational effects rather than by binding to interacting proteins. Consistent with this idea, there are only a very few cases where apical sorting information has been demonstrated to be saturable, or where similar peptide motifs have been identified in multiple proteins. Once proteins arrive at the apical surface, binding to the cytoskeleton via scaffolding proteins or PDZ-domain interactions can enhance their retention at this domain (see Sect. 5.4 for further details).

5.2.5 Apical Sorting Mechanisms

Clustering mediated by incorporation into glycolipid-enriched microdomains or by galectin-mediated crosslinking is implicated in the polarized sorting of lipid-raft associated and glycan-dependent apical proteins. The possibility that peptide-based sorting determinants regulate the conformation of polytopic proteins suggests that clustering may be an underlying mechanism for the apical targeting of this class of proteins as well (Bigay and Antonny 2012; Dehmelt and Bastiaens 2010). This is not to say that other features of polytopic proteins do not contribute to their clustering and/or apical sorting. There is certainly precedence for incorporation of ion channels and transporters into lipid rafts (Ares and Ortiz 2012; Bravo-Zehnder et al. 2000; Chen et al. 2011; Hill et al. 2007; Riquier et al. 2009). Moreover, glycan processing may enhance the oligomerization of polytopic proteins (Ozaslan et al. 2003). Thus, various features of a given protein may guide its incorporation into homomeric or heteromeric arrays that facilitate apical delivery.

How might cargo clustering lead to apical sorting? Cytosolic domains of some apical proteins may bind directly to Rab proteins, molecular motors, or adaptors. Proteins that may facilitate clustering and/or trafficking to the apical membrane include but are not limited to VIP17/MAL, caveolin-1, annexins, and the four-phosphate-adaptor protein 2 (FAPP2; reviewed in Fölsch 2008). Clustering will increase the avidity of these interactions to create a sorting "platform" that may nucleate the assembly of an apically destined transport carrier. Many apically destined proteins have very short (<10 amino acid) cytoplasmic tails, or in the case of GPI-anchored proteins, no domains accessible to the cytosol. Co-clustering of these proteins with others that can recruit sorting machinery may be essential for their sorting. Moreover, incorporation of transmembrane VAMPs into these clusters may be necessary for eventual fusion of transport carriers with the apical surface. VAMPs implied in fusion of vesicles with the apical membrane are VAMP2, VAMP7 (aka TI-VAMP), and VAMP8 (Caceres et al. 2014; Galli et al. 1998; Pocard et al. 2007). It should be noted that the concept of spatially regulated protein self-assembly or scaffolding into higher order clusters is not unique to apical sorting, and indeed models for self-organization have been demonstrated to drive numerous cellular functions (elegantly reviewed in Bigay and Antonny 2012; Dehmelt and Bastiaens 2010).

Another unresolved question is where along the biosynthetic pathway does apical sorting occur? Newly synthesized apical and basolateral proteins are segregated at least by the time they leave the TGN. Indeed, there is also evidence that raft-associated and raft-independent apical proteins are segregated when leaving this compartment (Fig. 5.1c; Guerriero et al. 2008; Jacob and Naim 2001). Moreover, it has become increasingly clear that these two classes of proteins take distinct routes to the apical surface, and that these routes intersect different endocytic compartments (Cresawn et al. 2007; Fölsch et al. 2009). Lipid raft associated proteins may pass through AEEs whereas glycan-dependent and some other apical cargos transit Rab11a-positive AREs. In addition, intestinal and kidney cells also express the Rab11b isoform, which appears to be localized to a subdomain of the ARE (Lapierre et al. 2003). Apically recycling CFTR and the epithelial sodium channel ENaC traffic through Rab11b-positive endosomes, although it is not known whether newly synthesized proteins also access this compartment (Butterworth et al. 2012; Silvis et al. 2009). Complicating matters further, there are multiple exit routes for proteins that traffic through the ARE (Mattila et al. 2012).

How can the concept of cluster-dependent sorting be reconciled with the presence of multiple apical sorting stations? If there is continual segregation of apical proteins into different export pathways as they move from one compartment to another, there may be different subsets of clusters that form at each site. Indeed, clustering mediated by distinct mechanisms may be initiated in different compartments. It is believed that proteins are incorporated into lipid rafts during their transit through the Golgi (Guerriero et al. 2008; Jacob and Naim 2001); and galectin 9 that binds to the raft-enriched glycolipid Forssman antigen (as well as to non-raft associated proteins) appears to access these structures at the TGN of MDCK cells (Mishra et al. 2010). In contrast, galectin 3 interaction with glycosylated proteins occurs after they leave the TGN (Delacour et al. 2007; Straube et al. 2013). Thus, non-raft associated proteins that traffic together to the ARE (for example, proteins with glycan-dependent and peptide-based sorting signals) might be segregated into distinct populations that differentially exit that compartment based on their ability to interact with galectin 3. Whether such hierarchical arrays of clustering actually occur remains to be determined.

5.3 Sorting to the Basolateral Membrane

Like apical targeting, there are multiple pathways that cargos may follow to reach the basolateral membrane during biosynthetic delivery and endocytic recycling (Arnspang et al. 2013; Farr et al. 2009; Fölsch et al. 2009). Many newly synthesized cargos travel from the TGN into REs during biosynthetic delivery, and cargos may be directed into the basolateral pathway from either of these sites (Ang et al. 2004; Fölsch et al. 2009; Nokes et al. 2008). Often, basolateral sorting signals are linear peptide motifs that are *cis*-dominant over apical sorting information such that mutation of the basolateral sorting signals results in a protein that is sorted to the apical membrane instead. In addition, basolateral sorting signals are often co-linear with lysosomal sorting signals and endocytic motifs. Indeed, low amounts of the lysosomal-associated membrane glycoprotein Lamp1 and the cation-dependent mannose 6-phosphate receptor (CD-MPR) that cycles between the TGN and late endosomes exclusively cycle through the basolateral membrane in MDCK cells (Distel et al. 1998; Hunziker et al. 1991; Nabi et al. 1991).

5.3.1 *Basolateral Sorting Signals and Adaptors*

Linear peptide sorting motifs that target proteins to the basolateral membrane are frequently either tyrosine-based [YxxØ, where Ø is a bulky hydrophobic amino acid (F, I, L, M, or V), and FxNPxY] or dileucine-based (LL, LI, or LM) often in the form of [D/E]xxxL[L/I]. These signals are generally recognized by cytosolic adaptor complexes, most commonly heterotetrameric clathrin adaptor protein (AP) complexes AP-1 (γ-β1-μ1-σ1), AP-2 (α-β2-μ2-σ2), AP-3 (δ-β3-μ3-σ3), and AP-4 (ε-β4-μ4-σ4) that does not associate with clathrin (Boehm and Bonifacino 2001; Hirst and Robinson 1998; Ohno et al. 1999). YxxØ signals are recognized by the AP medium subunits, whereas [D/E]xxxL[L/I]-signals are recognized simultaneously by σ1-γ subunits of AP-1 as well as σ2-α of AP-2, and σ3-δ of AP-3 (Doray et al. 2007; Janvier et al. 2003; Mattera et al. 2011). Another form of dileucine signal (DxxLL) is recognized at the TGN by monomeric clathrin adaptor proteins of the GGA (Golgi-localized, γ-ear containing, Arf-binding) protein family [discussed in (Kirchhausen 2002)], perhaps leading to incorporation into AP-1A vesicles (Doray et al. 2002). Moreover, the cytoplasmic tail of the transforming growth factor α (TGFα) encodes a specialized LL signal that is aided by a cluster of negative charges and recognized by the cytosolic adaptor Naked2 for sorting to the basolateral membrane (Li et al. 2004, 2007). Finally, FxNPxY signals are bridged to clathrin adaptor complexes by connector proteins such as numb, Dab2, and autosomal recessive hypercholesterolemia protein (ARH) (reviewed in Traub 2003).

Other peptide-based sorting motifs have been described that seem to be far less abundant. For example, the basolateral sorting signal of CD147 consists of only a single leucine residue (Deora et al. 2004). A different, non-related mono-leucine signal for basolateral sorting was also discovered in amphiregulin (Gephart et al. 2011). Moreover, a distinct VxxEED motif for basolateral delivery was identified in the type I transforming growth factor beta (TGF-β) receptor (Yin et al. 2017). No adaptor proteins have yet been described that recognize these unique sorting signals. In addition, proteins may be sorted to and selectively retained at the basolateral membrane by means of PDZ-binding motif interaction with PDZ domain containing proteins (reviewed in Brone and Eggermont 2005, see also Sect. 5.4). PDZ-directed sorting from endosomes may be facilitated by the sorting nexin 27 (SNX27; Lauffer et al. 2010). Furthermore, basolateral sorting from early endosomes may be facilitated by the 'retromer' complex and SNX17 (reviewed in Bonifacino 2014). SXN17 binds NPxY peptide motifs (Donoso et al. 2009; Stiegler et al. 2014). Finally, new motifs were recently discovered that interact specifically with μ1A of AP-1A. One type are acidic cluster motifs such as SDSEEDE in furin and similar clusters are also present in mannose 6-phosphate receptors (Navarro Negredo et al. 2017). The second new type is a cluster of three dibasic residues as identified in the cytoplasmic tail of L-selectin (Dib et al. 2017). It is currently unclear if these novel signals may help sorting to the basolateral membrane and if they are specific for μ1A or may also interact with μ1B of AP-1B.

Epithelial cells express two closely related AP-1 complexes, AP-1A and AP-1B, that differ only in the incorporation of their respective medium subunits μ1A or μ1B. Notably, μ1B expression is specific to columnar epithelial cells including MDCK and Caco-2 cell lines (Fölsch et al. 1999; Ohno et al. 1999). So far both AP-1A and AP-4 have been implicated in sorting of basolateral cargos at the TGN (Fig. 5.1c), and AP-1B was shown to facilitate cargo sorting in REs (reviewed in Ang and Fölsch 2012; Fölsch 2015b; Rodriguez-Boulan et al. 2013). Interestingly, whereas the connector proteins numb and Dab2 specifically interact with AP-2, ARH also interacts with AP-1B in REs to facilitate basolateral sorting of cargos with FxNPxY signals (Kang and Fölsch 2011).

5.3.2 AP-1B Expression in the Kidneys

AP-1B is expressed in many regions of the kidney including the medullary and cortical thick ascending limbs (TALs), and the cortical collecting duct (CCD) (Schreiner et al. 2010). However, AP-1B is not expressed in the proximal tubules and cell lines that were derived from that region (e.g. LLC-PK1 cells) (Ohno et al. 1999; Schreiner et al. 2010). Therefore, LLC-PK1 cells are a useful model system for analyzing AP-1B function and to determine if a protein depends on AP-1B for steady-state localization at the basolateral membrane (Fölsch 2015a). To this end, researchers use LLC-PK1 cells stably expressing exogenous copies of μ1B or μ1A as a control. Proteins that are apical in μ1A-expressing LLC-PK1 cells, but basolateral in μ1B-expressing cells are then identified as AP-1B-dependent cargos (Fields et al. 2007; Fölsch et al. 1999). Because AP-1B is the only adaptor complex that localizes in REs to facilitate basolateral sorting during endocytic recycling (Fölsch et al. 1999, 2001; Gan et al. 2002), its cargo proteins will be missorted to the apical membrane in the absence of functional AP-1B (Fields et al. 2007; Fölsch et al. 1999; Gravotta et al. 2007; Sugimoto et al. 2002). Proteins dependent on AP-1B for basolateral localization include many nutrient receptors like low-density lipoproteins receptor (LDLR) and TfnR, as well as the adhesion proteins E-cadherin and β1 integrin, and the growth hormone receptor EGFR (Fölsch et al. 1999; Gravotta et al. 2007; Ling et al. 2007; Ryan et al. 2010).

5.3.3 Autosomal Recessive Hypercholesterolemia Protein (ARH) Expression in the Kidneys

ARH expression is found predominantly in the distal nephron tubules including the distal convoluted tubule (DCT), the connecting tubule (CNT), as well as the cortical collecting duct (CCD), and to a lesser extent in the regions of the thick ascending limbs (TALs) (Fang et al. 2009). Thus, ARH expression largely overlaps with AP-1B expression in the kidneys ensuring AP-1B-dependent cargo sorting of proteins with FxNPxY sorting motifs.

5.3.4 Mechanisms of AP-1B-Mediated Basolateral Sorting

What is known so far about AP-1B function on a molecular level? It is thought that recruitment of AP-1B onto REs depends on $PI(3,4,5)P_3$ and possibly Arf6 (Fields et al. 2010; Shteyn et al. 2011). Strikingly, $PI(3,4,5)P_3$ accumulates in REs only in epithelial cells that express AP-1B in a reaction that may involve the phosphatidylinositol 4-phosphate 5-kinase PIPKIγ-90 (Fields et al. 2010). Indeed PIPKIγ-90 was shown to directly interact with AP-1B, but not AP-1A, and to play a role in basolateral sorting of E-cadherin (Ling et al. 2007). After membrane recruitment, AP-1B selects its cargos either directly or with the help of its co-adaptor ARH (Kang and Fölsch 2011). In addition, AP-1B recruits at least 2 subunits of the exocyst complex to REs for incorporation into AP-1B vesicles (Fölsch et al. 2003). AP-1B vesicles also contain the SNARE protein VAMP3 (aka cellubrevin) (Fields et al. 2007) and disruption of VAMP3 function in MDCK cells leads to dispersed REs to which AP-1B no longer localizes (Fields et al. 2007).

Tethering of AP-1B vesicles to the basolateral membrane is thought to depend on the exocyst complex (Grindstaff et al. 1998; Yeaman et al. 2001) and fusion of the vesicles may involve VAMP3 on the vesicles and syntaxin 4 at the basolateral membrane (Fields et al. 2007). Interestingly, basolateral sorting of syntaxin 4 itself depends on AP-1B expression, although how AP-1B influences basolateral sorting of syntaxin 4 is not yet entirely clear (Reales et al. 2011; Torres et al. 2011).

In addition to AP-1B-dependent sorting, syntaxin 4 is most likely also involved in fusion of vesicles that originated from BEEs. Indeed, syntaxin 4 was shown to directly interact with the early endosomal Rab4 protein (Li et al. 2001), and disruption of Rab4 function in MDCK cells resulted in apical missorting of basolateral membrane proteins (Mohrmann et al. 2002).

5.3.5 Basolateral Sorting of Transporters

Many transporters at the basolateral membrane contain tyrosine-based sorting motifs but are independent of AP-1B. This is perhaps due to low internalization rates and/or stable retention at the basolateral membrane. For example, the kidney anion exchanger 1 (kAE1) contains a YDEV signal in its cytoplasmic tail that is needed for AP-1A-mediated basolateral sorting at the TGN (Almomani et al. 2012; Junking et al. 2014; Toye et al. 2004). However, although kAE1 is initially sorted to the basolateral membrane independent of AP-1B expression, AP-1B is probably needed during endocytic recycling (Almomani et al. 2012). Tyrosine-based motifs have also been described in the chicken kAE1-4 splice isoform that contains these signals in a 63 amino acid N-terminal extension, which is missing in the apically sorted kAE1-3 splice isoform (Adair-Kirk et al. 1999). Interestingly, a disease-causing mutation in the C-terminal residue (M909T) of the human kAE1 results in the creation of a type I PDZ ligand and partial redirection of the protein to the apical surface (Fry et al.

2012). Because the wild type protein is targeted directly to the basolateral domain, it was concluded that the presence of this PDZ interacting motif causes apical misdirection of a subset of the mutant kAE1 within the biosynthetic pathway (Fry et al. 2012), although it is also possible that this mutation causes selective retention at the apical membrane.

Other transporters that do not rely on AP-1B for basolateral targeting are the homotetrameric water channels aquaporins 3 and 4 that are expressed in renal collecting ducts. Aquaporins have six transmembrane domains and both the N- and C-termini are exposed to the cytosol. Aquaporin 3 contains a tyrosine-based sorting motif that overlaps with a dileucine-based signal (YRLL) in its N-terminus (Rai et al. 2006). Mutation of either signal partially redirected aquaporin 3 to the apical membrane, and mutation of both motifs effectively disrupted surface delivery. Because aquaporin 3 is also localized to the basolateral membrane in LLC-PK1 cells, it was concluded that AP-1B is not needed for its basolateral sorting (Rai et al. 2006). Indeed, other receptors that contain dileucine-based motifs such as the Fc receptor FcRII-B2 likewise do not need AP-1B for basolateral targeting (Fields et al. 2007).

In contrast to aquaporin 3, aquaporin 4 contains basolateral sorting information in its C-terminus (Madrid et al. 2001). The first of two motifs is a tyrosine-based (S)YMEV(E) motif that also serves as an internalization motif and directly interacts with the medium subunits of AP-2 and AP-3, but not with AP-1A. The second motif is an acidic dileucine motif (ETEDLIL). Curiously, the YMEVE sequence may conform to the consensus sequence recognized by AP-4 (Burgos et al. 2010; Dell'Angelica et al. 1999), and it remains to be determined if it indeed interacts with this adaptor. Importantly, this motif overlaps with a casein kinase II (CKII) consensus site (SxxE/D), and phosphorylation by CKII introduces a negative charge directly upstream of the tyrosine leading to enhanced interaction with AP-3 and thus delivery of aquaporin 4 into lysosomes for degradation (Madrid et al. 2001). It is tempting to speculate that aquaporin 4 may be sorted at the TGN into AP-4 vesicles for basolateral targeting. After internalization, interaction with AP-3 may effectively remove the water channel from BEEs such that little if any aquaporin 4 may enter REs. Indeed, aquaporin 4 is also localized to the basolateral membrane in renal proximal tubules, indicating that AP-1B plays no role in its basolateral sorting (van Hoek et al. 2000).

5.4   Retention at the Cell Surface Through Interaction with PDZ Domains

Selective retention at the apical or basolateral membrane may be achieved through interaction with a specialized cytoskeleton. For example, ankyrin G and β2 spectrin specifically anchor proteins at the lateral membrane (Kizhatil et al. 2007). More common is retention through interaction of PDZ-binding motifs at the extreme C-terminus of transmembrane cargos with PDZ-domains in scaffolding proteins.

However, the presence of a PDZ-binding motif does not assure that it functions in retention, as disruption of the C-terminal PDZ-interacting sequences of hSMVT or ABCC2 has no effect on their localization (Nies et al. 2002; Subramanian et al. 2009). In other cases, retention at the apical membrane through a PDZ-binding motif may be overridden by a strong basolateral sorting signal as was the case for aquaporin 2 chimeras with aquaporin 3, in which the apical targeting signal in aquaporin 2, which is a PDZ-binding motif, was maintained and the basolateral sorting information of aquaporin 3 was added to the N-terminus effectively directing the chimera to the basolateral surface (Rai et al. 2006). Although PDZ binding motifs are thought to primarily facilitate selective retention at the plasma membrane, they are often also important for surface delivery (Lauffer et al. 2010; Maday et al. 2008).

PDZ-binding motifs may be present in addition to other sorting signals. For example, the basolateral potassium channel Kir2.3 contains a type I PDZ-binding motif that partially overlaps with a basolateral sorting signal; both motifs are needed for robust basolateral localization of Kir2.3 (Le Maout et al. 2001). Importantly, the PDZ-binding motif is necessary for stable expression at the plasma membrane and its deletion results in largely intracellular localization of the mutant channel. Retention at the basolateral membrane is most likely achieved through binding of Kir2.3 to the PDZ domain containing Lin-7/CASK protein complex (Olsen et al. 2002). Similarly, the epithelial γ-aminobutyric acid (GABA) transporter (BGT-1) is also retained at the basolateral membrane through interactions with Lin-7 (Perego et al. 1999). In MDCK cells, Lin-7 associates with adherens junctions through interaction with β-catenin (Perego et al. 2000).

Some transporters have tyrosine-based signals that are used for internalization, but not for basolateral targeting. For example, the renal outer medullary potassium secretory channel ROMK (aka Kir1.1) has an YDNPNFV motif that binds to ARH and serves as an endocytic signal (Fang et al. 2009). However, ROMK is localized to the apical membrane domain, although it is expressed in the kidney distal tubules that express both ARH and AP-1B (Fang et al. 2009; Schreiner et al. 2010). ROMK has a PDZ-binding motif at its C-terminus that is necessary for membrane expression of ROMK and interacts with the PDZ domain-containing proteins Na/H exchange regulatory factors NHERF1 and 2 (Yoo et al. 2004). Perhaps ROMK binding to NEHRF selectively stabilizes ROMK at the apical membrane effectively overriding basolateral sorting by AP-1B and ARH. PDZ-binding motifs have also been shown to contribute to the steady state apical localization of other ion transporters (Brone and Eggermont 2005) including CFTR (Milewski et al. 2000; Moyer et al. 1999), TRPV4 and TRPV5 (van de Graaf et al. 2006), and PMCA2b (Antalffy et al. 2012).

Interestingly, stabilizing proteins at the apical membrane through binding to NHERF may also ensure their exclusion from the primary cilium that extends from the apical membrane. Primary cilia are important appendages needed for correct cell signaling and homeostasis, and their membrane composition is distinct from the apical (and basolateral) membranes. Most apical membrane proteins are excluded from the primary cilium and it is thought that the base of the cilium contains a diffusion barrier that is formed by a member of the septin GTPase family, septin 2 (Hu et al. 2010). However, additional mechanisms exist. For example, it has been shown that interaction with NHERF1 prevents the apical membrane protein podocalyxin from entering the primary cilium (Francis et al. 2011). It is thought that NHERF1 immobilizes proteins in the apical membrane by cross-linking them to the apical actin network via ERM family members (Francis et al. 2011). It seems reasonable to assume that other apical proteins that interact with NHERF will be excluded from the primary cilium as well.

5.5 Sorting of Multi-subunit Transporters

Polarized sorting of heteromeric transporters may appear convoluted as each individual subunit may contain sorting information. Moreover, association with different subunit isoforms may drive the heteromers towards the apical or basolateral membrane. A well-studied example of this is the Na,K-ATPase, which in most tissues is localized basolaterally. The Na,K-ATPase consist of three subunits: the catalytic α subunit, the $\beta 1$ subunit that is thought to aid TGN exit of the complex, and the γ subunit that modulates ion channel activity. Multiple sorting signals have been described for these subunits. For example, both the α subunit and $\beta 1$ subunit contain basolateral sorting information (Marrs et al. 1995; Muth et al. 1998; Shoshani et al. 2005). However, the signal in the α subunit can be overturned by pairing with a different β subunit (Vagin et al. 2005). Whereas expression of $\beta 1$ directs the Na,K-ATPase α subunit to the basolateral surface, $\beta 2$ is responsible for apical localization of the Na,K-ATPase in gastric adenocarcinoma cells. Interestingly, this apical distribution of $\beta 2$ is apparently mediated by its prevalence of N-glycans, although the mechanism is unknown. Moreover, addition of new N-glycosylation consensus sequences to the more sparsely glycosylated $\beta 1$ subunit converts it to an apically localized protein (Vagin et al. 2005). Basolateral localization of the Na,K-ATPase has been further attributed to selective retention via interaction with the cytoskeleton (i.e., ankyrin G) (Kizhatil et al. 2007; Marrs et al. 1995), or through homotopic interaction of $\beta 1$ subunits across adjacent cells (Shoshani et al. 2005). Interestingly, basolateral localization of the $\beta 1$ subunit in sensory hair cells in zebrafish depends on AP-1 expression (Clemens Grisham et al. 2013), indicating that at least the $\beta 1$ subunit may have a tyrosine or dileucine-based motif that could be recognized by AP-1A at the TGN for basolateral sorting. Notably, AP-1B is not necessary for the basolateral localization of the Na,K-ATPase (Fölsch et al. 1999).

Both AP-1A and AP-1B have been implicated in the basolateral sorting of the H, K-ATPase β-subunit. This subunit has a reversed tyrosine-based signal, FRHY, that interacts with AP-1A and AP-1B in vitro (Duffield et al. 2004). The pump is localized to the basolateral membrane in MDCK cells and to the apical membrane in LLC-PK1 cells (Roush et al. 1998). Curiously, the pump is also localized to the apical membrane in LLC-PK1 cells stably transfected to express AP-1B, which normally reinstates basolateral sorting of proteins with tyrosine-based sorting motifs. This suggests that the FRHY signal may be masked in LLC-PK1 cells by an unknown mechanism (Duffield et al. 2004).

Caplan and colleagues have employed the sequence homology and structural similarities between the apically targeted gastric H,K-ATPase and the basolateral Na,K-ATPase to dissect the apical targeting information in the former. Interestingly, they found that replacing the fourth transmembrane domain (TM4) of the Na,K-ATPase α subunit with that of the closely related H,K-ATPase sequence was sufficient to redirect the Na,K-ATPase to the apical surface of polarized LLC-PK1 cells (Dunbar et al. 2000). In addition, apical targeting of the Na,K-ATPase was also observed when only the cytoplasmic and intracellular loop sequences surrounding this domain (but not TM4 itself) were replaced with the corresponding sequences from the H,K-ATPase. Similarly, a six amino acid sequence within the extracytoplasmic loop preceding TM4 of the related non-gastric H,K-ATPase was shown to be necessary for apical targeting in MDCK cells. In this case, a single point mutation in this region was sufficient to disrupt apical sorting (Lerner et al. 2006).

5.6 Challenges to the Field

Despite the many advances that were achieved over the years in determining polarized sorting mechanisms in epithelial cells, there remain many challenges to researchers in the field. For example, a confounding problem when working with ion transporters is that they are generally expressed at low levels in cells, and their folding, assembly, and surface delivery are often inefficient. Due to these technical challenges, polarized sorting determinants of polytopic proteins are often dissected by monitoring the steady-state distribution of heterologously expressed, and sometimes epitope tagged mutants. Unfortunately, by using this approach it is often difficult to gather information about the mechanisms that enable polarized localization of a given protein (e.g., polarized biosynthetic targeting vs. surface retention).

Another challenge for the field is the lack of available well-differentiated cell culture models to study the trafficking of ion transporters in their native cell type. As noted above, polarized delivery routes differ significantly between cells of liver, intestinal, and kidney origin. Moreover, there can be differences in the regulation of membrane traffic between distinct cell types in a single organ. As one example, cells in the proximal tubule of the kidney are highly specialized for apical endocytosis as opposed to cells further downstream along the nephron, and the organization of apical endocytic compartments in these cells reflects the functional elaboration of this pathway (Mattila et al. 2014). Similarly, the absence of μ1B expression in the proximal tubules has profound implications for the localization of many proteins. However, few cell culture models are available that maintain both the expression of endogenous transporters and the polarization of distinct kidney nephron cell types. While many challenges remain, recent advances in protein expression and knockdown strategies as well as the development of live animal imaging modalities have facilitated the study of ion transporter trafficking in situ. These techniques, in combination with the use of MDCK cells as an in vitro model, should lead to a better understanding of surface delivery of ion transporters in the future.

Acknowledgements We apologize to colleagues whose work we could not cite due to space constraints. Work in the Weisz lab was supported by NIH R01 DK118726 and DK125049, and work in the Fölsch lab was supported by NIH R01 GM0707036.

References

Adair-Kirk TL, Cox KH, Cox JV (1999) Intracellular trafficking of variant chicken kidney AE1 anion exchangers: role of alternative NH₂ termini in polarized sorting and Golgi recycling. J Cell Biol 147:1237–1248

Almomani EY, King JC, Netsawang J, Yenchitsomanus PT, Malasit P, Limjindaporn T, Alexander RT, Cordat E (2012) Adaptor protein 1 complexes regulate intracellular trafficking of the kidney anion exchanger 1 in epithelial cells. Am J Physiol Cell Physiol 303:C554–C566

Ang SF, Fölsch H (2012) The role of secretory and endocytic pathways in the maintenance of cell polarity. Essays Biochem 53:29–39

Ang AL, Taguchi T, Francis S, Fölsch H, Murrells LJ, Pypaert M, Warren G, Mellman I (2004) Recycling endosomes can serve as intermediates during transport from the Golgi to the plasma membrane of MDCK cells. J Cell Biol 167:531–543

Antalffy G, Mauer AS, Paszty K, Hegedus L, Padanyi R, Enyedi A, Strehler EE (2012) Plasma membrane calcium pump (PMCA) isoform 4 is targeted to the apical membrane by the w-splice insert from PMCA2. Cell Calcium 51:171–178

Ares GR, Ortiz PA (2012) Dynamin2, clathrin, and lipid rafts mediate endocytosis of the apical Na/K/2Cl cotransporter NKCC2 in thick ascending limbs. J Biol Chem 287:37824–37834

Arnspang EC, Sundbye S, Nelson WJ, Nejsum LN (2013) Aquaporin-3 and aquaporin-4 are sorted differently and separately in the trans-Golgi network. PLoS One 8:e73977

Bandler PE, Westlake CJ, Grant CE, Cole SP, Deeley RG (2008) Identification of regions required for apical membrane localization of human multidrug resistance protein 2. Mol Pharmacol 74:9–19

Bartles JR, Feracci HM, Stieger B, Hubbard AL (1987) Biogenesis of the rat hepatocyte plasma membrane in vivo: comparison of the pathways taken by apical and basolateral proteins using subcellular fractionation. J Cell Biol 105:1241–1251

Bibee KP, Augustin R, Gazit V, Moley KH (2013) The apical sorting signal for human GLUT9b resides in the N-terminus. Mol Cell Biochem 376:163–173

Bigay J, Antonny B (2012) Curvature, lipid packing, and electrostatics of membrane organelles: defining cellular territories in determining specificity. Dev Cell 23:886–895

Boehm M, Bonifacino JS (2001) Adaptins: the final recount. Mol Biol Cell 12:2907–2920

Bomsel M, Parton R, Kuznetsov SA, Schroer TA, Gruenberg J (1990) Microtubule- and motor-dependent fusion in vitro between apical and basolateral endocytic vesicles from MDCK cells. Cell 62:719–731

Bonifacino JS (2014) Adaptor proteins involved in polarized sorting. J Cell Biol 204:7–17

Bravo-Zehnder M, Orio P, Norambuena A, Wallner M, Meera P, Toro L, Latorre R, Gonzalez A (2000) Apical sorting of a voltage- and Ca_{2+} activated K_+ channel alpha-subunit in Madin-Darby canine kidney cells is independent of N-glycosylation. Proc Natl Acad Sci USA 97:13114–13119

Brone B, Eggermont J (2005) PDZ proteins retain and regulate membrane transporters in polarized epithelial cell membranes. Am J Physiol Cell Physiol 288:C20–C29

Burgos PV, Mardones GA, Rojas AL, daSilva LL, Prabhu Y, Hurley JH, Bonifacino JS (2010) Sorting of the Alzheimer's disease amyloid precursor protein mediated by the AP-4 complex. Dev Cell 18:425–436

Butterworth MB, Edinger RS, Silvis MR, Gallo LI, Liang X, Apodaca G, Frizzell RA, Johnson JP (2012) Rab11b regulates the trafficking and recycling of the epithelial sodium channel (ENaC). Am J Physiol Renal Physiol 302:F581–F590

Caceres PS, Mendez M, Ortiz PA (2014) Vesicle-associated membrane protein 2 (VAMP2) but not VAMP3 mediates cAMP-stimulated trafficking of the renal Na⁺K⁺2Cl co-transporter NKCC2 in thick ascending limbs. J Biol Chem 289:23951–23962

Casanova JE, Mishumi Y, Ikehara Y, Hubbard AL, Mostov KE (1991) Direct apical sorting of rat liver dipeptidylpeptidase IV expressed in Madin-Darby canine kidney cells. J Biol Chem 266:24428–24432

Chen G, Howe AG, Xu G, Frohlich O, Klein JD, Sands JM (2011) Mature N-linked glycans facilitate UT-A1 urea transporter lipid raft compartmentalization. FASEB J 25:4531–4539

Cheng CL, Glover G, Banker G, Amara SG (2002) A novel sorting motif in the glutamate transporter excitatory amino acid transporter 3 directs its targeting in Madin-Darby canine kidney cells and hippocampal neurons. J Neurosci 22:10643–10652

Chicka MC, Strehler EE (2003) Alternative splicing of the first intracellular loop of plasma membrane Ca²⁺ ATPase isoform 2 alters its membrane targeting. J Biol Chem 278:18464–18470

Clemens Grisham R, Kindt K, Finger-Baier K, Schmid B, Nicolson T (2013) Mutations in ap1b1 cause mistargeting of the Na⁺/K⁺-ATPase pump in sensory hair cells. PLoS One 8:e60866

Cresawn KO, Potter BA, Oztan A, Guerriero CJ, Ihrke G, Goldenring JR, Apodaca G, Weisz OA (2007) Differential involvement of endocytic compartments in the biosynthetic traffic of apical proteins. EMBO J 26:3737–3748

Dehmelt L, Bastiaens PI (2010) Spatial organization of intracellular communication: insights from imaging. Nat Rev 11:440–452

Delacour D, Gouyer V, Zanetta JP, Drobecq H, Leteurtre E, Grard G, Moreau-Hannedouche O, Maes E, Pons A, Andre S, Le Bivic A, Gabius HJ, Manninen A, Simons K, Huet G (2005) Galectin-4 and sulfatides in apical membrane trafficking in enterocyte-like cells. J Cell Biol 169:491–501

Delacour D, Cramm-Behrens CI, Drobecq H, Le Bivic A, Naim HY, Jacob R (2006) Requirement for galectin-3 in apical protein sorting. Curr Biol 16:408–414

Delacour D, Greb C, Koch A, Salomonsson E, Leffler H, Le Bivic A, Jacob R (2007) Apical sorting by galectin-3-dependent glycoprotein clustering. Traffic 8:379–388

Delacour D, Koch A, Jacob R (2009) The role of galectins in protein trafficking. Traffic 10:1405–1413

Dell'Angelica EC, Mullins C, Bonifacino JS (1999) AP-4, a novel protein complex related to clathrin adaptors. J Biol Chem 274:7278–7285

Deora AA, Gravotta D, Kreitzer G, Hu J, Bok D, Rodriguez-Boulan E (2004) The basolateral targeting signal of CD147 (EMMPRIN) consists of a single leucine and is not recognized by retinal pigment epithelium. Mol Biol Cell 15:4148–4165

Dib K, Tikhonova IG, Ivetic A, Schu P (2017) The cytoplasmic tail of L-selectin interacts with the adaptor-protein complex AP-1 subunit mu1A via a novel basic binding motif. J Biol Chem 292:6703–6714

Distel B, Bauer U, Le Borgne R, Hoflack B (1998) Basolateral sorting of the cation-dependent mannose 6-phosphate receptor in Madin-Darby canine kidney cells. Identification of a basolateral determinant unrelated to clathrin-coated pit localization signals. J Biol Chem 273:186–193

Donoso M, Cancino J, Lee J, van Kerkhof P, Retamal C, Bu G, Gonzalez A, Caceres A, Marzolo MP (2009) Polarized traffic of LRP1 involves AP1B and SNX17 operating on Y-dependent sorting motifs in different pathways. Mol Biol Cell 20:481–497

Doray B, Ghosh P, Griffith J, Geuze HJ, Kornfeld S (2002) Cooperation of GGAs and AP-1 in packaging MPRs at the trans-Golgi network. Science 297:1700–1703

Doray B, Lee I, Knisely J, Bu G, Kornfeld S (2007) The gamma/sigma1 and alpha/sigma2 hemicomplexes of clathrin adaptors AP-1 and AP-2 harbor the dileucine recognition site. Mol Biol Cell 18:1887–1896

Duffield A, Fölsch H, Mellman I, Caplan MJ (2004) Sorting of H,K-ATPase beta-subunit in MDCK and LLC-PK cells is independent of mu1B adaptin expression. Traffic 5:449–461

Dunbar LA, Aronson P, Caplan MJ (2000) A transmembrane segment determines the steady-state localization of an ion-transporting adenosine triphosphatase. J Cell Biol 148:769–778

Emi Y, Yasuda Y, Sakaguchi M (2012) A cis-acting five-amino-acid motif controls targeting of ABCC2 to the apical plasma membrane domain. J Cell Sci 125:3133–3143

Fang L, Garuti R, Kim BY, Wade JB, Welling PA (2009) The ARH adaptor protein regulates endocytosis of the ROMK potassium secretory channel in mouse kidney. J Clin Invest 119:3278–3289

Farr GA, Hull M, Mellman I, Caplan MJ (2009) Membrane proteins follow multiple pathways to the basolateral cell surface in polarized epithelial cells. J Cell Biol 186:269–282

Fields IC, Shteyn E, Pypaert M, Proux-Gillardeaux V, Kang RS, Galli T, Fölsch H (2007) v-SNARE cellubrevin is required for basolateral sorting of AP-1B-dependent cargo in polarized epithelial cells. J Cell Biol 177:477–488

Fields IC, King SM, Shteyn E, Kang RS, Fölsch H (2010) Phosphatidylinositol 3,4,5-trisphosphate localization in recycling endosomes is necessary for AP-1B-dependent sorting in polarized epithelial cells. Mol Biol Cell 21:95–105

Fölsch H (2008) Regulation of membrane trafficking in polarized epithelial cells. Curr Opin Cell Biol 20:208–213

Fölsch H (2015a) Analyzing the role of AP-1B in polarized sorting from recycling endosomes in epithelial cells. Method Cell Biol 130:289–305

Fölsch H (2015b) Role of the epithelial cell-specific clathrin adaptor complex AP-1B in cell polarity. Cell Logist 5:e1074331

Fölsch H, Ohno H, Bonifacino JS, Mellman I (1999) A novel clathrin adaptor complex mediates basolateral targeting in polarized epithelial cells. Cell 99:189–198

Fölsch H, Pypaert M, Schu P, Mellman I (2001) Distribution and function of AP-1 clathrin adaptor complexes in polarized epithelial cells. J Cell Biol 152:595–606

Fölsch H, Pypaert M, Maday S, Pelletier L, Mellman I (2003) The AP-1A and AP-1B clathrin adaptor complexes define biochemically and functionally distinct membrane domains. J Cell Biol 163:351–362

Fölsch H, Mattila PE, Weisz OA (2009) Taking the scenic route: biosynthetic traffic to the plasma membrane in polarized epithelial cells. Traffic 10:972–981

Francis SS, Sfakianos J, Lo B, Mellman I (2011) A hierarchy of signals regulates entry of membrane proteins into the ciliary membrane domain in epithelial cells. J Cell Biol 193:219–233

Fry AC, Su Y, Yiu V, Cuthbert AW, Trachtman H, Frankl FEK (2012) Mutation conferring apical-targeting motif on AE1 exchanger causes autosomal dominant distal RTA. J Am Soc Nephrol 23:1238–1249

Galli T, Zahraoui A, Vaidyanathan VV, Raposo G, Tian JM, Karin M, Niemann H, Louvard D (1998) A novel tetanus neurotoxin-insensitive vesicle-associated membrane protein in SNARE complexes of the apical plasma membrane of epithelial cells. Mol Biol Cell 9:1437–1448

Gan Y, McGraw TE, Rodriguez-Boulan E (2002) The epithelial-specific adaptor AP1B mediates post-endocytic recycling to the basolateral membrane. Nat Cell Biol 4:605–609

Gephart JD, Singh B, Higginbotham JN, Franklin JL, Gonzalez A, Fölsch H, Coffey RJ (2011) Identification of a novel mono-leucine basolateral sorting motif within the cytoplasmic domain of amphiregulin. Traffic 12:1793–1804

Gravotta D, Deora A, Perret E, Oyanadel C, Soza A, Schreiner R, Gonzalez A, Rodriguez-Boulan E (2007) AP1B sorts basolateral proteins in recycling and biosynthetic routes of MDCK cells. Proc Natl Acad Sci USA 104:1564–1569

Grindstaff KK, Yeaman C, Anandasabapathy N, Hsu SC, Rodriguez-Boulan E, Scheller RH, Nelson WJ (1998) Sec6/8 complex is recruited to cell-cell contacts and specifies transport vesicle delivery to the basal-lateral membrane in epithelial cells. Cell 93:731–740

Guerriero CJ, Lai Y, Weisz OA (2008) Differential sorting and Golgi export requirements for raft-associated and raft-independent apical proteins along the biosynthetic pathway. J Biol Chem 283:18040–18047

Hill WG, Butterworth MB, Wang H, Edinger RS, Lebowitz J, Peters KW, Frizzell RA, Johnson JP (2007) The epithelial sodium channel (ENaC) traffics to apical membrane in lipid rafts in mouse cortical collecting duct cells. J Biol Chem 282:37402–37411

Hirst J, Robinson MS (1998) Clathrin and adaptors. Biochim Biophys Acta 1404:173–193

Hu Q, Milenkovic L, Jin H, Scott MP, Nachury MV, Spiliotis ET, Nelson WJ (2010) A septin diffusion barrier at the base of the primary cilium maintains ciliary membrane protein distribution. Science 329:436–439

Hunziker W, Harter C, Matter K, Mellman I (1991) Basolateral sorting in MDCK cells requires a distinct cytoplasmic domain determinant. Cell 66:907–920

Imjeti NS, Lebreton S, Paladino S, de la Fuente E, Gonzalez A, Zurzolo C (2011) N-Glycosylation instead of cholesterol mediates oligomerization and apical sorting of GPI-APs in FRT cells. Mol Biol Cell 22:4621–4634

Ito M, Sakurai A, Hayashi K, Ohi A, Kangawa N, Nishiyama T, Sugino S, Uehata Y, Kamahara A, Sakata M, Tatsumi S, Kuwahata M, Taketani Y, Segawa H, Miyamoto K (2010) An apical expression signal of the renal type IIc Na+ dependent phosphate cotransporter in renal epithelial cells. Am J Physiol Renal Physiol 299:F243–F254

Jacob R, Naim HY (2001) Apical membrane proteins are transported in distinct vesicular carriers. Curr Biol 11:1444–1450

Janvier K, Kato Y, Boehm M, Rose JR, Martina JA, Kim BY, Venkatesan S, Bonifacino JS (2003) Recognition of dileucine-based sorting signals from HIV-1 Nef and LIMP-II by the AP-1 gamma-sigma1 and AP-3 delta-sigma3 hemicomplexes. J Cell Biol 163:1281–1290

Junking M, Sawasdee N, Duangtum N, Cheunsuchon B, Limjindaporn T, Yenchitsomanus PT (2014) Role of adaptor proteins and clathrin in the trafficking of human kidney anion exchanger 1 (kAE1) to the cell surface. Traffic 15:788–802

Kang RS, Fölsch H (2011) ARH cooperates with AP-1B in the exocytosis of LDLR in polarized epithelial cells. J Cell Biol 193:51–60

Kinlough CL, Poland PA, Gendler SJ, Mattila PE, Mo D, Weisz OA, Hughey RP (2011) Core-glycosylated mucin-like repeats from MUC1 are an apical targeting signal. J Biol Chem 286:39072–39081

Kirchhausen T (2002) Single-handed recognition of a sorting traffic motif by the GGA proteins. Nat Struct Biol 9:241–244

Kizhatil K, Yoon W, Mohler PJ, Davis LH, Hoffman JA, Bennett V (2007) Ankyrin-G and beta2-spectrin collaborate in biogenesis of lateral membrane of human bronchial epithelial cells. J Biol Chem 282:2029–2037

Kwon SH, Guggino WB (2004) Multiple sequences in the C terminus of MaxiK channels are involved in expression, movement to the cell surface, and apical localization. Proc Natl Acad Sci USA 101:15237–15242

Lapierre LA, Dorn MC, Zimmerman CF, Navarre J, Burnette JO, Goldenring JR (2003) Rab11b resides in a vesicular compartment distinct from Rab11a in parietal cells and other epithelial cells. Exp Cell Res 290:322–331

Lauffer BE, Melero C, Temkin P, Lei C, Hong W, Kortemme T, von Zastrow M (2010) SNX27 mediates PDZ-directed sorting from endosomes to the plasma membrane. J Cell Biol 190:565–574

Le Bivic A, Quaroni A, Nichols B, Rodriguez-Boulan E (1990) Biogenetic pathways of plasma membrane proteins in Caco-2, a human intestinal epithelial cell line. J Cell Biol 111:1351–1361

Le Maout S, Welling PA, Brejon M, Olsen O, Merot J (2001) Basolateral membrane expression of a K$^+$ channel, Kir 2.3, is directed by a cytoplasmic COOH-terminal domain. Proc Natl Acad Sci USA 98:10475–10480

Lerner M, Lemke D, Bertram H, Schillers H, Oberleithner H, Caplan MJ, Reinhardt J (2006) An extracellular loop of the human non-gastric H, K-ATPase alpha-subunit is involved in apical plasma membrane polarization. Cell Physiol Biochem 18:75–84

Li L, Omata W, Kojima I, Shibata H (2001) Direct interaction of Rab4 with syntaxin 4. J Biol Chem 276:5265–5273

Li C, Franklin JL, Graves-Deal R, Jerome WG, Cao Z, Coffey RJ (2004) Myristoylated Naked2 escorts transforming growth factor alpha to the basolateral plasma membrane of polarized epithelial cells. Proc Natl Acad Sci USA 101:5571–5576

Li C, Hao M, Cao Z, Ding W, Graves-Deal R, Hu J, Piston DW, Coffey RJ (2007) Naked2 acts as a cargo recognition and targeting protein to ensure proper delivery and fusion of TGF-alpha containing exocytic vesicles at the lower lateral membrane of polarized MDCK cells. Mol Biol Cell 18:3081–3093

Ling K, Bairstow SF, Carbonara C, Turbin DA, Huntsman DG, Anderson RA (2007) Type Igamma phosphatidylinositol phosphate kinase modulates adherens junction and E-cadherin trafficking via a direct interaction with mu1B adaptin. J Cell Biol 176:343–353

Lisanti MP, Caras IW, Davitz MA, Rodriguez-Boulan E (1989) A glycophospholipid membrane anchor acts as an apical targeting signal in polarized epithelial cells. J Cell Biol 109:2145–2156

Maday S, Anderson E, Chang HC, Shorter J, Satoh A, Sfakianos J, Fölsch H, Anderson JM, Walther Z, Mellman I (2008) A PDZ-binding motif controls basolateral targeting of syndecan-1 along the biosynthetic pathway in polarized epithelial cells. Traffic 9:1915–1924

Madrid R, Le Maout S, Barrault MB, Janvier K, Benichou S, Merot J (2001) Polarized trafficking and surface expression of the AQP4 water channel are coordinated by serial and regulated interactions with different clathrin-adaptor complexes. EMBO J 20:7008–7021

Marrs JA, Andersson-Fisone C, Jeong MC, Cohen-Gould L, Zurzolo C, Nabi IR, Rodriguez-Boulan E, Nelson WJ (1995) Plasticity in epithelial cell phenotype: modulation by expression of different cadherin cell adhesion molecules. J Cell Biol 129:507–519

Matter K, Brauchbar M, Bucher K, Hauri HP (1990) Sorting of endogenous plasma membrane proteins occurs from two sites in cultured human intestinal epithelial cells (Caco-2). Cell 60:429–437

Matter K, Whitney JA, Yamamoto EM, Mellman I (1993) Common signals control low density lipoprotein receptor sorting in endosomes and the Golgi complex of MDCK cells. Cell 74:1053–1064

Mattera R, Boehm M, Chaudhuri R, Prabhu Y, Bonifacino JS (2011) Conservation and diversification of dileucine signal recognition by adaptor protein (AP) complex variants. J Biol Chem 286:2022–2030

Mattila PE, Youker RT, Mo D, Bruns JR, Cresawn KO, Hughey RP, Ihrke G, Weisz OA (2012) Multiple biosynthetic trafficking routes for apically secreted proteins in MDCK cells. Traffic 13:433–442

Mattila PE, Raghavan V, Rbaibi Y, Baty CJ, Weisz OA (2014) Rab11a-positive compartments in proximal tubule cells sort fluid-phase and membrane cargo. Am J Physiol Cell Physiol 306: C441–C449

Milewski MI, Forrest JK, Mickle JE, Stanton B, Cutting GR (2000) The role of PDZ-binding motif in polarized distribution of CFTR and other apical or basolateral membrane proteins. Am J Hum Genet 67:198–198

Mishra R, Grzybek M, Niki T, Hirashima M, Simons K (2010) Galectin-9 trafficking regulates apical-basal polarity in Madin-Darby canine kidney epithelial cells. Proc Natl Acad Sci USA 107:17633–17638

Mo D, Costa SA, Ihrke G, Youker RT, Pastor-Soler N, Hughey RP, Weisz OA (2012) Sialylation of N-linked glycans mediates apical delivery of endolyn in MDCK cells via a galectin-9-dependent mechanism. Mol Biol Cell 23:3636–3646

Mohrmann K, Leijendekker R, Gerez L, van Der Sluijs P (2002) rab4 regulates transport to the apical plasma membrane in Madin-Darby canine kidney cells. J Biol Chem 277:10474–10481

Mostov K, Su T, ter Beest M (2003) Polarized epithelial membrane traffic: conservation and plasticity. Nat Cell Biol 5:287–293

Moyer BD, Denton J, Karlson KH, Reynolds D, Wang S, Mickle JE, Milewski M, Cutting GR, Guggino WB, Li M, Stanton BA (1999) A PDZ-interacting domain in CFTR is an apical membrane polarization signal. J Clin Invest 104:1353–1361

Muniz M, Zurzolo C (2014) Sorting of GPI-anchored proteins from yeast to mammals—common pathways at different sites? J Cell Sci 127:2793–2801

Muth TR, Gottardi CJ, Roush DL, Caplan MJ (1998) A basolateral sorting signal is encoded in the alpha-subunit of Na-K-ATPase. Am J Physiol Cell Physiol 274:C688–C696

Nabi IR, Le Bivic A, Fambrough D, Rodriguez-Boulan E (1991) An endogenous MDCK lysosomal membrane glycoprotein is targeted basolaterally before delivery to lysosomes. J Cell Biol 115:1573–1584

Navarro Negredo P, Edgar JR, Wrobel AG, Zaccai NR, Antrobus R, Owen DJ, Robinson MS (2017) Contribution of the clathrin adaptor AP-1 subunit micro1 to acidic cluster protein sorting. J Cell Biol 216:2927–2943

Nies AT, Konig J, Cui YH, Brom M, Spring H, Keppler D (2002) Structural requirements for the apical sorting of human multidrug resistance protein 2 (ABCC2). Eur J Biochem 269:1866–1876

Nokes RL, Fields IC, Collins RN, Fölsch H (2008) Rab13 regulates membrane trafficking between TGN and recycling endosomes in polarized epithelial cells. J Cell Biol 182:845–853

Ohno H, Tomemori T, Nakatsu F, Okazaki Y, Aguilar RC, Fölsch H, Mellman I, Saito T, Shirasawa T, Bonifacino JS (1999) Mu1B, a novel adaptor medium chain expressed in polarized epithelial cells. FEBS Lett 449:215–220

Olsen O, Liu H, Wade JB, Merot J, Welling PA (2002) Basolateral membrane expression of the Kir 2.3 channel is coordinated by PDZ interaction with Lin-7/CASK complex. Am J Physiol Cell Physiol 282:C183–C195

Ozaslan D, Wang S, Ahmed BA, Kocabas AM, McCastlain JC, Bene A, Kilic F (2003) Glycosyl modification facilitates homo- and hetero-oligomerization of the serotonin transporter. A specific role for sialic acid residues. J Biol Chem 278:43991–44000

Paladino S, Sarnataro D, Pillich R, Tivodar S, Nitsch L, Zurzolo C (2004) Protein oligomerization modulates raft partitioning and apical sorting of GPI-anchored proteins. J Cell Biol 167:699–709

Paladino S, Lebreton S, Tivodar S, Campana V, Tempre R, Zurzolo C (2008) Different GPI-attachment signals affect the oligomerisation of GPI-anchored proteins and their apical sorting. J Cell Sci 121:4001–4007

Perego C, Vanoni C, Villa A, Longhi R, Kaech SM, Frohli E, Hajnal A, Kim SK, Pietrini G (1999) PDZ-mediated interactions retain the epithelial GABA transporter on the basolateral surface of polarized epithelial cells. EMBO J 18:2384–2393

Perego C, Vanoni C, Massari S, Longhi R, Pietrini G (2000) Mammalian LIN-7 PDZ proteins associate with beta-catenin at the cell-cell junctions of epithelia and neurons. EMBO J 19:3978–3989

Pocard T, Le Bivic A, Galli T, Zurzolo C (2007) Distinct v-SNAREs regulate direct and indirect apical delivery in polarized epithelial cells. J Cell Sci 120:3309–3320

Potter BA, Hughey RP, Weisz OA (2006a) Role of N- and O-glycans in polarized biosynthetic sorting. Am J Physiol Cell Physiol 290:C1–C10

Potter BA, Weixel KM, Bruns JR, Ihrke G, Weisz OA (2006b) N-glycans mediate apical recycling of the sialomucin endolyn in polarized MDCK cells. Traffic 7:146–154

Rai T, Sasaki S, Uchida S (2006) Polarized trafficking of the aquaporin-3 water channel is mediated by an NH2-terminal sorting signal. Am J Physiol Cell Physiol 290:C298–C304

Reales E, Sharma N, Low SH, Fölsch H, Weimbs T (2011) Basolateral sorting of syntaxin 4 is dependent on its N-terminal domain and the AP1B clathrin adaptor, and required for the epithelial cell polarity. PLoS One 6:e21181

Riquier AD, Lee DH, McDonough AA (2009) Renal NHE3 and NaPi2 partition into distinct membrane domains. Am J Physiol Cell Physiol 296:C900–C910

Rodriguez-Boulan E, Kreitzer G, Musch A (2005) Organization of vesicular trafficking in epithelia. Nat Rev 6:233–247

Rodriguez-Boulan E, Perez-Bay A, Schreiner R, Gravotta D (2013) Response: the "tail" of the twin adaptors. Dev Cell 27:247–248

Roush DL, Gottardi CJ, Naim HY, Roth MG, Caplan MJ (1998) Tyrosine-based membrane protein sorting signals are differentially interpreted by polarized Madin-Darby canine kidney and LLC-PK1 epithelial cells. J Biol Chem 273:26862–26869

Ryan S, Verghese S, Cianciola NL, Cotton CU, Carlin CR (2010) Autosomal recessive polycystic kidney disease epithelial cell model reveals multiple basolateral epidermal growth factor receptor sorting pathways. Mol Biol Cell 21:2732–2745

Sai Y, Nies AT, Arias IM (1999) Bile acid secretion and direct targeting of mdr1-green fluorescent protein from Golgi to the canalicular membrane in polarized WIF-B cells. J Cell Sci 112:4535–4545

Schreiner R, Frindt G, Diaz F, Carvajal-Gonzalez JM, Perez Bay AE, Palmer LG, Marshansky V, Brown D, Philp NJ, Rodriguez-Boulan E (2010) The absence of a clathrin adapter confers unique polarity essential to proximal tubule function. Kidney Int 78:382–388

Sheff DR, Kroschewski R, Mellman I (2002) Actin dependence of polarized receptor recycling in Madin-Darby canine kidney cell endosomes. Mol Biol Cell 13:262–275

Shoshani L, Contreras RG, Roldan ML, Moreno J, Lazaro A, Balda MS, Matter K, Cereijido M (2005) The polarized expression of Na$^+$, K$^+$ ATPase in epithelia depends on the association between beta-subunits located in neighboring cells. Mol Biol Cell 16:1071–1081

Shteyn E, Pigati L, Fölsch H (2011) Arf6 regulates AP-1B-dependent sorting in polarized epithelial cells. J Cell Biol 194:873–887

Silvis MR, Bertrand CA, Ameen N, Golin-Bisello F, Butterworth MB, Frizzell RA, Bradbury NA (2009) Rab11b regulates the apical recycling of the cystic fibrosis transmembrane conductance regulator in polarized intestinal epithelial cells. Mol Biol Cell 20:2337–2350

Simons K, Ikonen E (1997) Functional rafts in cell membranes. Nature 387:569–572

Stechly L, Morelle W, Dessein AF, Andre S, Grard G, Trinel D, Dejonghe MJ, Leteurtre E, Drobecq H, Trugnan G, Gabius HJ, Huet G (2009) Galectin-4-regulated delivery of glycoproteins to the brush border membrane of enterocyte-like cells. Traffic 10:438–450

Stiegler AL, Zhang R, Liu W, Boggon TJ (2014) Structural determinants for binding of sorting nexin 17 (SNX17) to the cytoplasmic adaptor protein Krev interaction trapped 1 (KRIT1). J Biol Chem 289:25362–25373

Straube T, von Mach T, Honig E, Greb C, Schneider D, Jacob R (2013) pH-dependent recycling of galectin-3 at the apical membrane of epithelial cells. Traffic 14:1014–1027

Subramanian VS, Marchant JS, Boulware MJ, Said HM (2004) A C-terminal region dictates the apical plasma membrane targeting of the human sodium-dependent vitamin C transporter-1 in polarized epithelia. J Biol Chem 279:27719–27728

Subramanian VS, Marchant JS, Boulware MJ, Ma TY, Said HM (2009) Membrane targeting and intracellular trafficking of the human sodium-dependent multivitamin transporter in polarized epithelial cells. Am J Physiol Cell Physiol 296:C663–C671

Sugimoto H, Sugahara M, Fölsch H, Koide Y, Nakatsu F, Tanaka N, Nishimura T, Furukawa M, Mullins C, Nakamura N, Mellman I, Ohno H (2002) Differential recognition of tyrosine-based basolateral signals by AP-1B subunit mu1B in polarized epithelial cells. Mol Biol Cell 13:2374–2382

Sun AQ, Salkar R, Sachchidanand S, Xu L, Zeng MMZ, Suchy FJ (2003) A 14-amino acid sequence with a beta-turn structure is required for apical membrane sorting of the rat ileal bile acid transporter. J Biol Chem 278:4000–4009

Torres J, Funk HM, Zegers MM, ter Beest MB (2011) The syntaxin 4 N terminus regulates its basolateral targeting by munc18c-dependent and -independent mechanisms. J Biol Chem 286:10834–10846

Toye AM, Banting G, Tanner MJ (2004) Regions of human kidney anion exchanger 1 (kAE1) required for basolateral targeting of kAE1 in polarised kidney cells: mis-targeting explains dominant renal tubular acidosis (dRTA). J Cell Sci 117:1399–1410

Traub LM (2003) Sorting it out: AP-2 and alternate clathrin adaptors in endocytic cargo selection. J Cell Biol 163:203–208

Vagin O, Turdikulova S, Sachs G (2005) Recombinant addition of N-glycosylation sites to the basolateral Na, K-ATPase beta1 subunit results in its clustering in caveolae and apical sorting in HGT-1 cells. J Biol Chem 280:43159–43167

van de Graaf SF, Hoenderop JG, van der Kemp AW, Gisler SM, Bindels RJ (2006) Interaction of the epithelial Ca^{2+} channels TRPV5 and TRPV6 with the intestine- and kidney-enriched PDZ protein NHERF4. Pflugers Arch Eur J Physiol 452:407–417

van Hoek AN, Ma T, Yang B, Verkman AS, Brown D (2000) Aquaporin-4 is expressed in basolateral membranes of proximal tubule S3 segments in mouse kidney. Am J Physiol Renal Physiol 278:F310–F316

Weisz OA, Rodriguez-Boulan E (2009) Apical trafficking in epithelial cells: signals, clusters and motors. J Cell Sci 122:4253–4266

Yeaman C, Le Gall AH, Baldwin AN, Monlauzeur L, Le Bivic A, Rodriguez-Boulan E (1997) The O-glycosylated stalk domain is required for apical sorting of neurotrophin receptors in polarized MDCK cells. J Cell Biol 139:929–940

Yeaman C, Grindstaff KK, Wright JR, Nelson WJ (2001) Sec6/8 complexes on trans-Golgi network and plasma membrane regulate late stages of exocytosis in mammalian cells. J Cell Biol 155:593–604

Yin X, Kang JH, Andrianifahanana M, Wang Y, Jung MY, Hernandez DM, Leof EB (2017) Basolateral delivery of the type I transforming growth factor beta receptor is mediated by a dominant-acting cytoplasmic motif. Mol Biol Cell. 28:2701–2711

Yoo D, Flagg TP, Olsen O, Raghuram V, Foskett JK, Welling PA (2004) Assembly and trafficking of a multiprotein ROMK (Kir 1.1) channel complex by PDZ interactions. J Biol Chem 279:6863–6873

Youker RT, Bruns JR, Costa SA, Rbaibi Y, Lanni F, Kashlan OB, Teng H, Weisz OA (2013) Multiple motifs regulate apical sorting of p75 via a mechanism that involves dimerization and higher-order oligomerization. Mol Biol Cell 24:1996–2007

Chapter 6
Membrane Protein Structure and Folding

Aiping Zheng, Sophie C. Frizzell, Solomon M. Klombers, and Patrick H. Thibodeau

Abstract Seminal structural and biochemical studies demonstrated that proteins are organized by specific structural elements and that the information necessary to form these ordered structures is encoded in the polypeptide chain. Questions related to the mechanisms of protein folding for both soluble and transmembrane proteins remain a large field of research utilizing a combination of biochemical, biophysical, genetic, and large-scale computational approaches. Understanding these processes has broad implications for the functional identification of unknown and putative proteins found in sequenced genomes, the characterization of disease-causing mutations identified in genetic studies, and for rational design and protein engineering efforts to introduce novel biological function. This chapter will provide an overview of transmembrane protein structure determination and provide a survey of studies on CFTR folding and structure.

Keywords Protein folding · Protein biosynthesis · X-ray crystallography · Protein structure · Cryo-electron microscopy · ABC transporter · CFTR

6.1 Protein Folding and Biosynthesis

Seminal biochemical studies by Anfinsen demonstrated that the information for the specific structure and activity of a protein is contained in the polypeptide chain (Anfinsen and Redfield 1956). These observations, made initially on a soluble protein, RNaseA, established the field of protein folding and still underlie one of the great challenges in biology—understanding how the primary sequence of a protein facilitates folding into a compact and functional three-dimensional conformation. The fundamental observation was that a purified protein, when unfolded in

A. Zheng · S. C. Frizzell · S. M. Klombers · P. H. Thibodeau (✉)
Department of Microbiology and Molecular Genetics, University of Pittsburgh, School of Medicine, Pittsburgh, PA, USA
e-mail: aiz7@pitt.edu; smk144@pitt.edu; thibodea@pitt.edu

© The American Physiological Society 2020
K. L. Hamilton, D. C. Devor (eds.), *Basic Epithelial Ion Transport Principles and Function*, Physiology in Health and Disease,
https://doi.org/10.1007/978-3-030-52780-8_6

chemical denaturants, could adopt its native conformation upon dilution from the denaturant in vitro. These results demonstrated, for the first time, that the process of protein folding and the final protein structure are encoded by the primary sequence of the polypeptide chain (Anfinsen 1973).

6.1.1 Physical Regulation of Protein Folding and Structure

A variety of parameters regulate the formation of these compact, folded structures through both thermodynamic and kinetic mechanisms (Schlesinger and Barrett 1965; Friedland and Hastings 1967; Anfinsen 1973; Baker et al. 1992). A quantitative, first-principle understanding of the physical processes has not been sufficiently developed to generally predict protein structure de novo (Baker 2014). However, the basic physical processes underlying protein folding have been described and include specific interactions within the polypeptide chain, both local and long distance, as well as interactions between the polypeptide chain and its environment. At the heart of the protein folding reaction is the thermodynamic minimization of the free energy of the polypeptide chain (Anfinsen 1973). The energetic interactions between residues in the polypeptide and interactions between the polypeptide and solvent facilitate protein folding by providing a driving force for collapse and ordering.

The requirement for a driving force is illustrated by considering a random search in structure space, which was first formulated by Cyrus Levinthal (Levinthal 1968). A simple thought experiment considering only the polypeptide backbone and the two dihedral angles that describe its orientation, ϕ and ψ, illustrates the requirement for a protein folding "driving force." If one limits conformational space to three backbone positions for both dihedrals, the number of conformations is $3^{2(N-1)}$, where N is the number of amino acids in the polypeptide chain. If the polypeptide chain length is 100 amino acids, approximately 3^{198} (or 3×10^{94}) conformations would exist. Sampling each conformation at picosecond rates would require a timescale significantly longer than the lifespan of the known universe. Such sampling does not allow for the sub-millisecond folding seen in many experimental models. This formalization, while simplistic, illustrates the requirement for a driving force for protein folding and is known as the Levinthal paradox.

The initial physical process that drives protein folding is likely the exclusion of water from the hydrophobic amino acids (Nozaki and Tanford 1971; Moelbert et al. 2004). For soluble proteins in aqueous environments, the exposure of hydrophobic amino acids results in an increase in solvent ordering (i.e., a reduction in solvent entropy) (Dill 1990; Sun et al. 1995). This ordering results from the loss of polar interactions between solvent and solute and results in the formation of an ordered cage around the hydrophobic solute. This is energetically unfavorable and facilitates hydrophobic collapse wherein the apolar residues are buried within the protein, driving the formation of structure and increasing solvent entropy (Matsumura et al. 1988; Karpusas et al. 1989). For membrane proteins, the integration of hydrophobic spans into the lipid bilayer similarly excludes polar solvent from the hydrophobic residues and provides a major constraint for the protein structures spanning and proximal to the membrane (Trowbridge et al. 1991; Buchegger et al. 1996). The unique nature of the lipid bilayer places significant and specific constraints on the sequences of transmembrane proteins in comparison to soluble proteins, though both elements are generally enriched in densely packed hydrophobic amino acids.

Though these hydrophobic interactions provide a major energy source for the initial folding of the polypeptide chain, the native—or folded—state of a protein is stabilized by many small noncovalent interactions, including Van der Waals interactions, hydrogen bonding and electrostatic interactions (Dill 1990). The free energy associated with each of the individual interactions is small. However, the sum of thousands of these interactions is large for a folded protein. Hydrogen bonding contributes significantly to the stabilization of the native protein structure. In proteins, hydrogen bonds result from the differences in electronegativity between oxygen, nitrogen, and carbon atoms. The peptide backbone is regularly organized with hydrogen bond donors (amine N-H) and acceptors (carboxyl C=O), in addition to the donors and acceptors found in the polar amino acids. Hydrogen bonding is responsible for the organization of secondary structure elements based on the specific orientation of the amine and carbonyl groups in the peptide backbone.

Electrostatic interactions are associated with formal charges in ionizable groups—the side chains of charged amino acids and the N- and C-termini of the polypeptide. In addition, dipoles found in polar groups and secondary structure elements provide an additional source of electrostatic forces within proteins. The electrostatic interactions, in sum, contribute significantly to the free energy of the folded state. In addition, these forces are unique in their ability to propagate long distances through space and in response to the local environment. Van der Waals interactions are weak, short-range interactions that facilitate atom–atom packing. These optimal energetic interactions are maximized as two atoms make contact without interpenetrating. These small forces underlie the energetic basis of steric hindrance.

6.1.2 Thermodynamic and Kinetic Regulation of Protein Folding

From a thermodynamic perspective, the main resistance to the proper folding and formation of a compact structure is the entropic penalty associated with the constraint of the polypeptide chain (Dill 1990). In the absence of hydrophobic interactions, the unconstrained polypeptide chain would favor an extended, hydrated state with no regular structure. In an aqueous environment, the energies of solvent exclusion, electrostatics, hydrogen bonding, and Van der Waals interactions offset the entropic cost and favor the folded state (Fig. 6.1). Though the total energy associated with protein folding is large, the difference in free energy states is relatively small, with the free energies of folding often ranging between –5 and –

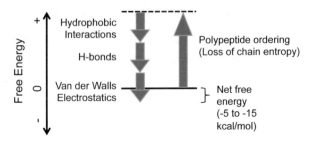

Fig. 6.1 Free energy of protein folding. A schematic of the free energies associated with protein folding is shown. The unfavorable entropic energy associated with polypeptide ordering is counterbalanced by hydrophobic interactions, hydrogen bonding, electrostatic, and Van der Waals interactions. The net free energy for most naturally occurring, folded proteins is relatively small, on the order of –5 to –15 kcal/mol

15 kcal/mol (Pace and Shaw 2000). Native state destabilization, resulting from changes to the primary sequence, is the molecular mechanism underlying many disease-causing protein mutations (Thomas et al. 1995).

The inherent complexity of the protein fold suggests that the kinetics of the folding reaction are similarly important. The formation of specific, individual structures occurs, in many cases, along a hierarchical pathway (Religa et al. 2005). For some proteins, folding pathways bifurcate into nonproductive branches wherein non-native structures are formed and trap the nascent chain (Baker et al. 1992; King et al. 1996). This has been conceptualized by the transformation of the classic two-state reaction into a three-dimensional funnel with "conformational space" plotted against free energy (Fig. 6.2) (Dill and Chan 1997). The energetic and structural transitions through this landscape are complicated by nonproductive, local energetic minima.

The kinetic regulation of folding provides another mechanism to conceptually "steer" the folding along productive transitions to the native state (Dill et al. 1993). This work suggests that proteins can follow some specific "folding pathways" that facilitate folding in a nonrandom fashion. These may follow a vectorial or hierarchical pathway or be more global and nonvectorial in nature (Jansens et al. 2002; Zhang et al. 2012). The hierarchical or vectorial pathway has previously been assumed for multi-domain proteins, however, single-molecule studies and high-resolution solution studies provide evidence that single-domain protein folding can proceed through specific ensembles of protein conformations and intermediate structures (Vendruscolo et al. 2001; Dobson 2003). While these events are driven by the thermodynamics of solvent-solute interactions and contacts with the folding protein, the nascent chain is thought to provide some control of the kinetics for both on- and off-pathway events (Baker et al. 1992). These controls may include rate-limited conformational changes associated with specific amino acids (i.e., proline isomerization or disulfide formation/breakage) (Deber et al. 1986; Ho and Brasseur 2005). Alternatively, they may be related to the formation of specific structural elements and their interactions in the folded structure.

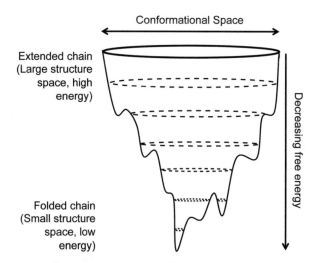

Fig. 6.2 Protein folding funnel. A schematic of a protein folding landscape is shown with structure space plotted against free energy as a "folding funnel." The thermodynamics of the protein fold drive the conformational constraints of the polypeptide into smaller and smaller ensembles of protein structures and into the native state. Local energy minima, represented as dips or valleys in the energy profile, may trap folding intermediates and require energy from chaperones or other sources to facilitate proper folding

6.1.3 Protein Folding in the Cellular Environment

The evolution of protein sequence and function occurs within the context of specific biological systems. The native state function of a protein evolves to regulate specific cellular processes within the context of the cell or tissue in which it is expressed. Similarly, the primary sequence of the protein evolves in the context of this cellular environment and is selected for proper folding in this context. As a result, efficient folding of many proteins relies not just on the primary sequence, but chaperones and other environmental factors associated with the cellular environment in which the protein evolved.

Soluble proteins rely on multiple chaperones systems to fold efficiently. These are too numerous to review in this context, but are well described in the primary and review literature, including Chap. 7 by Buck and Brodsky. Chaperones play several key roles in facilitating protein folding. First, they hold the extended polypeptide chain in order to keep it soluble during protein translation and folding (Slepenkov and Witt 2002; Cha et al. 2013; Powis et al. 2013; Roy et al. 2014). This "holdase" function is critical during protein synthesis, where unpaired protein sequences must be kept soluble while waiting for their partner sequences to be translated. Chaperones may also serve to facilitate folding of the nascent chain mechanically (Chan and Dill 1996; Wang and Tsou 1998; de Marco et al. 2005; Schroder and Kaufman 2005; Cha et al. 2013). This "foldase" function is often coupled to nucleotide hydrolysis and alters the nascent chain conformation. Additionally, chaperones may utilize an "unfoldase" function to reverse the process of protein misfolding and aggregation (Burton and Baker 2005; Sharma et al. 2009; Tamayo et al. 2011; Liu et al. 2013a). Finally, chaperone systems have increasingly been recognized as playing a key role in protein degradation, providing the initial recognition and signaling for ubiquitin modification (Guerriero et al. 2013; Mehnert et al. 2014; Zacchi et al. 2014).

While the basic concepts of protein folding and chaperones are the same for soluble and transmembrane proteins, the unique nature of the transmembrane protein—that is, its insertion into the membrane bilayer—requires the evolution of machinery to facilitate its integration into the membrane and of sequences that preferentially interact with the lipid bilayer.

6.1.4 Protein Insertion in the Biosynthetic Pathway

Transmembrane protein folding is intrinsically more complicated than the folding of soluble, cytosolic proteins (Do et al. 1996). While many of the soluble chaperone systems contribute to the folding of ER–lumenal or cytosolic domains of a transmembrane protein, a specific set of cellular systems have evolved to handle the insertion and assembly of transmembrane protein sequences (Krieg et al. 1989; Gorlich et al. 1990; Sanders and Schekman 1992). These include proteins that recognize transmembrane sequences and those that facilitate the integration of the transmembrane protein into the lipid bilayer (Moller et al. 1998; Wang and Dobberstein 1999).

Translation of the mRNA sequence occurs within the ribosome and involves multiple chaperones and accessory proteins that regulate translational rate and pausing, delivery to and entry into the secretory pathway, insertion in the membrane, and cytosolic and ER-lumenal domain folding (Nilsson et al. 2014). The regulation of protein translation and insertion is mediated by the signal recognition particle (SRP), the Sec61 translocon, and a variety of accessory proteins that target the ribosome-nascent chain complex to the ER insertion site (Wang and Dobberstein 1999). The function of SRP is to identify and bind hydrophobic stretches in the nascent chain in order to facilitate their targeting to the ER membrane (Jiang et al. 2008; Ataide et al. 2011; Nilsson et al. 2014). The Sec61 translocon, a complex of proteins that forms a channel spanning the ER membrane, is responsible for membrane insertion of the nascent chain. Together, these proteins serve as the "gate keepers" to membrane insertion and entry into the secretory pathway (Moller et al. 1998).

Transmembrane protein identification begins with interactions of the extending, nascent chain with SRP (Gorlich et al. 1990). SRP complexes with the ribosome near the exit tunnel and binds hydrophobic polypeptide signal sequences (Johnson et al. 2001; Flanagan et al. 2003; Bornemann et al. 2008). This binding results in translational pausing, through allosteric interactions in the ribosome, and directs the ribosome-nascent chain complex to the ER via binding to the SRP receptor and the translocon (Moller et al. 1998; Song et al. 2000; Lakkaraju et al. 2008; Mary et al.

2010; Noriega et al. 2014). Once bound to the translocon, SRP dissociates and the transmembrane sequences are transferred to the translocon. Translation inhibition is then relieved and the nascent chain continues its extension.

The translocon channel facilitates a complex series of events that includes the translocation of transmembrane sequence through the channel, assembly of these sequences into membrane-compatible structures and integration of these structures into the ER membrane (Crowley et al. 1993; Crowley et al. 1994; Sadlish et al. 2005; Pitonzo et al. 2009; Devaraneni et al. 2011). The first step involves interactions of the nascent chain with the aqueous translocon pore and residues lining the pore of the translocon (Crowley et al. 1994). Subsequent to this, the TM-spanning sequences are transferred through structural rearrangement of the translocon to either docking sites that can temporarily hold the TM spans in the translocon complex or directly into the membrane (Do et al. 1996). Transfer directly to the membrane or transient holding by the translocon depends on the specific sequence of the TM spans and the topology of the protein (Sadlish et al. 2005; Pitonzo et al. 2009; Devaraneni et al. 2011). Single TM span proteins are often transferred directly into the membrane environment. These spans tend to be formed almost exclusively by hydrophobic residues, so the free energy transfer into the lipid bilayer is favorable (Chan and Dill 1997). Proteins with multiple TM spans can be held transiently by the translocon while waiting for interacting sequences to be translated and inserted (Sadlish et al. 2005). This is often required to pair charged resides and/or to stabilize secondary structures before transfer to the lipid bilayer. The ability to neutralize and mask charged and polar residues in multispan transmembrane proteins allows for additional sequence complexity in the transmembrane sequences but requires the translocon to serve as a temporary holding site for unpaired residues (Sadlish et al. 2005; Pitonzo et al. 2009).

Extracellular and cytosolic domains are often found in transmembrane proteins and are also recognized by the ribosome–translocon complex. Extracellular domains are inserted through the translocon pore and into the ER (McCormick et al. 2003). The specific interactions between the translocon and these sequences vary from those of the TM-spanning sequences and facilitates their secretion into the lumen of the ER. In contrast, cytosolic domains require partial undocking of the ribosome from the translocon to allow those domains to be translated and fold in the cytosol. The mechanisms of this recognition are not fully elucidated, but are likely associated with changes in nascent chain hydrophobic properties and secondary structure.

6.1.5   Transmembrane Protein Sequences

The requirement that a protein span the lipid bilayer puts a unique and fundamental constraint on the evolution of the protein's primary sequence. Specifically, the favorable interaction of the polypeptide chain with the phospholipid bilayer generally restricts the amino acids that are found in transmembrane-spanning structures (Rost et al. 1995; Bigelow and Rost 2009). The majority of mammalian transmembrane proteins are thought to span the bilayer using helical structures (Reeb et al. 2014). These helices are enriched with hydrophobic amino acids and are generally composed of between 16 and 25 amino acids (Reeb et al. 2014). The shorter helices likely adopt a conformation that is normal to the plane of the bilayer. Longer helices may adopt conformations that cross the membrane at obtuse angles, requiring additional helical turns to span the bilayer.

The hydrophobicity of these sequences is likely the key determinant for their identification as membrane-spanning structures and their insertion into the lipid bilayer (Eisenberg et al. 1984). While any of the hydrophobic residues may be accommodated in the lipid bilayer, the distribution of amino acids in these sequences shows trends that provide additional specificity for the protein–membrane interaction. Two principle ideas emerge from statistical and energetic studies of model peptides and transmembrane proteins. First, the membrane bilayer is not uniform with respect to amino acid selection (Senes et al. 2001; Lehnert et al. 2004; Baeza-Delgado et al. 2013). The core of the bilayer, composed of the hydrocarbon tails of the lipids, preferentially interacts with small hydrophobic amino acids and the aromatics, Phe and Trp. These interactions likely are the result of strong positive nonpolar interactions between the hydrocarbon and aromatic sidechains with the lipid hydrocarbons. The interfacial regions of the membrane, those areas containing the phospho-head groups and bulk solvent, tend to be enriched in less apolar residues, though charged amino acids are generally disfavored. The subtle shift from hydrophobic to polar likely represents the need to interact favorably with both the hydrocarbon core and the electrostatics of the phospho-head groups. Second, charged residues that are found in the interfacial region of immediately flanking transmembrane spans show a propensity for basic amino acids to be located toward the cytoplasm (Engelman et al. 1986; Hartmann et al. 1989; Gafvelin et al. 1997; Wallin and von Heijne 1998). The enrichment of Lys and Arg residues on the cytoplasmic side of the membrane may facilitate proper topology of transmembrane helices when coupled with membrane potential. The membrane potential across the ER and plasma membranes is strongly negative, with increased negative charge distributed across the cytoplasmic leaves of these membranes. As a result, the propensity of transmembrane spans to include positive residues, likely provides a means to complement the negative charge and provide orientation and anchoring for multiple transmembrane helices.

6.2   Transmembrane Protein Structure

High-resolution structural biology of ion channels and transporters begins with critical discoveries related to X-rays and materials science in the late nineteenth and early twentieth centuries. Today, the growing number of high-resolution membrane protein structures aids in our understanding of protein function, the molecular basis of disease pathophysiologies, and drug discovery and refinement.

6.2.1 The History of Crystallography

In the late nineteenth century, Wilhelm Röntgen began studying what would be termed X-rays, a term that was established to describe their unknown ("X") qualities. For these observations, Röntgen won the Nobel Prize in 1901. The exact nature of this energy was not fully established until the twentieth century, however, experimentation with X-rays led to the imaging of multiple medical, biological, and chemical samples at the end of the nineteenth and early twentieth centuries (Keevil 1896; Glasser 1945). Though this technology was being explored in multiple applications, the real appreciation of its value—and its intrinsic danger—was not fully appreciated for decades after experimentation began.

In 1912, Max von Laue described the diffraction of X-rays by a solid matter crystal (Moffat et al. 1989). This seminal work established the basis for modern X-ray crystallography and is fundamentally unchanged today. Using salt crystals, Laue showed that an incoming light beam was diffracted, much like visible light through a diffraction grating, by a solid matter crystal. Moreover, when the diffracted X-rays were imaged, specific foci were generated with patterns that were dependent on the orientation and structure of the crystal itself. These observations led to the formulation of the Laue equations, which relate the diffraction pattern in reciprocal space to the real-world structure of the object diffracting the X-rays. These equations, which were subsequently reduced to Bragg's law, were central to solving the structures of small inorganic crystals in the early twentieth century and larger and more complicated macromolecules in the latter half of the century. In addition, this work demonstrated that X-rays behaved as waves with very short wavelengths, providing a partial description of the unknown properties of the X-rays described earlier by Röntgen.

In 1913, William Henry Bragg and William Lawrence Bragg, a father and son team, formalized a model that described the interactions between a solid matter crystal and X-rays (Bragg 1956). The physical interactions between the electrons with the incoming X-ray beam resulted in scattering and, depending on the wavelength of the incoming beam and the spacing between atoms in the crystal, produced constructive or destructive interference. The diffracted waves that were in phase to produce constructive interference could be detected as discrete "spots," and are the diffraction pattern that underlies the process of structure determination by crystallography (Fig. 6.3). The Laue equations and Bragg's law related the constructive interference seen to the spacing of the objects within the crystal. Thus, diffraction patterns collected from a sample could be used to back-calculate the foci of diffraction within the crystal.

Together, the efforts of the Braggs and Laue established the theoretical basis for the use of diffraction to determine special relationships in solid crystals. Laue was awarded the Nobel Prize in Physics in 1914 for his characterization of X-ray diffraction in solid matter crystals. For their theoretical contributions and the characterization of inorganic crystals and diamonds, the Braggs were awarded the Nobel Prize in Physics in 1915. Though several father–son pairs have separately won

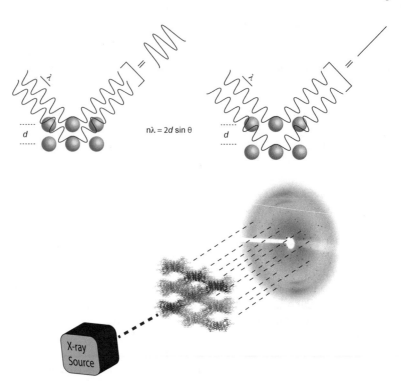

Fig. 6.3 X-ray diffraction by atoms in a crystal. A schematic showing the relationship between the light-source and diffracting atoms in a crystal is shown. The constructive interference from precisely spaced atoms results in "spots" in the diffraction image. Bragg's Law relates the incoming light beam and the space between atoms in an ordered crystal with the outgoing constructive interference as: n is an integer scalar, λ is the wavelength of the incoming light, d is the distance between planes in the crystal, and Θ is the angle between the incident ray and the diffracting crystal planes. The constructive interference is detected in diffraction experiments and the intensity and positions of the diffraction pattern are utilized to back-calculate the positions of atoms in the crystal. The generation of constructive and destructive interference based on the wavelength of incoming light and crystal planes is depicted, *top*. A schematic of the experimental setup is shown, *bottom*, with the experimental crystal lattice and diffraction image from the structure of ABCC6 NBD2 (REF)

Nobel Prizes, the Braggs are the only father and son to share a Noble Prize. William Lawrence Bragg, at the age of 25, remains the youngest winner of a scientific Nobel Prize to date.

6.2.2 Biological Application of X-Ray Diffraction

While these studies of the Braggs and Laue initially focused on small inorganic and organic compounds, the application of these approaches resulted in the first structures of biological macromolecules in the mid-twentieth century. The structure of DNA was solved using diffraction methodologies in 1953 by Watson and Crick and, in many ways, represents the beginning of modern molecular and structural biology (Watson and Crick 1953). In 1958 the first structure of a protein, whale myoglobin, was solved using X-ray crystallography (Kendrew et al. 1958). This structure provided the first detailed insights into the organization of secondary and tertiary structure in a folded protein and laid the groundwork for ~60 years of structural biology. Watson and Crick shared the 1962 Noble Prize in Physiology or Medicine with Maurice Wilkins. John Kendrew shared the Nobel Prize in Chemistry in 1962 with Max Perutz for structural studies on globular proteins and their organization.

The fundamental processes of protein structure determination remain largely unchanged today, though X-ray crystallography is now complemented by nuclear magnetic resonance (NMR), cryo-electron microscopy (cryo-EM) and other diffraction approaches (electron and neutron) for structure determination. Recent advances in cryo-EM technology have allowed for the rapid determination of many new membrane protein structures and remove some of the most significant limitations associated with crystallographic studies.

In 1985, the first structure of a membrane protein was solved by X-ray crystallography. Prior work had focused on soluble proteins, in large part, because membrane proteins were technically challenging to express and purify from their cellular membranes. In the early 1980s, Hartmut Michael demonstrated that the photosynthetic reaction complex from *Rhodopseudomonas viridis* could be purified and crystallized using a combination of detergents added during its extraction and purification (Michel 1982). Subsequent to this, Johann Deisenhofer, Robert Huber, and Hartmut Michael solved its structure—the first transmembrane crystal structure—at 3 Å resolution (Deisenhofer et al. 1985). This work showed the exquisite topology and orientation of the transmembrane and cytosolic components of the photosynthetic reaction center, providing insights into ATP generation and cellular energy homeostasis. This effort, which established the field of membrane protein crystallography, was acknowledged with the 1988 Nobel Prize in Chemistry awarded to Deisenhofer, Huber, and Michael.

6.2.3 Approaches to Membrane Protein Structure Determination

Since 1959, the number of crystal structures has grown tremendously. Currently, there are over 160,000 protein structures representing approximately 35,000 proteins in the Protein Data Bank (PDB), the central repository for protein structures (pdb. org). These include proteins from all kingdoms of life, representing a wide variety of enzymatic and structural functions. The rapid expansion of the PDB was in part the result of large-scale structural biology programs designed to define "fold-space," the range of protein folds that could be generated naturally (Skolnick et al. 2000). To date, there are roughly 1300–1500 different protein folds identified in the PDB, depending on classification methods used in the definition of these folds (Andreeva et al. 2008; Dawson et al. 2017). The majority of these are soluble, globular proteins, though multiple classifications of membrane protein topologies and fold organizations have been established with the growing number of available structures.

Since 1985, the number of transmembrane proteins has similarly grown (White 2009). There are over 950 unique transmembrane protein structures in the PDB, with a total of over 2800 structure files (pdb.org ; Structure). Until recently, the majority of high-resolution protein structures was solved by X-ray crystallography. Recent advances in cryo-EM provide alternatives for the determination of high-resolution membrane protein structures using diffraction methodologies (Cheng 2018).

6.2.4 X-Ray Determination of Membrane Protein Structures

Though technologies have improved, several fundamental challenges remain for membrane protein crystallographers. The chief difficulties are the expression and purification of multiple milligrams of homogenous, pure protein and the identification of conditions that produce crystals that diffract to high resolution.

The first challenge—that of protein expression and purification—has been addressed by the development of various recombinant technologies that allow for protein overexpression in prokaryotic and eukaryotic systems (Mus-Veteau 2010; Xiao et al. 2010; Young et al. 2012; Almo et al. 2013; Gileadi 2017; Baghban et al. 2019). Recombinant expression strategies and efficient site-directed mutagenesis have been critical for screening protein variants with specific thermodynamic and/or functional properties (Price and Nagai 1995; Lewis et al. 2005). In addition, the incorporation of purification tags facilitates the rapid capture of full-length proteins using efficient and economical purification technologies. Protein tags paired with specific and high-affinity column resins and fluorescent fusion reporters facilitate the screening and optimization of purification strategies (Kawate and Gouaux 2006; Xu et al. 2019).

Recombinant overexpression technologies have advanced to facilitate the generation of large quantities of biomass and expressed protein. Prokaryotic systems generally have distinct advantages over eukaryotic systems with respect to time and cost (Gileadi 2017). Cell mass can be readily generated in low-cost bacterial media, usually over the course of 1–2 days. However, particularly with eukaryotic transmembrane proteins, bacterial systems tend to exhibit problems with poor cell growth and protein expression (Mus-Veteau 2010). These problems may relate to differences in the membrane protein biosynthetic pathways, chaperone systems, membrane lipid composition, and functional regulation of a heterologous membrane protein.

Eukaryotic expression systems provide a more native-like environment for the production of mammalian membrane proteins (Baghban et al. 2019). Multiple yeast models, including *Saccharomyces cerevisiae* and *Pichia pastoris* have been utilized for transmembrane protein expression, as have mammalian cells grown in suspension. Large-scale yeast and mammalian cultures can be generated using fermenters and bioreactors, however, growth occurs over an extended period. In addition, medium costs, particularly for the growth of mammalian cells can be expensive when compared to the costs of bacterial medium. The added costs and time related to biomass generation in eukaryotic systems often limit their use.

The field of structural genomics has evolved, in part, as a result of difficulties in producing large quantities of transmembrane proteins (Skolnick et al. 2000). In this strategy, homologs of mammalian proteins of interest are expressed in prokaryotic systems and those that express well and can be purified efficiently are selected for downstream structural studies. While the primary sequences may differ between homologs, if the evolutionary similarity is sufficiently high, the structures of the homologs can serve as good models of the mammalian protein. Though atomic details may differ slightly between evolutionarily related proteins, homologous structures provide the basis for our understanding of many transmembrane channels, exchangers, and transporters.

The second major challenge is the identification of conditions that promote protein crystallization with adequate diffraction properties. At its heart, the process of crystallization remains empirical (Skarina et al. 2014). Often this involves screening through thousands of conditions, altering pH, salt, detergents, ligands, additives, temperature, and protein concentration to identify conditions that facilitate crystal formation and growth at high protein concentrations. This has traditionally required 10–100 s of milligrams of purified protein as proteins tend to crystallize when highly concentrated (5–10+ mg/ml purified protein). Liquid handling technologies have reduced the minimum quantities of protein required for crystal screening by 10–100-fold. Reducing the volumes of protein required from several microliters to 10–100 nanoliters per condition allows large-scale crystal trials to be attempted with smaller quantities of highly purified samples. Similarly, automated housing and handling robotics provide precise control of the conditions that support crystal growth and are capable of rapidly screening through thousands of crystallization conditions with minimal human intervention. However, the expression and purification of large quantities of membrane protein remain a significant barrier for diffraction studies.

Once crystal conditions have been identified and crystals produced with adequate size and diffraction qualities, the use of high-energy light sources and very high-quality detectors enables the acquisition of the highest resolution data. Using a combination of near ideal protein crystals and high-energy synchrotron X-ray sources, the highest resolution protein structures have been solved to sub-1 Å, sufficient to clearly visualize the sidechain rings of an aromatic amino acid sidechain and the void in the ring itself (Rosenbaum et al. 2015).

For pharmaceutical and academic applications, the data collection and acquisition systems are increasingly being automated with robotics controlling all aspects of crystal handling, diffraction screening, and data collection. Combining the increased quality of the source X-rays with high resolution and high sensitivity detection equipment allows for smaller crystals to be used for diffraction experiments, facilitating structure determination using less protein and for proteins that otherwise do not crystalize well (Rosenbaum et al. 2015).

New technologies have also emerged that allow for the collection of diffraction data from nanocrystals. Traditionally, protein crystals are grown such that the crystals are >10–50 μm in every dimension. With higher energy light sources and better detection systems, data from smaller crystals can be collected and used for structure determination. Recent work with electron free lasers has demonstrated that macroscopic crystals (>1–3 μm in any given dimension) may not be required for structure determination (Hunter and Fromme 2011; Boutet et al. 2012; Liu et al. 2013b). The first demonstration of this technology showed that microscopic crystals could be sprayed, in suspension, through a laser source and used for structure determination. These crystals were destroyed by the energy of the laser beam, but only after they diffracted a small portion of the incident beam. Data could be collected on the stream of crystals and integrated to determine the diffraction patterns. At the time of the first edition of this book, this approach was still in its infancy with only a small number of facilities existing with these light sources. While this methodology has continued to develop, advances in cryo-EM technologies have moved microscopy to the forefront of membrane protein structural biology with lower financial and physical costs for researchers.

6.2.5 Cryo-electron Microscopy Determination of Membrane Protein Structures

Major advances in cryo-EM technologies have propelled this technique to the forefront of membrane protein structure determination since the first edition of this book. These cryo-EM technologies have powered a so-called "Resolution Revolution," enabling the determination of protein structures to resolutions that have previously been attained only by using X-ray crystallography (<2 Å) (Kuhlbrandt 2014a, b). Prior to these advancements, single particle EM often led to maps that were limited to resolutions lower than 10 Å. These maps were sufficient to see gross topological features, domain organization, and oligomerization status, but failed to reveal fine-grain details usually needed to understand protein mechanochemistry. The development of new methods and technologies has dramatically increased cryo-EM structure resolution and opens the doors to structural studies otherwise limited by diffraction (X-ray) studies.

Fundamentally, the cryo-EM experiment follows the same workflow as in the past (Fig. 6.4). Highly purified protein samples are spotted onto small screens that are rapidly frozen to generate a thin layer of ice. The protein samples are effectively immobilized within the ice in more-or-less random orientations. Projection images of the protein samples are then acquired using a transmission electron microscope (TEM). These particles are subsequently binned based on their orientation. With a

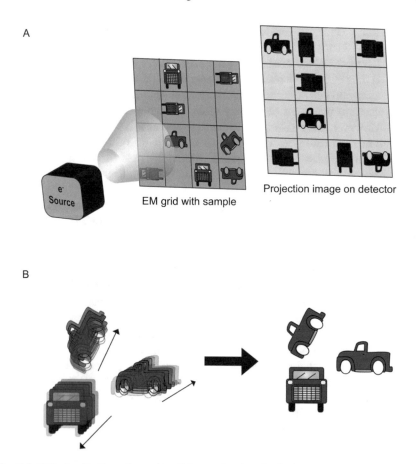

Fig. 6.4 EM visualization of proteins. Schema showing the basic concepts of EM structure determination are shown. (**a**) A cartoon showing image acquisition is shown. A TEM is used to acquire projection images of single particles frozen on a sample grid. If sufficient numbers and orientations of particles are collected, a three-dimensional reconstruction of the original object can be generated. (**b**) High-resolution reconstructions are built from movies collected on the TEM. Motion induced by the electron beam can be identified in the individual frames of the recorded movie. Software analysis identifies the vectors of motion and corrects for the motion in the movie frames. This method effectively increases the integration time of the images while simultaneously reducing the resolution loss associated with particle motion in the ice. The corrected images are binned according to their orientation and maps are electron density maps are then generated for model building

sufficient number of particle orientations in these two dimensions projection images, a three-dimensional reconstruction of the proteins particle can be generated (Scheres 2012a, b; Punjani et al. 2017).

Two major technological breakthroughs have powered this cryo-EM revolution. First, direct detector technology has been developed for the cameras used to capture the projection images of the sample (Kuhlbrandt 2014a, b). Prior to the development of these direct detectors, EM images were collected indirectly, using a process that converted the incoming electron to a photon that was then imaged using film or a camera. The transmitted electrons would first interact with a scintillating film that converted the electron to a photon. The generated light would then be transmitted to film or a conventional detector (camera). This conversion was required because high energy electrons could not be directly visualized without excessive damage to the camera. However, this conversion introduced noise and limited resolution during the scintillation conversion. Direct detectors are capable of imaging the incoming electron images directly onto the camera without the need for a scintillation screen or other electron-to-photon conversion, removing a significant source of noise and major resolution limit.

Second, computational tools for the acquisition and refinement of data have been developed to filter and process images for structure determination (Scheres 2012a, b; Punjani et al. 2017). Data acquisition is first improved by taking short movies of the EM grid, instead of long, single-exposure images (Zheng et al. 2017). By collecting movies, motion induced by the electron beam can be corrected. Short exposure images do not provide sufficient contrast to reveal the structural details of the sample. Long exposure images capture blur associated with particle movement in the ice over the time course of the exposure. To circumvent these problems, movies of the particles are captured. Software processing after image acquisition then identities the motion of the individual particles and corrects for this motion. This image processing technique removes the blur normally captured as particles move during a single long-exposure image and integrate enough signal to capture high-resolution structural data. The processed movies effectively provide a long exposure (integration) image that has been refocused to minimize the effects of particle movement within the ice (Fig. 6.4).

These images must then be binned based on particle orientation and alignment. Data processing tools for the identification and binning of 2D classes have also developed (Scheres 2012b; Punjani et al. 2017). Automated particle picking, which identifies the individual proteins on an image plate, and binning, which classifies the orientations based on discrete structural features, facilitates the analysis of thousands of images and hundreds of thousands of particles. Initially, these processes were curated manually and required a significant amount of human oversight. Software analysis tools have sufficiently improved such that particle picking and binning is accomplished automatically with limited curation. Iterative analyses facilitate picking particles in multiple structural conformations and oligomeric states, allowing, in some cases, for multiple structures to be solved from single sample preparation (i.e., open and closed states) (Scheres 2016; Hofmann et al. 2019). The advances in software and computing technologies have decreased the time required for these analysis and processing steps from hours to days to near-real time. Once 2D classification and binning are accomplished, 3D maps are generated and used to model the protein chain. These structure refinement and analyses approaches are still being developed based, in part, on software packages developed for crystallography.

The development of these cryo-EM technologies circumvents several of the major limitations associated with X-ray crystallography. First, because cryo-EM does not require empirical screening for conditions that promote protein crystallization, the quantity of protein required to determine an EM structure is generally reduced compared to that required for crystallographic studies. The generation of purified protein is often a major rate-limiting step in crystallographic experiments. Membrane protein expression is often limited by low yield and extraction into detergent-containing buffers is expensive. As most conditions (perhaps >99%) will not generate protein crystals, the majority of the protein is lost during screening. In contrast, cryo-EM does not require this screening step and is accomplished in a relatively dilute solution. Thus, even small quantities of purified protein, which would not be sufficient for a crystallization screen, have the potential to yield structures using cryo-EM.

Another exciting potential for cryo-EM is the direct observation of multiple protein conformations (Scheres 2016; Hofmann et al. 2019). By definition, a crystal is a regularly ordered matrix of its constituent parts. In most cases, the formation and propagation of a crystal nucleus require that all of the proteins be the same size and shape. This generally precludes the capture of multiple protein conformations (though in rare cases, these are observed in X-ray studies). Structural heterogeneity also limits diffraction resolution, as the constructive interference needed for detection is dependent on highly-ordered, regularly spaced repeating units within the crystal. With cryo-EM, the ability to collect data does not require this highly ordered assembly. Thus, multiple conformations of the protein may exist on a single grid. Data acquisition and processing can be tailored to identify these different conformations, effectively searching for multiple structures within the same experiment. The ability to distinguish and reconstruct multiple conformations has the potential to facilitate studies on protein function and/or the impact of mutations.

6.3 Biological Insights Derived from Transmembrane Protein Structures

Given the complexity and technical challenges associated with transmembrane protein structure determination, there are often only a few representative structures for any given transmembrane protein family. However, a growing number of membrane protein families have been solved in multiple structural states, providing insight into mechanochemical cycles and conformational dynamics. The growing list of transmembrane protein structures is too long to fully cover in the context of this work.

ATP-binding cassette transporter (ABC transporters) structures have been solved by both X-ray and cryo-EM methods. The first structures of these proteins' cytosolic domains provided mechanistic insight into the mechano-chemistry associated with ATP binding and hydrolysis, while structures of the full-length proteins now provide insight into the physical mechanisms by which solutes are pumped across membranes. This section will outline the structures that underlie our understanding of the

ABC family of proteins and contribute to our understanding of its biosynthesis, function, and disease development, with an emphasis on CFTR.

6.3.1 ABC Transporters

The ABC transporter family of proteins is an evolutionarily conserved family of proteins that facilitate the import and export of solutes across the membranes of cells and organelles (Dean et al. 2001; Albers et al. 2004; Davidson and Chen 2004) (Fig. 6.5). These proteins are abundant in archaea, eubacteria, and eukaryotes. ABC transporters are composed of at least two core domains: a nucleotide-binding

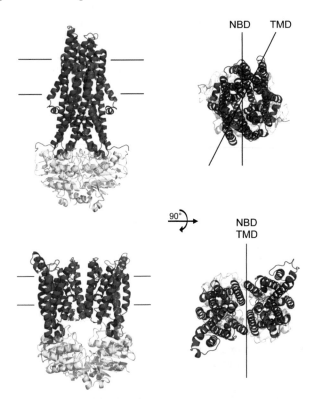

Fig. 6.5 Representative ABC transporter structures. A ribbon drawing showing the full-length structures of ABC transporters is shown. *Top*, the structure of Sav1866 is shown with the TMDs colored in blue and the NBDs colored in yellow. The TMDs are composed of six TM helices in each protomer. The axes of symmetry for both the TMD and NBD dimers are shown and are offset from one another. *Bottom*, the structure of BtuCD is shown and colored as for Sav1866. The TMDs of BtuC are composed of ten TM helices in each protomer and the axes of symmetry are roughly aligned in the BtuCD transporter. The independent evolution of the TMDs facilitates an expansion of function in prokaryotes. Mammalian ABC transporters putatively function exclusively as exporters and share the TMD organization seen in Sav1866

domain (NBD)—or ATP-binding cassette for which the family is named—and a transmembrane domain (TMD). These domains are expressed and function as dimers in the active protein, which is minimally composed of two NBDs and two TMDs. In addition, accessory domains are associated with a variety of ABC transporters and may include larger intracellular or extracellular globular domains or additional transmembrane domains (Dean et al. 2001).

While the basic protein architecture is well maintained in the ABC-transport proteins, the functional complex may be formed by association of one or more individual polypeptide chains (Dean et al. 2001; Albers et al. 2004; Davidson and Chen 2004). In many mammalian transporters, the complete ABC transporter is encoded by a single gene, which encodes both of the NBDs and TMDs. In contrast, multiple genes encode many bacterial transporters, each producing one or more of the individual domains that must assemble to form the functional tetrameric complex. Other transport systems are encoded as half-transporters with a single NBD and TMD on a polypeptide that then dimerizes to form the functional complex. These "half-transporter" molecules have been shown to homo- and hetero-dimerize, contributing to the evolutionary diversity of the family.

6.3.2 NBD Structure

The NBDs are functionally and structurally conserved domains across the ABC transporter protein family. Their function is highly conserved and couples the energy of ATP-binding and hydrolysis to domain dimerization and conformational movement within the TMDs (Carson and Welsh 1995; Bianchet et al. 1997; Senior and Gadsby 1997). The sequence and structural similarity are highly conserved within the NBDs, while the TMD components have evolved to recognize and transport a large number of solutes.

The core of the NBDs is composed of approximately 230 amino acids and contains the canonical nucleotide-binding sequences (Walker A and B) and the transport signature motif, sometimes referred to as the Walker C sequence (Schneider and Hunke 1998; Liu et al. 1999; Cui et al. 2001; Loo et al. 2002; Ambudkar et al. 2006; Buchaklian and Klug 2006). The Walker A (GxxGxKST) and Walker B (RX_{6-8}hydrophobic$_4$DE) sequences are associated with ATP binding and hydrolysis. These sequences are more generally conserved among other nucleotide-binding proteins. The transport signature motif (LSGGQ) is more specific to the ABC transporter family. These sequences represent the hallmark signatures of the NBDs, though other conserved elements (Q-loop, H-loop) are conserved throughout the family.

The structural organization of the NBDs is highly conserved (Fig. 6.6). Structures of multiple NBDs from a variety of bacterial and mammalian transporters have been solved and demonstrate that the core structural elements of these domains are highly similar (Hung et al. 1998; Diederichs et al. 2000; Yuan et al. 2001). The structures of the ABC transporter NBDs consist of two subdomains: a mixed α/β-core and an

A

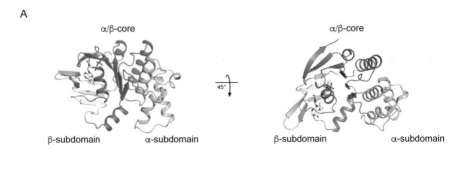

B

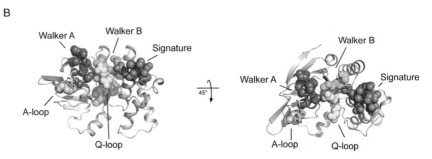

Fig. 6.6 ABC transporter NBD structure. A ribbon drawing of a bacterial nucleotide-binding domain from *M. jannaschii* is shown. The Mj0796 NBD fold is prototypical for the family of ABC transporters and is composed of three subdomains that undergo a small rigid body rotation upon nucleotide binding. Two views are shown, rotated by 45°. The nucleotide is shown in stick representation bound to the NBD monomer. (**a**) The NBD subdomains are colored: goldenrod, β-subdomain; α/β-core, slate; α-helical subdomain, green. (**b**) The critical ATP-associated sequences are shown in spheres and colored: A-loop, wheat; Walker A, slate; Q-loop, goldenrod; Signature sequence, magenta; Walker C, green

α-helical subdomain (Hung et al. 1998; Karpowich et al. 2001; Yuan et al. 2001). The α/β-core is a Rossman fold with high structural similarity to the RecA and F_1-core ATP-binding domains. The core contains both the Walker A and B sequences.

Nucleotides bind to the core of the domain *via* interactions between the Walker A and B sequences and the phosphate groups (Karpowich et al. 2001; Yuan et al. 2001). Additional ring–stacking interactions with an aromatic residue at the end of the first β-strand and the adenine ring further stabilize the bound ATP. ATP binding at these sites occurs on the surface of the NBD and is not followed by nucleotide hydrolysis (Hung et al. 1998; Moody et al. 2002). The ATP bound at this "half-site" requires active site completion by complementation with the LSGGQ from the second NBD protomer in the dimer (Smith et al. 2002). Only after NBD dimerization is the necessary elements present and arranged to support nucleotide hydrolysis. Hydrolysis and release of the γ-phosphate, alters the electrostatics of the nucleotides bound by the dimeric NBDs. This change in electrostatic potential is thought to cause dissociation of the dimer and allows for the release of the ADP.

The α-helical subdomain is divergent in both sequence and structural properties. This subdomain includes the LSGGQ sequence and is critical for the association of the NBDs with the TMDs in the functional transporter (Wilken et al. 1996; Locher et al. 2002; Pinkett et al. 2007). A flexible hinge connects the α-helical subdomain to the core domain and is termed the Q-loop for a highly conserved glutamine residue. ATP binding and hydrolysis induce a rigid body movement of the two subdomains via interactions with the Q-loop. This movement is thought to propagate into the TMDs and facilitate their structural rearrangement during the transport cycle (Karpowich et al. 2001; Locher 2004).

6.3.3   NBD–NBD Dimerization and Function

Initial studies of isolated NBDs suggested that ATPase activity required the dimerization of NBDs (Nikaido and Ames 1999; Hopfner et al. 2000; Moody et al. 2002; Smith et al. 2002; Chen et al. 2003; Zaitseva et al. 2005). Sequence analysis and mutagenesis suggested that both the Walker A and B sites, as well as the LSGGQ sequence were required for ATP hydrolysis and were physically close during the ATPase reaction cycle. However, the initial structures of monomeric NBDs revealed that these sites were separated by 15–25 Å and likely could not coordinate ATP in a monomeric state. The structure of the Mj0796 dimer, solved in 2002, was the first to show the biologically relevant dimerization events that underlie ABC-protein function (Fig. 6.7) (Smith et al. 2002). This structure was enabled by the mutation of the Walker B catalytic glutamate to a glutamine, rendering the NBD inactive and trapped in an ATP-bound, dimeric form. This NBD dimer showed the physical juxtaposition of the Walker A/B half-site from one monomer with the LSGGQ half-site of the opposing monomer. The ATP molecules were fully coordinated and buried within the NBD dimer, providing for the necessary chemistry for ATP hydrolysis.

Given the symmetry of the NBD dimers, two fully coordinated ATP molecules are bound by the juxtaposed Walker A and B sequences in the ATP-bound NBD dimer. Though ATP binding is required in at least one site to drive the formation of NBD dimers and transporter function, sequence analysis, and biochemical studies show that the hydrolytic function is not strictly required (Carson and Welsh 1995; Senior and Gadsby 1997; Nikaido and Ames 1999; Lewis et al. 2004, 2005). Multiple studies have shown that transport function minimally requires a single active ATP site and at least one family member, CFTR, has lost ATPase activity at one of the two NBD ATP-binding sites (Vergani et al. 2005).

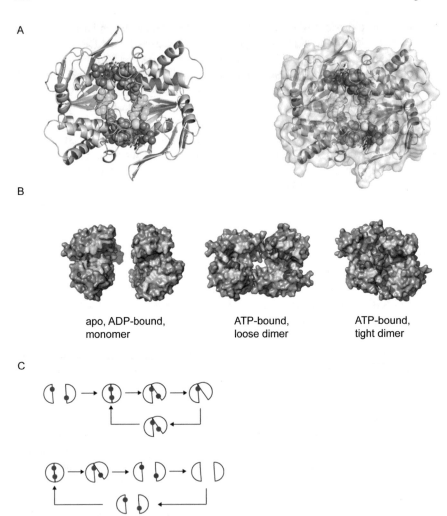

Fig. 6.7 ATP-driven NBD dimerization. A ribbon drawing of the ATP-driven NBD dimer is shown from Mj0796 in ribbon and surface renderings with the two bound nucleotides. The nucleotide-associated sequences are colored as: aromatic loop, *red*; Walker A, *green*; Walker B, *magenta*; and LSGGQ/Signature sequence, *blue*. Nucleotide-binding drives association of the NBD protomers and results in the formation of two complete active ATPase sites

6.3.4 TMD Structure

The structures of the TMDs are considerably more diverse than those of the NBDs. The TMD evolution likely results from the need to bind and transport a structurally diverse set of solutes. The TMDs are minimally composed of six transmembrane helices, though the numbers of helices increase in many bacterial transport systems (Dean et al. 2001). These domains dimerize in order to form the membrane-spanning

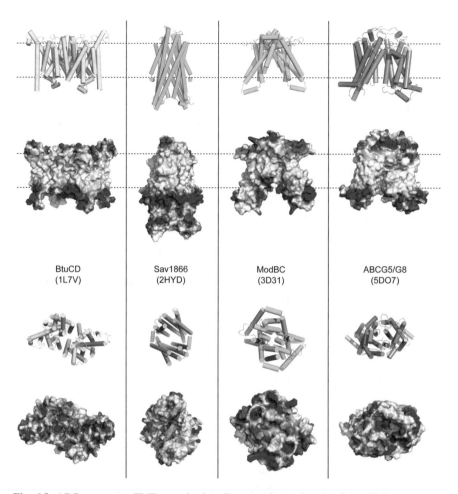

Fig. 6.8 ABC transporter TMD organization. Transmembrane domains from ABC transporters have evolved to accommodate and transport a large variety of substrates. Multiple structural classes of these domains have been identified with varying numbers of TM helices and topologies. Transmembrane domains from several representative structural classes are shown, including the vitamin B12 importer from *E. coli* (BtuCD), Sav1866 from *S. aureaus*, the molybdate importer from *M. acetivorans* (ModBC), and the human sterol transporter (ABCG5/G8). Views both from within the membrane, *top*, and from the extracellular side of the membrane, *bottom*, are shown as cartoon and surface electrostatics. The approximate position of the membrane is indicated by the dashed line in the top images. PDB codes for the structures are shown beneath the protein names

structures that facilitate solute transport. Less is known about the structural and functional rearrangements in the TMDs than in the NBDs. Structures of the full-length ABC transporters provided evidence that the dimerization of the NBDs leads to the reorganization of the TMDs (Fig. 6.8) (Dawson and Locher 2006, 2007; Jin et al. 2012; Shintre et al. 2013). The TMDs are generally found in one of two orientations. In an ATP-bound form, the NBDs form a tight dimer and the TMDs open into a vestibule on the outer leaflet of the membrane. This presumably would represent the low-affinity conformation associated with solute release in the exporters. In the second state, the NBDs are physically separated from one another and the TMDs have reoriented to face the inner leaflet or cytoplasm. These conformational changes are large and have not been established in a biological or in vivo system.

Though this alternating access scheme likely represents the physical basis of the mammalian ABC transporters, the magnitude of this movement may vary from that seen in vitro structurally. Similarly, the activity of the bacterial ABC importers likely follows a similar general alternating access scheme, but has not been fully elucidated structurally. This model for alternating access is likely similar for multiple mammalian transporters and cotransporters as the conformational changes within the protein shield the substrate from interactions with the hydrophobic membrane (Shimamura et al. 2010; Zhou et al. 2014).

Little is known about the nature of solute binding in these domains, though the recent structures of Pgp and molecular modeling of ABCC6 provide some insight into solute recognition and selection (Aller et al. 2009; Hosen et al. 2014). In both the structures and the molecular models, solute binding to the TMDs appears within the plane of the membrane. This could be readily accomplished by diffusion into the binding site by hydrophobic molecules found in the inner leaflet, as might be predicted for drug export, or through diffusion from the cytoplasm, as might be predicted for hydrophilic solutes. The exact mechanisms of solute binding are currently contentious and depend, in part, on the models used and the solutes chosen.

6.3.5 TMD–NBD Interactions

The interface between the TMDs and the NBDs is a critical region for the structures of the ABC proteins (Wilken et al. 1996; Lewis et al. 2004; Thibodeau et al. 2005). The NBDs contact the TMDs via extended intracellular loops, which have been named the intracellular domain (ICDs). The extension of the transmembrane helices includes short helices that are perpendicular to the transmembrane components, which have been named "coupling helices" (Fig. 6.9). Biochemical studies of chimeric ABC transporter molecules demonstrate that the α-helical subdomain in the NBD makes contacts with the loop sequences from the TMDs, providing specificity for the NBD–TMD interactions (Wilken et al. 1996). These contacts are critical for both biosynthesis and function (Du et al. 2005; Thibodeau et al. 2005, 2010; He et al. 2010).

Disruption of this interface by mutation is common across multiple ABC transporter family members (Fulop et al. 2009; Pomozi et al. 2014). The most studied mutant in this region is the F508del mutant in CFTR, which is found in >90% of all patients with cystic fibrosis (Rommens et al. 1989; Drumm et al. 1990; Cheng et al. 1990). This mutation is located in the N-terminal NBD and the sidechain of the Phe508 residue makes direct contact with intracellular loops from

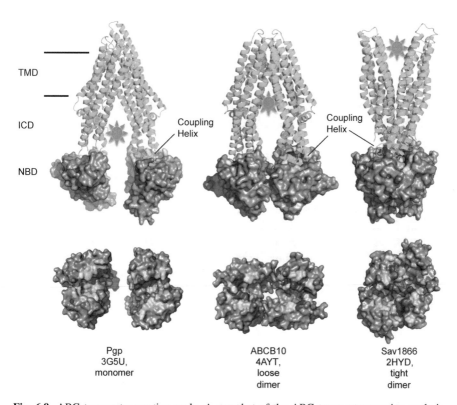

Fig. 6.9 ABC transporter reaction cycle. A snapshot of the ABC transporter reaction cycle is shown composed of the ATP-free Pgp, and ATP-bound ABCB10 and Sav1866 structures. The full-length transporters are shown with green cartoons for the transmembrane domains and surface representations of the nucleotide-binding domains. The region of substrate binding is shown with an orange star. The Sav1866 transporter is rotated 90° about the vertical axis to show the extracellular vestibule. A heat map, corresponding to the contacts in the NBD–TMD interaction is shown colored on the NBD surface and shows the dynamic structural changes associated with the catalytic and transport cycles

the TMDs. Alterations at this site affect the local structure of the NBD by reducing its thermodynamic stability and folding efficiency and alter the interactions of the NBD with the TMD (He et al. 2010; Thibodeau et al. 2010; Lukacs and Verkman 2012; Mendoza et al. 2012). The combined effect of NBD misfolding and a loss of appropriate NBD–TMD interactions lead to the misfolding of the full-length protein in vivo. Full rescue of CFTR misfolding has been proposed to require the restoration of NBD structure *and* NBD–TMD assembly.

Though only partially characterized, this conserved NBD–TMD interface appears to be a structural "hotspot" for other ABC transporter mutations (Fulop et al. 2009). The large number of disease-causing mutations found within this interface suggests that disruption of the NBD–TMD interaction is deleterious across the family of ABC proteins. These alterations may be manifest as biosynthetic or functional defects and the loss of functional protein likely explains the molecular pathophysiologies of these diseases.

6.4 Cystic Fibrosis, CFTR Folding and Structure, and Therapeutic Developments

Cystic fibrosis is caused by defects in the cystic fibrosis transmembrane conductance regulator (CFTR), an ABC transporter turned Cl⁻ channel (Riordan et al. 1989; Cheng et al. 1990). The CFTR protein serves to regulate fluid and ion movement across epithelial tissues, including the skin, airway, pancreas, and intestinal tract and is more thoroughly discussed by Sheppard and colleagues in Chapter 16, Volume 3. The loss of the functional protein at the plasma membrane, resulting from changes in protein expression, subcellular localization and trafficking, plasma membrane residence, or channel gating and conductance, results in CF. For further discussions of CFTR, cystic fibrosis and personalized medicine interested readers are referred to Bradbury's Chapter 16 in Volume 3.

6.4.1 CFTR Folding

A number of ABC transporter folding studies have focused on mutations associated with CFTR, most notably F508del. As might be expected, but not required, for a multi-domain protein, folding appears to follow a hierarchical pathway wherein the individual domains must fold and acquire native or native-like structure before assembly into a quaternary complex (Lukacs et al. 1994; Sharma et al. 2001; Du et al. 2005; Thibodeau et al. 2010). The folding and assembly steps may be energetically coupled, but mutations appear to show discrete changes in domain structure and/or assembly and thus act at multiple steps in a biosynthetic pathway.

6.4.1.1 Full-Length CFTR Folding Studies

The cloning of CFTR in 1989 (Riordan et al. 1989; Rommens et al. 1989) was followed closely by multiple reports indicating that the most common mutation, F508del, resulted in changes to protein biosynthesis and localization within the cell (Cheng et al. 1990). The F508del mutant failed to traffic to the plasma membrane and the loss of Cl⁻ channel activity was causative of disease. Cell biological and cell-based studies of full-length CFTR suggested the F508del variant reached a biosynthetic intermediate, but failed to adopt a fully native conformation. Limited proteolysis studies of CFTR demonstrated that the wild type and mutant proteins both acquired similar protease-resistant features but that the wild-type protein alters these proteolytic patterns in an ATP-dependent fashion (Lukacs et al. 1994; He et al. 2010). In contrast, the F508del protein is arrested in a structural intermediate and fails to adopt ATP-dependent, fully native conformation within the cell.

Studies of CFTR–chaperone interactions also suggested that several critical quality control systems showed altered association with the F508del protein, presumably by recognizing the nonnative conformation in which F508del was arrested (Amaral 2004; Riordan 2005; Nieddu et al. 2013; Ahner et al. 2013; Fukuda and Okiyoneda 2018). Multiple cytosolic and ER luminal chaperone systems playing key roles in CFTR biosynthesis have been identified, including Hsc/p70, 90, 40, and calnexin, as well as co-chaperones for these systems. Many of these systems show altered interactions between the wild type and F508del proteins, consistent with structural changes in F508del that are recognized within the cell. The various chaperone systems either facilitate CFTR maturation or detect its misfolding; failure to adopt a native conformation results in CFTR ubiquitylation and degradation by the proteasome via the ER-associated degradation pathway.

6.4.1.2 Folding Studies of NBD1

Studies of NBD1 in isolation were also developed to evaluate the local effects of F508del on protein structure and energetics (Thomas et al. 1992; Qu and Thomas 1996; Thibodeau et al. 2005, 2010). These studies were initially hampered by protein solubility and an inability to generate properly folded NBD1 protein. These obstacles were overcome, in part, by the dedicated structural biology efforts on NBD1 that lead to its crystallization (Lewis et al. 2004). Multi-milligram quantities of NBD1—wild type and mutant—were eventually generated for biophysical analyses, which revealed an increase in domain dynamics, a decrease in domain stability, and an increase in the formation of nonnative conformers with the F508del variant. These studies all suggested that the F508del mutation directly altered the folding and structure of NBD1. In addition, these changes in protein properties were largely restored by a number of substitutions at the Phe508 locus, suggesting the loss of the backbone was the critical alteration impacting NBD1 locally.

The studies of isolated NBD1, both with disease-causing mutations, engineered mutations, and second-site suppressors, lead to hierarchical folding models for CFTR. (Fig. 6.10) (Du et al. 2005; Thibodeau et al. 2005, 2010; Lukacs and Verkman 2012). The translation and membrane integration of TMD1 was followed by the translation and folding of NBD1. NBD1 adopted a native or near-native conformation, which then could dock onto the intracellular loops/coupling helices of the TMDs. Failure to adopt a native fold could potentially be recognized by cellular quality control, targeting the protein for degradation. Failure of the domain to assemble properly also lead to global changes in CFTR structure that were recognized by cellular quality control systems. Correction of these defects individually by specific second-site suppressors leads to partial restoration of channel biosynthesis and function (Thibodeau et al. 2010). Suppressors found within NBD1 increased its stability and partially restored F508del CFTR trafficking (Teem et al. 1993, 1996;

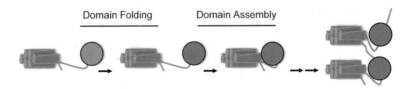

Domain Folding   Domain Assembly

Fig. 6.10 CFTR folding model. A model for the hierarchical folding of an ABC transporter is shown. Biosynthesis is initiated with the translation and membrane insertion of the first TMD. This is followed by the translation and folding of the first NBD. NBD folding to a native, or near-native, state is required to generate the appropriate NBD–TMD interaction surface. The folded NBD then docks against the intracellular loops from the transmembrane domains. Failure of the NBD to properly fold or to dock results in structural defects that propagate across the molecule and are recognized by cellular quality control. Models for the multi-step NBD1 folding and assembly are largely based on studies of CFTR. Two-step folding and assembly has been reported for NBD2 in ABCC6, suggesting late steps in may parallel those seen with NBD1

DeCarvalho et al. 2002; Pissarra et al. 2008). Likewise, suppressors found within the TMDs are thought to rescue the assembly defects between NBD1 and the intracellular loops of the TMDs and also partially restore F508del CFTR trafficking (Thibodeau et al. 2010; Mendoza et al. 2012). Correction of both the folding/stability and assembly defects lead to additively increased F508del trafficking, consistent with a multistep folding model and the correction of multiple steps in the biosynthetic pathway.

6.4.1.3 Folding Rescue and Therapeutic Strategies

Several critical experimental observations led to the idea that the conformational defects associated with F508del might be correctable therapeutically. First, expression of the F508del CFTR at low temperature (27–30 °C) resulted in partial restoration of its trafficking and function at the plasma membrane (Denning et al. 1992). These observations grew out of studies in *Xenopus* oocytes, which were cultured for electrophysiological experiments at room temperature. The partial restoration of F508del CFTR under these conditions demonstrated that the protein could be "rescued" at low temperatures. Moreover, these studies demonstrated this rescue led to the production of partially functional protein that retained some function at 37 °C after its biosynthesis and trafficking to the plasma membrane. These observations indicated the F508del defect was at least partially transient and was most impactful during CFTR biosynthesis.

Second, the addition of osmolytes into cell culture media partially restores CFTR trafficking and function (Sato et al. 1996; Fischer et al. 2001; Zaman et al. 2001; Zhang et al. 2003). These studies arose as mammalian osmolytes are natural products of the cell stress response and their presence is thought to stabilize proteins and facilitate biosynthesis in cells under chronic and acute physiological stress. These small molecule solutes likely induce global changes in protein–chaperone and protein–solvent interactions. Importantly, they provided evidence that an exogenous treatment with small molecules could partially restore the trafficking and function of F508del CFTR. Glycerol, myoinositol, trimethylamine N-oxide (TMAO) and S-nitrosoglutathione have all been shown to partially rescue F508del maturation, promoting limited trafficking, and function at the plasma membrane (Sato et al. 1996; Fischer et al. 2001; Zaman et al. 2001; Zhang et al. 2003).

Finally, second-site suppressors were identified that rescues F508del biosynthetic defects without restoring the Phe508 residue in CFTR (Teem et al. 1993, 1996). These observations provided evidence that the defects could be partially overcome, and channel function partially restored, without global changes to the cell and without the reintroduction of the Phe508 residue. The initial second site suppressors were isolated using a chimeric yeast transporter, Ste6, to genetically identify revertants of F508del (Teem et al. 1993). Subsequent studies demonstrated that many second-site mutations were capable of stabilizing the domain and rescuing the F508del defect (Teem et al. 1993, 1996; DeCarvalho et al. 2002; Lewis et al. 2005; Pissarra et al. 2008; Thibodeau et al. 2010). These suppressor mutations have been identified using protein engineering approaches and comparative evolutionary studies and all stabilize the NBD and partially rescue trafficking of the full-length F508del CFTR molecule (He et al. 2010, 2015; Yang et al. 2018).

These independent observations indicated a conformational defect was associated with F508del and, more importantly, the mutation could be corrected and protein activity could be restored by multiple different treatments. This led to therapeutic development efforts to identify CFTR "correctors," using cell-based high-throughput screening (HTS) electrophysiological assays initiated by the Cystic Fibrosis Foundation with Aurora Biosciences, which was acquired by Vertex Pharmaceuticals in 2001. These screens utilized a fluorescence-based reporter of CFTR channel activity to identify two functional sets of small molecules: potentiators and correctors. Potentiators were identified using the G551D CFTR variant, which trafficked normally to the plasma membrane but failed to open (Accurso et al. 2010). Potentiators restored the activity of this and other functional mutants, but do not facilitate trafficking of biosynthetic mutants. Correctors have been identified using F508del CFTR in the presence of potentiator compounds (Van Goor et al. 2011; Clancy et al. 2012). The design of these screens was such that the correctors would facilitate folding and trafficking, while the potentiator would serve to activate any channels that reached the plasma membrane.

In 2012, the FDA approved the potentiator ivacaftor (KALYDECO®) for use in G551D patients. Since its initial approval, the use of KALYDECO® has expanded to include multiple CFTR gating mutants (Yu et al. 2012; Van Goor et al. 2014). It was also approved by the FDA for use in combination with lumacaftor (as ORKAMBI®) as a dual therapy for CF patients with at least one copy of the F508del allele in 2015 (Lumacaftor/ivacaftor (Orkambi) for cystic fibrosis 2016). This dual therapy showed limited efficacy in many F508del patients and a second-gen combination with tezacaftor, a derivative of lumacaftor, was approved in 2018 and is sold as SYMD EKO® (Tezacaftor/Ivacaftor (Symdeko) for cystic fibrosis 2018). SYMDEKO® shows increased efficacy in some CF patients, though the levels of CFTR correction and physiological benefit are generally limited and rescue is not adequate to block disease progression.

The characterization of the hierarchical folding pathway and multiple discrete defects associated with the F508del mutation suggested that more than one conformational defect might need correcting. Stemming in part from these observations and a need for more efficacious therapies, next-gen corrector screens were initiated to identify new molecules in the presence of ivacaftor and lumacaftor/tezacaftor (Keating et al. 2018; Davies et al. 2018). The primary goal was to identify new corrector molecules that augmented the effects of existing correctors, potentially addressing the multiple F508del defects identified experimentally. The "triple combination" strategy has shown increased efficacy in Phase 3 clinical trials and Vertex filed an NDA for triple combination therapy (ivacaftor: potentiator; tezacaftor and elexacaftor: correctors) in summer 2019. Patients receiving these triple combination therapies show increase lung function (FEV1), decreased sweat chloride, and increased quality of life scores significantly beyond those reported for the dual therapies (ORKAMBI® or SYMDEKO®) (Keating et al. 2018; Davies et al. 2018). This second-generation therapy, Trikafta®, was approved by the FDA in October 2019 and promises to change the therapeutic landscape for the CF patient population (Keating et al. 2018; Davies et al. 2018; Heijerman et al. 2019).

6.4.2   Structural Biology of CFTR

Since its cloning in 1989, multiple groups have worked to characterize the structure of CFTR. Initially, structures of the isolated NBDs were experimentally determined by X-ray crystallography. Recent cryo-EM studies have provided the first looks at full-length CFTR structures.

6.4.2.1   Structures of NBD1

In 2001, a multimillion dollar collaboration was initiated between the Cystic Fibrosis Foundation and Structural GenomiX with the intent of solving structures of full-length CFTR for structure-based drug development. These efforts began with the expression and crystallization of the isolated soluble NBDs. The first domain structures of CFTR were solved and published in 2004, revealing the architecture of the *Mus musculus* NBD1, including the Phe508 position (Fig. 6.11) (Lewis et al. 2004).

The wild-type structures of NBD1 displayed the same domain architecture as other ABC-proteins. The NBD was composed of the three canonical subdomains (β-, α/β-, and α-) that showed no conformational changes in the absence or presence of nucleotide. Importantly, these structures showed the position of the Phe508 residue and suggested mechanisms by which its deletion or substitution might impact the NBD conformationally. Subsequent structures of Phe508 substitutions, including

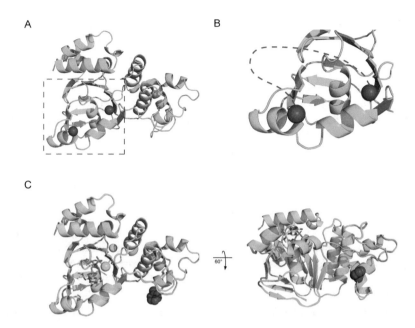

Fig. 6.11 Structure of CFTR NBD1. The structures of CFTR NBD1 were solved by X-ray crystallography and reveal features unique to the ABCC subfamily of ABC transporters, as well as the structural differences associated with the F508del mutation. (**a** and **b**) A cartoon representation of the murine NBD1 protein is shown. The architecture of NBD1 is similar to that of other NBDs. An insertion of ~15 amino acids in the β-sub-domain is not ordered or present in the electron density. Magenta spheres indicate the endpoints of the insertion in the domain. The close-up view presented in **b** is boxed for reference in **a**. (**c**) Wild type, *green*, and F508del, *cyan*, NBD1 structures are superposed with the Phe508 side chain shown as red spheres for reference. Only small local changes in NBD structure are seen surrounding the deletion site

F508C and F508R, and a F508del protein with numerous second-site suppressor mutations, suggested that the global fold of the NBD was not dramatically altered by mutation at the F508 position (Lewis et al. 2005; Thibodeau et al. 2005; Atwell et al. 2010a, b). Building from these, the structure of engineered human CFTR NBD1 proteins were also determined and used for structure-based ligand screens. Drug development efforts using X-ray crystallography failed to identify high-affinity NBD ligands, though some low-affinity molecules were shown to bind NBD1. In 2012, the structure of an unsuppressed F508del protein was published, showing no significant structural deviations from the previously published structures other than small local deviations associated with the backbone deletion at the 508 position (Mendoza et al. 2012).

Solution studies of WT and F508del NBD1 did reveal differences in the domain stability, solubility, and dynamics. Soluble expression of the F508del NBD is significantly reduced in multiple expression systems, consistent with a change in intrinsic domain properties (Lewis et al. 2005; Pissarra et al. 2008; Thibodeau et al. 2010; Mendoza et al. 2012). NMR studies of the purified NBDs reveal changes in protein dynamics across the domain when the F508del protein was analyzed (Hudson et al. 2012; Chong et al. 2015). The F508del mutation induces long-range intradomain changes that are evident in NMR spectra. These suggest that the effects of the F508del mutation propagate across the domain. Finally, multiple spectroscopic studies suggest that NBD unfolding proceeds through an intermediate, partially folded state (Protasevich et al. 2010; Wang et al. 2010). Under pseudo-equilibrium conditions, the F508del protein is more prone to occupy this nonnative conformational state. These data all suggest that the F508del mutation impacts that structure and energetics of NBD1 while inducing its complete unfolding globally.

6.4.2.2 Structures of NBD2

Structures of isolated CFTR NBD2 have now also been solved (Fig. 6.12) (Atwell et al. 2010a). Attempts to purify and crystalize NBD2 were initially hampered by poor protein solubility and stability. Significant protein engineering was required to obtain the first structure of human CFTR NBD2, which included multiple stabilizing and solubilizing mutations and a C-terminal fusion from the maltose transporter to improve stability and solubility. As with NBD1, the overall architecture of NBD2 includes the canonical subdomains and NBD features. However, the α-helical subdomain was displaced relative to the β- and α/β-core subdomains. This conformation would putatively not support NBD dimerization, requiring other interactions with CFTR to promote the ATP-induced dimerization of the NBDs and channel gating.

Few biophysical studies have been completed on NBD2 to date. Studies of NBD2 have been limited by the poor biochemical properties of the protein and the need to utilize a highly engineered domain. Unlike F508del in NBD1, only a handful of variants found in NBD2 are common in the patient population and are amenable to studies of the isolated domain. Thus, the progress in understanding NBD2 biophysical properties has remained slow.

6.4.2.3 Full-Length CFTR

While many groups have pursued the structure of CFTR since its cloning in 1989, the full-length structure was elusive until the most recent advances in cryo-EM technologies. Low-resolution single-particle and 2D crystal structures had been solved, but had limited utility owing to low map resolution (Rosenberg et al. 2011; Zhang et al. 2011). The structures of the bacterial ABC transporters BtuCD, published in 2002, and Sav1866, published in 2006, provided the first high-resolution structures of full-length ABC transporters and provided templates on which to build models of CFTR structures (see Fig. 6.5) (Locher et al. 2002; Dawson and Locher 2006, 2007). BtuCD functions as a vitamin B12 importer for nutrient acquisition and shows significant evolutionary divergence from the mammalian ABC-exporters, including CFTR. In contrast, Sav1866 shares a high degree of

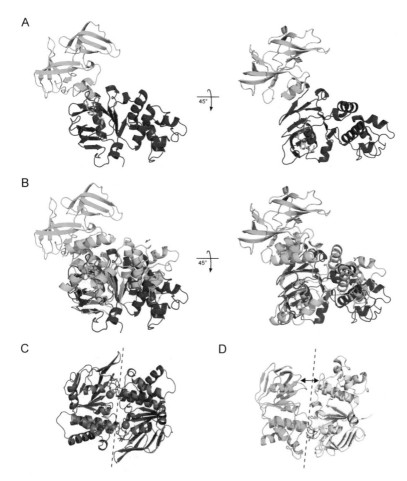

Fig. 6.12 Structure of CFTR NBD2. The structure of CFTR NBD2 was solved by X-ray crystallography and reveals alterations in subdomain orientation relative to other experimentally determined NBD structures. (**a**) Two orientations of NBD2 are show rotated about the *x*-axis by 45. The core NBD fold is colored purple with the C-terminal MalK fusion, required to maintain protein solubility and crystallization, shown in green. (**b**) superposition of NBD2, colored as in **a**, and NBD1, *cyan*. The conserved Walker A sequences were used to generate the structural alignment. The α/β-core domains superpose with only minimal local deviations. The α-helical subdomain is rotated away from the dimer interface in NBD2, while NBD1 shows a canonical conformation. (**c**) The structure of the Mj0796 homodimer is shown as a cartoon representation. The dimer interface is indicated by the dashed line. (**d**) The Mj0796 dimer was used as a template to model the CFTR heterodimer, with NBD1 shown in green and NBD2 in yellow. Alignments were made using the conserved Walker A sequence. The rotation of the α-helical sub-domain in NBD2 opens the dimer interface, separating the NBDs at the NBD1 composite site, indicated by the arrow. The flexible insertion shown in Fig. 6.11 is removed from NBD1 for this analysis

structural similarity to multiple subfamilies of mammalian ABC exporters, including CFTR. Despite their differences, both structures provided critical insight into conserved domain–domain interactions and channel architecture. Sav1866 was used extensively as a template for homology models of CFTR (Mornon et al. 2015) until its structure was solved and published in 2016 (Zhang and Chen 2016).

The first high-resolution structure of CFTR, from zebrafish (*Danio rerio*), was published in 2016, followed closely by that of human CFTR in 2017 (Fig. 6.13) (Zhang and Chen 2016; Liu et al. 2017). Both structures were solved in the channel closed conformation and had good local resolution in the transmembrane domains, but limited resolution in the nucleotide-binding domains. These structures have been supplemented with structures of both zebrafish and human CFTR in alternate conformations and in the presence of potentiator (activator) molecules with relatively high resolution (3.5–4.0 Å) across the entirety of the map (Zhang et al. 2017, 2018). In addition, structures of chicken CFTR have been published with marginally lower resolution (4.3–6.6 Å) (Fay et al. 2018).

The structures of CFTR indicate that it is similar in overall architecture to the bacterial Sav1866 transporter (Liu et al. 2017; Zhang et al. 2018). The domain organization and critical domain-domain interaction interfaces are preserved between Sav1866, and several other bacterial and eukaryotic ABC proteins, and CFTR. Importantly, two unique structural features are revealed in the EM studies of CFTR.

First, limited density for the R-domain is seen in the structures despite sample phosphorylation (Zhang et al. 2017, 2018). Short stretches of electron density have been attributed to residues within the R-domain, though its exact placement and the assignment of specific residues into this density has not been possible. This is consistent with in vitro studies of the R-region, which suggest it is highly disordered and fails to adopt regularly ordered structure (Ostedgaard et al. 2000; Baker et al. 2007). Much recent work suggests that disorder–order transitions underlie means of regulating protein structure and function by posttranslational modification (Bah and Forman-Kay 2016). Multiple phosphorylation sites have been identified biochemically and functionally in the R-region CFTR and these structural transitions underlie channel regulation (Cheng et al. 1991; Picciotto et al. 1992). Given its role in the phospho-regulation of CFTR and prior studies that demonstrate the R-domain was largely disordered in solution, it is not surprising that the R-region is not resolved in these structures.

Second, transmembrane helices seven and eight are altered when compared to the Sav1866 structures (Zhang and Chen 2016; Zhang et al. 2017). TM7 is displaced relative to its homologous TM helix in Sav1866. TM8 is non-continuous with a break in the lipid bilayer. In most ABC transporters, the TM helices span the membrane with no interruption in secondary structure. These continuous helices move as a rigid unit to transport substrates across the membrane. The breaks in CFTR's transmembrane helices are not observed, or predicted to occur, in other ABC proteins and may represent a critical evolutionary step away from an active transporter and toward a channel. It is currently not clear how the breaks in these helices contribute to CFTR's channel activity. Additional experiments will be

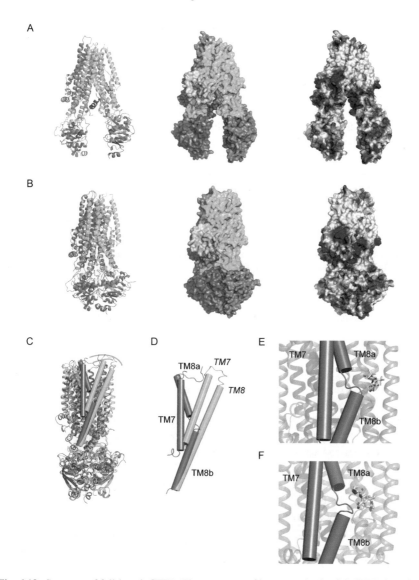

Fig. 6.13 Structure of full-length CFTR. The structures of human and zebrafish CFTR have been solved by cryo-EM and provide insight into the unique properties of this ABC transporter turned Cl⁻ channel. (**a**) Structures of CFTR in the putatively closed (non-conducting) state are shown as a cartoon, *left,* surface representation colored by domain, *center,* and electrostatic surface, *right.* (**b**) Structures of CFTR in a near-open state are shown as a cartoon, *left,* surface representation colored by domain, *center,* and electrostatic surface, *right.* Domains in **a** and **b** are colored as: N-terminal lasso domain, *goldenrod;* TMD1, *green*; NBD1, *blue*; R-region, *gray*; TMD2, *violet*; NBD2, *salmon.* (**c, d**) Differences in TM helices seven and eight, shown as cylinders, are observed when comparing CFTR to other ABC transporters in this structural family. TM7 and TM8 both show altered placement in CFTR, *blue,* when superposed onto Sav1866, *green.* TM7 is displaced in the helical bundle and TM8 is both displaced and interrupted (forming TM8a and TM8b) within the lipid bilayer. (**e, f**) Structural studies of potentiator binding by EM suggest ivacaftor and the GLPG1837 bind proximal to the break in TM8. Ivacaftor is shown as sticks in **e**; GLPG1837 is shown as sticks in **f**

required to evaluate how these changes in TM structure influence protein function and how they contribute to the channel pore. Curiously, the lower-resolution chicken structures deviate from the zebrafish and human CFTR structures in TM7 and TM8 (Fay et al. 2018). It is not clear how these observed changes in channel structure might alter its function.

The most recent structures of human CFTR also include channel potentiator (activator) compounds ivacaftor or an investigational potentiator from Galapagos (GLPG1837) bound to the transmembrane region (Yu et al. 2012; Van der Plas et al. 2018; Liu et al. 2019). Though both drugs potentiate CFTR, their structures and chemical scaffolds differ. Both bound drugs appear sandwiched between the breaks in transmembrane helix eight. Both drugs also appear to make significant interactions between the CFTR protein and the lipid bilayer. Mutagenic studies suggest that substitutions within the observed binding site decrease affinity for both drugs, consistent with the EM structural maps (Liu et al. 2019). Further structural studies will likely focus on identifying the binding sites of the corrector compounds and other clinically relevant small molecules, including nonclinical CFTR modulators and inhibitors.

6.5   Conclusions

Studies on protein folding and structure, though often very basic in nature, have provided key insights for understanding the molecular bases of disease and, perhaps more importantly, mechanisms by which defects in protein products can be corrected. Increasingly, pharmacological rescue of protein misfolding and instability is being recognized as a mechanism for therapeutic development (Bulawa et al. 2012; Gomes 2012; Suzuki 2013; Germain et al. 2016). The development of CFTR modulators as a means to treat CF arose from basic science studies of protein folding and channel structure. The translation of these basic science observations will provide clinical benefit to a majority of CF patients and serve as an example of the benefits derived from benchtop science. These clinical advances arise, in part, from the detailed structural and biophysical studies of proteins and grow from decades of work to understand how protein structures fold and function.

References

Accurso FJ, Rowe SM, Clancy JP, Boyle MP, Dunitz JM, Durie PR, Sagel SD, Hornick DB, Konstan MW, Donaldson SH, Moss RB, Pilewski JM, Rubenstein RC, Uluer AZ, Aitken ML, Freedman SD, Rose LM, Mayer-Hamblett N, Dong Q, Zha J, Stone AJ, Olson ER, Ordonez CL, Campbell PW, Ashlock MA, Ramsey BW (2010) Effect of VX-770 in persons with cystic fibrosis and the G551D-CFTR mutation. N Engl J Med 363(21):1991–2003. https://doi.org/10.1056/NEJMoa0909825

Ahner A, Gong X, Frizzell RA (2013) Cystic fibrosis transmembrane conductance regulator degradation: cross-talk between the ubiquitylation and SUMOylation pathways. FEBS J 280 (18):4430–4438. https://doi.org/10.1111/febs.12415

Albers SV, Koning SM, Konings WN, Driessen AJ (2004) Insights into ABC transport in archaea. J Bioenerg Biomembr 36(1):5–15

Aller SG, Yu J, Ward A, Weng Y, Chittaboina S, Zhuo R, Harrell PM, Trinh YT, Zhang Q, Urbatsch IL, Chang G (2009) Structure of P-glycoprotein reveals a molecular basis for poly-specific drug binding. Science 323(5922):1718–1722. https://doi.org/10.1126/science.1168750

Almo SC, Garforth SJ, Hillerich BS, Love JD, Seidel RD, Burley SK (2013) Protein production from the structural genomics perspective: achievements and future needs. Curr Opin Struct Biol 23(3):335–344. https://doi.org/10.1016/j.sbi.2013.02.014

Amaral MD (2004) CFTR and chaperones: processing and degradation. J Mol Neurosci 23 (1–2):41–48. https://doi.org/10.1385/JMN:23:1-2:041

Ambudkar SV, Kim IW, Xia D, Sauna ZE (2006) The A-loop, a novel conserved aromatic acid subdomain upstream of the Walker A motif in ABC transporters, is critical for ATP binding. FEBS Lett 580(4):1049–1055

Andreeva A, Howorth D, Chandonia JM, Brenner SE, Hubbard TJ, Chothia C, Murzin AG (2008) Data growth and its impact on the SCOP database: new developments. Nucleic Acids Res 36 (Database issue):D419–D425. https://doi.org/10.1093/nar/gkm993

Anfinsen CB (1973) Principles that govern the folding of protein chains. Science 181 (4096):223–230

Anfinsen CB, Redfield RR (1956) Protein structure in relation to function and biosynthesis. Adv Protein Chem 48(11):1–100

Ataide SF, Schmitz N, Shen K, Ke A, Shan SO, Doudna JA, Ban N (2011) The crystal structure of the signal recognition particle in complex with its receptor. Science 331(6019):881–886. https://doi.org/10.1126/science.1196473

Atwell S et al (2010a) 3GD7: crystal structure of human NBD2 complexed with N6-phenyl-ATP (PATP). Available via RCSB. http://www.rcsb/3GD7. Accessed 25 Aug 2020

Atwell S, Brouillette CG, Conners K, Emtage S, Gheyi T, Guggino WB, Hendle J, Hunt JF, Lewis HA, Lu F, Protasevich II, Rodgers LA, Romero R, Wasserman SR, Weber PC, Wetmore D, Zhang FF, Zhao X (2010b) Structures of a minimal human CFTR first nucleotide-binding domain as a monomer, head-to-tail homodimer, and pathogenic mutant. Protein Eng Des Sel 23(5):375–384. https://doi.org/10.1093/protein/gzq004

Baeza-Delgado C, Marti-Renom MA, Mingarro I (2013) Structure-based statistical analysis of transmembrane helices. Eur Biophys J EBJ 42(2–3):199–207. https://doi.org/10.1007/s00249-012-0813-9

Baghban R, Farajnia S, Rajabibazl M, Ghasemi Y, Mafi A, Hoseinpoor R, Rahbarnia L, Aria M (2019) Yeast expression systems: overview and recent advances. Mol Biotechnol 61 (5):365–384. https://doi.org/10.1007/s12033-019-00164-8

Bah A, Forman-Kay JD (2016) Modulation of intrinsically disordered protein function by post-translational modifications. J Biol Chem 291(13):6696–6705. https://doi.org/10.1074/jbc.R115.695056

Baker D (2014) Centenary Award and Sir Frederick Gowland Hopkins Memorial Lecture. Protein folding, structure prediction and design. Biochem Soc Trans 42(2):225–229. https://doi.org/10.1042/BST20130055

Baker D, Sohl JL, Agard DA (1992) A protein-folding reaction under kinetic control. Nature 356 (6366):263–265

Baker JM, Hudson RP, Kanelis V, Choy WY, Thibodeau PH, Thomas PJ, Forman-Kay JD (2007) CFTR regulatory region interacts with NBD1 predominantly via multiple transient helices. Nat Struct Mol Biol 14(8):738–745. https://doi.org/10.1038/nsmb1278

Bianchet MA, Ko YH, Amzel LM, Pedersen PL (1997) Modeling of nucleotide binding domains of ABC transporter proteins based on a F1-ATPase/recA topology: structural model of the nucleotide binding domains of the cystic fibrosis transmembrane conductance regulator (CFTR). J Bioenerg Biomembr 29(5):503–524

Bigelow H, Rost B (2009) Online tools for predicting integral membrane proteins. Methods Mol Biol 528:3–23. https://doi.org/10.1007/978-1-60327-310-7_1

Bornemann T, Jockel J, Rodnina MV, Wintermeyer W (2008) Signal sequence-independent membrane targeting of ribosomes containing short nascent peptides within the exit tunnel. Nat Struct Mol Biol 15(5):494–499. https://doi.org/10.1038/nsmb.1402

Boutet S, Lomb L, Williams GJ, Barends TR, Aquila A, Doak RB, Weierstall U, DePonte DP, Steinbrener J, Shoeman RL, Messerschmidt M, Barty A, White TA, Kassemeyer S, Kirian RA, Seibert MM, Montanez PA, Kenney C, Herbst R, Hart P, Pines J, Haller G, Gruner SM, Philipp HT, Tate MW, Hromalik M, Koerner LJ, van Bakel N, Morse J, Ghonsalves W, Arnlund D, Bogan MJ, Caleman C, Fromme R, Hampton CY, Hunter MS, Johansson LC, Katona G, Kupitz C, Liang M, Martin AV, Nass K, Redecke L, Stellato F, Timneanu N, Wang D, Zatsepin NA, Schafer D, Defever J, Neutze R, Fromme P, Spence JC, Chapman HN, Schlichting I (2012) High-resolution protein structure determination by serial femtosecond crystallography. Science 337(6092):362–364. https://doi.org/10.1126/science.1217737

Bragg L (1956) The diffraction of x-rays. Br J Radiol 29(339):121–126

Buchaklian AH, Klug CS (2006) Characterization of the LSGGQ and H motifs from the Escherichia coli lipid A transporter MsbA. Biochemistry 45(41):12539–12546

Buchegger F, Trowbridge IS, Liu LF, White S, Collawn JF (1996) Functional analysis of human/ chicken transferrin receptor chimeras indicates that the carboxy-terminal region is important for ligand binding. Eur J Biochem FEBS 235(1-2):9–17

Bulawa CE, Connelly S, Devit M, Wang L, Weigel C, Fleming JA, Packman J, Powers ET, Wiseman RL, Foss TR, Wilson IA, Kelly JW, Labaudiniere R (2012) Tafamidis, a potent and selective transthyretin kinetic stabilizer that inhibits the amyloid cascade. Proc Natl Acad Sci U S A 109(24):9629–9634. https://doi.org/10.1073/pnas.1121005109

Burton BM, Baker TA (2005) Remodeling protein complexes: insights from the AAA+ unfoldase ClpX and Mu transposase. Protein Sci 14(8):1945–1954. https://doi.org/10.1110/ps.051417505

Carson MR, Welsh MJ (1995) Structural and functional similarities between the nucleotide-binding domains of CFTR and GTP-binding proteins. Biophys J 69(6):2443–2448

Cha JY, Ahn G, Kim JY, Kang SB, Kim MR, Su'udi M, Kim WY, Son D (2013) Structural and functional differences of cytosolic 90-kDa heat-shock proteins (Hsp90s) in Arabidopsis thaliana. Plant Physiol Biochem 70:368–373. https://doi.org/10.1016/j.plaphy.2013.05.039

Chan HS, Dill KA (1996) A simple model of chaperonin-mediated protein folding. Proteins 24 (3):345–351. https://doi.org/10.1002/(SICI)1097-0134(199603)24:3<345::AID-PROT7>3.0. CO;2-F

Chan HS, Dill KA (1997) Solvation: how to obtain microscopic energies from partitioning and solvation experiments. Annu Rev Biophys Biomol Struct 26:425–459. https://doi.org/10.1146/annurev.biophys.26.1.425

Chen J, Lu G, Lin J, Davidson AL, Quiocho FA (2003) A tweezers-like motion of the ATP-binding cassette dimer in an ABC transport cycle. Mol Cell 12(3):651–661

Cheng Y (2018) Single-particle cryo-EM-How did it get here and where will it go. Science 361 (6405):876–880. https://doi.org/10.1126/science.aat4346

Cheng SH, Gregory RJ, Marshall J, Paul S, Souza DW, White GA, O'Riordan CR, Smith AE (1990) Defective intracellular transport and processing of CFTR is the molecular basis of most cystic fibrosis. Cell 63(4):827–834

Cheng SH, Rich DP, Marshall J, Gregory RJ, Welsh MJ, Smith AE (1991) Phosphorylation of the R domain by cAMP-dependent protein kinase regulates the CFTR chloride channel. Cell 66 (5):1027–1036. https://doi.org/10.1016/0092-8674(91)90446-6

Chong PA, Farber PJ, Vernon RM, Hudson RP, Mittermaier AK, Forman-Kay JD (2015) Deletion of phenylalanine 508 in the first nucleotide-binding domain of the cystic fibrosis transmembrane conductance regulator increases conformational exchange and inhibits dimerization. J Biol Chem 290(38):22862–22878. https://doi.org/10.1074/jbc.M115.641134

Clancy JP, Rowe SM, Accurso FJ, Aitken ML, Amin RS, Ashlock MA, Ballmann M, Boyle MP, Bronsveld I, Campbell PW, De Boeck K, Donaldson SH, Dorkin HL, Dunitz JM, Durie PR, Jain M, Leonard A, McCoy KS, Moss RB, Pilewski JM, Rosenbluth DB, Rubenstein RC, Schechter MS, Botfield M, Ordonez CL, Spencer-Green GT, Vernillet L, Wisseh S, Yen K, Konstan MW (2012) Results of a phase IIa study of VX-809, an investigational CFTR corrector compound, in subjects with cystic fibrosis homozygous for the F508del-CFTR mutation. Thorax 67(1):12–18. https://doi.org/10.1136/thoraxjnl-2011-200393

Crowley KS, Reinhart GD, Johnson AE (1993) The signal sequence moves through a ribosomal tunnel into a noncytoplasmic aqueous environment at the ER membrane early in translocation. Cell 73(6):1101–1115

Crowley KS, Liao S, Worrell VE, Reinhart GD, Johnson AE (1994) Secretory proteins move through the endoplasmic reticulum membrane via an aqueous, gated pore. Cell 78(3):461–471

Cui L, Hou YX, Riordan JR, Chang XB (2001) Mutations of the Walker B motif in the first nucleotide binding domain of multidrug resistance protein MRP1 prevent conformational maturation. Arch Biochem Biophys 392(1):153–161

Davidson AL, Chen J (2004) ATP-binding cassette transporters in bacteria. Annu Rev Biochem 73:241–268

Davies JC, Moskowitz SM, Brown C, Horsley A, Mall MA, EF MK, Plant BJ, Prais D, Ramsey BW, Taylor-Cousar JL, Tullis E, Uluer A, CM MK, Robertson S, Shilling RA, Simard C, Van Goor F, Waltz D, Xuan F, Young T, Rowe SM, VXS Group (2018) Structural and functional differences of cytosolic 90-kDa heat-shock proteins (Hsp90s) in Arabidopsis thaliana. N Engl J Med 379(17):1599–1611. https://doi.org/10.1056/NEJMoa1807119

Dawson RJ, Locher KP (2006) Structure of a bacterial multidrug ABC transporter. Nature 443 (7108):180–185

Dawson RJ, Locher KP (2007) Structure of the multidrug ABC transporter Sav1866 from Staphylococcus aureus in complex with AMP-PNP. FEBS Lett 581(5):935–938

Dawson NL, Lewis TE, Das S, Lees JG, Lee D, Ashford P, Orengo CA, Sillitoe I (2017) CATH: an expanded resource to predict protein function through structure and sequence. Nucleic Acids Res 45(D1):D289–D295. https://doi.org/10.1093/nar/gkw1098

Dean M, Rzhetsky A, Allikmets R (2001) The human ATP-binding cassette (ABC) transporter superfamily. Genome Res 11(7):1156–1166

Deber CM, Brandl CJ, Deber RB, Hsu LC, Young XK (1986) Amino acid composition of the membrane and aqueous domains of integral membrane proteins. Arch Biochem Biophys 251 (1):68–76

DeCarvalho AC, Gansheroff LJ, Teem JL (2002) Mutations in the nucleotide binding domain 1 signature motif region rescue processing and functional defects of cystic fibrosis transmembrane conductance regulator delta f508. J Biol Chem 277(39):35896–35905. https://doi.org/10.1074/jbc.M205644200

Deisenhofer J, Epp O, Miki K, Huber R, Michel H (1985) Structure of the protein subunits in the photosynthetic reaction centre of Rhodopseudomonas viridis at 3A resolution. Nature 318 (6047):618–624

de Marco A, Vigh L, Diamant S, Goloubinoff P (2005) Native folding of aggregation-prone recombinant proteins in Escherichia coli by osmolytes, plasmid- or benzyl alcohol-overexpressed molecular chaperones. Cell Stress Chaperones 10(4):329–339

Denning GM, Anderson MP, Amara JF, Marshall J, Smith AE, Welsh MJ (1992) Processing of mutant cystic fibrosis transmembrane conductance regulator is temperature-sensitive. Nature 358(6389):761–764. https://doi.org/10.1038/358761a0

Devaraneni PK, Conti B, Matsumura Y, Yang Z, Johnson AE, Skach WR (2011) Stepwise insertion and inversion of a type II signal anchor sequence in the ribosome-Sec61 translocon complex. Cell 146(1):134–147. https://doi.org/10.1016/j.cell.2011.06.004

Diederichs K, Diez J, Greller G, Muller C, Breed J, Schnell C, Vonrhein C, Boos W, Welte W (2000) Crystal structure of MalK, the ATPase subunit of the trehalose/maltose ABC transporter of the archaeon Thermococcus litoralis. EMBO J 19(22):5951–5961

Dill KA (1990) Dominant forces in protein folding. Biochemistry 29(31):7133

Dill KA, Chan HS (1997) From Levinthal to pathways to funnels. Nat Struct Biol 4(1):10–19

Dill KA, Fiebig KM, Chan HS (1993) Cooperativity in protein-folding kinetics. Proc Natl Acad Sci U S A 90(5):1942–1946

Do H, Falcone D, Lin J, Andrews DW, Johnson AE (1996) The cotranslational integration of membrane proteins into the phospholipid bilayer is a multistep process. Cell 85(3):369–378

Dobson CM (2003) Protein folding and misfolding. Nature 426(6968):884–890. https://doi.org/10.1038/nature02261

Drumm ML, Pope HA, Cliff WH, Rommens JM, Marvin SA, Tsui LC, Collins FS, Frizzell RA, Wilson JM (1990) Correction of the cystic fibrosis defect in vitro by retrovirus-mediated gene transfer. Cell 62(6):1227–1233

Du K, Sharma M, Lukacs GL (2005) The DeltaF508 cystic fibrosis mutation impairs domain-domain interactions and arrests post-translational folding of CFTR. Nat Struct Mol Biol 12 (1):17–25

Eisenberg D, Schwarz E, Komaromy M, Wall R (1984) Analysis of membrane and surface protein sequences with the hydrophobic moment plot. J Mol Biol 179(1):125–142

Engelman DM, Steitz TA, Goldman A (1986) Identifying nonpolar transbilayer helices in amino acid sequences of membrane proteins. Annu Rev Biophys Biophys Chem 15:321–353. https://doi.org/10.1146/annurev.bb.15.060186.001541

Fay JF, Aleksandrov LA, Jensen TJ, Cui LL, Kousouros JN, He L, Aleksandrov AA, Gingerich DS, Riordan JR, Chen JZ (2018) Cryo-EM visualization of an active high open probability CFTR anion channel. Biochemistry 57(43):6234–6246. https://doi.org/10.1021/acs.biochem.8b00763

Fischer H, Fukuda N, Barbry P, Illek B, Sartori C, Matthay MA (2001) Partial restoration of defective chloride conductance in DeltaF508 CF mice by trimethylamine oxide. Am J Physiol Lung Cell Mol Physiol 281(1):L52–L57. https://doi.org/10.1152/ajplung.2001.281.1.L52

Flanagan JJ, Chen JC, Miao Y, Shao Y, Lin J, Bock PE, Johnson AE (2003) Signal recognition particle binds to ribosome-bound signal sequences with fluorescence-detected subnanomolar affinity that does not diminish as the nascent chain lengthens. J Biol Chem 278 (20):18628–18637. https://doi.org/10.1074/jbc.M300173200

Friedland J, Hastings JW (1967) The reversibility of the denaturation of bacterial luciferase. Biochemistry 6(9):2893–2900

Fukuda R, Okiyoneda T (2018) Peripheral protein quality control as a novel drug target for CFTR stabilizer. Front Pharmacol 9:1100. https://doi.org/10.3389/fphar.2018.01100

Fulop K, Barna L, Symmons O, Zavodszky P, Varadi A (2009) Clustering of disease-causing mutations on the domain-domain interfaces of ABCC6. Biochemical and biophysical research communications 379(3):706–709. https://doi.org/10.1016/j.bbrc.2008.12.142

Gafvelin G, Sakaguchi M, Andersson H, von Heijne G (1997) Topological rules for membrane protein assembly in eukaryotic cells. J Biol Chem 272(10):6119–6127

Germain DP, Hughes DA, Nicholls K, Bichet DG, Giugliani R, Wilcox WR, Feliciani C, Shankar SP, Ezgu F, Amartino H, Bratkovic D, Feldt-Rasmussen U, Nedd K, Sharaf El Din U, Lourenco CM, Banikazemi M, Charrow J, Dasouki M, Finegold D, Giraldo P, Goker-Alpan O, Longo N, Scott CR, Torra R, Tuffaha A, Jovanovic A, Waldek S, Packman S, Ludington E, Viereck C, Kirk J, Yu J, Benjamin ER, Johnson F, Lockhart DJ, Skuban N, Castelli J, Barth J, Barlow C, Schiffmann R (2016) Treatment of Fabry's disease with the pharmacologic chaperone migalastat. N Engl J Med 375(6):545–555. https://doi.org/10.1056/NEJMoa1510198

Gileadi O (2017) Recombinant protein expression in E. coli : a historical perspective. Methods Mol Biol 1586:3–10. https://doi.org/10.1007/978-1-4939-6887-9_1

Glasser O (1945) Half a century of roentgen rays. Proc Rudolf Virchow Med Soc City New York 4:96–102

Gomes CM (2012) Protein misfolding in disease and small molecule therapies. Curr Topics Med Chem 12(22):2460–2469

Gorlich D, Prehn S, Hartmann E, Herz J, Otto A, Kraft R, Wiedmann M, Knespel S, Dobberstein B, Rapoport TA (1990) The signal sequence receptor has a second subunit and is part of a translocation complex in the endoplasmic reticulum as probed by bifunctional reagents. J Cell Biol 111(6 Pt 1):2283–2294

Guerriero CJ, Weiberth KF, Brodsky JL (2013) Hsp70 targets a cytoplasmic quality control substrate to the San1p ubiquitin ligase. J Biol Chem 288(25):18506–18520. https://doi.org/10.1074/jbc.M113.475905

Hartmann E, Rapoport TA, Lodish HF (1989) Predicting the orientation of eukaryotic membrane-spanning proteins. Proc Natl Acad Sci U S A 86(15):5786–5790

He L, Aleksandrov LA, Cui L, Jensen TJ, Nesbitt KL, Riordan JR (2010) Restoration of domain folding and interdomain assembly by second-site suppressors of the DeltaF508 mutation in CFTR. FASEB J 24(8):3103–3112. https://doi.org/10.1096/fj.09-141788

He L, Aleksandrov AA, An J, Cui L, Yang Z, Brouillette CG, Riordan JR (2015) Restoration of NBD1 thermal stability is necessary and sufficient to correct F508 CFTR folding and assembly. J Mol Biol 427(1):106–120. https://doi.org/10.1016/j.jmb.2014.07.026

Heijerman HGM, McKone EF, Downey DG, Van Braeckel E, Rowe SM, Tullis E, Mall MA, Welter JJ, Ramsey BW, McKee CM, Marigowda G, Moskowitz SM, Waltz D, Sosnay PR, Simard C, Ahluwalia N, Xuan F, Zhang Y, Taylor-Cousar JL, McCoy KS, VXT Group (2019) Efficacy and safety of the elexacaftor plus tezacaftor plus ivacaftor combination regimen in people with cystic fibrosis homozygous for the F508del mutation: a double-blind, randomised, phase 3 trial. Lancet. https://doi.org/10.1016/S0140-6736(19)32597-8

Ho BK, Brasseur R (2005) The Ramachandran plots of glycine and pre-proline. BMC Struct Biol 5:14

Hofmann S, Januliene D, Mehdipour AR, Thomas C, Stefan E, Bruchert S, Kuhn BT, Geertsma ER, Hummer G, Tampe R, Moeller A (2019) Conformation space of a heterodimeric ABC exporter under turnover conditions. Nature 571(7766):580–583. https://doi.org/10.1038/s41586-019-1391-0

Hopfner KP, Karcher A, Shin DS, Craig L, Arthur LM, Carney JP, Tainer JA (2000) Structural biology of Rad50 ATPase: ATP-driven conformational control in DNA double-strand break repair and the ABC-ATPase superfamily. Cell 101(7):789–800

Hosen MJ, Zubaer A, Thapa S, Khadka B, De Paepe A, Vanakker OM (2014) Molecular docking simulations provide insights in the substrate binding sites and possible substrates of the ABCC6 transporter. PloS one 9(7):e102779. https://doi.org/10.1371/journal.pone.0102779

Hudson RP, Chong PA, Protasevich II, Vernon R, Noy E, Bihler H, An JL, Kalid O, Sela-Culang I, Mense M, Senderowitz H, Brouillette CG, Forman-Kay JD (2012) Conformational changes relevant to channel activity and folding within the first nucleotide binding domain of the cystic fibrosis transmembrane conductance regulator. J Biol Chem 287(34):28480–28494. https://doi.org/10.1074/jbc.M112.371138

Hung LW, Wang IX, Nikaido K, Liu PQ, Ames GF, Kim SH (1998) Crystal structure of the ATP-binding subunit of an ABC transporter. Nature 396(6712):703–707

Hunter MS, Fromme P (2011) Toward structure determination using membrane-protein nanocrystals and microcrystals. Methods 55(4):387–404. https://doi.org/10.1016/j.ymeth.2011.12.006

Jansens A, van Duijn E, Braakman I (2002) Coordinated nonvectorial folding in a newly synthe-sized multidomain protein. Science 298(5602):2401–2403

Jiang Y, Cheng Z, Mandon EC, Gilmore R (2008) An interaction between the SRP receptor and the translocon is critical during cotranslational protein translocation. J Cell Biol 180(6):1149–1161. https://doi.org/10.1083/jcb.200707196

Jin MS, Oldham ML, Zhang Q, Chen J (2012) Crystal structure of the multidrug transporter P-glycoprotein from Caenorhabditis elegans. Nature 490(7421):566–569. https://doi.org/10.1038/nature11448

Johnson AE, Chen JC, Flanagan JJ, Miao Y, Shao Y, Lin J, Bock PE (2001) Structure, function, and regulation of free and membrane-bound ribosomes: the view from their substrates and products. Cold Spring Harbor Symp Quant Biol 66:531–541

Karpowich N, Martsinkevich O, Millen L, Yuan YR, Dai PL, MacVey K, Thomas PJ, Hunt JF (2001) Crystal structures of the MJ1267 ATP binding cassette reveal an induced-fit effect at the ATPase active site of an ABC transporter. Structure 9(7):571–586

Karpusas M, Baase WA, Matsumura M, Matthews BW (1989) Hydrophobic packing in T4 lysozyme probed by cavity-filling mutants. Proc Natl Acad Sci U S A 86(21):8237–8241

Kawate T, Gouaux E (2006) Fluorescence-detection size-exclusion chromatography for precrystallization screening of integral membrane proteins. Structure 14(4):673–681. https://doi.org/10.1016/j.str.2006.01.013

Keating D, Marigowda G, Burr L, Daines C, Mall MA, EF MK, Ramsey BW, Rowe SM, Sass LA, Tullis E, CM MK, Moskowitz SM, Robertson S, Savage J, Simard C, Van Goor F, Waltz D, Xuan F, Young T, Taylor-Cousar JL, VXS Group (2018) VX-445-tezacaftor-ivacaftor in patients with cystic fibrosis and one or two Phe508del alleles. N Engl J Med 379 (17):1612–1620. https://doi.org/10.1056/NEJMoa1807120

Keevil GM (1896) The Roentgen Rays. Br Med J 1(1833):433–434

Kendrew JC, Bodo G, Dintzis HM, Parrish RG, Wyckoff H, Phillips DC (1958) A three-dimensional model of the myoglobin molecule obtained by x-ray analysis. Nature 181 (4610):662–666

King J, Haase-Pettingell C, Robinson AS, Speed M, Mitraki A (1996) Thermolabile folding intermediates: inclusion body precursors and chaperonin substrates. FASEB J 10(1):57–66

Krieg UC, Johnson AE, Walter P (1989) Protein translocation across the endoplasmic reticulum membrane: identification by photocross-linking of a 39-kD integral membrane glycoprotein as part of a putative translocation tunnel. J Cell Biol 109(5):2033–2043

Kuhlbrandt W (2014a) Biochemistry. The resolution revolution. Science 343(6178):1443–1444. https://doi.org/10.1126/science.1251652

Kuhlbrandt W (2014b) Cryo-EM enters a new era. eLife 3:e03678. https://doi.org/10.7554/eLife.03678

Lakkaraju AK, Mary C, Scherrer A, Johnson AE, Strub K (2008) SRP keeps polypeptides translocation-competent by slowing translation to match limiting ER-targeting sites. Cell 133 (3):440–451. https://doi.org/10.1016/j.cell.2008.02.049

Lehnert U, Xia Y, Royce TE, Goh CS, Liu Y, Senes A, Yu H, Zhang ZL, Engelman DM, Gerstein M (2004) Computational analysis of membrane proteins: genomic occurrence, structure prediction and helix interactions. Q Rev Biophys 37(2):121–146

Levinthal C (1968) Are there pathways for protein folding? J Chim Phys 65:44–45

Lewis HA, Buchanan SG, Burley SK, Conners K, Dickey M, Dorwart M, Fowler R, Gao X, Guggino WB, Hendrickson WA, Hunt JF, Kearins MC, Lorimer D, Maloney PC, Post KW, Rajashankar KR, Rutter ME, Sauder JM, Shriver S, Thibodeau PH, Thomas PJ, Zhang M, Zhao X, Emtage S (2004) Structure of nucleotide-binding domain 1 of the cystic fibrosis transmembrane conductance regulator. EMBO J 23(2):282–293

Lewis HA, Zhao X, Wang C, Sauder JM, Rooney I, Noland BW, Lorimer D, Kearins MC, Conners K, Condon B, Maloney PC, Guggino WB, Hunt JF, Emtage S (2005) Impact of the deltaF508 mutation in first nucleotide-binding domain of human cystic fibrosis transmembrane conductance regulator on domain folding and structure. J Biol Chem 280(2):1346–1353

Liu PQ, Liu CE, Ames GF (1999) Modulation of ATPase activity by physical disengagement of the ATP-binding domains of an ABC transporter, the histidine permease. J Biol Chem 274 (26):18310–18318

Liu J, Mei Z, Li N, Qi Y, Xu Y, Shi Y, Wang F, Lei J, Gao N (2013a) Structural dynamics of the MecA-ClpC complex: a type II AAA+ protein unfolding machine. J Biol Chem 288 (24):17597–17608. https://doi.org/10.1074/jbc.M113.458752

Liu W, Wacker D, Gati C, Han GW, James D, Wang D, Nelson G, Weierstall U, Katritch V, Barty A, Zatsepin NA, Li D, Messerschmidt M, Boutet S, Williams GJ, Koglin JE, Seibert MM, Wang C, Shah ST, Basu S, Fromme R, Kupitz C, Rendek KN, Grotjohann I, Fromme P, Kirian RA, Beyerlein KR, White TA, Chapman HN, Caffrey M, Spence JC, Stevens RC, Cherezov V

(2013b) Serial femtosecond crystallography of G protein-coupled receptors. Science 342 (6165):1521–1524. https://doi.org/10.1126/science.1244142

Liu F, Zhang Z, Csanady L, Gadsby DC, Chen J (2017) Molecular structure of the human CFTR ion channel. Cell 169(1):85–95. e88. https://doi.org/10.1016/j.cell.2017.02.024

Liu F, Zhang Z, Levit A, Levring J, Touhara KK, Shoichet BK, Chen J (2019) Structural identification of a hotspot on CFTR for potentiation. Science 364(6446):1184–1188. https://doi.org/10.1126/science.aaw7611

Locher KP (2004) Structure and mechanism of ABC transporters. Curr Opin Struct Biol 14 (4):426–431

Locher KP, Lee AT, Rees DC (2002) The E. coli BtuCD structure: a framework for ABC transporter architecture and mechanism. Science 296(5570):1091–1098

Loo TW, Bartlett MC, Clarke DM (2002) The "LSGGQ" motif in each nucleotide-binding domain of human P-glycoprotein is adjacent to the opposing walker A sequence. J Biol Chem 277 (44):41303–41306

Lukacs GL, Verkman AS (2012) CFTR: folding, misfolding and correcting the DeltaF508 conformational defect. Trends Mol Med 18(2):81–91. https://doi.org/10.1016/j.molmed.2011.10.003

Lukacs GL, Mohamed A, Kartner N, Chang XB, Riordan JR, Grinstein S (1994) Conformational maturation of CFTR but not its mutant counterpart (delta F508) occurs in the endoplasmic reticulum and requires ATP. EMBO J 13(24):6076–6086

Lumacaftor/ivacaftor (Orkambi) for cystic fibrosis (2016) Med Lett Drugs Ther 58 (1491):41–42

Mary C, Scherrer A, Huck L, Lakkaraju AK, Thomas Y, Johnson AE, Strub K (2010) Residues in SRP9/14 essential for elongation arrest activity of the signal recognition particle define a positively charged functional domain on one side of the protein. RNA 16(5):969–979. https://doi.org/10.1261/rna.2040410

Matsumura M, Becktel WJ, Matthews BW (1988) Hydrophobic stabilization in T4 lysozyme determined directly by multiple substitutions of Ile 3. Nature 334(6181):406–410

McCormick PJ, Miao Y, Shao Y, Lin J, Johnson AE (2003) Cotranslational protein integration into the ER membrane is mediated by the binding of nascent chains to translocon proteins. Mol Cell 12(2):329–341

Mehnert M, Sommermeyer F, Berger M, Lakshmipathy SK, Gauss R, Aebi M, Jarosch E, Sommer T (2014) The interplay of Hrd3 and the molecular chaperone system ensures efficient degradation of malfolded secretory proteins. Mol Biol Cell. https://doi.org/10.1091/mbc.E14-07-1202

Mendoza JL, Schmidt A, Li Q, Nuvaga E, Barrett T, Bridges RJ, Feranchak AP, Brautigam CA, Thomas PJ (2012) Requirements for efficient correction of DeltaF508 CFTR revealed by analyses of evolved sequences. Cell 148(1–2):164–174. https://doi.org/10.1016/j.cell.2011.11.023

Michel H (1982) Three-dimensional crystals of a membrane protein complex. The photosynthetic reaction centre from Rhodopseudomonas viridis. J Mol Biol 158(3):567–572

Moelbert S, Emberly E, Tang C (2004) Correlation between sequence hydrophobicity and surface-exposure pattern of database proteins. Protein Sci 13(3):752–762

Moffat K, Bilderback D, Schildkamp W, Szebenyi D, Teng TY (1989) Laue photography from protein crystals. Basic Life Sci 51:325–330

Moller I, Jung M, Beatrix B, Levy R, Kreibich G, Zimmermann R, Wiedmann M, Lauring B (1998) A general mechanism for regulation of access to the translocon: competition for a membrane attachment site on ribosomes. Proc Natl Acad Sci U S A 95(23):13425–13430

Moody JE, Millen L, Binns D, Hunt JF, Thomas PJ (2002) Cooperative, ATP-dependent association of the nucleotide binding cassettes during the catalytic cycle of ATP-binding cassette transporters. J Biol Chem 277(24):21111–21114

Mornon JP, Hoffmann B, Jonic S, Lehn P, Callebaut I (2015) Full-open and closed CFTR channels, with lateral tunnels from the cytoplasm and an alternative position of the F508 region, as revealed by molecular dynamics. Cell Mol Life Sci 72(7):1377–1403. https://doi.org/10.1007/s00018-014-1749-2

Mus-Veteau I (2010) Heterologous expression of membrane proteins for structural analysis. Methods Mol Biol 601:1–16. https://doi.org/10.1007/978-1-60761-344-2_1

Nieddu E, Pollarolo B, Merello L, Schenone S, Mazzei M (2013) F508del-CFTR rescue: a matter of cell stress response. Curr Pharm Des 19(19):3476–3496. https://doi.org/10.2174/13816128113199990317

Nikaido K, Ames GF (1999) One intact ATP-binding subunit is sufficient to support ATP hydrolysis and translocation in an ABC transporter, the histidine permease. J Biol Chem 274 (38):26727–26735

Nilsson I, Lara P, Hessa T, Johnson AE, von Heijne G, Karamyshev AL (2014) The code for directing proteins for translocation across ER membrane: SRP cotranslationally recognizes specific features of a signal sequence. J Mol Biol. https://doi.org/10.1016/j.jmb.2014.06.014

Noriega TR, Chen J, Walter P, Puglisi JD (2014) Real-time observation of signal recognition particle binding to actively translating ribosomes. eLife 3. https://doi.org/10.7554/eLife.04418

Nozaki Y, Tanford C (1971) The solubility of amino acids and two glycine peptides in aqueous ethanol and dioxane solutions. Establishment of a hydrophobicity scale. J Biol Chem 246 (7):2211–2217

Ostedgaard LS, Baldursson O, Vermeer DW, Welsh MJ, Robertson AD (2000) A functional R domain from cystic fibrosis transmembrane conductance regulator is predominantly unstructured in solution. Proc Natl Acad Sci U S A 97(10):5657–5662. https://doi.org/10.1073/pnas.100588797

Pace CN, Shaw KL (2000) Linear extrapolation method of analyzing solvent denaturation curves. Proteins Suppl 4:1–7

pdb.org. http://www.pdb.org

Picciotto MR, Cohn JA, Bertuzzi G, Greengard P, Nairn AC (1992) Phosphorylation of the cystic fibrosis transmembrane conductance regulator. J Biol Chem 267(18):12742–12752

Pinkett HW, Lee AT, Lum P, Locher KP, Rees DC (2007) An inward-facing conformation of a putative metal-chelate-type ABC transporter. Science 315(5810):373–377

Pissarra LS, Farinha CM, Xu Z, Schmidt A, Thibodeau PH, Cai Z, Thomas PJ, Sheppard DN, Amaral MD (2008) Solubilizing mutations used to crystallize one CFTR domain attenuate the trafficking and channel defects caused by the major cystic fibrosis mutation. Chem Biol 15 (1):62–69. https://doi.org/10.1016/j.chembiol.2007.11.012

Pitonzo D, Yang Z, Matsumura Y, Johnson AE, Skach WR (2009) Sequence-specific retention and regulated integration of a nascent membrane protein by the endoplasmic reticulum Sec61 translocon. Mol Biol Cell 20(2):685–698. https://doi.org/10.1091/mbc.E08-09-0902

Pomozi V, Brampton C, Fulop K, Chen LH, Apana A, Li Q, Uitto J, Le Saux O, Varadi A (2014) Analysis of Pseudoxanthoma elasticum-causing missense mutants of ABCC6 in vivo; pharmacological correction of the mislocalized proteins. J Invest Dermatol 134(4):946–953. https://doi.org/10.1038/jid.2013.482

Powis K, Schrul B, Tienson H, Gostimskaya I, Breker M, High S, Schuldiner M, Jakob U, Schwappach B (2013) Get3 is a holdase chaperone and moves to deposition sites for aggregated proteins when membrane targeting is blocked. J Cell Sci 126(Pt 2):473–483. https://doi.org/10.1242/jcs.112151

Price SR, Nagai K (1995) Protein engineering as a tool for crystallography. Curr Opin Biotechnol 6 (4):425–430

Protasevich I, Yang Z, Wang C, Atwell S, Zhao X, Emtage S, Wetmore D, Hunt JF, Brouillette CG (2010) Thermal unfolding studies show the disease causing F508del mutation in CFTR thermodynamically destabilizes nucleotide-binding domain 1. Protein Sci 19(10):1917–1931. https://doi.org/10.1002/pro.479

Punjani A, Rubinstein JL, Fleet DJ, Brubaker MA (2017) cryoSPARC: algorithms for rapid unsupervised cryo-EM structure determination. Nat Methods 14(3):290–296. https://doi.org/10.1038/nmeth.4169

Qu BH, Thomas PJ (1996) Alteration of the cystic fibrosis transmembrane conductance regulator folding pathway. J Biol Chem 271(13):7261–7264. https://doi.org/10.1074/jbc.271.13.7261

Reeb J, Kloppmann E, Bernhofer M, Rost B (2014) Evaluation of transmembrane helix predictions in 2014. Proteins. https://doi.org/10.1002/prot.24749

Religa TL, Markson JS, Mayor U, Freund SM, Fersht AR (2005) Solution structure of a protein denatured state and folding intermediate. Nature 437(7061):1053–1056

Riordan JR (2005) Assembly of functional CFTR chloride channels. Annu Rev Physiol 67:701–718. https://doi.org/10.1146/annurev.physiol.67.032003.154107

Riordan JR, Rommens JM, Kerem B, Alon N, Rozmahel R, Grzelczak Z, Zielenski J, Lok S, Plavsic N, Chou JL et al (1989) Identification of the cystic fibrosis gene: cloning and characterization of complementary DNA. Science 245(4922):1066–1073. https://doi.org/10.1126/science.2475911

Rommens JM, Iannuzzi MC, Kerem B, Drumm ML, Melmer G, Dean M, Rozmahel R, Cole JL, Kennedy D, Hidaka N et al (1989) Identification of the cystic fibrosis gene: chromosome walking and jumping. Science 245(4922):1059–1065

Rosenbaum G, Ginell SL, Chen JC (2015) Energy optimization of a regular macromolecular crystallography beamline for ultra-high-resolution crystallography. J synchrotron Radiat 22 (Pt 1):172–174. https://doi.org/10.1107/S1600577514022619

Rosenberg MF, O'Ryan LP, Hughes G, Zhao Z, Aleksandrov LA, Riordan JR, Ford RC (2011) The cystic fibrosis transmembrane conductance regulator (CFTR): three-dimensional structure and localization of a channel gate. J Biol Chem 286(49):42647–42654. https://doi.org/10.1074/jbc.M111.292268

Rost B, Casadio R, Fariselli P, Sander C (1995) Transmembrane helices predicted at 95% accuracy. Protein Sci 4(3):521–533. https://doi.org/10.1002/pro.5560040318

Roy SS, Patra M, Nandy SK, Banik M, Dasgupta R, Basu T (2014) In vitro holdase activity of E. coli small heat-shock proteins IbpA, IbpB and IbpAB: a biophysical study with some unconventional techniques. Protein Pept Lett 21(6):564–571

Sadlish H, Pitonzo D, Johnson AE, Skach WR (2005) Sequential triage of transmembrane segments by Sec61alpha during biogenesis of a native multispanning membrane protein. Nat Struct Mol Biol 12(10):870–878. https://doi.org/10.1038/nsmb994

Sanders SL, Schekman R (1992) Polypeptide translocation across the endoplasmic reticulum membrane. J Biol Chem 267(20):13791–13794

Sato S, Ward CL, Krouse ME, Wine JJ, Kopito RR (1996) Glycerol reverses the misfolding phenotype of the most common cystic fibrosis mutation. J Biol Chem 271(2):635–638. https://doi.org/10.1074/jbc.271.2.635

Scheres SH (2012a) A Bayesian view on cryo-EM structure determination. J Mol Biol 415 (2):406–418. https://doi.org/10.1016/j.jmb.2011.11.010

Scheres SH (2012b) RELION: implementation of a Bayesian approach to cryo-EM structure determination. J Struct Biol 180(3):519–530. https://doi.org/10.1016/j.jsb.2012.09.006

Scheres SH (2016) Processing of structurally heterogeneous cryo-EM data in RELION. Methods Enzymol 579:125–157. https://doi.org/10.1016/bs.mie.2016.04.012

Schlesinger MJ, Barrett K (1965) The reversible dissociation of the alkaline phosphatase of Escherichia coli. I. Formation and reactivation of subunits. J Biol Chem 240(11):4284–4292

Schneider E, Hunke S (1998) ATP-binding-cassette (ABC) transport systems: functional and structural aspects of the ATP-hydrolyzing subunits/domains. FEMS Microbiol Rev 22(1):1–20

Schroder M, Kaufman RJ (2005) ER stress and the unfolded protein response. Mutat Res 569 (1–2):29–63. https://doi.org/10.1016/j.mrfmmm.2004.06.056

Senes A, Ubarretxena-Belandia I, Engelman DM (2001) The Calpha—H...O hydrogen bond: a determinant of stability and specificity in transmembrane helix interactions. Proc Natl Acad Sci U S A 98(16):9056–9061. https://doi.org/10.1073/pnas.161280798

Senior AE, Gadsby DC (1997) ATP hydrolysis cycles and mechanism in P-glycoprotein and CFTR. Semin Cancer Biol 8(3):143–150

Sharma M, Benharouga M, Hu W, Lukacs GL (2001) Conformational and temperature-sensitive stability defects of the delta F508 cystic fibrosis transmembrane conductance regulator in post-endoplasmic reticulum compartments. J Biol Chem 276(12):8942–8950

Sharma SK, Christen P, Goloubinoff P (2009) Disaggregating chaperones: an unfolding story. Curr Protein Pept Sci 10(5):432–446

Shimamura T, Weyand S, Beckstein O, Rutherford NG, Hadden JM, Sharples D, Sansom MS, Iwata S, Henderson PJ, Cameron AD (2010) Molecular basis of alternating access membrane transport by the sodium-hydantoin transporter Mhp1. Science 328(5977):470–473. https://doi.org/10.1126/science.1186303

Shintre CA, Pike AC, Li Q, Kim JI, Barr AJ, Goubin S, Shrestha L, Yang J, Berridge G, Ross J, Stansfeld PJ, Sansom MS, Edwards AM, Bountra C, Marsden BD, von Delft F, Bullock AN, Gileadi O, Burgess-Brown NA, Carpenter EP (2013) Structures of ABCB10, a human ATP-binding cassette transporter in apo- and nucleotide-bound states. Proc Natl Acad Sci U S A 110(24):9710–9715. https://doi.org/10.1073/pnas.1217042110

Skarina T, Xu X, Evdokimova E, Savchenko A (2014) High-throughput crystallization screening. Methods Mol Biol 1140:159–168. https://doi.org/10.1007/978-1-4939-0354-2_12

Skolnick J, Fetrow JS, Kolinski A (2000) Structural genomics and its importance for gene function analysis. Nat Biotechnol 18(3):283–287. https://doi.org/10.1038/73723

Slepenkov SV, Witt SN (2002) The unfolding story of the Escherichia coli Hsp70 DnaK: is DnaK a holdase or an unfoldase? Mol Microbiol 45(5):1197–1206

Smith PC, Karpowich N, Millen L, Moody JE, Rosen J, Thomas PJ, Hunt JF (2002) ATP binding to the motor domain from an ABC transporter drives formation of a nucleotide sandwich dimer. Mol Cell 10(1):139–149

Song W, Raden D, Mandon EC, Gilmore R (2000) Role of Sec61alpha in the regulated transfer of the ribosome-nascent chain complex from the signal recognition particle to the translocation channel. Cell 100(3):333–343

Structure MPoKD. http://blanco.biomol.uci.edu/mpstruc/

Sun S, Brem R, Chan HS, Dill KA (1995) Designing amino acid sequences to fold with good hydrophobic cores. Protein Eng 8(12):1205–1213

Suzuki Y (2013) Chaperone therapy update: Fabry disease, GM1-gangliosidosis and Gaucher disease. Brain Dev 35(6):515–523. https://doi.org/10.1016/j.braindev.2012.12.002

Tamayo AG, Slater L, Taylor-Parker J, Bharti A, Harrison R, Hung DT, Murphy JR (2011) GRP78 (BiP) facilitates the cytosolic delivery of anthrax lethal factor (LF) in vivo and functions as an unfoldase in vitro. Mol Microbiol 81(5):1390–1401. https://doi.org/10.1111/j.1365-2958.2011.07770.x

Teem JL, Berger HA, Ostedgaard LS, Rich DP, Tsui LC, Welsh MJ (1993) Identification of revertants for the cystic fibrosis delta F508 mutation using STE6-CFTR chimeras in yeast. Cell 73(2):335–346. https://doi.org/10.1016/0092-8674(93)90233-g

Teem JL, Carson MR, Welsh MJ (1996) Mutation of R555 in CFTR-delta F508 enhances function and partially corrects defective processing. Receptors Channels 4(1):63–72

Tezacaftor/Ivacaftor (Symdeko) for cystic fibrosis (2018) Med Lett Drugs Ther 60 (1558):174–176

Thibodeau PH, Brautigam CA, Machius M, Thomas PJ (2005) Side chain and backbone contributions of Phe508 to CFTR folding. Nat Struct Mol Biol 12(1):10–16

Thibodeau PH, Richardson JM 3rd, Wang W, Millen L, Watson J, Mendoza JL, Du K, Fischman S, Senderowitz H, Lukacs GL, Kirk K, Thomas PJ (2010) The cystic fibrosis-causing mutation deltaF508 affects multiple steps in cystic fibrosis transmembrane conductance regulator biogenesis. J Biol Chem 285(46):35825–35835. https://doi.org/10.1074/jbc.M110.131623

Thomas PJ, Shenbagamurthi P, Sondek J, Hullihen JM, Pedersen PL (1992) The cystic fibrosis transmembrane conductance regulator. Effects of the most common cystic fibrosis-causing mutation on the secondary structure and stability of a synthetic peptide. J Biol Chem 267 (9):5727–5730

Thomas PJ, Qu BH, Pedersen PL (1995) Defective protein folding as a basis of human disease. Trends Biochem Sci 20(11):456–459

Trowbridge IS, Collawn J, Jing S, White S, Esekogwu V, Stangel M (1991) Structure-function analysis of the human transferrin receptor: effects of anti-receptor monoclonal antibodies on tumor growth. Curr Stud Hematol Blood Transfus 58:139–147

Van der Plas SE, Kelgtermans H, De Munck T, Martina SLX, Dropsit S, Quinton E, De Blieck A, Joannesse C, Tomaskovic L, Jans M, Christophe T, van der Aar E, Borgonovi M, Nelles L, Gees M, Stouten P, Van Der Schueren J, Mammoliti O, Conrath K, Andrews M (2018) Discovery of N-(3-carbamoyl-5,5,7,7-tetramethyl-5,7-dihydro-4H-thieno[2,3-c]pyran-2-yl)-lH-pyr azole-5-carboxamide (GLPG1837), a novel potentiator which can open class III mutant cystic fibrosis transmembrane conductance regulator (CFTR) channels to a high extent. J Med Chem 61(4):1425–1435. https://doi.org/10.1021/acs.jmedchem.7b01288

Van Goor F, Hadida S, Grootenhuis PD, Burton B, Stack JH, Straley KS, Decker CJ, Miller M, McCartney J, Olson ER, Wine JJ, Frizzell RA, Ashlock M, Negulescu PA (2011) Correction of the F508del-CFTR protein processing defect in vitro by the investigational drug VX-809. Proc Natl Acad Sci U S A 108(46):18843–18848. https://doi.org/10.1073/pnas.1105787108

Van Goor F, Yu H, Burton B, Hoffman BJ (2014) Effect of ivacaftor on CFTR forms with missense mutations associated with defects in protein processing or function. J Cyst Fibros 13(1):29–36. https://doi.org/10.1016/j.jcf.2013.06.008

Vendruscolo M, Paci E, Dobson CM, Karplus M (2001) Three key residues form a critical contact network in a protein folding transition state. Nature 409(6820):641–645. https://doi.org/10.1038/35054591

Vergani P, Lockless SW, Nairn AC, Gadsby DC (2005) CFTR channel opening by ATP-driven tight dimerization of its nucleotide-binding domains. Nature 433(7028):876–880

Wallin E, von Heijne G (1998) Genome-wide analysis of integral membrane proteins from eubacterial, archaean, and eukaryotic organisms. Protein Sci 7(4):1029–1038. https://doi.org/10.1002/pro.5560070420

Wang L, Dobberstein B (1999) Oligomeric complexes involved in translocation of proteins across the membrane of the endoplasmic reticulum. FEBS Lett 457(3):316–322

Wang CC, Tsou CL (1998) Enzymes as chaperones and chaperones as enzymes. FEBS Lett 425 (3):382–384

Wang C, Protasevich I, Yang Z, Seehausen D, Skalak T, Zhao X, Atwell S, Spencer Emtage J, Wetmore DR, Brouillette CG, Hunt JF (2010) Integrated biophysical studies implicate partial unfolding of NBD1 of CFTR in the molecular pathogenesis of F508del cystic fibrosis. Protein Sci 19(10):1932–1947. https://doi.org/10.1002/pro.480

Watson JD, Crick FH (1953) Molecular structure of nucleic acids; a structure for deoxyribose nucleic acid. Nature 171(4356):737–738

White SH (2009) Biophysical dissection of membrane proteins. Nature 459(7245):344–346. https://doi.org/10.1038/nature08142

Wilken S, Schmees G, Schneider E (1996) A putative helical domain in the MalK subunit of the ATP-binding-cassette transport system for maltose of Salmonella typhimurium (MalFGK2) is crucial for interaction with MalF and MalG. A study using the LacK protein of Agrobacterium radiobacter as a tool. Mol Microbiol 22(4):655–666

Xiao R, Anderson S, Aramini J, Belote R, Buchwald WA, Ciccosanti C, Conover K, Everett JK, Hamilton K, Huang YJ, Janjua H, Jiang M, Kornhaber GJ, Lee DY, Locke JY, Ma LC, Maglaqui M, Mao L, Mitra S, Patel D, Rossi P, Sahdev S, Sharma S, Shastry R, Swapna GV, Tong SN, Wang D, Wang H, Zhao L, Montelione GT, Acton TB (2010) The high-throughput protein sample production platform of the Northeast Structural Genomics Consortium. J Struct Biol 172(1):21–33. https://doi.org/10.1016/j.jsb.2010.07.011

Xu H, Clairfeuille T, Jao CC, Ho H, Sweeney Z, Payandeh J, Koth CM (2019) A flexible and scalable high-throughput platform for recombinant membrane protein production. Methods Mol Biol 2025:389–402. https://doi.org/10.1007/978-1-4939-9624-7_18

Yang Z, Hildebrandt E, Jiang F, Aleksandrov AA, Khazanov N, Zhou Q, An J, Mezzell AT, Xavier BM, Ding H, Riordan JR, Senderowitz H, Kappes JC, Brouillette CG, Urbatsch IL (2018) Structural stability of purified human CFTR is systematically improved by mutations in nucleotide binding domain 1. Biochim Biophys Acta Biomembr 1860(5):1193–1204. https://doi.org/10.1016/j.bbamem.2018.02.006

Young CL, Britton ZT, Robinson AS (2012) Recombinant protein expression and purification: a comprehensive review of affinity tags and microbial applications. Biotechnol J 7(5):620–634. https://doi.org/10.1002/biot.201100155

Yu H, Burton B, Huang CJ, Worley J, Cao D, Johnson JP Jr, Urrutia A, Joubran J, Seepersaud S, Sussky K, Hoffman BJ, Van Goor F (2012) Ivacaftor potentiation of multiple CFTR channels with gating mutations. J Cyst Fibros 11(3):237–245. https://doi.org/10.1016/j.jcf.2011.12.005

Yuan YR, Blecker S, Martsinkevich O, Millen L, Thomas PJ, Hunt JF (2001) The crystal structure of the MJ0796 ATP-binding cassette. Implications for the structural consequences of ATP hydrolysis in the active site of an ABC transporter. J Biol Chem 276(34):32313–32321

Zacchi LF, Wu HC, Bell SL, Millen L, Paton AW, Paton JC, Thomas PJ, Zolkiewski M, Brodsky JL (2014) The BiP molecular chaperone plays multiple roles during the biogenesis of torsinA, an AAA+ ATPase associated with the neurological disease early-onset torsion dystonia. J Biol Chem 289(18):12727–12747. https://doi.org/10.1074/jbc.M113.529123

Zaitseva J, Jenewein S, Jumpertz T, Holland IB, Schmitt L (2005) H662 is the linchpin of ATP hydrolysis in the nucleotide-binding domain of the ABC transporter HlyB. EMBO J 24 (11):1901–1910

Zaman K, McPherson M, Vaughan J, Hunt J, Mendes F, Gaston B, Palmer LA (2001) S-nitrosoglutathione increases cystic fibrosis transmembrane regulator maturation. Biochem Biophys Res Commun 284(1):65–70. https://doi.org/10.1006/bbrc.2001.4935

Zhang Z, Chen J (2016) Atomic structure of the cystic fibrosis transmembrane conductance regulator. Cell 167(6):1586–1597. e1589. https://doi.org/10.1016/j.cell.2016.11.014

Zhang XM, Wang XT, Yue H, Leung SW, Thibodeau PH, Thomas PJ, Guggino SE (2003) Organic solutes rescue the functional defect in delta F508 cystic fibrosis transmembrane conductance regulator. J Biol Chem 278(51):51232–51242. https://doi.org/10.1074/jbc.M309076200

Zhang L, Aleksandrov LA, Riordan JR, Ford RC (2011) Domain location within the cystic fibrosis transmembrane conductance regulator protein investigated by electron microscopy and gold labelling. Biochim Biophys Acta 1808(1):399–404. https://doi.org/10.1016/j.bbamem.2010.08.012

Zhang L, Conway JF, Thibodeau PH (2012) Calcium-induced folding and stabilization of the Pseudomonas aeruginosa alkaline protease. J Biol Chem 287(6):4311–4322. https://doi.org/10.1074/jbc.M111.310300

Zhang Z, Liu F, Chen J (2017) Conformational changes of CFTR upon phosphorylation and ATP binding. Cell 170(3):483–491. e488. https://doi.org/10.1016/j.cell.2017.06.041

Zhang Z, Liu F, Chen J (2018) Molecular structure of the ATP-bound, phosphorylated human CFTR. Proc Natl Acad Sci U S A 115(50):12757–12762. https://doi.org/10.1073/pnas.1815287115

Zheng SQ, Palovcak E, Armache JP, Verba KA, Cheng Y, Agard DA (2017) MotionCor2: anisotropic correction of beam-induced motion for improved cryo-electron microscopy. Nat Methods 14(4):331–332. https://doi.org/10.1038/nmeth.4193

Zhou X, Levin EJ, Pan Y, McCoy JG, Sharma R, Kloss B, Bruni R, Quick M, Zhou M (2014) Structural basis of the alternating-access mechanism in a bile acid transporter. Nature 505 (7484):569–573. https://doi.org/10.1038/nature12811

Chapter 7
Epithelial Ion Channel Folding and ER-Associated Degradation (ERAD)

Teresa M. Buck and Jeffrey L. Brodsky

Abstract Epithelial ion channels and transporters play critical roles in salt and water homeostasis. The expression, assembly, and trafficking of these proteins are tightly regulated at both the transcriptional and posttranscriptional levels. For example, ion channel residence at the plasma membrane is regulated in response to intra- and extracellular signals that can alter anterograde trafficking, endocytosis, and/or degradation. While regulation at the plasma membrane has been thoroughly characterized for a number of ion channels and transporters, the mechanisms underlying early events in the endoplasmic reticulum (ER) have only more recently been defined. Notably, all epithelial ion channels and transporters are synthesized at and insert into the ER membrane, at which time posttranslational modifications are added, the protein begins to fold, and subunits of oligomeric proteins assemble. However, before epithelial ion channels and transporters can traffic from the ER, they are subject to protein quality control "decisions," which ensure that only properly modified, folded, and assembled subunits are directed to the Golgi apparatus and ultimately to the plasma membrane via vesicle intermediates. Here, we first describe the general steps to which ion channels and transporters are subject in the ER and highlight the factors that facilitate protein folding. We will also discuss the events that lead to the degradation of improperly modified, misfolded, and misassembled proteins in the ER. This pathway is known as ER-associated degradation, or ERAD. We will then provide specific examples of critical epithelial proteins that undergo these events in ER protein quality control.

Keywords Endoplasmic reticulum (ER) · ER-associated degradation (ERAD) · Kidney · Ion channels · Ion transporters · Molecular chaperones · ENaC · Na,K-ATPase · ROMK · V2R · NCC · Aquaporin-2

T. M. Buck (✉) · J. L. Brodsky
Department of Biological Sciences, University of Pittsburgh, Pittsburgh, PA, USA
e-mail: teb20@pitt.edu; jbrodsky@pitt.edu

© The American Physiological Society 2020

207

K. L. Hamilton, D. C. Devor (eds.), *Basic Epithelial Ion Transport Principles and Function*, Physiology in Health and Disease,
https://doi.org/10.1007/978-3-030-52780-8_7

7.1   Introduction

Approximately one-third of all cellular proteins, including all membrane proteins, transit through the secretory pathway (Ghaemmaghami et al. 2003). The translation of these secretory proteins begins in the cytosol. Signal recognition particle (SRP) binds to either an N-terminal signal sequence or the first transmembrane domain (TMD) of a nascent protein as it emerges from the ribosome and stalls translation (Walter and Blobel 1981; Halic et al. 2004, 2006). The SRP–ribosome nascent chain complex is transferred to the ER membrane by virtue of SRP binding to the SRP receptor, which facilitates ribosome docking to the Sec61 translocation channel in the ER membrane (Meyer and Dobberstein 1980; Gilmore et al. 1982). After the ribosome docks onto Sec61, translation resumes and secretory proteins are cotranslationally translocated into the ER lumen through Sec61 (Fig. 7.1) (Gorlich et al. 1992). For secretory proteins, the signal sequence may be cleaved by signal peptidase (Fig. 7.1a). For membrane proteins, such as epithelial ion channels and transporters, the TMDs move laterally through a Sec61 "gate" and are thus integrated into the lipid bilayer. Alternating TMDs act as either signal sequences or stop transfer sequences (Fig. 7.1b–c) (Egea and Stroud 2010). After membrane integration is complete, downstream events—such as protein folding, oligomeric protein assembly, and posttranslational modifications—occur. These events precede a quality control checkpoint which ensures that only properly folded proteins, and especially membrane proteins, can leave the ER and traffic to the plasma membrane (Barlowe and Helenius 2016).

7.2   Protein Folding in the Endoplasmic Reticulum

The integration of multi-pass ion-conducting membrane proteins is particularly challenging for the following reasons. First, ion-conducting pores are lined with polar residues that are embedded in the lipid bilayer. In fact, an analysis of all known TMDs predicts that ~30% are only marginally hydrophobic and may not integrate efficiently into the ER membrane (Hessa et al. 2007). Therefore, TMDs either from within the protein or from another subunit in an oligomeric complex are often required to stabilize polar TMDs (Beguin et al. 1998, 2000; Lu et al. 2000; Buck et al. 2007). Second, the final structure of a membrane protein requires the assembly of TMDs that might be synthesized early (i.e., at the N-terminus) and later (i.e., at the C-terminus). Thus, a TMD at the N-terminus might be orphaned for an extended time until it can assemble with a partner TMD. Third, the folding of a membrane protein takes place in three chemically distinct compartments: The ER lumen, which will ultimately be extracellular, the ER lipid bilayer, and the cytoplasm (Fig. 7.1). With few exceptions, it is generally unclear how the folding of a protein in these environments is coordinated. Finally, ribosomes synthesizing subunits of an oligomeric protein may not utilize adjacent Sec61 pores, so after membrane insertion each

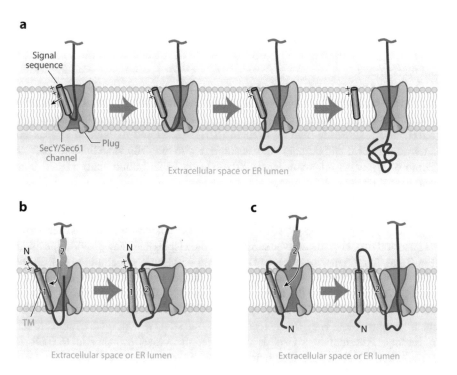

Fig. 7.1 Protein translocation. The ribosome nascent chain complex docks at the Sec61 channel. For soluble proteins (**a**) a signal sequence orients the polypeptide with positively charged amino acids facing the cytosol. The signal sequence is cleaved by signal peptidase prior to translocation of the nascent polypeptide chain into the ER lumen. For membrane proteins (**b** and **c**) the first transmembrane domain acts as a signal sequence to orient the N-terminus either toward the cytosol (**b**) or toward the ER lumen (extracellular) (**c**). Transmembrane domains move through a lateral gate in the Sec61 channel into the lipid bilayer. Subsequent transmembrane domains act as either stop transfers (**b**) or signal anchors (**c**) to position the more C-terminal transmembrane domains appropriately. Figure republished with permission of Rapoport et al. (2017). Permission conveyed through Copyright Clearance Center, Inc

subunit will need to diffuse in the plane of the ER to find one another. To overcome some of these hurdles, protein folding is aided by the contributions of other factors.

7.2.1 The Role of Molecular Chaperones in Protein Folding

Molecular chaperones bind nascent polypeptides chains as they emerge from the ribosome and facilitate folding, prevent aggregation, promote proper oligomerization, and/or trafficking from the ER. Chaperones continue to act often through cycles of binding and release until a mature protein conformation is reached. Chaperones are a structurally diverse group of proteins that include Hsp70s, Hsp90s, lectins, thiol oxidoreductases, and small heat shock proteins (Buchner 2019; Genest et al. 2019; Lang et al. 2017), each of which are required for the quality control of ion channels and transporters.

One of the most abundant and promiscuous chaperones are the Hsp70s (Fig. 7.2). Hsp70s are located in both the cytosol and the ER and participate in ATP-dependent protein folding. Hsp70s are composed of an N-terminal nucleotide-binding domain (NBD) and a substrate-binding domain (SBD) composed of eight β strands and C-terminal helical "lid domain" that is connected to the SBD via a linker (Fig. 7.2b) (Pobre et al. 2019; Alderson et al. 2016). In the ATP bound state, Hsp70 binds weakly to substrates. ATP hydrolysis, stimulated by the J-domain containing Hsp40s (Fig. 7.2a) (Misselwitz et al. 1998; Kityk et al. 2018), results in a conformational change whereby the "lid" closes over the SBD, which results in a tightly bound state (Fig. 7.2c) (Mayer and Kityk 2015; Mayer 2018). In addition to stimulating Hsp70 ATPase activity, Hsp40s deliver substrate to the Hsp70 and may provide specificity (Fig. 7.2a). In humans, there are 13 Hsp70s and ~50 Hsp40s (Kampinga et al. 2009). After ATP hydrolysis, a second class of co-chaperones, the nucleotide exchange factors (NEFs), stimulate the release of both the hydrolyzed nucleotide (i.e., ADP) and substrate (Fig. 7.2b) (Bracher and Verghese 2015). If the protein fails to reach a mature confirmation, Hsp70 may reengage the substrate to support additional rounds of folding. Alternatively, Hsp70s can target misfolded proteins in the ER for degradation, which is described below.

7.2.2   The Role of the Chaperone-Like Lectins in Protein Folding

Another class of ER chaperones, the glycan-binding lectins, recognize proteins containing N-linked glycans. Glycans are appended cotranslationally to polypeptides by the oligosaccharyl-transferase, which appends a core-glycan ($GlcNAc_2Man_9Glc_3$) to an Asn in the sequence Asn-X-Ser/Thr (where X can be any amino acid except Pro) (Fig. 7.3a). The glycan is then trimmed by glucosidases to yield a mono-glucosylated glycan ($GlcNAc_2Man_9Glc$), which enables the binding of two lectins, calnexin (CNX) and calreticulin (CRT) (Fig. 7.3b). CNX is a type I membrane protein, and CRT is a soluble paralog. While CNX/CRT binding was initially thought to be entirely glycan dependent, later evidence indicated that CNX/CRT can bind polypeptides in the absence of glycans (Danilczyk and Williams 2001; Sandhu et al. 2007; Brockmeier et al. 2009). CNX/CRT can also recruit a thioxidoreductase as well as a peptidyl-prolyl isomerase to respectively catalyze the formation of disulfide bonds in the ER and the interconversion of *trans–cis* Pro, both of which augment protein folding. In most cases, these events are sufficient to permit the mature protein to exit the ER after the action of another glucosidase. However, if folding is problematic, the UDP-glucose glycoprotein glucosyltransferase (UGGT1)

Fig. 7.2 Hsp70 cycle, cellular roles, and conformational changes. (**a**) Hsp40 J-domain proteins (JDPs) provide specificity for the diverse cellular roles and various substrates of Hsp70s. (**b**) Hsp70 ATPase/ substrate binding cycle as described in the text. Substrate (dark red) is targeted to Hps70 by JDPs. Structures of bacterial DnaK in the ATP bound open conformation (PDB 4B9Q) (Kityk et al. 2012) and the ADP-bound closed conformation (PDB 2KHO) (Bertelsen et al. 2009) bound to substrate (Zhu et al. 1996) are shown. Substrate-binding domain (SBD) is green. The nucleotide-binding domain (dark (lobe I) and light blue (lobe II)) linker between SBD and NBD is yellow. Nucleotide exchange factors (NEFs) promote release of ADP and ATP rebinding. (**c**) Model of Hsp70 conformational changes induced by ATP-binding and hydrolysis. Figure republished with permission of Mayer and Gierasch (2019). Permission conveyed through Copyright Clearance Center, Inc

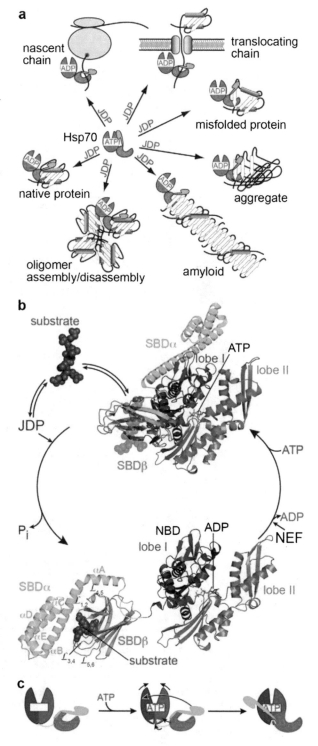

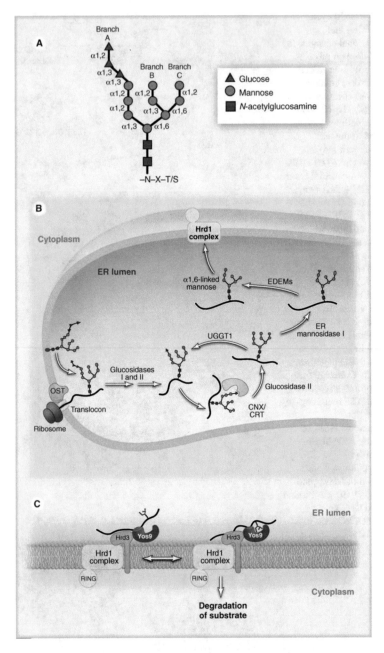

Fig. 7.3 N-linked glycosylation and the role of chaperone-like lectins in protein folding and degradation. (**a**) The core N-linked glycan unit is cotranslationally attached to nascent polypeptide chains as they enter the ER. (**b**) As described in the text, glycans are trimmed and glycoproteins undergo cycles of lectin binding and release prior to either degradation or reaching a mature confirmation. (**c**) Yos9 delivers the glycoprotein substrate to the Hrd1 complex by association with Hrd3. Figure republished with permission of Smith et al. (2011). Permission conveyed through Copyright Clearance Center, Inc

acts as a protein folding sensor. This enzyme adds a single glucose back to the core glycan, preventing ER exit and allowing for CNX/CRT reassociation (Fig. 7.3b) (Cherepanova et al. 2016). In principle, the cycle of glucose re-addition by the glucosyltransferase, favoring CNX/CRT binding, and glucose removal could continue indefinitely if the protein has failed to mature. However, proteins that exit the CNX/CRT cycle prior to reaching a mature state can be modified by EDEM (ER degradation enhancing α-mannosidase-like protein) (Ninagawa et al. 2014), thereby generating a substrate that is targeted for ERAD (Figs. 7.3b–c and 7.4, step

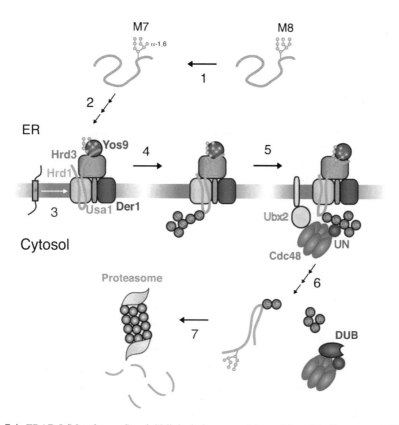

Fig. 7.4 ERAD-L/M pathway. *Step 1*: N-linked glycans are trimmed from 8 to 7 mannose residues (depicted as green circles). *Step 2*: The lectin Yos9 next binds to the terminal α1,6-linked mannose residues and the misfolded protein is inserted into Hrd1 channel with assistance from Hrd3 and Der1 (see discussion of retrotranslocation for alternative scenarios). *Step 3*: A single spanning membrane protein (ERAD-M substrate) interacts with Hrd1 through misfolded TMDs. *Step 4*: Substrate is ubiquitinated by the E3 ligase Hrd1. *Step 5*: Cdc48 (p97 in mammals) binds Ubx2 and the Cdc48 cofactors, Ufd1/Npl4 (UN) bind to the ubiquitinated substrate. *Step 6*: The substrate is retrotranslocated from the ER to the cytosol using the energy generated by Cdc48 ATP-hydrolysis. A deubiquitinating enzyme (DUB) trims the ubiquitin chain triggering substrate release from the Cdc48 complex. *Step 7*: The substrate is degraded by the 26S proteasome. Figure republished with permission of Wu and Rapoport (2018). Permission conveyed through Copyright Clearance Center, Inc

1–2). Although this cycle has been established for several model proteins, it remains unclear whether all proteins, especially membrane proteins, are subject to this cycle. In addition, some proteins are not glycosylated or the glycans may not be accessible to the enzymes that regulate protein retention and ERAD. Recent evidence indicates that other lectins, XTP-3B and Os9, help "decode" substrates for ERAD (van der Goot et al. 2018).

7.3   Endoplasmic Reticulum-Associated Degradation

During ERAD, substrates are first recognized by molecular chaperones (Ismail and Ng 2006; Meusser et al. 2005; Vembar and Brodsky 2008; Hebert and Molinari 2012; Rutkevich and Williams 2011; Young 2014). Following recognition, an integral membrane ERAD substrate—such as an ion channel or transporter—is next ubiquitinated by a series of ER-associated ubiquitin ligases (Fig. 7.4, step 3–5) (Preston and Brodsky 2017). The protein is then "retrotranslocated" or removed from the ER and transported to the cytosol through the action of an energy requiring complex that associates with ubiquitinated proteins (Fig. 7.4, step 6) (Bodnar and Rapoport 2017a). Finally, the ubiquitinated substrate is delivered to and degraded by the cytosolic 26S proteasome (Fig. 7.4, step 7) (Hiller et al. 1996; Hampton et al. 1996; Raasi and Wolf 2007).

7.3.1   Recognition of ERAD Substrates

Chaperones recognize folding lesions that include exposed hydrophobic patches, unpaired charged residues within membrane domains, or discrete sequences termed "degrons" that target proteins for degradation. Extensive experiments to characterize the degradation pathway of a diverse group of ERAD substrates have led to their placement in three categories based on the site of the folding lesion: ERAD-L (ER lumen), ERAD-M (membrane), or ERAD-C (cytosolic) (Vashist and Ng 2004; Taxis et al. 2003; Carvalho et al. 2006). When examined using model substrates, each class requires a unique set of factors for ERAD targeting. We will briefly discuss these requirements here, which were developed using yeast, but we note that the rules positioning substrates into these unique groups do not necessarily hold up for membrane proteins in mammalian cells (Bernasconi et al. 2010).

ERAD-L substrates are soluble proteins and are recognized by ER lumenal chaperones, including an Hsp70, BiP, along with Hsp40 and NEF cochaperones, select lectins, and thiol oxidoreductases. In addition to its role in protein folding, BiP identifies and targets ERAD-L substrates for ERAD and acts in conjunction with a subset of the seven Hsp40s (ERdj1-7) in the ER lumen (Pobre et al. 2019). For example, ERdj3 binds transiently to substrate through conserved, hydrophobic residues before delivering the substrate to BiP. Like most Hsp40s, ERdj3 is a dimer and interacts with BiP through the 70 amino acid J-domain. Once identified, BiP directs misfolded proteins to the mannose trimming EDEM proteins (Oda et al. 2003) and then on to a membrane complex for retrotranslocation and ubiquitination.

Due to the fact that the folding lesion resides within the membrane, ERAD-M substrates generally require fewer chaperones for substrate recognition. BiP, Hsp40s, and lectins are often dispensable to target ERAD-M substrates for degradation (Gardner et al. 2000; Sato et al. 2009). ERAD-M substrates rely instead on the Hrd1 complex to recognize misfolded lesions within the membrane (Figs. 7.4 and 7.5) (Sato et al. 2009). In one model, the exposure of hydrophilic or charged residues within the membrane may act as a folding lesion, which is in contrast to the exposure of hydrophobic residues in soluble proteins (e.g., ERAD-L substrates) or the cytoplasmic domain in ERAD-C proteins. To date, the mechanisms that lead to the selection of ion channels and transporters that are targeted for ERAD have not been defined.

These substrates are integral membrane proteins with folding lesions facing the cytoplasm and are recognized and targeted for degradation by cytosolic chaperones. For example, a well-characterized ion channel target of the ERAD pathway is the Cystic Fibrosis Transmembrane Conductance Regulator, CFTR. Mutations in CFTR are responsible for Cystic Fibrosis and the most prevalent CFTR mutant, F508del, results in ER retention and premature degradation by the ERAD pathway (Cheng et al. 1990). The bulk of the CFTR protein resides in the cytosol and consequently cytosolic chaperones are required to target CFTR for ERAD, whereas the ER lumenal chaperones are dispensable (Zhang et al. 2001). For example, cytosolic Hsp70 and sHsps are required for CFTR degradation (Rubenstein and Zeitlin 2000; Ahner et al. 2007) (also see Chap. 6 of this volume and Volume 3, Chaps. 15 and 16).

7.3.2 Ubiquitination of ERAD Substrates

The ubiquitination of an ERAD substrate occurs through a series of three enzymatic reactions carried out by an E1 ubiquitin-activating enzyme, an E2 ubiquitin-conjugating enzyme and an E3 ubiquitin ligase. The ubiquitin-activating enzyme catalyzes the formation of a high-energy thioester bond between the C-terminus of ubiquitin and an active site cysteine residue in an ATP-dependent reaction. The ubiquitin is then transferred to the active-site cysteine in an ubiquitin-conjugating enzyme. An E3 enzyme next facilitates the interaction between an E2 and the substrate and the transfer of ubiquitin to, most commonly, a Lys side chain. Ubiquitin conjugation onto Cys, Ser, Thr, or the N-terminus has also been observed (Shimizu et al. 2010; Swatek and Komander 2016). Polyubiquitin chains are formed when ubiquitin is appended to one of the seven Lys residues in ubiquitin (K6, K11, K27, K29, K33, K48, and K63). K48 polyubiquitin chains are the most common ubiquitin linkage for proteasomal degradation, and specificity of the linkage is provided by the E2. At least 4 ubiquitins in series are required for proteasome-dependent degradation. There are ~100 ubiquitin ligases in yeast and ~600 in humans, which reflects the role of these enzymes in substrate recognition, whereas there are only ~38 E2s and 2 E1s in humans (Mehrtash and Hochstrasser 2018). Hrd1 and Doa10 are the dedicated E3s for the ERAD pathway in yeast, and conserved homologs (HRD1 and gp78, and TEB4/MARCH6, respectively) function similarly in human cells. Each of these ligases is an ER membrane protein with the catalytic "RING" domains deposited in the cytosol. Other mammalian E3s associated with the ERAD pathway include TRC8, THEM129, Rfp2, RMA1, RNF145, RNF170, and RNF185 (Stagg et al. 2009; van den Boomen et al. 2014; Younger et al. 2006; Lu et al. 2011; El Khouri et al. 2013; Jiang et al. 2018; Lerner et al. 2007).

To facilitate substrate ubiquitination, each ligase requires somewhat different ubiquitin-conjugating enzymes. Hrd1 acts with Ubc7 in yeast to ubiquitinate ERAD-L or ERAD-M substrates. Ubc7 is tethered to the ER membrane by a membrane protein, Cue1, which directly activates conjugating activity (Fig. 7.5) (Metzger et al. 2013). Hrd1 also forms oligomers that are stabilized by another

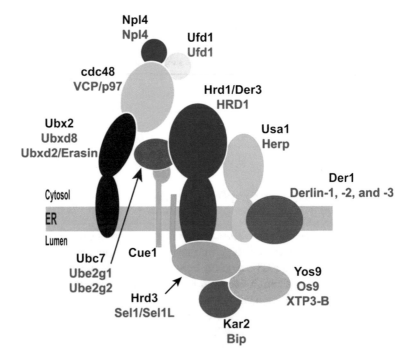

Fig. 7.5 Hrd1 E3 Ligase complex. Model of the *Saccharomyces cerevisiae* Hrd1 complex required for the degradation of ERAD-L substrates (black). Mammalian homologs are indicated in pink text. Key factors include the putative retrotranslocation channel, Hrd1, the adapter proteins, Usa1, Hrd3, and Der1, that complex with Hrd1, the Cdc48-Ufd1 complex (Npl4 is not shown) that retrotranslocates substrate, components of the ubiquitination machinery (Ubx2, Ubc7, Cue1) and chaperones, Kar2 and Yos9. Figure republished with permission of (Nguyen et al. 2009). Permission conveyed through Copyright Clearance Center, Inc

membrane protein, Usa1, which enhances ubiquitination activity (Horn et al. 2009; Carvalho et al. 2010). In addition, oligomerization increases the activity of gp78, the mammalian homolog of Hrd1, and gp78 can build polyubiquitin chains on the E2, Ube2g2, which in vitro can be transferred *en bloc* to an ERAD substrate, thereby increasing ubiquitination efficiency (Li et al. 2007). It is not clear whether this occurs in vivo. In contrast, Doa10 functions with both Ubc6 and Ubc7. Unlike Ubc7, Ubc6 is a tail-anchored protein in the ER membrane. Doa10 ubiquitinates ERAD-C substrates but exceptions to this rule have been noted (Habeck et al. 2015). Recently, the role of a conserved 16 amino acid, C-terminal element (CTE) in Doa10 was identified. Mutations in the CTE block the ubiquitination and degradation of select Doa10-dependent substrates. Therefore, the CTE may be important for substrate recognition. Mutations in the CTE of the mammalian homolog, MARCH6/TEB4, resulted in a phenotype similar to the Doa10 mutant (Zattas et al. 2016). Even though the action of these enzymes on the degradation of several ERAD substrates has been noted, the rules that define mechanisms underlying E3-dependent substrate selection are ill defined.

7.3.3   Retrotranslocation of ERAD Substrates

The driving force for the extraction or retrotranslocation of proteins from the ER into the cytosol is provided by the p97 complex (or Cdc48 complex in yeast) (Bays et al. 2001; Jarosch et al. 2002; Rabinovich et al. 2002; Ye et al. 2001). p97/Cdc48 is a hexameric AAA^+-ATPase and hydrolyzes ATP to spool proteins through a central aperture (Bodnar and Rapoport 2017b; Ripstein et al. 2017). Each p97/Cdc48 monomer contains two ATPase domains (D1 and D2) and an N-terminal domain (N). The two ATPase domains are stacked on top of one another. The N-terminal domains extend from the D1 ring, and movement of the N domains are coupled to ATP hydrolysis (Tang et al. 2010; Bodnar et al. 2018). p97/Cdc48 is involved in a variety of cellular activities that, in addition to ERAD, include mitochondrial associated degradation, removal of proteins from chromatin, membrane fusion, and vesicular trafficking (Xia et al. 2016). To date, ~38 p97 binding partners and cofactors have been identified, which dictate the cellular activity of p97 (Hanzelmann and Schindelin 2017).

During ERAD, the p97/Cdc48 cofactors—Ufd1 and Npl4—recognize polyubiquitinated proteins (Meyer et al. 2000, 2002). In addition to providing the driving force for retrotranslocation, the enzyme acts as an "unfoldase" (Bodnar et al. 2018). After recognition of a ubiquitinated substrate, ATP binds the D1 ring and the enzyme unfolds the protein using energy provided by ATP hydrolysis in the D2 ring. Partial trimming of the ubiquitin chain by a deubiquitinating enzyme may facilitate entry into the p97/Cdc48 central pore (Bodnar and Rapoport 2017b). ERAD remains proficient in a yeast strain lacking the deubiquitinating enzyme, suggesting that functional analogs are involved in retrotranslocation (Stein et al. 2014).

Does the retrotranslocation of an ERAD substrate require a dedicated channel? Early studies suggested that Sec61 (see, Fig. 7.1) is also required for the retrotranslocation and degradation of soluble ERAD substrates (Pilon et al. 1997, 1998; Wiertz et al. 1996; Plemper et al. 1997; Willer et al. 2008). However, later studies determined that Sec61 is dispensable for their retrotranslocation (Sato and Hampton 2006; Wahlman et al. 2007; Garza et al. 2009). In yeast, recent data strongly suggest that the E3 ubiquitin ligase, Hrd1, can also serve as a retrotranslocation channel for soluble substrates. For example, Hrd1 is necessary and sufficient for the retrotranslocation of an ER lumenal protein in vitro (Carvalho et al. 2010), and the structure of Hrd1 displays channel-like properties (Schoebel et al. 2017). In fact, Hrd1 links ERAD-requiring proteins both in the ER lumen and in the cytoplasm (Nakatsukasa et al. 2013). Nevertheless, the solubilization and extraction of membrane proteins require associated proteins, such as Der1 and Dfm1 (Gauss et al. 2006; Stein et al. 2014; Willer et al. 2008; Garza et al. 2009; Mehnert et al. 2014; Neal et al. 2018).

Yeast Der1 (Derlin in mammals) is an ER resident transmembrane protein required for the degradation of both soluble and membrane proteins, i.e., ERAD-L and ERAD-M substrates (Knop et al. 1996; Vashist et al. 2001; Neal et al. 2018; Hitt and Wolf 2004; Vashist and Ng 2004; Baldridge and Rapoport 2016). Der1 coprecipitates with other proteins involved in the ERAD pathway, and the association between Hrd1 and Der1 is mediated by Usa1 (Fig. 7.5) (Horn et al. 2009; Mehnert et al. 2014). p97 also binds to Der1 in an Hrd1-dependent manner (Gauss et al. 2006), consistent with Der1's role as an accessory during retrotranslocation (Fig. 7.5). The relationship between Der1 was further delineated via a site-specific cross-linking study in which it was established that the transmembrane domains of Der1 are in direct contact with both Hrd1 and Hrd3 and the lumenal domains and the transmembrane domains bind an ERAD substrate (Mehnert et al. 2014). Consistent with a direct role for Der1 in retrotranslocation, antibodies against Der1 inhibited the retrotranslocation of a soluble ERAD substrate that could be observed in real time (Wahlman et al. 2007). Together, Hrd3, which binds BiP (Mehnert et al. 2015), may recruit ERAD clients to Der1 (Fig. 7.5), and as the substrate emerges on the cytosolic face of the ER through the action of Hrd1, it is ubiquitinated by the cytoplasmic RING domain in Hrd1 (Mehnert et al. 2014).

Both Der1/Derlin and Dfm1 are members of the rhomboid protease family, which clip TMDs within membrane-spanning domains. However, both proteins lack critical catalytic residues and are therefore considered inactive rhomboids. A recent study indicated that rhomboids can distort and thin a lipid bilayer, suggesting that these proteins are unable to clip integral membrane ERAD substrates but instead help solubilize them prior to retrotranslocation (Kreutzberger et al. 2019). Consistent with this view, a genetic screen for ERAD effectors in yeast identified Dfm1 as being required for the degradation of integral membrane but not soluble ERAD substrates (Neal et al. 2018). Interestingly, yeast lacking Dfm1 rapidly acquire suppressor mutations that duplicate a region of the yeast chromosome that harbors the Hrd1 gene. Indeed, Hrd1 overexpression compensates for the deletion of Dfm1 (Neal et al. 2018). Together these data position a membrane complex, comprised of Hrd1 and inactive rhomboids, as playing a central role in retrotranslocation, assisted by the action of the p97/Cdc48 ATPase (Fig. 7.5). It is currently unknown whether these factors are involved in the degradation of misfolded epithelial ion channels and transporters.

7.3.4 Degradation by the 26S Proteasome

Prior to proteasome-dependent degradation, accessory factors facilitate the processing and targeting of substrates to this multi-catalytic protease. First, N-linked glycans may be removed by the cytosolic N-glycanase, Png1. Png1 is recruited to the ER membrane by Cdc48 (Li et al. 2006; Suzuki et al. 2000), suggesting that it acts immediately after or with retrotranslocation. Second, Dsk2 and Rad23 harbor both ubiquitin association domains as well as ubiquitin-like domains, which allows them to shuttle substrates to the proteasome (Richly et al. 2005; Verma et al. 2004; Bard et al. 2018). In yeast, deletion of the genes encoding these factors has a minimal effect on ERAD, suggesting that they function with only a subset of substrates. Consistent with this view, mutations in the human Png1 homolog result in a catastrophic disease, which may arise from impaired mitochondrial function (Suzuki 2015; Kong et al. 2018).

The 26S proteasome is composed of the 20S core particle (CP) (Finley et al. 2016) and the 19S regulatory particle (RP). The 20S CP contains four stacked heteroheptameric rings. Two rings of β subunits are flanked by two rings of α subunits. Three out of the seven β subunits contain proteolytic activity (six total active sites). The 20S CP contains three protease activities, chymotrypsin-like, trypsin-like, and caspase-like that cleave after hydrophobic, basic, and acidic residues, respectively. The 20S core of the proteasome forms a chamber that reduces nonspecific degradation of cellular proteins (Bard et al. 2018).

The 19S RP is responsible for unfolding and translocating polypeptide substrates into the 20S proteolytic chamber. The 19S RP is made of a base and a lid. The base consists of a six subunit AAA-ATPase and non-ATPase subunits that have multiple sites for binding ubiquitin and ubiquitin-like proteins, such as Rad23 and Dsk2 (see above). The hexameric AAA-ATPase ring is at the center of the base and provides the "motor" necessary to translocate proteins into the 20S proteolytic chamber. The lid also contains deubiquitinating enzymes that remove ubiquitin chains prior to polypeptide translocation into the 20S CP.

Recently, two cryo-EM structures of a substrate-engaged human 26S proteasome were published (Dong et al. 2019; de la Pena et al. 2018). Based on data from these studies, updated molecular models for substrate processing by the proteasome were proposed. It was reported that the 19S RP can be found in up to seven unique allosteric conformations (Dong et al. 2019; de la Pena et al. 2018; Bard et al. 2018). These include conformational rearrangements of both the base and the lid domain of 19S RP in response to the nucleotide-bound state, the presence of substrate, occupancy of the deubiquitinating enzyme, and the occupancy of ubiquitin receptors.

Therefore, substrates are initially targeted through ubiquitin chains by ubiquitin receptors prior to an unstructured region of the substrate initiating ATP-dependent translocation into the CP. Substrate engagement triggers major structural and conformational changes to augment 19S RP-dependent peptide translocation (Bard et al. 2018).

7.4 Epithelial Ion Channels and Transporters Subject to ER Protein Quality Control

7.4.1 The Na,K-ATPase

The Na,K-ATPase pump is ubiquitously expressed in epithelia and can reside on either the apical or basolateral membrane. The Na,K-ATPase facilitates the transport of sodium and potassium across the membrane and generates ion gradients to maintain membrane potential and drive vectorial sodium-dependent transepithelial transport by pumping three Na^+ ions into the extracellular space in exchange for two K^+ ions (Jorgensen 1986). The Na,K-ATPase pump is a heterodimer composed of an α-subunit that contains the ion-transport activity, and a β-subunit that is essential for channel maturation and trafficking (Geering 2008). The α-subunit is a ten TMD-containing protein with N- and C-termini facing the cytosol (Blanco 2005). The β-subunit is a single TMD-spanning glycoprotein with a short cytosolic tail and large extracellular domain (Geering 2001; Blanco 2005). While some studies suggested alternative subunit ratios (Yoshimura et al. 2008; Clifford and Kaplan 2008) analysis of the high-resolution structures of the Na,K-ATPase confirms a 1:1 α:β stoichiometry (Morth et al. 2007; Shinoda et al. 2009). There are four isomers of the α-subunit (α_1, α_2, α_3, and α_4) and three isoforms of the β-subunit (β_1, β_2, and β_3), and each show both tissue-specific and developmental differences in their expression patterns (Geering 2001; Blanco 2005). Each isoform also has unique kinetic properties and responds differently to substrate, ATP, and/or inhibitors (Blanco and Mercer 1998; Crambert et al. 2000). There is also evidence that the Na,K-ATPase is involved in cell adhesion and the maintenance of adherens junctions as the cytosolic domain of the pump binds E-cadherin and the ankyrin and spectrin cytoskeleton (Kizhatil et al. 2007; Nelson et al. 1990; Nelson and Veshnock 1987). The β-subunit dimerizes with β-subunits on opposing cells to maintain adherens junctions (Tokhtaeva et al. 2011; Padilla-Benavides et al. 2010). A more thorough discussion of the physiological roles and function of the Na,K-ATPase are provided in Volume 3, Chap. 1, of this series.

7.4.1.1   Assembly and ER-Associated Degradation of the Na,K-ATPase

In the absence of the β-subunit, the α-subunit is retained in the ER and targeted for ERAD (Beguin et al. 1998, 2000; Beggah et al. 1996, 1999). Several studies have provided insight into the mechanism by which the α-subunit is degraded (Beguin et al. 1998, 2000; Hasler et al. 2001). First, a series of truncated α-subunit TMD constructs were expressed in oocytes. Data indicated that the first four TMDs act alternatively as efficient "signal-anchor sequences" and "stop-transfer sequences" (see Fig. 7.1), thereby integrating TMD1-4 in the predicted topology (Beguin et al. 1998). However, efficient membrane integration of TMD6-10 required both C-terminal sequence information and the presence of the β-subunit. In the absence of the β-subunit, TMD7 was unable to act as a signal anchor sequence but remained in the cytosol. Therefore, the extracellular loop between TMD7-8, which contains a degradation signal (Beguin et al. 2000), is exposed to the cytosol, creating an unstable, immature protein (Beguin et al. 1998). Coexpressing the β-subunit promoted membrane integration of TMD7 and masked the degradation signal (Beguin et al. 2000). Membrane integration of the TMD7-8 pair resulted in proper localization of the TMD7-8 loop within the ER lumen and stabilized not only the full-length α-subunit but any construct containing at least TMD1-8. In the presence of the β-subunit, exposure of the degradation signal is only transient (Beguin et al. 1998, 2000). Thus, the β-subunit exhibits chaperone-like activity.

Although channel assembly is required for ER exit of the α_1-subunit (Beguin et al. 1998, 2000; Beggah et al. 1996, 1999), whether oligomerization is a prerequisite for trafficking of the β-subunit is less clear (Noguchi et al. 1990). However, later studies established that all mature β-subunits were bound to α-subunits (Clifford and Kaplan 2008), and *Xenopus* oocyte β-subunits were ER retained in the absence of α-subunits (Ackermann and Geering 1990; Jaunin et al. 1992; Beggah et al. 1996). The α-subunit also appears to be the limiting factor for α-β assembly (Tokhtaeva et al. 2009). As with α-subunits, ER retained unassembled β subunits are rapidly degraded, and mutations in the β-subunit TMD at residues predicted to interact with α-subunit TMD7 or the β-subunit extracellular domain results in ER-retention of the β-subunit and reduced colocalization with the α-subunit (Shinoda et al. 2009). Corresponding mutations in the α-subunit similarly resulted in impaired assembly and ER retention of the subunits (Tokhtaeva et al. 2009). These studies are all consistent with an inability of the β-subunit to traffic to the Golgi in the absence of the α-subunit.

7.4.1.2   The Roles of Chaperones in Na,K-ATPase Regulation

The role of molecular chaperones in Na,K-ATPase folding, assembly, quality control, and trafficking is not well characterized, but some evidence exists that BiP and calnexin regulate Na,K-ATPase biogenesis (Beggah et al. 1996; Tokhtaeva et al. 2010). Moreover, the least stable isoforms of the α- and β-subunits associated strongly with BiP, as did ER-retained subunits, whereas subunits with complex (Golgi appended) glycans did not coprecipitate with BiP. Consistent with these observations, coexpressing the α- and β-subunits reduced BiP association, presumably by promoting assembly, maturation, and trafficking of the Na,K-ATPase heterodimer (Beggah et al. 1996). These data strongly suggest that BiP is important either for Na,K-ATPase folding and assembly and/or for preferentially binding and targeting misfolded subunits for ERAD. These data are consistent with earlier data on yeast BiP, which is required for the folding and secretion of a wild-type version of a yeast glycoprotein; when mutated, BiP instead targets the substrate for ERAD (Simons et al. 1995; Plemper et al. 1997).

As described above, calnexin recognizes N-linked glycans and retains immature proteins in the ER. The three isoforms of the Na/K-ATPase β-subunit, β_1, β_2, and β_3, possess 3, 8, and 2 glycosylation sites, respectively. Surprisingly, β_1 and β_3 subunit functions are not impaired when glycosylation is blocked (Takeda et al. 1988; Lian et al. 2006; Beggah et al. 1997; Zamofing et al. 1989). However, when glycosylation of the β_2 isomer is blocked, the protein is retained in the ER and fails to assemble with the α-subunit. All eight glycosylation sites are not required for trafficking of β_2, but glycosylation sites two and seven are critical (Tokhtaeva et al. 2010). These data are consistent with studies in yeast that established a hierarchy of glycosylation sites in secreted proteins: Those that reside in hard-to-fold regions appear to be most vital for selecting the protein for ER retention and ERAD (Spear and Ng 2005). Consistent with these data, hydrophobic amino acids near glycosylation sites are required for calnexin binding and calnexin-mediated maturation of the β_2-subunit (Tokhtaeva et al. 2010).

7.4.2 The Epithelial Sodium Channel

Epithelial Sodium Channel (ENaC) is formed by homologous α-, β-, and γ-subunits, and each subunit contains two transmembrane domains (TMDs), a large extracellular loop (ECL) and cytosolic N- and C-termini (Canessa et al. 1994a). ENaC is highly expressed in the lung, colon, and kidney epithelia where it reabsorbs sodium. In the kidney, ENaC is expressed on the apical surface of the distal convoluted tubule, connecting tubule, and collecting duct, where ENaC is responsible for Na^+ reabsorption from urine and regulates salt and water homeostasis and therefore blood pressure (Rossier 2014; Bhalla and Hallows 2008; Snyder 2005; Palmer 2015). ENaC also provides the driving force for K^+ secretion by acting in conjunction with the ROMK potassium channel. Gain of function mutations in ENaC lead to hypertension and Liddle's syndrome (Staub et al. 1996), and loss of function mutations lead to hypotension, salt-wasting and Pseudohypoaldosteronism Type 1 (PHA1) (Grunder et al. 1997; Bhalla and Hallows 2008). In the lung, ENaC is negatively regulated by CFTR. In the absence of CFTR, ENaC hyperactively reabsorbs Na^+ which further dehydrates the airway mucosa and exacerbates cystic fibrosis symptoms (Mall et al. 2004; Hobbs et al. 2013; Gentzsch et al. 2010;

Ismailov et al. 1996; Yan et al. 2004). ENaC trafficking and surface expression are highly regulated and have been extensively studied, which is discussed in Volume 3, Chap. 18 of this series.

7.4.2.1 Posttranslational Modifications of ENaC

The ENaC subunits are modified by N-linked glycosylation, the formation of disulfide bonds, palmitoylation, acetylation, and proteolytic cleavage in the Golgi (Buck and Brodsky 2018). Interestingly, the α-, β-, and γENaC subunits are subject to different posttranslational modifications, even though they are ~30–40% identical (Bhalla and Hallows 2008; Canessa et al. 1994a; Mueller et al. 2010; Valentijn et al. 1998; Weisz et al. 2000; Snyder et al. 1994). For example, the α-, β-, and γENaC subunits contain 6, 13, and 7 sites for N-linked glycosylation, respectively (Buck and Brodsky 2018), and blocking glycosylation in any subunit altered channel conductance and Na^+-dependent inhibition (Kashlan et al. 2018). Expression of β-subunits lacking glycans led to the most dramatic effect, with an overall reduction in Na^+ current of ~50–80%. Interestingly, the α-subunit was more sensitive to protease under these conditions. Although glycosylation of the α-subunit was previously shown to be dispensable (Canessa et al. 1994a; Snyder et al. 1994), other studies established that tunicamycin treatment, which blocks the addition of N-linked sugars, inhibited ENaC current (Kieber-Emmons et al. 1999). ENaC trafficking is also subject to Na^+-dependent regulation: low Na^+ favors the maturation of glycans, consistent with trafficking to the later secretory pathway, whereas high Na^+ results in the acquisition of immature glycans (Heidrich et al. 2015). Because glycan maturation occurs in the Golgi under low Na^+ conditions (0 mM Na^+ in media), ENaC may follow an alternative trafficking pathway that bypasses the Golgi under conditions of high Na^+ (130 mM Na^+ in media) (Heidrich et al. 2015).

Members of the protein disulfide isomerase (PDI) family oxidize Cys residues within the ER to form disulfide bonds in nascent proteins (Galligan and Petersen 2012). The formation of disulfide bonds requires a thioredoxin motif (Cys-X-X-Cys), and to date 21 human PDIs are known (Galligan and Petersen 2012). Each ENaC subunit contains 16 conserved Cys residues, so perhaps not surprisingly each subunit in the channel forms several intrasubunit disulfide bonds, which are important for function (Buck and Brodsky 2018). For example, mutating the first or sixth Cys in α- or βENaC reduces surface expression, but mutating any of the other Cys residues results in an increase in sodium current. Disulfide bonds are also required for self-inhibition in response to Na^+ (Sheng et al. 2007; Firsov et al. 1999).

In contrast to most PDIs that harbor two thioredoxin motifs, some family members contain only a single Cys-X-X-Cys motif. These PDIs are unable to catalyze the formation of disulfide bonds in substrates but can form covalent bonds to single Cys residues and thus rearrange existing disulfide bonds. One of these family members is ERp29, which promotes ENaC trafficking from the ER to the Golgi (Grumbach et al. 2014). Specifically, ERp29 promotes the association of βENaC with a cargo recognition component of the coat complex II (COPII), which regulates ER exit and thus channel maturation.

A third posttranslational modification to which ENaC subunits are subject is the conjugation of palmitate, which is transferred to a Cys, Ser, or Thr near a hydrophobic amino acid patch by a palmitoyl transferase (De and Sadhukhan 2018; Mueller et al. 2010; Mukherjee et al. 2014). Palmitoylation increases the net hydrophobicity of a substrate, and ~50% of all human ion channels are palmitoylated. The addition of palmitate can inhibit protein–protein interactions, act as a signal for endocytosis and block recycling to the plasma membrane (Shipston 2011; Nadolski and Linder 2007; Linder and Deschenes 2007; Fukata et al. 2016). Both the βENaC (Cys43 and Cys557) and the γENaC (Cys33 and Cys41) subunits are modified (Mueller et al. 2010; Mukherjee et al. 2014). Treatment of cortical collecting duct cells with a palmitoyl transferase inhibitor reduced ENaC-specific Na^+ current (Mukherjee et al. 2017), and mutating the palmitoylation sites decreased ENaC channel open probability (P_o) and therefore reduced Na^+ current (Mueller et al. 2010; Mukherjee et al. 2014). Among the 23 mammalian palmitoyl transferases, coexpression of five (DHCC1, 2, 3, 7, and 14) increased Na^+ current, and ENaC coprecipitated with each of these enzymes (Mukherjee et al. 2017). All of the transferases are expressed in either the ER and/or the Golgi (Ohno et al. 2006), which raises the possibility that this modification affects early events during ENaC quality control.

Finally, each of the ENaC subunits is acetylated (Butler et al. 2015). Although protein acetylation is usually associated with histones, many other proteins are subject to this modification (Sadoul et al. 2008; Drazic et al. 2016). ENaC subunit acetylation increases both the cellular levels of the channel and the amount of ENaC present at the plasma membrane (Butler et al. 2015). Acetylation inhibitors increase ENaC current and expression, perhaps by inhibiting ubiquitination. Because ubiquitination of ENaC at the cell surface triggers endocytosis, these data suggest that acetylation and ubiquitination of Lys residues on the cytoplasmic tails of ENaC are antagonistic (Zhou et al. 2013; Oberfeld et al. 2011; Butler et al. 2015). In contrast, HDAC7 deacetylates ENaC early in the secretory pathway, because inhibition of HDAC7 increases the total ENaC pool, i.e., not just the population of surface-expressed ENaC (Butler et al. 2015). In the future, it will be important to determine if HDAC7-mediated deacetylation alters the targeting of ENaC for ERAD.

7.4.2.2 Regulation of ENaC by ERAD

Expression of all three ENaC subunits is required for efficient channel assembly and trafficking to the cell surface, and when the subunits are expressed individually they are rapidly targeted for ERAD (Snyder 2005; Staub et al. 1997; Valentijn et al. 1998; Asher et al. 1996; Masilamani et al. 1999). Moreover, even when all three subunits are present a significant percentage of the ENaC subunits continue to be degraded (Staub et al. 1997; Valentijn et al. 1998; Malik et al. 2001; Kashlan et al. 2007). To identify factors required for ENaC degradation, a yeast ENaC expression system was developed (Mueller et al. 2007; Buck et al. 2010, 2017). Initial studies using the yeast model confirmed that ENaC is targeted for ERAD, and more specifically that ENaC degradation requires the proteasome and the Cdc48 complex (Kashlan et al. 2007) (see above). The sHsps, which prevent protein aggregation in an ATP-independent manner, are also required for αENaC degradation (Kashlan et al. 2007). Importantly, overexpression of a mammalian sHsp in MDCK cells decreased ENaC expression, and co-expression of the sHsp with ENaC decreased Na^+ current and ENaC surface expression, consistent with results from the yeast model (Kashlan et al. 2007). Additionally, deletion of the ER lumenal Hsp40s, Jem1, and Scj1, which normally interact with BiP, stabilized ENaC in yeast, and co-expression of the human homologs with ENaC in *Xenopus* oocytes also reduced ENaC current and surface expression (Buck et al. 2010). Surprisingly, ENaC degradation showed no dependence on the yeast BiP homolog, even though the Hsp40s played an important role in degradation (Buck et al. 2010). Because ENaC ubiquitination was reduced in the absence of the Hsp40s, the Hsp40s—but not BiP—appear to be required to recognize or target ENaC for degradation.

Although BiP does not regulate ENaC degradation in yeast, another Hsp70-like molecule, Lhs1, is present in the ER. Lhs1 is a member of the Hsp110/GRP170 family of proteins, which act as a NEF for Hsp70 (Baxter et al. 1996; Craven et al. 1996; Steel et al. 2004; Tyson and Stirling 2000). Similar to Hsp70, mammalian Lhs1 can also bind misfolded proteins (Behnke and Hendershot 2014) and acts as a "holdase" to prevent misfolded protein aggregation (de Keyzer et al. 2009). In the yeast model, we found that αENaC degradation required this Hsp70-like NEF, and an unglycosylated species of αENaC was preferentially stabilized (Buck et al. 2013). Consistent with these results, the mammalian Lhs1 homolog, GRP170, also promoted αENaC degradation and preferentially bound to unglycosylated αENaC in HEK293 cells (Buck et al. 2013). In contrast to strains lacking the Hsp40s, ubiquitinated αENaC accumulated in the ER membrane when Lhs1 was deleted (Buck et al. 2017). These data suggest that Lhs1/GRP170 acts at a post-ubiquitination step and are consistent with this factor playing a role in αENaC retrotranslocation. Similarly, a role for GRP170 in the retrotranslocation of a viral protein and a misfolded glycoprotein, NHK, from the ER was observed, but retrotranslocation was BiP-dependent in both of these cases (Inoue and Tsai 2015, 2016).

Even though BiP is dispensable for ENaC biogenesis or quality control, the cytosolic Hsp70 promotes ENaC trafficking to the cell surface by enhancing ENaC association with COPII (Chanoux et al. 2012). Surprisingly, the closely related, constitutively expressed homolog of Hsp70, Hsc70, instead promotes ENaC degradation (Chanoux et al. 2013; Goldfarb et al. 2006). Although a mechanism for Hsc70-mediated degradation is unknown, the chaperone probably targets ENaC to the ERAD pathway. Consistent with these results, some stabilization of ENaC was noted when degradation was examined in yeast expressing a temperature-sensitive allele of a cytosolic Hsp70, Ssa1 (Buck et al. 2010).

Other modulators of ENaC stability and trafficking include PON2 (Paraoxonase-2) and Derlin-1 (Shi et al. 2017; You et al. 2017). PON2 is an ER-associated lactonase that is expressed in the ENaC-expressing principal cells of the distal nephron (Shi et al. 2017). The *Caenorhabditis elegans* PON2 homolog, MEC-6, promotes trafficking of an ENaC family member, the mechanosensitive degenerin, MEC-4 (Harel et al. 2004; Chelur et al. 2002). In transfected HEK293 cells, PON2 coimmunoprecipitates with ENaC, but in contrast to MEC-6, PON2 overexpression lowers ENaC currents and reduces ENaC surface expression (Shi et al. 2017). The reason that PON2 and MEC-6 appear to function antagonistically toward their respective sodium channels is unclear, but more work on these ill-defined chaperone-like proteins is essential. Similarly, overexpression of Derlin-1 promotes ENaC degradation in mammalian cells and associates with the channel (You et al. 2017). However, whether or not Derlin-1 serves as the ENaC retrotranslocation channel alone on in conjunction with other factors, perhaps gp78 or HRD1, must also be defined.

In yeast, ENaC degradation is dependent on both the ERAD-L/M E3 ubiquitin ligase, Hrd1, as well as the ERAD-C ubiquitin ligase, Doa10 (Buck et al. 2010). Whereas βENaC is nearly completely stabilized in a yeast strain lacking both Hrd1 and Doa10, a significant portion of both α- and γENaC continues to be degraded. This result suggests that a third unidentified ubiquitin ligase targets α- and γENaC for degradation. Because Hrd1 and Doa10 play a role in ENaC degradation, ENaC may contain folding lesions in multiple cellular compartments or Hrd1 and Doa10 may compensate for one another. To date, it remains mysterious which E3 ubiquitin ligases act on misfolded forms of ENaC in the mammalian ER.

7.4.2.3 ENaC Channel Assembly and ER Exit

Assembly of the α-, β-, and γENaC subunits likely occurs in the ER, and indirect evidence supports this assumption. First, in yeast, where nearly all ENaC is retained in the ER, the ENaC subunits coprecipitate (Buck et al. 2010, 2017). Second, in mammalian systems, immature channels lacking complex glycans or Golgi-associated protease activation also associate with one another, suggesting assembly occurs in a pre-Golgi compartment (Hughey et al. 2003). Third, when constructs containing only single αENaC domains were coexpressed with wild-type ENaC, channel trafficking was blocked (Bruns et al. 2003). These data imply that the single domain constructs assembled with wild-type ENaC subunits and prevented ENaC trafficking. While some heteromeric membrane proteins assemble in a defined order (Wanamaker et al. 2003; Adams et al. 1997; Feige et al. 2015), there is no evidence that ENaC assembly is also ordered. Nevertheless, intersubunit contacts between each of the ENaC domains is required for channel assembly, which was highlighted in the recent cryo-EM structure of the trimeric channel (Noreng et al. 2018). Assembly of ENaC TMDs is particularly important to reach a mature conformation. Evidence to support this view is derived from the expression of a series of chimeric α-βENaC constructs, which were then coexpressed with wild-type αENaC and γENaC. It was established that interactions between the TMDs alone, and not interactions between the large ER loops, are required to evade chaperone-dependent degradation (Buck et al. 2017).

What is the signal for the assembled channel to exit the ER after assembly is complete? Early data indicated that the αENaC subunit can form a homotrimer that inefficiently traffics to the plasma membrane in the absence of the β- and γENaC subunits, whereas the β- and γENaC subunits require the α-subunit for ER exit (Canessa et al. 1994b). Therefore, it was hypothesized that αENaC possesses an ER exit signal (Canessa et al. 1994b; Mueller et al. 2007). Indeed, using a reporter protein fused to cytoplasmic segments of ENaC, a five amino acid motif was identified that acted as a trafficking signal (Mueller et al. 2007). The mechanism by which these residues promote ER exit, or whether a competing ER retention sequence is also present, is unknown.

7.4.3 Other Epithelial Ion Channels and Transporters Regulated by ERAD

7.4.3.1 Renal Outer Medullary Potassium Channel

Renal Outer Medullary Potassium Channel (ROMK) (Kir1.1/*KCNJ1*), is a member of the inwardly rectifying potassium channel family. ROMK is a homotetramer, and each ROMK subunit contains two TMDs, a potassium selectivity filter, and large cytosolic N- and C-terminal tails (Ho et al. 1993). Like ENaC, ROMK is expressed on the apical surface of epithelial cells in the distal nephron of the kidney. ROMK is responsible for potassium secretion and also regulates salt-water homeostasis. ROMK physiology, trafficking, and regulation are discussed in depth in Volume 3, Chap. 19, of this series.

To date, over 60 mutations have been identified in ROMK that result in Type II Bartter syndrome, a disease that is associated with early developmental defects and lifelong hypokalemic alkalosis (Welling 2016; Ji et al. 2008). N-terminal frameshifts result in a nonfunctional channel, but some mutations alter ion selectivity, gating, or ligand binding (Fang et al. 2010; Peters et al. 2003). Although it was unknown whether other mutations instead altered protein folding, potentially targeting the channel for ERAD, four mutations that cluster in a C-terminal immunoglobulin-like domain result in trafficking defects in oocytes and/or HEK293 cells (Fallen et al. 2009; Peters et al. 2003). Later, via the development of a yeast model, the same mutants were found to be selected for proteasome-dependent degradation in a p97/Cdc48- and Hsp70-dependent manner (O'Donnell et al. 2017; Mackie et al. 2018). The same mutants were also degraded in a proteasome- and p97/Cdc48-dependent manner in mammalian cells. It will be important to next define which of the other uncharacterized disease-causing mutants are also targeted for ERAD.

7.4.3.2 Thiazide-Sensitive Sodium Chloride Cotransporter

Sodium Chloride Cotransporter (NCC) is a member of the SLC12 cation chloride cotransporter family of electroneutral transporters and is a 12 TMD protein with a glycosylated extracellular loop and large cytosolic N- and C-termini (Castaneda-Bueno et al. 2010). NCC is expressed on the apical surface of the distal convoluted tubule and also regulates sodium reabsorption, thereby playing a major role in salt-water homeostasis and blood pressure regulation. The transporter is also a target for the thiazide diuretics. Mutations in NCC result in Gitelman Syndrome, which is typified by salt-wasting, hypokalemic alkalosis, hypomagnesemia, and hypocalciura (Fulchiero and Seo-Mayer 2019). NCC is discussed in depth in Volume 3, Chap. 3, of this series.

Several characterized Gitelman Syndrome-causing mutations in NCC result in the production of an immature protein that contains only core N-linked glycans (Kunchaparty et al. 1999; De Jong et al. 2002; Sabath et al. 2004). To determine whether NCC is an ERAD substrate, and whether these mutations accelerated ERAD, a yeast expression system for NCC was developed (Needham et al. 2011). As anticipated, NCC was primarily ER localized, and degradation was slowed in the presence of a proteasome inhibitor. Further, NCC was ubiquitinated primarily by Hrd1 in an Hsp70-dependent manner. In mammalian cells, immature NCC was again ubiquitinated and targeted for proteasome-dependent degradation. Consistent with the data in yeast, Hsp70 overexpression increased the proportion of immature NCC and increased the degradation rate (Needham et al. 2011). A role for acceler-ated chaperone-dependent ERAD in Gitelman Syndrome was then verified via a mass spectrometry analysis. Major hits from this screen included Hsp90 and Hsp90 cochaperones, CHIP—an E3 ubiquitin ligase previously implicated in ERAD (McDonough and Patterson 2003)—an Hsp70/Hsp90 organizer protein, an Hsp40, as well as the known NCC effector, Hsp70 (Donnelly et al. 2013). Through the application of an Hsp90 inhibitor, it was established that NCC maturation depended on Hsp90 function (Donnelly et al. 2013). Notably, Gitelman syndrome mutants associated more strongly with Hsp70, Hsp40, and CHIP than wild-type NCC (Donnelly et al. 2013). The data are consistent with sequential NCC quality control in the ER, whereby the Hsp70 and Hsp90 chaperone complex acts sequentially to monitor NCC folding and ultimately promote channel trafficking (Donnelly et al. 2013).

7.4.3.3 V2 Vasopressin Receptor

The V2 vasopressin receptor (V2R) is a G-protein coupled receptor (GPCR) stim-ulated by arginine-vasopressin (AVP) (Birnbaumer et al. 1992). AVP responds to either an increase in NaCl or a decrease in blood pressure and binds V2R, which resides in the basolateral membrane of principal cells in the distal nephron. Activa-tion of V2R by AVP triggers AQP2 insertion in the plasma membrane and allows for water reabsorption from urine. In the absence of functional V2R, AQP2 remains embedded in intracellular vesicles. Mutations in the V2R (*AVPR2* gene) cause X-linked nephrogenic diabetes insipidus (NDI) (Hoekstra et al. 1996), and >270 NDI-causing mutations have been identified (Milano et al. 2017). These mutations are responsible for ~90% of the inherited cases of NDI, while the other ~10% are caused by mutations in the AQP2 gene, as described below. Patients with NDI are unable to respond to AVP and produce copious volumes of urine (polyuria) and exhibit excessive thirst (polydipsia).

NDI-causing V2R mutants are either retained in the ER or Golgi, or they traffic to the plasma membrane and are defective (Milano et al. 2017). Among nine NDI-causing mutants, most are ER-retained and subject to ERAD (Robben et al. 2005). Later, three V2R mutants with alterations in the cytosolic portion of the protein (L62P), within the membrane (V226E), and within the lumen (G201D) were characterized (Schwieger et al. 2008). G201D localized throughout the secretory pathway, whereas the L62P mutant was found exclusively in the ER. The V226E mutant was visualized in both the ER and ER-Golgi intermediate compartment. Proteasome inhibition stabilized the unglycosylated and core glycosylated species of both the wild type and the mutant V2Rs, and all of the proteins were ubiquitinated to varying degrees. Consistent with V2R being a substrate for ERAD, the unglycosylated and core glycosylated V2R mutants coimmunoprecipitated with p97 and Derlin-1. Although the chaperone and ubiquitin ligase requirements for degradation were not addressed, this will be an important future area of research.

7.4.3.4   Aquaporin-2

Aquaporin-2 (AQP2) is responsible for water reabsorption in the distal nephron and is expressed on the apical surface of principal cells, where ENaC, V2R, and ROMK also reside. AQP2 is a six TMD homotetramer (Bai et al. 1996). Over 60 NDI-causing mutations in AQP2 have been described (Milano et al. 2017), many of which lead to protein retention in the ER (Lin et al. 2002; Lloyd et al. 2005; Tamarappoo and Verkman 1998; Marr et al. 2002; Iolascon et al. 2007). Some of these mutants function if they reach the cell surface, suggesting the folding defect is minor (Leduc-Nadeau et al. 2010; Marr et al. 2002). Interestingly, 11 of the 65 mutants are dominant (Kuwahara et al. 2001; Sohara et al. 2006; Asai et al. 2003; Mulders et al. 1998) and form heterotrimers with wild-type subunits that are either misrouted to the lysosome (Marr et al. 2002) or the basolateral membrane (Kamsteeg et al. 2003) or are retained in the Golgi (Mulders et al. 1998; Procino et al. 2003).

Perhaps not surprisingly, several of the AQP2 NDI-associated mutants are ERAD substrates, but there is a dearth of information on the factors that recognize and target AQP2 mutants for degradation. One study observed that, in contrast to wild-type AQP2, the glycosylated form of four AQP2 mutants was significantly more stable than the unglycosylated species. Therefore, lectins may be involved in AQP2 folding and quality control, but this has not been investigated (Buck et al. 2004). There is also some evidence that the ERAD-associated E3 ubiquitin ligase, CHIP (see above), regulates AQP2 levels (Wu et al. 2018). AQP2 levels are higher in cells lacking CHIP, and CHIP seems to be required for both proteasomal and lysosomal AQP2 degradation. In an in vitro assay CHIP ubiquitinates AQP2, but Hsc70 blocks ubiquitination. AQP2 expression is also increased in a CHIP knockout mouse and water intake is lower, which is consistent with increased AQP2 activity (Wu et al. 2018). A more thorough discussion of AQPs is provided in Volume 3, Chap. 30, of this series.

7.4.3.5   Polycystin-2

Polycystin-2 (PC2) is a six TMD nonselective cation channel with cytosolic N- and C-termini (Mochizuki et al. 1996). The transmembrane domain of PC2 is homologous to polycystin-1 (PC1), which is a PC2 partner, and the TRP channels (Somlo and Ehrlich 2001). PC2 function is dependent on cellular residence. In the ER, PC2 acts as a Ca^{2+} release channel (Clapham 2003; Koulen et al. 2002), whereas at the plasma membrane PC2 may facilitate Ca^{2+} entry (Luo et al. 2003). In embryonic nodal cells and the primary cilia of renal tubular cells, PC2 acts as a flow sensor (McGrath et al. 2003), and when PC2 is inhibited the Ca^{2+} response to flow stimulation is absent (Nauli et al. 2003). Mutations in PC2—as well as PC1—are responsible for autosomal dominant polycystic kidney disease (ADPKD) (Torres and Harris 2006), which is the fourth leading cause of kidney failure.

Mounting evidence suggests that PC2 is targeted for ERAD. First, PC2 is ubiquitinated, and PC2 degradation is proteasome dependent (Liang et al. 2008). Second, PC2 degradation is dependent on p97, and PC2-GFP localizes to the ER and coprecipitates with p97 in MDCK cells. Liang et al. (2008) also determined that Herp, which may function similarly to the Hrd1-Der1 assembly factor, Usa1 (see above) and recruits substrates to this complex (Leitman et al. 2014), contributes to PC2 ERAD. It will be interesting to determine whether disease-causing mutations in PC2 enhance ERAD, as seen for several of the substrates discussed above, and which chaperones select PC2 for degradation.

7.4.3.6   The Sodium-Potassium Chloride Cotransporter-2

The sodium-potassium chloride cotransporter-2, NKCC2, is another renal protein that, when mutated, gives rise to a disease associated with altered salt-water balance (type I Bartter Syndrome). NKCC2, which is homologous to NCC (see above), resides in the thick ascending limb of the loop of Henle and plays a pivotal role in NaCl reabsorption. Prior work established that the protein is targeted for ERAD, which is consistent with its slow rate of ER exit (Zaarour et al. 2012). More recent evidence indicates that immature, ER glycosylated forms of the substrate are selected for degradation through the action of the ER lectin Os9 (Seaayfan et al. 2016). As in many of the examples outlined in this section, ongoing work on

NKCC2 will better define how the transporter is targeted for ERAD. A more thorough discussion of NKCC2 function is provided in Volume 3, Chap. 2, of this series.

7.5 Conclusions and Future Directions

In this chapter, we have provided an overview of the many hurdles that a nascent protein in the ER must overcome. We then outlined the chaperones, chaperone-like factors, and enzymes that impact protein folding and degradation, and focused on the mechanisms that lead to the selection, ubiquitination, and delivery of an ERAD substrate to the proteasome. As described in the preceding section, a growing body of literature indicates that TMD-containing proteins, many of which reside in epithelia, are targeted to this pathway. While disease-causing mutations may slow ER exit and hasten ERAD, even wild-type versions of these proteins fold inefficiently and are targeted for ERAD. Therefore, much has been learned from studying the biogenesis of both wild type and disease-associated mutants. As highlighted throughout the chapter the use of model systems has led to significant insights on ER quality control processes.

Although the field is in its infancy, there is growing interest in harnessing the knowledge obtained from studies on ER protein quality toward the development of therapeutics that correct misfolded epithelial proteins. Therefore, a better definition of how these varied ion channels and transporters assemble in the ER and traffic to the plasma membrane is imperative. Small molecules that target any of these steps might lead to the development of new therapeutic options.

For example, pharmacological chaperones are under development that target ER retained V2R mutants (Los et al. 2010). Like molecular chaperones, pharmacological chaperones are designed to assist protein folding. One class of these molecules are the nonpeptide vasopressin receptor antagonists, or vaptans (Francis and Tang 2004). Vaptans mimic the structure of AVP (Mouillac et al. 1995), are cell permeable, and thus bind the AVP binding site on V2R in the ER. Vaptans stabilize the mutant V2R allowing for maturation and trafficking while avoiding ERAD (Milano et al. 2017). Six vaptans can promote V2R trafficking, maturation, plasma membrane expression, and the AVP response (Robben et al. 2007; Morello et al. 2000; Wuller et al. 2004; Bernier et al. 2004). In small scale clinical trials for a vaptan, SR49059, patient urine volume was decreased while urine osmolality was significantly increased.

Perhaps the best example in which pharmacological approaches to target ERAD-associated diseases has been successful is Cystic Fibrosis. The F508del CFTR mutation, which targets this unfolded, ER-retained protein for ERAD, is significantly stabilized, can traffic to the apical membrane, and function in the presence of an FDA-approved drug combination (Lukacs and Verkman 2012). Current knowledge suggests these compounds bind directly to the mutated channel. One compound helps link two domains during channel synthesis, while the other compound binds to the cell surface population of F508del CFTR and increases its open probability (Clancy 2018). While improvements in the efficacy of this treatment are actively being pursued, these data provide hope that other drugs that target misfolded ion channels and transporters can be identified. In the case of Cystic Fibrosis, a molecular definition of the cause of the disease was instrumental in developing this therapy. For the other diseases discussed in this chapter, we similarly propose that basic research will pave the way for future cures.

Acknowledgments The authors would like to thank the editors, Kirk L. Hamilton and Daniel C. Devor, for the invitation to contribute to this edition. This work was supported by NIDDK DK109024 and DK117126 to T.M.B. and GM075061, GM131732 and DK079307 to J.L.B.

References

Ackermann U, Geering K (1990) Mutual dependence of Na,K-ATPase alpha- and beta-subunits for correct posttranslational processing and intracellular transport. FEBS Lett 269(1):105–108

Adams CM, Snyder PM, Welsh MJ (1997) Interactions between subunits of the human epithelial sodium channel. J Biol Chem 272(43):27295–27300

Ahner A, Nakatsukasa K, Zhang H, Frizzell RA, Brodsky JL (2007) Small heat-shock proteins select {delta}F508-CFTR for endoplasmic reticulum-associated degradation. Mol Biol Cell 18 (3):806–814

Alderson TR, Kim JH, Markley JL (2016) Dynamical structures of Hsp70 and Hsp70-Hsp40 complexes. Structure 24(7):1014–1030. https://doi.org/10.1016/j.str.2016.05.011

Asai T, Kuwahara M, Kurihara H, Sakai T, Terada Y, Marumo F, Sasaki S (2003) Pathogenesis of nephrogenic diabetes insipidus by aquaporin-2 C-terminus mutations. Kidney Int 64(1):2–10. https://doi.org/10.1046/j.1523-1755.2003.00049.x

Asher C, Wald H, Rossier BC, Garty H (1996) Aldosterone-induced increase in the abundance of Na$^+$ channel subunits. Am J Phys 271(2 Pt 1):C605–C611

Bai L, Fushimi K, Sasaki S, Marumo F (1996) Structure of aquaporin-2 vasopressin water channel. J Biol Chem 271(9):5171–5176

Baldridge RD, Rapoport TA (2016) Autoubiquitination of the Hrd1 ligase triggers protein retrotranslocation in ERAD. Cell 166(2):394–407. https://doi.org/10.1016/j.cell.2016.05.048

Bard JAM, Goodall EA, Greene ER, Jonsson E, Dong KC, Martin A (2018) Structure and function of the 26S proteasome. Annu Rev Biochem 87:697–724. https://doi.org/10.1146/annurev-biochem-062917-011931

Barlowe C, Helenius A (2016) Cargo capture and bulk flow in the early secretory pathway. Annu Rev Cell Dev Biol 32:197–222. https://doi.org/10.1146/annurev-cellbio-111315-125016

Baxter BK, James P, Evans T, Craig EA (1996) SSI1 encodes a novel Hsp70 of the *Saccharomyces cerevisiae* endoplasmic reticulum. Mol Cell Biol 16(11):6444–6456

Bays NW, Wilhovsky SK, Goradia A, Hodgkiss-Harlow K, Hampton RY (2001) HRD4/NPL4 is required for the proteasomal processing of ubiquitinated ER proteins. Mol Biol Cell 12 (12):4114–4128

Beggah A, Mathews P, Beguin P, Geering K (1996) Degradation and endoplasmic reticulum retention of unassembled alpha- and beta-subunits of Na,K-ATPase correlate with interaction of BiP. J Biol Chem 271(34):20895–20902

Beggah AT, Jaunin P, Geering K (1997) Role of glycosylation and disulfide bond formation in the beta subunit in the folding and functional expression of Na,K-ATPase. J Biol Chem 272 (15):10318–10326

Beggah AT, Beguin P, Bamberg K, Sachs G, Geering K (1999) beta-subunit assembly is essential for the correct packing and the stable membrane insertion of the H,K-ATPase alpha-subunit. J Biol Chem 274(12):8217–8223

Beguin P, Hasler U, Beggah A, Horisberger JD, Geering K (1998) Membrane integration of Na,K-ATPase alpha-subunits and beta-subunit assembly. J Biol Chem 273(38):24921–24931

Beguin P, Hasler U, Staub O, Geering K (2000) Endoplasmic reticulum quality control of oligomeric membrane proteins: topogenic determinants involved in the degradation of the unassembled Na,K-ATPase alpha subunit and in its stabilization by beta subunit assembly. Mol Biol Cell 11(5):1657–1672

Behnke J, Hendershot LM (2014) The large Hsp70 Grp170 binds to unfolded protein substrates in vivo with a regulation distinct from conventional Hsp70s. J Biol Chem 289(5):2899–2907. https://doi.org/10.1074/jbc.M113.507491

Bernasconi R, Galli C, Calanca V, Nakajima T, Molinari M (2010) Stringent requirement for HRD1, SEL1L, and OS-9/XTP3-B for disposal of ERAD-LS substrates. J Cell Biol 188 (2):223–235. https://doi.org/10.1083/jcb.200910042

Bernier V, Lagace M, Lonergan M, Arthus MF, Bichet DG, Bouvier M (2004) Functional rescue of the constitutively internalized V2 vasopressin receptor mutant R137H by the pharmacological chaperone action of SR49059. Mol Endocrinol 18(8):2074–2084. https://doi.org/10.1210/me.2004-0080

Bertelsen EB, Chang L, Gestwicki JE, Zuiderweg ER (2009) Solution conformation of wild-type E. coli Hsp70 (DnaK) chaperone complexed with ADP and substrate. Proc Natl Acad Sci U S A 106(21):8471–8476. https://doi.org/10.1073/pnas.0903503106

Bhalla V, Hallows KR (2008) Mechanisms of ENaC regulation and clinical implications. J Am Soc Nephrol 19(10):1845–1854

Birnbaumer M, Seibold A, Gilbert S, Ishido M, Barberis C, Antaramian A, Brabet P, Rosenthal W (1992) Molecular cloning of the receptor for human antidiuretic hormone. Nature 357 (6376):333–335. https://doi.org/10.1038/357333a0

Blanco G (2005) Na,K-ATPase subunit heterogeneity as a mechanism for tissue-specific ion regulation. Semin Nephrol 25(5):292–303. https://doi.org/10.1016/j.semnephrol.2005.03.004

Blanco G, Mercer RW (1998) Isozymes of the Na-K-ATPase: heterogeneity in structure, diversity in function. Am J Phys 275(5 Pt 2):F633–F650

Bodnar N, Rapoport T (2017a) Toward an understanding of the Cdc48/p97 ATPase. F1000Res 6:1318. https://doi.org/10.12688/f1000research.11683.1

Bodnar NO, Rapoport TA (2017b) Molecular mechanism of substrate processing by the Cdc48 ATPase complex. Cell 169(4):722–735.e729. https://doi.org/10.1016/j.cell.2017.04.020

Bodnar NO, Kim KH, Ji Z, Wales TE, Svetlov V, Nudler E, Engen JR, Walz T, Rapoport TA (2018) Structure of the Cdc48 ATPase with its ubiquitin-binding cofactor Ufd1-Npl4. Nat Struct Mol Biol 25(7):616–622. https://doi.org/10.1038/s41594-018-0085-x

Bracher A, Verghese J (2015) The nucleotide exchange factors of Hsp70 molecular chaperones. Front Mol Biosci 2:10. https://doi.org/10.3389/fmolb.2015.00010

Brockmeier A, Brockmeier U, Williams DB (2009) Distinct contributions of the lectin and arm domains of calnexin to its molecular chaperone function. J Biol Chem 284(6):3433–3444. https://doi.org/10.1074/jbc.M804866200

Bruns JB, Hu B, Ahn YJ, Sheng S, Hughey RP, Kleyman TR (2003) Multiple epithelial Na[+] channel domains participate in subunit assembly. Am J Physiol Renal Physiol 285(4):F600–F609

Buchner J (2019) Molecular chaperones and protein quality control: an introduction to the JBC Reviews thematic series. J Biol Chem 294(6):2074–2075. https://doi.org/10.1074/jbc.REV118.006739

Buck TM, Brodsky JL (2018) Epithelial sodium channel biogenesis and quality control in the early secretory pathway. Curr Opin Nephrol Hypertens 27(5):364–372. https://doi.org/10.1097/MNH.0000000000000438

Buck TM, Eledge J, Skach WR (2004) Evidence for stabilization of aquaporin-2 folding mutants by N-linked glycosylation in endoplasmic reticulum. Am J Physiol Cell Physiol 287(5):C1292–C1299. https://doi.org/10.1152/ajpcell.00561.2003

Buck TM, Wagner J, Grund S, Skach WR (2007) A novel tripartite motif involved in aquaporin topogenesis, monomer folding and tetramerization. Nat Struct Mol Biol 14(8):762–769

Buck TM, Kolb AR, Boyd C, Kleyman TR, Brodsky JL (2010) The ER associated degradation of the epithelial sodium channel requires a unique complement of molecular chaperones. Mol Biol Cell 21(6):1047–1058

Buck TM, Plavchak L, Roy A, Donnelly BF, Kashlan OB, Kleyman TR, Subramanya AR, Brodsky JL (2013) The Lhs1/GRP170 chaperones facilitate the endoplasmic reticulum-associated degradation of the epithelial sodium channel. J Biol Chem 288(25):18366–18380. https://doi.org/10.1074/jbc.M113.469882

Buck TM, Jordahl AS, Yates ME, Preston GM, Cook E, Kleyman TR, Brodsky JL (2017) Interactions between intersubunit transmembrane domains regulate the chaperone-dependent degradation of an oligomeric membrane protein. Biochem J 474(3):357–376. https://doi.org/10.1042/BCJ20160760

Butler PL, Staruschenko A, Snyder PM (2015) Acetylation stimulates the epithelial sodium channel by reducing its ubiquitination and degradation. J Biol Chem 290(20):12497–12503. https://doi.org/10.1074/jbc.M114.635540

Canessa CM, Merillat AM, Rossier BC (1994a) Membrane topology of the epithelial sodium channel in intact cells. Am J Phys 267(6 Pt 1):C1682–C1690

Canessa CM, Schild L, Buell G, Thorens B, Gautschi I, Horisberger JD, Rossier BC (1994b) Amiloride-sensitive epithelial Na⁺ channel is made of three homologous subunits. Nature 367 (6462):463–467

Carvalho P, Goder V, Rapoport TA (2006) Distinct ubiquitin-ligase complexes define convergent pathways for the degradation of ER proteins. Cell 126(2):361–373

Carvalho P, Stanley AM, Rapoport TA (2010) Retrotranslocation of a misfolded luminal ER protein by the ubiquitin-ligase Hrd1p. Cell 143(4):579–591. https://doi.org/10.1016/j.cell.2010.10.028

Castaneda-Bueno M, Vazquez N, Bustos-Jaimes I, Hernandez D, Rodriguez-Lobato E, Pacheco-Alvarez D, Carino-Cortes R, Moreno E, Bobadilla NA, Gamba G (2010) A single residue in transmembrane domain 11 defines the different affinity for thiazides between the mammalian and flounder NaCl transporters. Am J Physiol Renal Physiol 299(5):F1111–F1119. https://doi.org/10.1152/ajprenal.00412.2010

Chanoux RA, Robay A, Shubin CB, Kebler C, Suaud L, Rubenstein RC (2012) Hsp70 promotes epithelial sodium channel functional expression by increasing its association with coat complex II and its exit from endoplasmic reticulum. J Biol Chem 287(23):19255–19265. https://doi.org/10.1074/jbc.M112.357756

Chanoux RA, Shubin CB, Robay A, Suaud L, Rubenstein RC (2013) Hsc70 negatively regulates epithelial sodium channel trafficking at multiple sites in epithelial cells. Am J Physiol Cell Physiol 305(7):C776–C787. https://doi.org/10.1152/ajpcell.00059.2013

Chelur DS, Ernstrom GG, Goodman MB, Yao CA, Chen L, Hagan RO', Chalfie M (2002) The mechanosensory protein MEC-6 is a subunit of the C. elegans touch-cell degenerin channel. Nature 420(6916):669–673. https://doi.org/10.1038/nature01205

Cheng SH, Gregory RJ, Marshall J, Paul S, Souza DW, White GA, O'Riordan CR, Smith AE (1990) Defective intracellular transport and processing of CFTR is the molecular basis of most cystic fibrosis. Cell 63(4):827–834

Cherepanova N, Shrimal S, Gilmore R (2016) N-linked glycosylation and homeostasis of the endoplasmic reticulum. Curr Opin Cell Biol 41:57–65. https://doi.org/10.1016/j.ceb.2016.03.021

Clancy JP (2018) Rapid therapeutic advances in CFTR modulator science. Pediatr Pulmonol 53 (S3):S4–S11. https://doi.org/10.1002/ppul.24157

Clapham DE (2003) TRP channels as cellular sensors. Nature 426(6966):517–524. https://doi.org/10.1038/nature02196

Clifford RJ, Kaplan JH (2008) beta-Subunit overexpression alters the stoicheometry of assembled Na-K-ATPase subunits in MDCK cells. Am J Physiol Renal Physiol 295(5):F1314–F1323. https://doi.org/10.1152/ajprenal.90406.2008

Crambert G, Hasler U, Beggah AT, Yu C, Modyanov NN, Horisberger JD, Lelievre L, Geering K (2000) Transport and pharmacological properties of nine different human Na, K-ATPase isozymes. J Biol Chem 275(3):1976–1986

Craven RA, Egerton M, Stirling CJ (1996) A novel Hsp70 of the yeast ER lumen is required for the efficient translocation of a number of protein precursors. EMBO J 15(11):2640–2650

Danilczyk UG, Williams DB (2001) The lectin chaperone calnexin utilizes polypeptide-based interactions to associate with many of its substrates in vivo. J Biol Chem 276 (27):25532–25540. https://doi.org/10.1074/jbc.m100270200

De Jong JC, Van Der Vliet WA, Van Den Heuvel LP, Willems PH, Knoers NV, Bindels RJ (2002) Functional expression of mutations in the human NaCl cotransporter: evidence for impaired routing mechanisms in Gitelman's syndrome. J Am Soc Nephrol 13(6):1442–1448

de Keyzer J, Steel GJ, Hale SJ, Humphries D, Stirling CJ (2009) Nucleotide binding by Lhs1p is essential for its nucleotide exchange activity and for function in vivo. J Biol Chem 284 (46):31564–31571. https://doi.org/10.1074/jbc.M109.055160

de la Pena AH, Goodall EA, Gates SN, Lander GC, Martin A (2018) Substrate-engaged 26S proteasome structures reveal mechanisms for ATP-hydrolysis-driven translocation. Science 362 (6418):eaav0725. https://doi.org/10.1126/science.aav0725

De I, Sadhukhan S (2018) Emerging roles of DHHC-mediated protein S-palmitoylation in physiological and pathophysiological context. Eur J Cell Biol 97(5):319–338. https://doi.org/10.1016/j.ejcb.2018.03.005

Dong Y, Zhang S, Wu Z, Li X, Wang WL, Zhu Y, Stoilova-McPhie S, Lu Y, Finley D, Mao Y (2019) Cryo-EM structures and dynamics of substrate-engaged human 26S proteasome. Nature 565(7737):49–55. https://doi.org/10.1038/s41586-018-0736-4

Donnelly BF, Needham PG, Snyder AC, Roy A, Khadem S, Brodsky JL, Subramanya AR (2013) Hsp70 and Hsp90 multichaperone complexes sequentially regulate thiazide-sensitive cotransporter endoplasmic reticulum-associated degradation and biogenesis. J Biol Chem 288 (18):13124–13135. https://doi.org/10.1074/jbc.M113.455394

Drazic A, Myklebust LM, Ree R, Arnesen T (2016) The world of protein acetylation. Biochim Biophys Acta 1864(10):1372–1401. https://doi.org/10.1016/j.bbapap.2016.06.007

Egea PF, Stroud RM (2010) Lateral opening of a translocon upon entry of protein suggests the mechanism of insertion into membranes. Proc Natl Acad Sci U S A 107(40):17182–17187. https://doi.org/10.1073/pnas.1012556107

El Khouri E, Le Pavec G, Toledano MB, Delaunay-Moisan A (2013) RNF185 is a novel E3 ligase of endoplasmic reticulum-associated degradation (ERAD) that targets cystic fibrosis transmembrane conductance regulator (CFTR). J Biol Chem 288(43):31177–31191. https://doi.org/10.1074/jbc.M113.470500

Fallen K, Banerjee S, Sheehan J, Addison D, Lewis LM, Meiler J, Denton JS (2009) The Kir channel immunoglobulin domain is essential for Kir1.1 (ROMK) thermodynamic stability, trafficking and gating. Channels (Austin) 3(1):57–68

Fang L, Li D, Welling PA (2010) Hypertension resistance polymorphisms in ROMK (Kir1.1) alter channel function by different mechanisms. Am J Physiol Renal Physiol 299(6):F1359–F1364. https://doi.org/10.1152/ajprenal.00257.2010

Feige MJ, Behnke J, Mittag T, Hendershot LM (2015) Dimerization-dependent folding underlies assembly control of the clonotypic alphabetaT cell receptor chains. J Biol Chem 290 (44):26821–26831. https://doi.org/10.1074/jbc.M115.689471

Finley D, Chen X, Walters KJ (2016) Gates, channels, and switches: elements of the proteasome machine. Trends Biochem Sci 41(1):77–93. https://doi.org/10.1016/j.tibs.2015.10.009

Firsov D, Robert-Nicoud M, Gruender S, Schild L, Rossier BC (1999) Mutational analysis of cysteine-rich domains of the epithelium sodium channel (ENaC). Identification of cysteines essential for channel expression at the cell surface. J Biol Chem 274(5):2743–2749

Francis GS, Tang WH (2004) Vasopressin receptor antagonists: will the "vaptans" fulfill their promise? JAMA 291(16):2017–2018. https://doi.org/10.1001/jama.291.16.2017

Fukata Y, Murakami T, Yokoi N, Fukata M (2016) Local palmitoylation cycles and specialized membrane domain organization. Curr Top Membr 77:97–141. https://doi.org/10.1016/bs.ctm.2015.10.003

Fulchiero R, Seo-Mayer P (2019) Bartter syndrome and Gitelman syndrome. Pediatr Clin N Am 66 (1):121–134. https://doi.org/10.1016/j.pcl.2018.08.010

Galligan JJ, Petersen DR (2012) The human protein disulfide isomerase gene family. Hum Genomics 6:6. https://doi.org/10.1186/1479-7364-6-6

Gardner RG, Swarbrick GM, Bays NW, Cronin SR, Wilhovsky S, Seelig L, Kim C, Hampton RY (2000) Endoplasmic reticulum degradation requires lumen to cytosol signaling. Transmembrane control of Hrd1p by Hrd3p. J Cell Biol 151(1):69–82

Garza RM, Sato BK, Hampton RY (2009) In vitro analysis of Hrd1p-mediated retrotranslocation of its multispanning membrane substrate 3-hydroxy-3-methylglutaryl (HMG)-CoA reductase. J Biol Chem 284(22):14710–14722. https://doi.org/10.1074/jbc.M809607200

Gauss R, Sommer T, Jarosch E (2006) The Hrd1p ligase complex forms a linchpin between ER-lumenal substrate selection and Cdc48p recruitment. EMBO J 25(9):1827–1835. https://doi.org/10.1038/sj.emboj.7601088

Geering K (2001) The functional role of beta subunits in oligomeric P-type ATPases. J Bioenerg Biomembr 33(5):425–438

Geering K (2008) Functional roles of Na,K-ATPase subunits. Curr Opin Nephrol Hypertens 17 (5):526–532. https://doi.org/10.1097/MNH.0b013e3283036cbf

Genest O, Wickner S, Doyle SM (2019) Hsp90 and Hsp70 chaperones: collaborators in protein remodeling. J Biol Chem 294(6):2109–2120. https://doi.org/10.1074/jbc.REV118.002806

Gentzsch M, Dang H, Dang Y, Garcia-Caballero A, Suchindran H, Boucher RC, Stutts MJ (2010) The cystic fibrosis transmembrane conductance regulator impedes proteolytic stimulation of the epithelial Na$^+$ channel. J Biol Chem 285(42):32227–32232. https://doi.org/10.1074/jbc.M110.155259

Ghaemmaghami S, Huh WK, Bower K, Howson RW, Belle A, Dephoure N, O'Shea EK, Weissman JS (2003) Global analysis of protein expression in yeast. Nature 425(6959):737–741

Gilmore R, Blobel G, Walter P (1982) Protein translocation across the endoplasmic reticulum. I. Detection in the microsomal membrane of a receptor for the signal recognition particle. J Cell Biol 95(2 Pt 1):463–469

Goldfarb SB, Kashlan OB, Watkins JN, Suaud L, Yan W, Kleyman TR, Rubenstein RC (2006) Differential effects of Hsc70 and Hsp70 on the intracellular trafficking and functional expression of epithelial sodium channels. Proc Natl Acad Sci U S A 103(15):5817–5822

Gorlich D, Prehn S, Hartmann E, Kalies KU, Rapoport TA (1992) A mammalian homolog of SEC61p and SECYp is associated with ribosomes and nascent polypeptides during translocation. Cell 71(3):489–503

Grumbach Y, Bikard Y, Suaud L, Chanoux RA, Rubenstein RC (2014) ERp29 regulates epithelial sodium channel functional expression by promoting channel cleavage. Am J Physiol Cell Physiol 307(8):C701–C709. https://doi.org/10.1152/ajpcell.00134.2014

Grunder S, Firsov D, Chang SS, Jaeger NF, Gautschi I, Schild L, Lifton RP, Rossier BC (1997) A mutation causing pseudohypoaldosteronism type 1 identifies a conserved glycine that is involved in the gating of the epithelial sodium channel. EMBO J 16(5):899–907. https://doi.org/10.1093/emboj/16.5.899

Habeck G, Ebner FA, Shimada-Kreft H, Kreft SG (2015) The yeast ERAD-C ubiquitin ligase Doa10 recognizes an intramembrane degron. J Cell Biol 209(2):261–273. https://doi.org/10.1083/jcb.201408088

Halic M, Becker T, Pool MR, Spahn CM, Grassucci RA, Frank J, Beckmann R (2004) Structure of the signal recognition particle interacting with the elongation-arrested ribosome. Nature 427 (6977):808–814. https://doi.org/10.1038/nature02342

Halic M, Blau M, Becker T, Mielke T, Pool MR, Wild K, Sinning I, Beckmann R (2006) Following the signal sequence from ribosomal tunnel exit to signal recognition particle. Nature 444 (7118):507–511. https://doi.org/10.1038/nature05326

Hampton RY, Gardner RG, Rine J (1996) Role of 26S proteasome and HRD genes in the degradation of 3-hydroxy-3-methylglutaryl-CoA reductase, an integral endoplasmic reticulum membrane protein. Mol Biol Cell 7(12):2029–2044

Hanzelmann P, Schindelin H (2017) The interplay of cofactor interactions and post-translational modifications in the regulation of the AAA+ ATPase p97. Front Mol Biosci 4:21. https://doi.org/10.3389/fmolb.2017.00021

Harel M, Aharoni A, Gaidukov L, Brumshtein B, Khersonsky O, Meged R, Dvir H, Ravelli RB, McCarthy A, Toker L, Silman I, Sussman JL, Tawfik DS (2004) Structure and evolution of the serum paraoxonase family of detoxifying and anti-atherosclerotic enzymes. Nat Struct Mol Biol 11(5):412–419. https://doi.org/10.1038/nsmb767

Hasler U, Crambert G, Horisberger JD, Geering K (2001) Structural and functional features of the transmembrane domain of the Na,K-ATPase beta subunit revealed by tryptophan scanning. J Biol Chem 276(19):16356–16364. https://doi.org/10.1074/jbc.M008778200

Hebert DN, Molinari M (2012) Flagging and docking: dual roles for N-glycans in protein quality control and cellular proteostasis. Trends Biochem Sci 37(10):404–410. https://doi.org/10.1016/j.tibs.2012.07.005

Heidrich E, Carattino MD, Hughey RP, Pilewski JM, Kleyman TR, Myerburg MM (2015) Intracellular Na$^+$ regulates epithelial Na$^+$ channel maturation. J Biol Chem 290 (18):11569–11577. https://doi.org/10.1074/jbc.M115.640763

Hessa T, Meindl-Beinker NM, Bernsel A, Kim H, Sato Y, Lerch-Bader M, Nilsson I, White SH, von Heijne G (2007) Molecular code for transmembrane-helix recognition by the Sec61 translocon. Nature 450(7172):1026–1030. https://doi.org/10.1038/nature06387

Hiller MM, Finger A, Schweiger M, Wolf DH (1996) ER degradation of a misfolded luminal protein by the cytosolic ubiquitin-proteasome pathway. Science 273(5282):1725–1728

Hitt R, Wolf DH (2004) Der1p, a protein required for degradation of malfolded soluble proteins of the endoplasmic reticulum: topology and Der1-like proteins. FEMS Yeast Res 4(7):721–729

Ho K, Nichols CG, Lederer WJ, Lytton J, Vassilev PM, Kanazirska MV, Hebert SC (1993) Cloning and expression of an inwardly rectifying ATP-regulated potassium channel. Nature 362 (6415):31–38. https://doi.org/10.1038/362031a0

Hobbs CA, Da Tan C, Tarran R (2013) Does epithelial sodium channel hyperactivity contribute to cystic fibrosis lung disease? J Physiol 591(18):4377–4387. https://doi.org/10.1113/jphysiol.2012.240861

Hoekstra JA, van Lieburg AF, Monnens LA, Hulstijn-Dirkmaat GM, Knoers VV (1996) Cognitive and psychosocial functioning of patients with congenital nephrogenic diabetes insipidus. Am J Med Genet 61(1):81–88. https://doi.org/10.1002/(SICI)1096-8628(19960102)61:1<81::AID-AJMG17>3.0.CO;2-S

Horn SC, Hanna J, Hirsch C, Volkwein C, Schutz A, Heinemann U, Sommer T, Jarosch E (2009) Usa1 functions as a scaffold of the HRD-ubiquitin ligase. Mol Cell 36(5):782–793. https://doi.org/10.1016/j.molcel.2009.10.015

Hughey RP, Mueller GM, Bruns JB, Kinlough CL, Poland PA, Harkleroad KL, Carattino MD, Kleyman TR (2003) Maturation of the epithelial Na$^+$ channel involves proteolytic processing of the alpha- and gamma-subunits. J Biol Chem 278(39):37073–37082

Inoue T, Tsai B (2015) A nucleotide exchange factor promotes endoplasmic reticulum-to-cytosol membrane penetration of the nonenveloped virus simian virus 40. J Virol 89(8):4069–4079. https://doi.org/10.1128/JVI.03552-14

Inoue T, Tsai B (2016) The Grp170 nucleotide exchange factor executes a key role during ERAD of cellular misfolded clients. Mol Biol Cell 27(10):1650–1662. https://doi.org/10.1091/mbc.E16-01-0033

Iolascon A, Aglio V, Tamma G, D'Apolito M, Addabbo F, Procino G, Simonetti MC, Montini G, Gesualdo L, Debler EW, Svelto M, Valenti G (2007) Characterization of two novel missense mutations in the AQP2 gene causing nephrogenic diabetes insipidus. Nephron Physiol 105(3): p33–p41. https://doi.org/10.1159/000098136

Ismail N, Ng DT (2006) Have you HRD? Understanding ERAD is DOAble! Cell 126(2):237–239

Ismailov II, Awayda MS, Jovov B, Berdiev BK, Fuller CM, Dedman JR, Kaetzel M, Benos DJ (1996) Regulation of epithelial sodium channels by the cystic fibrosis transmembrane conductance regulator. J Biol Chem 271(9):4725–4732

Jarosch E, Taxis C, Volkwein C, Bordallo J, Finley D, Wolf DH, Sommer T (2002) Protein dislocation from the ER requires polyubiquitination and the AAA-ATPase Cdc48. Nat Cell Biol 4(2):134–139

Jaunin P, Horisberger JD, Richter K, Good PJ, Rossier BC, Geering K (1992) Processing, intracellular transport, and functional expression of endogenous and exogenous alpha-beta 3 Na,K-ATPase complexes in Xenopus oocytes. J Biol Chem 267(1):577–585

Ji W, Foo JN, O'Roak BJ, Zhao H, Larson MG, Simon DB, Newton-Cheh C, State MW, Levy D, Lifton RP (2008) Rare independent mutations in renal salt handling genes contribute to blood pressure variation. Nat Genet 40(5):592–599. https://doi.org/10.1038/ng.118

Jiang LY, Jiang W, Tian N, Xiong YN, Liu J, Wei J, Wu KY, Luo J, Shi XJ, Song BL (2018) Ring finger protein 145 (RNF145) is a ubiquitin ligase for sterol-induced degradation of HMG-CoA reductase. J Biol Chem 293(11):4047–4055. https://doi.org/10.1074/jbc.RA117.001260

Jorgensen PL (1986) Structure, function and regulation of Na,K-ATPase in the kidney. Kidney Int 29(1):10–20

Kampinga HH, Hageman J, Vos MJ, Kubota H, Tanguay RM, Bruford EA, Cheetham ME, Chen B, Hightower LE (2009) Guidelines for the nomenclature of the human heat shock proteins. Cell Stress Chaperones 14(1):105–111. https://doi.org/10.1007/s12192-008-0068-7

Kamsteeg EJ, Bichet DG, Konings IB, Nivet H, Lonergan M, Arthus MF, van Os CH, Deen PM (2003) Reversed polarized delivery of an aquaporin-2 mutant causes dominant nephrogenic diabetes insipidus. J Cell Biol 163(5):1099–1109

Kashlan OB, Mueller GM, Qamar MZ, Poland PA, Ahner A, Rubenstein RC, Hughey RP, Brodsky JL, Kleyman TR (2007) Small heat shock protein alphaA-crystallin regulates epithelial sodium channel expression. J Biol Chem 282(38):28149–28156

Kashlan OB, Kinlough CL, Myerburg MM, Shi S, Chen J, Blobner BM, Buck TM, Brodsky JL, Hughey RP, Kleyman TR (2018) N-linked glycans are required on epithelial Na^+ channel subunits for maturation and surface expression. Am J Physiol Renal Physiol 314(3):F483–F492. https://doi.org/10.1152/ajprenal.00195.2017

Kieber-Emmons T, Lin C, Foster MH, Kleyman TR (1999) Antiidiotypic antibody recognizes an amiloride binding domain within the alpha subunit of the epithelial Na^+ channel. J Biol Chem 274(14):9648–9655

Kityk R, Kopp J, Sinning I, Mayer MP (2012) Structure and dynamics of the ATP-bound open conformation of Hsp70 chaperones. Mol Cell 48(6):863–874. https://doi.org/10.1016/j.molcel.2012.09.023

Kityk R, Kopp J, Mayer MP (2018) Molecular mechanism of J-domain-TRIGGERED ATP hydrolysis by Hsp70 chaperones. Mol Cell 69(2):227–237.e224. https://doi.org/10.1016/j.molcel.2017.12.003

Kizhatil K, Davis JQ, Davis L, Hoffman J, Hogan BL, Bennett V (2007) Ankyrin-G is a molecular partner of E-cadherin in epithelial cells and early embryos. J Biol Chem 282(36):26552–26561. https://doi.org/10.1074/jbc.M703158200

Knop M, Finger A, Braun T, Hellmuth K, Wolf DH (1996) Der1, a novel protein specifically required for endoplasmic reticulum degradation in yeast. EMBO J 15(4):753–763

Kong J, Peng M, Ostrovsky J, Kwon YJ, Oretsky O, McCormick EM, He M, Argon Y, Falk MJ (2018) Mitochondrial function requires NGLY1. Mitochondrion 38:6–16. https://doi.org/10.1016/j.mito.2017.07.008

Koulen P, Cai Y, Geng L, Maeda Y, Nishimura S, Witzgall R, Ehrlich BE, Somlo S (2002) Polycystin-2 is an intracellular calcium release channel. Nat Cell Biol 4(3):191–197. https://doi.org/10.1038/ncb754

Kreutzberger AJB, Ji M, Aaron J, Mihaljevic L, Urban S (2019) Rhomboid distorts lipids to break the viscosity-imposed speed limit of membrane diffusion. Science 363(6426):eaao0076. https://doi.org/10.1126/science.aao0076

Kunchaparty S, Palcso M, Berkman J, Velazquez H, Desir GV, Bernstein P, Reilly RF, Ellison DH (1999) Defective processing and expression of thiazide-sensitive Na-Cl cotransporter as a cause of Gitelman's syndrome. Am J Phys 277(4):F643–F649. https://doi.org/10.1152/ajprenal.1999.277.4.F643

Kuwahara M, Iwai K, Ooeda T, Igarashi T, Ogawa E, Katsushima Y, Shinbo I, Uchida S, Terada Y, Arthus MF, Lonergan M, Fujiwara TM, Bichet DG, Marumo F, Sasaki S (2001) Three families with autosomal dominant nephrogenic diabetes insipidus caused by aquaporin-2 mutations in the C-terminus. Am J Hum Genet 69(4):738–748. https://doi.org/10.1086/323643

Lang S, Pfeffer S, Lee PH, Cavalie A, Helms V, Forster F, Zimmermann R (2017) An update on Sec61 channel functions, mechanisms, and related diseases. Front Physiol 8:887. https://doi.org/10.3389/fphys.2017.00887

Leduc-Nadeau A, Lussier Y, Arthus MF, Lonergan M, Martinez-Aguayo A, Riveira-Munoz E, Devuyst O, Bissonnette P, Bichet DG (2010) New autosomal recessive mutations in aquaporin-2 causing nephrogenic diabetes insipidus through deficient targeting display normal expression in Xenopus oocytes. J Physiol 588(Pt 12):2205–2218. https://doi.org/10.1113/jphysiol.2010.187674

Leitman J, Shenkman M, Gofman Y, Shtern NO, Ben-Tal N, Hendershot LM, Lederkremer GZ (2014) Herp coordinates compartmentalization and recruitment of HRD1 and misfolded proteins for ERAD. Mol Biol Cell 25(7):1050–1060. https://doi.org/10.1091/mbc.E13-06-0350

Lerner M, Corcoran M, Cepeda D, Nielsen ML, Zubarev R, Ponten F, Uhlen M, Hober S, Grander D, Sangfelt O (2007) The RBCC gene RFP2 (Leu5) encodes a novel transmembrane E3 ubiquitin ligase involved in ERAD. Mol Biol Cell 18(5):1670–1682. https://doi.org/10.1091/mbc.e06-03-0248

Li G, Zhao G, Zhou X, Schindelin H, Lennarz WJ (2006) The AAA ATPase p97 links peptide N-glycanase to the endoplasmic reticulum-associated E3 ligase autocrine motility factor receptor. Proc Natl Acad Sci U S A 103(22):8348–8353. https://doi.org/10.1073/pnas.0602747103

Li W, Tu D, Brunger AT, Ye Y (2007) A ubiquitin ligase transfers preformed polyubiquitin chains from a conjugating enzyme to a substrate. Nature 446(7133):333–337. https://doi.org/10.1038/nature05542

Lian WN, Wu TW, Dao RL, Chen YJ, Lin CH (2006) Deglycosylation of Na$^+$/K$^+$-ATPase causes the basolateral protein to undergo apical targeting in polarized hepatic cells. J Cell Sci 119 (Pt 1):11–22. https://doi.org/10.1242/jcs.02706

Liang G, Li Q, Tang Y, Kokame K, Kikuchi T, Wu G, Chen XZ (2008) Polycystin-2 is regulated by endoplasmic reticulum-associated degradation. Hum Mol Genet 17(8):1109–1119. https://doi.org/10.1093/hmg/ddm383

Lin SH, Bichet DG, Sasaki S, Kuwahara M, Arthus MF, Lonergan M, Lin YF (2002) Two novel aquaporin-2 mutations responsible for congenital nephrogenic diabetes insipidus in Chinese families. J Clin Endocrinol Metab 87(6):2694–2700. https://doi.org/10.1210/jcem.87.6.8617

Linder ME, Deschenes RJ (2007) Palmitoylation: policing protein stability and traffic. Nat Rev Mol Cell Biol 8(1):74–84. https://doi.org/10.1038/nrm2084

Lloyd DJ, Hall FW, Tarantino LM, Gekakis N (2005) Diabetes insipidus in mice with a mutation in aquaporin-2. PLoS Genet 1(2):e20. https://doi.org/10.1371/journal.pgen.0010020

Los EL, Deen PM, Robben JH (2010) Potential of nonpeptide (ant)agonists to rescue vasopressin V2 receptor mutants for the treatment of X-linked nephrogenic diabetes insipidus. J Neuroendocrinol 22(5):393–399. https://doi.org/10.1111/j.1365-2826.2010.01983.x

Lu Y, Turnbull IR, Bragin A, Carveth K, Verkman AS, Skach WR (2000) Reorientation of aquaporin-1 topology during maturation in the endoplasmic reticulum. Mol Biol Cell 11 (9):2973–2985

Lu JP, Wang Y, Sliter DA, Pearce MM, Wojcikiewicz RJ (2011) RNF170 protein, an endoplasmic reticulum membrane ubiquitin ligase, mediates inositol 1,4,5-trisphosphate receptor ubiquitination and degradation. J Biol Chem 286(27):24426–24433. https://doi.org/10.1074/jbc.M111.251983

Lukacs GL, Verkman AS (2012) CFTR: folding, misfolding and correcting the DeltaF508 conformational defect. Trends Mol Med 18(2):81–91. https://doi.org/10.1016/j.molmed.2011.144003

Luo Y, Vassilev PM, Li X, Kawanabe Y, Zhou J (2003) Native polycystin 2 functions as a plasma membrane Ca^{2+}-permeable cation channel in renal epithelia. Mol Cell Biol 23(7):2600–2607

Mackie TD, Kim BY, Subramanya AR, Bain DJ, O'Donnell AF, Welling PA, Brodsky JL (2018) The endosomal trafficking factors CORVET and ESCRT suppress plasma membrane residence of the renal outer medullary potassium channel (ROMK). J Biol Chem 293(9):3201–3217. https://doi.org/10.1074/jbc.M117.819086

Malik B, Schlanger L, Al-Khalili O, Bao HF, Yue G, Price SR, Mitch WE, Eaton DC (2001) Enac degradation in A6 cells by the ubiquitin-proteosome proteolytic pathway. J Biol Chem 276 (16):12903–12910

Mall M, Grubb BR, Harkema JR, O'Neal WK, Boucher RC (2004) Increased airway epithelial Na$^+$ absorption produces cystic fibrosis-like lung disease in mice. Nat Med 10(5):487–493

Marr N, Bichet DG, Hoefs S, Savelkoul PJ, Konings IB, De Mattia F, Graat MP, Arthus MF, Lonergan M, Fujiwara TM, Knoers NV, Landau D, Balfe WJ, Oksche A, Rosenthal W, Muller D, Van Os CH, Deen PM (2002) Cell-biologic and functional analyses of five new Aquaporin-2 missense mutations that cause recessive nephrogenic diabetes insipidus. J Am Soc Nephrol 13(9):2267–2277

Masilamani S, Kim GH, Mitchell C, Wade JB, Knepper MA (1999) Aldosterone-mediated regulation of ENaC alpha, beta, and gamma subunit proteins in rat kidney. J Clin Invest 104(7):R19–R23

Mayer MP (2018) Intra-molecular pathways of allosteric control in Hsp70s. Philos Trans R Soc Lond Ser B Biol Sci 373(1749):20170183. https://doi.org/10.1098/rstb.2017.0183

Mayer MP, Gierasch LM (2019) Recent advances in the structural and mechanistic aspects of Hsp70 molecular chaperones. J Biol Chem 294(6):2085–2097. https://doi.org/10.1074/jbc.REV118.002810

Mayer MP, Kityk R (2015) Insights into the molecular mechanism of allostery in Hsp70s. Front Mol Biosci 2:58. https://doi.org/10.3389/fmolb.2015.00058

McDonough H, Patterson C (2003) CHIP: a link between the chaperone and proteasome systems. Cell Stress Chaperones 8(4):303–308

McGrath J, Somlo S, Makova S, Tian X, Brueckner M (2003) Two populations of node monocilia initiate left-right asymmetry in the mouse. Cell 114(1):61–73

Mehnert M, Sommer T, Jarosch E (2014) Der1 promotes movement of misfolded proteins through the endoplasmic reticulum membrane. Nat Cell Biol 16(1):77–86. https://doi.org/10.1038/ncb2882

Mehnert M, Sommermeyer F, Berger M, Kumar Lakshmipathy S, Gauss R, Aebi M, Jarosch E, Sommer T (2015) The interplay of Hrd3 and the molecular chaperone system ensures efficient degradation of malfolded secretory proteins. Mol Biol Cell 26(2):185–194. https://doi.org/10.1091/mbc.E14-07-1202

Mehrtash AB, Hochstrasser M (2018) Ubiquitin-dependent protein degradation at the endoplasmic reticulum and nuclear envelope. Semin Cell Dev Biol 93:111–124. https://doi.org/10.1016/j.semcdb.2018.09.013

Metzger MB, Liang YH, Das R, Mariano J, Li S, Li J, Kostova Z, Byrd RA, Ji X, Weissman AM (2013) A structurally unique E2-binding domain activates ubiquitination by the ERAD E2, Ubc7p, through multiple mechanisms. Mol Cell 50(4):516–527. https://doi.org/10.1016/j.molcel.2013.04.004

Meusser B, Hirsch C, Jarosch E, Sommer T (2005) ERAD: the long road to destruction. Nat Cell Biol 7(8):766–772

Meyer DI, Dobberstein B (1980) A membrane component essential for vectorial translocation of nascent proteins across the endoplasmic reticulum: requirements for its extraction and reassociation with the membrane. J Cell Biol 87(2 Pt 1):498–502

Meyer HH, Shorter JG, Seemann J, Pappin D, Warren G (2000) A complex of mammalian ufd1 and npl4 links the AAA-ATPase, p97, to ubiquitin and nuclear transport pathways. EMBO J 19 (10):2181–2192

Meyer HH, Wang Y, Warren G (2002) Direct binding of ubiquitin conjugates by the mammalian p97 adaptor complexes, p47 and Ufd1-Npl4. EMBO J 21(21):5645–5652

Milano S, Carmosino M, Gerbino A, Svelto M, Procino G (2017) Hereditary nephrogenic diabetes insipidus: pathophysiology and possible treatment. An update. Int J Mol Sci 18(11):2385. https://doi.org/10.3390/ijms18112385

Misselwitz B, Staeck O, Rapoport TA (1998) J proteins catalytically activate Hsp70 molecules to trap a wide range of peptide sequences. Mol Cell 2(5):593–603

Mochizuki T, Wu G, Hayashi T, Xenophontos SL, Veldhuisen B, Saris JJ, Reynolds DM, Cai Y, Gabow PA, Pierides A, Kimberling WJ, Breuning MH, Deltas CC, Peters DJ, Somlo S (1996) PKD2, a gene for polycystic kidney disease that encodes an integral membrane protein. Science 272(5266):1339–1342

Morello JP, Salahpour A, Laperriere A, Bernier V, Arthus MF, Lonergan M, Petaja-Repo U, Angers S, Morin D, Bichet DG, Bouvier M (2000) Pharmacological chaperones rescue cell-surface expression and function of misfolded V2 vasopressin receptor mutants. J Clin Invest 105 (7):887–895. https://doi.org/10.1172/JCI8688

Morth JP, Pedersen BP, Toustrup-Jensen MS, Sorensen TL, Petersen J, Andersen JP, Vilsen B, Nissen P (2007) Crystal structure of the sodium-potassium pump. Nature 450 (7172):1043–1049. https://doi.org/10.1038/nature06419

Mouillac B, Chini B, Balestre MN, Elands J, Trumpp-Kallmeyer S, Hoflack J, Hibert M, Jard S, Barberis C (1995) The binding site of neuropeptide vasopressin V1a receptor. Evidence for a major localization within transmembrane regions. J Biol Chem 270(43):25771–25777

Mueller GM, Kashlan OB, Bruns JB, Maarouf AB, Aridor M, Kleyman TR, Hughey RP (2007) Epithelial sodium channel exit from the endoplasmic reticulum is regulated by a signal within the carboxyl cytoplasmic domain of the alpha subunit. J Biol Chem 282(46):33475–33483. https://doi.org/10.1074/jbc.M707339200

Mueller GM, Maarouf AB, Kinlough CL, Sheng N, Kashlan OB, Okumura S, Luthy S, Kleyman TR, Hughey RP (2010) Cys palmitoylation of the beta subunit modulates gating of the epithelial sodium channel. J Biol Chem 285(40):30453–30462. https://doi.org/10.1074/jbc.M110.151845

Mukherjee A, Mueller GM, Kinlough CL, Sheng N, Wang Z, Mustafa SA, Kashlan OB, Kleyman TR, Hughey RP (2014) Cysteine palmitoylation of the gamma subunit has a dominant role in modulating activity of the epithelial sodium channel. J Biol Chem 289(20):14351–14359. https://doi.org/10.1074/jbc.M113.526020

Mukherjee A, Wang Z, Kinlough CL, Poland PA, Marciszyn AL, Montalbetti N, Carattino MD, Butterworth MB, Kleyman TR, Hughey RP (2017) Specific palmitoyltransferases associate with and activate the epithelial sodium channel. J Biol Chem 292(10):4152–4163. https://doi.org/10. 1074/jbc.M117.776146

Mulders SM, Bichet DG, Rijss JP, Kamsteeg EJ, Arthus MF, Lonergan M, Fujiwara M, Morgan K, Leijendekker R, van der Sluijs P, van Os CH, Deen PM (1998) An aquaporin-2 water channel mutant which causes autosomal dominant nephrogenic diabetes insipidus is retained in the Golgi complex. J Clin Invest 102(1):57–66

Nadolski MJ, Linder ME (2007) Protein lipidation. FEBS J 274(20):5202–5210. https://doi.org/10. 1111/j.1742-4658.2007.06056.x

Nakatsukasa K, Brodsky JL, Kamura T (2013) A stalled retrotranslocation complex reveals physical linkage between substrate recognition and proteasomal degradation during ER-associated degradation. Mol Biol Cell 24(11):1765–1775, S1761–1768. https://doi.org/10. 1091/mbc.E12-12-0907

Nauli SM, Alenghat FJ, Luo Y, Williams E, Vassilev P, Li X, Elia AE, Lu W, Brown EM, Quinn SJ, Ingber DE, Zhou J (2003) Polycystins 1 and 2 mediate mechanosensation in the primary cilium of kidney cells. Nat Genet 33(2):129–137. https://doi.org/10.1038/ng1076

Neal S, Jaeger PA, Duttke SH, Benner C, Glass CK, Ideker T, Hampton RY (2018) The Dfm1 Derlin is required for ERAD retrotranslocation of integral membrane proteins. Mol Cell 69 (2):306–320.e304. https://doi.org/10.1016/j.molcel.2017.12.012

Needham PG, Mikoluk K, Dhakarwal P, Khadem S, Snyder AC, Subramanya AR, Brodsky JL (2011) The thiazide-sensitive NaCl cotransporter is targeted for chaperone-dependent endoplasmic reticulum-associated degradation. J Biol Chem 286(51):43611–43621. https://doi.org/10.1074/jbc.M111.288928

Nelson WJ, Veshnock PJ (1987) Ankyrin binding to (Na$^+$ + K$^+$)ATPase and implications for the organization of membrane domains in polarized cells. Nature 328(6130):533–536. https://doi.org/10.1038/328533a0

Nelson WJ, Shore EM, Wang AZ, Hammerton RW (1990) Identification of a membrane-cytoskeletal complex containing the cell adhesion molecule uvomorulin (E-cadherin), ankyrin, and fodrin in Madin-Darby canine kidney epithelial cells. J Cell Biol 110(2):349–357

Nguyen AD, Lee SH, DeBose-Boyd RA (2009) Insig-mediated, sterol-accelerated degradation of the membrane domain of hamster 3-hydroxy-3-methylglutaryl-coenzyme A reductase in insect cells. J Biol Chem 284(39):26778–26788. https://doi.org/10.1074/jbc.M109.032342

Ninagawa S, Okada T, Sumitomo Y, Kamiya Y, Kato K, Horimoto S, Ishikawa T, Takeda S, Sakuma T, Yamamoto T, Mori K (2014) EDEM2 initiates mammalian glycoprotein ERAD by catalyzing the first mannose trimming step. J Cell Biol 206(3):347–356. https://doi.org/10.1083/jcb.201404075

Noguchi S, Higashi K, Kawamura M (1990) A possible role of the beta-subunit of (Na,K)-ATPase in facilitating correct assembly of the alpha-subunit into the membrane. J Biol Chem 265 (26):15991–15995

Noreng S, Bharadwaj A, Posert R, Yoshioka C, Baconguis I (2018) Structure of the human epithelial sodium channel by cryo-electron microscopy. elife 7:e39340. https://doi.org/10.7554/eLife.39340

O'Donnell BM, Mackie TD, Subramanya AR, Brodsky JL (2017) Endoplasmic reticulum-associated degradation of the renal potassium channel, ROMK, leads to type II Bartter syndrome. J Biol Chem 292(31):12813–12827. https://doi.org/10.1074/jbc.M117.786376

Oberfeld B, Ruffieux-Daidie D, Vitagliano JJ, Pos KM, Verrey F, Staub O (2011) Ubiquitin-specific protease 2-45 (Usp2-45) binds to epithelial Na$^+$ channel (ENaC)-ubiquitylating enzyme Nedd4-2. Am J Physiol Renal Physiol 301(1):F189–F196. https://doi.org/10.1152/ajprenal.00487.2010

Oda Y, Hosokawa N, Wada I, Nagata K (2003) EDEM as an acceptor of terminally misfolded glycoproteins released from calnexin.[comment]. Science 299(5611):1394–1397

Ohno Y, Kihara A, Sano T, Igarashi Y (2006) Intracellular localization and tissue-specific distribution of human and yeast DHHC cysteine-rich domain-containing proteins. Biochim Biophys Acta 1761(4):474–483. https://doi.org/10.1016/j.bbalip.2006.03.010

Padilla-Benavides T, Roldan ML, Larre I, Flores-Benitez D, Villegas-Sepulveda N, Contreras RG, Cereijido M, Shoshani L (2010) The polarized distribution of Na$^+$,K$^+$-ATPase: role of the interaction between {beta} subunits. Mol Biol Cell 21(13):2217–2225. https://doi.org/10.1091/mbc.E10-01-0081

Palmer BF (2015) Regulation of potassium homeostasis. Clin J Am Soc Nephrol 10(6):1050–1060. https://doi.org/10.2215/CJN.08580813

Peters M, Ermert S, Jeck N, Derst C, Pechmann U, Weber S, Schlingmann KP, Seyberth HW, Waldegger S, Konrad M (2003) Classification and rescue of ROMK mutations underlying hyperprostaglandin E syndrome/antenatal Bartter syndrome. Kidney Int 64(3):923–932. https://doi.org/10.1046/j.1523-1755.2003.00153.x

Pilon M, Schekman R, Romisch K (1997) Sec61p mediates export of a misfolded secretory protein from the endoplasmic reticulum to the cytosol for degradation. EMBO J 16(15):4540–4548

Pilon M, Romisch K, Quach D, Schekman R (1998) Sec61p serves multiple roles in secretory precursor binding and translocation into the endoplasmic reticulum membrane. Mol Biol Cell 9 (12):3455–3473

Plemper RK, Bohmler S, Bordallo J, Sommer T, Wolf DH (1997) Mutant analysis links the translocon and BiP to retrograde protein transport for ER degradation. Nature 388 (6645):891–895

Pobre KFR, Poet GJ, Hendershot LM (2019) The endoplasmic reticulum (ER) chaperone BiP is a master regulator of ER functions: getting by with a little help from ERdj friends. J Biol Chem 294(6):2098–2108. https://doi.org/10.1074/jbc.REV118.002804

Preston GM, Brodsky JL (2017) The evolving role of ubiquitin modification in endoplasmic reticulum-associated degradation. Biochem J 474(4):445–469. https://doi.org/10.1042/BCJ20160582

Procino G, Carmosino M, Marin O, Brunati AM, Contri A, Pinna LA, Mannucci R, Nielsen S, Kwon TH, Svelto M, Valenti G (2003) Ser-256 phosphorylation dynamics of Aquaporin 2 during maturation from the ER to the vesicular compartment in renal cells. FASEB J 17 (13):1886–1888. https://doi.org/10.1096/fj.02-0870fje

Raasi S, Wolf DH (2007) Ubiquitin receptors and ERAD: a network of pathways to the proteasome. Semin Cell Dev Biol 18(6):780–791. https://doi.org/10.1016/j.semcdb.2007.09.008

Rabinovich E, Kerem A, Frohlich KU, Diamant N, Bar-Nun S (2002) AAA-ATPase p97/Cdc48p, a cytosolic chaperone required for endoplasmic reticulum-associated protein degradation. Mol Cell Biol 22(2):626–634

Rapoport TA, Li L, Park E (2017) Structural and mechanistic insights into protein translocation. Annu Rev Cell Dev Biol 33:369–390. https://doi.org/10.1146/annurev-cellbio-100616-060439

Richly H, Rape M, Braun S, Rumpf S, Hoege C, Jentsch S (2005) A series of ubiquitin binding factors connects CDC48/p97 to substrate multiubiquitylation and proteasomal targeting. Cell 120(1):73–84

Ripstein ZA, Huang R, Augustyniak R, Kay LE, Rubinstein JL (2017) Structure of a AAA+ unfoldase in the process of unfolding substrate. elife 6:e25754. https://doi.org/10.7554/eLife.25754

Robben JH, Knoers NV, Deen PM (2005) Characterization of vasopressin V2 receptor mutants in nephrogenic diabetes insipidus in a polarized cell model. Am J Physiol Renal Physiol 289(2): F265–F272. https://doi.org/10.1152/ajprenal.00404.2004

Robben JH, Sze M, Knoers NV, Deen PM (2007) Functional rescue of vasopressin V2 receptor mutants in MDCK cells by pharmacochaperones: relevance to therapy of nephrogenic diabetes insipidus. Am J Physiol Renal Physiol 292(1):F253–F260. https://doi.org/10.1152/ajprenal.00247.2006

Rossier BC (2014) Epithelial sodium channel (ENaC) and the control of blood pressure. Curr Opin Pharmacol 15:33–46. https://doi.org/10.1016/j.coph.2013.11.010

Rubenstein RC, Zeitlin PL (2000) Sodium 4-phenylbutyrate downregulates Hsc70: implications for intracellular trafficking of DeltaF508-CFTR. Am J Physiol Cell Physiol 278(2):C259–C267

Rutkevich LA, Williams DB (2011) Participation of lectin chaperones and thiol oxidoreductases in protein folding within the endoplasmic reticulum. Curr Opin Cell Biol 23(2):157–166. https://doi.org/10.1016/j.ceb.2010.10.011

Sabath E, Meade P, Berkman J, de los Heros P, Moreno E, Bobadilla NA, Vazquez N, Ellison DH, Gamba G (2004) Pathophysiology of functional mutations of the thiazide-sensitive Na-Cl cotransporter in Gitelman disease. Am J Physiol Renal Physiol 287(2):F195–F203. https://doi.org/10.1152/ajprenal.00044.2004

Sadoul K, Boyault C, Pabion M, Khochbin S (2008) Regulation of protein turnover by acetyltransferases and deacetylases. Biochimie 90(2):306–312. https://doi.org/10.1016/j.biochi.2007.06.009

Sandhu N, Duus K, Jorgensen CS, Hansen PR, Bruun SW, Pedersen LO, Hojrup P, Houen G (2007) Peptide binding specificity of the chaperone calreticulin. Biochim Biophys Acta 1774 (6):701–713. https://doi.org/10.1016/j.bbapap.2007.03.019

Sato BK, Hampton RY (2006) Yeast Derlin Dfm1 interacts with Cdc48 and functions in ER homeostasis. Yeast 23(14-15):1053–1064. https://doi.org/10.1002/yea.1407

Sato BK, Schulz D, Do PH, Hampton RY (2009) Misfolded membrane proteins are specifically recognized by the transmembrane domain of the Hrd1p ubiquitin ligase. Mol Cell 34 (2):212–222

Schoebel S, Mi W, Stein A, Ovchinnikov S, Pavlovicz R, DiMaio F, Baker D, Chambers MG, Su H, Li D, Rapoport TA, Liao M (2017) Cryo-EM structure of the protein-conducting ERAD channel Hrd1 in complex with Hrd3. Nature 548(7667):352–355. https://doi.org/10.1038/nature23314

Schwieger I, Lautz K, Krause E, Rosenthal W, Wiesner B, Hermosilla R (2008) Derlin-1 and p97/valosin-containing protein mediate the endoplasmic reticulum-associated degradation of human V2 vasopressin receptors. Mol Pharmacol 73(3):697–708. https://doi.org/10.1124/mol.107.040931

Seaayfan E, Defontaine N, Demaretz S, Zaarour N, Laghmani K (2016) OS9 protein interacts with Na-K-2Cl co-transporter (NKCC2) and targets its immature form for the endoplasmic reticulum-associated degradation pathway. J Biol Chem 291(9):4487–4502. https://doi.org/10.1074/jbc.M115.702514

Sheng S, Maarouf AB, Bruns JB, Hughey RP, Kleyman TR (2007) Functional role of extracellular loop cysteine residues of the epithelial Na$^+$ channel in Na+ self-inhibition. J Biol Chem 282 (28):20180–20190. https://doi.org/10.1074/jbc.M611761200

Shi S, Buck TM, Kinlough CL, Marciszyn AL, Hughey RP, Chalfie M, Brodsky JL, Kleyman TR (2017) Regulation of the epithelial Na$^+$ channel by paraoxonase-2. J Biol Chem 292 (38):15927–15938. https://doi.org/10.1074/jbc.M117.785253

Shimizu Y, Okuda-Shimizu Y, Hendershot LM (2010) Ubiquitylation of an ERAD substrate occurs on multiple types of amino acids. Mol Cell 40(6):917–926. https://doi.org/10.1016/j.molcel.2010.11.033

Shinoda T, Ogawa H, Cornelius F, Toyoshima C (2009) Crystal structure of the sodium-potassium pump at 2.4 A resolution. Nature 459(7245):446–450. https://doi.org/10.1038/nature07939

Shipston MJ (2011) Ion channel regulation by protein palmitoylation. J Biol Chem 286 (11):8709–8716. https://doi.org/10.1074/jbc.R110.210005

Simons JF, Ferro-Novick S, Rose MD, Helenius A (1995) BiP/Kar2p serves as a molecular chaperone during carboxypeptidase Y folding in yeast. J Cell Biol 130(1):41–49

Smith MH, Ploegh HL, Weissman JS (2011) Road to ruin: targeting proteins for degradation in the endoplasmic reticulum. Science 334(6059):1086–1090. https://doi.org/10.1126/science.1209235

Snyder PM (2005) Minireview: regulation of epithelial Na$^+$ channel trafficking. Endocrinology 146 (12):5079–5085

Snyder PM, McDonald FJ, Stokes JB, Welsh MJ (1994) Membrane topology of the amiloride-sensitive epithelial sodium channel. J Biol Chem 269(39):24379–24383

Sohara E, Rai T, Yang SS, Uchida K, Nitta K, Horita S, Ohno M, Harada A, Sasaki S, Uchida S (2006) Pathogenesis and treatment of autosomal-dominant nephrogenic diabetes insipidus caused by an aquaporin 2 mutation. Proc Natl Acad Sci U S A 103(38):14217–14222. https://doi.org/10.1073/pnas.0602331103

Somlo S, Ehrlich B (2001) Human disease: calcium signaling in polycystic kidney disease. Curr Biol 11(9):R356–R360

Spear ED, Ng DT (2005) Single, context-specific glycans can target misfolded glycoproteins for ER-associated degradation. J Cell Biol 169(1):73–82. https://doi.org/10.1083/jcb.200411136

Stagg HR, Thomas M, van den Boomen D, Wiertz EJ, Drabkin HA, Gemmill RM, Lehner PJ (2009) The TRC8 E3 ligase ubiquitinates MHC class I molecules before dislocation from the ER. J Cell Biol 186(5):685–692. https://doi.org/10.1083/jcb.200906110

Staub O, Dho S, Henry P, Correa J, Ishikawa T, McGlade J, Rotin D (1996) WW domains of Nedd4 bind to the proline-rich PY motifs in the epithelial Na$^+$ channel deleted in Liddle's syndrome. EMBO J 15(10):2371–2380

Staub O, Gautschi I, Ishikawa T, Breitschopf K, Ciechanover A, Schild L, Rotin D (1997) Regulation of stability and function of the epithelial Na$^+$ channel (ENaC) by ubiquitination. EMBO J 16(21):6325–6336

Steel GJ, Fullerton DM, Tyson JR, Stirling CJ (2004) Coordinated activation of Hsp70 chaperones. Science 303(5654):98–101

Stein A, Ruggiano A, Carvalho P, Rapoport TA (2014) Key steps in ERAD of luminal ER proteins reconstituted with purified components. Cell 158(6):1375–1388. https://doi.org/10.1016/j.cell. 2014.07.050

Suzuki T (2015) The cytoplasmic peptide:N-glycanase (Ngly1)-basic science encounters a human genetic disorder. J Biochem 157(1):23–34. https://doi.org/10.1093/jb/mvu068

Suzuki T, Park H, Hollingsworth NM, Sternglanz R, Lennarz WJ (2000) PNG1, a yeast gene encoding a highly conserved peptide:N-glycanase. J Cell Biol 149(5):1039–1052

Swatek KN, Komander D (2016) Ubiquitin modifications. Cell Res 26(4):399–422. https://doi.org/10.1038/cr.2016.39

Takeda K, Noguchi S, Sugino A, Kawamura M (1988) Functional activity of oligosaccharide-deficient (Na,K)ATPase expressed in Xenopus oocytes. FEBS Lett 238(1):201–204

Tamarappoo BK, Verkman AS (1998) Defective aquaporin-2 trafficking in nephrogenic diabetes insipidus and correction by chemical chaperones. J Clin Investig 101(10):2257–2267

Tang WK, Li D, Li CC, Esser L, Dai R, Guo L, Xia D (2010) A novel ATP-dependent conformation in p97 N-D1 fragment revealed by crystal structures of disease-related mutants. EMBO J 29 (13):2217–2229. https://doi.org/10.1038/emboj.2010.104

Taxis C, Hitt R, Park SH, Deak PM, Kostova Z, Wolf DH (2003) Use of modular substrates demonstrates mechanistic diversity and reveals differences in chaperone requirement of ERAD. J Biol Chem 278(38):35903–35913

Tokhtaeva E, Sachs G, Vagin O (2009) Assembly with the Na,K-ATPase alpha(1) subunit is required for export of beta(1) and beta(2) subunits from the endoplasmic reticulum. Biochemistry 48(48):11421–11431. https://doi.org/10.1021/bi901438z

Tokhtaeva E, Munson K, Sachs G, Vagin O (2010) N-glycan-dependent quality control of the Na, K-ATPase beta(2) subunit. Biochemistry 49(14):3116–3128. https://doi.org/10.1021/bi100115a

Tokhtaeva E, Sachs G, Souda P, Bassilian S, Whitelegge JP, Shoshani L, Vagin O (2011) Epithelial junctions depend on intercellular trans-interactions between the Na,K-ATPase beta(1) subunits. J Biol Chem 286(29):25801–25812. https://doi.org/10.1074/jbc.M111.252247

Torres VE, Harris PC (2006) Mechanisms of disease: autosomal dominant and recessive polycystic kidney diseases. Nat Clin Pract Nephrol 2(1):40–55. https://doi.org/10.1038/ncpneph0070

Tyson JR, Stirling CJ (2000) LHS1 and SIL1 provide a lumenal function that is essential for protein translocation into the endoplasmic reticulum. EMBO J 19(23):6440–6452. https://doi.org/10.1093/emboj/19.23.6440

Valentijn JA, Fyfe GK, Canessa CM (1998) Biosynthesis and processing of epithelial sodium channels in Xenopus oocytes. J Biol Chem 273(46):30344–30351

van den Boomen DJ, Timms RT, Grice GL, Stagg HR, Skodt K, Dougan G, Nathan JA, Lehner PJ (2014) TMEM129 is a Derlin-1 associated ERAD E3 ligase essential for virus-induced degradation of MHC-I. Proc Natl Acad Sci U S A 111(31):11425–11430. https://doi.org/10.1073/pnas.1409099111

van der Goot AT, Pearce MMP, Leto DE, Shaler TA, Kopito RR (2018) Redundant and antagonistic roles of XTP3B and OS9 in decoding glycan and non-glycan degrons in ER-associated degradation. Mol Cell 70(3):516–530.e516. https://doi.org/10.1016/j.molcel.2018.03.026

Vashist S, Ng DT (2004) Misfolded proteins are sorted by a sequential checkpoint mechanism of ER quality control. J Cell Biol 165(1):41–52

Vashist S, Kim W, Belden WJ, Spear ED, Barlowe C, Ng DT (2001) Distinct retrieval and retention mechanisms are required for the quality control of endoplasmic reticulum protein folding. J Cell Biol 155(3):355–368. https://doi.org/10.1083/jcb.200106123

Vembar SS, Brodsky JL (2008) One step at a time: endoplasmic reticulum-associated degradation. Nat Rev Mol Cell Biol 9(12):944–957

Verma R, Oania R, Graumann J, Deshaies RJ (2004) Multiubiquitin chain receptors define a layer of substrate selectivity in the ubiquitin-proteasome system. Cell 118(1):99–110

Wahlman J, DeMartino GN, Skach W, Bulleid NJ, Brodsky JL, Johnson AE (2007) Real-time fluorescence detection of ERAD substrate retro-translocation in a mammalian in vitro system. Cell 129(5):943–955

Walter P, Blobel G (1981) Translocation of proteins across the endoplasmic reticulum III. Signal recognition protein (SRP) causes signal sequence-dependent and site-specific arrest of chain elongation that is released by microsomal membranes. J Cell Biol 91(2 Pt 1):557–561

Wanamaker CP, Christianson JC, Green WN (2003) Regulation of nicotinic acetylcholine receptor assembly. Ann N Y Acad Sci 998:66–80

Weisz OA, Wang JM, Edinger RS, Johnson JP (2000) Non-coordinate regulation of endogenous epithelial sodium channel (ENaC) subunit expression at the apical membrane of A6 cells in response to various transporting conditions. J Biol Chem 275(51):39886–39893

Welling PA (2016) ROMK and Bartter syndrome type 2. Ion channels and transporters of epithelia in health and disease. Springer, New York

Wiertz EJ, Tortorella D, Bogyo M, Yu J, Mothes W, Jones TR, Rapoport TA, Ploegh HL (1996) Sec61-mediated transfer of a membrane protein from the endoplasmic reticulum to the proteasome for destruction. Nature 384(6608):432–438

Willer M, Forte GM, Stirling CJ (2008) Sec61p is required for ERAD-L: genetic dissection of the translocation and ERAD-L functions of Sec61P using novel derivatives of CPY. J Biol Chem 283(49):33883–33888. https://doi.org/10.1074/jbc.M803054200

Wu X, Rapoport TA (2018) Mechanistic insights into ER-associated protein degradation. Curr Opin Cell Biol 53:22–28. https://doi.org/10.1016/j.ceb.2018.04.004

Wu Q, Moeller HB, Stevens DA, Sanchez-Hodge R, Childers G, Kortenoeven MLA, Cheng L, Rosenbaek LL, Rubel C, Patterson C, Pisitkun T, Schisler JC, Fenton RA (2018) CHIP regulates aquaporin-2 quality control and body water homeostasis. J Am Soc Nephrol 29(3):936–948. https://doi.org/10.1681/ASN.2017050526

Wuller S, Wiesner B, Loffler A, Furkert J, Krause G, Hermosilla R, Schaefer M, Schulein R, Rosenthal W, Oksche A (2004) Pharmacochaperones post-translationally enhance cell surface expression by increasing conformational stability of wild-type and mutant vasopressin V2 receptors. J Biol Chem 279(45):47254–47263. https://doi.org/10.1074/jbc.M408154200

Xia D, Tang WK, Ye Y (2016) Structure and function of the AAA+ ATPase p97/Cdc48p. Gene 583 (1):64–77. https://doi.org/10.1016/j.gene.2016.02.042

Yan W, Samaha FF, Ramkumar M, Kleyman TR, Rubenstein RC (2004) Cystic fibrosis trans-membrane conductance regulator differentially regulates human and mouse epithelial sodium channels in Xenopus oocytes. J Biol Chem 279(22):23183–23192. https://doi.org/10.1074/jbc. M402373200

Ye Y, Meyer HH, Rapoport TA (2001) The AAA ATPase Cdc48/p97 and its partners transport proteins from the ER into the cytosol. Nature 414(6864):652–656

Yoshimura SH, Iwasaka S, Schwarz W, Takeyasu K (2008) Fast degradation of the auxiliary subunit of Na^+/K^+-ATPase in the plasma membrane of HeLa cells. J Cell Sci 121 (Pt 13):2159–2168. https://doi.org/10.1242/jcs.022905

You H, Ge Y, Zhang J, Cao Y, Xing J, Su D, Huang Y, Li M, Qu S, Sun F, Liang X (2017) Derlin-1 promotes ubiquitylation and degradation of the epithelial Na^+ channel, ENaC. J Cell Sci 130 (6):1027–1036. https://doi.org/10.1242/jcs.198242

Young JC (2014) The role of the cytosolic HSP70 chaperone system in diseases caused by misfolding and aberrant trafficking of ion channels. Dis Model Mech 7(3):319–329. https://doi.org/10.1242/dmm.014001

Younger JM, Chen L, Ren HY, Rosser MF, Turnbull EL, Fan CY, Patterson C, Cyr DM (2006) Sequential quality-control checkpoints triage misfolded cystic fibrosis transmembrane conductance regulator. Cell 126(3):571–582. https://doi.org/10.1016/j.cell.2006.06.041

Zaarour N, Demaretz S, Defontaine N, Zhu Y, Laghmani K (2012) Multiple evolutionarily conserved Di-leucine like motifs in the carboxyl terminus control the anterograde trafficking of NKCC2. J Biol Chem 287(51):42642–42653. https://doi.org/10.1074/jbc.M112.399162

Zamofing D, Rossier BC, Geering K (1989) Inhibition of N-glycosylation affects transepithelial Na$^+$ but not Na$^+$-K$^+$-ATPase transport. Am J Phys 256(5 Pt 1):C958–C966. https://doi.org/10.1152/ajpcell.1989.256.5.C958

Zattas D, Berk JM, Kreft SG, Hochstrasser M (2016) A conserved C-terminal element in the yeast Doa10 and human MARCH6 ubiquitin ligases required for selective substrate degradation. J Biol Chem 291(23):12105–12118. https://doi.org/10.1074/jbc.M116.726877

Zhang Y, Nijbroek G, Sullivan ML, McCracken AA, Watkins SC, Michaelis S, Brodsky JL (2001) Hsp70 molecular chaperone facilitates endoplasmic reticulum-associated protein degradation of cystic fibrosis transmembrane conductance regulator in yeast. Mol Biol Cell 12(5):1303–1314

Zhou R, Tomkovicz VR, Butler PL, Ochoa LA, Peterson ZJ, Snyder PM (2013) Ubiquitin-specific peptidase 8 (USP8) regulates endosomal trafficking of the epithelial Na$^+$ channel. J Biol Chem 288(8):5389–5397. https://doi.org/10.1074/jbc.M112.425272

Zhu X, Zhao X, Burkholder WF, Gragerov A, Ogata CM, Gottesman ME, Hendrickson WA (1996) Structural analysis of substrate binding by the molecular chaperone DnaK. Science 272 (5268):1606–1614

Chapter 8
Fundamentals of Epithelial Cl⁻ Transport

Bruce D. Schultz and Daniel C. Devor

Abstract While epithelial solute transport predates recorded history, our understanding of epithelial function has risen from the most basic level only recently. This chapter provides an historical perspective of epithelial electrophysiology and an initial foundation for much of the information contained in this three volume 2nd edition. Epithelial cell models are presented in their contemporary contexts to demonstrate the philosophical breakthroughs that they heralded along with the novel techniques that made them possible. The text touches on the roles that unique physiological systems such as eel gill, frog skin, rabbit intestine, and cultured cell lines have contributed to our understanding. Two examples of diseases associated with abnormal Cl⁻ transport, cholera diarrhea and cystic fibrosis, are discussed and underlying mechanisms that contribute to the pathology are identified. A hypothetical cell model with the minimal complement of transport proteins that are required for Cl⁻ secretion (Na^+/K^+-ATPase, $Na^+/K^+/2Cl^-$ cotransporter, K^+ channel, and Cl⁻ channel) along with their required localization to the mucosal (apical) or serosal (basolateral) membrane is presented. Selected examples of these transport mechanisms are presented and discussed in the light of their discoveries, biophysical characteristics, pharmacology, genetic identities, and their molecular partners. A recently published comprehensive cell model is presented as the climax of the chapter that sets the stage for distinct components that are presented in greater detail in various chapters of this volume and the other volumes of this 2nd edition.

Keywords Cl secretion · CFTR · K channels · Na/K-ATPAse · TMEM channels · Epithelia · Cotransporters

B. D. Schultz (✉)
Department of Anatomy and Physiology, Kansas State University, Manhattan, KS, USA
e-mail: bschultz@vet.ksu.edu

D. C. Devor
Department of Cell Biology, University of Pittsburgh, School of Medicine, Pittsburgh, PA, USA
e-mail: dd2@pitt.edu

© The American Physiological Society 2020
K. L. Hamilton, D. C. Devor (eds.), *Basic Epithelial Ion Transport Principles and Function*, Physiology in Health and Disease,
https://doi.org/10.1007/978-3-030-52780-8_8

8.1 Introduction

Fluid and electrolyte balance is critical for life whether viewed at the level of an individual cell, tissue, organ, or individual. The physiology or pathology associated with fluid intake or fluid loss has been appreciated since antiquity, although not always well-managed to achieve proper health—the contribution of active solute transport to fluid movement simply was not appreciated. Although Charles Darwin (1839) did not attribute solute transport functions to any particular tissues, he wrote in 1839 that Galapagos tortoises captured near lowland lagoons had urinary bladders that were distended with dilute fluid that was said to decrease in volume gradually and to become less pure (i.e., solutes became more concentrated) as animals were captured in highlands more distant from the lagoons. As early as the mid-nineteenth century, oral rehydration therapy was promoted as a treatment for cholera diarrhea along with the admonition that the "... water for drinking must be pure" (Johnson 1866) and especially free from sewage, providing a premonition to the role of bacteria and bacterial toxins in the regulation of epithelial solute and fluid transport (Snow 1849, 1855). Notable physiologists of the nineteenth century such as Ludwig Thiry, Luis Vella, Rudolf Heidenhain, and Ivan Pavlov studied epithelial fluid secretion and its regulation without appreciating the contributions of solute transport to fluid flux. Cellular mechanisms that account for solute (e.g., ions, sugars, and amino acids) transport were described initially in the mid-twentieth century and these mechanisms are a focus of ongoing research. Additional mechanisms continue to be discovered. New arrangements of well-defined and novel mechanisms, which account for solute and fluid flux in distinct tissues or situations, continue to be described. Their identities, orientations, distributions, regulation, and functions are the topics addressed in this book series.

All cells live in an extracellular fluid milieu and they maintain an intracellular environment that allows for cytosolic function. Within the vertebrate body, epithelial cells separate body compartments that can contain variable volumes and diverse solute compositions. The generation and maintenance of distinct fluid compositions in these extracellular body compartments are critical to the overall health of the individual and epithelial cells are responsible for moving solutes selectively from one compartment to another. Specialization is observed throughout the body. Gastric secretions are acid while pancreatic secretions are alkaline and have substantial buffering capacity. Airway surface fluid must be maintained with an appropriate thickness (i.e., volume), viscosity, and pH to allow for mucociliary clearance. Cells lining the inner ear generate the endolymph that is required for hearing (Wangemann 2002, 2006). Mammary epithelia generate milk for a period following gestation although there are wide variations in milk composition over time and across species. Abnormal epithelial function is associated with both male and female infertility (Denning et al. 1968; Gervais et al. 1996; Johannesson et al. 1998; Hoglund et al. 2006). The examples are nearly endless. Tightly regulated epithelial transport and barrier functions contribute to virtually all aspects of bodily function.

This overview of epithelial Cl⁻ transport will present information from an historical perspective and will focus on concepts and techniques that have played critical roles in developing the knowledge base. Often in this history, breakthroughs resulted from the development of novel techniques or novel applications of proven techniques. Examination of background material has not been exhaustive, although care is taken to identify many notable contributions. A portion of the pumps, transporters, and ion channels that are described in detail in this volume will be introduced and simple arrangements of these components that can account for observed phenomena will be shown.

8.2   Active Cl⁻ Transport

Research to examine the movement of Cl⁻ across an epithelium was initiated by August Krogh and his colleagues at the University of Copenhagen in the 1930s. Prior to this time, the characterization of epithelial transport was limited by analytical techniques that, by current standards, have very low precision. For example, gastric and pancreatic secretion characteristics were studied with the use of anionic or cationic dyes as surrogate markers (Ingraham and Visscher 1935). George de Hevesy, however, initiated the use of isotopes to study basic questions in physiology and metabolism (Hevesy and Hogfer 1934; Hahn et al. 1937, 1939). Krogh, who worked closely with de Hevesy, was quick to appreciate the advantages of these new tools and actively promoted the use of isotopes (which they obtained from Niels Bohr) to characterize metabolic pathways (Hevesy et al. 1935; Krogh 1937b). Work with frogs provided strong evidence that Cl⁻ could be taken up from dilute bathing solutions along with Na⁺ and K⁺, and could be concentrated to a 10,000-fold gradient across the skin (Krogh 1937a). Ca^{2+} uptake was not detectable with this preparation and in the presence of $CaCl_2$ the uptake of Cl⁻ was associated with bath alkalinization, which provided initial evidence for Cl⁻/base exchange. In retrospect, these observations are quite exciting, but at the time the author stated that "The present investigation raises the question . . . regarding the biological significance of this power [i.e., Cl⁻ uptake]" (Krogh 1937a). Nonetheless, Krogh extended his observations across phyla and reported active anion and cation transport across the skin or gills of frogs, goldfish, crayfish, and dragonfly larva (Krogh 1939). Ion transport was energy dependent and could be poisoned by cyanide. Importantly, the results showed that the anion uptake mechanism was selective for Cl⁻ or Br⁻, but virtually excluded I⁻, NO_3^-, and SCN⁻ (Krogh 1939). After comparing the uptake characteristics across this diverse selection of animals, Krogh concluded that cells accounting for transport are unknown, but reside in frog skin, in fish gills, and likely in the intestinal wall. He further speculated that these processes are likely regulated by nerves. Readers interested in further details of the use of radioisotopes in epithelia are directed to Chap. 1 of this volume.

Hans Ussing, a colleague of August Krogh in Copenhagen, provided the next major breakthrough in the field of epithelial ion transport. Through the 1940s,

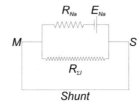

Fig. 8.1 Equivalent circuit representing short-circuited frog skin. E_{Na}: electromotive force of the sodium transporting mechanisms. R_{Na}: resistance to the sodium current. $R_{\Sigma I}$: resistance to passive ions. "*Shunt*" represents the net effect of the applied electromotive force between mucosa (M) and serosa (S). Modified from Ussing and Zerahn (1951)

isotopic tracers were used to characterize the distribution and turnover of ions in the body and the movement of ions into and out of cells. At the same time, measurements were being made to determine the electromotive forces across cell membranes in various physiological and nonphysiological buffers. Eventually, experiments were initiated to examine the movement of ions across isolated epithelial layers, specifically frog skin, and it was determined that some components of this transport could be modeled as passive diffusion while other aspects of transport required energy utilization to move ions against their electrochemical gradients. Importantly, Ussing built on earlier reports (Francis 1933) and introduced a means to separate the active transport of ions across epithelial cells from the passive movement driven by the naturally occurring transepithelial electrical potential—it was known that the isolated frog skin generated a transepithelial electrical potential, serosa positive (Ussing 1949a). Ussing's technique, the "Ussing chamber," coupled a standard flux chamber with external electrical circuitry to nullify the transepithelial voltage by the injection of electrical current (Ussing and Zerahn 1951). The system has great sensitivity, being able to resolve active transport of 1 μA on a minute-to-minute basis, or 0.62 nano equivalents/min. However, the system is blind to electroneutral processes and to charge carriers (e.g., Na^+ absorption cannot be differentiated from K^+ absorption or from Cl^- secretion). Thus, less sensitive complementary systems such as radiotracer flux must be used to define the charge carrier(s). Figure 8.1 shows that the frog skin transmural potential was attributed to net active Na^+ absorption and could contribute a driving force for the net flux of other ions (Ussing and Zerahn 1951). Ussing appreciated that there were significant mucosal-to-serosal and serosal-to-mucosal fluxes with a smaller net absorptive flux. Na^+ could be absorbed from the external environment, accumulate in the epithelial cells and then be transferred to the basolateral or blood-side of the epithelium, showing quite clearly that both the apical and the basal cell membranes play roles, and likely unique roles, in ion transport (i.e., the two membrane model) (Koefoed-Johnsen and Ussing 1958). Importantly, in their initial studies the authors fully accounted for short-circuit current, that is, the current required to maintain the transmural voltage at zero, with the absorption of Na^+. Only in the presence of adrenalin did the authors express a concern that the active transport of other ions might be implicated in the response (Ussing and Zerahn 1951). Indeed, in follow-up papers, the authors concluded that there was no net Cl^- flux in basal

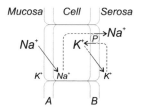

Fig. 8.2 A cell model illustrating the initial hypothesis that the frog skin potential resulted from the passive permeabilities of the apical (A) and basal (B) cell membranes to Na^+ and K^+, respectively, along with the Na^+/K^+ pump (P) in the basal membrane. Modified from Koefoed-Johnsen and Ussing (1958)

conditions, but that adrenalin stimulated net Cl⁻ secretion (Koefoed-Johnsen et al. 1952a, b). At the same time, Adrian Hogben, another of Ussing's colleagues, reported active Cl⁻ secretion by frog gastric epithelium (Hogben 1951). Additional evidence for active gastric Cl⁻ secretion was provided in a follow-up paper that also suggested the possibility of Cl^-/HCO_3^- exchange (Hogben 1955). Perhaps one of the most important conclusions to come from early studies of frog skin is an initial cell model that attributes distinct functions to the mucosal (facing the environment) and serosal (facing the blood) membranes. Figure 8.2 depicts a cell in which Na^+ enters passively across the apical membrane moving down its electrochemical gradient whereas K^+ passively exits the cell across the basal membrane. In this model, it is critical that the apical membrane is virtually impermeant to K^+ and that the basal membrane is virtually impermeant to Na^+. The authors invoked a Na^+/K^+ pump in the basal membrane to complete transepithelial Na^+ transport. That a Na^+/K^+ pump(s) exhibited stoichiometric linkage had been postulated for cephalopod nerves (Hodgkin and Keynes 1955) and it had been reported quite recently, at that time, that the enzyme derived from crab nerves required the presence of both Na^+ and K^+ for Mg^{2+}-dependent ATP hydrolysis to proceed (Skou 1957). This minimal cell model (Fig. 8.2) accounted for vectorial active transepithelial ion movement and set the stage for subsequent critical experiments to establish a role for anion secretion in health and disease. For a more thorough treatment of the fundamentals of Na^+ absorption across epithelia the reader is directed to Chap. 9 in this volume.

Initially, Ussing concluded that there was not a reason to suppose that Cl⁻ was actively transported by the epithelium—evidence supported the conclusion that Cl⁻ moved down the electrical potential established by active Na^+ transport, at least in basal conditions (Ussing 1949b). However, work by Hogben (1951, 1955) and other subsequent reports provided substantial evidence for both Cl⁻ absorption and Cl⁻ secretion by epithelia, although underlying mechanisms for active transport in either direction were not defined (Koefoed-Johnsen et al. 1952b; Curran and Solomon 1957; Zadunaisky et al. 1963; Zadunaisky 1966; Finn et al. 1967). Experiments were conducted to test for solvent drag that might accompany hydrostatically-driven fluid flux, a commonly hypothesized mechanism to account for solute flux. Hogben, however, was able to rule out this possibility and built a solid case for active electrogenic Cl⁻ secretion. Although rudimentary by today's standards, two important mechanisms were postulated to contribute to active transport in addition to Na^+/K^+ ATPase. The first is "exchange diffusion," which was introduced initially to account for Na^+/H^+ exchange across gastric mucosa (Teorell 1939). As stated above, exchange diffusion was implicated in the uptake of Cl^- across frog skin and to account for gastric secretion (Krogh 1937a; Ussing 1949b; Hogben 1955). A second salient concept, cell membrane pores that allow the movement of solutes and/or solvents down electrochemical gradients, was introduced in the 1930s (Ingraham and Visscher 1935; Curran and Solomon 1957). These fundamental mechanisms—channels, exchangers, and ATP-dependent pumps—are now well-recognized and myriad hypothetical arrangements of these (along with cotransporters that were described later) have been proposed to account for net flux across epithelia. However, cell models incorporating these components were not postulated until much later.

Although a variety of other techniques continued to be used, the Ussing-style flux chamber was adopted rapidly and was modified for the evaluation of tissues from a number of sources. In general, initial experiments focused on freshly excised epithelial sheets such as frog skin, toad bladder (Crabbe 1961), and rabbit ileum (Schultz et al. 1964). In each of these cases, Na^+ absorption could be demonstrated readily, but anion transport proved to be more elusive. Using open circuit conditions and other approaches, Cl^- absorption could be observed, but the possibility that this was secondary to active Na^+ transport, with Cl^- moving as a passive countervalent ion, could not be ruled out. By incorporating measurements of transepithelial electrical potential, rat ileal perfusion with radiotracer fluxes was used to conclude that both Na^+ and Cl^- were actively absorbed (Curran and Solomon 1957) in much the same way that Krogh had characterized uptake across the skin of a living frog two decades earlier (Krogh 1937a). Subsequently, Zadunaisky et al. (1963) provided compelling evidence for Cl^- absorption in Ussing chamber studies. Short circuit current across frog skin in standard Ringer's solution did not agree with measured $^{24}Na^+$ flux (i.e., there was a residual current), but in Cl^--free Ringer's solution the values for $^{24}Na^+$ flux and short circuit current concurred. Subsequent $^{36}Cl^-$ flux experiments were used to account for the initial discrepancy observed in standard Ringer's solution and thus to demonstrate active Cl^- absorption. The underlying reason for the inequality between Na^+ flux and short circuit current was that the concurrent absorptive flux of Na^+ and Cl^- is electroneutral, and therefore "invisible" to Ussing chamber-voltage clamp system. Importantly, ouabain, a glycoside ATPase inhibitor, reduced both Na^+ and Cl^- absorption, which suggested that these were active transport processes. Although it was not appreciated at the time, the outcomes revealed the activity of dual transport pathways. In addition to an ouabain-sensitive electrogenic Na^+ absorption mechanism, which had been observed in a number of tissues, the new results suggested that an additional process was expressed; either stoichiometric NaCl absorption or concurrent electrogenic Cl^- absorption. In this era, authors often speculated about the activity of a "Cl^- pump."

Other experiments examining basal transport properties revealed evidence for epithelial Na^+ and/or Cl^- secretion. Specialized tissues such as avian salt glands (Schmidt-Nielsen et al. 1958), spiny dogfish rectal glands (Burger and Hess 1960), and tear or nasal glands from loggerhead turtles and marine iguanas (Schmidt-Nielsen and Fange 1958) were reported to secrete fluids with Na^+ and Cl^- concentrations ranging from 500 to nearly 1000 mEq/l. Since these concentrations are substantially greater than plasma, one would conclude that an energy-consuming ion transport process must be involved for at least one of these ions. Since the lumen in an actively secreting gull nasal gland was positive relative to the interstitium, it was concluded that the active component of secretion was likely Na^+ and that Cl^- moved passively as a countervalent ion (Thesleff and Schmidt-Nielsen 1962). Such a conclusion was consistent with most previous reports that failed to gain evidence for active Cl^- transport in basal conditions. A distinct contrast, however, was shown when an Ussing-style flux chamber was developed to study frog cornea, which exhibited an aqueous humor-positive spontaneous potential. Secretory Cl^- fluxes were of the same magnitude as the short circuit current. Furthermore, the short circuit current was absent when Cl^- was removed, the magnitude of the current was a saturating function of the Cl^- concentration in the bathing medium and no net Na^+ flux was detected in these studies (Zadunaisky 1966). Thus, unequivocal evidence for active Cl^- secretion by an unstimulated epithelium was obtained.

The stage was set to characterize underlying mechanisms to account for active epithelial Cl^- transport. However, many obstacles remained—some conceptual and some practical. Ussing's early work introduced the "two membrane" model that attributed distinct functions to the mucosal (apical) and serosal (basal) cell membranes (Fig. 8.2) (Ussing 1949a; Ussing and Zerahn 1951). Yet, throughout the 1960s and sometimes beyond, conceptual models to describe ion transport were reduced to a single barrier, which confounded activities occurring at the serosal and mucosal membranes—a complete cell model to account for Cl^- movement would not be seen routinely until the 1970s (Linzell and Peaker 1971c; Schultz and Frizzell 1972; Wright 1972). Some experiments had been conducted to assess "sidedness" of buffer composition, but the concept of separate cell "loading" and "exit" mechanisms was seldom seen. Among the more pressing of the practical issues was to identify simple systems that expressed only one predominant transport process, which would allow it to be studied in relative isolation. Many of the early studies examining Cl^- transport were conducted in "basal" or unstimulated conditions and/or in complex tissues, which prevented the detection of some processes. This, however, was soon to change.

Readers interested in a historical perspective of Ussing, his work and the development of the short circuit current and the Ussing chamber technique are referred to Chap. 1 of this volume and Hamilton (2011).

8.3 Cl⁻ Transport Regulation

The first indication that active Cl^- transport could be regulated acutely came from early Ussing chamber studies. In frog skin, where electrogenic Na^+ absorption predominates, Koefoed-Johnsen et al. (1952a, b) reported adrenalin-induced Cl^-

secretion, supporting Krogh's prediction that ion transport was likely regulated by nerves (Krogh 1939). The authors concluded that mucus gland cells likely accounted for this phenomenon. This conclusion was built on earlier reports suggesting that subpopulations of cells in complex tissues could have specialized functions. For example, Keys and co-workers reported that specialized Cl^- secreting cells at the base of eel gill leaflets were responsible for Cl^- secretion in a marine environment, but not in fresh water (Keys 1931a; Keys and Willmer 1932). An in vitro system to assess eel gill permeability was developed with the goal of defining acute or chronic changes that occurred when animals transitioned between seawater and freshwater that showed high energy consumption and substantial Cl^- secretion in the marine, but not in the fresh water environment (Keys 1931b). Histological appearance was used to implicate this cell population (Keys and Willmer 1932). Although not focused on Cl^- secretion, Wright and co-workers (1940) [reviewed by Florey et al. (1941)] reported that sympathetic stimulation inhibited, and that cholinergic agonists stimulated feline intestinal fluid secretion and that atropine and some anesthetics such as barbiturates blocked these effects, which again highlighted the role of nerves in the regulation of epithelial functions. These observations reinforced the thought that epithelial cells were responsible for electrolyte and fluid transport and demonstrated that transport processes were induced, modified, or regulated by extracellular signals or conditions.

8.3.1 Cholera

The global cholera pandemic of 1961–1970 fueled substantial interest in the control of epithelial secretion and ultimately provided compelling evidence for the role of epithelial ion transport in the disease process (Pierce et al. 1971a). Asiatic cholera was first described by westerners in India in the late eighteenth century and first received general attention following the pandemic of 1817–1824, which reached from India to China and to the shores of the Mediterranean. In what proved to be one of the great epidemiological studies of all time, John Snow described the role played by contaminated water or food in the transmission of cholera (Snow 1849, 1855). Snow reported the rapid onset of explosive watery diarrhea that might be preceded by malaise and abdominal cramps, but typically was not accompanied by fever or other outward signs of illness. Cholera-induced intestinal fluid secretion is impressive by any standard. As a localized epidemic occurred in Bangkok in 1958, Watten et al. (1959) documented 24 h "rice water" stool volumes for adult patients of up to 17 L! Cholera diarrhea is a classic example of pathological Cl^- and/or HCO_3^- secretion. Importantly, delineation of cholera etiology provided both the impetus and the framework to define mechanisms regulating Cl^- secretion.

Electrophysiological studies examining intestinal anion transport initially were focused primarily on basal or unstimulated conditions and net Cl^- transport was difficult to detect in these conditions (Clarkson and Toole 1964; Schultz et al. 1964). It was known that nerve stimulation induced feline intestinal fluid secretion, but the underlying mechanism(s) were unknown (Wright et al. 1940; Florey et al. 1941). Likewise, inoculation with either *Vibrio cholerae* or media containing cholera toxin was known to induce intestinal fluid accumulation or diarrhea in rabbits, although fluid accumulation was not observed consistently in other species (De and Chatterje 1953; Dutta and Habbu 1955). Competing hypotheses included that cholera toxin inhibited the Na^+/K^+-ATPase or that it caused an increase in villus capillary permeability (De and Chatterje 1953; Finkelstein et al. 1966; Pierce et al. 1971b). Ultimately, rabbit ileal loop preparations (Burrows and Musteikis 1966; Leitch and Burrows 1968) and canine Thiry-Vella intestinal loops (Carpenter et al. 1968; Pierce et al. 1971a) were used to demonstrate that luminal inoculation led to fluid secretion. Seminal observations were reported by Field, Greenough, and co-workers, who showed that the dialyzed filtrate from a *Vibrio cholerae* culture induced an increase in short circuit current that was accompanied by an increase in Cl⁻ secretion across isolated rabbit ileum (Field et al. 1969; Greenough et al. 1969). The results showed quite clearly that cholera toxin caused intestinal Cl⁻ secretion and suggested that fluid secretion was osmotically driven. The authors built on an earlier report, which showed that exogenous cAMP or theophylline, a phosphodiesterase inhibitor that was used to block the breakdown of endogenously produced cAMP, caused a sustained increase in short circuit current (Field et al. 1968). Exposure to cholera toxin substantially reduced the effect of theophylline, suggesting that a similar mechanism may be involved in both pathways. At the same time, work with canine Thiry-Vella loops showed that exogenous cAMP, prostaglandin E1, theophylline, or cholera toxin could be used to stimulate fluid secretion of 0.5–1 ml cm^{-1} h^{-1}, the authors speculating that the key common mediator was cytosolic cAMP. Subsequently, Sharp and co-workers (Chen et al. 1971; Sharp and Hynie 1971; Sharp et al. 1971), using rabbit and human intestinal tissue homogenates, and Kimberg et al. (1971), using guinea pig and rabbit intestinal tissues, showed that cholera toxin caused a substantial and prolonged increase in adenylyl cyclase activity while having no detectable effect on phosphodiesterase. Importantly, it was shown that prostaglandins produced their effects when exposed to the serosal side of the cells whereas cholera toxin was effective only from the mucosal aspect, but that both were mediated by cytosolic adenylyl cyclase activity. Additional work would be required to fully describe the links between extracellular signaling molecules and the modulation of epithelial anion secretion. However, this flurry of reports in the late 1960s and early 1970s demonstrated that a variety of extracellular signaling molecules including vasopressin, prostaglandins, and microbial toxins could be used to upregulate anion secretion and it appeared that cAMP was a common cytosolic agent mediating the responses. The reader is directed to Chap. 2 of Volume 2 in this series for a more thorough description of the molecular mechanisms underlying the secretory diarrheas.

8.3.2 Crypts Are the Site of Intestinal Fluid Secretion

The studies outlined above demonstrate that, given appropriate stimulation, particular intestinal segments are capable of secreting vast quantities of fluids and electrolytes, however, these studies did not identify the cell type(s) that contribute to electrolyte secretion, which was a long-standing question. Florey et al. (1941) pointed to histological publications from the nineteenth century and indicated that "It has not, for instance, been settled whether the crypts secrete and the villi absorb ..." Early evidence that the crypts were the site of fluid secretion was obtained by showing that they dilated in response to pilocarpine injection into human subjects (Trier 1964). Isaacs et al. (1976) observed a greater density of nerve endings surrounding the crypts compared to the surface cells, suggesting they were the cells responding to neurotransmitters. Stronger evidence was provided by a number of studies showing that selectively damaging either surface or crypt cells with Na_2SO_4 or cycloheximide, respectively, resulted in an inhibition of cholera toxin- or acetylcholine-induced short circuit current and Cl^- secretion only when the crypts were damaged (Serebro et al. 1969; Roggin et al. 1972; Browning et al. 1978). Ultimately, Welsh et al. (1982b) demonstrated that the crypts were the site of fluid secretion by (1) directly observing the secretion of fluid using phase-contrast microscopy, and (2) using electrophysiological methods to directly demonstrate a change in the crypt cells potential profile.

8.4 Initial Cell Models for Cl^- Transport

The development of hypothetical cell models to account for ion fluxes was moved forward by the growing appreciation that the nature of particular tissues, or that the physiological or pathophysiological state of these tissues, impacts ion transport. Early studies with the Ussing-style flux chamber demonstrated net Cl^- secretion by frog gastric mucosa (Hogben 1955) and by frog skin that was exposed to adrenalin (Koefoed-Johnsen et al. 1952b). However, these observations predate the two-membrane epithelial model depicted in Fig. 8.2 and no attempt was made to place Cl^- secretion in the context of a cell system, per se. Throughout the period from the 1950s through the early 1970s a number of observations were made that ultimately were essential in developing a cell model to account for Cl^- secretion. Krogh (1937a) and Hogben (1955) suggested the likelihood of Cl^-/HCO_3^- exchange, which was supported by later observations (Hubel 1967, 1969) [Reviewed by Keynes (1969)]. Curran and Solomon (1957) suggested the presence of membrane pores—although the concept of ion channels had not yet appeared in this literature. Pharmacological agents began to be used to characterize overall transport processes. Ouabain became widely used to test for dependence on Na^+/K^+-ATPase activity (Bonting 1966). Ethacrynic acid, furosemide, and other sulfonamides began to be used as in vitro diagnostic tools, although the underlying mechanisms of action were, at best, speculative. Assessments with these agents tended to focus on changes in Na^+ transport that were attributed to Na^+/K^+-ATPase inhibition (Hook and Williamson 1965; Walker et al. 1967; Bank 1968), although this mechanism was soon ruled out (Bourgoignie et al. 1969). It was also learned that, in some tissues, Cl^- secretion was totally or partially dependent on the presence of Na^+, although Na^+ was not actively transported, which foreshadowed the discovery of electroneutral cotransporters that harnessed the electrochemical potential of one solute, typically Na^+, to drive the uptake of another solute (Zadunaisky 1966; Field et al. 1968) (see below and see Chaps. 1–4 of Volume 3 of this series). There was wide speculation regarding the expression of a "Cl^- pump(s)" as highlighted in the exhaustive review by Keynes (1969). The stage was set to build a cell diagram with components that could account for active Cl^- transport, but additional electrical driving forces still had to be defined.

Membrane potential and concentration gradients play critical roles in the movement of ions across mucosal or serosal cell membranes and transepithelial electrochemical potentials drive ions through the paracellular pathway. Numerous laboratories impaled epithelial cells with micropipettes to measure both transmembrane and transepithelial potentials in various conditions while also determining cytosolic ion concentrations (Giebisch 1958; Edmonds and Nielsen 1968; Shoemaker et al. 1970; Rose and Schultz 1971). It was noted that the predominant monovalent ions, Na^+, K^+, and Cl^- were not distributed across the cell membranes in accordance with the measured membrane potentials. Typically, there was a substantial inwardly directed electrochemical gradient for Na^+ while both K^+ and Cl^- exhibited outwardly directed electrochemical gradients of lesser magnitudes [for reviews see Schultz and Frizzell (1972) and Frizzell and Duffey (1980)]. Rose and Schultz (1971) proposed an electrical equivalent circuit that showed electrical coupling between the mucosal and serosal membranes and implicated a conductive pathway for Na^+ in the mucosal membrane. Serosal Na^+/K^+-ATPase activity could account for the outward movement of Na^+ and the inward movement of K^+ against their electrochemical gradients and for the distribution of these cations across the cell membrane. A critical aspect of this electrical model was to incorporate a "shunt" resistance representing the paracellular pathway. The model was particularly robust because the inclusion of the shunt resistance allowed the model to account for differences in serosal membrane potential between "tight" epithelia, such as frog skin, and "leaky" epithelia, such as rabbit ileum, as was demonstrated nicely by mathematical derivation in a subsequent letter (Schultz 1972).

In a series of papers and an elegant review, Linzell and Peaker (1971a, b, c, d) determined serosal, cytosolic, and mucosal ion composition for guinea pig mammary epithelium and the electrical potential across the mucosal and serosal membranes. Based on these observations, ion transport mechanisms were predicted, as shown in Fig. 8.3. The authors speculated that the mucosal membrane was freely permeable to both Na^+ and K^+—both cations were distributed such that their Nernst potential was equal to the measured potential across this membrane. The authors surmised that activity of the serosal Na^+/K^+-ATPase could account for the concentration of these cations in the cytosol and by extension in the milk. Chloride,

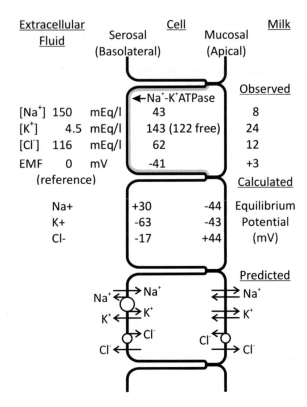

Fig. 8.3 Distribution of ions in mammary tissue of guinea pig. Top: measured concentrations in extracellular fluid and milk and calculated intracellular concentrations from an analysis of whole tissue and extracellular and milk spaces (Linzell and Peaker 1971b) and measured membrane potentials (Evans et al. 1971). Distribution of Na⁺-K⁺-ATPase is taken from Kimura (1969). Middle: equilibrium potentials for each ion across apical and basal membranes calculated from the Nernst equation. Bottom: a scheme that could account for observed distribution of ions and membrane potentials. Active transport mechanisms probably exist for Na⁺, K⁺, and possibly for Cl⁻ on basolateral membrane, but possibly for Cl⁻ only on the apical membrane. Modified from Linzell and Peaker (1971c)

however, was not distributed according to its Nernst potential at either membrane and mechanisms had to be envisioned that could account for Cl⁻ entry into the cell against an electrochemical gradient. The authors speculated, as others had in reference to other epithelia (Zadunaisky and Lande 1971), that an active pump for Cl⁻ uptake might be present in either or both membranes. The model was both novel and groundbreaking because it was able to use biophysical principles and known or predicted transporters to account for differences in mucosal and serosal fluid compositions. A schematic model to account for Cl⁻ secretion by choroid plexus was published a year later (Wright 1972). This model included new mechanisms such as a putative Na⁺/H⁺ antiporter in the serosal membrane working in concert with the carbonic acid reaction to account for HCO_3^- entry into the cell. An ion transporter working in conjunction with the carbonic acid reaction was not new, per se. Turnberg et al. (1970) previously published a model for human ileal mucosa that linked Na^+/H^+ exchange and Cl^-/HCO_3^- exchange to account for electroneutral uptake of Na^+, Cl^-, and/or HCO_3^- from various ion substituted buffers. The cell model proposed by Wright (1972) made a philosophical breakthrough by including a subset of known transport mechanisms and by placing the Na^+/K^+-ATPase in the mucosal membrane, demonstrating that components could be rearranged to account for differences in overall transport function between tissues. These models, however, included fundamental shortcomings. The authors included arrows across the serosal and/or mucosal membranes to indicate the movement of each ion, Na^+, K^+, and Cl^-, down (or sometimes against!) their respective electrochemical gradients. However, no permeation pathway is depicted and certainly no regulated pathway is envisioned by their presentations. Furthermore, these models did not extend to address one more recurring theme in epithelial anion transport, the movement of HCO_3^-. Krogh (1937a) showed that Cl^- uptake through frog skin was associated with bath alkalinization, suggesting Cl^-/HCO_3^- exchange, a mechanism that was also suggested for frog gastric mucosa (Hogben 1955). Cholera diarrhea is alkaline (Watten et al. 1959; Leitch and Burrows 1968) and Turnberg et al. (1970) suggested that a Cl^-/HCO_3^- exchange mechanism was expressed in small intestine. Nonetheless, the unveiling of these cell models to account for Cl^- transport was a substantial step forward.

8.5 Cl⁻ Conductances

Ion permeability pathways across cell membranes were undefined in early schematic models to account for ion transport. That cell membranes exhibited "conductances" to certain ions was long known (Cole and Curtis 1941) and it was demonstrated subsequently that these conductive pathways exhibited selectivity for one ion over another, at least in excitable cells (Hodgkin and Huxley 1952a, b; Hodgkin et al. 1952). It was speculated that epithelial cells, too, included membrane pores that allowed the permeation of solutes (Ingraham and Visscher 1935; Curran and Solomon 1957). However, the structure and mechanics of these pores remained virtually undefined. Ultimately, three distinct electrophysiology-based techniques were developed that contributed substantially to the discovery, kinetic description, and localization of permeation pathways through mucosal and/or serosal membranes.

Fluctuation analysis or noise analysis, which was enabled substantially by the development of computer algorithms (Cooley and Tukey 1965), was used initially to describe K^+-selective conductances in nodes of Ranvier (Derksen and Verveen 1966; Verveen et al. 1967). Lindemann and Van Driessche (Lindemann and Van Driessche 1977, 1978; Van Driessche and Zeiske 1980) used a similar approach, albeit with stationary kinetics, to demonstrate the presence of both K^+- and Na^+-conductive pathways across cell membranes in frog skin. The authors demonstrated that amiloride could be used to block Na^+ channels and introduce a new component in the power spectrum that shifted in a concentration-dependent manner (see Chaps. 8 and 9 of this volume and Chap. 18 of Volume 3 of this series). Amiloride was without effect on the K^+ channels identified in this system. An exhaustive review (Van Driessche and Zeiske 1985) demonstrated that this technique was embraced widely to define cation channels in a variety of epithelia. Not until 20 years later was the technique used to describe epithelial Cl^- channels in the mucosal membrane of an immortalized colonic epithelial cell line, T84 cells (Dharmsathaphorn et al. 1984), and the modulation of their activity by novel diarylsulfonylureas (Singh et al. 2004), as will be described further in subsequent paragraphs.

For further information about fluctuation (noise) analysis, interested readers are directed to Chap. 1 of this volume.

The membrane patch-voltage clamp (i.e., patch-clamp) technique provided a second independent method to define the biophysical characteristics of ion channels. In their Nobel Prize in Physiology or Medicine (1991) winning work, Neher and Sakmann (1976) made an electrically tight seal between a small aperture (i.e., ~10 μm^2) glass pipette and a frog muscle cell membrane to record stochastic deflections in the current records that were consistent with the inward flow of cations. Critical criteria that the authors required to define a membrane channel were that: (1) the amplitude was dependent on membrane potential; (2) the channel amplitude and kinetic rates should be consistent with predictions from fluctuation analysis; and (3) the pharmacology should be consistent with other assay systems. Surprisingly, the authors did not articulate a requirement for ion selectivity. Nonetheless, the authors reported a cholinergically stimulated 22 pS channel (at 8°C), which met the fundamental requirements that they had set. The patch-clamp technique was enhanced greatly by a subsequent report from the same laboratory (Hamill et al. 1981). Smaller pipettes (~1 μm^2) that were fire polished and electrically isolated with Sylgard were used a single time to achieve a gigaohm seal with a cell membrane. The authors showed well-resolved openings and closings of single channels in membranes of cells derived from different tissues, demonstrated voltage dependence of channel amplitude, and showed that current could be carried through a particular class of channels by different ions (e.g., Cs^+ and Na^+). Importantly, the report introduced four distinct recording configurations: cell attached; whole cell; excised inside out; and excised outside out patches, each with its own applications, strengths, and weaknesses. The technique was embraced rapidly for epithelial ion transport studies. Petersen and colleagues (Maruyama and Peterson 1982; Maruyama et al. 1983) reported Ca^{2+}- and voltage-regulated cation channels in pancreatic and salivary acinar cells. Apical membrane Cl^- channels were first reported by Nelson et al. (1984) in A6 cells, which were originally derived from *Xenopus* kidney (Perkins and Handler 1981). The authors reported a very large conductance (360 pS), stilbene-sensitive channel that was ~10-fold selective for Cl^- over Na^+ and remained active over a narrow voltage range (±20 mV).

The perfused tubule as a third powerful technique developed to define epithelial ion transport and ultimately this technique played a key role in identifying an apical Cl^- conductance. In 1966, Maurice "Mo" Burg and co-workers (Burg et al. 1966) at the National Institutes of Health reported that they had developed a method to dissect viable rabbit nephron segments and, using concentric pipettes, to perfuse them while measuring net fluid loss from the lumen. A companion paper (Grantham and Burg 1966) showed that the technique could be used to test for the effects of vasopressin and cAMP on water and urea permeability. The technique was enhanced to include electrical measurements and cable theory was introduced as a means to assess the outcomes (Burg et al. 1968; Lutz et al. 1973). In subsequent studies, the authors reported collecting duct transepithelial resistance of nearly 1 kΩ cm^2 (Helman et al. 1971), approximately 40 times greater than the resistance reported for proximal convoluted tubules (Giebisch et al. 1964). The technique was used extensively to define ion transport characteristics in various nephron segments, including active Cl⁻ absorption in the thick ascending limb of Henle's loop, and to localize the action of furosemide (Burg et al. 1973), a sulfonamide diuretic that was later shown to selectively inhibit a Na⁺/K⁺/2Cl⁻ cotransporter (Geck et al. 1980) (see Chap. 2 of Volume 3 of this series). The assay system was used to screen anthracene-9-carboxylate derivatives, which led to the discovery of diphenylamine-2'-carboxylate (DPC) as a Cl⁻ conductance blocker. DPC immediately was characterized further using the patch-clamp technique on the apical membrane of cells derived from shark (*Squalus acanthias*) rectal gland tubule, which showed that it blocked inwardly directed current carried by Cl⁻. Thus, a first pharmacological tool that selectively targeted epithelial Cl⁻ channels was identified. The study of carboxylates was expanded to screen more than 200 derivatives in this assay system, which led to the identification of 5-nitro-2-(3-phenylpropylammino)-benzoate (NPPB) as a high affinity (half maximal inhibitory concentration = 80 nM), selective epithelial Cl⁻ channel blocker (Wangemann et al. 1986). Greger (1985) provided an extensive review focused on the thick ascending limb, which included a comprehensive model to account for Cl⁻ reabsorption from the nephron lumen (Fig. 8.4). The model is similar to that described by others to account for Cl⁻ secretion by airway gland cells and shark rectal glands (Nadel and Davis 1978; Frizzell et al. 1981; Welsh et al. 1982a). Importantly, this schematic model accounts for both the entry and the exit of all pertinent ions (i.e., no ion is either accumulated or depleted by the depicted mechanisms running in concert), it accounts for the energy source that is required to move ions against their electrochemical gradients and it shows pharmacological agents that can be used to selectively block each component—the inhibition of any one component will disable the entire system. The ouabain-sensitive ATPase pumps Na⁺ out of and K⁺ into the cell with a 3:2 stoichiometry, which reduces cytosolic Na⁺ concentration and makes the cytosol electronegative relative to the bath (see Chap. 1 of Volume 3). The Na⁺ electrochemical gradient is harnessed by an electrically silent Na⁺/K⁺/2Cl⁻ cotransporter, which is inhibited by furosemide, to accumulate Cl⁻ in the cytosol above its equilibrium potential (see Chap. 2 for Volume 3). Both Cl⁻ and K⁺ exit the cell via facilitated diffusion through regulated ion-selective channels. DPC can be used to block Cl⁻ exit and Ba^{2+} was shown to block K⁺ channels. Pharmacological agents such as these have proven to be immeasurably important when defining contributions of particular components in complex tissues. The

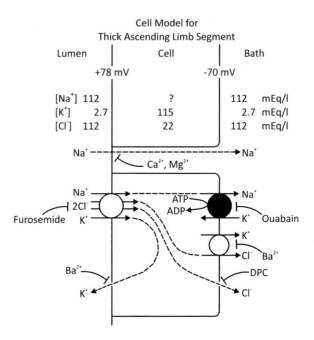

Fig. 8.4 Cell model for thick ascending limb segment. —•, diffusion; ○•, carrier; .●•, active pump; DPC, diphenylamine-2′-carboxylate. Modified from Greger (1985)

identification of additional selective blockers remains a high priority (see Chaps. 16, 21, and 24 of Volume 3).

The perfused tubule system was used to make the seminal observation that cystic fibrosis (CF), the most common life-shortening recessive genetic disease in Caucasians, is characterized by the apparent absence of an epithelial Cl⁻ conductive pathway. Cystic fibrosis of the pancreas, first described in the medical literature by Dorothy Andersen (1938), is characterized by steatorrhea, failure to thrive, pancreatic insufficiency, meconium ileus, intestinal obstruction, recurrent pulmonary infections, male infertility, and elevated concentrations of salt in sweat. Using sweat ducts derived from CF patients and from healthy controls, Quinton and co-workers (Quinton 1983, 1986; Quinton and Bijman 1983; Bijman and Quinton 1984) showed that patient ducts exhibited a greater lumen-negative transepithelial potential when compared to their unaffected counterparts and that perfusion with solutions having reduced Cl⁻ concentrations caused transepithelial hyperpolarization in healthy ducts, but was without effect in patient ducts (Quinton 1983). The authors concluded that the fundamental defect in CF was an abnormally low epithelial Cl⁻ permeability. This conclusion was supported by contemporary observations using freshly excised nasal epithelium and cultured nasal or tracheal epithelial cells, which showed normal expression of amiloride-sensitive Na⁺ absorption, but relative impermeability to Cl⁻ (Knowles et al. 1983; Widdicombe et al. 1985; Yankaskas et al. 1985).

For further discussions of Burg's isolated perfused tubule technique and the impact that his technique has had on understanding renal tubular physiology, the reader is directed to Chap. 1 of this volume and Hamilton and Moore (2016).

8.5.1 Apical Cl⁻ Conductances

The stage was set to determine the molecular or biophysical identity of the Cl⁻ conductance that was apparently absent in CF epithelia. The possibilities included that a Cl⁻ channel was absent or non-functional, or that a critical regulatory component was genetically inactivated (Frizzell et al. 1986). Both Welsh (1986) and Frizzell et al. (1986) employed airway epithelial cells with cell attached and excised inside out patch-clamp configurations to show the presence of an apical outwardly rectifying Cl⁻ channel (ORCC) that was observed frequently in cell-attached recordings from healthy individuals, but was not observed in cell-attached patches from patients. Following excision, an outwardly rectifying channel was observed in both CF and healthy tissues. While there were some similarities in the channels defined by these authors, there were substantial differences regarding the biophysical characteristics: e.g., channel size as defined by slope conductance at the reversal potential (20 or 50 pS) and sensitivity to Ca^{2+}. Additionally, Frizzell et al. (1986) reported that the channel displayed subconductance states while Welsh (1986) reported partial block by anthracene-9-carboxylate and a $Cl^-:Na^+$ selectivity ratio of 6.7:1. These reports set the tone for CF research for several years and numerous other reports describing ORCCs in epithelial apical membranes were published, especially focusing on the activation of these channels by protein kinases and the lack of kinase-dependent activation in CF cells (Hwang et al. 1989; Kunzelmann et al. 1989; Li et al. 1989) [reviewed by Guggino (1993)]. As was shown previously with ouabain for Na^+/K^+-ATPase, furosemide for the $Na^+/K^+/2Cl^-$ cotransporter, and amiloride to block a Na^+ conductance, effective, and selective blockers can provide compelling evidence for contributions of distinct transporters in different and/or complex ion transport systems. However, few Cl⁻ channel blockers were available. It was reported that the ORCC was inhibited by NPPB (Kunzelmann et al. 1989) and by 4,4′-dinitrostilbene-2,2′-disulfonic acid (DNDS) (Bridges et al. 1989), although others had reported previously that a related stilbene, 4-acetamido-4′-isothiocyanotostilbene-2,2′-sulfonic acid (SITS) was without effect on Cl⁻ uptake by normal and CF airway epithelia (Stutts et al. 1985). Bridges and co-workers (Singh et al. 1995) employed voltage-clamped synthetic lipid bilayers with membrane vesicles isolated from rat colon epithelial to develop para-sulfonated calixarenes as high-affinity selective blockers of the ORCC. While the discovery of the ORCC provided much excitement, it was determined ultimately that ORCC was likely not the primary determinant of the anion transport defect in CF.

Anion conductances with distinct biophysical characteristics were observed in colonic and pancreatic epithelial cells. Gray et al. (1989) reported that the dominant channel type observed in cell-attached pancreatic cell membrane patches was a 4–7 pS channel that was Cl^- selective and activated by forskolin. The small conductance channel was not active in excised membrane patches, but the ORCC was observed frequently in this configuration. The authors suggested that the predominant small conductance channel might work in concert with a Cl^-/HCO_3^- exchanger to account for high concentrations of HCO_3^- in pancreatic secretions (Gray et al. 1989). Frizzell and co-workers (Worrell et al. 1989; Cliff and Frizzell 1990) employed the whole-cell membrane patch to identify three distinct types of Cl^- conductances in T84 cells. Hypo-osmotic shifts were used to induce cell swelling that activated both a K^+ conductance and an outwardly rectifying Cl^- conductance in the whole-cell configuration with time-dependent activation at negative potentials and time-dependent inactivation at positive potentials. The authors speculated that this Cl^- conductance played a major role in colonic cell volume regulation (Worrell et al. 1989). Forskolin or cAMP was used to activate a Cl^- conductance, without an effect on K^+ conductance, that was linear and showed no time-dependent activation or inactivation (Cliff and Frizzell 1990). Ca^{2+} iono-phores, ionomycin, and A23187, activated a Cl^- conductance that exhibited an outward rectification, time-dependent inactivation at negative potentials, and time-dependent activation at positive potentials (Cliff and Frizzell 1990). Taken together, the observations from the Frizzell laboratory suggest that there are at least three classes of Cl^- channels in T84 cells that have distinctly different activation mechanisms and biophysical fingerprints. The outcomes, however, do not inform the reader regarding the identity or localization (apical or basolateral) of the corresponding Cl^- channels. An often-stated goal for the treatment of CF is to activate an alternative Cl^- channel that can circumvent the disease pathology. These outcomes suggested the potential for this treatment scheme.

8.5.2 Cystic Fibrosis Transmembrane Conductance Regulator

The cystic fibrosis transmembrane conductance regulator, CFTR, provided the first opportunity to link a molecular identity (i.e., a gene) with a particular Cl^- conductance. The gene coding for the protein that is defective in CF tissues was identified in 1989 and named CFTR (Kerem et al. 1989; Riordan et al. 1989; Rommens et al. 1989). The sequences of approximately one half million base pairs on the long arm of human chromosome seven were evaluated to identify a span of approximately 250 KB that contains the coding regions for CFTR (Rommens et al. 1989). Twenty-four exons were identified that coded for a protein that includes 1480 amino acids (Riordan et al. 1989). Haplotype analysis suggested that a single three-base deletion accounted for approximately 70% of disease-associated alleles and that the remaining 30% included a variety of different mutations (Kerem et al. 1989) (see Chaps. 15 and 16 of Volume 3). Sequence analysis suggested that the resulting protein included two membrane-spanning domains with six membrane-spanning segments in each, two nucleotide-binding domains, and a large regulatory domain that contained numerous charged residues and 16 consensus sites for phosphorylation by either PKC or PKA (Riordan et al. 1989). Access to the cDNA coding for CFTR provided an unprecedented opportunity to define the molecular basis of CF and to develop therapies targeted to circumvent or cure the disease.

Biophysical characteristics of CFTR were examined closely using a variety of artificial expression systems. Within a year after the sequence of CFTR was determined, the protein was expressed in CFPAC-1 cells (Drumm et al. 1990), a cell line derived from a pancreatic tumor in a CF patient (Schoumacher et al. 1990). Whole-cell patch-clamp recordings revealed a forskolin- or cAMP-stimulated conductance that had high $Cl^-:Na^+$ selectivity and a linear current–voltage relationship (Drumm et al. 1990), similar to that reported by Cliff and Frizzell (1990). Similar observations were made when CFTR was expressed in HeLa, CHO, or NIH 3T3 cells (Anderson et al. 1991c) and when expressed with the baculovirus expression system in Sf9 cells (Kartner et al. 1991), which were derived from *Spodoptera frugiperda* ovary (Vaughn et al. 1977). Cell attached recordings following CFTR expression in either Sf9 cells or *Xenopus* oocytes revealed a small (7–10 pS) linear Cl⁻ channel that was activated by cAMP (Bear et al. 1991; Kartner et al. 1991) and exhibited substantial similarities to channels that had been reported in pancreatic duct cells (Gray et al. 1989). The most compelling evidence that CFTR was a Cl⁻ channel and not a Cl⁻ channel regulator came from work by Anderson et al. (1991b) showing that mutations in the putative transmembrane segments were associated with differences in whole-cell anion selectivity when expressed in HeLa cells. The evidence was compelling that CFTR coded for the small linear channel and that mutations in this channel accounted for CF-associated Cl⁻ transport deficiencies. The relationship of CFTR to other ion channels and the potential to harness alternative Cl⁻ channels remained undetermined. These topics are covered in detail elsewhere in this volume (see Chaps. 15 and 16 of Volume 3). Potential alternative anion channels are also covered in detail in this volume (Chaps. 13 and 17 of Volume 3).

8.5.2.1 CFTR Pharmacology and Potential Therapeutic Applications

Access to CFTR expression systems provided unprecedented opportunities to define physiological and pharmacological regulators of the CFTR anion channel (see Chaps. 15 and 16 of Volume 3). The deduced protein structure suggested that kinases and nucleotides could affect protein function (Riordan et al. 1989) and this was certainly shown to be the case by a variety of reports. Welsh and colleagues (Anderson et al. 1991a; Anderson and Welsh 1992) provided initial compelling evidence that nucleotides played a critical role in channel gating. Venglarik and Schultz (Venglarik et al. 1994; Schultz et al. 1995b) employed the excised membrane patch configuration and fluctuation analysis to derive the underlying rate constants for the interaction of nucleotides with the CFTR channel and to determine the relationship with channel gating. Sheppard and Welsh (1992) reported that glibenclamide and tolbutamide, ATP-dependent K^+ channel blockers, inhibited the CFTR-associated conductance in whole-cell recordings. Excised membrane patches were used in conjunction with fluctuation analysis to demonstrate that these sulfonamides directly interacted with the CFTR channel (Schultz et al. 1996; Venglarik et al. 1996) and that structurally related compounds, diarylsulfonylureas, which have no detectable effect on K^+ channels, inhibit CFTR, as well (Schultz et al. 1995a). This structure was exploited through medicinal chemistry to produce analogs that blocked CFTR in excised membrane patches and in intact epithelial monolayers (Singh et al. 2004). DASU-01 was used as a probe in fluctuation analysis and based on the outcomes, estimates of unitary current, channel number, and open probability could be made for CFTR channels in a native cell environment (Singh et al. 2004). DASU-02 has been used in combination with other selective channel blockers to demonstrate that CFTR plays a critical role in intestinal secretion induced by *E. coli* heat-stable and heat-labile toxins and by extension to cholera toxin (O'Donnell et al. 2000; Veilleux et al. 2008), making it a potential lead compound for the treatment of secretory diarrhea. Alternatively, Cuppoletti et al. (2014) employed DASU-02 to demonstrate a contrast between CFTR-mediated whole-cell current and current that was stimulated by lubiprostone and inhibited by methadone, which the authors attribute to ClC-2 Cl^- channels. CLC-2, anoctamin and CFTR channels, their pharmacologies and the general topic of secretory diarrheas are handled in detail in Chap. 2 of Volume 2 and Chaps. 13–17, and 25 of Volume 3.

8.6   Apical Cl^-/HCO_3^- Exchangers

As indicated in previous paragraphs, a functional linkage between Cl^- and HCO_3^- transport has been observed or postulated frequently. Among the examples presented were frog skin (Krogh 1937a), and frog gastric mucosa (Hogben 1955) while other contemporary examples could be presented [e.g., mosquito gut (Keynes 1969)], rabbit small intestine (Schultz and Frizzell 1972), rabbit colon (Frizzell et al. 1976), human ileum (Turnberg et al. 1970). Research concerning HCO_3^- transporters, and especially concerning anion exchangers, has distinct challenges associated with it. First, the carbonic acid reaction can be catalyzed both in the cytosol and in the extracellular fluid by carbonic anhydrases, which can produce or consume HCO_3^- in either compartment. Second, CO_2, which is another component of the carbonic acid reaction, is volatile, is present in large quantities in the environment, and can cross cell membranes more easily than HCO_3^-. Thus, it is challenging to "clamp" the HCO_3^- concentration at a particular level or to measure the production or disappearance of HCO_3^- during an experiment. Finally, many of the HCO_3^- transport mechanisms are electroneutral, which greatly reduces the utility of electrophysiological techniques. Nonetheless, useful techniques have been developed and two major classes of HCO_3^- transporters have been identified.

The route to the discovery of anion exchangers was initiated in 1971 when proteins from human erythrocyte ghosts were resolved on polyacrylamide gels

(Fairbanks et al. 1971). Six major protein bands were observed and "Band 3" (as it became known) accounted for more than 25% of the total intensity on the gel. Calculations suggested that the Band 3 protein was present in two- to five-fold molar excess when compared to the intensity of other bands (Fairbanks et al. 1971). Further research showed that Band 3 allowed for anion permeability across the membrane and that permeation could be inhibited by a number of compounds including pyridoxal phosphate and disulfonic stilbenes (Zaki et al. 1975; Rothstein et al. 1976). Band 3, which later became known as anion exchanger 1 (AE1), was the first transport protein to be cloned and expressed (Kopito and Lodish 1985). Subsequently, a number of related proteins were identified by homology cloning and a gene family, SLC4, containing ten members has been described (Romero et al. 2004; Alper 2006). The gene family includes not only classical anion exchangers (SLC4A1-4), but also electrogenic and electroneutral Na^+/HCO_3^- cotransporters (SLC4A4-8). The latter proteins play important roles in systemic acid–base balance by mediating Na^+-coupled HCO_3^- recovery in the kidney and in loading pancreatic cells for HCO_3^- secretion that ultimately buffers the proximal small intestine. The reader is directed to Chap. 4 of Volume 3 for a more extensive description of SLC4 HCO_3^- transporters.

A second family of HCO_3^- transporters, the SLC26 gene family, has been described more recently, although certain family members are associated with medical conditions that were described more than 50 years ago. The founding member of the gene family, Sat-1 (SLC26A1), is a sulfate transporter that was cloned from rat hepatocytes (Bissig et al. 1994). Like many of the anion transporters that have been described, SLC26A1 is inhibited by disulfonic stilbenes. SLC26A2 is a sulfate transporter that is predominantly expressed in cartilage and in the intestine. Naturally occurring mutations are associated with diastrophic dysplasia, which was first described by Lamy and Maroteaux in 1960 (Everett and Green 1999). The gene was cloned in 1994 (Hastbacka et al. 1994). SLC26A3 is a Cl^-/HCO_3^- exchanger that is known also as DRA (downregulated in adenoma), which is highly expressed in normal colonic epithelium, but expressed at lower levels in polyps and in adenocarcinomas of the colon (Schweinfest et al. 1993). The same gene has been associated with Cl^- losing diarrhea (CLD), which was first described in 1945 as congenital alkalosis with diarrhea (Darrow 1945; Gamble et al. 1945). More recently, Hoglund and colleagues (Hihnala et al. 2006; Hoglund et al. 2006; Kujala et al. 2007) reported that mutations in SLC26A3 are associated also with male infertility. Mutations in SLC26A4 are associated with deafness and goiter, a condition first described by Vaughan Pendred (1896). The gene associated with Pendred syndrome was speculated initially to be a sulfate transporter based on sequence homology (Everett et al. 1997), but it is now known to function as a Cl^-/HCO_3^- exchanger. Recently it was reported by Wangemann and colleagues (Choi et al. 2011; Li et al. 2013) that SLC26A4 expression is required during fetal development in the endolymphatic sac, but not in the cochlea or vestibular sensory organs, for hearing and balance to develop, and that it is not required during postnatal life for hearing to be maintained. SLC26A6 is an anion exchanger that is expressed in many tissues and at higher levels in kidney, where it may have a role in formate or oxalate transport. SLC26A6 null mice appear to be healthy, but have some renal transport abnormalities (Wang et al. 2005). There has been speculation that there are physical or functional interactions between CFTR and SLC26A6 or SLC26A3 that can account for CFTR-dependent HCO_3^- secretion, especially in the pancreas where segregated expression was suggested (Lee et al. 1999; Ko et al. 2004; Singh et al. 2010). More recently it was shown that both SLC26A3 and SLC26A6 are expressed at high levels in or near the mucosal membrane of all epithelial cells lining porcine vas deferens with no detectable gradient in expression over the length of the duct (Pierucci-Alves et al. 2015). Although protein localization was not assessed, there were no detectable differences in the expression of either SLC26A3 or SLC26A6 in the vas deferens of pigs lacking CFTR or expressing the most common mutant form of CFTR, $\Delta F508$ (Pierucci-Alves et al. 2011). Obviously, additional work is required to demine the roles that these transport proteins play in normal physiology of different tissues. Excellent reviews of this transporter family have been published (Everett and Green 1999; Mount and Romero 2004) and the reader is directed to Chap. 12 of Volume 3 for an expanded discussion focused on SLC26 anion transporters.

8.7   Evidence for a $Na^+/K^+/2Cl^-$ Cotransporter

As pointed out by Tidball (1961), Cl^- typically is secreted against an electrochemical potential gradient such that the process must include an active energy-consuming process. This active step could occur at either the basolateral membrane where Cl^- uptake occurs, or Cl^- could be actively extruded across the apical membrane. Measurements of intracellular Cl^- from numerous laboratories demonstrated that Cl^- is accumulated above electrochemical equilibrium in the cell (Reuss et al. 1983; Welsh 1983b; Greger and Schlatter 1984b; Shorofsky et al. 1984), indicative of Cl^- being actively transported into the cell across the basolateral membrane. Whether this active step involved a pump or a cotransport process remained an open question, however.

This question began to be addressed when Al-Bazzaz and Al-Awqati (1979) demonstrated that Cl^- secretion from dog trachea could be blocked by simply removing Na^+ from the serosal solution; being consistent with Na^+-coupled entry of Cl^- into the cell. More direct evidence for Na^+-coupled Cl^- entry was provided by measuring intracellular Cl^- with ion-selective microelectrodes in the presence and absence of bath Na^+ (Welsh 1983b). Initially, intracellular Cl^- was found to be accumulated above electrochemical equilibrium. However, when Na^+ was removed from the bath, intracellular Cl^- activity declined to a value very near that expected for an equilibrium distribution (Welsh 1983b). Identical results were reported by Reuss et al. (1983) in bullfrog corneal epithelium. By carrying out Na^+ and Cl^- influx studies in dog trachea, Widdicombe et al. (1983) provided additional evidence for coupled Na^+-Cl^- cotransport. That is, these authors demonstrated that removal of Na^+ reduced Cl^- influx, while removal of Cl^- reduced Na^+ influx into the cells.

Demonstrating a dependence on K^+ in the uptake of Cl^- across the basolateral membrane proved to be more difficult. This stems from the fact that simply removing K^+ from the serosal solution will inhibit the Na^+/K^+-ATPase, thereby confounding the results. Greger and Schlatter (1984b) proposed that 2 Cl^- ions were transported for each Na^+ in the spiny dogfish (*Squalus acanthias*) rectal gland, based on the Hill coefficients for Na^+-dependent Cl^- transport and Cl^--dependent Na^+ transport. This notion was extended by showing that furosemide, a known blocker of the cotransporter (Geck et al. 1980), caused the concentration of Cl^- to drop in the cell twice as fast as Na^+ (Greger and Schlatter 1984a). These authors also demonstrated that removal of serosal K^+ caused a drop in intracellular Cl^- that was not due to an effect on the Na^+/K^+-ATPase and, therefore, proposed a $1Na^+:1K^+:2Cl^-$ stoichiometry for the carrier (Greger and Schlatter 1984b). This stoichiometry was similarly proposed for this carrier in the MDCK cell line when it was found that the uptake of radiolabeled Na^+ only occurred in the presence of both Cl^- and K^+, while the uptake of radiolabeled Rb^+ (a tracer typically used in place of K^+) required both Na^+ and Cl^- (McRoberts et al. 1982). Finally, O'Grady et al. (1986), working in winter flounder intestine, carried out simultaneous NaCl and RbCl influx measurements and demonstrated a ratio of 2.2 for $Cl^-:Na^+$ and 1.8 for $Cl^-:Rb^+$; confirming a $1Na^+:1K^+:2Cl^-$ stoichiometry.

One of the consequences of proposing a $1Na^+:1K^+:2Cl^-$ stoichiometry for the cotransporter is that it must be electroneutral and therefore insensitive to membrane voltage. Early evidence that this was the case came from duck red blood cells (Haas et al. 1982). In these studies, the duck red cells were exposed to the K^+ ionophore valinomycin, thereby increasing membrane K^+ permeability such that the membrane voltage was dominated by the K^+ reversal potential (E_K). Thus, the uptake of Na^+ could be studied while bath K^+, and therefore membrane voltage, was varied. Under these conditions, the uptake of Na^+ was not affected by changes in membrane voltage; demonstrating the electroneutrality of the cotransporter (Haas et al. 1982). Similar experiments were carried out on MDCK cells using valinomycin with a similar lack of dependence on membrane voltage demonstrated (McRoberts et al. 1982). Finally, electrophysiological methods in canine tracheal epithelium confirmed that the cotransporter is electroneutral by confirming that the potential across the basolateral membrane was unaffected when the cotransporter was blocked by either Cl^- removal (Welsh 1983b) or furosemide block (Welsh 1983a).

As noted above, Cl^- entry into the cell requires that it moves against an electrochemical potential. Thus, the coupling of Cl^- entry to Na^+ and K^+, such that this process is electroneutral, is an energetically efficient way of accumulating Cl^- above electrochemical equilibrium. Indeed, using values for intracellular Na^+, K^+, and Cl^- a net driving force of 40–70 mV can be calculated, favoring Cl^- entry. This indicates that there is sufficient energy in the established ion gradients to drive Cl^- into the cell above electrochemical equilibrium using a $1Na^+:1K^+:2Cl^-$ transport process.

A complete account of the regulation, cell biology and molecular biology of the $Na^+/K^+/2Cl^-$ cotransporters are discussed in Chap. 2 of Volume 3 of this series and the reader is directed there for additional details.

8.8 Evidence for a Basolateral Membrane K$^+$ Channel

As detailed above, it was recognized early on that both Na$^+$ absorption and Cl$^-$ secretion were associated with the exit of K$^+$ across the basolateral membrane of epithelia. Indeed, as can be appreciated from the above discussion, activation of an apical Cl$^-$ conductance alone would induce an initial Cl$^-$ secretory response, followed by a return to steady state. This is because Cl$^-$ is initially above its electrochemical equilibrium, due to the Na$^+$/K$^+$/2Cl$^-$ cotransporter, such that Cl$^-$ can flow down its electrochemical potential gradient upon opening of an Cl$^-$ channel exit pathway in the apical membrane. Subsequently, however, activation of the Cl$^-$ conductance alone would result in the apical membrane depolarizing until it approaches the reversal potential for Cl$^-$ (E_{Cl}), after which there would be no net driving force favoring Cl$^-$ exit from the cell, resulting in an attenuation of Cl$^-$ secretion. Clearly then, an additional conductive pathway, capable of hyperpolarizing the cell such that Cl$^-$ remains above electrochemical equilibrium is required; the aforementioned K$^+$ exit pathway provides a critical force to maintain membrane polarity.

Early electrophysiological studies demonstrated, in both canine trachea (Welsh et al. 1982a) and spiny dogfish rectal gland (Greger and Schlatter 1984b), that cAMP-mediated agonists induced activation of a basolateral membrane K$^+$ conductance. Using microelectrodes, Welsh et al. (1982a) showed that epinephrine caused an initial depolarization of membrane potential, due to the activation of an apical Cl$^-$ conductance, followed approximately 20 sec later by a repolarization that was associated with a decrease in the basolateral membrane fractional resistance; consistent with activation of a basolateral K$^+$ conductive pathway. Further studies in both canine trachea (Smith and Frizzell 1984) and shark rectal gland (Greger and Schlatter 1984a) preparations demonstrated that cAMP-mediated agonist-induced Cl$^-$ secretion could be inhibited by either raising serosal K$^+$ or by adding a nonspecific K$^+$ channel blocker, e.g., Ba^{2+}, to the serosal bath. Both of these maneuvers would depolarize the cell; decreasing the electrochemical driving force for Cl$^-$ exit across the apical membrane. These early studies confirmed a role for a basolateral membrane K$^+$ conductance in the Cl$^-$ secretory response.

While activation of a basolateral membrane K$^+$ conductance hyperpolarizes the cell and thus maintains the electrochemical driving force for Cl$^-$ exit, activation of this conductive pathway plays an additional important role that can be overlooked. That is, early investigators noted that when cells are stimulated to secrete Cl$^-$ there is an increase in intracellular Na$^+$ (Greger et al. 1984; Shorofsky et al. 1986). This is due to Na$^+$ entering on the Na$^+$/K$^+$/2Cl$^-$ cotransporter with Cl$^-$, as detailed above. This Na$^+$ is removed from the cell via the Na$^+$/K$^+$-ATPase, which results in a potential increase in intracellular K$^+$. However, it has been shown that intracellular K$^+$ does not change during agonist stimulation (Greger and Schlatter 1984b; Smith and Frizzell 1984). Thus, K$^+$ entry via both the Na$^+$/K$^+$-ATPase and the Na$^+$/K$^+$/2Cl$^-$ cotransporter must be equaled by K$^+$ exit across the basolateral membrane. Indeed, in the absence of an increase in K$^+$ permeability there would be a requirement for the driving force for K^+ exit to increase. Since potential across the basolateral membrane is either not changed during stimulation (Smith and Frizzell 1984) or slightly decreased (Greger et al. 1984) this would require an increase in cellular K^+ to levels 100–600 mM above normal (Welsh et al. 1982a; Smith and Frizzell 1984), which, as stated above, does not happen. Although a role for a basolateral membrane K^+ conductive pathway in the Cl^- secretory response was confirmed by these studies, the identities of these K^+ channels would remain elusive for many more years until selective blockers and molecular cloning techniques could be applied.

8.8.1 Identification of the cAMP- and Ca^{2+}-Activated Basolateral Membrane K^+ Channels

The original studies of Frizzell and colleagues (Welsh et al. 1982a, 1983; Heintze et al. 1983; Smith and Frizzell 1984), Greger and colleagues (Greger and Schlatter 1984a, b; Greger et al. 1984) and Shorofsky et al. (1983) clearly suggested that cAMP was regulating a basolateral membrane K^+ conductance. Additional microelectrode studies in locust hindgut similarly showed cAMP-activation of a basolateral K^+ channel (Hanrahan and Phillips 1984) and patch-clamp studies confirmed activation of a basolateral K^+ channel in *Necturus* oxyntic cells (Ueda et al. 1987). Dharmsathaphorn and colleagues (Weymer et al. 1985; Mandel et al. 1986) demonstrated that the colonic cell line, T84, also exhibited a cAMP-activated K^+ conductance that could be blocked by Ba^{2+}.

In addition to cAMP, Ca^{2+}-mediated agonists have been shown to stimulate transepithelial Cl^- secretion via a similar model. Frizzell (1977) demonstrated that a Ca^{2+} ionophore, A23187, could induce Cl^- secretion from rabbit colon. Similarly, Bolton and Field (1977) demonstrated that A23187 increased serosa-to-mucosa Cl^- flux, resulting in Cl^- secretion in rabbit ileal mucosa. These authors further demonstrated that carbachol stimulated Cl^- secretion in the absence of any detectable change in cytosolic cAMP (Bolton and Field 1977). In addition, Zimmerman et al. (1982) demonstrated that bethanechol induced Cl^- secretion from rat intestine in a Ca^{2+}-dependent, cAMP-independent manner. Finally, using the T84 cell line, Dharmsathaphorn and Pandol (1986) demonstrated that carbachol-induced Cl^- secretion by increasing intracellular Ca^{2+} with no detectable change in cAMP, identical to what had been described in ex vivo tissue. Indeed, these authors went on to show that carbachol stimulated K^+ efflux across the basolateral membrane, indicative of an increase in basolateral K^+ conductance as shown for cAMP-mediated agonists. While these data confirm that both cAMP- and Ca^{2+}-mediated agonists activate a basolateral membrane K^+ conductance, in the absence of specific blockers and molecular techniques little additional evidence was forthcoming to differentiate amongst these conductances.

McRoberts et al. (1985) initially demonstrated in T84 cells that the cAMP- and Ca^{2+}-activated basolateral membrane K^+ conductances were differentially inhibited by Ba^{2+}, consistent with these two second messengers activating unique K^+ channels. Unfortunately, there were no blockers available at that time that could allow the investigator to conclusively identify Ca^{2+}- and cAMP-dependent K^+ channels. The first breakthrough in this line of investigation came when McCann et al. (1990) and Devor and Frizzell (1993) characterized the Ca^{2+}-activated basolateral membrane K^+ channel in airway and colonic epithelia, respectively; subsequently demonstrating it was blocked with high affinity by charybdotoxin, but was insensitive to block by the Maxi-K channel blockers, paxilline and iberiotoxin (Devor et al. 1996b). A second critical breakthrough in the pharmacology of these channels came when Lohrmann et al. (1995) characterized the 1,4-sulfonylaminochromanoles as a novel class of inhibitors of the cAMP-mediated K^+ channel in secretory epithelia, with 293B being the most potent compound described. Third, Devor and coworkers (Devor et al. 1996a, b) described the first class of openers of the Ca^{2+}-activated K^+ channels, 1-ethyl-2-benzimidazolinone (1-EBIO). These authors demonstrated that 1-EBIO activated a charybdotoxin-sensitive basolateral K^+ conductance that was insensitive to 293B, whereas cAMP-mediated Cl^- secretion was insensitive to charybdotoxin, but blocked by 293B. Finally, Rufo et al. (1996) characterized clotrimazole as a blocker of Ca^{2+}-mediated Cl^- secretion and that this effect was due to block of a basolateral K^+ conductance. Devor et al. (1997) subsequently confirmed that this was due to a direct inhibition of the charybdotoxin-sensitive basolateral K^+ channel (see Chap. 22 of Volume 3 for additional details). The use of these pharmacological tools unequivocally demonstrated that cAMP- and Ca^{2+}-mediated agonists activate unique basolateral membrane K^+ channels and played a pivotal role in confirming their molecular identities.

8.8.2 *KCa3.1 Is the Ca^{2+}-Activated Basolateral Membrane K^+ Channel*

The molecular identity of the Ca^{2+}-activated, basolateral membrane K^+ channel was revealed in 1997 when Joiner et al. (1997) and Ishii et al. (1997) nearly simultaneously reported the cloning of this channel. This channel was the fourth member of the KCNN gene family that contains the SK1-3 family members previously reported by Kohler et al. (1996). SK1-3 encode small conductance, Ca^{2+}-activated and apamin-sensitive K^+ channels expressed primarily in nerve and smooth muscle (Adelman et al. 2012). Given the ~40% similarity in amino acid sequence, Joiner et al. (1997) referred to the novel clone as SK4. However, given the intermediate conductance of the channel, based on historical precedence, Ishii et al. (1997) referred to this channel as IK1. Based on the standard nomenclature derived from the International Union of Pharmacology [IUPHAR; (Gutman et al. 2003)] this channel is now referred to as KCa3.1. Joiner et al. (1997) and Ishii et al. (1997)

demonstrated KCa3.1 was activated by Ca^{2+} and blocked by charybdotoxin and clotrimazole while being insensitive to blockers of Maxi-K and small conductance K^+ channels, similar to what had been reported for the endogenously expressed basolateral membrane, Ca^{2+}-activated K^+ channel of epithelia. Further studies confirmed KCa3.1 was activated by 1-EBIO (Pedersen et al. 1999; Syme et al. 2000; Jensen et al. 2001), consistent with data in both T84 cells (Devor et al. 1996a, b) as well as murine tracheal epithelia (Devor et al. 1996a) and human bronchial epithelia (Devor et al. 2000). Subsequent to the cloning of KCa3.1 and the realization that this channel displayed identical biophysical and pharmacological characteristics to the basolateral membrane Ca^{2+}-activated K^+ conductance of secretory epithelia, Warth et al. (1999) and Gerlach et al. (2000) confirmed by RT-PCR and Northern blot, respectively, this was the channel expressed in T84 cells. Finally, Flores et al. (2007) carried out studies on KCa3.1 knockout mice and demonstrated that Ca^{2+}-mediated Cl⁻ secretion was completely eliminated in both distal colon and small intestinal epithelium, which resulted in a marked reduction in water content in the stools. In total, these studies unequivocally demonstrate that KCa3.1 is the basolateral membrane K^+ channel activated by Ca^{2+}-mediated agonists as a means of maintaining transepithelial Cl⁻ secretion across intestinal epithelia. For a complete account of KCa3.1 in epithelial transport, the reader is directed to Chap. 22 of Volume 3 of this series.

8.8.3 KCNQ1 (Kv7.1)/KCNE3 (Mirp2) Is the cAMP-Activated Basolateral Membrane K⁺ Channel

The demonstration that 293B blocked cAMP-mediated transepithelial Cl⁻ secretion by inhibiting the basolateral membrane K^+ conductance, with no effect on the apical membrane Cl⁻ conductance, was a critical turning point in the identification of this K^+ channel (Lohrmann et al. 1995). The next step came when it was discovered that expression of minK (for minimal K^+ channel; now called KCNE1), a protein first cloned in 1988 (Takumi et al. 1988), in *Xenopus laevis* oocytes resulted in a slowly activating, voltage-dependent K^+ current that was blocked by 293B [reviewed in Suessbrich and Busch (1999)]. However, it was difficult to imagine how a protein of 130 amino acids and a single transmembrane domain could itself form a K^+ channel. In addition, KCNE1 is expressed at very low levels in colonic epithelia, the 293B-sensitive current does not display the same biophysical characteristics of *Xenopus laevis*-expressed KCNE1 and expression of KCNE1 in mammalian expression systems failed to induce an 293B-sensitive current. This conundrum was solved with the discovery that KCNE1 co-assembles with KvLQT1 (Kv7.1, KCNQ1) in *Xenopus* oocytes to form this slowly activating current (Barhanin et al. 1996; Sanguinetti et al. 1996). KvLQT1 was identified as the cardiac K^+ channel responsible for Long QT Syndrome (Type 1; LQT1; (Wang et al. 1996) and was subsequently shown to be expressed in various tissues, including intestine. Coexpression of KCNE1 and KCNQ1 resulted in a dramatically increased macroscopic current that displayed slowed activation and almost no inactivation kinetics (Splawski et al. 1997) and was blocked by various chromanol derivatives (including 293B) with a similar profile to that described for the basolateral K^+ conductance in distal colon (Suessbrich et al. 1996). These results suggested KCNQ1 was the basolateral membrane K^+ conductance responsible for maintaining a favorable electrochemical driving force for transepithelial Cl^- secretion (Bleich et al. 1997). Readers interested in addition information about KCNQ channels are directed to Chap. 25 of Volume 3.

While the above results argued for KCNQ1/KCNE1 channels being the basolateral membrane, cAMP-activated K^+ conductance in epithelia an additional puzzle remained to be solved. That is, KCNQ1/KCNE1 channels are slowly activating and voltage-dependent, whereas the cAMP-activated basolateral membrane K^+ conductance is voltage-independent and constitutively active. A possible solution to this puzzle came with the cloning of KCNE3, another small, single transmembrane domain subunit, which, when co-expressed with KCNQ1 in *Xenopus* oocytes resulted in a near-instantaneous activation of the channel complex resulting in a constitutively active, 293B-blockable K^+ conductance similar to what has been described in colonic epithelia (Schroeder et al. 2000). These results suggest KCNQ1 may be obligatorily associated with KCNE3 in certain epithelia resulting in a linear current–voltage relationship for this K^+ channel as described in colonic epithelia (Schroeder et al. 2000). However, expression of KCNQ1/KCNE3 in mammalian cells does not reproduce these biophysical characteristics (Mazhari et al. 2002; Jespersen et al. 2004). Thus, while there is clear evidence demonstrating that KCNQ1 is the basolateral membrane K^+ channel α-subunit activated by cAMP in secretory epithelia, there may be additional components to the channel complex required to produce the biophysical properties observed endogenously in epithelia. For a detailed account of KCNQ/KCNE channel gating and regulation, the reader is directed to Chap. 25 of Volume 3 in this series, as well as several excellent reviews (Jespersen et al. 2005; Zaydman and Cui 2014; Liin et al. 2015; Nakajo and Kubo 2015).

8.9   Conclusion: An Extensive Cell Model

At this juncture, it seems appropriate to respond to the statement published in 1937 by Nobel laureate August Krogh regarding epithelial Cl^- uptake, "The present investigation raises the question ... regarding the biological significance of this power" (Krogh 1937a). There is now little doubt regarding the significance of Cl^- transport in light of the many pathologies that can be attributed to changes in the expression or activity of individual proteins such as CFTR, SLC26A3, or SLC26A4. Pavlov, Krogh, Ussing, Burg, Neher, Sackman, Mullis, Agre, Venter, Collins, and others have provided groundbreaking techniques and insights that allowed generations of scientists to build a progressively better understanding of epithelial function. There is no longer a question regarding the importance of Cl^- transport. Rather, one necessarily asks, "What epithelial mechanisms are required to accomplish Cl^- transport in health and which molecular targets can be modulated to remedy disease?" Minimal models can account for secretion with as few as four components: Na⁺/K⁺-ATPase, Na⁺/K⁺/2Cl⁻ cotransporter, and a K⁺ channel in the serosal membrane and a Cl⁻ channel in the mucosal membrane. This "classical model" is

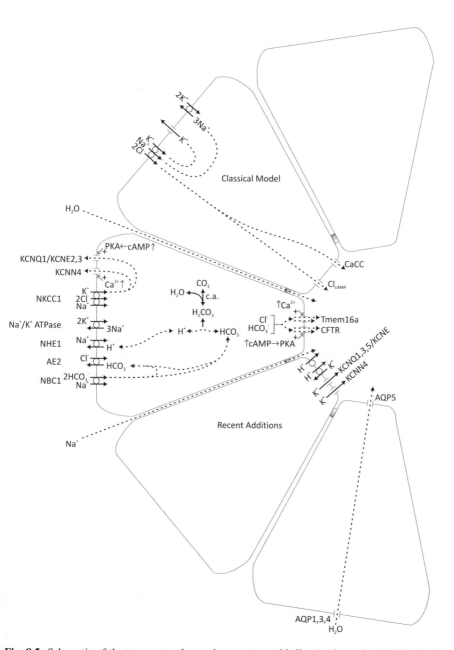

Fig. 8.5 Schematic of the transport pathways in secretory epithelia. As shown in the "Classical model," Cl⁻ entry across the basolateral membrane is driven primarily via Na⁺ cotransport. Apical

presented in a cell near the top of Fig. 8.5 as depicted by Frizzell and Hanrahan (2012) in an excellent review. Much of this chapter shows that unique transporters fulfill these general roles in different cells and tissues. Additionally, the components can be paired with different partners or arranged in different configurations to account for differences in transport that are observed across tissues (e.g., see Fig. 8.4). The lower cells in Fig. 8.5 present a few recent additions to the classical model including distinct Cl^- and K^+ channels, Na^+/H^+ exchangers (e.g., NHE), anion exchangers (e.g., AE2), Na^+/HCO_3^- cotransporters (e.g., NBC1), and an H^+-ATPase. The diagram is not exhaustive, but provides a new, fuller framework as a foundation for ongoing studies that are designed to better understand the mechanisms of ion transport. Many of the chapters in this volume describe these components in great detail and build on this simple model with the addition of other transport proteins. The authors characterize the structures, functions, and regulation of these epithelial transporters in health and disease.

References

Adelman JP, Maylie J, Sah P (2012) Small-conductance Ca^{2+}-activated K^+ channels: form and function. Annu Rev Physiol 74:245–269. https://doi.org/10.1146/annurev-physiol-020911-153336

Al-Bazzaz FJ, Al-Awqati Q (1979) Interaction between sodium and chloride transport in canine tracheal mucosa. J Appl Physiol Respir Environ Exerc Physiol 46:111–119

Alper SL (2006) Molecular physiology of SLC4 anion exchangers. Exp Physiol 91:153–161. https://doi.org/10.1113/expphysiol.2005.031765

Andersen DH (1938) Cystic fibrosis of the pancreas and its relation to celiac disease: a clinical and pathological study. Am J Dis Child 56:344–399

Anderson MP, Welsh MJ (1992) Regulation by ATP and ADP of CFTR chloride channels that contain mutant nucleotide-binding domains. Science 257:1701–1704

Fig. 8.5 (continued) Cl^- efflux is mediated by cAMP- and Ca^{2+}-activated anion conductances. The scheme labeled "Recent additions" identifies the numerous transporters and channels proposed to mediate ion and fluid secretion in various epithelia. Accordingly, basolateral Cl^- loading may occur via both NKCC1 and AE2. Cl^- and HCO_3^- exit involves apical CFTR, Tmem16A, and probably other Cl^- channels/transporters, such as members of the SLC26A family (not shown). NBC1-mediated HCO_3^- entry during cAMP stimulation maintains intracellular pH and generates electrogenic transepithelial HCO_3^- secretion. NHE1 maintains intracellular pH during stimulation by Ca^{2+} secretagogues when cells are hyperpolarized and net Cl^- secretion is favored. There is evidence for apical H^+ secretion mediated by vacuolar proton pumps, H^+/K^+-ATPase, and also apical K^+ channels, which contribute to apical hyperpolarization and thus enhance the driving force for anion secretion. Most transepithelial Na^+ flux occurs passively through the paracellular pathway driven by the lumen-negative voltage arising from electrogenic anion secretion. In both schemes, regulated apical anion conductance is rate-limiting at low/medium levels of anion secretion, but basolateral K^+ conductance becomes increasingly limiting at higher secretory rates, explaining the additive nature of secretagogues that operate via different second-messenger pathways. Modified from Frizzell and Hanrahan (2012)

Anderson MP, Berger HA, Rich DP, Gregory RJ, Smith AE, Welsh MJ (1991a) Nucleoside triphosphates are required to open the CFTR chloride channel. Cell 67:775–784

Anderson MP, Gregory RJ, Thompson S, Souza DW, Paul S, Mulligan RC, Smith AE, Welsh MJ (1991b) Demonstration that CFTR is a chloride channel by alteration of its anion selectivity. Science 253:202–205

Anderson MP, Rich DP, Gregory RJ, Smith AE, Welsh MJ (1991c) Generation of cAMP-activated chloride currents by expression of CFTR. Science 251:679–682

Bank N (1968) Physiological basis of diuretic action. Annu Rev Med 19:103–118. https://doi.org/10.1146/annurev.me.19.020168.000535

Barhanin J, Lesage F, Guillemare E, Fink M, Lazdunski M, Romey G (1996) K(V)LQT1 and IsK (minK) proteins associate to form the I(Ks) cardiac potassium current. Nature 384:78–80. https://doi.org/10.1038/384078a0

Bear CE, Duguay F, Naismith AL, Kartner N, Hanrahan JW, Riordan JR (1991) Cl⁻ channel activity in Xenopus oocytes expressing the cystic fibrosis gene. J Biol Chem 266:19142–19145

Bijman J, Quinton PM (1984) Influence of abnormal Cl⁻ impermeability on sweating in cystic fibrosis. Am J Physiol Cell Physiol 247:C3–C9

Bissig M, Hagenbuch B, Stieger B, Koller T, Meier PJ (1994) Functional expression cloning of the canalicular sulfate transport system of rat hepatocytes. J Biol Chem 269:3017–3021

Bleich M, Briel M, Busch AE, Lang HJ, Gerlach U, Gogelein H, Greger R, Kunzelmann K (1997) KVLQT channels are inhibited by the K⁺ channel blocker 293B. Pflugers Arch 434:499–501

Bolton JE, Field M (1977) Ca ionophore-stimulated ion secretion in rabbit ileal mucosa: relation to actions of cyclic 3′,5′-AMP and carbamylcholine. J Membr Biol 35:159–173

Bonting SL (1966) Studies on sodium-potassium-activated adenosinetriphosphatase. XV. The rectal gland of the elasmobranchs. Comp Biochem Physiol 17:953–966

Bourgoignie J, Klahr S, Yates J, Guerra L, Bricker NS (1969) Characteristics of ATPase system of turtle bladder epithelium. Am J Phys 217:1496–1503

Bridges RJ, Worrell RT, Frizzell RA, Benos DJ (1989) Stilbene disulfonate blockade of colonic secretory Cl⁻ channels in planar lipid bilayers. Am J Physiol Cell Physiol 256:C902–C912

Browning JG, Hardcastle J, Hardcastle PT, Redfern JS (1978) Localization of the effect of acetylcholine in regulating intestinal ion transport. J Physiol 281:15–27

Burg M, Grantham J, Abramow M, Orloff J (1966) Preparation and study of fragments of single rabbit nephrons. Am J Phys 210:1293–1298

Burg MB, Issaacson L, Grantham J, Orloff J (1968) Electrical properties of isolated perfused rabbit renal tubules. Am J Phys 215:788–794

Burg M, Stoner L, Cardinal J, Green N (1973) Furosemide effect on isolated perfused tubules. Am J Phys 225:119–124

Burger JW, Hess WN (1960) Function of the rectal gland in the spiny dogfish. Science 131:670–671

Burrows W, Musteikis GM (1966) Cholera infection and toxin in the rabbit ileal loop. J Infect Dis 116:183–190

Carpenter CC, Sack RB, Feeley JC, Steenberg RW (1968) Site and characteristics of electrolyte loss and effect of intraluminal glucose in experimental canine cholera. J Clin Invest 47:1210–1220. https://doi.org/10.1172/JCI105810

Chen LC, Rohde JE, Sharp GW (1971) Intestinal adenyl-cyclase activity in human cholera. Lancet 1:939–941

Choi BY, Kim HM, Ito T, Lee KY, Li X, Monahan K, Wen Y, Wilson E, Kurima K, Saunders TL, Petralia RS, Wangemann P, Friedman TB, Griffith AJ (2011) Mouse model of enlarged vestibular aqueducts defines temporal requirement of Slc26a4 expression for hearing acquisition. J Clin Invest 121:4516–4525. https://doi.org/10.1172/JCI59353

Clarkson TW, Toole SR (1964) Measurement of short-circuit current and ion transport across the ileum. Am J Phys 206:658–668

Cliff WH, Frizzell RA (1990) Separate Cl⁻ conductances activated by cAMP and Ca²⁺ in Cl⁻-secreting epithelial cells. Proc Natl Acad Sci U S A 87:4956–4960

Cole KS, Curtis HJ (1941) Membrane potential of the squid giant axon during current flow. J Gen Physiol 24:551–563

Cooley JW, Tukey JW (1965) An algorithm for the machine calculation of complex Fourier series. Math Comput 19:297–301

Crabbe J (1961) Stimulation of active sodium transport by the isolated toad bladder with aldosterone in vitro. J Clin Invest 40:2103–2110. https://doi.org/10.1172/JCI104436

Cuppoletti J, Chakrabarti J, Tewari KP, Malinowska DH (2014) Differentiation between human ClC-2 and CFTR Cl⁻ channels with pharmacological agents. Am J Physiol Cell Physiol 307: C479–C492. https://doi.org/10.1152/ajpcell.00077.2014

Curran PF, Solomon AK (1957) Ion and water fluxes in the ileum of rats. J Gen Physiol 41:143–168

Darrow CW (1945) Congenital alkalosis with diarrhea. J Pediatr 26:519–532. https://doi.org/10. 1016/S0022-3476(45)80079-3

Darwin C (1839) Journal of researches into the geology and natural history of the various countries visited by H.M.S. Beagle [under the command of Captain Fitzroy, R.N. from 1832 to 1836. Colburn, London

De SN, Chatterje DN (1953) An experimental study of the mechanism of action of *Vibrio cholerae* on the intestinal mucous membrane. J Pathol Bacteriol 66:559–562

Denning CR, Sommers SC, Quigley HJ Jr (1968) Infertility in male patients with cystic fibrosis. Pediatrics 41:7–17

Derksen HE, Verveen AA (1966) Fluctuations of resting neural membrane potential. Science 151:1388–1389

Devor DC, Frizzell RA (1993) Calcium-mediated agonists activate an inwardly rectified K⁺ channel in colonic secretory cells. Am J Physiol Cell Physiol 265:C1271–C1280

Devor DC, Singh AK, Bridges RJ, Frizzell RA (1996a) Modulation of Cl⁻ secretion by benzimidazolones. II. Coordinate regulation of apical G_{Cl} and basolateral G_K. Am J Physiol Lung Cell Mol Physiol 271:L785–L795

Devor DC, Singh AK, Frizzell RA, Bridges RJ (1996b) Modulation of Cl⁻ secretion by benzimidazolones. I. Direct activation of a Ca^{2+}-dependent K⁺ channel. Am J Physiol Lung Cell Mol Physiol 271:L775–L784

Devor DC, Singh AK, Gerlach AC, Frizzell RA, Bridges RJ (1997) Inhibition of intestinal Cl⁻ secretion by clotrimazole: direct effect on basolateral membrane K⁺ channels. Am J Physiol Cell Physiol 273:C531–C540

Devor DC, Bridges RJ, Pilewski JM (2000) Pharmacological modulation of ion transport across wild-type and ΔF508 CFTR-expressing human bronchial epithelia. Am J Physiol Cell Physiol 279:C461–C479

Dharmsathaphorn K, Pandol SJ (1986) Mechanism of chloride secretion induced by carbachol in a colonic epithelial cell line. J Clin Invest 77:348–354. https://doi.org/10.1172/JCI112311

Dharmsathaphorn K, McRoberts JA, Mandel KG, Tisdale LD, Masui H (1984) A human colonic tumor cell line that maintains vectorial electrolyte transport. Am J Physiol Gastrointest Liver Physiol 246:G204–G208

Drumm ML, Pope HA, Cliff WH, Rommens JM, Marvin SA, Tsui L-C, Collins FS, Frizzell RA, Wilson JM (1990) Correction of the cystic fibrosis defect in vitro by retrovirus-mediated gene transfer. Cell 62:1227–1233

Dutta NK, Habbu MK (1955) Experimental cholera in infant rabbits: a method for chemotherapeutic investigation. Br J Pharmacol Chemother 10:153–159

Edmonds CJ, Nielsen OE (1968) Transmembrane electrical potential differences and ionic composition of mucosal cells of rat colon. Acta Physiol Scand 72:338–349. https://doi.org/10.1111/j. 1748-1716.1968.tb03856.x

Evans MH, Linzell JL, Peaker M (1971) Membrane potentials in the mammary gland of the lactating rat. J Physiol 213:49P–50P

Everett LA, Green ED (1999) A family of mammalian anion transporters and their involvement in human genetic diseases. Hum Mol Genet 8:1883–1891

Everett LA, Glaser B, Beck JC, Idol JR, Buchs A, Heyman M, Adawi F, Hazani E, Nassir E, Baxevanis AD, Sheffield VC, Green ED (1997) Pendred syndrome is caused by mutations in a putative sulphate transporter gene (PDS). Nat Genet 17:411–422. https://doi.org/10.1038/ng1297-411

Fairbanks G, Steck TL, Wallach DF (1971) Electrophoretic analysis of the major polypeptides of the human erythrocyte membrane. Biochemistry 10:2606–2617

Field M, Plotkin GR, Silen W (1968) Effects of vasopressin, theophylline and cyclic adenosine monophosphate on short-circuit current across isolated rabbit ileal mucosa. Nature 217:469–471

Field M, Fromm D, Wallace CK, Greenough WBI (1969) Stimulation of active chloride secretion in small intestine by cholera exotoxin. J Clin Invest 48:24a

Finkelstein RA, Atthasampunna P, Chulasamaya M, Charunmethee P (1966) Pathogenesis of experimental cholera: biologic activities of purified procholeragen A. J Immunol 96:440–449

Finn AL, Handler JS, Orloff J (1967) Active chloride transport in the isolated toad bladder. Am J Phys 213:179–184

Flores CA, Melvin JE, Figueroa CD, Sepulveda FV (2007) Abolition of Ca^{2+}-mediated intestinal anion secretion and increased stool dehydration in mice lacking the intermediate conductance Ca^{2+}-dependent K^+ channel Kcnn4. J Physiol 583:705–717. https://doi.org/10.1113/jphysiol.2007.134387

Florey HW, Wright RD, Jennings MA (1941) The secretions of the intestine. Physiol Rev 21:36–69

Francis WL (1933) Output of electrical energy by frog-skin. Nature 131:805

Frizzell RA (1977) Active chloride secretion by rabbit colon: calcium-dependent stimulation by ionophore A23187. J Membr Biol 35:175–187

Frizzell RA, Duffey ME (1980) Chloride activities in epithelia. Fed Proc 39:2860–2864

Frizzell RA, Hanrahan JW (2012) Physiology of epithelial chloride and fluid secretion. Cold Spring Harb Perspect Med 2:a009563. https://doi.org/10.1101/cshperspect.a009563

Frizzell RA, Koch MJ, Schultz SG (1976) Ion transport by rabbit colon. I Active and passive components. J Membr Biol 27:297–316

Frizzell RA, Welsh MJ, Smith PL (1981) Electrophysiology of chloride-secreting epithelia. Soc Gen Physiol Ser 36:137–145

Frizzell RA, Rechkemmer G, Shoemaker RL (1986) Altered regulation of airway epithelial cell chloride channels in cystic fibrosis. Science 233:558–560

Gamble JL, Fahey KR, Appleton J, MacLachlan E (1945) Congenital alkalosis with diarrhea. J Pediatr 26:509–518. https://doi.org/10.1016/S0022-3476(45)80078-1

Geck P, Pietrzyk C, Burckhardt BC, Pfeiffer B, Heinz E (1980) Electrically silent cotransport on Na^+, K^+ and Cl⁻ in Ehrlich cells. Biochim Biophys Acta 600:432–447

Gerlach AC, Gangopadhyay NN, Devor DC (2000) Kinase-dependent regulation of the intermediate conductance, calcium-dependent potassium channel, hIK1. J Biol Chem 275:585–598

Gervais R, Dumur V, Letombe B, Larde A, Rigot JM, Roussel P, Lafitte JJ (1996) Hypofertility with thick cervical mucus: another mild form of cystic fibrosis? JAMA 276:1638

Giebisch G (1958) Electrical potential measurements on single nephrons of Necturus. J Cell Physiol 51:221–239

Giebisch G, Klose RM, Malnic G, Sullivan WJ, Windhager EE (1964) Sodium movement across single perfused proximal tubules of rat kidneys. J Gen Physiol 47:1175–1194

Grantham JJ, Burg MB (1966) Effect of vasopressin and cyclic AMP on permeability of isolated collecting tubules. Am J Phys 211:255–259

Gray MA, Harris A, Coleman L, Greenwell JR, Argent BE (1989) Two types of chloride channel on duct cells cultured from human fetal pancreas. Am J Physiol Cell Physiol 257:C240–C251

Greenough WBI, Pierce NF, Al Awqati Q, Carpenter CCJ (1969) Stimulation of gut electrolyte secretion by prostaglandins, theophylline, and cholera exotoxin. J Clin Invest 48:32a

Greger R (1985) Ion transport mechanisms in thick ascending limb of Henle's loop of mammalian nephron. Physiol Rev 65:760–797

Greger R, Schlatter E (1984a) Mechanism of NaCl secretion in rectal gland tubules of spiny dogfish (*Squalus acanthias*). II. Effects of inhibitors. Pflugers Arch 402:364–375

Greger R, Schlatter E (1984b) Mechanism of NaCl secretion in the rectal gland of spiny dogfish (*Squalus acanthias*). I. Experiments in isolated in vitro perfused rectal gland tubules. Pflugers Arch 402:63–75

Greger R, Schlatter E, Wang F, Forrest JN Jr (1984) Mechanism of NaCl secretion in rectal gland tubules of spiny dogfish (*Squalus acanthias*). III. Effects of stimulation of secretion by cyclic AMP. Pflugers Arch 402:376–384

Guggino WB (1993) Outwardly rectifying chloride channels and CF: a divorce and remarriage. J Bioenerg Biomembr 25:27–35

Gutman GA, Chandy KG, Adelman JP, Aiyar J, Bayliss DA, Clapham DE, Covarriubias M, Desir GV, Furuichi K, Ganetzky B, Garcia ML, Grissmer S, Jan LY, Karschin A, Kim D, Kuperschmidt S, Kurachi Y, Lazdunski M, Lesage F, Lester HA, McKinnon D, Nichols CG, O'Kelly I, Robbins J, Robertson GA, Rudy B, Sanguinetti M, Seino S, Stuehmer W, Tamkun MM, Vandenberg CA, Wei A, Wulff H, Wymore RS (2003) International Union of Pharmacology. XLI. Compendium of voltage-gated ion channels: potassium channels. Pharmacol Rev 55:583–586. https://doi.org/10.1124/pr.55.4.9

Haas M, Schmidt WF 3rd, McManus TJ (1982) Catecholamine-stimulated ion transport in duck red cells. Gradient effects in electrically neutral [Na + K + 2Cl] Co-transport. J Gen Physiol 80:125–147

Hahn LA, Hevesy GC, Lundsgaard EC (1937) The circulation of phosphorus in the body revealed by application of radioactive phosphorus as indicator. Biochem J 31:1705–1709

Hahn LA, Hevesy GC, Rebbe OH (1939) Do the potassium ions inside the muscle cells and blood corpuscles exchange with those present in the plasma? Biochem J 33:1549–1558

Hamill OP, Marty A, Neher E, Sakmann B, Sigworth FJ (1981) Improved patch-clamp techniques for high-resolution current recording from cells and cell-free membrane patches. Pflugers Arch 391:85–100

Hamilton KL (2011) Ussing's 'Little Chamber': 60 years+ old and counting. Front Physiol 2:6. https://doi.org/10.3389/fphys.2011.0000

Hamilton KL, Moore AB (2016) 50 years of renal physiology from one man and the perfused tubule: Maurice B. Burg. Am J Physiol Renal Physiol 311:F291–F304

Hanrahan JW, Phillips JE (1984) KCl Transport across an insect epithelium: II. electrochemical potentials and electrophysiology. J Membr Biol 80:27–47

Hastbacka J, de la Chapelle A, Mahtani MM, Clines G, Reeve-Daly MP, Daly M, Hamilton BA, Kusumi K, Trivedi B, Weaver A, Coloma A, Lovett M, Buckler A, Kaitila I, Lander ES (1994) The diastrophic dysplasia gene encodes a novel sulfate transporter: positional cloning by fine-structure linkage disequilibrium mapping. Cell 78:1073–1087

Heintze K, Stewart CP, Frizzell RA (1983) Sodium-dependent chloride secretion across rabbit descending colon. Am J Physiol Gastrointest Liver Physiol 244:G357–G365

Helman SI, Grantham JJ, Burg MB (1971) Effect of vasopressin on electrical resistance of renal cortical collecting tubules. Am J Phys 220:1825–1832

Hevesy G, Hogfer E (1934) Elimination of water from the human body. Nature 134:879

Hevesy GV, Hofer E, Krogh A (1935) The permeability of the skin of frogs to water as determined by D_2O and H_2O. Skandinavisches Archiv Für Physiologie 72:199–214

Hihnala S, Kujala M, Toppari J, Kere J, Holmberg C, Hoglund P (2006) Expression of SLC26A3, CFTR and NHE3 in the human male reproductive tract: role in male subfertility caused by congenital chloride diarrhoea. Mol Hum Reprod 12:107–111. https://doi.org/10.1093/molehr/gal009

Hodgkin AL, Huxley AF (1952a) The components of membrane conductance in the giant axon of *Loligo*. J Physiol 116:473–496

Hodgkin AL, Huxley AF (1952b) Currents carried by sodium and potassium ions through the membrane of the giant axon of *Loligo*. J Physiol 116:449–472

Hodgkin AL, Keynes RD (1955) Active transport of cations in giant axons from *Sepia* and *Loligo*. J Physiol 128:28–60

Hodgkin AL, Huxley AF, Katz B (1952) Measurement of current-voltage relations in the membrane of the giant axon of *Loligo*. J Physiol 116:424–448

Hogben CA (1951) The chloride transport system of the gastric mucosa. Proc Natl Acad Sci U S A 37:393–395

Hogben CA (1955) Active transport of chloride by isolated frog gastric epithelium: origin of the gastric mucosal potential. Am J Phys 180:641–649

Hoglund P, Hihnala S, Kujala M, Tiitinen A, Dunkel L, Holmberg C (2006) Disruption of the SLC26A3-mediated anion transport is associated with male subfertility. Fertil Steril 85:232–235

Hook JB, Williamson HE (1965) Lack of correlation between natriuretic activity and inhibition of renal NaK-activated ATPase. Proc Soc Exp Biol Med 120:358–360

Hubel KA (1967) Bicarbonate secretion in rat ileum and its dependence on intraluminal chloride. Am J Phys 213:1409–1413

Hubel KA (1969) Effect of luminal chloride concentration on bicarbonate secretion in rat ileum. Am J Phys 217:40–45

Hwang TC, Lu L, Zeitlin PL, Gruenert DC, Huganir R, Guggino WB (1989) Cl⁻ channels in CF: lack of activation by protein kinase C and cAMP-dependent protein kinase. Science 244:1351–1353

Ingraham RC, Visscher MB (1935) Studies on the elimination of dyes in the gastric and pancreatic secretions, and inferences there from concerning the mechanisms of secretion of acid and base. J Gen Physiol 18:695–716

Isaacs PE, Corbett CL, Riley AK, Hawker PC, Turnberg LA (1976) In vitro behavior of human intestinal mucosa. The influence of acetyl choline on ion transport. J Clin Invest 58:535–542. https://doi.org/10.1172/JCI108498

Ishii TM, Silvia C, Hirschberg B, Bond CT, Adelman JP, Maylie J (1997) A human intermediate conductance calcium-activated potassium channel. Proc Natl Acad Sci U S A 94:11651–11656

Jensen BS, Strobaek D, Olesen SP, Christophersen P (2001) The Ca²⁺-activated K⁺ channel of intermediate conductance: a molecular target for novel treatments? Curr Drug Targets 2:401–422

Jespersen T, Rasmussen HB, Grunnet M, Jensen HS, Angelo K, Dupuis DS, Vogel LK, Jorgensen NK, Klaerke DA, Olesen SP (2004) Basolateral localisation of KCNQ1 potassium channels in MDCK cells: molecular identification of an N-terminal targeting motif. J Cell Sci 117:4517–4526. https://doi.org/10.1242/jcs.01318

Jespersen T, Grunnet M, Olesen SP (2005) The KCNQ1 potassium channel: from gene to physiological function. Physiology (Bethesda) 20:408–416. https://doi.org/10.1152/physiol. 00031.2005

Johannesson M, Landgren BM, Csemiczky G, Hjelte L, Gottlieb C (1998) Female patients with cystic fibrosis suffer from reproductive endocrinological disorders despite good clinical status. Hum Reprod 13:2092–2097

Johnson G (1866) Rules for the treatment of epidemic diarrhoea and cholera. Br Med J 2:63–65

Joiner WJ, Wang LY, Tang MD, Kaczmarek LK (1997) hSK4, a member of a novel subfamily of calcium-activated potassium channels. Proc Natl Acad Sci U S A 94:11013–11018

Kartner N, Hanrahan JW, Jensen TJ, Naismith AL, Sun SZ, Ackerley CA, Reyes EF, Tsui LC, Rommens JM, Bear CE, Riordan R (1991) Expression of the cystic fibrosis gene in non-epithelial invertebrate cells produces a regulated anion conductance. Cell 64:681–691

Kerem B, Rommens JM, Buchanan JA, Markiewicz D, Cox TK, Chakravarti A, Buchwald M, Tsui LC (1989) Identification of the cystic fibrosis gene: genetic analysis. Science 245:1073–1080

Keynes RD (1969) From frog skin to sheep rumen: a survey of transport of salts and water across multicellular structures. Q Rev Biophys 2:177–281

Keys AB (1931a) Chloride and water secretion and absorption by the gills of the eel. Z vgl Physiol 15:364–388. https://doi.org/10.1007/bf00339115

Keys AB (1931b) The heart-gill preparation of the eel and its perfusion for the study of a natural membrane in situ. Z vgl Physiol 15:352–363. https://doi.org/10.1007/bf00339114

Keys A, Willmer EN (1932) "Chloride secreting cells" in the gills of fishes, with special reference to the common eel. J Physiol 76(368-378):362

Kimberg DV, Field M, Johnson J, Henderson A, Gershon E (1971) Stimulation of intestinal mucosal adenyl cyclase by cholera enterotoxin and prostaglandins. J Clin Invest 50:1218–1230. https://doi.org/10.1172/JCI106599

Kimura T (1969) [Electron microscopic study of the mechanism of secretion of milk]. Nihon Sanka Fujinka Gakkai Zasshi 21:301–308

Knowles MR, Stutts MJ, Spock A, Fischer N, Gatzy JT, Boucher RC (1983) Abnormal ion permeation through cystic fibrosis respiratory epithelium. Science 221:1067–1070

Ko SB, Zeng W, Dorwart MR, Luo X, Kim KH, Millen L, Goto H, Naruse S, Soyombo A, Thomas PJ, Muallem S (2004) Gating of CFTR by the STAS domain of SLC26 transporters. Nat Cell Biol 6:343–350

Koefoed-Johnsen V, Ussing HH (1958) The nature of the frog skin potential. Acta Physiol Scand 42:298–308. https://doi.org/10.1111/j.1748-1716.1958.tb01563.x

Koefoed-Johnsen V, Levi H, Ussing HH (1952a) The mode of passage of chloride ions through the isolated frog skin. Acta Physiol Scand 25:150–163. https://doi.org/10.1111/j.1748-1716.1952.tb00866.x

Koefoed-Johnsen V, Ussing HH, Zerahn K (1952b) The origin of the short-circuit current in the adrenaline stimulated frog skin. Acta Physiol Scand 27:38–48. https://doi.org/10.1111/j.1748-1716.1953.tb00922.x

Kohler M, Hirschberg B, Bond CT, Kinzie JM, Marrion NV, Maylie J, Adelman JP (1996) Small-conductance, calcium-activated potassium channels from mammalian brain. Science 273:1709–1714

Kopito RR, Lodish HF (1985) Primary structure and transmembrane orientation of the murine anion exchange protein. Nature 316:234–238

Krogh A (1937a) Osmotic regulation in the frog (*R. esculenta*) by active absorption of chloride ion. Skandinavisches Archiv Für Physiologie 76:60–74

Krogh A (1937b) The use of isotopes as indicators in biological research. Science 85:187–191. https://doi.org/10.1126/science.85.2199.187

Krogh A (1939) The active uptake of ions into cells and organisms. Proc Natl Acad Sci U S A 25:275–277

Kujala M, Hihnala S, Tienari J, Kaunisto K, Hastbacka J, Holmberg C, Kere J, Hoglund P (2007) Expression of ion transport-associated proteins in human efferent and epididymal ducts. Reproduction 133:775–784. https://doi.org/10.1530/rep.1.00964

Kunzelmann K, Pavenstadt H, Greger R (1989) Properties and regulation of chloride channels in cystic fibrosis and normal airway cells. Pflugers Arch 415:172–182

Lee MG, Wigley WC, Zeng W, Noel LE, Marino CR, Thomas PJ, Muallem S (1999) Regulation of Cl^-/HCO_3^- exchange by cystic fibrosis transmembrane conductance regulator expressed in NIH 3T3 and HEK 293 cells. J Biol Chem 274:3414–3421

Leitch GJ, Burrows W (1968) Experimental cholera in the rabbit ligated intestine: ion and water accumulation in the duodenum, ileum and colon. J Infect Dis 118:349–359

Li M, McCann JD, Anderson MP, Clancy JP, Liedtke CM, Nairn AC, Greengard P, Welsch MJ (1989) Regulation of chloride channels by protein kinase C in normal and cystic fibrosis airway epithelia. Science 244:1353–1356

Li X, Sanneman JD, Harbidge DG, Zhou F, Ito T, Nelson R, Picard N, Chambrey R, Eladari D, Miesner T, Griffith AJ, Marcus DC, Wangemann P (2013) SLC26A4 targeted to the endolymphatic sac rescues hearing and balance in Slc26a4 mutant mice. PLoS Genet 9(7):e1003641. https://doi.org/10.1371/journal.pgen.1003641

Liin SI, Barro-Soria R, Larsson HP (2015) The KCNQ1 channel—remarkable flexibility in gating allows for functional versatility. J Physiol 593(12):2605–2615. https://doi.org/10.1113/jphysiol.2014.287607

Lindemann B, Van Driessche W (1977) Sodium-specific membrane channels of frog skin are pores: current fluctuations reveal high turnover. Science 195:292–294

Lindemann B, Van Driessche W (1978) The mechanism of Na-uptake through Na-selective channels in the epithelium of frog skin. In: Hoffman JF (ed) Membrane transport processes, vol 1. Raven Press, New York, pp 155–178

Linzell JL, Peaker M (1971a) The effects of oxytocin and milk removal on milk secretion in the goat. J Physiol Lond 216:717–734

Linzell JL, Peaker M (1971b) Intracellular concentrations of sodium, potassium and chloride in the lactating mammary gland and their relation to the secretory mechanism. J Physiol Lond 216:683–700

Linzell JL, Peaker M (1971c) Mechanism of milk secretion. Physiol Rev 51:564–597

Linzell JL, Peaker M (1971d) The permeability of mammary ducts. J Physiol Lond 216:701–716

Lohrmann E, Burhoff I, Nitschke RB, Lang HJ, Mania D, Englert HC, Hropot M, Warth R, Rohm W, Bleich M, Greger R (1995) A new class of inhibitors of cAMP-mediated Cl⁻ secretion in rabbit colon, acting by the reduction of cAMP-activated K⁺ conductance. Pflugers Arch 429:517–530

Lutz MD, Cardinal J, Burg MB (1973) Electrical resistance of renal proximal tubule perfused *in vitro*. Am J Phys 225:729–734

Mandel KG, Dharmsathaphorn K, McRoberts JA (1986) Characterization of a cyclic AMP-activated Cl-transport pathway in the apical membrane of a human colonic epithelial cell line. J Biol Chem 261:704–712

Maruyama Y, Peterson OH (1982) Single-channel currents in isolated patches of plasma membrane from basal surface of pancreatic acini. Nature 299:159–161

Maruyama Y, Gallacher DV, Petersen OH (1983) Voltage and Ca²⁺-activated K⁺ channel in basolateral acinar cell membranes of mammalian salivary glands. Nature 302:827–829

Mazhari R, Nuss HB, Armoundas AA, Winslow RL, Marban E (2002) Ectopic expression of KCNE3 accelerates cardiac repolarization and abbreviates the QT interval. J Clin Invest 109:1083–1090. https://doi.org/10.1172/JCI15062

McCann JD, Matsuda J, Garcia M, Kaczorowski G, Welsh MJ (1990) Basolateral K⁺ channels in airway epithelia. I. Regulation by Ca²⁺ and block by charybdotoxin. Am J Physiol Lung Cell Mol Physiol 258:L334–L342

McRoberts JA, Erlinger S, Rindler MJ, Saier MH Jr (1982) Furosemide-sensitive salt transport in the Madin-Darby canine kidney cell line. Evidence for the cotransport of Na⁺, K⁺, and Cl⁻. J Biol Chem 257:2260–2266

McRoberts JA, Beuerlein G, Dharmsathaphorn K (1985) Cyclic AMP and Ca²⁺-activated K⁺ transport in a human colonic epithelial cell line. J Biol Chem 260:14163–14172

Mount DB, Romero MF (2004) The SLC26 gene family of multifunctional anion exchangers. Pflugers Arch 447:710–721

Nadel JA, Davis B (1978) Regulation of Na⁺ and Cl⁻ transport and mucous gland secretion in airway epithelium. CIBA Found Symp 54:133–147

Nakajo K, Kubo Y (2015) KCNQ1 channel modulation by KCNE proteins via the voltage-sensing domain. J Physiol 593(12):2617–2625. https://doi.org/10.1113/jphysiol.2014.287672

Neher E, Sakmann B (1976) Single-channel currents recorded from membrane of denervated frog muscle fibres. Nature 260:799–802

Nelson DJ, Tang JM, Palmer LG (1984) Single-channel recordings of apical membrane chloride conductance in A6 epithelial cells. J Membr Biol 80:81–89

O'Donnell EK, Sedlacek RL, Singh AK, Schultz BD (2000) Inhibition of enterotoxin-induced porcine colonic secretion by diarylsulfonylureas in vitro. Am J Physiol Gastrointest Liver Physiol 279:G1104–G1112

O'Grady SM, Musch MW, Field M (1986) Stoichiometry and ion affinities of the Na-K-Cl cotransport system in the intestine of the winter flounder (*Pseudopleuronectes americanus*). J Membr Biol 91:33–41

Pedersen KA, Schroder RL, Skaaning-Jensen B, Strobaek D, Olesen SP, Christophersen P (1999) Activation of the human intermediate-conductance Ca²⁺-activated K⁺ channel by 1-ethyl-2-benzimidazolinone is strongly Ca²⁺-dependent. Biochim Biophys Acta 1420:231–240

Pendred V (1896) Deaf-mutism and goitre. Lancet 2:532. https://doi.org/10.1016/S0140-6736(01) 74403-0

Perkins FM, Handler JS (1981) Transport properties of toad kidney epithelia in culture. Am J Physiol Cell Physiol 241:C154–C159

Pierce NF, Carpenter CC Jr, Elliott HL, Greenough WB 3rd (1971a) Effects of prostaglandins, theophylline, and cholera exotoxin upon transmucosal water and electrolyte movement in the canine jejunum. Gastroenterology 60:22–32

Pierce NF, Greenough WB 3rd, Carpenter CC Jr (1971b) *Vibrio cholerae* enterotoxin and its mode of action. Bacteriol Rev 35:1–13

Pierucci-Alves F, Akoyev V, Stewart JC 3rd, Wang LH, Janardhan KS, Schultz BD (2011) Swine models of cystic fibrosis reveal male reproductive tract phenotype at birth. Biol Reprod 85:442–451. https://doi.org/10.1095/biolreprod.111.090860

Pierucci-Alves F, Akoyev V, Schultz BD (2015) Bicarbonate exchangers SLC26A3 and SLC26A6 are localized at the apical membrane of porcine vas deferens epithelium. Physiol Rep 3(4): e12380. https://doi.org/10.14814/phy2.12380

Quinton PM (1983) Chloride impermeability in cystic fibrosis. Nature 301:421–422

Quinton PM (1986) Missing Cl conductance in cystic fibrosis. Am J Physiol Cell Physiol 251: C649–C652

Quinton PM, Bijman J (1983) Higher bioelectric potentials due to decreased chloride absorption in the sweat glands of patients with cystic fibrosis. N Engl J Med 308:1185–1189. https://doi.org/10.1056/NEJM198305193082002

Reuss L, Reinach P, Weinman SA, Grady TP (1983) Intracellular ion activities and Cl-transport mechanisms in bullfrog corneal epithelium. Am J Physiol Cell Physiol 244:C336–C347

Riordan JR, Rommens JM, Kerem B, Alon N, Rozmahel R, Grzelczak Z, Zielenski J, Lok S, Plavsic N, Chou JL, Drumm ML, Iannuzzi MC, Collins FS, Tsui LC (1989) Identification of the cystic fibrosis gene: cloning and characterization of complementary DNA. Science 245:1066–1073

Roggin GM, Banwell JG, Yardley JH, Hendrix TR (1972) Unimpaired response of rabbit jejunum to cholera toxin after selective damage to villus epithelium. Gastroenterology 63:981–989

Romero MF, Fulton CM, Boron WF (2004) The SLC4 family of HCO$_3^-$ transporters. Pflugers Arch 447:495–509. https://doi.org/10.1007/s00424-003-1180-2

Rommens JM, Iannuzzi MC, Kerem B, Drumm ML, Melmer G, Dean M, Rozmahel R, Cole JL, Kennedy D, Hidaka N, Zsiga M, Buchwald M, Riordan JR, Tsui LC, Collins FS (1989) Identification of the cystic fibrosis gene: chromosome walking and jumping. Science 245:1059–1065

Rose RC, Schultz SG (1971) Studies on the electrical potential profile across rabbit ileum. Effects of sugars and amino acids on transmural and transmucosal electrical potential differences. J Gen Physiol 57:639–663

Rothstein A, Cabantchik ZI, Knauf P (1976) Mechanism of anion transport in red blood cells: role of membrane proteins. Fed Proc 35:3–10

Rufo PA, Jiang L, Moe SJ, Brugnara C, Alper SL, Lencer WI (1996) The antifungal antibiotic, clotrimazole, inhibits Cl secretion by polarized monolayers of human colonic epithelial cells. J Clin Invest 98:2066–2075. https://doi.org/10.1172/JCI119012

Sanguinetti MC, Curran ME, Zou A, Shen J, Spector PS, Atkinson DL, Keating MT (1996) Coassembly of K$_V$LQT1 and minK (IsK) proteins to form cardiac I$_{Ks}$ potassium channel. Nature 384:80–83. https://doi.org/10.1038/384080a0

Schmidt-Nielsen K, Fange R (1958) Salt glands in marine reptiles. Nature 182:783–784

Schmidt-Nielsen K, Jorgensen CB, Osaki H (1958) Extrarenal salt excretion in birds. Am J Phys 193:101–107

Schoumacher RA, Ram J, Iannuzzi MC, Bradbury NA, Wallace RW, Hon CT, Kelly DR, Schmid SM, Gelder FB, Rado TA, Frizzell RA (1990) A cystic fibrosis pancreatic adenocarcinoma cell line. Proc Natl Acad Sci U S A 87:4012–4016

Schroeder BC, Waldegger S, Fehr S, Bleich M, Warth R, Greger R, Jentsch TJ (2000) A constitutively open potassium channel formed by KCNQ1 and KCNE3. Nature 403:196–199. https://doi.org/10.1038/35003200

Schultz SG (1972) Electrical potential differences and electromotive forces in epithelial tissues. J Gen Physiol 59:794–798

Schultz SG, Frizzell RA (1972) An overview of intestinal absorptive and secretory processes. Gastroenterology 63:161–170

Schultz SG, Zalusky R, Gass AE Jr (1964) Ion transport in isolated rabbit ileum. 3. Chloride fluxes. J Gen Physiol 48:375–378

Schultz BD, Singh AK, Aguilar-Bryan L, Frizzell RA, Bridges RJ (1995a) LY295501; a sulfonylurea that blocks CFTR Cl⁻ channels, but does not alter pancreatic β-cell function. Pediatr Pulmonol Suppl 12:200

Schultz BD, Venglarik CJ, Bridges RJ, Frizzell RA (1995b) Regulation of CFTR Cl⁻ channel gating by ADP and ATP analogues. J Gen Physiol 105:329–361

Schultz BD, DeRoos AD, Venglarik CJ, Singh AK, Frizzell RA, Bridges RJ (1996) Glibenclamide blockade of CFTR chloride channels. Am J Physiol Lung Cell Mol Physiol 271:L192–L200

Schweinfest CW, Henderson KW, Suster S, Kondoh N, Papas TS (1993) Identification of a colon mucosa gene that is down-regulated in colon adenomas and adenocarcinomas. Proc Natl Acad Sci U S A 90:4166–4170

Serebro HA, Iber FL, Yardley JH, Hendrix TR (1969) Inhibition of cholera toxin action in the rabbit by cycloheximide. Gastroenterology 56:506–511

Sharp GW, Hynie S (1971) Stimulation of intestinal adenyl cyclase by cholera toxin. Nature 229:266–269

Sharp GW, Hynie S, Lipson LC, Parkinson DK (1971) Action of cholera toxin to stimulate adenyl cyclase. Trans Assoc Am Phys 84:200–211

Sheppard DN, Welsh MJ (1992) Effect of ATP-sensitive K⁺ channel regulators on cystic fibrosis transmembrane conductance regulator chloride currents. J Gen Physiol 100:573–591

Shoemaker RL, Makhlouf GM, Sachs G (1970) Action of cholinergic drugs on *Necturus* gastric mucosa. Am J Phys 219:1056–1060

Shorofsky SR, Field M, Fozzard HA (1983) Electrophysiology of Cl secretion in canine trachea. J Membr Biol 72:105–115

Shorofsky SR, Field M, Fozzard HA (1984) Mechanism of Cl secretion in canine trachea: changes in intracellular chloride activity with secretion. J Membr Biol 81:1–8

Shorofsky SR, Field M, Fozzard HA (1986) Changes in intracellular sodium with chloride secretion in dog tracheal epithelium. Am J Physiol Cell Physiol 250:C646–C650

Singh AK, Venglarik CJ, Bridges RJ (1995) Development of chloride channel modulators. Kidney Int 48:985–993

Singh AK, Schultz BD, van Driessche W, Bridges RJ (2004) Transepithelial fluctuation analysis of chloride secretion. J Cyst Fibros 3(Suppl 2):127–132. https://doi.org/10.1016/j.jcf.2004.05.027

Singh AK, Riederer B, Chen M, Xiao F, Krabbenhoft A, Engelhardt R, Nylander O, Soleimani M, Seidler U (2010) The switch of intestinal Slc26 exchangers from anion absorptive to HCO₃⁻ secretory mode is dependent on CFTR anion channel function. Am J Physiol Cell Physiol 298: C1057–C1065. https://doi.org/10.1152/ajpcell.00454.2009

Skou JC (1957) The influence of some cations on an adenosine triphosphatase from peripheral nerves. Biochim Biophys Acta 23:394–401

Smith PL, Frizzell RA (1984) Chloride secretion by canine tracheal epithelium: IV. Basolateral membrane K permeability parallels secretion rate. J Membr Biol 77:187–199

Snow J (1849) On the mode of the communication of cholera. John Churchill, London

Snow J (1855) On the mode of the communication of cholera (second edition), 2nd edn. John Churchill, London

Splawski I, Timothy KW, Vincent GM, Atkinson DL, Keating MT (1997) Molecular basis of the long-QT syndrome associated with deafness. N Engl J Med 336:1562–1567. https://doi.org/10.1056/NEJM199705293362204

Stutts MJ, Cotton CU, Yankaskas JR, Cheng E, Knowles MR, Gatzy JT, Boucher RC (1985) Chloride uptake into cultured airway epithelial cells from cystic fibrosis patients and normal individuals. Proc Natl Acad Sci U S A 82:6677–6681

Suessbrich H, Busch AE (1999) The I_{Ks} channel: coassembly of I_{sK} (minK) and KvLQT1 proteins. Rev Physiol Biochem Pharmacol 137:191–226

Suessbrich H, Bleich M, Ecke D, Rizzo M, Waldegger S, Lang F, Szabo I, Lang HJ, Kunzelmann K, Greger R, Busch AE (1996) Specific blockade of slowly activating I_{sK} channels by chromanols—impact on the role of I_{sK} channels in epithelia. FEBS Lett 396:271–275

Syme CA, Gerlach AC, Singh AK, Devor DC (2000) Pharmacological activation of cloned intermediate- and small-conductance Ca^{2+}-activated K^+ channels. Am J Physiol Cell Physiol 278:C570–C581

Takumi T, Ohkubo H, and Nakanishi S (1988) Cloning of a membrane protein that induces a slow voltage-gated potassium current. Science 242:1042–1045

Teorell T (1939) On the permeability of the stomach mucosa for acids and sme other substances. J Gen Physiol 23:263–274

Thesleff S, Schmidt-Nielsen K (1962) An electrophysiological study of the salt gland of the herring gull. Am J Phys 202:597–600

Tidball CS (1961) Active chloride transport during intestinal secretion. Am J Phys 200:309–312

Trier JS (1964) Studies on small intestinal crypt epithelium. I Evidence for the mechanisms of secretory activity by undifferentiated crypt cells of the human small intestine. Gastroenterology 47:480–495

Turnberg LA, Bieberdorf FA, Morawski SG, Fordtran JS (1970) Interrelationships of chloride, bicarbonate, sodium, and hydrogen transport in the human ileum. J Clin Invest 49:557–567. https://doi.org/10.1172/JCI106266

Ueda S, Loo DD, Sachs G (1987) Regulation of K^+ channels in the basolateral membrane of *Necturus* oxyntic cells. J Membr Biol 97:31–41

Ussing HH (1949a) The active ion transport through the isolated frog skin in the light of tracer studies. Acta Physiol Scand 17:1–37. https://doi.org/10.1111/j.1748-1716.1949.tb00550.x

Ussing HH (1949b) Transport of ions across cellular membranes. Physiol Rev 29:127–155

Ussing HH, Zerahn K (1951) Active transport of sodium as the source of electric current in the short-circuited isolated frog skin. Acta Physiol Scand 23:110–127. https://doi.org/10.1111/j.1748-1716.1951.tb00800.x

Van Driessche W, Zeiske W (1980) Spontaneous fluctuations of potassium channels in the apical membrane of frog skin. J Physiol 299:101–116

Van Driessche W, Zeiske W (1985) Ionic channels in epithelial cell membranes. Physiol Rev 65:833–903

Vaughn JL, Goodwin RH, Tompkins GJ, McCawley P (1977) The establishment of two cell lines from the insect *Spodoptera frugiperda* (Lepidoptera; Noctuidae). In Vitro 13:213–217

Veilleux S, Holt N, Schultz BD, Dubreuil JD (2008) *Escherichia coli* EAST1 toxin toxicity of variants 17-2 and O 42. Comp Immunol Microbiol Infect Dis 31:567–578. https://doi.org/10.1016/j.cimid.2007.10.003

Venglarik CJ, Schultz BD, Frizzell RA, Bridges RJ (1994) ATP alters current fluctuations of cystic fibrosis transmembrane conductance regulator: evidence for a three-state activation mechanism. J Gen Physiol 104:123–146

Venglarik CJ, Schultz BD, de Roos AD, Singh AK, Bridges RJ (1996) Tolbutamide causes open channel blockade of cystic fibrosis transmembrane conductance regulator Cl^- channels. Biophys J 70:2696–2703

Verveen AA, Derksen HE, Schick KL (1967) Voltage fluctuations of neural membrane. Nature 216:588–589

Walker WG, Cooke CR, Iber FL, Lesch M, Caranasos GJ (1967) Topics in clinical medicine. A symposium Uses and complications of diuretic therapy. Johns Hopkins Med J 121:194–216

Wang Q, Curran ME, Splawski I, Burn TC, Millholland JM, VanRaay TJ, Shen J, Timothy KW, Vincent GM, de Jager T, Schwartz PJ, Toubin JA, Moss AJ, Atkinson DL, Landes GM,

Connors TD, Keating MT (1996) Positional cloning of a novel potassium channel gene: KVLQT1 mutations cause cardiac arrhythmias. Nat Genet 12:17–23. https://doi.org/10.1038/ng0196-17

Wang Z, Wang T, Petrovic S, Tuo B, Riederer B, Barone S, Lorenz JN, Seidler U, Aronson PS, Soleimani M (2005) Renal and intestinal transport defects in Slc26a6-null mice. Am J Physiol Cell Physiol 288:C957–C965

Wangemann P (2002) Adrenergic and muscarinic control of cochlear endolymph production. Adv Otorhinolaryngol 59:42–50

Wangemann P (2006) Supporting sensory transduction: cochlear fluid homeostasis and the endocochlear potential. J Physiol 576:11–21. https://doi.org/10.1113/jphysiol.2006.112888

Wangemann P, Wittner M, Di Stefano A, Englert HC, Lang HJ, Schlatter E, Greger R (1986) Cl⁻ channel blockers in the thick ascending limb of the loop of Henle. Structure activity relationship. Pflugers Arch 407(Suppl 2):S128–S141

Warth R, Hamm K, Bleich M, Kunzelmann K, von Hahn T, Schreiber R, Ullrich E, Mengel M, Trautmann N, Kindle P, Schwab A, Greger R (1999) Molecular and functional characterization of the small Ca^{2+}-regulated K^+ channel (rSK4) of colonic crypts. Pflugers Arch 438(4):437–444

Watten RH, Morgan FM, Yachai Na S, Vanikiati B, Phillips RA (1959) Water and electrolyte studies in cholera. J Clin Invest 38:1879–1889. https://doi.org/10.1172/JCI103965

Welsh MJ (1983a) Inhibition of chloride secretion by furosemide in canine tracheal epithelium. J Membr Biol 71:219–226

Welsh MJ (1983b) Intracellular chloride activities in canine tracheal epithelium. Direct evidence for sodium-coupled intracellular chloride accumulation in a chloride-secreting epithelium. J Clin Invest 71:1392–1401

Welsh MJ (1986) An apical-membrane chloride channel in human tracheal epithelium. Science 232:1648–1650

Welsh MJ, Smith PL, Frizzell RA (1982a) Chloride secretion by canine tracheal epithelium: II. The cellular electrical potential profile. J Membr Biol 70:227–238

Welsh MJ, Smith PL, Fromm M, Frizzell RA (1982b) Crypts are the site of intestinal fluid and electrolyte secretion. Science 218:1219–1221

Welsh MJ, Smith PL, Frizzell RA (1983) Chloride secretion by canine tracheal epithelium: III. Membrane resistances and electromotive forces. J Membr Biol 71:209–218

Weymer A, Huott P, Liu W, McRoberts JA, Dharmsathaphorn K (1985) Chloride secretory mechanism induced by prostaglandin E1 in a colonic epithelial cell line. J Clin Invest 7:1828–1836. https://doi.org/10.1172/JCI112175

Widdicombe JH, Nathanson IT, Highland E (1983) Effects of "loop" diuretics on ion transport by dog tracheal epithelium. Am J Physiol Cell Physiol 245:C388–C396

Widdicombe JH, Welsh MJ, Finkbeiner WE (1985) Cystic fibrosis decreases the apical membrane chloride permeability of monolayers cultured from cells of tracheal epithelium. Proc Natl Acad Sci U S A 82:6167–6171

Worrell RT, Butt AG, Cliff WH, Frizzell RA (1989) A volume-sensitive chloride conductance in human colonic cell line T84. Am J Physiol Cell Physiol 256:C1111–C1119

Wright EM (1972) Mechanisms of ion transport across the choroid plexus. J Physiol 226:545–571

Wright RD, Jennings MA, Florey HW, Lium R (1940) The influence of nerves and drugs on secretion by the small intestine and an investigation of the enzymes in intestinal juice. Exp Physiol 30:73–120

Yankaskas JR, Cotton CU, Knowles MR, Gatzy JT, Boucher RC (1985) Culture of human nasal epithelial cells on collagen matrix supports. A comparison of bioelectric properties of normal and cystic fibrosis epithelia. Am Rev Respir Dis 132:1281–1287

Zadunaisky JA (1966) Active transport of chloride in frog cornea. Am J Phys 211:506–512

Zadunaisky JA, Lande MA (1971) Active chloride transport and control of corneal transparency. Am J Phys 221:1837–1844

Zadunaisky JA, Candia OA, Chiarandini DJ (1963) The origin of the short-circuit current in the isolated skin of the South American frog *Leptodactylus Ocellatus*. J Gen Physiol 47:393–402

Zaki L, Fasold H, Schuhmann B, Passow H (1975) Chemical modification of membrane proteins in relation to inhibition of anion exchange in human red blood cells. J Cell Physiol 86:471–494. https://doi.org/10.1002/jcp.1040860305

Zaydman MA, Cui J (2014) PIP2 regulation of KCNQ channels: biophysical and molecular mechanisms for lipid modulation of voltage-dependent gating. Front Physiol 5:195. https://doi.org/10.3389/fphys.2014.00195

Zimmerman TW, Dobbins JW, Binder HJ (1982) Mechanism of cholinergic regulation of electrolyte transport in rat colon in vitro. Am J Physiol Gastrointest Liver Physiol 242:G116–G123

Chapter 9
Fundamentals of Epithelial Na⁺ Absorption

Alexander Staruschenko, Daria V. Ilatovskaya, and Kenneth R. Hallows

Abstract The maintenance of electrolyte balance is essential for the control of many functions in the human body. Na^+, K^+, and Cl^- are key electrolytes that contribute to a variety of processes ranging from the maintenance of cellular membrane potential to the regulation of cell volume and extracellular fluid. Fundamentals of epithelial Cl^- and K^+ transport are discussed in the preceding and following chapters and will be only briefly touched upon here. Na^+ absorption occurs across the epithelial barriers of many organs, including the lung, gastrointestinal tract, exocrine glands, and the kidney. Total body Na^+ content is the primary determinant of blood volume, and a number of physiological mechanisms that control blood pressure mediate their effects by adjusting Na^+ balance in the kidney. This chapter describes the classical understanding of epithelial Na^+ absorption and highlights some recently described mechanisms involved in the physiological regulation of Na^+ transport in specific epithelia with a particular emphasis on the kidney.

Keywords Kidney · Hypertension · Na^+/K^+ pump · NHE · SGLT · NKCC · NCC · ENaC · NBC · NPT

A. Staruschenko (✉)
Department of Physiology, Medical College of Wisconsin, Milwaukee, WI, USA
e-mail: staruschenko@mcw.edu

D. V. Ilatovskaya
Division of Nephrology, Department of Medicine, Medical University of South Carolina, Charleston, SC, USA
e-mail: ilatovskaya@musc.edu

K. R. Hallows (✉)
Division of Nephrology and Hypertension, Department of Medicine, USC/UKRO Kidney Research Center, University of Southern California Keck School of Medicine, Los Angeles, CA, USA
e-mail: hallows@usc.edu

© The American Physiological Society 2020
K. L. Hamilton, D. C. Devor (eds.), *Basic Epithelial Ion Transport Principles and Function*, Physiology in Health and Disease, https://doi.org/10.1007/978-3-030-52780-8_9

9.1 Introduction

In this book, specific chapters deal with all aspects of the transport of the individual solutes and electrolytes. In the present chapter, we focus on general principles of epithelial Na^+ absorption and mechanisms controlling this process. As summarized in an excellent review by Kotchen and colleagues, Na^+ consumption in modern society generally exceeds total body Na^+ needs and is associated with adverse clinical outcomes (Kotchen et al. 2013). High Na^+ intake as found in Western diets is associated with hypertension and increased rates of cardiovascular disease. Of note, it has become recognized that the dietary K^+ to Na^+ ratio plays a critical role in blood pressure control, so increasing the dietary K^+/Na^+ ratio (e.g., through higher fruit and vegetable intake) may be highly beneficial in reducing hypertension (McDonough et al. 2017; Staruschenko 2018). Thus, despite the critical role that Na^+ plays in the maintenance of multiple mechanisms in the human body, many individuals, especially those on the Western diet, might need to limit Na^+ consumption and increase their K^+ consumption. When we eat too much NaCl, several organs, especially the kidneys, act to help us excrete the excess salt load. Sodium absorption in the kidney is precisely regulated and controlled by numerous mechanisms, many of which are reviewed in this chapter. Multiple studies have focused on transport processes in the kidney. As a consequence, much of our current knowledge of the principles of epithelial transport has been derived from studies on renal epithelia. Thus, we provide here the fundamentals of epithelial Na^+ absorption, focusing primarily on the mechanisms involved in the transport of this important ion in the kidney. In addition, modulation of transepithelial Na^+ reabsorption in the lung plays a key role in determining airway surface liquid volume and airway clearance, which is dysregulated in certain diseases like cystic fibrosis. The following section of this chapter will briefly describe sodium transport in the lung and other organs.

Although our understanding of the specific details and mechanisms of epithelial Na^+ absorption has greatly expanded recently, the main concepts remain remarkably similar to those described 50–60 years ago. Several excellent review articles and book chapters have focused on these mechanisms, and we briefly summarize them here. In addition, we recommend the reader to peruse renal physiology textbooks, including those edited by Schrier, Brenner, Boron, and Guyton/Hall.

Na^+ is the principal extracellular cation that is present within cells at a much lower concentration than outside. Typically, Na^+ concentrations are 10–20 mM and 135–145 mM in the intracellular and extracellular compartments in humans, respectively. The glomeruli filter approximately 25,000 mEq of Na^+ per day. As sodium is freely filtered by glomeruli, the kidneys must reabsorb the vast majority of Na^+ as the filtrate flows along the nephron. Na^+ absorption in the kidney is co-regulated with the transport of other ions (including chloride, phosphate, bicarbonate, potassium, protons, calcium, and magnesium), neutral solutes (e.g., glucose and amino acids) and water via directly or functionally coupled transport processes at both the apical and basolateral membranes. This is accomplished by the integrated function of consecutive sodium transporters and channels along the nephron.

9.2 General Concepts of Sodium Absorption in Epithelia

9.2.1 Basic Principles of Sodium Transport

In general, epithelial sodium (Na^+) absorption represents reabsorption of sodium ions from the tubules back into the blood. In this case, it is called reabsorption rather than absorption because sodium returns to the circulation. The classical concept of transepithelial NaCl absorption, which was formulated for the frog skin more than 50 years ago by Koefoed-Johnsen and Ussing (Koefoed-Johnsen and Ussing 1958) is still essentially valid for sodium reabsorption in the kidney. Palmer and Andersen briefly discussed the historical background and wide-ranging impact and implications of this seminal work (Palmer and Andersen 2008; Hamilton 2011).

In the kidney, Na^+, along with other ions such as chloride (Cl^-) and bicarbonate (HCO_3^-), is extensively reabsorbed. However, the rate of reabsorption depends on the current needs of the body and is precisely controlled to maintain homeostatic balance. To be reabsorbed, Na^+ must be transported across the tubular epithelial membranes in complex processes that will be discussed in this chapter. Reabsorption of Na^+ across the tubular epithelium into the interstitium includes active transport, passive diffusion, or both. The rate of total transport across a cellular membrane (dM/dt) is the sum of passive (PT) and active (carrier-mediated; CM) transport, and can be described by using the Michaelis–Menten equation (Eq. 9.1) (Sugano et al. 2010).

$$\frac{dM}{dt} = A \times (P_{PT} + P_{CM}) \times C = A \times P_{PT} \times C + \frac{V_{max} \times C}{K_m + C} \tag{9.1}$$

where A is the surface area of a membrane (length²); P is the permeability (length/time); C is the concentration of a permeant species (amount/length³); K_m is the Michaelis constant (amount/length³); and V_{max} is the maximum carrier-mediated transport (amount/time) (Sugano et al. 2010).

As shown in Fig. 9.1, there are two basic mechanisms of epithelial Na^+ reabsorption: (1) the transcellular route whereby sodium is transported through the cell across both the apical and basolateral plasma membranes; and (2) the paracellular route, where sodium ions traverse the spaces between cell–cell junctions. After Na^+ is absorbed across the tubular epithelial cells into the interstitial fluid, sodium is then transported through peritubular capillary walls into the blood by ultrafiltration. This process is mediated by Starling forces (Levick and Michel 2010; Starling 1896), a combination of hydrostatic pressure and colloid osmotic forces, which will not be covered further in this chapter.

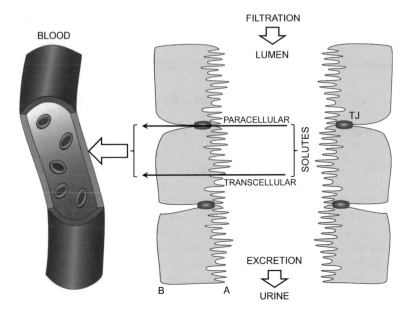

Fig. 9.1 General concept of active and passive Na$^+$ transport through the cell. Reabsorption of filtered Na$^+$ from the tubular lumen across the epithelial cells back into the blood. Na$^+$ is transported either by active transport through the cells (transcellular pathway) or between the cells (paracellular pathway) by diffusion. Tight junctions (TJ) are shown

9.2.2 Cytosolic Diffusion

As described in Einstein's theory of Brownian motion, molecular diffusion (or diffusion) consists of the random movements of a molecule with respect to adjacent molecules and occurs as the consequence of thermal energy. Since the diffusional movement of an individual molecule is random, a concentration gradient is required for any net transfer of molecules to occur across a membrane. Thus, the concentration gradient represents a major driving force for net transport. In addition, the movements of charged solutes (ions) are governed by electrical gradients created by potential differences between compartments, and thus net ion fluxes are dictated by their electrochemical gradients across membranes.

9.2.3 Maintenance of Membrane Potential

In addition to its critical role in the maintenance of electrolyte homeostasis, Na$^+$ concentrations play an important role in the maintenance of cellular membrane potential, as originally formulated by Goldman, Hodgkin, and Katz (Goldman 1943; Hodgkin and Katz 1949). Studies of various segments of the nephron have revealed heterogeneity of the electrical properties along these segments. The electrophysiological properties, such as transepithelial voltage and resistance (V_t and R_T, respectively), vary widely. Na$^+$, along with K$^+$ and Cl$^-$ contributes to the maintenance of concentration and charge differences across cell membranes. The intracellular and extracellular concentration differences between K$^+$ and Na$^+$ create an electrochemical gradient across the plasma membrane. In kidney epithelial cells, the cellular membrane potential is maintained mostly by ionic gradients generated by the Na$^+$/K$^+$-ATPase and governed by the relative permeability of the major ions, K$^+$, Cl$^-$, and Na$^+$ through the ensemble of channels and transporters present in each cell type. Tight control of the cellular membrane potential is recognized as being critical in other organs such as brain, heart, and muscle; the maintenance of membrane potential in renal epithelial cells is also important for the coordinated transport of solutes and overall kidney function.

For charged solutes, the driving force for transport is the sum of the chemical and electrical potential gradients. The Nernst equation (Eq. 9.2) describes the equilibrium condition for a membrane permeable only to a single ionic species:

$$V_m = V_2 - V_1 = \frac{RT}{ZF} \ln \frac{C_o}{C_i} \tag{9.2}$$

where R is the gas constant, T is the absolute temperature, Z is the valence of the solute, F is the Faraday constant, and C and V are concentration and electrical potential terms, respectively. At equilibrium, then, the voltage (V_m) across an ideally selective membrane is defined by the concentrations of the permeant ion on the inside and outside of the membrane, C_i and C_o, respectively.

For systems containing more than one permeant ion, i.e., almost all real physiological systems, the equilibrium membrane potential (V_m) can be described by the Goldman–Hodgkin–Katz (GHK) equation (Eq. 9.3; the equation is shown for Na$^+$, K$^+$, and Cl$^-$):

$$V_m = \frac{RT}{F} \ln \left[\frac{\left(P_{Na}C_{Na}^o + P_K C_K^o + P_{Cl}C_{Cl}^i\right)}{\left(P_{Na}C_{Na}^i + P_K C_K^i + P_{Cl}C_{Cl}^o\right)} \right] \tag{9.3}$$

where P_x is the permeability of the respective solutes, such as Na$^+$, K$^+$, and Cl$^-$. Thus, in a polarized epithelial system containing multiple charged solutes, one can estimate the transmembrane voltages at both membranes as a function of the relative transmembrane concentrations and permeabilities of each solute on both sides of the apical and basolateral membranes.

9.2.4 Mechanisms of Na⁺ Transport Across the Plasma Membrane

Na$^+$ absorption in the kidney includes a series of transport mechanisms. Active transcellular transport is the main mechanism for Na$^+$ absorption along the nephron. However, passive paracellular transport of Na$^+$ at least partially mediates sodium reabsorption in the kidney. For instance, in the proximal convoluted tubules (PCT) and thick ascending limb (TAL) of Henle's loop, sodium reabsorption takes both transcellular and paracellular routes (Olinger et al. 2018; Pei et al. 2016; Plain and Alexander 2018; Yu 2017). Passive paracellular transport is the predominant route for ion flows in the PCT, where the paracellular resistance is low. In the case of active transport, Na$^+$ ions have to cross both the apical and basolateral membranes (or vice versa) in series. Due to the hydrophobic properties of the membrane lipids, electrolytes such as Na$^+$ cannot cross the membrane freely and have to interact with specific transport proteins, which will be discussed in corresponding sections. The molecular basis of many of these transport proteins have been identified, and our knowledge about these transport mechanisms has significantly improved in recent years.

9.2.4.1 Active Transcellular Transport

The coupling of solute transport to an energy source can take two forms: primary active transport, where solute transport is linked directly to an energy-yielding reaction, and secondary active transport, where the solute movement against its electrochemical gradient is energized by the movement of another solute down its own gradient.

The most common example of primary active transport is the transport of Na$^+$ and K$^+$ by the Na$^+$/K$^+$-ATPase (also often referred to as the sodium pump or Na$^+$/K$^+$ pump), which is a transmembrane ATPase mediating the exchange of 3 Na$^+$ ions outward for 2 K$^+$ ions inward at the expense of ATP hydrolysis with each cycle. Therefore, the pump is electrogenic. The Na$^+$/K$^+$ pump is localized at the basolateral membrane of epithelial cells in different segments of the kidney, but it is especially abundant in the TAL of Henle's loop, cells of the distal convoluted tubule (DCT), connecting tubule (CNT) cells, and principal cells of the collecting duct (CD). In general, segments with high rates of active Na$^+$ transport have high Na$^+$/K$^+$ pump activity (Feraille and Doucet 2001). Na$^+$/K$^+$ pump functions as a heterodimeric protein complex comprised of catalytic α- and auxiliary β-subunits. The Na$^+$/K$^+$ pump is responsible for maintaining the intracellular Na$^+$ at a low level, which provides the gradient for the Na$^+$-coupled transport of many other solutes into the cell. More specific details on the structure and function of the Na$^+$/K$^+$ pump are provided in Chap. 1 of Volume 3.

In contrast to primary active transport, in secondary active transport, also known as coupled transport, the stored energy is used to transport molecules across a

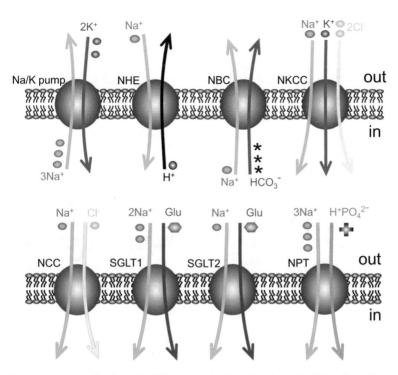

Fig. 9.2 Ion movement in the main Na⁺ transporters involved in epithelial sodium absorption. Shown are primary transport pathways for Na⁺/K⁺-ATPase, Na⁺/H⁺-exchanger (NHE), Na⁺-HCO⁻₃ cotransporter (NBC), Na⁺-K⁺-Cl⁻ cotransporter (NKCC), Na⁺-Cl⁻ cotransporter (NCC), sodium-glucose cotransporters (SGLT1 and SGLT2), and Na⁺-phosphate cotransporter (NPT). Arrows demonstrate the direction of ion movement inside or outside of the cell

membrane without its direct coupling to ATP hydrolysis. Na⁺, because of its steep inward electrochemical gradient maintained by the Na⁺/K⁺ pump, often participates in the transport of other solutes. Depending on the direction of the movement of coupled solutes (same or opposite), secondary active transporters are classified either as cotransporters (or symporters) or exchangers (or antiporters). Robert K. Crane first proposed flux coupling in 1960 suggesting a model for the sodium-glucose cotransporter (SGLT; described below in Sect. 9.3.2.1) (Crane 1960). Multiple cotransporters and exchangers involved in Na⁺ absorption are expressed in epithelial cells. Specific transporters mediating Na⁺ influx in exchange for efflux of other solutes are discussed below in sections describing Na⁺ transport in defined segments.

There are multiple electrogenic Na⁺-driven cotransporters expressed in epithelial cells in the kidney, such as Na⁺-coupled HCO₃⁻ transporters (Parker and Boron 2013). In addition to electrogenic transporters, Na⁺ entry is also mediated by electroneutral (i.e., when no net charge is transported across the membrane) Na⁺-H⁺ exchangers (e.g., NHE3) and Na⁺-coupled cotransporters (e.g., NCC and NKCC). Shown in Fig. 9.2 are examples of both primary and secondary active transporters mediating epithelial Na⁺ absorption. As seen from this figure,

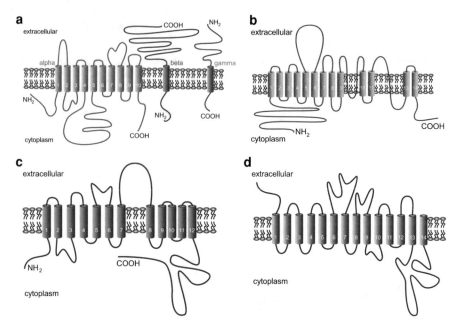

Fig. 9.3 Membrane topology of several Na⁺ transporters involved in epithelial sodium absorption. (**a**) Organization of the α- and β-subunits of Na⁺/K⁺-ATPases in the plasma membrane. The γ-subunit, which is found associated with some isoforms of Na⁺/K⁺-ATPase is also shown. (**b**) Presumed topology of the electrogenic Na⁺-HCO₃⁻ cotransporter. The model shows the extended cytosolic amino- and carboxy-terminal domains (NH₂ and C termini) linked via a transmembrane domain that includes 14 transmembrane spans, one of which is thought to be an extended region rather than an α-helix [modified from Parker and Boron (2013)]. (**c**) The topology proposed for Na⁺-coupled chloride cotransporters (NKCC) [modified from Gamba (2005)]. (**d**) Secondary structure of Na⁺-glucose cotransporter SGLT. Shown are 14 transmembrane helices with both the NH₂ and COOH termini facing the extracellular solution [modified from Turk et al. (1996) and Wright et al. (2011)]

ion/substrate coupling stoichiometry of the transporters varies. For instance, the coupling stoichiometry of the SGLT1 and SGLT2 cotransporters is 2 Na⁺ ions to 1 glucose molecule and 1 Na⁺:1 glucose per transport cycle for SGLT1 and SGLT2, respectively; for NKCC 1 Na⁺ ion is transported in the same direction as 1 K⁺ and 2 Cl⁻ ions. Importantly, due to their diverse nature and function, they have different plasma membrane topologies. Figure 9.3 illustrates some examples of various transporters' topologies.

9.2.4.2 Passive Paracellular Transport

The paracellular pathway in renal tubular epithelia such as the PCT, which reabsorbs the largest fraction of filtered NaCl, is important for the transport of electrolytes and water (Olinger et al. 2018; Plain and Alexander 2018; Yu 2017). The gatekeeper of the paracellular pathway is the tight junction, which is located at apical cell–cell interactions of adjacent epithelial cells (see Fig. 9.1). The tight junction separates the apical domain from the basolateral domain and provides a barrier to paracellular movement of water and ions. Claudins are key integral proteins that provide the barrier function and permit selective paracellular transport (Yu 2014, 2017). Claudin-2 is the main claudin responsible for reabsorption of Na⁺ (Kiuchi-Saishin et al. 2002). In addition to the claudin family members, there are several other tight junction transmembrane proteins that might directly influence the adhesive barrier, including occludins, junctional adhesion molecules, and tricellulin (Anderson and Van Itallie 2009). Interestingly, approximately 30% of the Na⁺ that is transported in the proximal tubule from lumen to blood by the transcellular pathway diffuses back to the urine by the paracellular pathway.

9.2.5 Methods of Na⁺ Transport Measurement

To a large degree, advancement in our understanding of the epithelial transport physiology of sodium and other electrolytes has been afforded by the development of key measurement methods and tools. Stockand and colleagues have provided a clear overview and description of traditional and contemporary in vivo and ex vivo tools used to study renal tubule transport physiology (Stockand et al. 2012). These include isotopic flux measurements, micropuncture, and microperfusion techniques, and the patch-clamp technique performed ex vivo in various configurations on isolated kidney tubules.

9.3 Sodium Homeostasis and Its Role in the Kidney

9.3.1 Role of Sodium Reabsorption in the Passive Diffusion of Water, Urea, and Other Solutes

Na⁺ reabsorption is critical for the transport of other solutes, and overall water homeostasis. When sodium is transported out of the kidney tubule, its concentration decreases inside the tubular lumen while increasing in the interstitial space. Various active and passive transport processes are involved in the maintenance of water and solute homeostasis in different nephron segments. Shown in Fig. 9.4 are basic mechanisms by which water, chloride, and urea reabsorption are coupled with sodium transport in the kidney. The generation of transmembrane concentration differences of sodium and other solutes induces osmotic water flow from the lumen to the renal interstitium. Certain segments of the nephron, such as the proximal tubule, are highly permeable to water and small ions via the tight junctions. In addition, AQP1 is abundant in the apical and basolateral membranes of the proximal

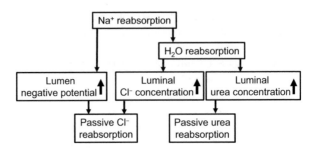

Fig. 9.4 Mechanisms of coupling of sodium and water, chloride, and urea reabsorption. When sodium is reabsorbed, anions such as chloride and bicarbonate are transported along with it; as Na^+ is positively charged, its transport from the lumen creates a lumen-negative potential, which promotes the diffusion of chloride through the paracellular pathway. Furthermore, transmembrane concentration differences of sodium facilitate water reabsorption from the lumen to the renal interstitium, which in turn can promote additional reabsorption of chloride ions and urea due to the increased concentration of these solutes present in the tubular lumen resulting from the water loss

tubules and descending thin limbs of Henle's loop, where this water channel provides the transcellular pathway for water reabsorption following small osmotic gradients at the apical and basolateral surfaces of the cell (Agre 2000; Nielsen et al. 1993). In contrast to the proximal tubules, less leaky epithelia such as the loop of Henle, distal tubules, and collecting duct prevent the osmotic flow of water across the tight junctions of the plasma membrane. However, vasopressin, also known as arginine vasopressin (AVP) or antidiuretic hormone (ADH), greatly increases water permeability in the collecting tubules, as discussed later (see Sect. 9.3.2.4). Therefore, changes in sodium reabsorption significantly influence the reabsorption of water and many other solutes.

In addition to water, sodium reabsorption plays an important role in the transport of solutes. When sodium is reabsorbed across epithelial cells, anions such as chloride and bicarbonate are transported along with sodium through coupled transporters. As Na^+ is positively charged, its transport from the lumen creates a negative lumen potential, which promotes the diffusion of anions such as chloride through the paracellular pathway in certain nephron segments (e.g., PCT). Osmotic water flow from the lumen to the interstitium also promotes the additional reabsorption of chloride ions due to the increased chloride concentration present in the tubular lumen. Thus, sodium reabsorption is coupled to the passive reabsorption of chloride by changes in the electrical potential and chloride concentration gradient. Moreover, sodium and chloride transport are tightly linked since these ions can also be reabsorbed by secondary active transport through coupled cotransporters (e.g., NCC and NKCC, as discussed below). This lumen-negative potential caused by Na^+ reabsorption in the collecting duct promotes K^+ secretion (via ROMK, discussed below).

Urea is mainly transported by UT-A and UT-B urea transporters (Klein et al. 2011, 2012; Klein and Sands 2016; Nawata and Pannabecker 2018). Besides, urea is also passively reabsorbed from the tubule. As described above, water is reabsorbed from the lumen by osmosis. Thus, urea concentration in the lumen increases, consequently, and urea is also reabsorbed by the concentration gradient. However, urea permeates the tubule to a much lesser extent than water or chloride. In addition to these facilitated urea transporters, sodium-dependent, secondary active urea transport mechanisms have been characterized. Thus, it was shown that removing sodium from the perfusate completely inhibited net urea reabsorption, demonstrating that active urea transport is dependent upon the presence of sodium in the tubular lumen (Isozaki et al. 1994; Sands et al. 1996). Recent studies in both human and animal models revealed that the renal concentration mechanism couples natriuresis with correspondent renal water reabsorption and results in concurrent extracellular volume conservation and concentration of salt excreted into the urine. This water-conserving mechanism of dietary salt excretion is dependent on urea transporter-driven urea recycling by the kidneys and urea production by the liver and skeletal muscle (Kitada et al. 2017; Rakova et al. 2017).

9.3.2 Sodium Absorption in Different Nephron Segments

Precise regulation of Na$^+$ absorption by the kidneys relies on sequential actions of the various nephron segments, each with highly specialized transport capabilities. Each human kidney contains about one million nephrons capable of forming urine (Hoy et al. 2005). Figure 9.5 provides an overview of consecutive segments of the nephron and corresponding Na$^+$ transport proteins along the nephron. Every nephron is composed of a renal corpuscle containing the glomerulus and Bowman's capsule, a proximal tubule (proximal convoluted and straight tubules), loop of Henle, a distal convoluted tubule (segments DCT1 and DCT2), connecting tubule (CNT), and the collecting duct system, which includes the initial collecting tubule (ICT), the cortical collecting duct (CCD), the outer medullary collecting duct (OMCD) and the inner medullary collecting duct (IMCD). All of the components of the nephron, including the collecting duct system, which collects urine from several nephrons, are functionally interconnected. In general, the absolute rates of Na$^+$ reabsorption are greatest in the proximal tubule and fall as the tubular fluid proceeds from proximal to distal segments. The proximal tubule reabsorbs the majority of the filtered Na$^+$ load (up to 60–70%). However, as discussed below, the proximal tubule has a rather limited ability to alter Na$^+$ transport. In contrast, the DCT and collecting ducts reabsorb only a minor fraction (<5–10%) of the filtered Na$^+$ but are finely regulated by various physiological factors.

9.3.2.1 Proximal Tubule (PT)

The proximal tubule is the major site for Na$^+$ absorption in the nephron. There is considerable heterogeneity of both morphologic and functional characteristics along

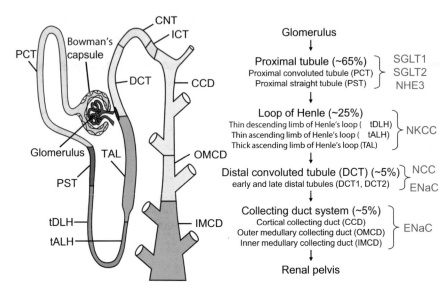

Fig. 9.5 Structure and consecutive segments of the nephron. Proximal convoluted tubule (PCT), proximal straight tubule (PST), thin descending limb of Henle's loop (tDLH), thin ascending limb of Henle's loop (tALH), thick ascending limb of Henle's loop (TAL), distal convoluted tubule (with segments DCT1 and DCT2), connecting tubule (CNT), initial collecting tubule (ICT), cortical collecting duct (CCD), outer medullary collecting duct (OMCD), and inner medullary collecting duct (IMCD) segments of the nephron are shown. Approximate percentage of sodium reabsorption in different segments and major sodium transporters mediating these processes are also depicted [modified from Staruschenko (2012)]

the proximal tubule. Typically, the proximal tubule consists of convoluted and straight tubules (PCT and PST, respectively). Based on the ultrastructure, it can be further divided into S1, S2, and S3 segments. The first two segments are located in the kidney cortex, and the straight segments (S3) descend into the outer medulla. The net rates of the Na^+ transport in the late proximal segment are, in general, lower than in the initial proximal convoluted tubule (Jacobson 1982). Importantly, the absorption of sodium in the proximal tubule provides the driving force for the absorption of other solutes, such as bicarbonate, glucose, phosphate, and amino acids (Parker and Boron 2013; Skelton et al. 2010). Under most circumstances, fluid along the proximal tubule has virtually the same Na^+ concentration and osmolality as plasma (Ullrich et al. 1963). The isosmotic nature of proximal tubule fluid absorption derives from the high water permeability of this segment (Andreoli et al. 1978), which effectively clamps the osmolality of the tubular fluid as that of plasma. Although Na^+ transport in the proximal tubule occurs in the absence of large electrical or chemical gradients, the bulk of Na^+ absorption in the proximal tubule relies on active transport. Earlier studies demonstrated that Na^+ can be reabsorbed in this tubule segment against both concentration (Giebisch et al. 1964; Ullrich et al. 1963) and electrical (Barratt et al. 1974) gradients. A significant amount of Na^+ transport in the proximal tubules also occurs passively (Schafer et al. 1975). For

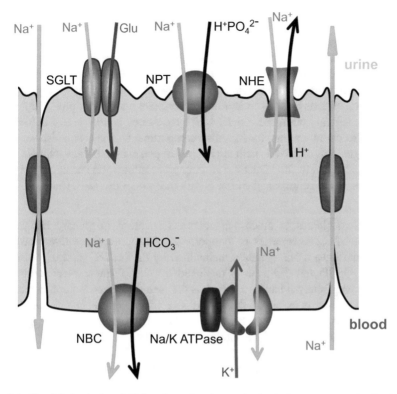

Fig. 9.6 Simplified scheme of Na^+ reabsorption in proximal tubules. In the proximal tubules, sodium can be reabsorbed against both concentration and electrical gradients; however, a significant amount of Na^+ transport also occurs passively. The majority of sodium ions are reabsorbed via the Na^+/H^+ exchanger (NHE); sodium-glucose cotransporter (SGLT) cotransports Na^+ and glucose into the cells, and so does the Na^+-phosphate cotransporter (NPT) along with phosphate ions. Importantly, the absorption of sodium in the proximal tubule provides the driving force for the absorption of other solutes, such as bicarbonate, glucose, phosphate, and amino acids. On the basolateral side, sodium is removed from the cells with the help of Na/HCO_3 cotransporter (NBC) and Na^+/K^+-ATPase

example, in the late convoluted and straight proximal tubules, Na^+ diffuses passively out of the tubule driven by the lumen-positive electrical potential difference. This difference is driven by a Cl^- concentration gradient across the tubule wall. Shown in Fig. 9.6 is a simplified scheme of Na^+ reabsorption in proximal tubules.

NHE3

NHE3 is a Na^+/H^+ exchanger involved in large amounts of neutral Na^+ absorption in the PCT. NHE3 (in human encoded by the *SLC9A3* gene) is one of nine isoforms of the mammalian Na^+/H^+ exchanger (NHE) gene family (Donowitz and Li 2007; Fenton et al. 2017). NHE3 is localized at the apical plasma membrane of the proximal tubules and is responsible for the majority of sodium absorption in the kidney. NHE3 performs an electroneutral exchange of one extracellular Na^+ ion with one extracellular proton (see Fig. 9.2). NHE3 is functionally linked to a Cl^-/HCO_3^- exchanger and takes part in neutral NaCl absorption (Soleimani 2013). Recent studies using genetic deletion of NHE3 revealed that NHE3 in the proximal tubules plays an essential role in the maintenance of basal blood pressure (Li et al. 2018).

The activity of NHE3 is regulated allosterically by intracellular pH, stimulated by cellular acidification. NHE3-mediated neutral NaCl absorption is highly regulated by dietary conditions and the endocrine environment under both physiological and pathophysiological conditions. Under basal conditions NHE3 cycles between the apical plasma membrane and the recycling compartments. The mechanisms involved in acute regulation of NHE3 include changes in plasma membrane expression and turnover, which can be modulated by altering both endocytosis and exocytosis rates, and changes in the cellular half-life of NHE3 (see also Volume 3, Chap. 5).

SGLT

There are two classes of glucose transporters involved in glucose homeostasis. Sodium-glucose cotransporters or symporters (SGLTs) are the active transporters driven by the inward Na^+ gradient maintained by the Na^+/K^+ pump. Transport of glucose via facilitated diffusion is mediated by GLUT transporters (uniporters). SGLTs are exquisitely selective for Na^+ as the cation required for cotransport with glucose. One interpretation of the cation selectivity data is that cation binding initiates a change in the conformation of the sugar-binding site (Wright et al. 2011). Interestingly, SGLT1 (in humans is encoded by the *SLC5A1* gene) and SGLT3 (*SLC5A4*) have two strongly interacting Na^+ binding sites while SGLT2 (*SLC5A2*) only has one Na^+ binding site (Díez-Sampedro et al. 2001; Hummel et al. 2011), as shown in Fig. 9.2.

SGLT1 cotransports Na^+ and glucose in the PCT. The activity of this transporter depolarizes the plasma membrane, which can serve as a trigger for further signaling mechanisms. Interestingly, in the absence of glucose, SGLT1 can still conduct Na^+ currents. The first evidence that SGLT1 worked as a uniporter was reported in 1990 (Umbach et al. 1990), soon after SGLT1 was cloned in 1987 (Hediger et al. 1987). Thus, it was shown that the SGLT inhibitor phlorizin blocked current in the absence of glucose. This current accounted for approximately 8% of the total Na^+-glucose cotransporter current and was not observed in control cells (Umbach et al. 1990). Moreover, this current was saturated with increasing external Na^+ concentration (Loo et al. 1999). In addition, it was shown that SGLT1 transports water and urea along with Na^+ and glucose (Loo et al. 1999; Wright et al. 2011).

There are 12 members of the human SGLT (SLC5) gene family, including cotransporters for sugars, anions, vitamins, and fatty acids. However, only four members of the SLC5 family are Na^+-glucose cotransporters. Two of these are responsible for glucose transport in the kidney: SGLT1 and SGLT2. SGLT1 is expressed mainly in the S2 and S3 segments of the PT, and SGLT2 is highly expressed in the S1 segment of the PT. SGLT3 was also found to be expressed in the kidney (Sotak et al. 2017); however, in contrast to SGLT1/2, SGLT3 functions as a glucose sensor, not a Na^+-glucose cotransporter (Díez-Sampedro et al. 2003). Early micropuncture studies identified that the glomerular filtrate is glucose-free when it reaches the end of the PCT. Further microperfusion experiments with isolated nephrons revealed that active glucose reabsorption occurs in the early segments of PCT via a low-affinity high-capacity system, whereas active reabsorption in the late PST occurred by a high-affinity, low-capacity system. As it was later identified, the majority of glucose (and Na$^+$, respectively) is transported in the PCT by SGLT2 and the rest in PST by SGLT1 (Wright et al. 2011). However, although the critical role of SGLT2 has been demonstrated, the functional properties of SGLT2 in the kidney are less thoroughly studied than SGLT1 due to its poor expression in heterologous systems. Nevertheless, SGLT2 inhibitors are emerging to be very important clinically in the treatment of type 2 diabetes mellitus. They also have beneficial effects on the cardiovascular system and the kidney (Nadkarni et al. 2017; Nespoux and Vallon 2018; Rieg and Vallon 2018; Thomas and Cherney 2018).

9.3.2.2 The Loop of Henle

The loop of Henle consists of structurally and functionally distinct thin limbs [thin ascending limb of Henle (tALH), thin descending limb of Henle (tDLH)], and the thick ascending limb (TAL) (refer to Fig. 9.5). Compared to the proximal tubule, the loop of Henle reabsorbs a smaller amount of filtered Na$^+$ [approximately 25% of the total load (Bennett et al. 1968)]. The key difference between the PCT and the loop of Henle is that the loop as a whole reabsorbs more salt than water, whereas in the proximal tubule this occurs in essentially equal proportions. Interestingly, the reabsorption of salt and water is performed in different places along the loop. The descending limb reabsorbs very little Na$^+$ or Cl$^-$, but allows transport of water; both ascending limbs (thick and thin) are impermeable to water, but extensively reabsorb sodium and chloride (which is why they are often referred to as "diluting segments"). As a result, the fluid leaving the loop is hypoosmotic relative to plasma, indicating that more salt than water is reabsorbed. Ultimately, the loop of Henle is a nephron segment that allows water and salt excretion to be regulated independently.

Thin Ascending and Thin Descending Loops of Henle
Intense water extraction via aquaporin water channels (Halperin et al. 2008; Nielsen et al. 2002; Verkman 2006) and the near absence of sodium reabsorption in the tDLH concentrate tubular fluid (Kokko 1970; Kokko and Rector 1972). High interstitial levels of NaCl and urea provide additional osmotic energy for water reabsorption by tDLH. Therefore, when the fluid exits the tDLH, it possesses a favorable gradient of luminal sodium, which facilitates further passive sodium reabsorption in the water-impermeable tALH (Imai and Kokko 1974). The formation of dilute urine by the loop of Henle begins at the tALH. Transepithelial movement of sodium from the lumen occurs via the paracellular pathway, whereas chloride diffuses transcellularly via ClC-K1 chloride channels, which are selectively localized to the tALH and appear to be the major mediators of chloride movement there (Reeves et al. 2001). At large, the osmolarity of the fluid leaving the tALH is decreased due to a fall in NaCl content, which occurs almost entirely via passive and paracellular pathways.

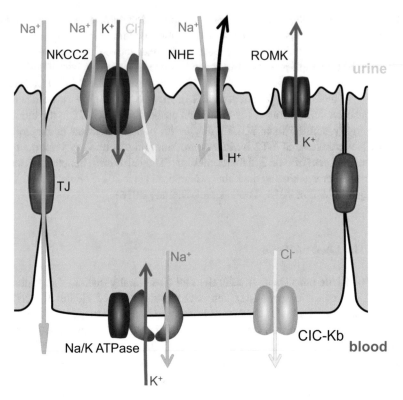

Fig. 9.7 Simplified scheme of Na⁺ reabsorption in the loop of Henle. Transepithelial movement of sodium from the lumen in the tDLH and tALH occurs via the paracellular pathway, whereas chloride diffuses transcellularly via ClC-K1 chloride channels. In contrast to the tDLH and tALH, the major mechanisms of sodium and chloride reabsorption in the TAL are active. The dominant step of luminal sodium and chloride uptake in this segment occurs via the sodium–potassium–chloride symporter (NKCC2 isoform); an alternative mechanism for the uptake of Na⁺ from the lumen is provided by the apical the Na⁺/H⁺ antiporter (NHE). ClC-K1 channels and Na⁺/K⁺-ATPase localized at the basolateral membrane allow Cl⁻ and Na⁺ transepithelial resorption, respectively. K⁺ enters the cell as a result of Na⁺/K⁺-ATPase function and then leaks back into the urine through apical potassium (ROMK) channels

Thick Ascending Loop of Henle

As the luminal fluid moves into the TAL tubules, the reabsorbing properties of the epithelium change. In contrast to the tDLH and tALH, the major mechanisms of sodium and chloride reabsorption in the TAL are active. As with the tALH, the TAL consists of a tight epithelium, which is impermeable to water (Hebert 1986).

As shown in Fig. 9.7, the dominant step of luminal sodium and chloride uptake in this segment occurs via the sodium–potassium–chloride symporter (NKCC2 isoform) (Greger 1981; Koenig et al. 1983; Markadieu and Delpire 2014). ClC-K1 (hClC-Ka in human) localized at the basolateral membrane allows passive Cl⁻ transepithelial resorption in the TAL (Fahlke and Fischer 2010; Imbrici et al.

2014). An accessory ClC-K channel subunit, barttin, encoded by the gene product of *BSND*, is required to form a functional channel (Birkenhager et al. 2001; Estevez et al. 2001). It was reported that ClC-K1 is upregulated in water-restricted animals (Fushimi et al. 1993; Uchida et al. 1993). Further studies revealed that the ClC-K1 knockout mice can maintain salt and water balance under normal conditions but are unable to concentrate urine under water restriction; deletion of ClC-K1 results in diabetes insipidus (Matsumura et al. 1999). The electroneutral process of Na^+-K^+-$2Cl^-$ cotransport is driven largely by the favorable electrochemical gradient for Na^+ entry, which is maintained by the continuous operation of the Na^+/K^+ pump in the basolateral membrane. Na^+-K^+-$2Cl^-$ transport is selectively inhibited by so-called loop diuretics (e.g., furosemide and bumetanide), as discussed further below in Sect. 9.3.5.1 (Russell 2000). The apical membrane of TAL cells also expresses a Na^+/H^+ antiporter isoform [major isoform being NHE3 (Amemiya et al. 1995; Borensztein et al. 1995; Capasso et al. 2002; Laghmani et al. 1997)], which provides an alternative mechanism for the uptake of Na^+ and regulates intracellular pH.

Interestingly, NKCC2 requires equal amounts of Na^+ and K^+ to enter the cell. However, as there is much less K^+ than Na^+ in the lumen, the tubular fluid would seem to get depleted of K^+ long before enough Na^+ was reabsorbed. This issue is resolved by K^+ recycling via the apical K^+-conductive pathway. K^+ leaks back into the lumen through apical potassium channels (mostly through ROMK channels, but the involvement of other potassium channels has also been reported) (Giebisch 2001; Guggino et al. 1987; Hebert et al. 2005; Wang 1994, 2012; Wang et al. 1990), and this makes K^+ available for Na^+-K^+-$2Cl^-$ cotransport again. Furthermore, K^+ channels actually dominate the apical membrane conductance of the TAL, creating a lumen-positive transepithelial voltage, which provides the driving force for the paracellular movement of Na^+. This passive diffusion of Na^+ accounts for as much as 50% of total Na^+ reabsorption in this segment. The Cl^- that enters TAL cells apically via NKCC2 exits via ClC-Kb Cl^- channels on the basolateral membrane (Zaika et al. 2016) (Fig. 9.7).

NKCC2

The uptake of Na^+ in the loop of Henle is coupled to that of Cl^- and K^+. The coupling ratio is fixed at 2 Cl^- per 1 Na^+ and 1 K^+ (see Fig. 9.2). The renal-specific apical form of this cotransporter encoded by *SLC12A1* gene was denoted NKCC2 (also known as BSC1, or bumetanide-sensitive cotransporter), based on high homology with the basolateral Na^+-K^+-$2Cl^-$ cotransporter NKCC1 (Gamba 2005). After its initial cloning from rat and rabbit (Gamba et al. 1994; Payne and Forbush 1994), NKCC2 was identified in the mouse exclusively in the kidney (Igarashi et al. 1995), and later in humans (Simon et al. 1996a).

NKCC2 belongs to a family of electroneutral cation–chloride cotransporters, which are all members of the amino acid polyamine cotransporter (APC) superfamily. As resolved from similarity to other crystallized APC family members, NKCC2 contains 12 transmembrane domains (see Fig. 9.3c) with a large extracellular loop between membrane segments 7 and 8; the *SLC12A1* gene encodes the protein with a molecular weight of 115–120 kDa (approximately 1100 a.a.) (Gamba et al. 1994). At least six different isoforms of NKCC2 have been identified in the mouse kidney (Ares et al. 2011; Gamba 2001; Igarashi et al. 1995; Mount et al. 1999; Payne and Forbush 1994; Plata et al. 2001; Simon et al. 1996a; Yang et al. 1996); such molecular diversity results from alternative splicing variants. Interestingly, NKCC2 isoforms are differentially expressed along the TAL. In situ RNA hybridization and RT-PCR studies have revealed that isoform A is expressed in the renal cortex and medulla outer stripe, F in the outer medulla with higher density in the inner stripe, and B in the renal cortex. In the macula densa, both isoforms A and B were found (Castrop and Schnermann 2008). Furthermore, A, B, and F isoforms have different transport capabilities; A and B possess a higher affinity to Na^+, K^+, and Cl^- than the F isoform (Plata et al. 2002).

The role of apical NKCC2-mediated Na^+-K^+-$2Cl^-$ contransport in the TAL was initially demonstrated by mutations in *SLC12A1*, which are associated with the Bartter's syndrome (Simon et al. 1996a), a disorder characterized by hypokalemia, metabolic alkalosis, and hyperaldosteronism (see Table 9.1). NKCC2-knockout mice reproduce a clinical phenotype characteristic of this illness (clinical features included severe renal failure, high plasma potassium, and metabolic acidosis) (Takahashi et al. 2000). Furthermore, NKCC2 has been shown to play a role in cardiovascular diseases, including hypertension. This is particularly important as this cotransporter is the main target of loop diuretics (e.g., furosemide and bumetanide), which are potent pharmacological agents currently in wide clinical use (Gamba 2005). Inhibition of NKCC2 results in reduced salt reabsorption in the TAL, which increases salt delivery to the distal nephron and produces substantial diuresis and natriuresis. It has been shown that NKCC2 along the TAL is activated in the Milan hypertensive rats, and this significantly contributes to the increase in systemic blood pressure in this model (Carmosino et al. 2011). Several rare functionally impairing mutations in human NKCC2 were associated with lower blood pressure (Monette et al. 2011). In Dahl salt-sensitive (SS) rats, high salt intake increases apical NKCC2 expression and phosphorylation in the TAL as compared with salt-resistant animals, and this may contribute to enhanced NaCl reabsorption in SS rats during high salt intake (Ares et al. 2012; Haque et al. 2011). In general, NKCC2 constitutes the major salt transporting pathway in mammalian TAL, and its proper function is fundamental not only for salt reabsorption but for the kidney's ability to produce urine that is more dilute or concentrated than plasma, which is essential for the survival of the land mammals (Gamba 2005).

NKCC2 activity is regulated by several hormones and factors, including vasopressin, via the arginine vasopressin 2 receptor (V2R), prostaglandins, other peptide hormones, angiotensin II, adrenergic agents, nitric oxide, and osmolality (Bachmann and Mutig 2017; Gonzalez-Vicente et al. 2019). Downstream of vasopressin activation, in addition to cAMP-dependent signaling cascades that are activated, the Ste20 SPS1-related proline alanine-rich kinase (SPAK) and oxidative stress-responsive kinase 1 (OSR1) are related kinases that bind to the N-terminus of NKCC2, directly phosphorylate stimulatory residues on NKCC2 and regulate its activity (Cheng et al. 2015). Further details on NKCC structure and function are provided in Chap. 2 of volume 3.

Table 9.1 Inherited disorders of renal sodium transport

Disorder	Affected gene product	OMIM #	Nephron distribution	Functional consequences	Clinical features
Hypotensive disorders					
Bartter, Type 1 (antenatal)	NKCC2 (*SLC12A1*)	600839	TAL	Decreased NKCC cotransport	Hypotension, hypokalemia, metabolic alkalosis
Bartter, Type 2 (antenatal)	ROMK (*KCNJ1*)	600359	TAL, CCD	Decreased apical K^+ recycling	Same as above
Bartter, Type 3 (infantile)	ClC-Kb (*CLCNKB*)	602023	TAL	Decreased basolateral Cl^- flux	Hypokalemia, alkalosis, hyperreninemia
Bartter, Type 4	Barttin (*BSND*)	606412	tAL, TAL, inner ear	Decreased basolateral Cl^- flux	As above, with sensorineural deafness
Bartter, Type 5	CaSR (*CASR*)	601199	PT, TAL, parathyroid, thyroid, brain	Activating mutation of Ca^{2+}-sensing receptor with decreased K^+ recycling and NaCl reabsorption	Hypokalemia, alkalosis, hyperreninemia, hypocalcemia
Gitelman	NCCT (*SLC12A3*)	600968	DCT	Decreased NaCl cotransport	Hypokalemia, metabolic alkalosis, hypocalciuria
Pseudohypoaldosteronism type 1 (autosomal recessive)	ENaC subunits (α, β, γ) (*SCNN1A, SCNN1B, SCNN1G*)	264350	Collecting duct	Decreased amiloride-sensitive sodium transport	Hypotension, hyperkalemia, metabolic acidosis
Pseudohypoaldosteronism type 1 (autosomal dominant)	Mineralocorticoid receptor (*NR3C2*)	177735	Collecting duct	Decreased response to mineralocorticoids	Hypotension, hyperkalemia (less severe than recessive)
Hypertensive disorders					
Liddle	β or γ ENaC (*SCNN1B, SCNN1G*)	177200	Collecting duct	Increased cell surface expression and activity of ENaC channels	Hypertension, hypokalemia, metabolic alkalosis, responsive to amiloride
Pseudohypoaldosteronism type 2 (Gordon)	WNK1 or WNK4 (*WNK1, WNK4*)	145260	DCT, collecting duct	Increased NCC cotransport, increased paracellular Cl^- permeability	Hypertension, hyperkalemia, metabolic acidosis, responsive to thiazide diuretics

(continued)

Table 9.1 (continued)

Disorder	Affected gene product	OMIM #	Nephron distribution	Functional consequences	Clinical features
Apparent mineralocorticoid excess	11β-HSD2 (*HSD11B2*)	218030	DCT, collecting duct	Decreased oxidation of glucocorticoids causing activation of mineralocorticoid receptor	Hypertension, hypokalemia, metabolic alkalosis, responsive to dexamethasone
Other disorders					
Renal glucosuria	SGLT2 (*SLC5A2*)	233100	Proximal tubule	Decreased sodium-coupled glucose absorption	Decreased threshold for glucosuria
Proximal RTA	NBC1 (*SLC4A4*)	603345	Proximal tubule	Decreased sodium-coupled HCO_3^- transport across basolateral membrane	Metabolic acidosis, ocular and dental abnormalities, growth and mental retardation

TAL thick ascending limb, *tAL* thin ascending limb, *DCT* distal convoluted tubule, *OMIM #* accession number for Online Mendelian Inheritance in Man. Updated references available at http://www.ncbi.nlm.nih.gov/entrez/query.fcgi?db=OMIM. Data provided are based on Subramanya et al. (2012)

9.3.2.3 Distal Convoluted Tubule

The distal convoluted tubule (DCT) is the initial segment of the aldosterone-sensitive distal nephron (ASDN), which consists of DCT, the connecting tubule (CNT), and the collecting ducts (CD). DCT can be further divided into two functionally different segments, denominated "early" and "late" DCT (Reilly and Ellison 2000) by their sensitivity to aldosterone. Distal tubules reabsorb much smaller fractions of filtered Na^+ load than PCT or the Loop of Henle (5–10%) (Hierholzer and Wiederholt 1976), and this process occurs almost exclusively by the transcellular pathway. NaCl concentrations generated along the loop of Henle cannot be maintained by the DCT, and within the initial 20% of the DCT, salt concentration almost doubles (Schnermann et al. 1982). Later, luminal Na^+ concentration decreases to a value of approximately 30 mM at the end of DCT (Khuri et al. 1975). This occurs to a large extent via active Na^+ transport and because the transepithelial voltage becomes progressively more negative moving from the early to late DCT (Allen and Barratt 1985; Barratt et al. 1975; Hayslett et al. 1977; Wright 1971). The absorption of Na^+ in the DCT is dependent on the load. Higher tubular flow rates result in the increased delivery of Na^+ to the normally unsaturated later parts of the distal tubule (Khuri et al. 1975; Shimizu et al. 1989).

The major step of apical Na^+ intake in the early DCT is via the Na^+-Cl^- cotransporter (NCC) (Ostrosky-Frid et al. 2019; Velazquez et al. 1987), which is characteristically different from the NKCC2 in the TAL, independent of K^+ and highly sensitive to a different class of drugs—thiazide diuretics (see Sect. 9.3.5.2 and Fig. 9.8). As mentioned above, lumen transepithelial electrical potential is more negative in the late rather than early DCT, and this is probably caused by differences in the Na^+ transporters. The late DCT, as distinct from the early DCT, also has an amiloride-sensitive pathway of Na^+ reabsorption (Reilly and Ellison 2000), which is mediated by the epithelial Na^+ channel (ENaC; see Sect. 9.3.3.4) (Campean et al. 2001; Ciampolillo et al. 1996; Costanzo 1984; Loffing et al. 2000, 2001; Schmitt et al. 1999). However, species differences have been reported for ENaC expression in the late DCT. ENaC has been found in the DCT of mouse and rat renal tissues (Loffing et al. 2000; Schmitt et al. 1999) but not in human or rabbit kidney (Biner et al. 2002; Velazquez et al. 2001). It was also proposed that ENaC and NCC might interact with each other in the late DCT (Mistry et al. 2016; Wynne et al. 2017).

The removal of Na^+ from the basolateral side of the DCT cells is controlled by the very intense activity of the Na^+/K^+ pump in both early and late segments (Doucet 1988). Basolateral K^+ is then removed via the channels which belong to the K_{ir} family (Hamilton and Devor 2012; Lourdel et al. 2002; Ookata et al. 2000; Zaika et al. 2013; Zhang et al. 2013), either Kir4.1 homomer (known as KCNJ10) or the Kir4.1/Kir5.1 (KCNJ10/KCNJ16) heteromer (Palygin et al. 2017a), which help maintain the activity of the Na^+/K^+ pump and, thus, salt reabsorption capacity (Bockenhauer et al. 2009; Palygin et al. 2017b; Paulais et al. 2011; Reichold et al. 2010; Staruschenko 2018). Apically, K^+ is secreted via ROMK channels (Wang and Giebisch 2009; Wang et al. 2010; Xu et al. 1997). Similar to the TAL, basolateral Cl

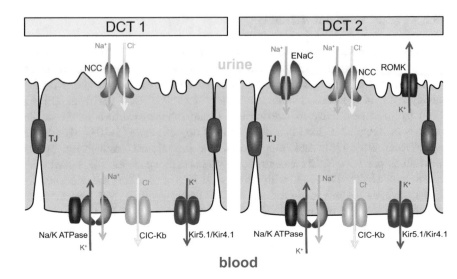

Fig. 9.8 Simplified scheme of Na^+ reabsorption in the distal convoluted tubules. Apical Na^+ intake is mediated via the Na^+-Cl^- cotransporter (NCC) (DCT1) or both NCC and ENaC (DCT2). The removal of Na^+ from the basolateral side of the DCT cells is controlled by the very intense activity of the Na^+/K^+ pump. K^+ is secreted apically via ROMK channels in DCT2. Basolaterally, K^+ is extruded through inwardly rectifying $K_{ir}5.1/K_{ir}4.1$ channels. Similar to the TAL, basolateral Cl efflux is mediated via the ClC-K channels, which significantly ensure the maintenance of the gradient for the entry of Na^+ and mitigate against intracellular Cl^- concentration increases resulting from apical NaCl entry via NCC

efflux is mediated via the ClC-K (mostly through ClC-K2/ClC-Kb) channels (Fahlke and Fischer 2010). These chloride channels ensure the maintenance of the gradient for the entry of Na^+ and mitigate against intracellular Cl^- concentration increases resulting from apical NaCl entry via Na^+-Cl^- cotransport (Estevez et al. 2001; Lourdel et al. 2003). The apical membranes and tight junctions of DCT have almost no water permeability.

NCC

The Na^+-Cl^- cotransporter (NCC, or *SLC12A3*) is a cation-coupled Cl^- cotransporter that, like NKCC2, belongs to the SLC12 family (Gamba 2005). NCC is dose-dependently inhibited by thiazide diuretics (e.g., chlorthalidone, hydrochlorothiazide, bendroflumethiazide, and metolazone) rather than loop diuretics (Costanzo 1985; Ellison et al. 1987; Gamba et al. 1994; Stokes 1984), and is thus also known as the thiazide-sensitive cotransporter (TSC). Moreover, it was also shown to be insensitive to barium, acetazolamide, furosemide, amiloride, 4,4′-diisothiocyano-stilbene-2,2′-disulfonic acid (DIDS), and ouabain (Gamba 2005). NCC mediates coupled absorption of Na^+ and Cl^-; this mechanism provides the means for maintaining intracellular Cl^- levels above the electrochemical equilibrium (Boron and Sackin 1983). Na^+ and Cl^- absorption in DCT are interdependent, and mammalian TSC has a high affinity for both ions (half-maximal concentrations for these ions are ~10 mM) (Monroy et al. 2000; Velazquez et al. 1984). The presence of NCC was first discovered in the urinary bladder of the winter flounder (Renfro 1975, 1977). Subsequent studies demonstrated that it is expressed in the renal cortex of rat, mouse, rabbit, and humans (Bostanjoglo et al. 1998; Gamba et al. 1994; Kunchaparty et al. 1999; Mastroianni et al. 1996; Simon et al. 1996c). The basic structure of NCC shares similarity with NKCC2 (shown in Fig. 9.3c): the protein has 12 transmembrane domains flanked by a short amino-terminal domain and a long predominantly hydrophilic C-terminal domain, both located in the cytoplasm; a long hydrophilic extracellular loop connects segments 7 and 8.

Loss-of-function mutations in the *SLC12A3* gene encoding NCC lead to the development of the Gitelman syndrome, which is clinically characterized by hypokalemia, hypomagnesemia, metabolic alkalosis, arterial hypotension, and hypocalciuria (see Table 9.1). On the other hand, increased NCC activity (caused by a gain-of-function mutation) results in the opposite: hyperkalemic metabolic acidosis, arterial hypertension, and hypercalciuria, known as Gordon syndrome, pseudohypoaldosteronism type II, or familial hyperkalemic hypertension (see Table 9.1) (Gamba 2005; Ostrosky-Frid et al. 2019). New roles for NCC in sodium and potassium handling and blood pressure regulation have been unraveled. For instance, the discoveries that NCC is activated by angiotensin II but inhibited by dietary potassium shed light on how the kidney handles sodium during hypovolemia (accompanied with high angiotensin II) and hyperkalemia (Moes et al. 2014; Su et al. 2019; Wang et al. 2018). In addition, much of the complex molecular machinery controlling the transporter's activity has been recently revealed. For example, modification of multiple kinases or ubiquitin ligases (including WNKs, SGK1, SPAK, Nedd4-2, Cullin-3, Kelch-like 3 and others), alteration of intracellular calcium signaling, changes in hormonal status (aldosterone, vasopressin, insulin, angiotensin II), and circadian rhythms all modulate the transporter's activity and may contribute to disease states such as hypo- or hypertension, and have detrimental effects on the ability of the kidneys to excrete sodium, potassium, calcium, and protons (Chavez-Canales et al. 2013; Chiga et al. 2008; Eladari et al. 2014; Gailly et al. 2014; Gamba 2012; Lagnaz et al. 2014; Lee et al. 2013; Moes et al. 2014; Mutig et al. 2010; Richards et al. 2014; Rieg et al. 2013; Saritas et al. 2013; Terker et al. 2014; Vallon et al. 2009). Of note, recent studies have highlighted the role of [K⁺] in the regulation of NCC via changes in intracellular [Cl⁻] and regulation of WNK1 and SPAK with NCC phosphorylation at key regulatory residues (Cuevas et al. 2017; Terker et al. 2015; Wang et al. 2018). Further details on the structure, function, and regulation of NCC are provided in Chap. 3 of Volume 3.

9.3.2.4 Connecting Tubule and Collecting Duct

Heterogenic vectorial transport in this segment is tightly controlled. This segment of the nephron is the most sensitive to hormones, which oversee the fine control of plasma Na⁺ and K⁺ concentrations. A variety of hormones (e.g., aldosterone, angiotensin II, vasopressin, atrial natriuretic peptide, and insulin) are involved in this tight regulation of CNT and CCD ion transport. The late DCT, CNT, and CCD are often referred to as the aldosterone-sensitive distal nephron (ASDN). Aldosterone, which is the final effector of the renin–angiotensin–aldosterone system (RAAS), increases the reabsorption of Na^+ and water and the secretion of K^+ ions.

The apical entry of Na^+ in the CNT and the CCD is regulated by the epithelial Na^+ channel (ENaC), which is composed of three subunits: α, β, and γ. All three subunits are required to form a functional trimeric channel (Canessa et al. 1994; Noreng et al. 2018; Staruschenko et al. 2005). The critical role of ENaC is highlighted by its mutations leading to Liddle syndrome and pseudohypoaldosteronism type 1 (PHA1) (Hanukoglu and Hanukoglu 2016; Lifton et al. 2001). Within the scope of this overview, only some aspects of ENaC regulation will be provided. Different aspects of ENaC regulation are highlighted in Chap. 18 of Volume 3, and are subjects of several excellent reviews (Bhalla and Hallows 2008; Butterworth et al. 2009; Hanukoglu and Hanukoglu 2016; Loffing and Korbmacher 2009; Mutchler and Kleyman 2019; Pavlov and Staruschenko 2017; Rossier 2014; Soundararajan et al. 2010).

As described in Sect. 9.2.4.1, there are different transporters and exchangers involved in the maintenance of sodium homeostasis in the kidney. Interestingly, ENaC is the only "classical" channel mediating sodium reabsorption in the renal tubules. Considering that ENaC is responsible for the fine-tuning of sodium reabsorption in the collecting ducts, the role of this channel in sodium reabsorption in the kidney in critical and unique. ENaC plays a key role in the final Na^+ composition of urine. and is controlled by different factors. Evolutionarily, the ENaC/Degenerin (Deg) superfamily members appeared in the Metazoan ancestor (Studer et al. 2011), along with the Na^+/K^+ pump α- and β-subunits. The emergence of the ENaC/Deg superfamily and the complete formation of the active Na^+/K^+ pump allowed for the stricter control of salt balance with changes in the environment. Like with regards to other transporters involved in membrane transport and osmolarity regulation, the hypothesis that these proteins helped regulate the transition to multicellularity is appealing. The evolution of ENaC and the Na^+/K^+ pump as limiting factors of aldosterone action on Na^+ transport has been recently discussed by Rossier and colleagues (Studer et al. 2011).

Since ENaC is an electrogenic channel that generally conducts Na^+ ions in only one direction, it is possible to apply the patch-clamp electrophysiological approach to directly measure inward ENaC-mediated Na^+ flux in native cells (Mironova et al. 2013; Stockand et al. 2012) and in overexpression mammalian systems (Staruschenko et al. 2006). Expression of ENaC subunits in *Xenopus* oocytes is also successfully used to assess ENaC function (Chen et al. 2014; Krappitz et al. 2014; Malik et al. 2005; Zhou et al. 2013). In combination with contemporary molecular genetics and biochemical tools, electrophysiology enables us to identify the function and specific mechanisms of renal ENaC channels' action. This has led to an explosive growth in the investigation and subsequent understanding of many physiological and pathophysiological processes performed by ENaC in the aldosterone-sensitive distal nephron. A more thorough description of ENaC structure, function, and regulation can be found in Chap. 18 of Volume 3.

9.3.3 Physiological Regulation of Na^+ Absorption

The tight regulation of transcellular Na^+ concentrations is so important that multiple mechanisms work in a coordinated manner to control them. The expression and activity of Na^+ channels and transporters are regulated by specific hormones and different extra- and intracellular signaling mechanisms. Recently, significant progress has been made in our understanding of the cellular and molecular mechanisms responsible for Na^+ absorption in the kidney. Multiple laboratories have employed electrophysiological, biochemical, microscopical, molecular, and genetic methods to study ion channels and transporters mediating sodium transport in both normal and pathological conditions. The development and application of various tools to study proteins mediating renal function have revealed some of the intriguing physiological functions of the kidney. Research advances naturally resulted in the cloning of multiple genes that are involved in water and electrolyte transport in different renal tubular segments. Moreover, new knowledge about the function of particular segments of the kidney has accrued as a result of the development and application of gene deletion techniques, both conventional and cell-specific gene knockouts (Lang and Shumilina 2013; Rubera et al. 2009; Wen et al. 2014).

Importantly, mechanisms similar to those controlling Na^+ absorption are implicated in the maintenance of other ions and solutes. Thus, multiple studies have demonstrated that the same mechanisms controlling activation of ENaC results in the coordinate regulation of ROMK K^+ channels (Frindt et al. 2011). Similar findings are reported for integrated control of sodium, chloride, bicarbonate, phosphate, and other ions. Recent papers have also revealed cross talk between ENaC and Na^+/K^+ pump that may play a role in the correlation between Na^+ delivery and reabsorption independently of hormonal influence (Wang et al. 2014).

9.3.4 Tubulo-Glomerular Feedback (TGF) Mechanisms

Changes in sodium reabsorption and glomerular filtration are closely coordinated to avoid fluctuations in urinary excretion. Autoregulation of renal blood flow and thus glomerular filtration rate (GFR) maintains constant delivery of NaCl to the distal nephron, where fine control of sodium and chloride transport by hormones, such as aldosterone and arginine vasopressin (AVP), governs water and electrolyte balance. The TGF mechanism operates at the single nephron level to maintain distal delivery within the narrow limits of the reabsorptive capacity of distal tubules (Braam et al. 1993). The mechanism for autoregulation during an acute increase in blood pressure is a reduction in afferent arteriolar radius that normalizes flow to the glomerulus.

This arteriolar constriction is driven both by a myogenic mechanism and TGF, a response that senses NaCl delivery and transport by the apical NKCC cotransporter in the macula densa (MD) (McDonough et al. 2003). NKCC was proposed to be essential for the sensing step since the intraluminal application of furosemide abolished the TGF-mediated reduction in single-nephron GFR (Franco et al. 1988; Wright and Schnermann 1974). MD is a region formed by about 20 cells in the tubular epithelium between the TAL and DCT, where it establishes contact with its parent glomerulus and afferent arteriole. MD cell volume increases when increments are isosmotic, and shrinks if osmolality increases. The MD cells are different from TAL and DCT epithelial cells both anatomically and functionally. For instance, in contrast to the water-impermeable TAL and DCT segments, the apical membrane of MD cells is permeable to water (Komlosi et al. 2006). Furthermore, MD cells have lower Na^+/K^+ pump activity than the adjacent tubular epithelial cells (Komlosi et al. 2009; Schnermann and Marver 1986).

TGF stabilizes nephron function through negative feedback by establishing an inverse relationship between the tubular NaCl load and the GFR of the same nephron. Although the molecular mechanisms of TGF signaling are still not completely uncovered, it is generally accepted that purinergic signaling via ATP or adenosine is at least partially responsible for the tubular signal-dependent regulation of glomerular hemodynamics (Braam et al. 1993; Komlosi et al. 2009; Peti-Peterdi 2006; Schnermann 2011). The sensor function is also coupled with other intracellular signaling events including changes in Cl^- concentrations, intracellular calcium, pH, membrane depolarization, and cell volume. Furthermore, in addition to purines, the MD produces and releases prostaglandin E_2 and nitric oxide (Komlosi et al. 2009). MD also plays a critical role in the secretion of renin, which is the major component of the renin–angiotensin–aldosterone system (RAAS). Products of the RAAS formed downstream from the renin act to conserve salt by causing glomerular vasoconstriction and increasing reabsorption of salt and water from the proximal and distal nephrons.

In addition to the MD contacts with afferent arteriole, it was previously reported that the superficial nephrons of the renal cortex also come into close proximity to the corresponding afferent arteriole through the CNT (Barajas et al. 1986; Capasso 2007; Dorup et al. 1992). Studies by Ren et al. provided evidence of the functional connection between the CNT and afferent arteriole (Ren et al. 2007). This cross talk was called "connecting tubule glomerular feedback" (CTGF) to differentiate it from TGF. Inhibition of ENaC with amiloride blocked the dilatation induced by CTGF, but inhibition of NCC with hydrochlorothiazide failed to prevent renal afferent arterioles dilatation (Ren et al. 2007). Thus, in addition to its critical role in collecting ducts, ENaC is important for sodium reabsorption in the upstream segment, CNT (Nesterov et al. 2012; Rubera et al. 2003). Recent studies provided some further details about specific mechanisms involved in CTGF (Ren et al. 2013, 2014; Romero and Carretero 2019; Wang et al. 2013). The CTGF response has not been widely studied and this mechanism needs to be more fully explored. However, these studies emphasize the role of the connecting tubules and collecting ducts, where the final control of urinary electrolytes takes place.

9.3.5   Pharmacological Control of Na$^+$ Absorption

9.3.5.1   Loop Diuretics

As suggested earlier, the apical membranes of TAL epithelial cells are hyperpolarized because of high ROMK K$^+$ channel activity. Thus, the transmembrane potential of these apical membranes is strongly dependent on the equilibrium potential for K$^+$ (V_K). In contrast, the basolateral membrane has many Cl$^-$ channels, so the basolateral transmembrane potential is less negative than V_K (i.e., the Cl$^-$ conductance depolarizes the basolateral membrane). As a result of the hyperpolarization of the luminal membrane and depolarization of the basolateral membrane, there exists a transepithelial potential difference of ~10 mV, with the lumen being positive with respect to the interstitial space. The lumen-positive potential provides an important driving force for the paracellular flux of Na$^+$, Ca^{2+}, and Mg^{2+} into the interstitial space. Thus, inhibitors of Na$^+$-K$^+$-2Cl$^-$ cotransport also attenuate Ca^{2+} and Mg^{2+} reabsorption in the TAL by abolishing the transepithelial potential difference. Mutations in genes encoding the Na$^+$-K$^+$-2Cl$^-$ cotransporter, the apical K$^+$ channel, the basolateral Cl$^-$ channel, the Barttin subunit of this Cl$^-$ channel, or the calcium-sensing receptor on the basolateral membrane of these cells can all cause Bartter Syndrome (inherited hypokalemic alkalosis with salt wasting and hypotension; see Table 9.1) (Lifton et al. 2001; Simon et al. 1996a, b, c).

Na$^+$-K$^+$-2Cl$^-$ cotransport inhibitors are chemically diverse. The drugs currently available in this group are: furosemide and bumetanide, which contain a sulfonamide moiety, ethacrynic acid, which is a phenoxyacetic acid derivative, and torsemide, which is a sulfonylurea (Wargo and Banta 2009). These drugs inhibit NKCC2 at the apical membrane of the TAL of Henle's loop, so these diuretics are often called "loop diuretics." Diuretics acting at sites distal to the TAL have lower efficacy because a much smaller percentage of the filtered Na$^+$ reaches these segments. In contrast, NKCC inhibitors are highly efficacious because ~25% of the filtered Na$^+$ normally is reabsorbed by the TAL, and nephron segments distal to the TAL do not possess the reabsorptive capacity to rescue the flood of unreabsorbed salt from the TAL.

9.3.5.2   Thiazide Diuretics

Figure 9.8 illustrates the current model of electrolyte transport in the DCT. As with other nephron segments, transport is powered by a Na$^+$/K$^+$ pump in the basolateral membrane. Energy of the electrochemical gradient for Na$^+$ is harnessed by the NCC symporter in the luminal membrane, which moves Cl$^-$ into the epithelial cell against its electrochemical gradient. Cl$^-$ then exits the basolateral membrane passively via a Cl$^-$ channel. Thiazide diuretics inhibit the Na$^+$-Cl$^-$ symporter. Mutations in the *SLC12A3* gene, encoding NCC cause a form of inherited hypokalemic alkalosis called Gitelman Syndrome (see Table 9.1) (Gamba 2005, 2012).

Since the first inhibitors of the NCC were 1,2,4-benzothiazide-1,1-dioxides, this class of diuretics became known as thiazide diuretics. Later, drugs that are pharmacologically similar to thiazide diuretics, but are not chemically thiazides, were developed and called thiazide-like diuretics. There are many thiazide and thiazide-like diuretics available in the USA, including hydrochlorothiazide, chlorothiazide, metolazone, and chlorthalidone. These drugs (especially hydrochlorothiazide and chlorthalidone) are often the first-line treatments for patients with hypertension. Moreover, because some resistance to the effects of high-dose loop diuretics may occur over time or in patients with chronic kidney disease or heart failure due to compensatory upregulation of DCT NCC expression and function, thiazide diuretics (esp. metolazone or chlorothiazide) are often used in this setting to potentiate diuresis in such cases (Fliser et al. 1994; Jentzer et al. 2010).

9.3.5.3 Amiloride and Its Analogs

Principal cells in the late DCT, CNT, and CD regulate ENaC expression in their apical membranes, which provides a conductive pathway for Na^+ entry into the cell down the electrochemical gradient created by the basolateral Na^+ pump (Staruschenko 2012). The higher permeability of the luminal membrane for Na^+ depolarizes the luminal membrane but not the basolateral membrane, creating a lumen-negative transepithelial potential difference (TEPD). This provides an important driving force for the secretion of K^+ into the lumen via K^+ channels (primarily ROMK and the Ca^{2+}-activated K^+ channel (BK); see Fig. 9.9) in the apical membrane. Loop and thiazide diuretics increase distal Na^+ delivery, a situation that is often associated with increased K^+ and H^+ excretion. The increased luminal Na^+ concentration in the distal nephron induced by this type of diuretics augments depolarization of the apical membrane and thereby enhances the lumen-negative TEPD, which facilitates K^+ excretion. In addition to principal cells, the CD also contains Type A intercalated cells (Al-Awqati 2013) that mediate H^+ secretion into the tubular lumen via the apical vacuolar H^+-ATPase (V-ATPase or proton pump), and this pump is aided by partial depolarization of the apical membrane. However, increased distal Na^+ delivery is not the only mechanism by which diuretics increase K^+ and H^+ excretion. Activation of the renin–angiotensin–aldosterone system by diuretics also contributes to diuretic-induced K^+ and H^+ excretion, as discussed in the next section on mineralocorticoid antagonists.

Considerable evidence indicates that amiloride blocks ENaC in the luminal membrane of principal cells in the late distal tubule and CD (Warnock et al. 2014). Importantly, amiloride blocks ENaC in the low nM concentration range. Used in higher concentrations, it is also able to inhibit other sodium channels and transporters, such as NHE (Kleyman and Cragoe 1988; Teiwes and Toto 2007). Appropriate modification of amiloride has produced analogs that are several 100-fold more active than amiloride against specific transporters

Triamterene and amiloride are the only two drugs of this class in clinical use. Amiloride is a pyrazinoylguanidine derivative, and triamterene is a pteridine

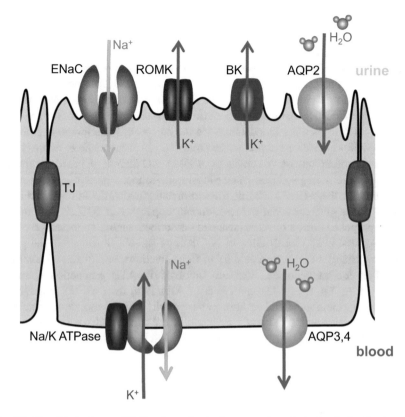

Fig. 9.9 Simplified scheme of Na⁺ reabsorption in the connecting tubules/cortical collecting duct. The apical entry of Na⁺ in the CNT and the CCD is regulated by the epithelial Na⁺ channel (ENaC); water follows the concentration gradient created by sodium transport and is taken up from the urine via AQP2 (aquaporin 2) channels. Basolaterally, sodium is reabsorbed via the Na⁺/K⁺-ATPase, and water is removed by aquaporins 3 and 4 (AQP3 and 4). Secretion of K⁺ into the lumen occurs primarily via ROMK and the Ca²⁺-activated K⁺ channel (BK)

(Tamargo et al. 2014). Thus, both drugs are organic bases and are transported by the organic base secretory mechanism in the proximal tubule (McKinney 1984). Both drugs cause small increases in NaCl excretion and are often employed for their antikaliuretic actions to offset the effects of other diuretics that increase K⁺ excretion and may cause hypokalemia. Consequently, triamterene and amiloride, along with spironolactone (see next section), are classified as potassium-sparing diuretics. These drugs are also often used in the setting of ongoing lithium treatment to limit the potential damage to CD epithelial cells caused by lithium entry via ENaC, which over time may result in nephrogenic diabetes insipidus (Kishore and Ecelbarger 2013; Kortenoeven et al. 2009).

9.3.5.4 Mineralocorticoid Receptors (MR) Antagonists

Mineralocorticoids induce salt and water retention and increase K^+ and H^+ excretion by binding to specific MRs. Currently, two MR antagonists are available in the USA, spironolactone and eplerenone. These MR antagonists are also called K^+-sparing diuretics or aldosterone antagonists. Epithelial cells in the late distal tubule and CD contain cytosolic MRs that have a high affinity for aldosterone. Aldosterone enters the epithelial cell from the bloodstream via the basolateral membrane and binds to MRs. The MR–aldosterone complex then translocates to the nucleus where it binds to specific hormone-responsive elements of DNA and thereby regulates the expression of multiple gene products called aldosterone-induced proteins (AIPs), which notably include the α-ENaC subunit, the serum and glucocorticoid-regulated kinase (SGK1) and the glucocorticoid-induced leucine zipper protein (GILZ). A number of different proposed effects of AIPs have been described, including increased cellular protein expression and localization of Na^+ pumps and Na^+ channels in the plasma membrane, changes in the permeability of tight junctions, and increased activity of enzymes in the mitochondria that are involved in ATP production (Law and Edelman 1978; Yu 2014). The net effect of AIPs is to increase Na^+ conductance of the luminal membrane and sodium pump activity in the basolateral membrane. Consequently, transepithelial NaCl transport is enhanced and the lumen-negative transepithelial voltage is increased. The latter effect increases the driving force for the secretion of K^+ and H^+ into the tubular lumen.

9.3.5.5 SGLT Inhibitors

Sodium-glucose cotransporters SGLT1, and especially SGLT2, have become the major targets for regulation of blood glucose levels in diabetes (De Nicola et al. 2014; Oliva and Bakris 2014; Rieg and Vallon 2018; Vallon et al. 2011, 2014). The classic competitive inhibitor of SGLTs is phlorizin (Ehrenkranz et al. 2005). Phlorizin has a higher affinity for hSGLT2 (K_i 40 nM) than hSGLT1 (K_i 200 nM). Several pharmaceutical companies have attempted to modify the phlorizin structure to further enhance selectivity for SGLT2 over SGLT1 and develop oral SGLT2 inhibitors. Examples of these new FDA-approved SGLT2 inhibitors include dapagliflozin, empagliflozin, and canagliflozin (Rieg and Vallon 2018; Riser Taylor and Harris 2013; Thomas and Cherney 2018). Please see Chap. 6 of Volume 3 for details.

9.4   Sodium Balance and Its Role in Other Organs

9.4.1   Sodium Absorption in the Lung

Alveoli
The alveoli of the lung promote the exchange of oxygen and CO_2 between the air spaces and the blood and contain two main cell types, termed alveolar type I and type II cells. Efficient gas exchange requires the presence of a thin liquid layer on the apical surface of the alveolar epithelium, which is finely regulated by a balance between passive secretion of fluid from the vasculature by a paracellular route governed by permeability of claudins and active reabsorption governed by the presence of apical Na$^+$ channels [reviewed in (Eaton et al. 2009)]. Although more is known about the surfactant-secreting type II cells, both alveolar cell types appear to contain a variety of ion transport proteins, including two forms of amiloride-sensitive channels. One form has the electrophysiological signature of ENaC expressed in the kidney and colon, as it is highly selective for Na$^+$ over other cations and has a low single-channel conductance of ~5 pS and strong sensitivity to amiloride. The other form is relatively nonselective for Na$^+$ over K$^+$, has a higher single-channel conductance (~20 pS) and a weaker sensitivity to amiloride and may contain primarily α-ENaC subunits rather than all three subunits. Na$^+$ entering at the apical membrane exits at the basolateral membrane via the Na$^+$/K$^+$ pump. ENaC in the airways is highly regulated by agonists interacting with G protein-coupled receptors (e.g., purinergic and adrenergic agents), circulating hormones (e.g., glucocorticoids), chemokines and inflammatory mediators (e.g., TNF-α and interleukins), and reactive oxygen and nitrogen species. These regulatory factors may differentially modulate the function of highly selective ENaC channels versus nonselective cation channels (Eaton et al. 2009). Further details about the role of ENaC and other cation channels in the lung are provided in Chap. 18 of Volume 3.

Airways
Under normal conditions in the conducting airways, a balance is maintained between fluid reabsorption and fluid secretion to preserve airway surface liquid (ASL) height, cilia function, and thus normal mucociliary clearance. This balance appears to be regulated by ASL height itself and signaling factors such as secreted nucleotides, although the mechanisms that underlie this absorption-secretion coupling are still under active investigation. Apical membrane ENaC and possibly cyclic nucleotide-gated cation channels contribute to active transcellular Na$^+$ reabsorption, which occurs in concert with basolateral Na$^+$/K$^+$ pump activity and paracellular Cl$^-$ reabsorption (Hollenhorst et al. 2011). This active salt absorption is coupled with passive transcellular water flow mediated by aquaporin water channels (see Fig. 9.10) (Donaldson and Boucher 2003). On the other hand, fluid secretion in the conducting airways occurs via transcellular Cl$^-$ secretion predominantly through apical CFTR channels, but also through Ca^{2+}-activated Cl$^-$ channels like TMEM16A (Rock et al. 2009), coupled with basolateral NKCC-mediated Cl$^-$

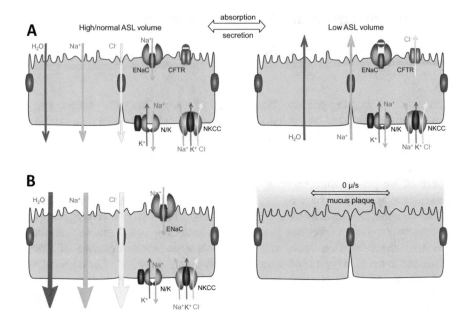

Fig. 9.10 Balance is maintained between airway surface liquid absorption and secretion mediated by transcellular and paracellular Na^+, Cl^- and water transport pathways under normal conditions (**a**), but this balance gets disturbed in patients with cystic fibrosis, who have excessive ENaC-mediated Na^+ (and thus fluid) absorption (**b**). See text for further details

entry, paracellular Na^+ secretion (with ENaC inhibited), and transcellular water flow from the basolateral to the apical compartment (Fig. 9.10).

Disease Correlations

The importance of ENaC in the regulation of alveolar and airway surface liquid volume is evidenced by the fact that loss-of-function ENaC mutations underlie pseudohypoaldosteronism type I, which is associated with neonatal respiratory distress syndrome caused by pulmonary edema (Keszler and Sivasubramanian 1983; Malagon-Rogers 1999). Hypoxia decreases ENaC expression in nongenetic forms of this disorder among premature infants. In addition to inducing increased surfactant production, dexamethasone may improve this condition by counteracting hypoxia-induced decreases in ENaC-mediated sodium reabsorption. Decreased ENaC activity in the lung is also seen in patients susceptible to high-altitude pulmonary edema (Scherrer et al. 1999). A potential therapeutic role for ENaC modulation in acute respiratory distress syndrome has also been explored (Matthay et al. 2005).

Perhaps the most thoroughly studied role for ENaC in pulmonary disease is in patients with cystic fibrosis (CF), caused by mutations in the CFTR Cl^- channel (see Chaps. 15 and 16 of Volume 3). In patients with CF, the normal balance between fluid secretion and reabsorption in the airway gets disrupted due to a loss of CFTR function. As CFTR normally inhibits ENaC activity (by yet unclear mechanisms), a constitutive gain of ENaC function appears to ensue in the airways of patients with CF, which might be a major contributing factor in the pathogenesis of CF, causing low ASL volume, mucous plugging, infections and inflammation and chronic bronchiectasis with lung parenchymal destruction (Fig. 9.10) (Donaldson and Boucher 2007). Of note, a β-ENaC transgenic lung mouse model revealed that overexpression of ENaC appears to be sufficient to recapitulate the CF lung phenotype of decreased airway surface liquid volume and increased airway inflammation in mice (Mall et al. 2004). However, this view that low ASL volume, per se, underlies much of the pathogenesis of CF lung disease has been challenged, and the actual situation that exists in human CF lung disease is still far from settled (Song et al. 2009).

9.4.2 Sodium Absorption in the Gastrointestinal and Endocrine Systems

Small Intestine

There are regional differences in the mechanisms of Na$^+$ transport and salt absorption versus secretion in the gastrointestinal tract. Electroneutral NaCl absorption is predominant in the small intestine and proximal colon and occurs at the apical membrane via parallel Na$^+$/H$^+$ exchange (mediated by NHE2 and NHE3) and Cl$^-$/HCO$_3^-$ exchanger (Kato and Romero 2011). These transport proteins are regulated by intracellular pH and intracellular [HCO$_3^-$] directly, as H$^+$ and HCO$_3^-$ are substrates, and indirectly through pH-dependent allosteric effects. In addition, there is nutrient-coupled Na$^+$ absorption, which occurs via SGLTs for Na$^+$-glucose cotransport, along with several Na$^+$-amino acid cotransporters and Na$^+$-coupled solute carriers. This salt and fluid absorption is balanced by electrogenic Cl$^-$ secretion mediated by apical CFTR, which is stimulated under conditions that increase cellular cAMP, cGMP, or Ca^{2+} levels. These same signaling mediators concomitantly inhibit electroneutral NaCl. Once Na$^+$ and Cl$^-$ are apically absorbed, the combination of a basolateral Cl$^-$ channel (ClC-2), the Na$^+$/K$^+$-ATPase, and the Kir7.1 K$^+$ channel, which allows K$^+$ pumped in by the Na$^+$/K$^+$ pump to recycle back out of the cell, afford the net transfer of NaCl across the basolateral membrane (Kato and Romero 2011).

Functional interactions among these transporters, along with a network of kinase signaling cascades, coordinate the regulation of salt absorption from the small intestine. The endocrine, autonomic, and immune systems can regulate epithelial NaCl transport function via the enteric nervous system and gene expression of the various transport proteins (Kato and Romero 2011). Glucocorticoids stimulate apical membrane NHE3 via SGK1 and stimulate SGLT1 transporters (Grahammer et al. 2006). The mineralocorticoid aldosterone also weakly stimulates electroneutral NaCl absorption in the small intestine.

Colon

In the colon, there is electroneutral NaCl reabsorption under basal conditions; aldosterone, however, is a key stimulator of both electroneutral NaCl absorption in the proximal colon and ENaC-mediated electrogenic Na^+ reabsorption in the distal colon, where it works to enhance Na^+ (salt) reabsorption from these segments of the gut (Harvey et al. 2008). Thus, in the distal colon aldosterone causes a switch from electroneutral NaCl absorption to stimulated electrogenic Na^+ absorption by inducing expression of apical ENaC and basolateral Na^+/K^+-ATPase (Kunzelmann and Mall 2002). In parallel, net K^+ absorption is converted to net K^+ secretion by induction of apical K^+ channels (Sweiry and Binder 1989). ENaC expression at the apical membrane is generally more near the tips of the villi, whereas CFTR Cl^- channel expression is more pronounced at the apical membrane near the crypts of the villi. Like in the small intestine, CFTR is activated during secretion.

Disease Correlations

Whereas enhanced apical Cl^- secretion via CFTR in the small intestine plays a key role in several secretory diarrheal diseases, such as cholera, where cholera toxin-induced cAMP formation promotes PKA-dependent CFTR-mediated fluid secretion, inhibition of sodium and chloride absorption in the colon is important for diarrheal disorders such as ulcerative colitis (Sandle 2005). Reduced expression/activity of apical Na^+ channels and basolateral $Na^{+/}K^+$-ATPase, leading to a loss of electrogenic Na^+ absorption in the distal colon and rectum was also reported (Greig et al. 2004). There is also likely to be a decrease in electroneutral NaCl cotransport, which is present throughout the colon. Preliminary work on basolateral K^+ channel abundance and activity in colonic epithelial cells suggests that whole-cell K^+ conductance is decreased in ulcerative colitis, leading to epithelial cell depolarization, and further limitation of Na^+ absorption. In addition, there is a marked reduction in colonic epithelial resistance, which reflects a decrease in the integrity of intercellular tight junctions and the presence of apoptotic foci (Sandle 2005).

9.5 Sodium Transport in Epithelia and Human Diseases

In the kidney, the primary role of renal Na^+ transport is the control of extracellular fluid volume. Thus, a number of genetic disorders are characterized by either hypo- or hypertension. Many Na^+ transport proteins have been linked to specific genetic disorders in the kidney (Table 9.1). However, mutations in Na^+ channels and transporters may cause disease manifestations in other organs such as lung and colon (described above) and in other tissues. Importantly, Na^+ channels and transporters are highly critical for human diseases not only due to certain mutations identified in those proteins. Mutations in various genes associated with human diseases can trigger abnormal signaling cascades leading to Na^+ transport disturbance as the main endpoint effector. Thus, ion channels and transporters are attractive targets for pharmaceutical interventions even when diseases are caused by mutations in other proteins.

9.6 Final Conclusions

Epithelial Na⁺ reabsorption in the kidneys and other tissues plays the key role in regulating total body salt and water balance, airway function, and fluid and electrolyte reabsorption. In addition to focusing on individual Na⁺ channels and other transporters, future work will be required to provide us with detailed information about the coordinated function of various components of sodium absorption and their tight regulation.

Acknowledgments Research from the authors' laboratories is supported by NIH grants R35 HL135749, P01 HL116264 (to A.S.), and R00 DK105160, PKD Foundation Research grant 221G18a (to D.V.I.), and U.S. Dept. of Defense grants W81XWH-15-1-0420 and W81XWH-15-1-0663 (to K.R.H.).

References

Agre P (2000) Homer W. Smith award lecture Aquaporin water channels in kidney. J Am Soc Nephrol 11:764–777

Al-Awqati Q (2013) Cell biology of the intercalated cell in the kidney. FEBS Lett 587:1911–1914

Allen GG, Barratt LJ (1985) Origin of positive transepithelial potential difference in early distal segments of rat kidney. Kidney Int 27:622–629

Amemiya M, Loffing J, Lotscher M, Kaissling B, Alpern RJ, Moe OW (1995) Expression of NHE-3 in the apical membrane of rat renal proximal tubule and thick ascending limb. Kidney Int 48:1206–1215

Anderson JM, Van Itallie CM (2009) Physiology and function of the tight junction. Cold Spring Harb Perspect Biol 1:a002584

Andreoli TE, Schafer JA, Troutman SL (1978) Perfusion rate-dependence of transepithelial osmosis in isolated proximal convoluted tubules: estimation of the hydraulic conductance. Kidney Int 14:263–269

Ares GR, Caceres PS, Ortiz PA (2011) Molecular regulation of NKCC2 in the thick ascending limb. Am J Physiol Renal Physiol 301:F1143–F1159

Ares GR, Haque MZ, Delpire E, Ortiz PA (2012) Hyperphosphorylation of Na-K-2Cl cotransporter in thick ascending limbs of Dahl salt-sensitive rats. Hypertension 60:1464–1470

Bachmann S, Mutig K (2017) Regulation of renal Na-(K)-Cl cotransporters by vasopressin. Pflugers Arch 469:889–897

Barajas L, Powers K, Carretero O, Scicli AG, Inagami T (1986) Immunocytochemical localization of renin and kallikrein in the rat renal cortex. Kidney Int 29:965–970

Barratt LJ, Rector FC Jr, Kokko JP, Seldin DW (1974) Factors governing the transepithelial potential difference across the proximal tubule of the rat kidney. J Clin Invest 53:454–464

Barratt LJ, Rector FC Jr, Kokko JP, Tisher CC, Seldin DW (1975) Transepithelial potential difference profile of the distal tubule of the rat kidney. Kidney Int 8:368–375

Bennett CM, Brenner BM, Berliner RW (1968) Micropuncture study of nephron function in the rhesus monkey. J Clin Invest 47:203–216

Bhalla V, Hallows KR (2008) Mechanisms of ENaC regulation and clinical implications. J Am Soc Nephrol 19:1845–1854

Biner HL, Arpin-Bott MP, Loffing J, Wang X, Knepper M, Hebert SC, Kaissling B (2002) Human cortical distal nephron: distribution of electrolyte and water transport pathways. J Am Soc Nephrol 13:836–847

Birkenhager R et al (2001) Mutation of BSND causes Bartter syndrome with sensorineural deafness and kidney failure. Nat Genet 29:310–314

Bockenhauer D et al (2009) Epilepsy, ataxia, sensorineural deafness, tubulopathy, and KCNJ10 mutations. N Engl J Med 360:1960–1970

Borensztein P, Froissart M, Laghmani K, Bichara M, Paillard M (1995) RT-PCR analysis of Na+/H + exchanger mRNAs in rat medullary thick ascending limb. Am J Phys 268:F1224–F1228

Boron WF, Sackin H (1983) Measurement of intracellular ionic composition and activities in renal tubules. Annu Rev Physiol 45:483–496

Bostanjoglo M et al (1998) 11Beta-hydroxysteroid dehydrogenase, mineralocorticoid receptor, and thiazide-sensitive Na-Cl cotransporter expression by distal tubules. J Am Soc Nephrol 9:1347–1358

Braam B, Mitchell KD, Koomans HA, Navar LG (1993) Relevance of the tubuloglomerular feedback mechanism in pathophysiology. J Am Soc Nephrol 4:1257–1274

Butterworth MB, Edinger RS, Frizzell RA, Johnson JP (2009) Regulation of the epithelial sodium channel by membrane trafficking. Am J Physiol Renal Physiol 296:F10–F24

Campean V, Kricke J, Ellison D, Luft FC, Bachmann S (2001) Localization of thiazide-sensitive Na$^+$-Cl$^-$ cotransport and associated gene products in mouse DCT. Am J Physiol Renal Physiol 281:F1028–F1035

Canessa CM, Schild L, Buell G, Thorens B, Gautschi I, Horisberger JD, Rossier BC (1994) Amiloride-sensitive epithelial Na$^+$ channel is made of three homologous subunits. Nature 367:463–467

Capasso G (2007) A new cross-talk pathway between the renal tubule and its own glomerulus. Kidney Int 71:1087–1089

Capasso G, Rizzo M, Pica A, Di Maio FS, Moe OW, Alpern RJ, De Santo NG (2002) Bicarbonate reabsorption and NHE-3 expression: abundance and activity are increased in Henle's loop of remnant rats. Kidney Int 62:2126–2135

Carmosino M et al (2011) NKCC2 is activated in Milan hypertensive rats contributing to the maintenance of salt-sensitive hypertension. Pflugers Arch 462:281–291

Castrop H, Schnermann J (2008) Isoforms of renal Na-K-2Cl cotransporter NKCC2: expression and functional significance. Am J Physiol Renal Physiol 295:F859–F866

Chavez-Canales M et al (2013) Insulin increases the functional activity of the renal NaCl cotransporter. J Hypertens 31:303–311

Chen J, Kleyman TR, Sheng S (2014) Deletion of alpha-subunit exon 11 of the epithelial Na+ channel reveals a regulatory module. Am J Physiol Renal Physiol 306:F561–F567

Cheng CJ, Yoon J, Baum M, Huang CL (2015) STE20/SPS1-related proline/alanine-rich kinase (SPAK) is critical for sodium reabsorption in isolated, perfused thick ascending limb. Am J Physiol Renal 308:F437–F443

Chiga M, Rai T, Yang SS, Ohta A, Takizawa T, Sasaki S, Uchida S (2008) Dietary salt regulates the phosphorylation of OSR1/SPAK kinases and the sodium chloride cotransporter through aldosterone. Kidney Int 74:1403–1409

Ciampolillo F, McCoy DE, Green RB, Karlson KH, Dagenais A, Molday RS, Stanton BA (1996) Cell-specific expression of amiloride-sensitive, Na$^+$-conducting ion channels in the kidney. Am J Phys 271:C1303–C1315

Costanzo LS (1984) Comparison of calcium and sodium transport in early and late rat distal tubules: effect of amiloride. Am J Phys 246:F937–F945

Costanzo LS (1985) Localization of diuretic action in microperfused rat distal tubules: Ca and Na transport. Am J Phys 248:F527–F535

Crane RK (1960) Intestinal absorption of sugars. Physiol Rev 40:789–825

Cuevas CA et al (2017) Potassium sensing by renal distal tubules requires Kir4.1. J Am Soc Nephrol 28:1814–1825

De Nicola L, Gabbai FB, Liberti ME, Sagliocca A, Conte G, Minutolo R (2014) Sodium/Glucose cotransporter 2 inhibitors and prevention of diabetic nephropathy: targeting the renal tubule in diabetes. Am J Kidney Dis 64:16–24

Díez-Sampedro A, Eskandari S, Wright EM, Hirayama BA (2001) Na+-to-sugar stoichiometry of SGLT3. Am J Physiol Renal Physiol 280:F278–F282

Díez-Sampedro A et al (2003) A glucose sensor hiding in a family of transporters. Proc Natl Acad Sci U S A 100:11753–11758

Donaldson SH, Boucher RC (2003) Update on pathogenesis of cystic fibrosis lung disease. Curr Opin Pulm Med 9:486–491

Donaldson SH, Boucher RC (2007) Sodium channels and cystic fibrosis. Chest 132:1631–1636

Donowitz M, Li X (2007) Regulatory Binding Partners and Complexes of NHE3. Physiol Rev 87:825–872

Dorup J, Morsing P, Rasch R (1992) Tubule-tubule and tubule-arteriole contacts in rat kidney distal nephrons. A morphologic study based on computer-assisted three-dimensional reconstructions. Lab Investig 67:761–769

Doucet A (1988) Function and control of Na-K-ATPase in single nephron segments of the mammalian kidney. Kidney Int 34:749–760

Eaton DC, Helms MN, Koval M, Bao HF, Jain L (2009) The contribution of epithelial sodium channels to alveolar function in health and disease. Annu Rev Physiol 71:403–423

Ehrenkranz JR, Lewis NG, Kahn CR, Roth J (2005) Phlorizin: a review. Diabetes Metab Res Rev 21:31–38

Eladari D, Chambrey R, Picard N, Hadchouel J (2014) Electroneutral absorption of NaCl by the aldosterone-sensitive distal nephron: implication for normal electrolytes homeostasis and blood pressure regulation. Cell Mol Life Sci 71:2879–2895

Ellison DH, Velazquez H, Wright FS (1987) Thiazide-sensitive sodium chloride cotransport in early distal tubule. Am J Phys 253:F546–F554

Estevez R, Boettger T, Stein V, Birkenhager R, Otto E, Hildebrandt F, Jentsch TJ (2001) Barttin is a Cl⁻ channel beta-subunit crucial for renal Cl⁻ reabsorption and inner ear K⁺ secretion. Nature 414:558–561

Fahlke C, Fischer M (2010) Physiology and pathophysiology of ClC-K/barttin channels. Front Physiol 1:155

Fenton RA, Poulsen SB, de la Mora CS, Soleimani M, Dominguez Rieg JA, Rieg T (2017) Renal tubular NHE3 is required in the maintenance of water and sodium chloride homeostasis. Kidney Int 92:397–414

Feraille E, Doucet A (2001) Sodium-potassium-adenosinetriphosphatase-dependent sodium transport in the kidney: hormonal control. Physiol Rev 81:345–418

Fliser D, Schroter M, Neubeck M, Ritz E (1994) Coadministration of thiazides increases the efficacy of loop diuretics even in patients with advanced renal failure. Kidney Int 46:482–488

Franco M, Bell PD, Navar LG (1988) Evaluation of prostaglandins as mediators of tubuloglomerular feedback. Am J Phys 254:F642–F649

Frindt G, Houde V, Palmer LG (2011) Conservation of Na⁺ versus K⁺ by the rat cortical collecting duct. Am J Physiol Renal Physiol 301:F14–F20

Fushimi K, Uchida S, Hara Y, Hirata Y, Marumo F, Sasaki S (1993) Cloning and expression of apical membrane water channel of rat kidney collecting tubule. Nature 361:549–552

Gailly P et al (2014) P2Y receptor activation inhibits the expression of the sodium-chloride cotransporter NCC in distal convoluted tubule cells. Pflugers Arch 466(11):2035–2047

Gamba G (2001) Alternative splicing and diversity of renal transporters. Am J Physiol Renal Physiol 281:F781–F794

Gamba G (2005) Molecular physiology and pathophysiology of electroneutral cation-chloride cotransporters. Physiol Rev 85:423–493

Gamba G (2012) Regulation of the renal Na+-Cl- cotransporter by phosphorylation and ubiquitylation. Am J Physiol Renal Physiol 303:F1573–F1583

Gamba G, Miyanoshita A, Lombardi M, Lytton J, Lee WS, Hediger MA, Hebert SC (1994) Molecular cloning, primary structure, and characterization of two members of the mammalian electroneutral sodium-(potassium)-chloride cotransporter family expressed in kidney. J Biol Chem 269:17713–17722

Giebisch G (2001) Renal potassium channels: function, regulation, and structure. Kidney Int 60:436–445

Giebisch G, Klose RM, Malnic G, Sullivan WJ, Windhager EE (1964) Sodium movement across single perfused proximal tubules of rat kidneys. J Gen Physiol 47:1175–1194

Goldman DE (1943) Potential, impedance, and rectification in membranes. J Gen Physiol 27:37–60

Gonzalez-Vicente A, Saez F, Monzon CM, Asirwatham J, Garvin JL (2019) Thick ascending limb sodium transport in the pathogenesis of hypertension. Physiol Rev 99:235–309

Grahammer F et al (2006) Intestinal function of gene-targeted mice lacking serum- and glucocorticoid-inducible kinase 1. Am J Physiol Gastrointest Liver Physiol 290:G1114–G1123

Greger R (1981) Chloride reabsorption in the rabbit cortical thick ascending limb of the loop of Henle. A sodium dependent process. Pflugers Arch 390:38–43

Greig ER, Boot-Handford RP, Mani V, Sandle GI (2004) Decreased expression of apical Na$^+$ channels and basolateral Na$^+$, K$^+$-ATPase in ulcerative colitis. J Pathol 204:84-92.

Guggino SE, Guggino WB, Green N, Sacktor B (1987) Ca^{2+}-activated K$^+$ channels in cultured medullary thick ascending limb cells. Am J Phys 252:C121–C127

Halperin ML, Kamel KS, Oh MS (2008) Mechanisms to concentrate the urine: an opinion. Curr Opin Nephrol Hypertens 17:416–422

Hamilton KL (2011) Ussing's "little chamber": 60 years+ old and counting. Front Physiol 2:6

Hamilton KL, Devor DC (2012) Basolateral membrane K+ channels in renal epithelial cells. Am J Physiol Renal Physiol 302:F1069–F1081

Hanukoglu I, Hanukoglu A (2016) Epithelial sodium channel (ENaC) family: phylogeny, structure-function, tissue distribution, and associated inherited diseases. Gene 579:95–132

Haque MZ, Ares GR, Caceres PS, Ortiz PA (2011) High salt differentially regulates surface NKCC2 expression in thick ascending limbs of Dahl salt-sensitive and salt-resistant rats. Am J Physiol Renal Physiol 300:F1096–F1104

Harvey BJ, Alzamora R, Stubbs AK, Irnaten M, McEneaney V, Thomas W (2008) Rapid responses to aldosterone in the kidney and colon. J Steroid Biochem Mol Biol 108:310–317

Hayslett JP, Boulpaep EL, Kashgarian M, Giebisch GH (1977) Electrical characteristics of the mammalian distal tubule: comparison of Ling-Gerard and macroelectrodes. Kidney Int 12:324–331

Hebert SC (1986) Hypertonic cell volume regulation in mouse thick limbs. II. Na+-H+ and Cl(-)-HCO3- exchange in basolateral membranes. Am J Phys 250:C920–C931

Hebert SC, Desir G, Giebisch G, Wang W (2005) Molecular diversity and regulation of renal potassium channels. Physiol Rev 85:319–371

Hediger MA, Coady MJ, Ikeda TS, Wright EM (1987) Expression cloning and cDNA sequencing of the Na+/glucose co-transporter. Nature 330:379–381

Hierholzer K, Wiederholt M (1976) Some aspects of distal tubular solute and water transport. Kidney Int 9:198–213

Hodgkin AL, Katz B (1949) The effect of sodium ions on the electrical activity of giant axon of the squid. J Physiol 108:37–77

Hollenhorst MI, Richter K, Fronius M (2011) Ion transport by pulmonary epithelia. J Biomed Biotechnol 2011:174306

Hoy WE, Hughson MD, Bertram JF, Douglas-Denton R, Amann K (2005) Nephron number, hypertension, renal disease, and renal failure. J Am Soc Nephrol 16:2557–2564

Hummel CS, Lu C, Loo DD, Hirayama BA, Voss AA, Wright EM (2011) Glucose transport by human renal Na+/D-glucose cotransporters SGLT1 and SGLT2. Am J Physol Cell Physiol 300: C14–C21

Igarashi P, Vanden Heuvel GB, Payne JA, Forbush B 3rd (1995) Cloning, embryonic expression, and alternative splicing of a murine kidney-specific Na-K-Cl cotransporter. Am J Phys 269: F405–F418

Imai M, Kokko JP (1974) Sodium chloride, urea, and water transport in the thin ascending limb of Henle. Generation of osmotic gradients by passive diffusion of solutes. J Clin Invest 53:393–402

Imbrici P, Liantonio A, Gradogna A, Pusch M, Camerino DC (2014) Targeting kidney CLC-K channels: pharmacological profile in a human cell line versus Xenopus oocytes. Biochim Biophys Acta 1838:2484–2491

Isozaki T, Lea JP, Tumlin JA, Sands JM (1994) Sodium-dependent net urea transport in rat initial inner medullary collecting ducts. J Clin Invest 94:1513–1517

Jacobson HR (1982) Transport characteristics of in vitro perfused proximal convoluted tubules. Kidney Int 22:425–433

Jentzer JC, DeWald TA, Hernandez AF (2010) Combination of loop diuretics with thiazide-type diuretics in heart failure. J Am Coll Cardiol 56:1527–1534

Kato A, Romero MF (2011) Regulation of electroneutral NaCl absorption by the small intestine. Annu Rev Physiol 73:261–281

Keszler M, Sivasubramanian KN (1983) Pseudohypoaldosteronism. Am J Dis Child 137:738–740

Khuri RN, Strieder N, Wiederholt M, Giebisch G (1975) Effects of graded solute diuresis on renal tubular sodium transport in the rat. Am J Phys 228:1262–1268

Kishore BK, Ecelbarger CM (2013) Lithium: a versatile tool for understanding renal physiology. Am J Physiol Renal Physiol 304(9):F1139–F1149

Kitada K et al (2017) High salt intake reprioritizes osmolyte and energy metabolism for body fluid conservation. J Clin Invest 127:1944–1959

Kiuchi-Saishin Y, Gotoh S, Furuse M, Takasuga A, Tano Y, Tsukita S (2002) Differential expression patterns of claudins, tight junction membrane proteins, in mouse nephron segments. J Am Soc Nephrol 13:875–886

Klein JD, Sands JM (2016) Urea transport and clinical potential of urearetics. Curr Opin Nephrol Hypertens 25:444–451

Klein JD, Blount MA, Sands JM (2011) Urea transport in the kidney. Compr Physiol 1:699–729

Klein JD, Blount MA, Sands JM (2012) Molecular mechanisms of urea transport in health and disease. Pflugers Arch 464:561–572

Kleyman TR, Cragoe EJ Jr (1988) Amiloride and its analogs as tools in the study of ion transport. J Membr Biol 105:1–21

Koefoed-Johnsen V, Ussing HH (1958) The nature of the frog skin potential. Acta Physiol Scand 42:298–308

Koenig B, Ricapito S, Kinne R (1983) Chloride transport in the thick ascending limb of Henle's loop: potassium dependence and stoichiometry of the NaCl cotransport system in plasma membrane vesicles. Pflugers Arch 399:173–179

Kokko JP (1970) Sodium chloride and water transport in the descending limb of Henle. J Clin Invest 49:1838–1846

Kokko JP, Rector FC Jr (1972) Countercurrent multiplication system without active transport in inner medulla. Kidney Int 2:214–223

Komlosi P, Fintha A, Bell PD (2006) Unraveling the relationship between macula densa cell volume and luminal solute concentration/osmolality. Kidney Int 70:865–871

Komlosi P, Bell PD, Zhang ZR (2009) Tubuloglomerular feedback mechanisms in nephron segments beyond the macula densa. Curr Opin Nephrol Hypertens 18:57–62

Kortenoeven ML, Li Y, Shaw S, Gaeggeler HP, Rossier BC, Wetzels JF, Deen PM (2009) Amiloride blocks lithium entry through the sodium channel thereby attenuating the resultant nephrogenic diabetes insipidus. Kidney Int 76:44–53

Kotchen TA, Cowley AW Jr, Frohlich ED (2013) Salt in health and disease-a delicate balance. N Engl J Med 368:1229–1237

Krappitz M, Korbmacher C, Haerteis S (2014) Demonstration of proteolytic activation of the epithelial sodium channel (ENaC) by combining current measurements with detection of cleavage fragments. J Vis Exp 89:51582

Kunchaparty S et al (1999) Defective processing and expression of thiazide-sensitive Na-Cl cotransporter as a cause of Gitelman's syndrome. Am J Phys 277:F643–F649

Kunzelmann K, Mall M (2002) Electrolyte transport in the mammalian colon: mechanisms and implications for disease. Physiol Rev 82:245–289

Laghmani K et al (1997) Chronic metabolic acidosis enhances NHE-3 protein abundance and transport activity in the rat thick ascending limb by increasing NHE-3 mRNA. J Clin Invest 99:24–30

Lagnaz D et al (2014) WNK3 abrogates the NEDD4-2-mediated inhibition of the renal Na^+:Cl^- Cotransporter. Am J Physiol Renal Physiol 307(3):F275–F286

Lang F, Shumilina E (2013) Regulation of ion channels by the serum- and glucocorticoid-inducible kinase SGK1. FASEB J 27:3–12

Law PY, Edelman IS (1978) Induction of citrate synthase by aldosterone in the rat kidney. J Membr Biol 41:41–64

Lee DH et al (2013) Effects of ACE inhibition and ANG II stimulation on renal Na-Cl cotransporter distribution, phosphorylation, and membrane complex properties. Am J Physiol Cell Physiol 304:C147–C163

Levick JR, Michel CC (2010) Microvascular fluid exchange and the revised Starling principle. Cardiovasc Res 87:198–210

Li XC et al (2018) Proximal Tubule-Specific Deletion of the NHE3 (Na+/H+ Exchanger 3) Promotes the Pressure-Natriuresis Response and Lowers Blood Pressure in Mice. Hypertension 72:1328–1336

Lifton RP, Gharavi AG, Geller DS (2001) Molecular mechanisms of human hypertension. Cell 104:545–556

Loffing J, Korbmacher C (2009) Regulated sodium transport in the renal connecting tubule (CNT) via the epithelial sodium channel (ENaC). Pflugers Arch 458:111–135

Loffing J et al (2000) Differential subcellular localization of ENaC subunits in mouse kidney in response to high- and low-Na diets. Am J Physiol Renal Physiol 279:F252–F258

Loffing J et al (2001) Distribution of transcellular calcium and sodium transport pathways along mouse distal nephron. Am J Physiol Renal Physiol 281:F1021–F1027

Loo DDF, Hirayama BA, Meinild A-K, Chandy G, Zeuthen T, Wright EM (1999) Passive water and ion transport by cotransporters. J Physiol 518:195–202

Lourdel S et al (2002) An inward rectifier K(+) channel at the basolateral membrane of the mouse distal convoluted tubule: similarities with Kir4-Kir5.1 heteromeric channels. J Physiol 538:391–404

Lourdel S, Paulais M, Marvao P, Nissant A, Teulon J (2003) A chloride channel at the basolateral membrane of the distal-convoluted tubule: a candidate ClC-K channel. J Gen Physiol 121:287–300

Malagon-Rogers M (1999) A patient with pseudohypoaldosteronism type 1 and respiratory distress syndrome. Pediatr Nephrol 13:484–486

Malik B, Yue Q, Yue G, Chen XJ, Price SR, Mitch WE, Eaton DC (2005) Role of Nedd4-2 and polyubiquitination in epithelial sodium channel degradation in untransfected renal A6 cells expressing endogenous ENaC subunits. Am J Physiol Renal Physiol 289:F107–F116

Mall M, Grubb BR, Harkema JR, O'Neal WK, Boucher RC (2004) Increased airway epithelial Na+ absorption produces cystic fibrosis-like lung disease in mice. Nat Med 10:487–493

Markadieu N, Delpire E (2014) Physiology and pathophysiology of SLC12A1/2 transporters. Pflugers Arch 466:91–105

Mastroianni N et al (1996) Molecular cloning, expression pattern, and chromosomal localization of the human Na-Cl thiazide-sensitive cotransporter (SLC12A3). Genomics 35:486–493

Matsumura Y et al (1999) Overt nephrogenic diabetes insipidus in mice lacking the CLC-K1 chloride channel. Nat Genet 21:95–98

Matthay MA, Robriquet L, Fang X (2005) Alveolar epithelium: role in lung fluid balance and acute lung injury. Proc Am Thorac Soc 2:206–213

McDonough AA, Leong PKK, Yang LE (2003) Mechanisms of Pressure Natriuresis. Ann N Y Acad Sci 986:669–677

McDonough AA, Veiras LC, Guevara CA, Ralph DL (2017) Cardiovascular benefits associated with higher dietary K^+ vs. lower dietary Na^+: evidence from population and mechanistic studies. Am J Physiol Endocrinol Metab 312:E348–E356

McKinney TD (1984) Further studies of organic base secretion by rabbit proximal tubules. Am J Phys 246:F282–F289

Mironova E, Bugay V, Pochynyuk O, Staruschenko A, Stockand JD (2013) Recording ion channels in isolated, split-opened tubules. Methods Mol Biol 998:341–353

Mistry AC et al (2016) The sodium chloride cotransporter (NCC) and epithelial sodium channel (ENaC) associate. Biochem J 473:3237–3252

Moes AD, van der Lubbe N, Zietse R, Loffing J, Hoorn EJ (2014) The sodium chloride cotransporter SLC12A3: new roles in sodium, potassium, and blood pressure regulation. Pflugers Arch 466:107–118

Monette MY, Rinehart J, Lifton RP, Forbush B (2011) Rare mutations in the human Na-K-Cl cotransporter (NKCC2) associated with lower blood pressure exhibit impaired processing and transport function. Am J Physiol Renal Physiol 300:F840–F847

Monroy A, Plata C, Hebert SC, Gamba G (2000) Characterization of the thiazide-sensitive Na(+)-Cl (-) cotransporter: a new model for ions and diuretics interaction. Am J Physiol Renal Physiol 279:F161–F169

Mount DB et al (1999) Isoforms of the Na-K-2Cl cotransporter in murine TAL I. Molecular characterization and intrarenal localization. Am J Phys 276:F347–F358

Mutchler SM, Kleyman TR (2019) New insights regarding epithelial Na+ channel regulation and its role in the kidney, immune system and vasculature. Curr Opin Nephrol Hypertens 28:113–119

Mutig K et al (2010) Short-term stimulation of the thiazide-sensitive Na+-Cl- cotransporter by vasopressin involves phosphorylation and membrane translocation. Am J Physiol Renal Physiol 298:F502–F509

Nadkarni GN et al (2017) Acute kidney injury in patients on SGLT2 inhibitors: a propensity-matched analysis. Diabetes Care 40:1479–1485

Nawata CM, Pannabecker TL (2018) Mammalian urine concentration: a review of renal medullary architecture and membrane transporters. J Comp Physiol B 188:899–918

Nespoux J, Vallon V (2018) SGLT2 inhibition and kidney protection. Clin Sci 132:1329–1339

Nesterov V, Dahlmann A, Krueger B, Bertog M, Loffing J, Korbmacher C (2012) Aldosterone-dependent and -independent regulation of the epithelial sodium channel (ENaC) in mouse distal nephron. Am J Physiol Renal Physiol 303:F1289–F1299

Nielsen S, Smith BL, Christensen EI, Knepper MA, Agre P (1993) CHIP28 water channels are localized in constitutively water-permeable segments of the nephron. J Cell Biol 120:371–383

Nielsen S, Frokiaer J, Marples D, Kwon TH, Agre P, Knepper MA (2002) Aquaporins in the kidney: from molecules to medicine. Physiol Rev 82:205–244

Noreng S, Bharadwaj A, Posert R, Yoshioka C, Baconguis I (2018) Structure of the human epithelial sodium channel by cryo-electron microscopy. Elife 7:e39340

Olinger E, Houillier P, Devuyst O (2018) Claudins: a tale of interactions in the thick ascending limb. Kidney Int 93:535–537

Oliva RV, Bakris GL (2014) Blood pressure effects of sodium-glucose co-transport 2 (SGLT2) inhibitors. J Am Soc Hypertens 8:330–339

Ookata K, Tojo A, Suzuki Y, Nakamura N, Kimura K, Wilcox CS, Hirose S (2000) Localization of inward rectifier potassium channel Kir7.1 in the basolateral membrane of distal nephron and collecting duct. J Am Soc Nephrol 11:1987–1994

Ostrosky-Frid M, Castañeda-Bueno M, Gamba G (2019) Regulation of the renal NaCl cotransporter by the WNK/SPAK pathway: lessons learned from genetically altered animals. Am J Physiol Renal Physiol 316:F146–F158

Palmer LG, Andersen OS (2008) The two-membrane model of epithelial transport: Koefoed-Johnsen and Ussing (1958). J Gen Physiol 132:607–612

Palygin O et al (2017a) Essential role of $K_{ir}5.1$ channels in renal salt handling and blood pressure control. JCI. Insight 2:e92331

Palygin O, Pochynyuk O, Staruschenko A (2017b) Role and mechanisms of regulation of the basolateral $K_{ir}4.1/K_{ir}5.1$ K^+ channels in the distal tubules. Acta Physiol 219:260–273

Parker MD, Boron WF (2013) The divergence, actions, roles, and relatives of sodium-coupled bicarbonate transporters. Physiol Rev 93:803–959

Paulais M et al (2011) Renal phenotype in mice lacking the Kir5.1 (Kcnj16) K+ channel subunit contrasts with that observed in SeSAME/EAST syndrome. Proc Natl Acad Sci U S A 108:10361–10366

Pavlov TS, Staruschenko A (2017) Involvement of ENaC in the development of salt-sensitive hypertension. Am J Physiol Renal Physiol 313:F135–F140

Payne JA, Forbush B 3rd (1994) Alternatively spliced isoforms of the putative renal Na-K-Cl cotransporter are differentially distributed within the rabbit kidney. Proc Natl Acad Sci U S A 91:4544–4548

Pei L et al (2016) Paracellular epithelial sodium transport maximizes energy efficiency in the kidney. J Clin Invest 126:2509–2518

Peti-Peterdi J (2006) Calcium wave of tubuloglomerular feedback. Am J Physiol Renal Physiol 291:F473–F480

Plain A, Alexander RT (2018) Claudins and nephrolithiasis. Curr Opin Nephrol Hypertens 27:268–276

Plata C, Meade P, Hall A, Welch RC, Vazquez N, Hebert SC, Gamba G (2001) Alternatively spliced isoform of apical Na^+-K^+-Cl^- cotransporter gene encodes a furosemide-sensitive Na^+-Cl^-cotransporter. Am J Physiol Renal Physiol 280:F574–F582

Plata C, Meade P, Vazquez N, Hebert SC, Gamba G (2002) Functional properties of the apical Na^+-K^+-$2Cl^-$—cotransporter isoforms. J Biol Chem 277:11004–11012

Rakova N et al (2017) Increased salt consumption induces body water conservation and decreases fluid intake. J Clin Invest 127:1932–1943

Reeves WB, Winters CJ, Andreoli TE (2001) Chloride channels in the loop of Henle. Annu Rev Physiol 63:631–645

Reichold M et al (2010) KCNJ10 gene mutations causing EAST syndrome (epilepsy, ataxia, sensorineural deafness, and tubulopathy) disrupt channel function. Proc Natl Acad Sci U S A 107:14490–14495

Reilly RF, Ellison DH (2000) Mammalian distal tubule: physiology, pathophysiology, and molecular anatomy. Physiol Rev 80:277–313

Ren Y, Garvin JL, Liu R, Carretero OA (2007) Crosstalk between the connecting tubule and the afferent arteriole regulates renal microcirculation. Kidney Int 71:1116–1121

Ren Y, D'Ambrosio MA, Garvin JL, Wang H, Carretero OA (2013) Prostaglandin E2 mediates connecting tubule glomerular feedback. Hypertension 62:1123–1128

Ren Y et al (2014) Aldosterone sensitizes connecting tubule glomerular feedback via the aldosterone receptor GPR30. Am J Physiol Renal Physiol 307(4):F427–F434

Renfro JL (1975) Water and ion transport by the urinary bladder of the teleost *Pseudopleuronectes americanus*. Am J Phys 228:52–61

Renfro JL (1977) Interdependence of Active Na+ and Cl- transport by the isolated urinary bladder of the teleost, Pseudopleuronectes americanus. J Exp Zool 199:383–390

Richards J, Ko B, All S, Cheng KY, Hoover RS, Gumz ML (2014) A role for the circadian clock protein Per1 in the regulation of the NaCl co-transporter (NCC) and the with-no-lysine kinase (WNK) cascade in mouse distal convoluted tubule cells. J Biol Chem 289:11791–11806

Rieg T, Vallon V (2018) Development of SGLT1 and SGLT2 inhibitors. Diabetologia 61:2079–2086

Rieg T, Tang T, Uchida S, Hammond HK, Fenton RA, Vallon V (2013) Adenylyl cyclase 6 enhances NKCC2 expression and mediates vasopressin-induced phosphorylation of NKCC2 and NCC. Am J Pathol 182:96–106

Riser Taylor S, Harris KB (2013) The clinical efficacy and safety of sodium glucose cotransporter-2 inhibitors in adults with type 2 diabetes mellitus. Pharmacotherapy 33:984–999

Rock JR, O'Neal WK, Gabriel SE, Randell SH, Harfe BD, Boucher RC, Grubb BR (2009) Transmembrane protein 16A (TMEM16A) is a Ca2+-regulated Cl- secretory channel in mouse airways. J Biol Chem 284:14875–14880

Romero CA, Carretero OA (2019) A novel mechanism of renal microcirculation regulation: connecting tubule-glomerular feedback. Curr Hypertens Rep 21:8

Rossier BC (2014) Epithelial sodium channel (ENaC) and the control of blood pressure. Curr Opin Pharmacol 15:33–46

Rubera I et al (2003) Collecting duct-specific gene inactivation of {alpha}ENaC in the mouse kidney does not impair sodium and potassium balance. J Clin Invest 112:554–565

Rubera I, Hummler E, Beermann F (2009) Transgenic mice and their impact on kidney research. Pflugers Arch 458:211–222

Russell JM (2000) Sodium-potassium-chloride cotransport. Physiol Rev 80:211–276

Sandle GI (2005) Pathogenesis of diarrhea in ulcerative colitis: new views on an old problem. J Clin Gastroenterol 39:S49–S52

Sands JM, Martial S, Isozaki T (1996) Active urea transport in the rat inner medullary collecting duct: functional characterization and initial expression cloning. Kidney Int 49:1611–1614

Saritas T et al (2013) SPAK differentially mediates vasopressin effects on sodium cotransporters. J Am Soc Nephrol 24:407–418

Schafer JA, Patlak CS, Andreoli TE (1975) A component of fluid absorption linked to passive ion flows in the superficial pars recta. J Gen Physiol 66:445–471

Scherrer U, Sartori C, Lepori M, Allemann Y, Duplain H, Trueb L, Nicod P (1999) High-altitude pulmonary edema: from exaggerated pulmonary hypertension to a defect in transepithelial sodium transport. Adv Exp Med Biol 474:93–107

Schmitt R et al (1999) Developmental expression of sodium entry pathways in rat nephron. Am J Phys 276:F367–F381

Schnermann J (2011) Maintained tubuloglomerular feedback responses during acute inhibition of P2 purinergic receptors in mice. Am J Physiol Renal Physiol 300:F339–F344

Schnermann J, Marver D (1986) ATPase activity in macula densa cells of the rabbit kidney. Pflugers Arch 407:82–86

Schnermann J, Briggs J, Schubert G (1982) In situ studies of the distal convoluted tubule in the rat. I. Evidence for NaCl secretion. Am J Phys 243:F160–F166

Shimizu T, Yoshitomi K, Taniguchi J, Imai M (1989) Effect of high NaCl intake on Na+ and K+ transport in the rabbit distal convoluted tubule. Pflugers Arch 414:500–508

Simon DB, Karet FE, Hamdan JM, DiPietro A, Sanjad SA, Lifton RP (1996a) Bartter's syndrome, hypokalaemic alkalosis with hypercalciuria, is caused by mutations in the Na-K-2Cl cotransporter NKCC2. Nat Genet 13:183–188

Simon DB et al (1996b) Genetic heterogeneity of Bartter's syndrome revealed by mutations in the K + channel, ROMK. Nat Genet 14:152–156

Simon DB et al (1996c) Gitelman's variant of Bartter's syndrome, inherited hypokalaemic alkalosis, is caused by mutations in the thiazide-sensitive Na-Cl cotransporter. Nat Genet 12:24–30

Skelton LA, Boron WF, Zhou Y (2010) Acid-base transport by the renal proximal tubule. J Nephrol 23(Suppl 16):S4–S18

Soleimani M (2013) SLC26 Cl-/HCO3- exchangers in the kidney: roles in health and disease. Kidney Int 84:657–666

Song Y, Namkung W, Nielson DW, Lee JW, Finkbeiner WE, Verkman AS (2009) Airway surface liquid depth measured in ex vivo fragments of pig and human trachea: dependence on Na+ and Cl- channel function. Am J Physiol Lung Cell Mol Physiol 297:L1131–L1140

Sotak M, Marks J, Unwin RJ (2017) Putative tissue location and function of the SLC5 family member SGLT3. Exp Physiol 102:5–13

Soundararajan R, Pearce D, Hughey RP, Kleyman TR (2010) Role of epithelial sodium channels and their regulators in hypertension. J Biol Chem 285:30363–30369

Starling EH (1896) On the absorption of fluids from the connective tissue spaces. J Physiol 19:312–326

Staruschenko A (2012) Regulation of transport in the connecting tubule and cortical collecting duct. Compr Physiol 2:1541–1584

Staruschenko A (2018) Beneficial effects of high potassium: contribution of renal basolateral K$^+$ channels. Hypertension 71:1015–1022

Staruschenko A, Adams E, Booth RE, Stockand JD (2005) Epithelial Na+ channel subunit stoichiometry. Biophys J 88:3966–3975

Staruschenko A, Booth RE, Pochynyuk O, Stockand JD, Tong Q (2006) Functional reconstitution of the human epithelial Na+ channel in a mammalian expression system. Methods Mol Biol 337:3–13

Stockand JD, Vallon V, Ortiz P (2012) In vivo and ex vivo analysis of tubule function. Compr Physiol 2:2495–2525

Stokes JB (1984) Sodium chloride absorption by the urinary bladder of the winter flounder. A thiazide-sensitive, electrically neutral transport system. J Clin Invest 74:7–16

Studer RA, Person E, Robinson-Rechavi M, Rossier BC (2011) Evolution of the epithelial sodium channel and the sodium pump as limiting factors of aldosterone action on sodium transport. Physiol Genomics 43:844–854

Su X-T, Ellison DH, Wang W-H (2019) Kir4.1/5.1 in the DCT plays a role in the regulation of renal K+ excretion. Am J Physiol Renal Physiol 316(3):F582–F586. https://doi.org/10.1152/ajprenal. 00412.2018

Subramanya AR, Reeves WB, Hallows KR (2012) Tubular sodium transport. In: Falk RJ, Schrier RW, Coffman TM, Molitoris BA (eds) Schrier's diseases of the kidney. Wolters Kluwer, Philadelphia, pp 159–193

Sugano K et al (2010) Coexistence of passive and carrier-mediated processes in drug transport. Nat Rev Drug Discov 9:597–614

Sweiry JH, Binder HJ (1989) Characterization of aldosterone-induced potassium secretion in rat distal colon. J Clin Invest 83:844–851

Takahashi N, Chernavvsky DR, Gomez RA, Igarashi P, Gitelman HJ, Smithies O (2000) Uncompensated polyuria in a mouse model of Bartter's syndrome. Proc Natl Acad Sci U S A 97:5434–5439

Tamargo J, Solini A, Ruilope LM (2014) Comparison of agents that affect aldosterone action. Semin Nephrol 34:285–306

Teiwes J, Toto RD (2007) Epithelial sodium channel inhibition in cardiovascular disease. A potential role for amiloride. Am J Hypertens 20:109–117

Terker AS et al (2014) Sympathetic stimulation of thiazide-sensitive sodium chloride cotransport in the generation of salt-sensitive hypertension. Hypertension 64:178–184

Terker AS et al (2015) Potassium modulates electrolyte balance and blood pressure through effects on distal cell voltage and chloride. Cell Metab 21:39–50

Thomas MC, Cherney DZI (2018) The actions of SGLT2 inhibitors on metabolism, renal function and blood pressure. Diabetologia 61:2098–2107

Turk E, Kerner CJ, Lostao MP, Wright EM (1996) Membrane topology of the human Na+/glucose cotransporter SGLT1. J Biol Chem 271:1925–1934

Uchida S, Sasaki S, Furukawa T, Hiraoka M, Imai T, Hirata Y, Marumo F (1993) Molecular cloning of a chloride channel that is regulated by dehydration and expressed predominantly in kidney medulla. J Biol Chem 268:3821–3824

Ullrich KJ, Schmidt-Nielson B, O'Dell R, Pehling G, Gottschalk CW, Lassiter WE, Mylle M (1963) Micropuncture study of composition of proximal and distal tubular fluid in rat kidney. Am J Phys 204:527–531

Umbach JA, Coady MJ, Wright EM (1990) Intestinal Na+/glucose cotransporter expressed in Xenopus oocytes is electrogenic. Biophys J 57:1217–1224

Vallon V, Schroth J, Lang F, Kuhl D, Uchida S (2009) Expression and phosphorylation of the Na+-Cl- cotransporter NCC in vivo is regulated by dietary salt, potassium, and SGK1. Am J Physiol Renal Physiol 297:F704–F712

Vallon V et al (2011) SGLT2 mediates glucose reabsorption in the early proximal tubule. J Am Soc Nephrol 22:104–112

Vallon V et al (2014) SGLT2 inhibitor empagliflozin reduces renal growth and albuminuria in proportion to hyperglycemia and prevents glomerular hyperfiltration in diabetic Akita mice. Am J Physiol Renal Physiol 306:F194–F204

Velazquez H, Good DW, Wright FS (1984) Mutual dependence of sodium and chloride absorption by renal distal tubule. Am J Phys 247:F904–F911

Velazquez H, Ellison DH, Wright FS (1987) Chloride-dependent potassium secretion in early and late renal distal tubules. Am J Phys 253:F555–F562

Velazquez H, Silva T, Andujar E, Desir GV, Ellison DH, Greger R (2001) The distal convoluted tubule of rabbit kidney does not express a functional sodium channel. Am J Physiol Renal Physiol 280:F530–F539

Verkman AS (2006) Roles of aquaporins in kidney revealed by transgenic mice. Semin Nephrol 26:200–208

Wang WH (1994) Two types of K+ channel in thick ascending limb of rat kidney. Am J Phys 267:F599–F605

Wang T (2012) Renal outer medullary potassium channel knockout models reveal thick ascending limb function and dysfunction. Clin Exp Nephrol 16:49–54

Wang WH, Giebisch G (2009) Regulation of potassium (K) handling in the renal collecting duct. Pflugers Arch 458:157–168

Wang WH, White S, Geibel J, Giebisch G (1990) A potassium channel in the apical membrane of rabbit thick ascending limb of Henle's loop. Am J Phys 258:F244–F253

Wang WH, Yue P, Sun P, Lin DH (2010) Regulation and function of potassium channels in aldosterone-sensitive distal nephron. Curr Opin Nephrol Hypertens 19:463–470

Wang H, D'Ambrosio MA, Garvin JL, Ren Y, Carretero OA (2013) Connecting tubule glomerular feedback in hypertension. Hypertension 62:738–745

Wang Y-B et al (2014) Sodium transport is modulated by p38 kinase–dependent cross-talk between ENaC and Na,K-ATPase in collecting duct principal cells. J Am Soc Nephrol 25:250–259

Wang MX et al (2018) Potassium intake modulates the thiazide-sensitive sodium-chloride cotransporter (NCC) activity via the Kir4.1 potassium channel. Kidney Int 93:893–902

Wargo KA, Banta WM (2009) A comprehensive review of the loop diuretics: should furosemide be first line? Ann Pharmacother 43:1836–1847

Warnock DG, Kusche-Vihrog K, Tarjus A, Sheng S, Oberleithner H, Kleyman TR, Jaisser F (2014) Blood pressure and amiloride-sensitive sodium channels in vascular and renal cells. Nat Rev Nephrol 10:146–157

Wen D, Cornelius RJ, Sansom SC (2014) Interacting influence of diuretics and diet on BK channel-regulated K homeostasis. Curr Opin Pharmacol 15:28–32

Wright FS (1971) Increasing magnitude of electrical potential along the renal distal tubule. Am J Phys 220:624–638

Wright FS, Schnermann J (1974) Interference with feedback control of glomerular filtration rate by furosemide, triflocin, and cyanide. J Clin Invest 53:1695–1708

Wright EM, Loo DDF, Hirayama BA (2011) Biology of human sodium glucose transporters. Physiol Rev 91:733–794

Wynne BM, Mistry AC, Al-Khalili O, Mallick R, Theilig F, Eaton DC, Hoover RS (2017) Aldosterone Modulates the Association between NCC and ENaC. Sci Rep 7:4149

Xu JZ, Hall AE, Peterson LN, Bienkowski MJ, Eessalu TE, Hebert SC (1997) Localization of the ROMK protein on apical membranes of rat kidney nephron segments. Am J Phys 273:F739–F748

Yang T, Huang YG, Singh I, Schnermann J, Briggs JP (1996) Localization of bumetanide- and thiazide-sensitive Na-K-Cl cotransporters along the rat nephron. Am J Phys 271:F931–F939

Yu AS (2014) Claudins and the kidney. J Am Soc Nephrol 26(1):11–19

Yu ASL (2017) Paracellular transport and energy utilization in the renal tubule. Curr Opin Nephrol Hypertens 26:398–404

Zaika OL, Mamenko M, Palygin O, Boukelmoune N, Staruschenko A, Pochynyuk O (2013) Direct inhibition of basolateral Kir4.1/5.1 and Kir4.1 channels in the cortical collecting duct by dopamine. Am J Physiol Renal Physiol 305:F1277–F1287

Zaika O, Tomilin V, Mamenko M, Bhalla V, Pochynyuk O (2016) New perspective of ClC-Kb/2 Cl- channel physiology in the distal renal tubule. Am J Physiol Renal Physiol 310:F923–F930

Zhang C, Wang L, Thomas S, Wang K, Lin DH, Rinehart J, Wang WH (2013) Src family protein tyrosine kinase regulates the basolateral K channel in the distal convoluted tubule (DCT) by phosphorylation of KCNJ10 protein. J Biol Chem 288:26135–26146

Zhou R, Tomkovicz VR, Butler PL, Ochoa LA, Peterson ZJ, Snyder PM (2013) Ubiquitin-specific peptidase 8 (USP8) regulates endosomal trafficking of the epithelial Na$^+$ channel. J Biol Chem 288:5389–5397

Chapter 10
Physiologic Influences of Transepithelial K⁺ Secretion

Dan R. Halm

Abstract Cellular ionic balance relies on ion channels and coupled transporters to maintain and use the transmembrane electrochemical gradients of the cations Na^+ and K^+. High intracellular K^+ concentration provides a ready reserve within the body allowing epithelia to secrete K^+ into the fluid covering the apical membrane in the service of numerous physiologic activities. A major role for transepithelial K^+ secretion concerns the balance of total body K^+ such that excretion of excess K^+ in the diet safeguards against disturbances to cellular balance. Accomplishing this transepithelial flow involves two archetypical cellular mechanisms, Na^+ absorption and Cl^- secretion. Ion channels for K^+, Na^+, and Cl^-, as well as cotransporters, exchangers, and pumps contribute to produce transepithelial flow by coupling electrochemical gradients such that K^+ flow enters across the basolateral membrane and exits through the apical membrane. Beyond excretion, transepithelial K^+ secretion serves to create the high K^+ concentration of endolymph in the inner ear that supports the sensation of sound and body orientation. For several epithelia such as those in airways and gastric mucosa, the elevated K^+ concentration of apical fluid may occur largely as a consequence of supporting the secretion of other ions such as Cl^- or H^+. Less well-appreciated consequences of K^+ secretion may result as in saliva and colonic luminal fluid where a high K^+ concentration likely influences interactions with the resident microbiome. Independent control of K^+ secretion also allows for specific adjustments in rate that serve the physiology of organs large and small.

Keywords Airway fluid · Ammonium transport · Ectothelium · Gastric acid · Inner ear · Intestine · Microbiome · Nutrient absorption · Olfaction · Pancreatic acini and duct · Renal excretion · Saliva · Sweat

D. R. Halm (✉)
Department of Neuroscience, Cell Biology and Physiology, Wright State University Boonshoft School of Medicine, Dayton, OH, USA
e-mail: dan.halm@wright.edu

© The American Physiological Society 2020
K. L. Hamilton, D. C. Devor (eds.), *Basic Epithelial Ion Transport Principles and Function*, Physiology in Health and Disease,
https://doi.org/10.1007/978-3-030-52780-8_10

337

10.1 Introduction

Potassium and sodium comprise the major cations in body fluids and thereby command a pervasive influence on cellular ionic balance, from establishing the electrical potential difference across the cell's plasma membrane to nutrient uptake that supports cellular metabolism (Williams and Dawson 2006). The high Na^+ and low K^+ concentrations of extracellular fluid likely mimics the relative abundance of these ions in the ocean during early stages of development for this cellular balance (Armstrong 2015; Cereijido et al. 2004; Knauth 2005; Mulkidjanian et al. 2012; Pinti 2005; Schrum et al. 2010; Wilson and Lin 1980).

A provocative hypothesis proposes that life originated within submarine hydrothermal vent systems using the geochemical differences between the alkaline vent fluid and more acidic ocean water (Branscomb and Russell 2018; Corliss et al. 1981; Kitadai et al. 2019). A further elaboration extends the proposal to include an epithelioid stage in which protocells occluded apertures in the porous rocky vent structure (Lane 2017; Sojo et al. 2016). The resulting transcellular disparity in pH then would have allowed the continued evolution of ATP synthesis using this persistent H^+ electrochemical gradient.

Sequestering K^+ inside cells and excluding Na^+ to the extracellular space became a cornerstone of ionic balance that permits maintenance of near-constant cellular volume (Vol. 1, Chap. 11). Constancy of these extracellular concentrations maintained by the kidneys for vertebrate animals (Malnic et al. 2013; Palmer and Clegg 2016) permits all cells in the body to operate under reliable conditions of the milieu intérieur.

Key components of the cellular process for maintaining ionic and osmotic balance are the Na^+/K^+-pump (Vol. 3, Chap. 1) and K^+ channels (Vol. 3, Chaps. 19–25). These proteinaceous nanomachines operating in the plasma membrane guide flow of Na^+ and K^+ into and out of cells. The combined action of these two types of nanomachines lowers Na^+ and raises K^+ concentrations intracellularly such that a plasma membrane electrical potential difference (ψ_m) develops that is negative inside the cell compared with the extracellular fluid. As simple as this system may seem, it underlies the orientation of transmembrane flow for most other small solutes a cell encounters.

Using membrane nanomachines with specific solute coupling stoichiometries allows the inward orientation of the Na^+ electrochemical gradient to influence the flow of various solutes (Vol. 3, Chaps. 1–8) such that the intracellular concentrations of these solutes deviate from the expectations governed by the electrochemical energy of that solute alone. Similarly, the dominance of K^+ channels in contributing to plasma membrane conductance (G_m) sets the polarity of ψ_m, thereby influencing flow through all the types of ion channels and other electrogenic transporters present (Vol. 1, Chaps. 1, 2, 4; Vol. 3, Chaps. 1, 4, 6–8, and 11–29).

Epithelial cells make up the barrier that defines the inside of the body (Cereijido et al. 2004; Le Bivic 2013; Palmer 2017; Shashikanth et al. 2017; Vol. 1, Chaps. 3 and 5). In general, epithelia also form the lining for various body compartments as well as covering the body surface. Specifically, the epithelium forming the boundary between blood plasma and the interstitium has the histologic designation of the endothelium (Jackson 2017; Levick and Michel 2010), and body cavities are lined by epithelia designated as mesothelium (Flessner 2005; Ji and Nie 2008; Markov and Amasheh 2014). A similar histological designation apparently has not been assigned to the specific epithelium separating the inside of the body and the external environment, but "*ectothelium*" may suffice in making that distinction. Fluid contiguous with the external environment covers the apical membrane of the ectothelium whether in the lungs, the kidneys, or the gut, such that this contact with the external environment makes the ectothelium the sole site of absorption into and excretion from the body.

By bounding the extracellular space, epithelia permit a stable internal environment via selective entry and exit of substances into and from the body. These absorptive and secretory events occur across epithelial regions located primarily within specialized organs including the lungs, kidneys, and gastrointestinal tract. The often-intricate tubular geometry of these epithelial regions leads to a functional distinction between secretion and excretion. In the broadest biologic sense, secretion denotes the exit from a cell, but in the context of ectothelial cells, the focus is on exit across the apical membrane and thereby out of the body. The clearest phrasing for this physiologic focus is to consider transepithelial secretion, the net movement from the extracellular fluid of the interstitium into the cell across the basolateral membrane and then out across the apical membrane. As for the distinction between excretion and transepithelial secretion, accomplishing excretion requires that no further opportunity exists to retrieve the transepithelially secreted solute. Considering the variation in epithelial transport along renal tubules, for example, the simple transepithelial secretion of K$^+$ at a proximal site could be reversed by transepithelial absorption from the tubular fluid at any more distal site. Ultimate renal excretion of K$^+$ (or any other solute) therefore occurs when urine passes away from contact with the body. This type of serial processing of tubular fluid K$^+$ not only allows renal K$^+$ excretion to be highly regulated but also illustrates that transepithelial K$^+$ secretion can lead to high K$^+$ concentration, [K$^+$], on the apical surface that supports a range of physiologic functions beyond the simple need for excretion of excess K$^+$.

Any transepithelial K$^+$ secretion leads to increased [K$^+$] in the aqueous fluid covering the apical membrane with the volume of this space determining how rapidly that increase occurs. These exo-epithelial apical compartments often are the lumens of small-bore epithelial tubules (5–30 μm diameter) such as in the kidney, salivary gland, or intestinal crypts of Leiberkühn, but also occur as a thin film on the airways or skin surface. Slow fluid flow along these tubules or over the surface allows high [K$^+$] to result, whereas rapid fluid flow limits the buildup. In the case of excretion, attaining a higher luminal [K$^+$] allows the least water elimination to remove a specified amount of K$^+$, with the only limitations being the requirements for handling other solutes at the same time. The renal tubule uses an elaborate set of regulatory schemes to produce appropriate excretion of multiple solutes nearly simultaneously, but each epithelial location includes a suitable anion to accompany the transepithelially secreted K$^+$.

Many occurrences of K^+ secretion operate primarily to support the absorption/ secretion of other ions without leading to eventual K^+ excretion. These situations can be seen as extensions of the regulated cellular mechanisms that lead to K^+ excretion from the kidney (see Sect. 10.3.1), but instead use K^+ as a recycling cofactor to promote another epithelial action (see Sect. 10.3.3). The stomach provides an example for this type of balance in the gastric parietal cell where apical K^+ channels support acid secretion via an H^+/K^+-pump without creating an impact on the ultimate gastrointestinal excretion of K^+. Transepithelial K^+ secretion also can act primarily to make K^+ the dominant cation within an extracellular compartment, instead of Na^+ (see Sect. 10.3.4). The endolymph of the inner ear provides the most dramatic example of changing the dominant cation, where the high luminal $[K^+]$ serves in the sensation of sound and body orientation (see Sect. 10.3.2.1). These contributions of K^+ secretion to the physiology of numerous organs require precise coupling of transmembrane electrochemical gradients using a multitude of nanomachines assembled from various membrane transport proteins.

10.2   Pathways for Transepithelial K^+ Secretion

Transepithelial transport of solute drives fluid flow across epithelia as well as altering the composition of the aqueous fluid covering the apical membrane. Electrogenic Na^+ absorption and Cl^- secretion (Vol. 1, Chaps. 8 and 9) typify the transport mechanisms responsible for flow via the cellular pathway across epithelia. Paracellular flow in comparison occurs through the tight junctions and along the lateral spaces between epithelial cells (Anderson and Van Itallie 2009; DiBona 1985; Frömter and Diamond 1972; Furuse 2010; Günzel and Fromm 2012; Lane et al. 1992; Lee et al. 2008a; Pinto da Silva and Kachar 1982). Selectivity for this paracellular shunt pathway is governed primarily by characteristics of tight junction proteins (Günzel and Yu 2013; Günzel 2017). Together these parallel cellular and paracellular routes combine to determine the outcomes of epithelial actions.

Insertion of ion channels, cotransporters, exchangers, and pumps into cell membranes creates the pathways for the solute flow observed across epithelia, but the electrochemical energy gradients of these solutes sets the direction for the flow Eq. 10.1a.[1] Chemical gradients developed by pump (transport ATPase) activity constitute the ultimate electromotive forces (*EMF* or simply *E*) for ions, with the combination of all the conductive membrane components determining the size and orientation of ψ_m (Sackin and Palmer 2013; Williams and Dawson 2006; Vol. 1, Chaps. 1, 2, 4). Transformation from energy gradient to the potential gradient (Eq. 10.1b) gives the familiar driving force for ions flowing through channels,

---

[1]μ is electrochemical energy (joules), j is the solute of interest, T is absolute temperature (degrees Kelvin), R is the universal gas constant (joules/mole·°K), F is Faraday's constant (coulombs/mole), z_j is the electrical valence of the solute, ln is natural log.

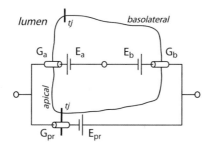

Fig. 10.1 Epithelial equivalent circuit. The pathways for ion flow are indicated by conductors (G) and batteries (E) at the apical (a) and basolateral (b) membranes as well as in the paracellular route (pr). The apical and basolateral conductors represent channels or other electrogenic transporters, and the conductor for the paracellular route represents a combination of tight junction proteins (tj) and lateral space fluid conductivity. The distributed nature of the lateral space conductance along the lateral membrane (Clausen et al. 1979; Frömter and Diamond 1972; Sackin and Palmer 2013) is lumped together as part of this simplified G_{pr}. The EMF's (E) are represented as batteries for the conductive ions, as in Eqs. 10.1a, 10.1b, 10.2a, 10.2b, and 10.2c

indicating whether that ion type will flow into or out of the cell. Conveniently, the same formalism allows a calculation of the driving forces for other transporters as well by including "$\Delta\mu_j/F$" terms as appropriate for cotransport, exchange, or pump activity. A determination of the direction for transepithelial flow whether absorptive or secretory, therefore, results from the cellular control of ion composition and ψ_m as expressed through the transmembrane nanomachines chosen.

$$\Delta^{21}\mu_j = \left(\mu_{j2} - \mu_{j1}\right) = RT \ln\left(\frac{C_{j2}}{C_{j1}}\right) + z_j F(\psi_2 - \psi_1) \tag{10.1a}$$

$$\frac{\Delta^{21}\mu_j}{F} = \frac{RT}{F} \ln\left(\frac{C_{j2}}{C_{j1}}\right) + z_j(\psi_2 - \psi_1) = z_j\left(\psi_m - E_j\right);$$

$$E_j = -\frac{RT}{z_j F} \ln\left(\frac{C_{j2}}{C_{j1}}\right) \tag{10.1b}$$

Epithelial ion flow can be analyzed using the schematic approach of an equivalent circuit to represent the transport elements (Lewis et al. 1996; Schultz et al. 1981; Thompson 1986; Vol. 1, Chaps. 1, 2, 4). The conductance (G) of each membrane is the sum of the individual conductances from all the ion channels and other electrogenic transporters, and each membrane EMF is the weighted average of the E_j's (Nernst potentials from concentration gradients as in Eq. 10.1b) for all of the conductors present (Fig. 10.1). This circuit illustrates the influence of these E_j's on the transepithelial electrical potential difference (ψ_t) as well as on the apical (ψ_a) and basolateral (ψ_b) membranes. During steady-state transepithelial flow, cell influx clearly must equal efflux, which is accomplished through regulatory adjustments to conductances that lead as well to altered driving forces. As conductance decreases or increases, consequent increases or decreases in driving force ultimately feed back to maintain the flow rate steady across the membrane. Any epithelial cell using a secretory mechanism with apical membrane K^+ channels maintains the size and orientation of ψ_a such that cell K^+ exits into the lumen, $z_K\left(\psi_a - E_K^a\right) > 0$.

The dependence of electrogenic K^+ secretion on ψ_a allows the flow of other ions to influence K^+ exit through the apical membrane into the lumen. Equations 10.2a, 10.2b, and 10.2c provide this relationship of ψ_a to conductive ion flow, derived from the equivalent circuit in Fig. 10.1 (Halm and Frizzell 1991). Similar relations can be derived for ψ_b and ψ_t. For the case with conductance through the paracellular route (G_{pr}) much smaller than via the cellular route (G_{cr}), the E_j's of the apical membrane alone determine ψ_a (Eq. 10.2b). This situation raises a crucial point, that the most conductive ions exert the largest influence on determining the size and orientation of ψ_m. As long as the E_j's for some of the conductive ions are more positive than E_K^a, then $z_K\left(\psi_a - E_K^a\right)$ will be positive and K^+ will exit into the lumen. If only K^+ channels open, then ψ_a equals E_K with K^+ exit stopping after an astonishingly brief period of K^+ flow that charges the membrane capacitance. This stoppage of flow results directly from the impediment imposed by the physical constraint of macroscopic electroneutrality, where nanoscopic charge flow generates a ψ_m opposing further ion flow.

$$\psi_a = \left(\frac{G_{pr}}{G_{cr} + G_{pr}}\right)\left\{\left(\frac{G_a}{G_a + G_b} + \frac{G_{cr}}{G_{pr}}\right)E_a + \left(\frac{G_b}{G_a + G_b}\right)(E_b - E_{pr})\right\};$$

$$G_{cr} = G_a\left(1 + \frac{G_a}{G_b}\right)^{-1} \tag{10.2a}$$

$$G_{pr} \ll G_{cr} \rightarrow \psi_a \approx E_a = \sum\left(\frac{G_j^a}{G_a}\right)E_j^a \tag{10.2b}$$

$$G_{pr} \gg G_{cr} \rightarrow \psi_a \approx \left(\frac{G_a}{G_a + G_b}\right)E_a + \left(\frac{G_b}{G_a + G_b}\right)(E_b - E_{pr})$$

$$= \sum\left(\frac{G_j}{G_{cell}}\right)E_j - \left(\frac{G_b}{G_a + G_b}\right)E_{pr}; G_{cell} = \sum G_j \tag{10.2c}$$

Epithelial ion flows within various organs illustrate the many inventive strategies evolved for promoting counter charge flow to deal with this stricture of electroneutrality. For epithelia with even modest G_{pr}, the E_j's of the basolateral membrane and paracellular pathway also contribute to ψ_a, via the electrical shunting of G_{pr}. In this more typical situation (Eq. 10.2a), a large driving force for apical K^+ exit occurs when basolateral channels open for ions with E_j's more positive than E_K^a (Dawson and Richards 1990; Li et al. 2003). This combining of specific transporter types in the apical and basolateral membranes harnesses the available electrochemical energy (generated by ion pumps) to support the ultimate goal of transepithelial fluid movement and/or alterations in apical fluid composition.

Several transport features emanate from the influence of the paracellular pathway on transcellular ion flow. As just noted, the electrical connection between the lumen and the interstitial compartment allows basolateral membrane E_j's to alter ψ_a and thereby the driving forces of electrogenic solute flow across the apical membrane. Since active transcellular ion flow often produces a non-zero ψ_t, net ion flows occur through the G_{pr} driven by ψ_t (Eq. 10.3). These net flows lead to changes in luminal composition such that the concentrations of the ions (solutes) relative to one another differ over time, further altering transepithelial driving forces (E_{pr}). Any resulting increase or decrease in the total number of solutes in the luminal space also leads to an osmotic gradient across the epithelium that drives water flow. In this way, paracellular features complement cellular events allowing epithelial actions to serve body demands for transepithelial absorption or secretion of solutes and water.

$$\psi_t = \left(\frac{G_{cr}}{G_{cr} + G_{pr}}\right)(-E_a + E_b) + \left(\frac{G_{pr}}{G_{cr} + G_{pr}}\right)E_{pr} \qquad (10.3)$$

Additionally, similar electrical shunting paths occur if more than one cell type exists in proximity within an epithelial region. The principal and intercalated cells of the renal distal tubule and collecting duct provide an example (Kriz and Kaissling 2013; Malnic et al. 2013). Differences in the ion selectivity between the cell types alter driving forces as the expanded number of equivalent circuit components contribute to ψ_a and ψ_b, as well as ψ_t. The shunting action may occur via active transepithelial ion flow as for intercalated cells (H⁺ and K⁺ secretion) or more simply as a transcellular conductive route (G_{Cl}^a and G_{Cl}^b) occurring in the mitochondrial-rich cells of amphibian skin and minority cell types of other epithelia (Fan et al. 2012; Jakab et al. 2013; Larsen 2011; Montoro et al. 2018; Plasschaert et al. 2018).

10.2.1 Cellular Mechanisms for Transepithelial K⁺ Flow

The cellular model proposed by Koefoed-Johnsen and Ussing (1958) for electrogenic Na⁺ absorption formalized the concepts necessary to describe transepithelial flow via the cellular pathway, by including the balance of influx and efflux that maintains cellular homeostasis while accomplishing movement across an epithelial cell layer (Vol. 1, Chap. 9). Subsequent studies of epithelial solute transport built directly on these ideas. The concept of Na⁺-coupled glucose transport (Crane 1965; Wright et al. 2011; Vol. 3, Chap. 6) fits with this cellular mechanism such that Na⁺-dependent glucose absorption contributes to the osmotic gradient driving water absorption in the small intestine and renal proximal tubule. Based on these insights, fluid secretion can be envisioned to occur as a simple switching of the membrane localization for these transport elements such that solute and fluid flow would be guided into the lumen (external environment) rather than into the body. The cellular model that best fits observations in many epithelia only relocates the Na^+-dependent uptake component to the basolateral membrane and additionally includes apical membrane Cl^- channels (Degnan et al. 1977; DiBona and Mills 1979; Ernst and Mills 1977; Field 1993; Frizzell et al. 1979; Heintze et al. 1983; Karnaky et al. 1977; Shorofsky et al. 1983, 1984; Silva et al. 1977; Smith and Frizzell 1984; Welsh et al. 1982, 1983b), which leads directly to electrogenic Cl^- secretion (Vol. 1, Chap. 8). K^+ plays a central role in these epithelial archetypes for absorption and secretion.

10.2.1.1 Electrogenic Na^+ Absorption

Producing K^+ secretion coupled to Na^+ absorption only requires the additional presence of apical membrane K^+ channels in the Koefoed-Johnsen–Ussing model (Fig. 10.2). The first proposals of this coupling accounted for the Na^+-dependence of renal K^+ excretion occurring in cells of the distal tubule and collecting duct (Berliner 1961; Giebisch et al. 1967). Opening of K^+ channels in both apical and basolateral membranes allows for a variable rate of K^+ exit into the lumen as the K^+ taken up by the Na^+/K^+-pump exits from the cell according to the size of the G_K and driving forces at apical and basolateral membranes. When all K^+ exit occurs through apical channels, the Na^+/K^+ coupling stoichiometry of the Na^+/K^+-pump sets an upper limit on the maximal rate for K^+ secretion. The generally opposing orientation of the Na^+

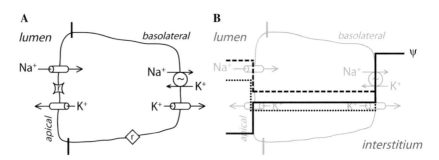

Fig. 10.2 Electrogenic Na^+ absorption and K^+ secretion. The Koefoed-Johnsen and Ussing model for Na^+ absorption (Koefoed-Johnsen and Ussing 1958) includes apical membrane Na^+ channels together with Na^+/K^+-pumps and K^+ channels in the basolateral membrane that constitute the minimal set of transporters required for absorption. (**a**) The addition of apical membrane K^+ channels to this cellular mechanism produces K^+ secretion dependent on the rate of Na^+ absorption. Variation in the opening of basolateral K^+ channels adjusts the K^+ secretory rate. Receptors (r) in the basolateral and apical membranes transduce extracellular signals into activation or inhibition of transepithelial Na^+ and K^+ flow. (**b**) The E_j's of the conductive ions determine the electrical potential profile across this epithelium (ψ, referenced to the interstitial space). For the idealized case of $G_{pr} = 0$ and $G_K^a = 0$ (Eqs. 10.2a, 10.2b, and 10.2c), a staircase-type profile occurs (solid line) with $\psi_a = E_{Na}^a$ and $\psi_b = E_K^b$. A well-type profile occurs with increases in G_{pr} such that E_K^b contributes to ψ_a as well as E_{Na}^a contributing to ψ_b (dashed line), or during increases in G_K^a that alter the polarization of ψ_a via E_K^a. Mineralocorticoid stimulation hyperpolarizes ψ_b (dotted line), apparently via the action of Na^+/K^+-pumps (Koeppen and Giebisch 1985)

and K$^+$ concentration gradients at the apical membrane, created by the Na$^+$/K$^+$-pump, assures that Na$^+$ enters and K$^+$ exits, since ψ_a resides between the value of E_{Na}^a and E_K^a, such that $z_{Na}(\psi_a - E_{Na}^a) < 0$ and $z_K(\psi_a - E_K^a) > 0$. When cellular regulation dramatically decreases G_K^a relative to G_{Na}^a thereby reducing K$^+$ secretion, the rate of Na$^+$ absorption becomes more dependent upon the electrical properties of the paracellular pathway and basolateral membrane to maintain an inwardly directed driving force for Na$^+$ (Eq. 10.2a). The critical dependence of K$^+$ secretion on the rate of Na$^+$ absorption becomes most apparent as ongoing Na$^+$ absorption depletes Na$^+$ in the apical fluid layer thereby leading to decreased E_{Na}^a as well as limited Na$^+$/K$^+$-pump turnover which together attenuate K$^+$ secretion.

10.2.1.2 Electrogenic Cl$^-$ Secretion

Opening apical membrane K$^+$ channels in a cell already capable of Cl$^-$ secretion produces a coupling of K$^+$ secretion with Cl$^-$ secretion (Fig. 10.3). This proposed combination of ion secretions accounts for the dependence of electrogenic K$^+$ secretion in mammalian colon on Na$^+$ uptake across the basolateral membrane (Frizzell et al. 1984a; Halm 1984; Halm and Frizzell 1986). As noted for the Na$^+$ absorbing cell, the presence of K$^+$ channels in the apical and basolateral membranes readily sets up a regulatory scheme for K$^+$ secretion to occur via adjustments in the relative conductance at the two membranes (Li and Halm 2002). Activating both Cl$^-$ and K$^+$ channels in the apical membrane nearly assures exit into the lumen for both ions since ψ_a will reside between E_{Cl}^a and E_K^a, with both $z_{Cl}(\psi_a - E_{Cl}^a)$ and $z_K(\psi_a - E_K^a)$ positive. The low rates of K$^+$ secretion associated with few open apical K$^+$ channels make Cl$^-$ secretion more dependent on the properties of the paracellular pathway and basolateral membrane (Eq. 10.2a). Higher rates of K$^+$ secretion produced by opening apical K$^+$ channels readily enhance the driving force for apical Cl$^-$ exit by moving ψ_a away from E_{Cl}^a (Fig. 10.3b). This possibility of adjusting G_K^a to maximize the rate of Cl$^-$ secretion (Cook and Young 1989; Palk et al. 2010) allows for a separate regulation that optimizes G_{pr} to slow paracellular flow, with the consequence of sizable K$^+$ secretion and an increase in luminal [K$^+$].

An extra feature of this cellular mechanism for K$^+$ and Cl$^-$ secretion emerges from a need for a basolateral membrane Cl$^-$ exit path during K$^+$ secretion occurring with low Cl$^-$ secretory rates. In the mammalian colon, renal intercalated cells, and the inner ear, K$^+$ secretion often occurs with negligible Cl$^-$ secretion (Fig. 10.3c), and basolateral membrane Cl$^-$ channels appear to account for the Cl$^-$ exit that balances Cl$^-$ entry via Na$^+$\K$^+$\2Cl$^-$-cotransporters (He et al. 2011a; Li et al. 2003; Liu et al. 2011; Marcus and Wangemann 2009). In the absence of Cl$^-$ secretion, apical K$^+$ exit depends completely on the depolarizing influence of E_{Cl}^b (and possibly E_{pr}) on ψ_a so that $z_K(\psi_a - E_K^a) > 0$. Whether these basolateral Cl$^-$ channels open during Cl$^-$ secretion remains to be fully explored, but Cl$^-$ channel inhibitors acting on the basolateral side increase Cl$^-$ secretion suggesting a general applicability of a cellular mechanism that includes Cl$^-$ channels opening in both apical and

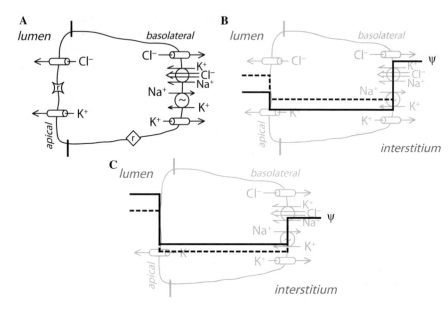

Fig. 10.3 Electrogenic Cl⁻ and K⁺ secretion. The cellular model for Cl⁻ secretion includes apical membrane Cl⁻ channels together with Na⁺/K⁺-pumps, Na⁺\K⁺\2Cl⁻-cotransporters, and K⁺ channels in the basolateral membrane. (**a**) Apical membrane K⁺ channels in this cellular mechanism produce K⁺ secretion dependent on the rate of Cl⁻ secretion. Opening basolateral Cl⁻ channels allows the K⁺ secretory rate to exceed the rate of Cl⁻ secretion. Receptors (r) in the basolateral and apical membranes transform extracellular signals into activation or inhibition of transepithelial Cl⁻ and K⁺ flow. (**b**) The E_j's of the conductive ions determine the electrical potential profile across this epithelium (ψ, referenced to the interstitial space). For the idealized case of $G_{pr} = 0$ and $G_K^a = 0$ (Eqs. 10.2a, 10.2b, and 10.2c), a well-type profile occurs (solid line) with $\psi_a = E_{Cl}^a$ and $\psi_b = E_K^b$. Increasing G_{pr} and G_K^a alters the profile (dashed line) with E_K^b and E_K^a hyperpolarizing ψ_a as well as E_{Cl}^a depolarizing ψ_b. (**c**) For the case of $G_{Cl}^a = 0$ with Cl⁻ secretion thereby absent, a well-type profile occurs (solid line) with $\psi_a = E_K^a$. Increasing G_{pr} to a value greater than zero alters the profile (dashed line) with E_{Cl}^b depolarizing ψ_a away from E_K^a

basolateral locations (Duta et al. 2006; Fischer et al. 2007; He et al. 2011a). Adjustments of G_{Cl}^b also allow optimization of the driving force for basolateral K⁺ exit, $z_K(\psi_b - E_K^b)$, without resorting to large increases in G_{pr}. With this insight, both Cl⁻ and K⁺ enter the cell via basolateral membrane Na⁺-dependent uptake and exit across either the apical or basolateral membrane according to the regulatory influences that govern the open status of the respective channels (Halm 2004).

10.2.2   Transport Proteins Supporting Transcellular K$^+$ Secretion

The specific transport proteins found in K$^+$ secretory epithelia provide the pathways for flow and contribute to producing appropriate driving forces for maintaining the directionality of movement. Both Na$^+$ absorptive and Cl$^-$ secretory epithelia require Na$^+$/K$^+$-pumps (Vol. 3, Chap. 1) in the basolateral membrane to establish the electrochemical gradients needed for transepithelial flow. After K$^+$ uptake across the basolateral membrane via Na$^+$/K$^+$-pumps (*ATP1A*, Na/K-ATPase, EC3.6.3.9) as well as Na$^+$\K$^+$\2Cl$^-$-cotransporters (*SLC12A2*, NKCC1; Vol. 3, Chap. 2) in the Cl$^-$ secretory scheme, K$^+$ exit across the apical membrane generally occurs through K$^+$ channels.

Seventy some genes encode the family of K$^+$ channel proteins, many of which undergo alternative splicing such that the number of different functional channel types is much larger (González et al. 2012; Gutman et al. 2003; Yu and Catterall 2004). Although all conduct K$^+$, differences in control of channel opening allow each K$^+$ channel type to mesh with the specific signaling cascades used in various cells. For cells absorbing Na$^+$ (Fig. 10.2), K$_{ir}$1.1 (*KCNJ1*, RomK), and K$_{Ca}$1.1 (*KCNMA1*, BK, Maxi-K) commonly contribute to the G_K^a permitting simultaneous K$^+$ secretion (Nakamoto et al. 2008; Wang and Huang 2013). K$^+$ secreting cells using a Cl$^-$ secretory type mechanism (Fig. 10.3) often have K$_{Ca}$1.1, K$_{Ca}$3.1 (*KCNN4*, IK1), and K$_V$7.1 (*KCNQ1*, KVLQT) in the apical membrane (Barmeyer et al. 2010; Estilo et al. 2008; Grahammer et al. 2001; Joiner et al. 2003; Marcus and Wangemann 2009; Sørensen et al. 2010b; Zhang et al. 2012). Auxillary β-subunits further increase the functional diversity of K$^+$ channels by altering gating kinetics and membrane localization, particularly for K$_V$7.1 and K$_{Ca}$1.1 (Abbott 2012, 2016; Pongs and Schwarz 2010; Gonzalez-Perez and Lingle 2019; Vol. 3, Chap. 25). In gastric parietal cells, *KCNE2* enhances K$_V$7.1 targeting to the apical membrane while also minimizing voltage-dependent gating and conferring acid activation, all of which promote gastric acid secretion (Abbott 2012). The β-subunits for K$_{Ca}$1.1 alter Ca^{2+}-sensitivity and voltage dependence, with K$_{Ca}$1.1β1 (*KCNMB1*) present in the Na$^+$-absorbing principal cells and K$_{Ca}$1.1β4 (*KCNMB4*) in the acid/base-secreting intercalated cells of the kidney (Grimm et al. 2007; Holtzclaw et al. 2011). At the low physiologic extracellular [K$^+$], K$_{Ca}$1.1β4 shifts voltage activation of K$_{Ca}$1.1 more negative by ~40 mV (Jaffe et al. 2011) promoting increased G_K^a at physiologic ψ_a for intercalated cells. Contributions from K$_{Ca}$1.1γ-subunits (leucine-rich repeat-containing proteins, *LRRC26*, *LRRC52*, *LRRC55*, *LRRC38*) also shift the voltage dependence of K$_{Ca}$1.1 by 20–140 mV, such that activation occurs at negative ψ_m thereby allowing greater influence from Ca^{2+} stimulation and other regulatory pathways (Gonzalez-Perez and Lingle 2019; Latorre et al. 2017; Li and Yan 2016; Yang et al. 2017). Additional K$^+$ channel types have been observed in the apical membrane (Hayashi and Novak 2013; Heitzmann and Warth 2008; O'Grady and Lee 2005), but K$_{ir}$1.1, K$_{Ca}$1.1, K$_{Ca}$3.1, and K$_V$7.1 (Vol. 3, Chaps. 19, 22, 23, and 25)

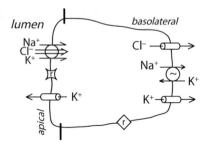

Fig. 10.4 Electroneutral Na^+Cl^- absorption and K^+ secretion. The cellular model for Na^+Cl^- absorption in the thick ascending limb of Henle's loop of the kidney (TALH) and in the fish intestine includes apical membrane $Na^+\backslash K^+\backslash 2Cl^-$-cotransporters and K^+ channels together with basolateral membrane Na^+/K^+-pumps, Cl^- channels, and K^+ channels. Apical membrane K^+ channels in this cellular mechanism produce K^+ secretion dependent on the rate of Na^+Cl^- absorption. Basolateral and apical membrane receptors (r) activate or inhibit transepithelial Cl^-, Na^+, and K^+ flow

may represent a group particularly suited to promote the apical K^+ exit needed for K^+ secretion.

Additional transporter proteins often reside in the apical membrane along with K^+ channels, including other ion channel types, cotransporters, exchangers, and pumps. The epithelial Na^+ channel ENaC (*SCNN1*) supports Na^+ absorption (Fig. 10.2; Vol. 1, Chap. 9; Vol. 3, Chap. 18, Kleyman et al. 2018), and along with CFTR (*ABCC7*) various Cl^- channels such as the Ca^{2+}-activated TMEM16A (*ANO1*) contribute to Cl^- secretion (Fig. 10.3; Huang et al. 2012; Woodward and Guggino 2013; Vol. 1, Chap. 8; Vol. 3, Chaps. 12, 13, and 15–17). Na^+ uptake across the apical membrane also occurs via $Na^+\backslash K^+\backslash 2Cl^-$-cotransporters (*SLC12A1*, NKCC2; Vol. 3, Chap. 2) and $Na^+\backslash Cl^-$-cotransporters (*SLC12A3*, NCC; Vol. 3, Chap. 3) in cells of the renal thick ascending limb of Henle's loop (TALH) and distal tubule (Delpire and Gagnon 2018; Gamba 2013; Vol. 3, Chaps. 2 and 3) as well as fish intestine (Frizzell et al. 1984b; Musch et al. 1982) and urinary bladder (Dawson and Frizzell 1989; Gamba et al. 1993; Stokes et al. 1984). The presence of either $Na^+\backslash K^+\backslash 2Cl^-$-cotransporters or $Na^+\backslash Cl^-$-cotransporters illustrates that the net rate and direction of K^+ transport depends on the combined action of all influx and efflux pathways (Fig. 10.4). Coupling of apical K^+ influx with acid secretion occurs via H^+/K^+-pump (H/K-ATPase, EC3.6.3.10; Vol. 3, Chap. 10), most dramatically (*ATP4A*, gHKA) in gastric parietal cells (Fig. 10.5; Heitzmann and Warth 2007; Kopic et al. 2010; Vol. 3, Chap. 10) but also (*ATP12A*, cHKA) in the distal colon and the kidney intercalated cell (Crowson and Shull 1992; Gumz et al. 2010; Suzuki and Kaneko 1987). Intercalated cells also operate electrogenic H^+-pumps (*ATP6V*, V-ATPase, EC3.6.3.14) in the apical membrane together with $K_{Ca}1.1$ (Wagner et al. 2004). In addition to K^+ channels, K^+ exit across the apical membrane into the lumen could occur via any transporter with an appropriate driving force (Eqs. 10.1a and 10.1b). Renal distal tubule cells employ $K^+\backslash Cl^-$-cotransporters (*SLC12A3*, KCC) for that purpose, $-\left(z_K E_K^a + z_{Cl} E_{Cl}^a\right) > 0$, together with apical membrane

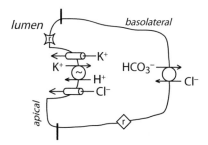

Fig. 10.5 Gastric acid secretion. The cellular model for acid secretion in gastric parietal cells includes apical domain H^+/K^+-pumps, Cl^- channels, and K^+ channels that reside in tubulovesicles prior to stimulated fusion with the apical membrane. Together with basolateral membrane Cl^-/HCO_3^--exchangers, these transporters suffice to generate activated secretory rates for H^+ and Cl^-. Basolateral and apical membrane receptors (r) activate or inhibit transepithelial H^+ and Cl^- flow

$Na^+\backslash Cl^-$-cotransporters supplying Na^+ influx for Na^+/K^+-ATPase turnover (Ellison et al. 1987; Malnic et al. 2013). In combination, these apical transporters influence the fluid covering the apical membrane by decreasing apical fluid pH or by contributing to fluid absorption while K^+ secretion increases the $[K^+]$ of the remaining fluid.

Basolateral membrane transporters must complement apical K^+ exit by providing appropriate pathways and driving forces for K^+ influx into the K^+ secretory epithelial cell (Bachmann et al. 2011). The Na^+/K^+-ATPase establishes the primary energy for this K^+ influx, and in cells using the Cl^- secretory mechanism $Na^+\backslash K^+\backslash 2Cl^-$-cotransporters provide Na^+ influx for maintaining Na^+/K^+-ATPase turnover as well as contributing to K^+ influx, $-\left(z_{Na}E_{Na}^b + z_K E_K^b + 2z_{Cl}E_{Cl}^b\right) < 0$. K^+ channels and Cl^- channels make up the other key components of basolateral ion transport, but the molecular identity of the channel types remains to be fully defined. In Na^+ absorbing cells (Fig. 10.2; Vol. 1, Chap. 9; Vol. 3, Chap. 20), K^+ channels provide a basolateral exit path such that the K^+ secretory rate can be lower than defined by the Na^+ absorption rate and the Na^+/K^+ coupling stoichiometry of the Na^+/K^+-pump. Basolateral Cl^- channels provide a necessary exit pathway in K^+ secretory cells relying on basolateral $Na^+\backslash K^+\backslash 2Cl^-$-cotransporters for maintaining Na^+/K^+-pump turnover and also for those that absorb Cl^-. Cells of the stria marginalia in the inner ear use CLC-Ka and CLC-Kb along with the β-subunit barttin (*CLCNKA*, *CLCNKB*, and *BSND*; Vol. 2, Chap. 8) to support primary electrogenic K^+ secretion (Hibino and Kurachi 2006; Marcus and Wangemann 2009). Colonic epithelial cells likely employ TMEM16A and CLC-2 (Vol. 3, Chaps. 13 and 17) to support K^+ secretion in Cl^- secretory and absorptive cells, respectively (Catalán et al. 2002; He et al. 2011a; Zdebik et al. 2004). The range of biophysical behavior exhibited by Cl^- currents in colon epithelial cells suggests that other Cl^- channel types likely contribute (Halm 2004; He et al. 2011a; Li et al. 2003; Mignen et al. 2000). Renal tubule cells use CLC-Kb in Cl^- absorbing segments, TALH and distal tubule, as well as intercalated cells (Gamba et al. 2013; Stauber et al. 2012). The combined action of these basolateral transport proteins assures continued K^+ secretion during stimulation by physiologic agonists.

10.2.3 Paracellular K⁺ Flow

Epithelia absorbing Na^+ or secreting Cl^- by electrogenic cellular mechanisms develop a lumen-negative ψ_t (Eq. 10.3) with an immediate consequence of electrically driven ion flow through the paracellular pathway. Distinguishing cellular mechanisms for transepithelial ion flow from this paracellular route (Figs. 10.2 and 10.3) can be accomplished experimentally by maintaining $\psi_t = 0$ and the ionic composition of the fluid on the two sides of the epithelium identical (all $E_j^{pr} = 0$), generally in an ex vivo or in vitro setting (Sackin and Palmer 2013; Vol. 1, Chaps. 1, 2, 4). The presence in vivo of any sizable G_{pr} for K^+ leads inexorably to higher luminal $[K^+]$ due to this lumen-negative ψ_t acting on paracellular K^+ flow, $z_K(\psi_t - E_K^{pr})$. Given sufficient time, the $[K^+]$ of this exo-epithelial fluid residing over the apical membrane reaches a value determined by $\psi_t(E_K^{pr} = \psi_t)$. For $\psi_t = -30$ mV, the luminal $[K^+]$ rises to 15 mM (assuming interstitial $[K^+]$ of 5 mM), and thereby lacking a transepithelial driving force greater than zero, net paracellular K^+ flow ceases; but the transcellular flow would continue, governed by the cellular constraints noted above. Electrogenic K^+ secretion in the absence of either Na^+ absorption or Cl^- secretion (Fig. 10.3c) leads to a lumen-positive ψ_t along with increased luminal $[K^+]$, such that both E_K^{pr} and ψ_t act to drive paracellular K^+ flow out of the lumen. The net amount of K^+ secreted into the lumen often depends on whether the overlying fluid remains stationary, allowing the luminal $[K^+]$ to attain thermodynamic equilibrium with ψ_t, or flows away to a location distant from the site of secretion.

Paracellular permeability generally compromises any transcellular ion flow whether absorptive or secretory, since the concentration and electrical gradients (E_j^{pr}, ψ_t) generated by that transcellular movement drives paracellular flow for that ion in the opposing direction. Tight junctions specifically limit this flow through the paracellular route primarily via the claudin proteins (Günzel and Fromm 2012; Günzel and Yu 2013; Tsukita et al. 2019; Zhao et al. 2018). Simply halting any ion backflow by resorting to barrier claudins for achieving a maximal reduction in paracellular permeability uncovers the already noted electrical difficulties for creating sufficient driving forces to sustain net transcellular ion flow. Instead, the presence of the paracellular pathway, with a context-appropriate size for G_{pr}, supports both maintenance of transcellular driving forces and paracellular flow driven by ψ_t and E_j^{pr}. These paracellular flows often promote energetic efficiencies particularly in renal tubules where transepithelial fluid absorption alters luminal composition thereby driving further absorption of physiologic benefit (Muto 2017; Yu 2017). The ultimate energy efficiency of transepithelial transport includes both transcellular and paracellular routes such that driving forces at the apical and basolateral membranes as well as the tight junction optimize the energy required to generate the range of solute flows.

Interestingly, the range of ion selectivity exhibited by the family of claudin proteins (*CLDN*) provides opportunities for maintaining the electrical shunting of the paracellular route without undue ion backflow (Günzel 2017). Ideally, a Cl^-

selective claudin would accompany an electrogenic Na$^+$ absorptive cell (Fig. 10.2) such that net paracellular Cl$^-$ absorption occurs producing net Na$^+$Cl$^-$ absorption, while also providing electrical shunting to generate a substantial driving force for apical Na$^+$ entry but without paracellular Na$^+$ backflow. For an electrogenic Cl$^-$ secretory cell (Fig. 10.3), a Na$^+$ selective claudin, in contrast, would provide the ideal electrical shunt for the paracellular pathway by promoting net Na$^+$Cl$^-$ secretion without paracellular Cl$^-$ backflow.

Although the various claudin isoforms do not exhibit these ideal ion selectivity characteristics, heterogeneity of epithelial claudin expression generally follows the concept of matching transcellular ion flow with a claudin of the opposing charge polarity, thus providing an optimal electrical counterflow (Günzel and Yu 2013; Hou et al. 2013; Krug et al. 2014). The aqueous mobilities for Na$^+$ and Cl$^-$, $(P_{Na}/P_{Cl})_{aq} = 0.656$, provide the most illustrative point of comparison for paracellular ion selectivity because claudins restrict an otherwise aqueous route for ion flow. Of the presently designated Cl$^-$ conductive claudins (*CLDN4, CLDN10a, CLDN11, CLDN17*) most exhibit at best an aqueous selectivity during expression in model epithelia (Günzel et al. 2009; Hou et al. 2010; Krug et al. 2012; Van Itallie et al. 2003). An absence of dramatic selection for anions by claudins may indicate a fundamental difficulty with discriminating by charge polarity, or more likely an experimental problem with finding a suitable model epithelium to accentuate any intrinsic claudin anion preference. Apparently, Cl$^-$ conductive claudins act primarily by limiting cation conductance, whereas cation conductive claudins act by enhancing cation conductance. A plausible interpretation of P_{Cl} measurements for *CLDN4* in cultured renal collecting duct epithelia (Hou et al. 2010) suggests $(P_{Na}/P_{Cl})_{CLDN4} \approx 0.22$ indicating significant anion preference. The cells absorbing Na$^+$ electrogenically within both the renal collecting duct and the distal colon surface epithelium express *CLDN4* together with *CLDN8* supporting the pairing of an anion-selective paracellular route with transcellular electrogenic Na$^+$ absorption (Günzel and Yu 2013; Holmes et al. 2006; Rahner et al. 2001; Yu 2015). Notably, cation conductive claudins (*CLDN2, CLDN10b, CLDN15*) exhibit $(P_{Na}/P_{Cl})_{CLDN} = 6–10$ when expressed in model epithelia (Amasheh et al. 2002; Günzel et al. 2009; Milatz et al. 2017; Tamura et al. 2011; Van Itallie et al. 2003). The cells electrogenically secreting Cl$^-$ within distal colonic crypts express *CLDN2, CLDN15*, and likely *CLDN10b*, supporting the pairing of a cation-selective paracellular route with transcellular electrogenic Cl$^-$ secretion (Fujita et al. 2006; Gumber et al. 2014; Holmes et al. 2006; Rahner et al. 2001; Tamura et al. 2011).

Association of paracellular K$^+$ secretion with either electrogenic Na$^+$ absorption or electrogenic Cl$^-$ secretion results directly from the lumen-negative ψ_t generated by these cellular transport pathways. During electrogenic Na$^+$ absorption, the likelihood of an anion-selective tight junction as observed in the distal colon and renal collecting duct (Holmes et al. 2006; Rahner et al. 2001; Yu 2015) limits not only paracellular backflow of absorbed Na$^+$ but also paracellular K$^+$ secretion. Instead, the primary route for K$^+$ secretion occurs via the cellular mechanism in Fig. 10.2, together with restricted paracellular K$^+$ backflow. The anion selectivity of the tight junction thus ideally constrains luminal [K$^+$] to the level governed by K$^+$ exit across the apical membrane. Electrogenic Cl$^-$ secretion presents a different issue in that the likely cation-selective tight junction potentially allows both Na^+ and K^+ to flow into the lumen. Since the interstitial concentration of Na^+ exceeds that for K^+ by ~30-fold, Na^+ dominates net salt flow except in the unlikely case of high K^+ selectivity at the tight junction, $(P_K/P_{Na})_{tj} \geq 30$. The cation selectivity of claudins ranges from $(P_K/P_{Na})_{CLDN} = 0.70$ for *CLDN10b*, 1.03 for *CLDN2*, to 1.22 for *CLDN15* (Amasheh et al. 2002; Günzel et al. 2009; Tamura et al. 2011), compared with $(P_K/P_{Na})_{aq} = 1.47$, which assures that Na^+ and Cl^- constitute the major ions in the secreted fluid when G_K^a remains minimal. Importantly, the same cells that secrete Cl^- in the distal colon likely also can activate to produce K^+ secretion alone (Fig. 10.3c; Halm and Frizzell 1986; Rechkemmer et al. 1996; Sørensen et al. 2010a) such that a lumen-positive ψ_t occurs (Frizzell et al. 1984a). Unless the activating secretagogues also rapidly regulate tight junction ion selectivity, this primary transcellular K^+ secretion will occur together with the cation-selective claudins that support Na^+Cl^- secretion. Although paracellular Cl^- flow into the lumen may appear the logical complement to transcellular K^+ secretion, perhaps cation-favoring claudins such as *CLDN10b* or *CLDN2* serve to promote cation exchange resulting from luminal $[K^+]$ rising via transcellular secretion and luminal $[Na^+]$ falling via paracellular absorption, with the added consequence of minimal fluid secretion.

10.3   Physiologic Contributions of Transepithelial K^+ Secretion

10.3.1   Potassium Excretion

Excretion constitutes a paramount motivation for transepithelial K^+ secretion. Maintaining cell volume and electrical activity requires sufficient K^+ for it to be the major intracellular cation. Since dietary intake of either animal or plant components includes cell contents, K^+ in the diet generally exceeds that needed to maintain the balance of total body composition. Excretion of any excess K^+ occurs primarily via the kidneys (~90%), with the remainder via the gut (Agarwal et al. 1994; Gumz et al. 2015; McDonough and Youn 2017; Malnic et al. 2013; Palmer 2015; Palmer and Clegg 2016; Wrong et al. 1981). The amount excreted is the product of the $[K^+]$ and the volume of exo-epithelial fluid ejected from the body proper. While the typical $[K^+]$ in urine (20–70 mM) and fecal water (~75 mM) is similar for humans, daily urine volume ranges from 0.5 to 18 L (typical ~1.5 L) and fecal water remains below 0.2 L. Any increase in fecal water loss constitutes diarrhea with attendant pH disturbances and dehydration, because unlike the kidney, colonic transport that adjusts luminal fluid pH is not governed by requirements for maintaining systemic pH. Even though higher gut loss of K^+ can be beneficial during renal failure (Epstein and Lifschitz 2016; Hayes et al. 1967; Sandle and Hunter 2010), this limit on fecal volume loss restricts the possibility of large K^+ losses via physiologic compensation by the gut. In contrast to terrestrial mammals, seawater teleost fish produce a low urine volume, which aids in water conservation but also limits renal K^+ excretion. The apparent route of K^+ excretion proceeds via cells in the gill and larval skin that include apical K^+ channels ($K_{ir}1.1$, *KCNJ1*) and basolateral $Na^+\backslash K^+\backslash 2Cl^-$-cotransporters (Furukawa et al. 2014; Guh and Hwang 2017; Horng et al. 2017), similar to Fig. 10.3c. Salt glands of desert iguanid lizards also secrete a fluid with high $[K^+]$ that relieves the kidneys of producing urine to manage excess K^+ (Hazard 2001; Shuttleworth et al. 1987). Achieving total body K^+ balance ultimately occurs by regulating the increases and decreases in the activity of K^+ transporting ectothelial cells.

10.3.1.1 Cellular K⁺ Secretory Mechanisms

Potassium secretion in both the kidney and gut occurs in cells using the Na^+ absorptive type (Fig. 10.2) and Cl^- secretory type (Fig. 10.3) mechanisms. Positioning these cells in distal locations allows a definitive influence on the final excreted fluid. In the kidney, these cells reside in the distal tubule and collecting duct, with principal cells absorbing Na^+ via apical membrane ENaC and intercalated cells using Na^+ entry via basolateral membrane $Na^+\backslash K^+\backslash 2Cl^-$-cotransporters (Carrisoza-Gaytan et al. 2016; Frindt and Palmer 2009; Liu et al. 2011; Malnic et al. 2013). Although transport localization in the colon remains unsettled (Frizzell et al. 1984a; Grotjohann et al. 1998; Halm et al. 1993; Halm and Rick 1992; Köckerling and Fromm 1993; Köckerling et al. 1993; Rechkemmer et al. 1996; Sørensen et al. 2010a, 2011; Zhang et al. 2012), Na^+ absorbing cells likely reside in the surface epithelium and Cl^- secretory cells in the crypts of Lieberkühn. The presence of these distinct K^+ secretory cell types in proximity at distal locations suggests a physiologic advantage for coupling K^+ secretion to different cellular driving forces. At the least, establishing both mechanisms provides for flexible control of luminal $[K^+]$ in combination with other solute and fluid transport requirements.

Dependence on Na⁺ Absorption

The $K_{ir}1.1$ K^+ channel (*KCNJ1*, RomK) constitutes the primary component of G_K^a in principal cells with $K_{Ca}1.1$ (*KCNMA1*, BK) and $K_{Ca}2.3$ (*KCNN3*, SK3) also contributing in some conditions (Berrout et al. 2014; Hebert et al. 2005; Welling 2016). Coupling K^+ secretion to Na^+ absorption (Fig. 10.2) allows the requirements for K^+ excretion to work together with the need for Na^+ conservation in the typical primordial environment of Na^+ scarcity, producing a low $[Na^+]$ in the final urine and fecal fluid along with higher $[K^+]$. As noted earlier (see Sect. 10.2.1), cellular conditions with substantial G_K^a aid in maintaining a sizable driving force for apical Na^+ entry via Na^+ channels (ENaC; Vol. 1, Chap. 9; Vol. 3, Chap. 18, Kleyman et al. 2018) that sustains electrogenic Na^+ absorption without resorting to the higher G_{pr}

required to harness E_K^b for assuring Na^+ entry (Eqs. 10.2a, 10.2b, and 10.2c). This lower G_{pr} aids in limiting paracellular backflow of actively transported ions (see Sect. 10.2.3), thereby preserving the ion concentrations of the fluid destined for excretion.

Mineralocorticoid receptor activation via aldosterone increases Na^+ absorption and K^+ secretion in the distal segments of the renal tubule and in the distal colon, increasing Na^+ retention and K^+ excretion (Malnic et al. 2013; Palmer and Clegg 2016). When electrogenic Na^+ absorption substantially exceeds K^+ secretion, paracellular electrical shunting acts to maintain the driving force for basolateral K^+ exit by depolarizing ψ_b via the influence of E_{Na}^a such that $z_K(\psi_b - E_K^b) > 0$. Sustained mineralocorticoid stimulation over many days leads to a renal K^+ secretory rate near to or above two-thirds the rate of Na^+ absorption (O'Neil and Helman 1977; Schwartz and Burg 1978; Stokes et al. 1981). Since the coupling ratio of the Na^+/K^+-pump ($3Na^+/2K^+$) sets the limit on the direct dependence of K^+ secretion upon Na^+ absorption in principal cells, basolateral K^+ exit via G_K^b may cease in this condition, $z_K(\psi_b - E_K^b) \approx 0$, such that all of the K^+ taken in by the Na^+/K^+-pump exits across the apical membrane. However, intercalated cells contribute substantially to renal K^+ secretion without contributing to electrogenic Na^+ absorption, so the relative rates of renal K^+ secretion and Na^+ absorption are not strictly constrained by the coupling stoichiometry of the Na^+/K^+-pump.

The lumen-negative ψ_t produced by electrogenic Na^+ absorption drives Na^+ and K^+ passively into the renal tubule lumen thereby enhancing K^+ secretion and diminishing net Na^+ absorption (see Sect. 10.2.3). The presumed anion selectivity of the tight junctions conferred by the presence of *CLDN4* and *CLDN8* in collecting ducts likely limits this paracellular K^+ and Na^+ flow (Hou et al. 2013; Yu 2015). During mineralocorticoid stimulation, G_{pr} decreases approximately threefold (Koeppen and Giebisch 1985) presumably restricting further the paracellular Na^+ and K^+ flow. If this decrease in G_{pr} occurs via regulated remodeling of the tight junction by specific modification or recycling of claudins (Shigetomi and Ikenouchi 2018; Stamatovic et al. 2017; Turner et al. 2014) that enhances *CLDN4* and *CLDN8*, then aldosterone may produce profound restrictions on paracellular cation flow with Cl^- dominating counter ion flow.

The series arrangement of G_K^b and G_K^a suggests another possible mechanism for coupling K^+ secretion to Na^+ absorption (Fig. 10.2), a cellular conductive shunt pathway for K^+ (Malnic et al. 2013). Mineralocorticoid activation of cortical collecting duct principal cells hyperpolarizes ψ_b to around -103 mV, seemingly larger than E_K^b such that $z_K(\psi_b - E_K^b) < 0$ and K^+ enters principal cells via G_K^b (Koeppen and Giebisch 1985; Sansom and O'Neil 1985, 1986). The concurrent decrease in G_{pr} (Koeppen and Giebisch 1985) reduces the depolarizing influence of E_{Na}^a on ψ_b, thereby favoring this basolateral K^+ entry. With apical K^+ exit maintained by the depolarizing influence of E_{Na}^a on ψ_a, a conductive transcellular K^+ secretory shunt would be established, increasing the rate of K^+ secretion through principal cells to a value above two-thirds the rate of Na^+ absorption. The observed hyperpolarization of ψ_b to values more negative than E_K^b was proposed to occur via high turnover of Na^+/K^+-pumps (with an electrogenic $3Na^+/2K^+$ coupling ratio) that supports the observed increase of Na^+ flow, thus producing an apparent reversal of conductive K^+ flow across the basolateral membrane. The large *EMF* of Na^+/K^+-pumps, more negative than -120 mV (Rakowski and Paxson 1988; Rakowski et al. 1989), could support such an action if the effective conductance of the Na^+/K^+-pumps becomes comparable to the other components of G_b (mostly G_K^b). Complicating this interpretation would be any changes to $[K^+]$ resulting from the high Na^+/K^+-pump turnover. A local depletion of interstitial and lateral space $[K^+]$ due to this high pump activity, from the typical ~5 mM to as low as 3 mM, would make E_K^b more negative by up to 14 mV and likely stop secretory flow through this transcellular K^+ shunt, $z_K\left(\psi_b - E_K^b\right) \geq 0$.

Mammalian distal colon exhibits K^+ secretion in vivo that relies on ENaC-dependent electrogenic Na^+ absorption (Edmonds 1981; Wrong et al. 1981). But, during ex vivo measurements with the transepithelial electrochemical gradients nullified by short-circuiting and identical bath composition, K^+ secretion generally fails to depend on apical Na^+ entry via ENaC (Frizzell et al. 1976; Halm and Dawson 1984a; McCabe et al. 1984; Rechkemmer and Halm 1989; Yorio and Bentley 1977). An ex vivo demonstration of K^+ secretion sensitivity to the ENaC inhibitor amiloride suggestive of coupling to Na^+ absorption occurs under the extraordinary condition of a nominal absence of Na^+ only on the basolateral side of the colonic mucosa (Sweiry and Binder 1989). The possibility exists in this unusual condition that amiloride-sensitive Na^+ absorption (Fig. 10.2) increases interstitial $[Na^+]$ that supports K^+ secretion via basolateral Na^+ uptake (Fig. 10.3). Colon from turtle exhibits amiloride-sensitive transcellular K^+ secretion during ex vivo current-clamping that maintains ψ_t near the spontaneous level, around -100 mV (Halm and Dawson 1984b; Wilkinson et al. 1993). Voltage-clamping ψ_t to zero eliminates this transcellular K^+ secretion, supporting the concept of voltage-dependent G_K^a that limits K^+ secretion via this colonic cell type to conditions with lumen-negative ψ_t and consequently a relatively depolarized ψ_a, as occurs during high rates of electrogenic Na^+ absorption. This dependence of K^+ secretion on depolarized ψ_a, together with sensitivity to block by tetraethylammonium (Wilkinson et al. 1993), makes $K_{Ca}1.1$ a likely candidate for G_K^a in these Na^+ absorbing colonic cells. In the mammalian colon, single-channel recordings support the presence of $K_{Ca}1.1$ in the apical membrane of Na^+ absorption generating surface cells based on single-channel conductance and block by tetraethylammonium (Butterfield et al. 1997; Sandle et al. 2007).

Colonic epithelial cells turnover in 5–7 days, regenerating from stem cell division at the crypt base, such that surface cells constitute a brief and terminally differentiated state of Na^+ absorptive cells (Barker 2014; Chang and Leblond 1971; Chang and Nadler 1975). Localization using antibody and Iberiotoxin (IbTx) labeling supports the presence of $K_{Ca}1.1$ in the apical membrane of surface cells (Flores et al. 2007; Hay-Schmidt et al. 2003; Sausbier et al. 2006; Zhang et al. 2012), and targeting to the apical membrane (Vol. 1, Chap. 5) of surface cells may involve the $K_{Ca}1.1$ C-terminal splice variant KYVQEERL (Zhang et al. 2012). The low $K_{Ca}1.1$

mRNA expression for surface cells compared with crypt cells suggests an altered protein retention plan for maintaining $K_{Ca}1.1$ activity during terminal differentiation (Sørensen et al. 2011; Zhang et al. 2012) possibly by preferential translation or delaying protein degradation. The Ca^{2+}-activated $K_{Ca}3.1$ channel (*KCNN4*, IK1) also may contribute to surface cell G_K^a based on antibody localization (Barmeyer et al. 2010; Furness et al. 2003).

Colon surface cells which absorb Na^+ electrogenically also express *CLDN4* and *CLDN8* (Amasheh et al. 2009; Fujita et al. 2006; Günzel and Yu 2013; Holmes et al. 2006; Rahner et al. 2001) similar to renal principal cells, thus establishing a tight junction preferential to anion counter flow such that ψ_t-driven paracellular Cl^- absorption accompanies transcellular Na^+ absorption. Ex vivo stimulation of distal colon with aldosterone increases ENaC-dependent Na^+ absorption along with a approximately threefold increase in *CLDN8* expression, without altering the *CLDN4* already abundantly present (Amasheh et al. 2009). Similarly, the colonic cell line HT29/B6 responds to steroid stimulation with an approximately twofold increase in *CLDN8* expression together with an approximately threefold decrease in G_{pr} (Amasheh et al. 2009), supporting the concept that *CLDN8* enhances the capacity of *CLDN4* to limit cation permeability thereby leading to tight junction anion selectivity (Günzel and Yu 2013). Furthermore, inhibition of ENaC with amiloride blocks both this decrease in G_{pr} and the increase in *CLDN8* expression indicating a direct dependence on apical Na^+ entry rather than via steroid action targeted specifically toward the tight junction. These results suggest the possibility of rapid alteration in tight junction ion selectivity that matches changes in apical Na^+ entry. Whether these actions occur via signaling-mediated claudin modifications or membrane recycling of *CLDN8* (Shigetomi and Ikenouchi 2018; Stamatovic et al. 2017; Turner et al. 2014) remains to be determined.

Dependence on Basolateral Membrane $Na^+\backslash K^+\backslash 2Cl^-$-Cotransporter

Secreting K^+ by a mechanism independent of Na^+ absorption frees the epithelium from direct dependence on changes in luminal $[Na^+]$. Turning to basolateral uptake of Na^+ via $Na^+\backslash K^+\backslash 2Cl^-$-cotransporters to permit continued turnover of Na^+/K^+-pumps establishes a steady driving force for basolateral K^+ influx (Fig. 10.3). Secretagogue activation of K^+ secretion in the colon occurs together with Cl^- secretion that ranges from small and transient, to large and sustained (Halm and Frizzell 1986; Hosoda et al. 2002; Rechkemmer et al. 1996; Zhang et al. 2009b). Several of these secretagogues, adrenergic, cholinergic, peptidergic, and prostanoid activate electrogenic K^+ secretion without accompanying sustained Cl^- secretion (Carew and Thorn 2000; Halm and Halm 2001; Liao et al. 2005; Matos et al. 2005; Traynor and O'Grady 1996; Zhang et al. 2009b) supporting the concept that Cl^- exits conductively via basolateral Cl^- channels (Halm 2004; Halm and Frizzell 1986; He et al. 2011a; Li et al. 2003). Renal intercalated cells also use this secretory mechanism with $Na^+\backslash K^+\backslash 2Cl^-$-cotransporters operating together with conductive

Cl$^-$ exit only across the basolateral membrane (Liu et al. 2011; Sabolić et al. 1999). A consequence of secreting K$^+$ via apical K$^+$ channels without either Na$^+$ channels or Cl$^-$ channels in the apical membrane (Fig. 10.3c) is a need to maintain sufficient G_{pr} so that basolateral E_j's, in this case E^b_{Cl}, depolarize ψ_a away from E^a_K to support apical K$^+$ exit, $z_K(\psi_a - E^a_K) > 0$. The relatively high luminal [K$^+$] in the renal tubule and colon decreases the magnitude of E^a_K constraining the required size for E^b_{Cl}, which needs to be more positive than E^a_K (likely through high intracellular [Cl$^-$]) in order to sustain K$^+$ secretion. Proximity of cells using this K$^+$-alone secretory mechanism (Fig. 10.3c) to the Na$^+$ absorptive type cells (Fig. 10.2) in the kidney and the colon leads to interactions that alter the rates of all the conductively transported ions. The lumen-positive influence on the orientation of ψ_t promoted by the K$^+$-alone secretory cell enhances the driving force for electrogenic Na$^+$ absorption. Similarly, the Na$^+$-absorptive cells generate a lumen-negative influence on ψ_t that contributes to apical K$^+$ exit for these K$^+$-alone secretory cells thereby lessening the demands on tight control of E^b_{Cl}.

The K$^+$ secretion generated by intercalated cells proceeds via apical K$^+$ exit through K$_{Ca}$1.1 and K$_V$1.3 (*KCNA3*) imparting G^a_K with sensitivity to intracellular Ca^{2+} and ψ_a (Carrisoza-Gaytán et al. 2010, 2016; Hebert et al. 2005; Rodan et al. 2011; Welling 2016). Increasing shear forces associated with more rapid tubular fluid flow activates K$_{Ca}$1.1 (Carrisoza-Gaytan et al. 2016; Satlin et al. 2006), and a high K$^+$ diet likely leads to apical insertion of K$_V$1.3 (Carrisoza-Gaytán et al. 2010). Opening of K$_{Ca}$2.3 (*KCNN3*, SK3) during TRPV4 activated Ca^{2+} entry supports an involvement in K$^+$ secretion (Berrout et al. 2014), but likely not as part of flow-induced stimulation (Woda et al. 2001). Whether flow-induced K$^+$ secretion occurs solely from intercalated cells or also includes a contribution from principal cells remains unsettled (Carrisoza-Gaytan et al. 2016, 2017; Welling 2016).

Colonic crypt columnar cells also exhibit apical membrane K$_{Ca}$1.1 (Sandle et al. 2007; Sausbier et al. 2006; Simon et al. 2008; Zhang et al. 2012) and K$_V$1.3 (Grunnet et al. 2003) as well as K$_{Ca}$3.1 (*KCNN4*, IK; Barmeyer et al. 2010; Furness et al. 2003; Halm et al. 2006), with additional appearances in the apical pole suggestive of storage within vesicular compartments. K$_{Ca}$1.1 plays a dominant role as indicated by the complete loss of electrogenic K$^+$ secretion with genetic ablation of K$_{Ca}$1.1 (Sausbier et al. 2006; Sørensen et al. 2010b). However, the functional composition of G^a_K remains for elucidation, since ~50% of secretagogue-activated K$^+$ secretion persists in the presence of K$_{Ca}$1.1 inhibitors (Zhang et al. 2012). Although a small K$_{Ca}$3.1-dependent K$^+$ secretion occurs in the proximal colon (Joiner et al. 2003), secretagogue-activated K$^+$ secretion in distal colon exhibits only marginal sensitivity to K$_{Ca}$3.1-inhibitors (<4%) at concentrations capable of strongly blocking multiple K$^+$ channel types (Halm et al. 2006; Zhang et al. 2012). The low sensitivity to K$_{Ca}$3.1-inhibitors for the apical splice variant of K$_{Ca}$3.1 (Barmeyer et al. 2010; Basalingappa et al. 2011) only compounds the difficulty of assigning the other channel types contributing to G^a_K. The relatively low IbTx sensitivity of secretagogue-activated K$^+$ secretion (Zhang et al. 2012) supports the assembly of ~3 K$_{Ca}$1.1β4 subunits together with the tetrameric pore-forming K$_{Ca}$1.1α subunits

(Gonzalez-Perez and Lingle 2019), as well as $K_{Ca}1.1\beta1$ possibly occupying the fourth regulatory position in the channel complex. Furthermore, cells in the crypt express $K_{Ca}1.1\alpha$ mRNA at high levels (Sørensen et al. 2011; Zhang et al. 2012) perhaps as part of the cell proliferation occurring in this epithelial location. Specificity for targeting to the apical membrane (Vol. 1, Chap. 5) differs in colonic cells during residence in the crypt compared to their arrival in the surface as indicated by the selective lack of the $K_{Ca}1.1\alpha$ C-terminal splice variant KYVQEERL in crypt apical membrane and expression instead of either or both the C-terminal splice variants NRKEMVYR and KKEMVYR (Zhang et al. 2012).

The K^+-alone secretory cells of the kidney and colon concurrently extrude H^+ via apical membrane pumps. Renal α-intercalated cells use electrogenic H^+-pumps (*ATP6V*, V-ATPase) to support systemic acid/base balance, which also contributes a lumen-positive influence on ψ_t that enhances Na^+ absorption in neighboring principal cells (Andersen et al. 1985; Malnic et al. 2013; Wagner et al. 2004). Mammalian colonic mucosa during ex vivo measurements exhibits net K^+ secretion in the proximal region and net K^+ absorption in the distal region (Foster et al. 1984; Rechkemmer and von Engelhardt 1993; Sullivan and Smith 1986), with stimulation by secretagogues leading toward net K^+ secretion in the distal region (Halm and Frizzell 1986; Rechkemmer et al. 1996). These K^+ secretory cells in the distal colon secrete H^+ using H^+/K^+-pumps (*ATP12A*, chKA; Binder et al. 1999; Shao et al. 2010; Suzuki and Kaneko 1987) accounting for the observed net K^+ absorption. Activity of these apical H^+/K^+-pumps exerts control over conductive apical K^+ exit such that complete inhibition of these apical pumps limits K^+ secretion, supporting the combined presence of electroneutral K^+ absorption and electrogenic K^+ secretion in the same cell type (Rechkemmer et al. 1996). Genetic deletion of the colonic H^+/K^+-pump (chKA) results in higher fecal K^+ excretion and leads to more severe hypokalemia during a K^+ restricted diet (Meneton et al. 1998). Renal compensation during dietary K^+ restriction lowers K^+ excretion from the kidney likely via activation of both chKA and gHKA (Crambert 2014; Gumz et al. 2010; Malnic et al. 2013). Switching to H^+/K^+-pumps for renal acid secretion during K^+ restriction introduces the consequence of possibly reducing the enhancement of Na^+ absorption produced by the lumen-positive influence on ψ_t generated via electrogenic H^+-pumps (V-ATPase).

Intercalated cell use of H^+-pumps (V-ATPase) in the apical membrane of α-type for acid secretion and in the basolateral membrane of β-type for HCO_3^- secretion (Breton and Brown 2013; Roy et al. 2015; Vol. 1, Chap. 12) likely extends to powering K^+ secretion. The general mechanism for K^+-alone secretion (Fig. 10.3c) uses basolateral membrane Na^+/K^+-pumps to maintain gradients of $[Na^+]$ and $[K^+]$ that set the driving forces for K^+ secretion, but the apparent low density of Na^+/K^+-pumps for intercalated cells suggests the need for other Na^+ extrusion pathways (Palmer and Frindt 2007; Sabolić et al. 1999). Likely Na^+ exit mechanisms involve the abundant H^+-pumps together with transporters coupled to HCO_3^-, which also contribute to the maintenance of acid/base balance via directional exit of H^+ and HCO_3^- (Wieczorek et al. 1999). The basolateral location of $Na^+\backslash HCO_3^-$-cotransporters (AE4, *SLC4A9*; Parker and Boron 2013) provides a pathway and driving force for Na$^+$ exit $\left(-2\psi_b - E_{Na}^b + 3E_{HCO3}^b > 0\right)$ that returns Na$^+$ entering via Na$^+$\K$^+$\2Cl$^-$-cotransporters back to the peritubular interstitium (Carrisoza-Gaytan et al. 2016). For β-intercalated cells, this basolateral Na$^+$\HCO$_3$$^-$-cotransporter-(AE4)-driven HCO$_3$$^-$ exit decreases HCO$_3$$^-$ secretion, but for α-intercalated cells, this attendant basolateral HCO$_3$$^-$ exit via an electrogenic Na$^+$\HCO$_3$$^-$-cotransporter assists basolateral Cl$^-$/HCO$_3$$^-$-exchangers ($E_{HCO3}^b - E_{Cl}^b > 0$; AE1, *SLC4A1*) in maintaining acid secretion. Non-α-intercalated cells also absorb Na$^+$ and Cl$^-$ via the combined action of apical Cl$^-$/HCO$_3$$^-$-exchangers ($E_{Cl}^a - E_{HCO3}^a < 0$; pendrin, *SLC26A4*; Vol. 3, Chap. 12) and Na$^+$-driven-Cl$^-$/HCO$_3$$^-$-exchangers (NDCBE, *SLC4A8*; $-E_{Na}^a + 2E_{HCO3}^a - E_{Cl}^a < 0$), such that HCO$_3$$^-$ secretion occurs whenever net apical Cl$^-$ entry exceeds net apical Na$^+$ entry (Eladari et al. 2014). The involvement of these apical Na$^+$-driven-Cl$^-$/HCO$_3$$^-$-exchangers in promoting K$^+$ secretion has been proposed for the renal adjustments occurring during a low Na$^+$ and high K$^+$ alkaline diet (Cornelius et al. 2016; Vol. 3, Chap. 23). With the expected low luminal [Na$^+$] due to principal cell Na$^+$ absorption, the driving force for intercalated cell apical Na$^+$-driven-Cl$^-$/HCO$_3$$^-$-exchangers likely reverses ($-E_{Na}^a + 2E_{HCO3}^a - E_{Cl}^a > 0$) allowing apical exit for a portion of the Na$^+$ entering via basolateral Na$^+$\K$^+$\2Cl$^-$-cotransporters, together with increased HCO$_3$$^-$ secretion. Principal cells reabsorb this Na$^+$ exiting into the lumen creating an adjacent-cell coupling mechanism between ENaC-dependent Na$^+$ absorption and intercalated cell K$^+$ secretion. In each of these conditions, intercalated cell K$^+$ secretion requires basolateral Na$^+$\K$^+$\2Cl$^-$-cotransporters as the sole source of K$^+$ entry.

Colon crypt cells that secrete Cl$^-$ electrogenically also express *CLDN2*, *CLDN10b*, and *CLDN15* (Fujita et al. 2006; Gumber et al. 2014; Holmes et al. 2006; Rahner et al. 2001; Tamura et al. 2011), establishing tight junctions preferential to cation counterflow such that ψ_t-driven paracellular Na$^+$ secretion accompanies transcellular Cl$^-$ secretion. Luminal cation composition of distal colonic crypts exhibits a switch toward higher [K$^+$] and lower [Na$^+$] during stimulation of K$^+$ secretion (Halm and Rick 1992), and the crypt cells respond to aldosterone using the K$^+$-alone mechanism (Fig. 10.3c) to increase electrogenic K$^+$ secretion without attendant Cl$^-$ secretion (Halm and Halm 1994; Rechkemmer and Halm 1989). Long-term aldosterone elevation during low Na$^+$ or high K$^+$ diet intake also promotes the expression of K$_{Ca}$1.1α and K$_{Ca}$3.1 in the distal colon (Furukawa et al. 2017; Singh et al. 2012; Sørensen et al. 2008) as well as *CLDN2* in the lower portion of distal colonic crypts (Furukawa et al. 2017). Thus, aldosterone aids in maintaining transcellular K$^+$ secretion together with attendant paracellular Na$^+$ absorption driven by the lumen-positive ψ_t, such that luminal cation composition shifts toward high [K$^+$], low [Na$^+$] without large changes in luminal volume.

10.3.1.2 Interactions with Ammonium

Systemic acid/base balance imposes further conditions on K^+ excretion since the kidney generates ammonium (NH_4^+) to support acid excretion (Weiner and Verlander 2013), and the enteric microbiome produces NH_4^+ while fermenting the remnants of the host diet (Davila et al. 2013; Wrong et al. 1981). The charge and size similarity of K^+ and NH_4^+ makes NH_4^+ an important physiologic competitor leading to potential influences on K^+ transport, with Na^+/K^+-pumps (Halm and Dawson 1983; Skou and Esmann 1992; Wall and Koger 1994), H^+/K^+-pumps (Codina et al. 1999; Cougnon et al. 1999; Swarts et al. 2005), $Na^+\backslash K^+\backslash 2Cl^-$-cotransporters (Kinne et al. 1986), and the K^+ channels, $K_{ir}1.1$ (Yang et al. 2012), $K_{Ca}1.1$ (Eisenman et al. 1986), $K_{Ca}3.1$ (Christophersen 1991), K_V1 (Heginbotham and MacKinnon 1993) having significant NH_4^+ permeability via interaction at K^+ binding sites. This multitude of cellular transport routes constitutes a major cause of toxicity as increasing plasma $[NH_4^+]$ leads to the disruption of cellular physiology.

Ammonium excretion proceeds via selective transport proteins (Weiner and Verlander 2014; Worrell et al. 2008), but the presence of apical K^+ transporters complicates the task of trapping NH_4^+ in the lumen (Nakamura et al. 1999; Weiner and Verlander 2013). Apical membrane K^+ channels, in particular, could dissipate this NH_4^+ gradient, since the cell-negative ψ_a contributes a substantial driving force for NH_4^+ entry from the lumen into the epithelial cells of renal tubules and colon, $z_{NH4}(\psi_a - E_{NH4}^a) < 0$. The physiologic range of ψ_a and the opposing orientation for E_K^a and E_{NH4}^a maintain the driving forces for K^+ exit and NH_4^+ entry. However, several features of K^+ channels act to limit apical NH_4^+ uptake, including ion competition at binding sites and a reduced probability of opening during NH_4^+ passage (Bennekou and Christophersen 1990; Eisenman et al. 1986; Heginbotham and MacKinnon 1993; Yang et al. 2012). The characteristics of $K_{ir}1.1$ (Yang et al. 2012), conducting K^+ in renal TALH cells and principal cells, illustrate how the multisite permeation pathway in K^+ channels likely blunts NH_4^+ entry. Competition with Mg^{2+} for an extracellularly facing site limits NH_4^+ access to the conduction path, and stronger K^+ binding within the selectivity filter blocks inward progress of NH_4^+ at physiologic ψ_a. Impeding NH_4^+ entry may have been a key stressor for driving the evolution of the multisite selectivity filter in the K^+ channel family of proteins long before the inception of epithelia.

10.3.2 Epithelial K^+ Gradients Supporting Sensory Physiology

10.3.2.1 Balance and Hearing in the Inner Ear

The inner ear senses sound waves (cochlea) and body orientation (vestibular labyrinth) using transepithelial electrochemical gradients to transduce these physical forces into electrical signals for nerve conduction (Hibino and Kurachi 2006; Marcus and Wangemann 2009; Vol. 2, Chap. 8). The sensory cells reside in the epithelia bounding the endolymph compartment with the high endolymph [K⁺] of ~150 mM occurring from a steady-state balance in K⁺ secretion and subsequent absorption by the sensory cells. The vestibular dark cell epithelium secretes K⁺ into the endolymph using the K⁺-alone secretory mechanism (Fig. 10.3c), with $K_V7.1$/kcne1 (*KCNQ1/ KCNE1*) as the apical K⁺ channel and CLC-K/barttin (*CLCNK/BSND*) as the basolateral Cl⁻ channel. High endolymph [K⁺] reduces E_K^a to low levels requiring a depolarized ψ_a in the secreting cells so that K⁺ exits from the cell into the endolymph, $z_K(\psi_a - E_K^a) > 0$. This condition implies a high intracellular [Cl⁻] that produces a low E_{Cl}^b for maintaining this outward apical driving force for K⁺. The high endolymph [K⁺] depolarizes vestibular hair cells during the activation of sensory mechano-electrical transducer channels.

In the cochlea, the combination of two epithelia in series comprise the stria vascularis that maintains the high endolymph [K⁺] in this region, generating an endocochlear potential of +80 to +100 mV (Hibino and Kurachi 2006; Marcus and Wangemann 2009; Wangemann 2006; Nin et al. 2016; Vol. 2, Chap. 8). The marginal cell epithelium of the stria vascularis secretes K⁺ directly into the endo-lymph using essentially same transport proteins as in the vestibular dark cell epithelium. A syncytium of intermediate and basal cells together with fibrocytes forms the other stria vascularis epithelium that secretes K⁺, by the K⁺-alone mech-anism, into the intra-strial space facing the basolateral membrane of the marginal cell epithelium. Restricted paracellular access to this intra-strial space occurs via anion-selective tight junctions of marginal cells due to claudin-4 (*CLDN4*; Florian et al. 2003) and of syncytial basal cell claudin-11 (*CLDN11*; Gow et al. 2004; Kitajiri et al. 2004; Liu et al. 2017). The apical K⁺ channel ($K_{ir}4.1$, *KCNJ10*) for the syncytial epithelium resides only in the intermediate cells, such that the generation of the stria vascularis electrochemical gradient occurs with the syncytial epithelium producing a ψ_t of +90 mV and an intra-strial [K⁺] of ~2 mM. At the apical membrane of the syncytial epithelium, the low intra-strial [K⁺] increases E_K^a thereby hyperpolarizing ψ_a. A low E_{Cl}^b aids in maintaining the large ψ_t and the outward K⁺ electrochemical gradient at the apical membrane. The marginal epithelium then secretes K⁺ from this intra-strial space into the endolymph with a ψ_t near 0 mV. This tandem epithelial arrangement creates the endocochlear potential that allows the depolarization of hair cells in the adjacent organ of Corti resulting in sensory nerve activity and ultimately in hearing.

10.3.2.2 Olfactory Sensation

Terrestrial vertebrates sense olfactory cues via the main olfactory epithelium and the vomeronasal organ (Lucero 2013; Tirindelli et al. 2009). Detection of odorants and pheromones occurs through receptors on olfactory sensory neurons projecting into the apical surface fluid. This exo-epithelial fluid on the olfactory epithelium and the vomeronasal organ exhibits a [K⁺] of ~70 mM (Kim et al. 2012; Reuter et al. 1998).

Similar to the inner ear, the resulting low value for E_K^a, near -25 mV, contributes to depolarization of the sensory neurons, likely modulating signal generation and the sensory response. The transport character of the columnar (support/sustentacular) cells in the olfactory epithelium resembles the tracheal mucosa with capacities for electrogenic Na^+ absorption using ENaC and Cl^- secretion via CFTR and/or TMEM16A (Dauner et al. 2012; Grubb et al. 2009; Henriques et al. 2019; Maurya et al. 2015; Merigo et al. 2011). These columnar epithelial cells open apical membrane $K_{Ca}1.1$ channels during purinergic activation, generating a lumen-positive ψ_t consistent with K^+ secretion that likely maintains the high exo-epithelial $[K^+]$ (Grubb et al. 2009; Hegg et al. 2009; Vogalis et al. 2005). Coordination between these epithelial sensory neurons and columnar cells serves to match olfactory sensitivity to the external odorant environment (Lucero 2013).

10.3.3 Transport Cofactor

Many epithelia generate K^+ secretion using apical membrane K^+ channels that results in only modest increases of apical fluid $[K^+]$. In these cases, the K^+ secretion often serves as part of a K^+ circulation that promotes the transepithelial flow of other ions and solutes. Generally, retrieval of the secreted K^+ occurs by absorptive pathways in the secretory cell (parietal cell, TALH cell) or by fluid absorption at a distant site. The primary example of this second type of retrieval consists of gut handling of saliva and airway fluid. Removing the consequence of excretion allows these epithelia to focus K^+ secretion on enhancing other transport activities.

10.3.3.1 Gastric Acid Secretion

Parietal cells secrete H^+ at high rates leading to a luminal pH as low as 1–2 and $[K^+]$ of 10–20 mM (Heitzmann and Warth 2007; Kopic et al. 2010; Watt and Wilson 1958). The secretory mechanism involves H^+/K^+-pumps and K^+ channels (Fig. 10.5) trafficked into the apical membrane by stimulated generation of a canaliculus from tubulovesicles residing in the apical pole of the parietal cell (Forte and Zhu 2010). During lower rates of secretion, the transport mechanism resembles the K^+Cl^- secretory cell (Fig. 10.3) found in the distal colon that includes apical H^+/K^+-pumps (Halm and Frizzell 1986; Rechkemmer et al. 1996; Suzuki and Kaneko 1987). The $K_V7.1/kcne2$ (*KCNQ1/KCNE2*) channel provides the dominant contribution to stimulated parietal cell G_K^a (Heitzmann et al. 2004; Roepke et al. 2006; Song et al. 2009), with $K_{ir}4.1$ (*KCNJ10*; Fujita et al. 2002; Song et al. 2011), $K_{ir}4.2$ (*KCNJ15*; He et al. 2011b; Yuan et al. 2015), $K_{ir}2.1$ (*KCNJ2*; Malinowska et al. 2004), as well as $K_{ir}1.1$ (*KCNJ1*; Vucic et al. 2015) also involved. Although mistakenly dismissed by the investigators, an mRNA-screen of parietal cell K^+ channel expression supports a strong involvement of $K_{ir}5.1$, as well as the expected contribution by $K_V7.1$ (*KCNQ1*) and *KCNE2* (Lambrecht et al. 2005). Regulation of K^+ channel opening by pH likely serves to match conductive apical K^+ exit to increased turnover of the H^+/K^+-pumps that produces extracellular acidification and intracellular alkalinization. Low extracellular pH increases the open probability of $K_{ir}2.1$ from parietal cells (Malinowska et al. 2004) and *KCNE2* confers $K_V7.1$ with activation by low extracellular pH (Heitzmann et al. 2007). Interestingly, the presence of $K_{ir}5.1$ in parietal cells (He et al. 2011b; Lambrecht et al. 2005; Song et al. 2011) may confer $K_{ir}4.1$ and $K_{ir}4.2$ with activation by increased intracellular pH through the formation of heteromeric K_{ir} channels (Hibino et al. 2010; Sepúlveda et al. 2015). This sidedness of pH sensitivity likely imposes distinct regulatory demands, such that the more tightly controlled intracellular environment requires a close proximity between H^+/K^+-pumps and those K^+ channels sensitive to internal pH whereas K^+ channels sensitive to external pH could reside further away from the H^+/K^+-pumps and still respond to increased activity.

Localization studies place $K_V7.1/kcne2$, $K_{ir}4.1$, $K_{ir}4.2$, and $K_{ir}2.1$ in proximity with H^+/K^+-pumps in tubulovesicular and canalicular membranes (He et al. 2011b; Heitzmann et al. 2004; Lambrecht et al. 2005; Malinowska et al. 2004). $K_V7.1/kcne2$ and $K_{ir}4.2$ traffic to the canaliculus in tubulovesicles distinct from those containing H^+/K^+-pumps (He et al. 2011b; Nguyen et al. 2013), while $K_{ir}4.1$ resides in the same tubulovesicles as the H^+/K^+-pumps (Kaufhold et al. 2008). Fluid secretion driven osmotically by apical exit of K^+ and Cl^- provides flow that propels secreted acid out of the canaliculus and gastric pit (Heitzmann et al. 2004). In addition to providing K^+ in the apical fluid for turnover of H^+/K^+-pumps, a large G_K^a moves ψ_a toward E_K^a which aids in maintaining an outward Cl^- driving force at the apical membrane, $z_{Cl}\left(\psi_a - E_{Cl}^a\right) > 0$, for the Cl^- secretion needed to support continued acid secretion. Since the 0.08 mM K_m^K of the gastric H^+/K^+-pump (Shao et al. 2010) more than assures maximal turnover at the typical canalicular $[K^+]$, the luminal level of K^+ likely derives more as a consequence of the G_K^a needed for maintaining adequate rates of Cl^- secretion. A computational model (Crothers et al. 2016) provides further insight into how canalicular $[K^+]$ alters driving forces for apical K^+ and Cl^- exit during the onset and steady-state of H^+ secretion indicating that decreases in canalicular $[K^+]$ due to increased pump activity increases the driving force for apical K^+ exit via G_K^a, $z_K\left(\psi_a - E_K^a\right) > 0$.

10.3.3.2 Pancreatic Acinar Enzyme Release

Digestion of nutrients in the intestinal lumen requires enzymes released from pancreatic acinar cells. These digestive enzymes reside in zymogen granules within the apical pole of acinar cells such that the release of the stored contents into the lumen involves the fusion of the granule membrane with the apical membrane followed by fluid flow that propels the contents out (Thévenod 2002). Vesicular Cl^- channels and K^+ channels promote ion secretion that drives water flow via aquaporins into to the widening fusion pit. The presence of these channels in the

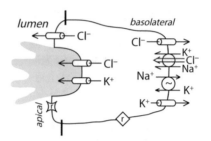

Fig. 10.6 Exocytotic apical release. The cellular model for enzyme release from pancreatic acinar cells includes vesicular Cl⁻ channels and K⁺ channels together with basolateral membrane transporters supporting Cl⁻ secretion (Fig 10.3). A local osmotic gradient drives water flow into granules fused with the apical membrane. Basolateral and apical membrane receptors (r) activate or inhibit acinar secretions

vesicular membrane focuses the osmotic gradient and water flow on the points of release (Fig. 10.6). Both $K_V7.1$ (*KCNQ1*) and $K_{ir}6.1$ (*KCNJ8*) likely contribute to this vesicular G_K (Kelly et al. 2005; Lee et al. 2008b). Since acinar cell stimulation primarily activates Cl⁻ channels in the apical membrane (Heitzmann and Warth 2008), the Cl⁻ and K⁺ secretion from the fusion events likely contributes only a minor amount to the ionic composition of the secreted acinar fluid.

10.3.3.3 Na⁺Cl⁻ Absorption

The thick ascending limb of Henle's loop (TALH) in the kidney absorbs Na⁺ and Cl⁻ using Na⁺\K⁺\2Cl⁻-cotransporters (NKCC2) as the apical step in the cellular mechanism (Fig. 10.4). During transit of tubular fluid along the TALH, absorption decreases [Na⁺], [Cl⁻], and [K⁺], with [K⁺] dropping from ~8 to ~1 mM at the distal end (Vallon et al. 1997). Longitudinal differences in splice variants of NKCC2 promote absorption in the face of these decreasing substrate concentrations, with the variant in the distal half of TALH exhibiting an approximately fourfold higher affinity for K⁺ than the variant in the proximal half (Giménez et al. 2002). K⁺ secretion via apical $K_{ir}1.1$ assists in maintaining Na⁺ and Cl⁻ absorption by adding K⁺ to the tubular fluid such that luminal [K⁺] permits uptake via Na⁺\K⁺\2Cl⁻-cotransporters. The presence of NH_4^+ in the tubular fluid as part of acid/base management also likely contributes to Na⁺ and Cl⁻ absorption (Weinstein 2010), since Na⁺\K⁺\2Cl⁻-cotransporters also accept NH_4^+ at the K⁺ transport site (Kinne et al. 1986). In addition, the lumen-positive ψ_t generated by this electrogenic K⁺ secretion drives absorption of Na⁺ as well as Mg^{2+} and Ca^{2+} via the cation-selective (*CLDN10b, CLDN3, CLDN16, CLDN19*) paracellular pathway (Bleich et al. 2017; Milatz et al. 2017).

The balance between apical K⁺ uptake via Na⁺\K⁺\2Cl⁻-cotransporters and exit through $K_{ir}1.1$ determines whether net K⁺ absorption or secretion occurs in the TALH. High Na⁺ intake typical of a contemporary western diet generally favors

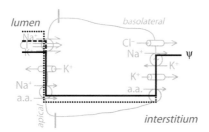

Fig. 10.7 Intestinal nutrient absorption. The cellular model for Na^+Cl^- absorption in fish intestine includes apical membrane $Na^+\backslash K^+\backslash 2Cl^-$-cotransporters, K^+ channels, and Na^+-coupled amino acid transporters together with basolateral membrane Na^+/K^+-pumps, Cl^- channels, and amino acid transporters (Halm et al. 1985). At high luminal $[Na^+]$ and $[Cl^-]$, ψ_t is small (solid line); decreases in luminal $[Na^+]$ and $[Cl^-]$ during absorption hyperpolarizes ψ_a (dashed line); associated decreases in luminal $[K^+]$ to 1 mM further hyperpolarizes ψ_a (dotted line)

absorption as illustrated by the low $[K^+]$ of TALH tubular fluid (Vallon et al. 1997). In stark contrast, the low Na^+ and high K^+ alkaline composition mimicking a plant-dominated diet expected during more primordial conditions shifts the TALH to net K^+ secretion, such that ~40% of renal K^+ excretion depends on TALH output (Wang et al. 2017, 2019).

Intestines of fish absorb fluid driven by Na^+ and Cl^- absorption using a cellular mechanism similar to the TALH (Fig. 10.4). Marine fish ingest seawater during feeding and to replace water lost to the external environment. Because intestinal water uptake requires Na^+Cl^- absorption, systemic balance depends on the gills to excrete the excess Na^+ and Cl^-. Progressive intestinal fluid absorption leads to a luminal composition of low $[Na^+]$, $[Cl^-]$, and $[K^+]$ dominated instead by Mg^{2+} and SO_4^{2-} from the ingested seawater, with Ca^{2+} precipitating as $CaCO_3$ (Whittamore et al. 2010). As in TALH, intestinal K^+ secretion maintains luminal $[K^+]$ for continued uptake via $Na^+\backslash K^+\backslash 2Cl^-$-cotransporters and supports a lumen-positive ψ_t for paracellular Na^+ absorption (Frizzell et al. 1984b). The G_K^a responsible for K^+ secretion also creates a ψ_a supporting Na^+-coupled nutrient absorption (Fig. 10.7; Halm et al. 1985). Apical uptake of glucose and some amino-acids occurs via Na^+-coupled transporters using the Na^+ electrochemical gradient, $z_{Na}(\psi_a - E_{Na}^a)$, that drives luminal nutrient concentrations to low levels (Bröer 2008; Musch et al. 1987; Thompson and Kleinzeller 1985; Wright et al. 2011; Vol. 3, Chaps. 6 and 7). As luminal $[Na^+]$ drops due to absorption, the Na^+ driving force for this nutrient uptake also decreases. Assuming a near-constant intracellular $[Na^+]$ (Hudson and Schultz 1984), E_{Na}^a would decrease by ~43 mV. Also, G_K^a increases in response to lower luminal $[Na^+]$, possibly via reduced ion uptake by $Na^+\backslash K^+\backslash 2Cl^-$-cotransporters, such that ψ_a hyperpolarizes by ~42−106 mV (Halm et al. 1985), which compensates for the decreased E_{Na}^a with $z_{Na}(\psi_a - E_{Na}^a)$ remaining at ~120 mV inward. The presence of electrogenic K^+ secretion in these intestinal cells contributes to the driving forces for $Na^+\backslash K^+\backslash 2Cl^-$-cotransporters and paracellular Na^+ absorption as well as nutrient absorption.

Endothelia of the brain vasculature contribute an essential layer to the blood–brain barrier (Daneman and Prat 2015; Greene et al. 2019; Hladky and Barrand 2016). Many of the key features found in this endothelium resemble the ectothelial specializations already discussed including tight junction limited paracellular flow as well as transcellular flow guided by various pumps, cotransporters, exchangers, and channels. Exit from the brain becomes tantamount to excretion given the privileged status of this interstitial space. A major component of trans-endothelial flow occurs as Na^+Cl^- uptake from the capillary lumen by a mechanism similar to that in the thick ascending limb of Henle's loop in the renal tubule (Fig. 10.4; Hladky and Barrand 2016; O'Donnell et al. 2017). Fluid influx into the brain interstitium driven by the attendant osmotic gradient generated by solute uptake provides fluid replenishment together with a means to carry excess metabolites through interstitial spaces for eventual efflux back to the circulation. The $K_{Ca}3.1$ channel (*KCNN4*, IK1) contributes to G_K^a such that net K^+ efflux from the brain may occur concurrently with fluid entry (Chen et al. 2015). As in K^+ transporting ectothelia, regulation of K^+ exit across the luminal (apical) versus the abluminal (basolateral) membrane constitutes a means to either remove K^+ from the brain during excess or return K^+ to the brain during episodes of depletion. Particularly during periods of ischemia, a local transendothelial route for K^+ exit would benefit the need for reducing the buildup of extracellular K^+. Whether the brain endothelium exhibits episodic regional heterogeneity that creates a regulated flow pathway of entry and exit for fluid as well as for K^+ remains for further investigation.

10.3.3.4 Cl^- Secretion

The driving force for apical Cl^- exit during Cl^- secretion (Fig. 10.3) requires hyperpolarization of ψ_a away from E_{Cl}^a with $z_{Cl}(\psi_a - E_{Cl}^a) > 0$. Opening basolateral K^+ channels provides this hyperpolarization (Frizzell and Hanrahan 2012), but only in conjunction with a sizable G_{pr} that electrically couples apical and basolateral membranes (Eqs. 10.2a, 10.2b, and 10.2c) producing an added complication of possible nonselective paracellular flow of ions that may partially dissipate the active transcellular Cl^- flow. Simulations of salivary acinar Cl^- secretion indicate that modest activation of G_K^a increases Cl^- secretion by 10–25% (Cook and Young 1989; Palk et al. 2010), lessening the dependence on G_{pr}. Supporting this concept of E_K^a contributing to the driving force for apical Cl^- exit, stimulation of parotid acinar cells activates apical K^+ currents consistent with the presence of $K_{Ca}1.1$ and $K_{Ca}3.1$ (Almassy et al. 2012). As this primary secreted acinar fluid flows along into salivary ducts, these epithelial cells alter the ion composition by a mechanism similar to the Na^+ absorptive model (Fig. 10.2). Submandibular ducts secrete K^+ using apical $K_{Ca}1.1$, leading to a saliva $[K^+]$ of ~35 mM (Mangos et al. 1973; Nakamoto et al. 2008). In these ductal cells, the G_K^a producing K^+ secretion also assists in setting the driving force for apical Na^+ entry.

Airway epithelia secrete Cl$^-$ such that the resulting exo-epithelial apical fluid layer supports mucociliary clearance (Hoegger et al. 2014). An associated K$^+$ secretion, only ~3% of the Cl$^-$ secretory rate (Clarke et al. 1997; Welsh et al. 1983a), leads to apical fluid [K$^+$] of 9–30 mM (Joris et al. 1993; Knowles et al. 1997; Namkung et al. 2009; Robinson et al. 1989). The K$_{Ca}$1.1 K$^+$ channel (Manzanares et al. 2011), and possibly K$_V$7 (Moser et al. 2008; Namkung et al. 2009), supports secretagogue-activated K$^+$ secretion occurring together with ciliary beating in this apical fluid layer. Since the physiologic stimulus via ATP release followed by P2Y receptor activation produces a transient response (Kis et al. 2016; Manzanares et al. 2011; Zaidman et al. 2017), the augmentation of Cl$^-$ secretion and thereby fluid secretion likely occurs as short bursts spatially dispersed to locally activated ATP release sites along the airway. These focal sites of fluid secretion likely contribute to the detachment of mucus strands and threads, which serves mucociliary clearance of airway debris (Hoegger et al. 2014; Quinton 2017; Xie et al. 2018).

Alveolar type-2 cells absorb Na$^+$ electrogenically (Fig. 10.2) with G_{Cl}^a providing a pathway for electrically coupled transcellular Cl$^-$ absorption (O'Grady and Lee 2003). The observed localization of several K$_V$-type K$^+$ channels in the apical membrane (Lee et al. 2003) supports the possibility for K$^+$ secretion (Nielson 1986; Saumon and Basset 1993) that also alters Cl$^-$ flow via this transcellular route. In the absence of activated G_K^a, Na$^+$ and Cl$^-$ channels determine ψ_a such that both ions enter the cell, $z_{Na}(\psi_a - E_{Na}^a) < 0$ and $z_{Cl}(\psi_a - E_{Cl}^a) < 0$, as the first step in transcellular absorption. Opening apical K$^+$ channels increases apical fluid [K$^+$] while hyperpolarizing ψ_a toward E_K^a such that the Cl$^-$ driving force may reach zero or reverse direction, $z_{Cl}(\psi_a - E_{Cl}^a) \geq 0$. Inhibition of these K$_V$ channels during hypoxic episodes (Bardou et al. 2009; O'Grady and Lee 2005) would increase an inward Cl$^-$ driving force thus enhancing apical Cl$^-$ entry via G_{Cl}^a, thereby contributing to maximal thinning of alveolar surface liquid that promotes gas exchange (Matthay 2014).

Sweat glands and mammary glands elaborate an exo-epithelial fluid onto the skin surface by producing a primary fluid secretion in the secretory coil followed by ductal modification (Quinton et al. 1999; Sato 2001; Shennan and Peaker 2000; Vol. 2, Chaps. 5 and 6). Mammary epithelial cell lines recapitulate milk production by exhibiting Cl$^-$ secretion (Blaug et al. 2003) and Na$^+$ absorption (Palmer et al. 2011). K$^+$ secretion accompanies Na$^+$ absorption (Palmer et al. 2011), supporting a process in which ductal modification elevates [K$^+$] to 15–25 mM and lowers [Na$^+$] to ~8 mM (Shennan and Peaker 2000). Apical K$^+$ exit proceeds through K$_{Ca}$3.1 and several K$_{2P}$ (K$_{2P}$1, K$_{2P}$2, K$_{2P}$3, K$_{2P}$9, K$_{2P}$10; KCNK) channel types, which activate further via P2Y-purinergic receptors (Palmer et al. 2011; Srisomboon et al. 2018). Persistent P2Y stimulation diminishes K$_{2P}$ activity thereby leaving primarily the K$_{Ca}$3.1 exit route. Independence of control occurs through distinct receptor locations with Na$^+$ absorption activated via basolateral P2Y receptors and K$^+$ secretion via apical P2Y receptors.

Human eccrine sweat [K$^+$] reaches 7–8 mM via actions of the secretory coil and duct cells (Bijman and Quinton 1984; Saint-Criq and Gray 2017; Sato and Sato

1990). Frog skin produces a surface liquid layer by a similar process driven via exocrine gland Cl^- secretion followed by electrogenic Na^+ absorption across the skin surface epithelium (Larsen 2011; Larsen and Ramløv 2013; Watlington and Huf 1971). Recirculation of Na^+ allows evaporative water loss without major Na^+ loss. A Ca^{2+}-activated K^+ channel consistent with $K_{Ca}1.1$ opens during Cl^- secretion in these exocrine glands (Sørensen et al. 2001) leading to $[K^+]$ in this fluid of 15–20 mM (Larsen and Ramløv 2013). In contrast, rat and mouse paw sweat glands produce sweat with $[K^+]$ of 140–190 mM and $[Na^+]$ of 60–10 mM, with a primary fluid from the secretory coil of 120–160 mM $[K^+]$ and 60–35 mM $[Na^+]$ (Brusilow et al. 1968; Quatrale and Laden 1968; Sato and Sato 1978; Sato et al. 1994). Compared with the inner ear that produces endolymph with a high $[K^+]$, secretory coil cells secrete K^+ with a ψ_a ranging from -50 to -40 mV so that apical exit of K^+ via channels would require unreasonably high intracellular $[K^+]$ to assure $z_K(\psi_a - E_K^a) > 0$ (Sato 1980). Instead, K^+ secretion apparently relies on an apical transporter that couples K^+ flow to a large outwardly directed energy gradient, perhaps apical exit of lactate$^-$ (see Sect 10.3.4, K^+\citrate^{3-}-cotransporter), unless K^+ secretion occurs in a minority cell type with a ψ_a near 0 mV.

Bicarbonate secretion proceeds along the pancreatic duct modifying the primary fluid secreted from the acinar cells by increasing both the volume and luminal $[HCO_3^-]$, reaching $\sim$140 mM in human pancreas (Argent et al. 2012; Lee et al. 2012; Vol. 1, Chap. 12; Vol. 2, Chap. 4). The presence of Cl^-/HCO_3^--exchangers (*SLC26A6*; Vol. 3, Chap. 12) in the apical membrane replaces luminal Cl^- with HCO_3^- producing net HCO_3^- secretion. Opening apical membrane K^+ channels acts to hyperpolarize ψ_a toward E_K^a potentially augmenting the driving force for Cl^- and HCO_3^- exit via the apical CFTR channel $(z_{Cl}(\psi_a - E_{Cl}^a) > 0;\ z_{HCO3}(\psi_a - E_{HCO3}^a) > 0)$ as well as the electrogenic exchanger *SLC26A6* $(2E_{HCO3}^a - E_{Cl}^a - \psi_a > 0)$ (Saint-Criq and Gray 2017). Increases in G_K^a occur through Ca^{2+}-dependent activation of apical $K_{Ca}1.1$ during luminal bile acid stimulated HCO_3^- secretion, but not during secretin stimulation (Venglovecz et al. 2011). A computational model of pancreatic HCO_3^- secretion supports the relative lack of augmentation by increasing G_K^a during the production of high luminal $[HCO_3^-]$ (Yamaguchi et al. 2017). The constant $[K^+]$ of pancreatic fluid ($\sim$4 mM) during secretin stimulation of an ultimately high $[HCO_3^-]$, $\sim$140 mM, also indicates a lack of substantial K^+ secretion expected with increased G_K^a (Case 2006; Case et al. 1969; Padfield et al. 1989). However, in the rat, which produces a lower pancreatic fluid $[HCO_3^-]$ of $\sim$70 mM, secretin stimulation elevates luminal $[K^+]$ from $\sim$5 mM to $\sim$8 mM (Sewell and Young 1975). This K^+ secretion may occur either via paracellular flow due to the lumen-negative ψ_t generated by Cl^- and HCO_3^- secretion or via a cellular route through G_K^a. Claudin expression patterns during pancreas development support the presence of cation-selective tight junctions (Westmoreland et al. 2012; Vol. 2, Chap. 4), and localization of K^+ channel types supports a presence in the apical membrane for several types in addition to $K_{Ca}1.1$ (Hayashi and Novak 2013; Venglovecz et al. 2015). Potential roles for G_K^a augmentation of the driving force for apical anion exit appear limited to particular conditions with lower luminal [HCO$_3$$^-$] (Venglovecz et al. 2011). Specifically, E^a_{HCO3} at the point of apical flow reversal for HCO$_3$$^-$ depends critically on ψ_a for both CFTR-dependent $\left(^{rev}E^a_{HCO3} = \psi_a\right)$ and SLC26A6-dependent $\left(^{rev}E^a_{HCO3} = \frac{1}{2}\left(\psi_a + E^a_{Cl}\right)\right)$ exit (Argent et al. 2012; Case 2006; Ishiguro et al. 2002). Conditions with low luminal [Cl$^-$] decrease E^a_{Cl} thereby disadvantaging the use of SLC26A6. A major constraint imposed by employing increases in G^a_K results from the attendant rise in [K$^+$] that decreases E^a_K, thus limiting the size of ψ_a and the ultimate attainment of a high luminal [HCO$_3$$^-$].

10.3.4   Apical Fluid Composition

Several epithelia produce an exo-epithelial fluid covering the apical membrane with K$^+$ as a major cation component. Aside from the excretory action by the kidney and the high endolymph [K$^+$] of the inner ear that serves sensory transduction, the utility of other high [K$^+$] exo-epithelial fluids often remains obscure. In those cases that G^a_K aids in potentiating Cl$^-$ secretion, luminal [K$^+$] generally remains lower than [Na$^+$]. In contrast, the colonic lumen exhibits a high [K$^+$] fluid space from which nutrient and fluid absorption occur. Although colonic K$^+$ secretion has been suggested to operate in support of Cl$^-$ secretion (Nanda Kumar et al. 2010), many physiologic secretagogues stimulate K$^+$ secretion largely independent of Cl$^-$ secretion (Fig. 10.3c, see Sect. 10.2.1.2). The K$^+$ secreted into the colonic lumen may serve instead primarily as an ionic osmolyte for maintaining a substantial luminal volume in the face of continual absorption of Na$^+$ together with short-chain fatty acids. This augmentation of luminal fluid volume provides space for the large community of resident bacteria (a major component of the overall microbiome) that ferments poorly digestible meal components into short-chain fatty acids as well as other metabolites (Davila et al. 2013; Preidis and Versalovic 2009). Using K$^+$ rather than Na$^+$ for sustaining this microbiome space allows the final excretion of gut fluid to be more nearly devoid of Na$^+$, as well as contributing a modest amount to total body K$^+$ excretion, ~10%.

The presence of K$^+$ in these exo-epithelial fluids also may confer other advantages, such as maintaining the epithelial barrier against infection. Saliva suppresses microbial growth using secreted proteins and specific ionic composition (Carpenter 2013; van't Hof et al. 2014; Viejo-Díaz et al. 2004; Wang and Germaine 1993). By secreting K$^+$ and reabsorbing Na$^+$, salivary ducts generate saliva that restricts bacterial use of Na$^+$-dependent transporters and diminishes bacterial K$^+$ gradients, E^{bac}_K. The high [K$^+$] sweat of rodent paws also may serve an antimicrobial role, as well as possibly enhancing the sensation of pheromones by adding K$^+$ to inhaled aerosols (see Sect. 10.3.2.2). Similarly, the bioenergetics enforced by a low Na$^+$ and high K$^+$ environment may contribute to the stability of the colonic microbiome and resistance to overpopulation by virulent strains (Diskowski et al. 2015; Epstein and Schultz 1965; Häse et al. 2001; Vol. 2, Chap. 2). Flagella constitute an important virulence factor for *Vibrio cholera*, with the Na^+-driven flagellar motor of *Vibrio* rotating ~5 times faster than H^+-driven flagellar motors (Häse and Barquera 2001; Minamino and Imada 2015). A quorum-sensing receptor of *Escherichia coli* (including the enterohemorrhagic strain O157:H7) responds to norepinephrine/epinephrine as well as its own aromatic amine, leading to biofilm formation (Rossi et al. 2018; Strandwitz 2018). The high [norepinephrine] in the colonic lumen (~0.3 μM; Asano et al. 2012) could result from the overflow of enteric nerves stimulating K^+ secretion (Green et al. 2004; Zhang et al. 2009b; see Sect. 10.4) and/or production by lumen resident bacteria (Strandwitz 2018) that leads to alterations in microbiota composition (Moreira et al. 2016). Bacterial biofilms also likely contribute to luminal $[K^+]$, which recruits motile bacteria for incorporation into the biofilm (Humphries et al. 2017). In general, exo-epithelial fluids high in K^+ likely limit bacterial growth bioenergetically while modulating the virulence of biofilm formation thereby easing the burden on other components of immune system surveillance.

The tubular lumen of the epididymis contains an exo-epithelial fluid maintaining spermatozoa in a quiescent state during their temporary residence. Together with low luminal pH and secreted signaling factors, a high luminal $[K^+]$ aids in maturation of spermatozoa as well as limiting their motility (Breton and Brown 2013; Dacheux and Dacheux 2013; Gao et al. 2019; Shum et al. 2011). The resulting decrease in E_K^{sperm} depolarizes ψ_{sperm} likely influencing Ca^{2+}-dependent regulation of the flagellar dynein motor (Lishko et al. 2012; Miller et al. 2015; Mizuno et al. 2012; Yoshida and Yoshida 2011). The extent of this depolarization alters the balance between activating voltage-gated Ca^{2+} channels and decreasing the driving force for Ca^{2+} entry into the spermatazoan, $z_{Ca}(\psi_{sperm} - E_{Ca}^{sperm})$. Particularly for spermatozoa of certain fishes, the high $[K^+]$ of seminal fluid provides the key restriction on motility until release into the external environment dilutes the $[K^+]$ (Alavi and Cosson 2006). Mammalian epididymal fluid exhibits a luminal $[K^+]$ ranging among species from 18 to 55 mM with a median of ~35 mM (Jones 1978; Levine and Marsh 1971; Turner et al. 1977). Ion absorption and secretion contributes to adjustments of luminal fluid volume and composition along the epididymis (Breton and Brown 2013; Shum et al. 2011; Wong and Yeung 1978), such that $[K^+]$ increases or decreases as a direct result. Measurement of current (I_{sc}) flowing across epithelial monolayers from primary cultures of epididymal cells supports the capacity for electrogenic K^+ secretion (Fig. 10.3c; Gao et al. 2019). Apical membrane K^+ exit likely occurs via $K_{Ca}1.1$ and K_{ATP} (possibly $K_{ir}6.2/SUR2$) activated by H_2S signaling pathways (Chan et al. 1995; Gao et al. 2019; Huang et al. 1999; Lybaert et al. 2008). This capacity for transcellular electrogenic K^+ secretion in the principal cells together with the segregation of H^+ secretion to the clear cells allows independent regulation of $[K^+]$ and pH in the epididymal lumen (Breton and Brown 2013; Gao et al. 2019).

Addition of prostatic fluid, ~65 mM K^+, to seminal fluid during spermatozoan transit into the urethra maintains a relatively high luminal $[K^+]$ while contributing [citrate^{3-}] ranging from 7 to 200 mM (median 94 mM) into the exo-epithelial fluid carrying the spermatozoa (Kavanagh 1985). Citrate^{3-} uptake provides a metabolic substrate for spermatozoan ATP synthesis as motile capability increases (Mycielska et al. 2015). The proposed prostate epithelial mechanism for K$^+$ secretion involves basolateral membrane uptake via Na$^+$/K$^+$-ATPase and apical exit via K$^+$\citrate^{3-}-cotransport which links the exit into the lumen of K$^+$ and citrate^{3-} (Mobasheri et al. 2003), unlike the more common channel mechanism for apical K$^+$ exit (see Sect. 10.2.1, Figs. 10.2 and 10.3). The 4 to 1 coupling stoichiometry of the K$^+$\citrate^{3-}-cotransporter (*SLC25A1*; Mazurek et al. 2010; Mycielska and Djamgoz 2004) sets an outward driving force for citrate^{3-} maintaining exit across the apical membrane into the prostate gland lumen, $\psi_a - 4E_K^a + 3E_{cit}^a > 0$. Uptake of K$^+$ across the apical membrane via H$^+$/K$^+$-ATPase allows luminal [citrate^{3-}] to exceed luminal [K$^+$] while also adjusting luminal pH (Pestov et al. 2002).

Variations in the [K$^+$] of uterine luminal fluid (15–30 mM) likely contributes to spermatozoan motility as well as blastocyst implantation (Borland et al. 1977; Clemetson et al. 1972; Casslén and Nilsson 1984; Iritani et al. 1974; Wang and Dey 2006; Zhang et al. 2017). Production of this uterine fluid results from Cl$^-$ secretion (Fig. 10.3) across the endometrial gland epithelium with an accompanying K$^+$ secretion via apical membrane K$^+$ channels (K$_{Ca}$2.1, *KCNN1*; and K$_{Ca}$2.3, *KCNN3*) occurring at a rate ~25% that of Cl$^-$ secretion (Palmer et al. 2008; Ruan et al. 2014). The estradiol stimulated switching of expression from K$_{Ca}$2.1 to K$_{Ca}$2.3 (Palmer et al. 2008) may relate to estrous cycle-dependent decreases in luminal [K$^+$] during mid-cycle that enhances spermatozoan motility (Casslén and Nilsson 1984). The ultimate destination of the spermatozoa in the fallopian tube exhibits an elevated luminal [K$^+$], ~25 mM, which may contribute to the fertilization process (Borland et al. 1977, 1980; Leese et al. 2001; Li and Winuthayanon 2017; Maillo et al. 2016). The cellular mechanism that secretes this K$^+$ into the fallopian tube lumen likely also involves Cl$^-$ secretion (Fig. 10.3; Brunton and Brinster 1971; Downing et al. 1997; Keating and Quinlan 2008, 2012; Keating et al. 2019) together with apical membrane K$_{Ca}$1.1 (James and Okada 1994). The hyperpolarization of ψ_{sperm} necessary for spermatazoan capacitation that culminates in oocyte fertilization (Aitken and Nixon 2013; Lishko et al. 2012; Puga Molina et al. 2018) likely requires a lowering of luminal [K$^+$] via a pulse of fluid secretion driven by Cl$^-$ secretion without accompanying K$^+$ secretion. Movement along the fallopian tube takes the embryo to the uterus, where, as a blastocyst, implantation within the endometrium occurs (Wang and Dey 2006). Elevation of luminal [K$^+$] to ~45 mM during the window of receptivity for implantation enhances blastocyst contact with the endometrial surface possibly by Ca^{2+} signals resulting from depolarizing ψ_a of the trophectoderm and endometrium (Clemetson et al. 1972; Ruan et al. 2014).

10.4   Signaling Pathways for Transepithelial K⁺ Secretion

Regulation of K^+ secretion often follows the control of Na^+ absorption and Cl^- secretion (Vol. 1, Chaps. 8 and 9), since these cellular mechanisms (Figs. 10.2 and 10.3) depend on basolateral Na^+/K^+-pumps that provide K^+ uptake, and since opening either Na^+ or Cl^- channels in the apical membrane generally enhances the driving force for K^+ exit via apical channels (see Sect. 10.2.1). Understanding of the control mechanisms maintaining the steady-state balance of K^+ uptake and exit during K^+ secretion remain incomplete, but responses to physiologic stimuli provide insights into epithelial regulation (Heitzmann and Warth 2008, Malnic et al. 2013, Welling 2016; Vol. 3, Chaps. 12–14, 16–20, 22, 23, and 25). Stimulation of electrogenic Na^+ absorption by aldosterone acts via increasing the number of Na^+ channels (ENaC) in the apical membrane by slowing endocytosis (Butterworth et al. 2009). Direct control of apical K^+ channel activity also occurs in the renal principal cells and intercalated cells responsible for K^+ excretion via cellular signaling pathways leading to channel phosphorylation and interaction with membrane lipids (Carrisoza-Gaytan et al. 2016; Satlin et al. 2006; Wang and Huang 2013; Welling 2013, 2016; Vol. 3, Chaps. 19 and 23). Mirroring the regulation of ENaC, $K_{ir}1.1$ activity depends on the kinetics of membrane insertion and endocytosis thus providing major control points for adjusting G_K^a. Increases in tubular fluid flow activate $K_{Ca}1.1$ via increases in cytosolic Ca^{2+} produced by mechanical deformation of the primary cilium and paracrine activation of purinergic receptors. A gut-derived factor also stimulates renal K^+ excretion, as well as extrarenal K^+ sequestration, such that plasma $[K^+]$ remains nearly constant during ingestion of K^+ containing meals (Epstein and Lifschitz 2016; Lee et al. 2007; McDonough and Youn 2017; Oh et al. 2011; Youn 2013). Acting through inhibition of distal tubule $Na^+\backslash Cl^-$-cotransporters (NCC) via rapid dephosphorylation, this gut-derived factor leads to increased fluid flow and Na^+ delivery to the collecting duct that activates electrogenic Na^+ absorption along with K^+ secretion (Penton et al. 2015; Sørensen et al. 2013).

Colonic K^+ secretion often occurs without activation of Cl^- secretion, indicating the presence of signaling pathways allowing independent control of apical K^+ and Cl^- channels. Since both K^+ secretion and Cl^- secretion use cAMP production as a key signaling event, activating K^+ secretion without Cl^- secretion requires a separation between these signaling cascades either within distinct cell types or via functional compartmentalization within a common secretory cell. Support for compartmentalization comes from β-adrenergic activation that produces transient Cl^- secretion and sustained K^+ secretion (Halm et al. 2010; Zhang et al. 2009a, b). The transient Cl^- secretion results from cAMP production via a transmembrane adenylyl cyclase activated by β2-adrenergic receptors. In contrast, activation of sustained K^+ secretion proceeds via soluble adenylyl cyclase activated by a β1\β2-adrenergic receptor complex. Phosphodiesterases likely contribute to the functional compartmentalization by restricting cAMP spread within the cell (Pozdniakova and Ladilov 2018; Torres-Quesada et al. 2017). These secretory cells also respond to aldosterone with K$^+$ secretion independent of Na$^+$ absorption and Cl$^-$ secretion (Halm and Halm 1994; Rechkemmer and Halm 1989). This activation requires gene transcription and protein translation, occurring with a delay of ~15 min and half-time of ~40 min, suggesting a mechanism critically dependent on a protein within the signaling cascade with a relatively short half-life. Paracrine signaling also activates apical K$^+$ channels in Cl$^-$ secretory cells, as illustrated by stimulation of K$^+$ secretion in the airway and colon by the release of ATP into the apical fluid (Clarke et al. 1997; Manzanares et al. 2011; Sausbier et al. 2006; Zhang et al. 2009b). This range of control for colonic K$^+$ secretion allows for an adaptable maintenance of luminal [K$^+$] that serves both excretion and interactions with the enteric microbiome.

10.5 Summary

Transepithelial K$^+$ secretion contributes to a wide range of physiologic actions by increasing the [K$^+$] of exo-epithelial spaces. This K$^+$ secretion generally occurs in cells also capable of absorbing Na$^+$ or secreting Cl$^-$ such that the electrochemical gradients for these ions couple to maintain the rate and orientation of transepithelial flow. Several epithelia, notably from the inner ear, kidney, and colon, modify the Cl$^-$ secretory mechanism to produce K$^+$ secretion without transcellular Cl$^-$ flow, by redirecting Cl$^-$ taken up via Na$^+$\K$^+$\2Cl$^-$-cotransporters back across the basolateral membrane through Cl$^-$ channels. Renal and colonic K$^+$ secretion together produce K$^+$ excretion into a fluid with high [NH$_4^+$], which competes for passage through many K$^+$ transport proteins including the apical K$^+$ channels. Sensation in the inner ear depends crucially on the high [K$^+$] of endolymph, while olfaction likely modulates sensory signals via high apical fluid [K$^+$]. Apical K$^+$ channels involved in secretion also contribute to maintaining Cl$^-$ exit in several Cl$^-$ secretory epithelia and to intestinal nutrient uptake. The elevated [K$^+$] in fluid on several apical surfaces, including saliva and sweat as well as airway and colonic fluid, likely influences the activity of the resident microbiome living in those exo-epithelial fluids. The physiologic impact of K$^+$ secretion clearly ranges from the macroscopic issues of total body balance attended by excretion, to the microscopic manipulations that adjust transport for sensation and cellular metabolism.

References

Abbott GW (2012) KCNE2 and the K$^+$ channel: the tail wagging the dog. Channels (Austin) 6:1–10
Abbott GW (2016) KCNE1 and KCNE3: the yin and yang of voltage-gated K(+) channel regulation. Gene 576:1–13
Agarwal R, Afzalpurkar R, Fordtran JS (1994) Pathophysiology of potassium absorption and secretion by the human intestine. Gastroenterology 107:548–571
Aitken RJ, Nixon B (2013) Sperm capacitation: a distant landscape glimpsed but unexplored. Mol Hum Reprod 19:785–793

Alavi SM, Cosson J (2006) Sperm motility in fishes. (II) effects of ions and osmolality: a review. Cell Biol Int 30:1–14

Almassy J, Won JH, Begenisich TB, Yule DI (2012) Apical Ca^{2+}-activated potassium channels in mouse parotid acinar cells. J Gen Physiol 139:121–133

Amasheh S, Meiri N, Gitter AH, Schöneberg T, Mankertz J, Schulzke JD, Fromm M (2002) Claudin-2 expression induces cation-selective channels in tight junctions of epithelial cells. J Cell Sci 115:4969–4976

Amasheh S, Milatz S, Krug SM, Bergs M, Amasheh M, Schulzke JD, Fromm M (2009) Na^+ absorption defends from paracellular back-leakage by claudin-8 upregulation. Biochem Biophys Res Commun 378:45–50

Andersen OS, Silveira JE, Steinmetz PR (1985) Intrinsic characteristics of the proton pump in the luminal membrane of a tight urinary epithelium. The relation between transport rate and $\Delta\mu_H$. J Gen Physiol 86:215–234

Anderson JM, Van Itallie CM (2009) Physiology and function of the tight junction. Cold Spring Harb Perspect Biol 1:a002584

Argent BE, Gray MA, Steward MC, Case RM (2012) Secretory control of basolateral membrane K^+ and Cl^- channels in colonic crypt cells. In: Johnson LR (ed) Physiology of the gastrointestinal tract, 5th edn. Academic Press, Boston, pp 1399–1423

Armstrong CM (2015) Packaging life: the origin of ion-selective channels. Biophys J 109:173–177

Asano Y, Hiramoto T, Nishino R, Aiba Y, Kimura T, Yoshihara K, Koga Y, Sudo N (2012) Critical role of gut microbiota in the production of biologically active, free catecholamines in the gut lumen of mice. Am J Physiol Gastrointest Liver Physiol 303:G1288–G1295

Bachmann O, Juric M, Seidler U, Manns MP, Yu H (2011) Basolateral ion transporters involved in colonic epithelial electrolyte absorption, anion secretion and cellular homeostasis. Acta Physiol (Oxford) 201:33–46

Bardou O, Trinh NT, Brochiero E (2009) Molecular diversity and function of K^+ channels in airway and alveolar epithelial cells. Am J Physiol Lung Cell Mol Physiol 296:L145–L155

Barker N (2014) Adult intestinal stem cells: critical drivers of epithelial homeostasis and regeneration. Nat Rev Mol Cell Biol 15:19–33

Barmeyer C, Rahner C, Yang Y, Sigworth FJ, Binder HJ, Rajendran VM (2010) Cloning and identification of tissue-specific expression of KCNN4 splice variants in rat colon. Am J Physiol Cell Physiol 299:C251–C263

Basalingappa KM, Rajendran VM, Wonderlin WF (2011) Characteristics of Kcnn4 channels in the apical membranes of an intestinal epithelial cell line. Am J Physiol Gastrointest Liver Physiol 301:G905–G911

Bennekou P, Christophersen P (1990) The gating of human red cell Ca^{2+}-activated K^+-channels is strongly affected by the permeant cation species. Biochim Biophys Acta 1030:183–187

Berliner RW (1961) Renal mechanisms for potassium excretion. In: The harvey lectures, series 55. Academic Press, New York, pp 141–171

Berrout J, Mamenko M, Zaika OL, Chen L, Zhang W, Pochynyuk O, O'Neil RG (2014) Emerging role of the calcium-activated, small conductance, SK3 K^+ channel in distal tubule function: regulation by TRPV4. PLoS One 9:e95149

Bijman J, Quinton PM (1984) Influence of abnormal Cl^- impermeability on sweating in cystic fibrosis. Am J Physiol Cell Physiol 247:C3–C9

Binder HJ, Sangan P, Rajendran VM (1999) Physiological and molecular studies of colonic H^+,K^+-ATPase. Semin Nephrol 19:405–414

Blaug S, Rymer J, Jalickee S, Miller SS (2003) P2 purinoceptors regulate calcium-activated chloride and fluid transport in 31EG4 mammary epithelia. Am J Physiol Cell Physiol 284:C897–C909

Bleich M, Wulfmeyer VC, Himmerkus N, Milatz S (2017) Heterogeneity of tight junctions in the thick ascending limb. Ann N Y Acad Sci 1405:5–15

Borland RM, Hazra S, Biggers JD, Lechene CP (1977) The elemental composition of the environments of the gametes and preimplantation embryo during the initiation of pregnancy. Biol Reprod 16:147–157

Borland RM, Biggers JD, Lechene CP, Taymor ML (1980) Elemental composition of fluid in the human fallopian tube. J Reprod Fertil 58:479–482

Branscomb E, Russell MJ (2018) Frankenstein or a submarine alkaline vent: who is responsible for abiogenesis? Part 2: as life is now, so it must have been in the beginning. Bioessays 40: e1700182

Breton S, Brown D (2013) Regulation of luminal acidification by the V-ATPase. Physiology (Bethesda) 28:318–329

Bröer S (2008) Amino acid transport across mammalian intestinal and renal epithelia. Physiol Rev 88:249–286

Brunton WJ, Brinster RL (1971) Active chloride transport in the isolated rabbit oviduct. Am J Phys 221:658–661

Brusilow SW, Ikai K, Gordes E (1968) Comparative physiological aspects of solute secretion by the eccrine sweat gland of the rat. Proc Soc Exp Biol Med 129:731–732

Butterfield I, Warhurst G, Jones MN, Sandle GI (1997) Characterization of apical potassium channels induced in rat distal colon during potassium adaptation. J Physiol 501:537–547

Butterworth MB, Edinger RS, Frizzell RA, Johnson JP (2009) Regulation of the epithelial sodium channel by membrane trafficking. Am J Physiol Renal Physiol 296:F10–F24

Carew MA, Thorn P (2000) Carbachol-stimulated Cl⁻ secretion in mouse colon: evidence of a role for autocrine prostaglandin-E₂ release. Exp Physiol 85:67–72

Carpenter GH (2013) The secretion, components, and properties of saliva. Annu Rev Food Sci Technol 4:267–276

Carrisoza-Gaytán R, Salvador C, Satlin LM, Liu W, Zavilowitz B, Bobadilla NA, Trujillo J, Escobar LI (2010) Potassium secretion by voltage-gated potassium channel $K_V1.3$ in the rat kidney. Am J Physiol Renal Physiol 299:F255–F264

Carrisoza-Gaytan R, Carattino MD, Kleyman TR, Satlin LM (2016) An unexpected journey: conceptual evolution of mechanoregulated potassium transport in the distal nephron. Am J Physiol Cell Physiol 310:C243–C259

Carrisoza-Gaytán R, Wang L, Schreck C, Kleyman TR, Wang WH, Satlin LM (2017) The mechanosensitive BKα/β1 channel localizes to cilia of principal cells in rabbit cortical collecting duct (CCD). Am J Physiol Renal Physiol 312:F143–F156

Case RM (2006) Is the rat pancreas an appropriate model of the human pancreas? Pancreatology 6:180–190

Case RM, Harper AA, Scratcherd T (1969) The secretion of electrolytes and enzymes by the pancreas of the anaesthetized cat. J Physiol 201:335–348

Casslén B, Nilsson B (1984) Human uterine fluid, examined in undiluted samples for osmolarity and the concentrations of inorganic ions, albumin, glucose, and urea. Am J Obstet Gynecol 150:877–881

Catalán M, Cornejo I, Figueroa CD, Niemeyer MI, Sepúlveda FV, Cid LP (2002) CLC-2 in guinea pig colon: mRNA, immunolabeling, and functional evidence for surface epithelium localization. Am J Physiol Gastrointest Liver Physiol 283:G1004–G1013

Cereijido M, Contreras RG, Shoshani L (2004) Cell adhesion, polarity, and epithelia in the dawn of metazoans. Physiol Rev 84:1229–1262

Chan HC, Fu WO, Chung YW, Chan PS, Wong PY (1995) An ATP-activated cation conductance in human epididymal cells. Biol Reprod 52:645–652

Chang WW, Leblond CP (1971) Renewal of the epithelium in the descending colon of the mouse. I. Presence of three cell populations: vacuolated-columnar, mucous and argentaffin. Am J Anat 131:73–99

Chang WW, Nadler NJ (1975) Renewal of the epithelium in the descending colon of the mouse. IV. Cell population kinetics of vacuolated-columnar and mucous cells. Am J Anat 144:39–56

Chen YJ, Wallace BK, Yuen N, Jenkins DP, Wulff H, O'Donnell ME (2015) Blood-brain barrier KCa3.1 channels: evidence for a role in brain Na^+ uptake and edema in ischemic stroke. Stroke 46:237–244

Christophersen P (1991) Ca^{2+}-activated K^+ channel from human erythrocyte membranes: single channel rectification and selectivity. J Membr Biol 119:75–83

Clarke LL, Chinet T, Boucher RC (1997) Extracellular ATP stimulates K^+ secretion across cultured human airway epithelium. Am J Physiol Lung Cell Mol Physiol 272:L1084–L1091

Clausen C, Lewis SA, Diamond JM (1979) Impedance analysis of a tight epithelium using a distributed resistance model. Biophys J 26:291–317

Clemetson CA, Kim JK, Mallikarjuneswara VR, Wilds JH (1972) The sodium and potassium concentrations in the uterine fluid of the rat at the time of implantation. J Endocrinol 54:417–423

Codina J, Pressley TA, DuBose TD Jr (1999) The colonic H^+,K^+-ATPase functions as a Na^+-dependent K^+(NH_4^+)-ATPase in apical membranes from rat distal colon. J Biol Chem 274:19693–19698

Cook DI, Young JA (1989) Effect of K^+ channels in the apical plasma membrane on epithelial secretion based on secondary active Cl^- transport. J Membr Biol 110:139–146

Corliss JB, Baross JA, Hoffman SE (1981) An hypothesis concerning the relationships between submarine hot springs and the origin of life on earth. In: Proceedings 26th International Geological Congress, Geology of Oceans Symposium, Paris, July 1980. Oceanologica Acta Special Issue:59–69

Cornelius RJ, Wang B, Wang-France J, Sansom SC (2016) Maintaining K^+ balance on the low-Na^+, high-K^+ diet. Am J Physiol Renal Physiol 310:F581–F595

Cougnon M, Bouyer P, Jaisser F, Edelman A, Planelles G (1999) Ammonium transport by the colonic H^+-K^+-ATPase expressed in Xenopus oocytes. Am J Physiol Cell Physiol 277:C280–C287

Crambert G (2014) H-K-ATPase type 2: relevance for renal physiology and beyond. Am J Physiol Renal Physiol 306:F693–F700

Crane RK (1965) Na^+-dependent transport in the intestine and other animal tissues. Fed Proc 24:1000–1006

Crothers JM Jr, Forte JG, Machen TE (2016) Computer modeling of gastric parietal cell: significance of canalicular space, gland lumen, and variable canalicular [K^+]. Am J Physiol Gastrointest Liver Physiol 310:G671–G681. Corrigendum 312:G535

Crowson MS, Shull GE (1992) Isolation and characterization of a cDNA encoding the putative distal colon H,K-ATPase. Similarity of deduced amino acid sequence to gastric H^+,K^+-ATPase and Na^+,K^+-ATPase and mRNA expression in distal colon, kidney, and uterus. J Biol Chem 267:13740–13748

Dacheux JL, Dacheux F (2013) New insights into epididymal function in relation to sperm maturation. Reproduction 147:R27–R42

Daneman R, Prat A (2015) The blood-brain barrier. Cold Spring Harb Perspect Biol 7:a020412

Dauner K, Lissmann J, Jeridi S, Frings S, Möhrlen F (2012) Expression patterns of anoctamin 1 and anoctamin 2 chloride channels in the mammalian nose. Cell Tissue Res 347:327–341

Davila AM, Blachier F, Gotteland M, Andriamihaja M, Benetti PH, Sanz Y, Tomé D (2013) Intestinal luminal nitrogen metabolism: role of the gut microbiota and consequences for the host. Pharmacol Res 68:95–107

Dawson DC, Frizzell RA (1989) Mechanism of active K^+ secretion by flounder urinary bladder. Pflugers Arch 414:393–400

Dawson DC, Richards NW (1990) Basolateral K^+ conductance: role in regulation of Na^+Cl^- absorption and secretion. Am J Physiol Cell Physiol 259:C181–C195

Degnan KJ, Karnaky KJ Jr, Zadunaisky JA (1977) Active chloride transport in the in vitro opercular skin of a teleost (Fundulus heteroclitus), a gill-like epithelium rich in chloride cells. J Physiol 271:155–191

Delpire E, Gagnon KB (2018) Na^+-K^+-$2Cl^-$ Cotransporter (NKCC) Physiological Function in Nonpolarized Cells and Transporting Epithelia. Compr Physiol 8:871–901

DiBona DR (1985) Functional analysis of tight junction organization. Pflugers Arch 405(Suppl 1): S59–S66

DiBona DR, Mills JW (1979) Distribution of Na$^+$-pump sites in transporting epithelia. Fed Proc 38:134–143

Diskowski M, Mikusevic V, Stock C, Hänelt I (2015) Functional diversity of the superfamily of K$^+$ transporters to meet various requirements. Biol Chem 396:1003–1014

Downing SJ, Maguiness SD, Watson A, Leese HJ (1997) Electrophysiological basis of human fallopian tubal fluid formation. J Reprod Fertil 111:29–34

Duta V, Duta F, Puttagunta L, Befus AD, Duszyk M (2006) Regulation of basolateral Cl$^-$ channels in airway epithelial cells: the role of nitric oxide. J Membr Biol 213:165–174

Edmonds CJ (1981) Amiloride sensitivity of the transepithelial electrical potential and of sodium and potassium transport in rat distal colon in vivo. J Physiol 313:547–559

Eisenman G, Latorre R, Miller C (1986) Multi-ion conduction and selectivity in the high-conductance Ca^{2+}-activated K$^+$ channel from skeletal muscle. Biophys J 50:1025–1034

Eladari D, Chambrey R, Picard N, Hadchouel J (2014) Electroneutral absorption of NaCl by the aldosterone-sensitive distal nephron: implication for normal electrolytes homeostasis and blood pressure regulation. Cell Mol Life Sci 71:2879–2895

Ellison DH, Velázquez H, Wright FS (1987) Mechanisms of sodium, potassium and chloride transport by the renal distal tubule. Miner Electrolyte Metab 13:422–432

Epstein M, Lifschitz MD (2016) The unappreciated role of extrarenal and gut sensors in modulating renal potassium handling: implications for diagnosis of dyskalemias and interpreting clinical trials. Kidney Int Rep 1:43–56

Epstein W, Schultz SG (1965) Cation Transport in Escherichia coli: V. Regulation of cation content. J Gen Physiol 49:221–234

Ernst SA, Mills JW (1977) Basolateral plasma membrane localiztion of ouabain-sensitive sodium transport sites in the secretory epithelium of the avian salt gland. J Cell Biol 75:74–94

Estilo G, Liu W, Pastor-Soler N, Mitchell P, Carattino MD, Kleyman TR, Satlin LM (2008) Effect of aldosterone on BK channel expression in mammalian cortical collecting duct. Am J Physiol Renal Physiol 295:F780–F788

Fan S, Harfoot N, Bartolo RC, Butt AG (2012) CFTR is restricted to a small population of high expresser cells that provide a forskolin-sensitive transepithelial Cl$^-$ conductance in the proximal colon of the possum, Trichosurus vulpecula. J Exp Biol 215:1218–1230

Field M (1993) Intestinal electrolyte secretion. History of a paradigm. Arch Surg 128:273–278

Fischer H, Illek B, Finkbeiner WE, Widdicombe JH (2007) Basolateral Cl$^-$ channels in primary airway epithelial cultures. Am J Physiol Lung Cell Mol Physiol 292:L1432–L1443

Flessner MF (2005) The transport barrier in intraperitoneal therapy. Am J Physiol Renal Physiol 288:F433–F442

Flores CA, Melvin JE, Figueroa CD, Sepúlveda FV (2007) Abolition of Ca^{2+}-mediated intestinal anion secretion and increased stool dehydration in mice lacking the intermediate conductance Ca^{2+}-dependent K$^+$ channel Kcnn4. J Physiol 583:705–717

Florian P, Amasheh S, Lessidrensky M, Todt I, Bloedow A, Ernst A, Fromm M, Gitter AH (2003) Claudins in the tight junctions of stria vascularis marginal cells. Biochem Biophys Res Commun 304:5–10

Forte JG, Zhu L (2010) Apical recycling of the gastric parietal cell H$^+$,K$^+$-ATPase. Annu Rev Physiol 72:273–296

Foster ES, Hayslett JP, Binder HJ (1984) Mechanism of active potassium absorption and secretion in the rat colon. Am J Physiol Gastrointest Liver Physiol 246:G611–G617

Frindt G, Palmer LG (2009) K$^+$ secretion in the rat kidney: Na$^+$ channel-dependent and -independent mechanisms. Am J Physiol Renal Physiol 297:F389–F396

Frizzell RA, Hanrahan JW (2012) Physiology of epithelial chloride and fluid secretion. Cold Spring Harb Perspect Med 2:a009563

Frizzell RA, Koch MJ, Schultz SG (1976) Ion transport by rabbit colon. I. Active and passive components. J Membr Biol 27:297–316

Frizzell RA, Field M, Schultz SG (1979) Sodium-coupled chloride transport by epithelial tissues. Am J Physiol Renal Physiol 236:F1–F8

Frizzell RA, Halm DR, Krasny EJ Jr (1984a) Active secretion of chloride and potassium across the large intestine. In: Skadhauge E, Heintze K (eds) Intestinal absorption and secretion. International conference on intestinal absorption and secretion, Titisee, West Germany, June 1983. Falk Symposium 36. MTP Press, Boston, pp 313–324

Frizzell RA, Halm DR, Musch MW, Stewart CP, Field M (1984b) Potassium transport by flounder intestinal mucosa. Am J Physiol Renal Physiol 246:F946–F951

Frömter E, Diamond J (1972) Route of passive ion permeation in epithelia. Nat New Biol 235:9–13

Fujita A, Horio Y, Higashi K, Mouri T, Hata F, Takeguchi N, Kurachi Y (2002) Specific localization of an inwardly rectifying K^+ channel, Kir4.1, at the apical membrane of rat gastric parietal cells; its possible involvement in K^+ recycling for the H^+-K^+-pump. J Physiol 540:85–92

Fujita H, Chiba H, Yokozaki H, Sakai N, Sugimoto K, Wada T, Kojima T, Yamashita T, Sawada N (2006) Differential expression and subcellular localization of claudin-7, -8, -12, -13, and -15 along the mouse intestine. J Histochem Cytochem 54:933–944

Furness JB, Robbins HL, Selmer IS, Hunne B, Chen MX, Hicks GA, Moore S, Neylon CB (2003) Expression of intermediate conductance potassium channel immunoreactivity in neurons and epithelial cells of the rat gastrointestinal tract. Cell Tissue Res 314:179–189

Furukawa F, Watanabe S, Kakumura K, Hiroi J, Kaneko T (2014) Gene expression and cellular localization of ROMKs in the gills and kidney of Mozambique tilapia acclimated to fresh water with high potassium concentration. Am J Physiol Regul Integr Comp Physiol 307:R1303–R1312

Furukawa C, Ishizuka N, Hayashi H, Fujii N, Manabe A, Tabuchi Y, Matsunaga T, Endo S, Ikari A (2017) Up-regulation of claudin-2 expression by aldosterone in colonic epithelial cells of mice fed with NaCl-depleted diets. Sci Rep 7:12223

Furuse M (2010) Claudins, tight junctions, and the paracellular barrier. In: Yu ASL (ed) Claudins, Current topics in membranes, vol 65. Elsevier, Boston, pp 1–19

Gamba G (2013) Physiology and pathophysiology of the Na^+Cl^- co-transporters in the kidney. In: Alpern RJ, Moe OW, Caplan M (eds) Seldin and Giebisch's the kidney: physiology and pathophysiology, 5th edn. Elsevier, Boston, pp 1047–1080

Gamba G, Saltzberg SN, Lombardi M, Miyanoshita A, Lytton J, Hediger MA, Brenner BM, Hebert SC (1993) Primary structure and functional expression of a cDNA encoding the thiazide-sensitive, electroneutral sodium-chloride cotransporter. Proc Natl Acad Sci U S A 90:2749–2753

Gamba G, Wang W, Schild L (2013) Sodium chloride transport in the loop of Henle, distal convoluted tubule, and collecting duct. In: Alpern RJ, Moe OW, Caplan M (eds) Seldin and Giebisch's the kidney: physiology and pathophysiology, 5th edn. Elsevier, Boston, pp 1143–1179

Gao D-D, Xu J-W, Qin W-B, Peng L, Qiu Z-E, Wang L-L, Lan C-F, Cao X-N, Xu J-B, Zhu Y-X, Tang Y-G, Zhang Y-L, Zhou W-L (2019) Cellular mechanism underlying hydrogen sulfide mediated epithelial K^+ secretion in rat epididymis. Front Physiol 9:1886

Giebisch G, Klose RM, Malnic G (1967) Renal tubular potassium transport. Bull Schweiz Akad Med Wiss 23:287–312

Giménez I, Isenring P, Forbush B (2002) Spatially distributed alternative splice variants of the renal Na^+-K^+-Cl^- cotransporter exhibit dramatically different affinities for the transported ions. J Biol Chem 277:8767–8770

González C, Baez-Nieto D, Valencia I, Oyarzún I, Rojas P, Naranjo D, Latorre R (2012) K^+ channels: function-structural overview. Compr Physiol 2:2087–2149

Gonzalez-Perez V, Lingle CJ (2019) Regulation of BK channels by beta and gamma subunits. Annu Rev Physiol 81:113–137

Gow A, Davies C, Southwood CM, Frolenkov G, Chrustowski M, Ng L, Yamauchi D, Marcus DC, Kachar B (2004) Deafness in Claudin 11-null mice reveals the critical contribution of basal cell tight junctions to stria vascularis function. J Neurosci 24:7051–7062

Grahammer F, Herling AW, Lang HJ, Schmitt-Gräff A, Wittekindt OH, Nitschke R, Bleich M, Barhanin J, Warth R (2001) The cardiac K⁺ channel KCNQ1 is essential for gastric acid secretion. Gastroenterology 120:1363–1371

Green BT, Lyte M, Chen C, Xie Y, Casey MA, Kulkarni-Narla A, Vulchanova L, Brown DR (2004) Adrenergic modulation of *Escherichia coli* O157:H7 adherence to the colonic mucosa. Am J Physiol Gastrointest Liver Physiol 287:G1238–G1246

Greene C, Hanley N, Campbell M (2019) Claudin-5: gatekeeper of neurological function. Fluids Barriers CNS 16:3

Grimm PR, Foutz RM, Brenner R, Sansom SC (2007) Identification and localization of BK-β subunits in the distal nephron of the mouse kidney. Am J Physiol Renal Physiol 293:F350–F359

Grotjohann I, Gitter AH, Köckerling A, Bertog M, Schulzke JD, Fromm M (1998) Localization of cAMP- and aldosterone-induced K⁺ secretion in rat distal colon by conductance scanning. J Physiol 507:561–570

Grubb BR, Rogers TD, Boucher RC, Ostrowski LE (2009) Ion transport across CF and normal murine olfactory and ciliated epithelium. Am J Physiol Cell Physiol 296:C1301–C1309

Grunnet M, Rasmussen HB, Hay-Schmidt A, Klærke DA (2003) The voltage-gated potassium channel subunit, Kᵥ1.3, is expressed in epithelia. Biochim Biophys Acta 1616:85–94

Guh YJ, Hwang PP (2017) Insights into molecular and cellular mechanisms of hormonal actions on fish ion regulation derived from the zebrafish model. Gen Comp Endocrinol 251:12–20

Gumber S, Nusrat A, Villinger F (2014) Immunohistological characterization of intercellular junction proteins in rhesus macaque intestine. Exp Toxicol Pathol 66:437–444

Gumz ML, Lynch IJ, Greenlee MM, Cain BD, Wingo CS (2010) The renal H⁺-K⁺-ATPases: physiology, regulation, and structure. Am J Physiol Renal Physiol 298:F12–F21

Gumz ML, Rabinowitz L, Wingo CS (2015) An integrated view of potassium homeostasis. N Engl J Med 373:60–72. erratum 373:1281

Günzel D (2017) Claudins: vital partners in transcellular and paracellular transport coupling. Pflugers Arch 469:35–44

Günzel D, Fromm M (2012) Claudins and other tight junction proteins. Compr Physiol 2:1819–1852

Günzel D, Yu AS (2013) Claudins and the modulation of tight junction permeability. Physiol Rev 93:525–569

Günzel D, Stuiver M, Kausalya PJ, Haisch L, Krug SM, Rosenthal R, Meij IC, Hunziker W, Fromm M, Müller D (2009) Claudin-10 exists in six alternatively spliced isoforms that exhibit distinct localization and function. J Cell Sci 122:1507–1517

Gutman GA, Chandy KG, Adelman JP, Aiyar J, Bayliss DA, Clapham DE, Covarriubias M, Desir GV, Furuichi K, Ganetzky B, Garcia ML, Grissmer S, Jan LY, Karschin A, Kim D, Kuperschmidt S, Kurachi Y, Lazdunski M, Lesage F, Lester HA, McKinnon D, Nichols CG, O'Kelly I, Robbins J, Robertson GA, Rudy B, Sanguinetti M, Seino S, Stuehmer W, Tamkun MM, Vandenberg CA, Wei A, Wulff H, Wymore RS (2003) International Union of Pharmacology. XLI. Compendium of voltage-gated ion channels: Potassium channels. Pharm Rev 55:583–586

Halm DR (1984) β-adrenergic effects on transport. In: Gaginella TS (ed) Neuromodulation of intestinal ion transport. Symposium on neuromodulation of intestial ion transport, 34th annual fall meeeting, American Society for Pharmacology and Experimental Therapeutics, Philadelphia, August 1983. Fed Proc 43:2931–2932

Halm DR (2004) Secretory control of basolateral membrane K⁺ and Cl⁻ channels in colonic crypt cells. In: Lauf PK, Adragna NC (eds) Cell volume and signaling. international symposium on cell volume and signal transduction, Dayton, September 2003, Advances in experimental medicine and biology, vol 559. Springer, New York, pp 119–129

Halm DR, Dawson DC (1983) Cation activation of the basolateral sodium-potassium pump in turtle colon. J Gen Physiol 82:315–329

Halm DR, Dawson DC (1984a) Potassium transport by turtle colon: active secretion and active absorption. Am J Physiol Cell Physiol 246:C315–C322

Halm DR, Dawson DC (1984b) Control of potassium transport by turtle colon: role of membrane potential. Am J Physiol Cell Physiol 247:C26–C32

Halm DR, Frizzell RA (1986) Active K$^+$ transport across rabbit distal colon: relation to Na$^+$ absorption and Cl$^-$ secretion. Am J Physiol Cell Physiol 251:C252–C267

Halm DR, Frizzell RA (1991) Ion transport across the large intestine. In: Field M, Frizzell RA (eds) Intestinal absorption and secretion. Schultz SG (ed) The gastrointestinal system. Rauner BB (ed) Handbook of physiology, sect 6, vol IV, Chap. 8. American Physiological Society, Bethesda, MD, pp 257–273

Halm DR, Halm ST (1994) Aldosterone stimulates K$^+$ secretion prior to onset of Na$^+$ absorption in guinea pig distal colon. Am J Physiol Cell Physiol 266:C552–C558

Halm DR, Halm ST (2001) Prostanoids stimulate K$^+$ secretion and Cl$^-$ secretion in guinea pig distal colon via distinct pathways. Am. J. Physiol Gastrointest Liver Physiol 281:G984–G996

Halm DR, Rick R (1992) Secretion of K$^+$ and Cl$^-$ across colonic epithelium: cellular localization using electron microprobe analysis. Am J Physiol Cell Physiol 262:C1392–C1402

Halm DR, Krasny EJ Jr, Frizzell RA (1985) Electrophysiology of flounder intestinal mucosa. II. Relation of the electrical potential profile to coupled NaCl absorption. J Gen Physiol 85:865–883

Halm DR, Kirk KL, Sathiakumar KC (1993) Stimulation of Cl permeability in colonic crypts of Lieberkühn measured with a fluorescent indicator. Am J Phys 265:G423–G431

Halm ST, Liao T, Halm DR (2006) Distinct K$^+$ conductive pathways are required for Cl$^-$ and K$^+$ secretion across distal colonic epithelium. Am J Physiol Cell Physiol 291:C636–C648

Halm ST, Zhang J, Halm DR (2010) β-adrenergic activation of K$^+$ and Cl$^-$ secretion in guinea pig distal colonic epithelium proceeds via separate cAMP signaling pathways. Am J Physiol Gastrointest Liver Physiol 299:G81–G95

Häse CC, Barquera B (2001) Role of sodium bioenergetics in Vibrio cholerae. Biochim Biophys Acta 1505:169–178

Häse CC, Fedorova ND, Galperin MY, Dibrov PA (2001) Sodium ion cycle in bacterial pathogens: evidence from cross-genome comparisons. Microbiol Mol Biol Rev 65:353–370

Hayashi M, Novak I (2013) Molecular basis of potassium channels in pancreatic duct epithelial cells. Channels (Austin) 7:432–441

Hayes CP Jr, McLeod ME, Robinson RR (1967) An extravenal mechanism for the maintenance of potassium balance in severe chronic renal failure. Trans Assoc Am Phys 80:207–216

Hay-Schmidt A, Grunnet M, Abrahamse SL, Knaus HG, Klærke DA (2003) Localization of Ca^{2+}-activated big-conductance K$^+$ channels in rabbit distal colon. Pflugers Arch 446:61–68

Hazard LC (2001) Ion secretion by salt glands of desert iguanas (Dipsosaurus dorsalis). Physiol Biochem Zool 74:22–31

He Q, Halm ST, Zhang J, Halm DR (2011a) Activation of the basolateral membrane Cl$^-$ conductance essential for electrogenic K$^+$ secretion suppresses electrogenic Cl$^-$ secretion. Exp Physiol 96:305–316

He W, Liu W, Chew CS, Baker SS, Baker RD, Forte JG, Zhu L (2011b) Acid secretion-associated translocation of KCNJ15 in gastric parietal cells. Am J Physiol Gastrointest Liver Physiol 301: G591–G600

Hebert SC, Desir G, Giebisch G, Wang W (2005) Molecular diversity and regulation of renal potassium channels. Physiol Rev 85:319–371

Hegg CC, Irwin M, Lucero MT (2009) Calcium store-mediated signaling in sustentacular cells of the mouse olfactory epithelium. Glia 57:634–644

Heginbotham L, MacKinnon R (1993) Conduction properties of the cloned Shaker K$^+$ channel. Biophys J 65:2089–2096

Heintze K, Stewart CP, Frizzell RA (1983) Sodium-dependent chloride secretion across rabbit descending colon. Am J Phys 244:G357–G365

Heitzmann D, Warth R (2007) No potassium, no acid: K⁺ channels and gastric acid secretion. Physiology (Bethesda) 22:335–341

Heitzmann D, Warth R (2008) Physiology and pathophysiology of potassium channels in gastrointestinal epithelia. Physiol Rev 88:1119–1182

Heitzmann D, Grahammer F, von Hahn T, Schmitt-Gräff A, Romeo E, Nitschke R, Gerlach U, Lang HJ, Verrey F, Barhanin J, Warth R (2004) Heteromeric KCNE2/KCNQ1 potassium channels in the luminal membrane of gastric parietal cells. J Physiol 561:547–557

Heitzmann D, Koren V, Wagner M, Sterner C, Reichold M, Tegtmeier I, Volk T, Warth R (2007) KCNE beta subunits determine pH sensitivity of KCNQ1 potassium channels. Cell Physiol Biochem 19:21–32

Henriques T, Agostinelli E, Hernandez-Clavijo A, Maurya DK, Rock JR, Harfe BD, Menini A, Pifferi S (2019) TMEM16A calcium-activated chloride currents in supporting cells of the mouse olfactory epithelium. J Gen Physiol May 2: jgp.201812310.

Hibino H, Kurachi Y (2006) Molecular and physiological bases of the K⁺ circulation in the mammalian inner ear. Physiology (Bethesda) 21:336–345

Hibino H, Inanobe A, Furutani K, Murakami S, Findlay I, Kurachi Y (2010) Inwardly rectifying potassium channels: their structure, function, and physiological roles. Physiol Rev 90:291–366

Hladky SB, Barrand MA (2016) Fluid and ion transfer across the blood-brain and blood-cerebrospinal fluid barriers; a comparative account of mechanisms and roles. Fluids Barriers CNS 13:19

Hoegger MJ, Fischer AJ, McMenimen JD, Ostedgaard LS, Tucker AJ, Awadalla MA, Moninger TO, Michalski AS, Hoffman EA, Zabner J, Stoltz DA, Welsh MJ (2014) Impaired mucus detachment disrupts mucociliary transport in a piglet model of cystic fibrosis. Science 345:818–822

Holmes JL, Van Itallie CM, Rasmussen JE, Anderson JM (2006) Claudin profiling in the mouse during postnatal intestinal development and along the gastrointestinal tract reveals complex expression patterns. Gene Expr Patterns 6:581–588

Holtzclaw JD, Grimm PR, Sansom SC (2011) Role of BK channels in hypertension and potassium secretion. Curr Opin Nephrol Hypertens 20:512–517

Horng JL, Yu LL, Liu ST, Chen PY, Lin LY (2017) Potassium regulation in Medaka (Oryzias latipes) larvae acclimated to fresh water: passive uptake and active secretion by the skin cells. Sci Rep 7:16215

Hosoda Y, Karaki S, Shimoda Y, Kuwahara A (2002) Substance P-evoked Cl⁻ secretion in guinea pig distal colonic epithelia: interaction with PGE₂. Am J Physiol Gastrointest Liver Physiol 283: G347–G356

Hou J, Renigunta A, Yang J, Waldegger S (2010) Claudin-4 forms paracellular chloride channel in the kidney and requires claudin-8 for tight junction localization. Proc Natl Acad Sci U S A 107:18010–18015

Hou J, Rajagopal M, Yu AS (2013) Claudins and the kidney. Annu Rev Physiol 75:479–501

Huang Y, Chung YW, Wong PY (1999) Potassium channel activity recorded from the apical membrane of freshly isolated epithelial cells in rat caudal epididymis. Biol Reprod 60:1509–1514

Huang F, Wong X, Jan LY (2012) International union of basic and clinical pharmacology. LXXXV: calcium-activated chloride channels. Pharmacol Rev 64:1–15

Hudson RL, Schultz SG (1984) Sodium-coupled sugar transport: effects on intracellular sodium activities and sodium-pump activity. Science 224:1237–1239

Humphries J, Xiong L, Liu J, Prindle A, Yuan F, Arjes HA, Tsimring L, Süel GM (2017) Species-independent attraction to biofilms through electrical signaling. Cell 168:200–209

Iritani A, Sato E, Nishikawa Y (1974) Secretion rates and chemical composition of oviduct and uterine fluids in sows. J Anim Sci 39:582–588

Ishiguro H, Steward MC, Sohma Y, Kubota T, Kitagawa M, Kondo T, Case RM, Hayakawa T, Naruse S (2002) Membrane potential and bicarbonate secretion in isolated interlobular ducts from guinea-pig pancreas. J Gen Physiol 120:617–628

Jackson WF (2017) Boosting the signal: endothelial inward rectifier K$^+$ channels. Microcirculation 24(3). https://doi.org/10.1111/micc.12319

Jaffe DB, Wang B, Brenner R (2011) Shaping of action potentials by type I and type II large-conductance Ca^{2+}-activated K$^+$ channels. Neuroscience 192:205–218

Jakab RL, Collaco AM, Ameen NA (2013) Characterization of CFTR high expresser cells in the intestine. Am J Physiol Gastrointest Liver Physiol 305:G453–G465

James AF, Okada Y (1994) Maxi K$^+$ channels from the apical membranes of rabbit oviduct epithelial cells. J Membr Biol 137:109–118

Ji HL, Nie HG (2008) Electrolyte and fluid transport in mesothelial cells. J Epithel Biol Pharmacol 1:1–7

Joiner WJ, Basavappa S, Vidyasagar S, Nehrke K, Krishnan S, Binder HJ, Boulpaep EL, Rajendran VM (2003) Active K$^+$ secretion through multiple K$_{Ca}$-type channels and regulation by IK$_{Ca}$ channels in rat proximal colon. Am J Physiol Gastrointest Liver Physiol 285:G185–G196

Jones R (1978) Comparative biochemistry of mammalian epididymal plasma. Comp Biochem Physiol B 61:365–370

Joris L, Dab I, Quinton PM (1993) Elemental composition of human airway surface fluid in healthy and diseased airways. Am Rev Respir Dis 148:1633–1637

Karnaky KJ Jr, Degnan KJ, Zadunaisky JA (1977) Chloride transport across isolated opercular epithelium of killifish: a membrane rich in chloride cells. Science 195:203–205

Kaufhold MA, Krabbenhöft A, Song P, Engelhardt R, Riederer B, Fährmann M, Klöcker N, Beil W, Manns M, Hagen SJ, Seidler U (2008) Localization, trafficking, and significance for acid secretion of parietal cell Kir4.1 and KCNQ1 K$^+$ channels. Gastroenterology 134:1058–1069

Kavanagh JP (1985) Sodium, potassium, calcium, magnesium, zinc, citrate and chloride content of human prostatic and seminal fluid. J Reprod Fertil 75:35–41

Keating N, Quinlan LR (2008) Effect of basolateral adenosine triphosphate on chloride secretion by bovine oviductal epithelium. Biol Reprod 78:1119–1126

Keating N, Quinlan LR (2012) Small conductance potassium channels drive ATP-activated chloride secretion in the oviduct. Am J Physiol Cell Physiol 302:C100–C109

Keating N, Dev K, Hynes AC, Quinlan LR (2019) Mechanism of luminal ATP activated chloride secretion in a polarized epithelium. J Physiol Sci 69:85–95

Kelly ML, Abu-Hamdah R, Jeremic A, Cho SJ, Ilie AE, Jena BP (2005) Patch clamped single pancreatic zymogen granules: direct measurements of ion channel activities at the granule membrane. Pancreatology 5:443–449

Kim S, Ma L, Jensen KL, Kim MM, Bond CT, Adelman JP, Yu CR (2012) Paradoxical contribution of SK3 and GIRK channels to the activation of mouse vomeronasal organ. Nat Neurosci 15:1236–1244

Kinne R, Kinne-Saffran E, Schütz H, Schölermann B (1986) Ammonium transport in medullary thick ascending limb of rabbit kidney: involvement of the Na$^+$,K$^+$,Cl$^-$-cotransporter. J Membr Biol 94:279–284

Kis A, Krick S, Baumlin N, Salathe M (2016) Airway hydration, apical K$^+$ secretion, and the large-conductance, Ca^{2+}-activated and voltage-dependent potassium (BK) Channel. Ann Am Thorac Soc 13(Suppl 2):S163–S168

Kitadai N, Nakamura R, Yamamoto M, Takai K, Yoshida N, Oono Y (2019) Metals likely promoted protometabolism in early ocean alkaline hydrothermal systems. Sci Adv 5:eaav7848

Kitajiri S, Miyamoto T, Mineharu A, Sonoda N, Furuse K, Hata M, Sasaki H, Mori Y, Kubota T, Ito J, Furuse M, Tsukita S (2004) Compartmentalization established by claudin-11-based tight junctions in stria vascularis is required for hearing through generation of endocochlear potential. J Cell Sci 117:5087–5096

Kleyman TR, Kashlan OB, Hughey RP (2018) Epithelial Na$^+$ Channel Regulation by Extracellular and Intracellular Factors. Annu Rev Physiol 80:263–281

Knauth LP (2005) Temperature and salinity history of the Precambrian ocean: implications for the course of microbial evolution. Palaeogeogr Palaeoclimatol Palaeoecol 219:53–69

Knowles MR, Robinson JM, Wood RE, Pue CA, Mentz WM, Wager GC, Gatzy JT, Boucher RC (1997) Ion composition of airway surface liquid of patients with cystic fibrosis as compared with normal and disease-control subjects. J Clin Invest 100:2588–2595

Köckerling A, Fromm M (1993) Origin of cAMP-dependent Cl⁻ secretion from both crypts and surface epithelia of rat intestine. Am J Physiol Cell Physiol 264:C1294–C1301

Köckerling A, Sorgenfrei D, Fromm M (1993) Electrogenic Na⁺ absorption of rat distal colon is confined to surface epithelium: a voltage-scanning study. Am J Physiol Cell Physiol 264: C1285–C1293

Koefoed-Johnsen V, Ussing HH (1958) The nature of the frog skin potential. Acta Physiol Scand 42:298–308

Koeppen BM, Giebisch GH (1985) Mineralocorticoid regulation of sodium and potassium transport by the cortical collecting duct. In: Graves S (ed) Regulation and development of membrane transport processes. 37th annual symposium of the society of general physiologists, Woods Hole, September 1983, Society of general physiologists series, vol 39. Wiley, New York, pp 89–104

Kopic S, Murek M, Geibel JP (2010) Revisiting the parietal cell. Am J Physiol Cell Physiol 298: C1–C10

Kriz W, Kaissling B (2013) Structural organization of the mammalian kidney. In: Alpern RJ, Moe OW, Caplan M (eds) Seldin and Giebisch's the kidney: physiology and pathophysiology, 5th edn. Elsevier, Boston, pp 595–691

Krug SM, Günzel D, Conrad MP, Rosenthal R, Fromm A, Amasheh S, Schulzke JD, Fromm M (2012) Claudin-17 forms tight junction channels with distinct anion selectivity. Cell Mol Life Sci 69:2765–2778

Krug SM, Schulzke JD, Fromm M (2014) Tight junction, selective permeability, and related diseases. Semin Cell Dev Biol 36:166–176

Lambrecht NW, Yakubov I, Scott D, Sachs G (2005) Identification of the K⁺ efflux channel coupled to the gastric H⁺-K⁺-ATPase during acid secretion. Physiol Genomics 21:81–91

Lane N (2017) Proton gradients at the origin of life. Bioessays 39:e1600217

Lane NJ, Reese TS, Kachar B (1992) Structural domains of the tight junctional intramembrane fibrils. Tissue Cell 24:291–300

Larsen EH (2011) Reconciling the Krogh and Ussing interpretations of epithelial chloride transport—presenting a novel hypothesis for the physiological significance of the passive cellular chloride uptake. Acta Physiol (Oxford) 202:435–464

Larsen EH, Ramløv H (2013) Role of cutaneous surface fluid in frog osmoregulation. Comp Biochem Physiol A Mol Integr Physiol 165:365–370

Latorre R, Castillo K, Carrasquel-Ursulaez W, Sepulveda RV, Gonzalez-Nilo F, Gonzalez C, Alvarez O (2017) Molecular Determinants of BK Channel Functional Diversity and Functioning. Physiol Rev 97:39–87

Le Bivic A (2013) Evolution and cell physiology. 4. Why invent yet another protein complex to build junctions in epithelial cells? Am J Physiol Cell Physiol 305:C1193–C1201

Lee SY, Maniak PJ, Ingbar DH, O'Grady SM (2003) Adult alveolar epithelial cells express multiple subtypes of voltage-gated K⁺ channels that are located in apical membrane. Am J Physiol Cell Physiol 284:C1614–C1624

Lee FN, Oh G, McDonough AA, Youn JH (2007) Evidence for gut factor in K⁺ homeostasis. Am J Physiol Renal Physiol 293:F541–F547

Lee DB, Jamgotchian N, Allen SG, Abeles MB, Ward HJ (2008a) A lipid-protein hybrid model for tight junction. Am J Physiol Renal Physiol 295:F1601–F1612

Lee WK, Torchalski B, Roussa E, Thévenod F (2008b) Evidence for KCNQ1 K⁺ channel expression in rat zymogen granule membranes and involvement in cholecystokinin-induced pancreatic acinar secretion. Am J Physiol Cell Physiol 294:C879–C892

Lee MG, Ohana E, Park HW, Yang D, Muallem S (2012) Molecular mechanism of pancreatic and salivary gland fluid and HCO_3 secretion. Physiol Rev 92:39–74

Leese HJ, Tay JI, Reischl J, Downing SJ (2001) Formation of Fallopian tubal fluid: role of a neglected epithelium. Reproduction 121:339–346

Levick JR, Michel CC (2010) Microvascular fluid exchange and the revised Starling principle. Cardiovasc Res 87:198–210

Levine N, Marsh DJ (1971) Micropuncture studies of the electrochemical aspects of fluid and electrolyte transport in individual seminiferous tubules, the epididymis and the vas deferens in rats. J Physiol 213:557–570

Lewis SA, Clausen C, Wills NK (1996) Impedance analysis of epithelia. In: Wills NK, Reuss L, Lewis SA (eds) Epithelial transport: a guide to methods and experimental analysis. Chapman & Hall, New York, pp 118–145

Li Y, Halm DR (2002) Secretory modulation of basolateral membrane inwardly rectified K^+ channel in guinea pig distal colonic crypts. Am J Physiol Cell Physiol 282:C719–C735

Li S, Winuthayanon W (2017) Oviduct: roles in fertilization and early embryo development. J Endocrinol 232:R1–R26

Li Q, Yan J (2016) Modulation of BK channel function by auxiliary beta and gamma subunits. Int Rev Neurobiol 128:51–90

Li Y, Halm ST, Halm DR (2003) Secretory activation of basolateral membrane Cl^- channels in guinea pig distal colonic crypts. Am J Physiol Cell Physiol 284:C918–C933

Liao T, Wang L, Halm ST, Lu L, Fyffe RE, Halm DR (2005) K^+ channel K_VLQT located in the basolateral membrane of distal colonic epithelium is not essential for activating Cl^- secretion. Am J Physiol Cell Physiol 289:C564–C575

Lishko PV, Kirichok Y, Ren D, Navarro B, Chung JJ, Clapham DE (2012) The control of male fertility by spermatozoan ion channels. Annu Rev Physiol 74:453–475

Liu W, Schreck C, Coleman RA, Wade JB, Hernandez Y, Zavilowitz B, Warth R, Kleyman TR, Satlin LM (2011) Role of NKCC in BK channel-mediated net K^+ secretion in the CCD. Am J Physiol Renal Physiol 301:F1088–F1097

Liu W, Schrott-Fischer A, Glueckert R, Benav H, Rask-Andersen H (2017) The human "cochlear battery"—claudin-11 barrier and ion transport proteins in the lateral wall of the cochlea. Front Mol Neurosci 10:239

Lucero MT (2013) Peripheral modulation of smell: fact or fiction? Semin Cell Dev Biol 24:58–70

Lybaert P, Vanbellinghen AM, Quertinmont E, Petein M, Meuris S, Lebrun P (2008) K_{ATP} channel subunits are expressed in the epididymal epithelium in several mammalian species. Biol Reprod 79:253–261

Maillo V, Sánchez-Calabuig MJ, Lopera-Vasquez R, Hamdi M, Gutierrez-Adan A, Lonergan P, Rizos D (2016) Oviductal response to gametes and early embryos in mammals. Reproduction 152:R127–R141

Malinowska DH, Sherry AM, Tewari KP, Cuppoletti J (2004) Gastric parietal cell secretory membrane contains PKA- and acid-activated Kir2.1 K^+ channels. Am J Physiol Cell Physiol 286:C495–C506

Malnic G, Giebisch G, Muto S, Wang WH, Bailey MA, Satlin LM (2013) Regulation of K^+ excretion. In: Alpern RJ, Moe OW, Caplan M (eds) Seldin and Giebisch's the kidney: physiology and pathophysiology, 5th edn. Elsevier, Boston, pp 1659–1715

Mangos JA, McSherry NR, Nousia-Arvanitakis S, Irwin K (1973) Secretion and transductal fluxes of ions in exocrine glands of the mouse. Am J Phys 225:18–24

Manzanares D, Gonzalez C, Ivonnet P, Chen RS, Valencia-Gattas M, Conner GE, Larsson HP, Salathe M (2011) Functional apical large conductance, Ca^{2+}-activated, and voltage-dependent K^+ channels are required for maintenance of airway surface liquid volume. J Biol Chem 286:19830–19839

Marcus DC, Wangemann P (2009) Cochlear and vestibular function and dysfunction. In: Alvarez-Leefmans FJ, Delpire E (eds) Physiology and pathology of chloride transporters and channels in the nervous system—from molecules to diseases. Elsevier, Boston, pp 425–437

Markov AG, Amasheh S (2014) Tight junction physiology of pleural mesothelium. Front Physiol 5:221

Matos JE, Robaye B, Boeynaems JM, Beauwens R, Leipziger J (2005) K⁺ secretion activated by luminal $P2Y_2$ and $P2Y_4$ receptors in mouse colon. J Physiol 564:269–279

Matthay MA (2014) Resolution of pulmonary edema. Thirty years of progress. Am J Respir Crit Care Med 189:1301–1308

Maurya DK, Henriques T, Marini M, Pedemonte N, Galietta LJ, Rock JR, Harfe BD, Menini A (2015) Development of the olfactory epithelium and nasal glands in TMEM16A-/- and TMEM16A+/+ mice. PLoS One 10:e0129171

Mazurek MP, Prasad PD, Gopal E, Fraser SP, Bolt L, Rizaner N, Palmer CP, Foster CS, Palmieri F, Ganapathy V, Stühmer W, Djamgoz MB, Mycielska ME (2010) Molecular origin of plasma membrane citrate transporter in human prostate epithelial cells. EMBO Rep 11:431–437

McCabe RD, Smith PL, Sullivan LP (1984) Ion transport by rabbit descending colon: mechanisms of transepithelial potassium transport. Am J Physiol Gastrointest Liver Physiol 246:G594–G602

McDonough AA, Youn JH (2017) Potassium homeostasis: the knowns, the unknowns, and the health benefits. Physiology (Bethesda) 32:100–111

Meneton P, Schultheis PJ, Greeb J, Nieman ML, Liu LH, Clarke LL, Duffy JJ, Doetschman T, Lorenz JN, Shull GE (1998) Increased sensitivity to K⁺ deprivation in colonic H⁺,K⁺-ATPase-deficient mice. J Clin Invest 101:536–542

Merigo F, Mucignat-Caretta C, Cristofoletti M, Zancanaro C (2011) Epithelial membrane transporters expression in the developing to adult mouse vomeronasal organ and olfactory mucosa. Dev Neurobiol 71:854–869

Mignen O, Egee S, Liberge M, Harvey BJ (2000) Basolateral outward rectifier chloride channel in isolated crypts of mouse colon. Am J Physiol Gastrointest Liver Physiol 279:G277–G287

Milatz S, Himmerkus N, Wulfmeyer VC, Drewell H, Mutig K, Hou J, Breiderhoff T, Müller D, Fromm M, Bleich M, Günzel D (2017) Mosaic expression of claudins in thick ascending limbs of Henle results in spatial separation of paracellular Na⁺ and Mg^{2+} transport. Proc Natl Acad Sci U S A 114:E219–E227

Miller MR, Mansell SA, Meyers SA, Lishko PV (2015) Flagellar ion channels of sperm: similarities and differences between species. Cell Calcium 58:105–113

Minamino T, Imada K (2015) The bacterial flagellar motor and its structural diversity. Trends Microbiol 23:267–274

Mizuno K, Shiba K, Okai M, Takahashi Y, Shitaka Y, Oiwa K, Tanokura M, Inaba K (2012) Calaxin drives sperm chemotaxis by Ca^{2+}-mediated direct modulation of a dynein motor. Proc Natl Acad Sci U S A 109:20497–20502

Mobasheri A, Pestov NB, Papanicolaou S, Kajee R, Cózar-Castellano I, Avila J, Martín-Vasallo P, Foster CS, Modyanov NN, Djamgoz MB (2003) Expression and cellular localization of Na,K-ATPase isoforms in the rat ventral prostate. BJU Int 92:793–802

Montoro DT, Haber AL, Biton M, Vinarsky V, Lin B, Birket SE, Yuan F, Chen S, Leung HM, Villoria J, Rogel N, Burgin G, Tsankov AM, Waghray A, Slyper M, Waldman J, Nguyen L, Dionne D, Rozenblatt-Rosen O, Tata PR, Mou H, Shivaraju M, Bihler H, Mense M, Tearney GJ, Rowe SM, Engelhardt JF, Regev A, Rajagopal J (2018) A revised airway epithelial hierarchy includes CFTR-expressing ionocytes. Nature 560:319–324

Moreira CG, Russell R, Mishra AA, Narayanan S, Ritchie JM, Waldor MK, Curtis MM, Winter SE, Weinshenker D, Sperandio V (2016) Bacterial adrenergic sensors regulate virulence of enteric pathogens in the gut. MBio 7:e00826–e00816

Moser SL, Harron SA, Crack J, Fawcett JP, Cowley EA (2008) Multiple KCNQ potassium channel subtypes mediate basal anion secretion from the human airway epithelial cell line Calu-3. J Membr Biol 221:153–163

Mulkidjanian AY, Bychkov AY, Dibrova DV, Galperin MY, Koonin EV (2012) Origin of first cells at terrestrial, anoxic geothermal fields. Proc Natl Acad Sci U S A 109:E821–E830

Musch MW, Orellana SA, Kimberg LS, Field M, Halm DR, Krasny EJ Jr, Frizzell RA (1982) Na⁺-K⁺-Cl⁻ co-transport in the intestine of a marine teleost. Nature 300:351–353

Musch MW, McConnell FM, Goldstein L, Field M (1987) Tyrosine transport in winter flounder intestine: interaction with Na^+-K^+-$2Cl^-$ cotransport. Am J Physiol Regul Integr Comp Physiol 253:R264–R269

Muto S (2017) Physiological roles of claudins in kidney tubule paracellular transport. Am J Physiol Renal Physiol 312:F9–F24

Mycielska ME, Djamgoz MB (2004) Citrate transport in the human prostate epithelial PNT2-C2 cell line: electrophysiological analyses. J Physiol 559:821–833

Mycielska ME, Milenkovic VM, Wetzel CH, Rümmele P, Geissler EK (2015) Extracellular citrate in health and disease. Curr Mol Med 15:884–891

Nakamoto T, Romanenko VG, Takahashi A, Begenisich T, Melvin JE (2008) Apical maxi-K ($K_{Ca}1.1$) channels mediate K^+ secretion by the mouse submandibular exocrine gland. Am J Physiol Cell Physiol 294:C810–C819

Nakamura S, Amlal H, Galla JH, Soleimani M (1999) NH_4^+ secretion in inner medullary collecting duct in potassium deprivation: role of colonic H^+-K^+-ATPase. Kidney Int 56:2160–2167

Namkung W, Song Y, Mills AD, Padmawar P, Finkbeiner WE, Verkman AS (2009) In situ measurement of airway surface liquid [K^+] using a ratioable K^+-sensitive fluorescent dye. J Biol Chem 284:15916–15926

Nanda Kumar NS, Singh SK, Rajendran VM (2010) Mucosal potassium efflux mediated via Kcnn4 channels provides the driving force for electrogenic anion secretion in colon. Am J Physiol Gastrointest Liver Physiol 299:G707–G714

Nguyen N, Kozer-Gorevich N, Gliddon BL, Smolka AJ, Clayton AH, Gleeson PA, van Driel IR (2013) Independent trafficking of the KCNQ1 K^+ channel and H^+-K^+-ATPase in gastric parietal cells from mice. Am J Physiol Gastrointest Liver Physiol 304:G157–G166

Nielson DW (1986) Electrolyte composition of pulmonary alveolar subphase in anesthetized rabbits. J Appl Physiol 60:972–979

Nin H, Yoshida T, Sawamura S, Ogata G, Ota T, Higuchi T, Murakami S, Doi K, Kurachi Y, Hibino H (2016) The unique electrical properties in an extracellular fluid of the mammalian cochlea; their functional roles, homeostatic processes, and pathological significance. Pflugers Arch 468:1637–1649

O'Donnell ME, Wulff H, Chen YJ (2017) Blood-brain barrier mechanisms of edema formation: the role of ion transporters and channels. In: Badaut J, Plesnila N (eds) Brain edema: from Molecular mechanisms to clinical practice. Academic Press, Cambridge MA, pp 130–149

O'Grady SM, Lee SY (2003) Chloride and potassium channel function in alveolar epithelial cells. Am J Physiol Lung Cell Mol Physiol 284:L689–L700

O'Grady SM, Lee SY (2005) Molecular diversity and function of voltage-gated (K_V) potassium channels in epithelial cells. Int J Biochem Cell Biol 37:1578–1594

O'Neil RG, Helman SI (1977) Transport characteristics of renal collecting tubules: influences of DOCA and diet. Am J Physiol Renal Physiol 233:F544–F558

Oh KS, Oh YT, Kim SW, Kita T, Kang I, Youn JH (2011) Gut sensing of dietary K^+ intake increases renal K^+ excretion. Am J Physiol Regul Integr Comp Physiol 301:R421–R429

Padfield PJ, Garner A, Case RM (1989) Patterns of pancreatic secretion in the anaesthetised guinea pig following stimulation with secretin, cholecystokinin octapeptide, or bombesin. Pancreas 4:204–209

Palk L, Sneyd J, Shuttleworth TJ, Yule DI, Crampin EJ (2010) A dynamic model of saliva secretion. J Theor Biol 266:625–640

Palmer BF (2015) Regulation of Potassium Homeostasis. Clin J Am Soc Nephrol 10:1050–1060

Palmer LG (2017) Epithelial transport in The Journal of General Physiology. J Gen Physiol 149:897–909

Palmer BF, Clegg DJ (2016) Physiology and pathophysiology of potassium homeostasis. Adv Physiol Educ 40:480–490

Palmer LG, Frindt G (2007) High-conductance K channels in intercalated cells of the rat distal nephron. Am J Physiol Renal Physiol 292:F966–F973

Palmer ML, Schiller KR, O'Grady SM (2008) Apical SK potassium channels and Ca^{2+}-dependent anion secretion in endometrial epithelial cells. J Physiol 586:717–726

Palmer ML, Peitzman ER, Maniak PJ, Sieck GC, Prakash YS, O'Grady SM (2011) K$_{Ca}$3.1 channels facilitate K$^+$ secretion or Na$^+$ absorption depending on apical or basolateral P2Y receptor stimulation. J Physiol 589:3483–3394

Parker MD, Boron WF (2013) The divergence, actions, roles, and relatives of sodium-coupled bicarbonate transporters. Physiol Rev 93:803–959

Penton D, Czogalla J, Loffing J (2015) Dietary potassium and the renal control of salt balance and blood pressure. Pflugers Arch 467:513–530

Pestov NB, Korneenko TV, Adams G, Tillekeratne M, Shakhparonov MI, Modyanov NN (2002) Nongastric H-K-ATPase in rodent prostate: lobe-specific expression and apical localization. Am J Physiol Cell Physiol 282:C907–C916

Pinti DL (2005) The origin and evolution of the oceans. In: Gargaud M, Barbier B, Martin H, Reisse J (eds) Lectures in astrobiology, Advances in astrobiology and biogeophysics, vol 1. Springer, Berlin, pp 83–112

Pinto da Silva P, Kachar B (1982) On tight-junction structure. Cell 28:441–450

Plasschaert LW, Žilionis R, Choo-Wing R, Savova V, Knehr J, Roma G, Klein AM, Jaffe AB (2018) A single-cell atlas of the airway epithelium reveals the CFTR-rich pulmonary ionocyte. Nature 560:377–381

Pongs O, Schwarz JR (2010) Ancillary subunits associated with voltage-dependent K$^+$ channels. Physiol Rev 90:755–796

Pozdniakova S, Ladilov Y (2018) Functional Significance of the Adcy10-Dependent Intracellular cAMP Compartments. J Cardiovasc Dev Dis 5:29

Preidis GA, Versalovic J (2009) Targeting the human microbiome with antibiotics, probiotics, and prebiotics: gastroenterology enters the metagenomics era. Gastroenterology 136:2015–2031

Puga Molina LC, Luque GM, Balestrini PA, Marín-Briggiler CI, Romarowski A, Buffone MG (2018) Molecular basis of human sperm capacitation. Front Cell Dev Biol 6:72

Quatrale RP, Laden K (1968) Solute and water secretion by the eccrine sweat glands of the rat. J Invest Dermatol 51:502–504

Quinton PM (2017) Both ways at once: keeping small airways clean. Physiology (Bethesda) 32:380–390

Quinton PM, Elder HY, Jenkinson DM, Bovell DL (1999) Structure and function of human sweat glands. In: Laden K (ed) Antiperspirants and deodorants, 2nd edn. Marcel Dekker, New York, pp 17–57

Rahner C, Mitic LL, Anderson JM (2001) Heterogeneity in expression and subcellular localization of claudins 2, 3, 4, and 5 in the rat liver, pancreas, and gut. Gastroenterology 120:411–422

Rakowski RF, Paxson CL (1988) Voltage dependence of Na$^+$/K$^+$ pump current in Xenopus oocytes. J Membr Biol 106:173–182

Rakowski RF, Gadsby DC, De Weer P (1989) Stoichiometry and voltage dependence of the sodium pump in voltage-clamped, internally dialyzed squid giant axon. J Gen Physiol 93:903–941

Rechkemmer G, Halm DR (1989) Aldosterone stimulates K$^+$ secretion across mammalian colon independent of Na$^+$ absorption. Proc Natl Acad Sci U S A 86:397–401

Rechkemmer G, von Engelhardt W (1993) Absorption and secretion of electrolytes and short-chain fatty acids in the guinea pig large intestine. In: Clauss W (ed) Ion transport in vertebrate colon, Advances in comparative and environmental physiology, vol 16. Springer, New York, pp 139–167

Rechkemmer G, Frizzell RA, Halm DR (1996) Active K$^+$ transport across guinea pig distal colon: action of secretagogues. J Physiol 493:485–502

Reuter D, Zierold K, Schröder WH, Frings S (1998) A depolarizing chloride current contributes to chemoelectrical transduction in olfactory sensory neurons in situ. J Neurosci 18:6623–6630

Robinson NP, Kyle H, Webber SE, Widdicombe JG (1989) Electrolyte and other chemical concentrations in tracheal airway surface liquid and mucus. J Appl Physiol 66:2129–2135

Rodan AR, Cheng CJ, Huang CL (2011) Recent advances in distal tubular potassium handling. Am J Physiol Renal Physiol 300:F821–F827

Roepke TK, Anantharam A, Kirchhoff P, Busque SM, Young JB, Geibel JP, Lerner DJ, Abbott GW (2006) The KCNE2 potassium channel ancillary subunit is essential for gastric acid secretion. J Biol Chem 281:23740–23747

Rossi E, Cimdins A, Lüthje P, Brauner A, Sjöling Å, Landini P, Römling U (2018) "It's a gut feeling"—*Escherichia coli* biofilm formation in the gastrointestinal tract environment. Crit Rev Microbiol 44:1–30

Roy A, Al-bataineh MM, Pastor-Soler NM (2015) Collecting duct intercalated cell function and regulation. Clin J Am Soc Nephrol 10:305–324

Ruan YC, Chen H, Chan HC (2014) Ion channels in the endometrium: regulation of endometrial receptivity and embryo implantation. Hum Reprod Update 20:517–529

Sabolić I, Herak-Kramberger CM, Breton S, Brown D (1999) Na/K-ATPase in intercalated cells along the rat nephron revealed by antigen retrieval. J Am Soc Nephrol 10:913–922

Sackin H, Palmer LG (2013) Electrophysiological analysis of transepithelial transport. In: Alpern RJ, Moe OW, Caplan M (eds) Seldin and Giebisch's the kidney: physiology and pathophysiology, 5th edn. Elsevier, Boston, pp 177–216

Saint-Criq V, Gray MA (2017) Role of CFTR in epithelial physiology. Cell Mol Life Sci 74:93–115

Sandle GI, Hunter M (2010) Apical potassium (BK) channels and enhanced potassium secretion in human colon. Q J Med 103:85–89

Sandle GI, Perry MD, Mathialahan T, Linley JE, Robinson P, Hunter M, MacLennan KA (2007) Altered cryptal expression of luminal potassium (BK) channels in ulcerative colitis. J Pathol 212:66–73

Sansom SC, O'Neil RG (1985) Mineralocorticoid regulation of apical cell membrane Na^+ and K^+ transport of the cortical collecting duct. Am J Physiol Renal Physiol 248:F858–F868

Sansom SC, O'Neil RG (1986) Effects of mineralocorticoids on transport properties of cortical collecting duct basolateral membrane. Am J Physiol Renal Physiol 251:F743–F757

Satlin LM, Carattino MD, Liu W, Kleyman TR (2006) Regulation of cation transport in the distal nephron by mechanical forces. Am J Physiol Renal Physiol 291:F923–F931

Sato K (1980) Electrochemical driving forces for K^+ secretion by rat paw eccrine sweat gland. Am J Physiol Cell Physiol 239:C90–C97

Sato K (2001) The mechanism of eccrine sweat secretion. In: Gisolfi CV, Lamb DR, Nadel ER (eds) Exercise, heat, and thermoregulation, Perspectives in exercise science and sports medicine, vol 6. Cooper Publishing Group, Traverse City, MI, pp 85–117

Sato F, Sato K (1978) Secretion of a potassium-rich fluid by the secretory coil of the rat paw eccrine sweat gland. J Physiol 274:37–50

Sato K, Sato F (1990) Na^+, K^+, H^+, Cl^-, and Ca^{2+} concentrations in cystic fibrosis eccrine sweat in vivo and in vitro. J Lab Clin Med 115:504–511

Sato K, Cavallin S, Sato KT, Sato F (1994) Secretion of ions and pharmacological responsiveness in the mouse paw sweat gland. Clin Sci (Lond) 86:133–139

Saumon G, Basset G (1993) Electrolyte and fluid transport across the mature alveolar epithelium. J Appl Physiol 74:1–15

Sausbier M, Matos JE, Sausbier U, Beranek G, Arntz C, Neuhuber W, Ruth P, Leipziger J (2006) Distal colonic K^+ secretion occurs via BK channels. J Am Soc Nephrol 17:1275–1282

Schrum JP, Zhu TF, Szostak JW (2010) The origins of cellular life. Cold Spring Harb Perspect Biol 2:a002212

Schultz SG, Thompson SM, Suzuki Y (1981) Equivalent electrical circuit models and the study of Na transport across epithelia: nonsteady-state current-voltage relations. Fed Proc 40:2443–2449

Schwartz GJ, Burg MB (1978) Mineralocorticoid effects on cation transport by cortical collecting tubules in vitro. Am J Physiol Renal Physiol 235:F576–F585

Sepúlveda FV, Pablo Cid L, Teulon J, Niemeyer MI (2015) Molecular aspects of structure, gating, and physiology of pH-sensitive background K_{2P} and K_{ir} K^+-transport channels. Physiol Rev 95:179–217. Corrigendum 96:1665

Sewell WA, Young JA (1975) Secretion of electrolytes by the pancreas of the anaestetized rat. J Physiol 252:379–396

Shao J, Gumz ML, Cain BD, Xia SL, Shull GE, van Driel IR, Wingo CS (2010) Pharmacological profiles of the murine gastric and colonic H,K-ATPases. Biochim Biophys Acta 1800:906–911

Shashikanth N, Yeruva S, Ong MLDM, Odenwald MA, Pavlyuk R, Turner JR (2017) Epithelial organization: the gut and beyond. Compr Physiol 7:1497–1518

Shennan DB, Peaker M (2000) Transport of milk constituents by the mammary gland. Physiol Rev 80:925–951

Shigetomi K, Ikenouchi J (2018) Regulation of the epithelial barrier by post-translational modifications of tight junction membrane proteins. J Biochem 163:265–272

Shorofsky SR, Field M, Fozzard HA (1983) Electrophysiology of Cl secretion in canine trachea. J Membr Biol 72:105–115

Shorofsky SR, Field M, Fozzard HA (1984) Mechanism of Cl secretion in canine trachea: changes in intracellular chloride activity with secretion. J Membr Biol 81:1–8

Shum WW, Ruan YC, Da Silva N, Breton S (2011) Establishment of cell-cell cross talk in the epididymis: control of luminal acidification. J Androl 32:576–586

Shuttleworth TJ, Thompson JL, Dantzler WH (1987) Potassium secretion by nasal salt glands of desert lizard Sauromalus obesus. Am J Physiol Regul Integr Comp Physiol 253:R83–R90

Silva P, Stoff J, Field M, Fine L, Forrest JN, Epstein FH (1977) Mechanism of active chloride secretion by shark rectal gland: role of Na⁺-K⁺-ATPase in chloride transport. Am J Physiol Renal Physiol 233:F298–F306

Simon M, Duong JP, Mallet V, Jian R, MacLennan KA, Sandle GI, Marteau P (2008) Over-expression of colonic K⁺ channels associated with severe potassium secretory diarrhoea after haemorrhagic shock. Nephrol Dial Transplant 23:3350–3352

Singh SK, O'Hara B, Talukder JR, Rajendran VM (2012) Aldosterone induces active K⁺ secretion by enhancing mucosal expression of Kcnn4c and Kcnma1 channels in rat distal colon. Am J Physiol Cell Physiol 302:C1353–C1360

Skou JC, Esmann M (1992) The Na⁺,K⁺-ATPase. J Bioenerg Biomembr 24:249–261

Smith PL, Frizzell RA (1984) Chloride secretion by canine tracheal epithelium: IV. Basolateral membrane K permeability parallels secretion rate. J Membr Biol 77:187–199

Sojo V, Herschy B, Whicher A, Camprubí E, Lane N (2016) The origin of life in alkaline hydrothermal vents. Astrobiology 16:181–197

Song P, Groos S, Riederer B, Feng Z, Krabbenhöft A, Smolka A, Seidler U (2009) KCNQ1 is the luminal K⁺ recycling channel during stimulation of gastric acid secretion. J Physiol 587:3955–3965

Song P, Groos S, Riederer B, Feng Z, Krabbenhöft A, Manns MP, Smolka A, Hagen SJ, Neusch C, Seidler U (2011) Kir4.1 channel expression is essential for parietal cell control of acid secretion. J Biol Chem 286:14120–14128

Sørensen JB, Nielsen MS, Gudme CN, Larsen EH, Nielsen R (2001) Maxi K⁺ channels co-localised with CFTR in the apical membrane of an exocrine gland acinus: possible involvement in secretion. Pflugers Arch 442:1–11

Sørensen MV, Matos JE, Sausbier M, Sausbier U, Ruth P, Praetorius HA, Leipziger J (2008) Aldosterone increases $K_{Ca}1.1$ (BK) channel-mediated colonic K⁺ secretion. J Physiol 586:4251–4264

Sørensen MV, Matos JE, Praetorius HA, Leipziger J (2010a) Colonic potassium handling. Pflugers Arch 459:645–656

Sørensen MV, Sausbier M, Ruth P, Seidler U, Riederer B, Praetorius HA, Leipziger J (2010b) Adrenaline-induced colonic K⁺ secretion is mediated by $K_{Ca}1.1$ (BK) channels. J Physiol 588:1763–1777

Sørensen MV, Strandsby AB, Larsen CK, Praetorius HA, Leipziger J (2011) The secretory $K_{Ca}1.1$ channel localises to crypts of distal mouse colon: functional and molecular evidence. Pflugers Arch 462:745–752

Sørensen MV, Grossmann S, Roesinger M, Gresko N, Todkar AP, Barmettler G, Ziegler U, Odermatt A, Loffing-Cueni D, Loffing J (2013) Rapid dephosphorylation of the renal sodium chloride cotransporter in response to oral potassium intake in mice. Kidney Int 83:811–824

Srisomboon Y, Zaidman NA, Maniak PJ, Deachapunya C, O'Grady SM (2018) P2Y receptor regulation of K2P channels that facilitate K$^+$ secretion by human mammary epithelial cells. Am J Physiol Cell Physiol 314:C627–C639

Stamatovic SM, Johnson AM, Sladojevic N, Keep RF, Andjelkovic AV (2017) Endocytosis of tight junction proteins and the regulation of degradation and recycling. Ann N Y Acad Sci 1397:54–65

Stauber T, Weinert S, Jentsch TJ (2012) Cell biology and physiology of CLC chloride channels and transporters. Compr Physiol 2:1701–1744

Stokes JB, Lee I, Williams A (1981) Potassium secretion by cortical collecting tubule: relation to sodium absorption, luminal sodium concentration, and transepithelial voltage. Am J Physiol Renal Physiol 241:F395–F402

Stokes JB, Lee I, D'Amico M (1984) Sodium chloride absorption by the urinary bladder of the winter flounder. A thiazide-sensitive, electrically neutral transport system. J Clin Invest 74:7–16

Strandwitz P (2018) Neurotransmitter modulation by the gut microbiota. Brain Res 1693:128–133

Sullivan SK, Smith PL (1986) Active potassium secretion by rabbit proximal colon. Am J Physiol Gastrointest Liver Physiol 250:G475–G483

Suzuki Y, Kaneko K (1987) Acid secretion in isolated guinea pig colon. Am J Physiol Gastrointest Liver Physiol 253:G155–G164

Swarts HG, Koenderink JB, Willems PH, De Pont JJ (2005) The non-gastric H,K-ATPase is oligomycin-sensitive and can function as an H$^+$,NH$_4$$^+$-ATPase. J Biol Chem 280:33115–33122

Sweiry JH, Binder HJ (1989) Characterization of aldosterone-induced potassium secretion in rat distal colon. J Clin Invest 83:844–851

Tamura A, Hayashi H, Imasato M, Yamazaki Y, Hagiwara A, Wada M, Noda T, Watanabe M, Suzuki Y, Tsukita S (2011) Loss of claudin-15, but not claudin-2, causes Na$^+$ deficiency and glucose malabsorption in mouse small intestine. Gastroenterology 140:913–923

Thévenod F (2002) Ion channels in secretory granules of the pancreas and their role in exocytosis and release of secretory proteins. Am J Physiol Cell Physiol 283:C651–C672

Thompson SM (1986) Relations between chord and slope conductances and equivalent electromotive forces. Am J Physiol Cell Physiol 250:C333–C339

Thompson KA, Kleinzeller A (1985) Glucose transport in intestinal epithelia of winter flounder. Am J Physiol Regul Integr Comp Physiol 248:R573–R577

Tirindelli R, Dibattista M, Pifferi S, Menini A (2009) From pheromones to behavior. Physiol Rev 89:921–956

Torres-Quesada O, Mayrhofer JE, Stefan E (2017) The many faces of compartmentalized PKA signalosomes. Cell Signal 37:1–11

Traynor TR, O'Grady SM (1996) Regulation of colonic ion transport by GRP. I. GRP stimulates transepithelial K$^+$ and Na$^+$ secretion. Am J Physiol Cell Physiol 270:C848–C858

Tsukita S, Tanaka H, Tamura A (2019) The claudins: from tight junctions to biological systems. Trends Biochem Sci 44:141–152

Turner TT, Hartmann PK, Howards SS (1977) In vivo sodium, potassium, and sperm concentrations in the rat epididymis. Fertil Steril 28:191–194

Turner JR, Buschmann MM, Romero-Calvo I, Sailer A, Shen L (2014) The role of molecular remodeling in differential regulation of tight junction permeability. Semin Cell Dev Biol 36:204–212

Vallon V, Osswald H, Blantz RC, Thomson S (1997) Potential role of luminal potassium in tubuloglomerular feedback. J Am Soc Nephrol 8:1831–1837

Van Itallie CM, Fanning AS, Anderson JM (2003) Reversal of charge selectivity in cation or anion-selective epithelial lines by expression of different claudins. Am J Physiol Renal Physiol 285:F1078–F1084 van't Hof W, Veerman EC, Nieuw Amerongen AV, Ligtenberg AJ (2014) Antimicrobial defense systems in saliva. Monogr Oral Sci 24:40–51

Venglovecz V, Hegyi P, Rakonczay Z Jr, Tiszlavicz L, Nardi A, Grunnet M, Gray MA (2011) Pathophysiological relevance of apical large-conductance Ca^{2+}-activated potassium channels in pancreatic duct epithelial cells. Gut 60:361–369

Venglovecz V, Rakonczay Z Jr, Gray MA, Hegyi P (2015) Potassium channels in pancreatic duct epithelial cells: their role, function and pathophysiological relevance. Pflugers Arch 467:625–640

Viejo-Díaz M, Andrés MT, Fierro JF (2004) Modulation of in vitro fungicidal activity of human lactoferrin against Candida albicans by extracellular cation concentration and target cell metabolic activity. Antimicrob Agents Chemother 48:1242–1248

Vogalis F, Hegg CC, Lucero MT (2005) Electrical coupling in sustentacular cells of the mouse olfactory epithelium. J Neurophysiol 94:1001–1012

Vucic E, Alfadda T, MacGregor GG, Dong K, Wang T, Geibel JP (2015) Kir1.1 (ROMK) and $K_V7.1$ (KCNQ1/K_VLQT1) are essential for normal gastric acid secretion: importance of functional Kir1.1. Pflugers Arch 467:1457–1468

Wagner CA, Finberg KE, Breton S, Marshansky V, Brown D, Geibel JP (2004) Renal vacuolar H⁺-ATPase. Physiol Rev 84:1263–1314

Wall SM, Koger LM (1994) NH_4^+ transport mediated by Na^+-K^+-ATPase in rat inner medullary collecting duct. Am J Physiol Renal Physiol 267:F660–F670

Wang H, Dey SK (2006) Roadmap to embryo implantation: clues from mouse models. Nat Rev Genet 7:185–199

Wang YB, Germaine GR (1993) Effects of pH, potassium, magnesium, and bacterial growth phase on lysozyme inhibition of glucose fermentation by Streptococcus mutans 10449. J Dent Res 72:907–911

Wang WH, Huang CL (2013) The molecular biology of renal K⁺ channels. In: Alpern RJ, Moe OW, Caplan M (eds) Seldin and Giebisch's the kidney: physiology and pathophysiology, 5th edn. Elsevier, Boston, pp 1601–1627

Wang B, Wen D, Li H, Wang-France J, Sansom SC (2017) Net K⁺ secretion in the thick ascending limb of mice on a low-Na, high-K diet. Kidney Int 92:864–875

Wang B, Wang-France J, Li H, Sansom SC (2019) Furosemide reduces BK-αβ4-mediated K⁺ secretion in mice on an alkaline high-K⁺ diet. Am J Physiol Renal Physiol 316:F341–F350

Wangemann P (2006) Supporting sensory transduction: cochlear fluid homeostasis and the endocochlear potential. J Physiol 576:11–21

Watlington CO, Huf EG (1971) β-adrenergic stimulation of frog skin mucous glands: non-specific inhibition by adrenergic blocking agents. Comp Gen Pharmacol 2:295–305

Watt J, Wilson CW (1958) Gastric secretion in the normal guinea-pig. J Physiol 142:233–241

Weiner ID, Verlander JW (2013) Renal ammonia metabolism and transport. Compr Physiol 3:201–220

Weiner ID, Verlander JW (2014) Ammonia transport in the kidney by Rhesus glycoproteins. Am J Physiol Renal Physiol 306:F1107–F1120

Weinstein AM (2010) A mathematical model of rat ascending Henle limb. I. Cotransporter function. Am J Physiol Renal Physiol 298:F512–F524

Welling PA (2013) Regulation of renal potassium secretion: molecular mechanisms. Semin Nephrol 33:215–228

Welling PA (2016) Roles and Regulation of Renal K Channels. Annu Rev Physiol 78:415–435

Welsh MJ, Smith PL, Frizzell RA (1982) Chloride secretion by canine tracheal epithelium: II. The cellular electrical potential profile. J Membr Biol 70(3):227–238

Welsh MJ, Karp P, Ruppert TR (1983a) Evidence for basolateral membrane potassium conductance in canine tracheal epithelium. Am J Physiol Cell Physiol 244:C377–C384

Welsh MJ, Smith PL, Frizzell RA (1983b) Chloride secretion by canine tracheal epithelium: III. Membrane resistances and electromotive forces. J Membr Biol 71:209–218

Westmoreland JJ, Drosos Y, Kelly J, Ye J, Means AL, Washington MK, Sosa-Pineda B (2012) Dynamic distribution of claudin proteins in pancreatic epithelia undergoing morphogenesis or neoplastic transformation. Dev Dyn 241:583–594

Whittamore JM, Cooper CA, Wilson RW (2010) HCO_3^- secretion and $CaCO_3$ precipitation play major roles in intestinal water absorption in marine teleost fish in vivo. Am J Physiol Regul Integr Comp Physiol 298:R877–R886

Wieczorek H, Brown D, Grinstein S, Ehrenfeld J, Harvey WR (1999) Animal plasma membrane energization by proton-motive V-ATPases. Bioessays 21:637–648

Wilkinson DJ, Kushman NL, Dawson DC (1993) Tetraethylammonium-sensitive apical K^+ channels mediating K^+ secretion by turtle colon. J Physiol 462:697–714

Williams JA, Dawson DC (2006) Cell structure and function. In: Mulholland MW, Lillemoe KD, Doherty GM, Maier RV, Upchurch GR Jr (eds) Greenfield's surgery: scientific principles and practice, 4th edn. Lippincott Williams & Wilkins, Philadelphia, pp 2–42

Wilson TH, Lin EC (1980) Evolution of membrane bioenergetics. J Supramol Struct 13:421–446

Woda CB, Bragin A, Kleyman TR, Satlin LM (2001) Flow-dependent K^+ secretion in the cortical collecting duct is mediated by a maxi-K channel. Am J Physiol Renal Physiol 280:F786–F793

Wong PY, Yeung CH (1978) Absorptive and secretory functions of the perfused rat cauda epididymidis. J Physiol 275:13–26

Woodward OM, Guggino WB (2013) Anion channels. In: Alpern RJ, Moe OW, Caplan M (eds) Seldin and Giebisch's the kidney: physiology and pathophysiology, 5th edn. Elsevier, Boston, pp 1019–1045

Worrell RT, Merk L, Matthews JB (2008) Ammonium transport in the colonic crypt cell line, T84: role for Rhesus glycoproteins and NKCC1. Am J Physiol Gastrointest Liver Physiol 294:G429–G440

Wright EM, Loo DD, Hirayama BA (2011) Biology of human sodium glucose transporters. Physiol Rev 91:733–794

Wrong OM, Edmonds CJ, Chadwick VS (eds) (1981) The large intestine: it's role in mammalian nutrition and homeostasis. MTP Press, Lancaster, p 217

Xie Y, Ostedgaard L, Abou Alaiwa MH, Lu L, Fischer AJ, Stoltz DA (2018) Mucociliary transport in healthy and cystic fibrosis pig airways. Ann Am Thorac Soc 15(Suppl 3):S171–S176

Yamaguchi M, Steward MC, Smallbone K, Sohma Y, Yamamoto A, Ko SB, Kondo T, Ishiguro H (2017) Bicarbonate-rich fluid secretion predicted by a computational model of guinea-pig pancreatic duct epithelium. J Physiol 595:1947–1972

Yang L, Edvinsson J, Sackin H, Palmer LG (2012) Ion selectivity and current saturation in inward-rectifier K^+ channels. J Gen Physiol 139:145–157

Yang C, Gonzalez-Perez V, Mukaibo T, Melvin JE, Xia XM, Lingle CJ (2017) Knockout of the LRRC26 subunit reveals a primary role of LRRC26-containing BK channels in secretory epithelial cells. Proc Natl Acad Sci U S A 114:E3739–E3747

Yorio T, Bentley PJ (1977) Permeability of the rabbit colon in vitro. Am J Phys 232:F5–F9

Yoshida M, Yoshida K (2011) Sperm chemotaxis and regulation of flagellar movement by Ca^{2+}. Mol Hum Reprod 17:457–465

Youn JH (2013) Gut sensing of potassium intake and its role in potassium homeostasis. Semin Nephrol 33:248–256

Yu AS (2015) Claudins and the kidney. J Am Soc Nephrol 26:11–19

Yu ASL (2017) Paracellular transport and energy utilization in the renal tubule. Curr Opin Nephrol Hypertens 26:398–404

Yu FH, Catterall WA (2004) The VGL-Chanome: a protein superfamily specialized for electrical signaling and ionic homeostasis. Sci STKE 2004(253):re15

Yuan J, Liu W, Karvar S, Baker SS, He W, Baker RD, Ji G, Xie J, Zhu L (2015) Potassium channel KCNJ15 is required for histamine-stimulated gastric acid secretion. Am J Physiol Cell Physiol 309:C264–C270

Zaidman NA, Panoskaltsis-Mortari A, O'Grady SM (2017) Large-conductance Ca^{2+}-activated K^+ channel activation by apical P2Y receptor agonists requires hydrocortisone in differentiated airway epithelium. J Physiol 595:4631–4645

Zdebik AA, Cuffe JE, Bertog M, Korbmacher C, Jentsch TJ (2004) Additional disruption of the CLC-2 Cl^- channel does not exacerbate the cystic fibrosis phenotype of cystic fibrosis transmembrane conductance regulator mouse models. J Biol Chem 279:22276–22283

Zhang J, Halm ST, Halm DR (2009a) Adrenergic activation of electrogenic K^+ secretion in guinea pig distal colonic epithelium: involvement of β1- and β2-adrenergic receptors. Am J Physiol Gastrointest Liver Physiol 297:G269–G277

Zhang J, Halm ST, Halm DR (2009b) Adrenergic activation of electrogenic K^+ secretion in guinea pig distal colonic epithelium: desensitization via the Y2-neuropeptide receptor. Am J Physiol Gastrointest Liver Physiol 297:G278–G291

Zhang J, Halm ST, Halm DR (2012) Role of the BK channel ($K_{Ca}1.1$) during activation of electrogenic K^+ secretion in guinea pig distal colon. Am J Physiol Gastrointest Liver Physiol 303:G1322–G1334

Zhang Y, Wang Q, Wang H, Duan E (2017) Uterine fluid in pregnancy: a biological and clinical outlook. Trends Mol Med 23:604–614

Zhao J, Krystofiak ES, Ballesteros A, Cui R, Van Itallie CM, Anderson JM, Fenollar-Ferrer C, Kachar B (2018) Multiple claudin-claudin cis interfaces are required for tight junction strand formation and inherent flexibility. Commun Biol 1:50

Chapter 11
Volume Regulation in Epithelia

Erik Hviid Larsen and Else Kay Hoffmann

Abstract Polarized epithelia generate regulated water flows and solute fluxes for serving extracellular homeostasis, which imposes changes in epithelial water volume and osmolyte concentrations that would be critical to normal function if not regulated. The studies have been challenged by the fact that epithelia may contain more than one cell type, and by the large osmotic permeability of some epithelial membranes that presupposes successful elimination of unstirred layer effects following aniso-osmotic perturbations.

Illustrated by several examples applying a range of methods, we review studies showing that volume regulation is governed by principles similar to non-polarized cells by having acquired well-developed regulatory volume decrease (RVD) and regulatory volume increase (RVI). RVI may not be seen unless the cell has undergone a prior RVD. The rate of RVD and RVI is faster in cells of high osmotic permeability like amphibian gallbladder and mammalian proximal tubule as compared to amphibian skin and mammalian cortical collecting tubule of low and intermediate osmotic permeability. Cross talk between entrance and exit mechanisms interferes with volume regulation both at aniso-osmotic and isosmotic volume perturbations. For example, the inevitable volume increase resulting from Na^+/K^+ pump arrest is delayed by inhibition of Na^+ and K^+ leak permeabilities. This may even be preceded by a transient volume decrease associated with a reduction of the cytosolic Cl^- pool if kept above equilibrium by downhill Na^+ entrance. It has been proposed that cell volume regulation is an intrinsic function of isoosmotic fluid transport that depends on Na^+ recirculation. The causative relationship is discussed for amphibian skin epithelium and submucosal glands in which all major ion transporters and channels including the Na^+ recirculation mechanisms have been identified.

A large number of transporters and ion channels involved in volume regulation have been cloned. The volume-regulated anion channel (VRAC) exhibiting specific electrophysiological characteristics seems exclusively to serve volume regulation.

E. H. Larsen · E. K. Hoffmann (✉)
Department of Biology, University of Copenhagen, Copenhagen, Denmark
e-mail: ehlarsen@bio.ku.dk; ekhoffmann@bio.ku.dk

© The American Physiological Society 2020
K. L. Hamilton, D. C. Devor (eds.), *Basic Epithelial Ion Transport Principles and Function*, Physiology in Health and Disease,
https://doi.org/10.1007/978-3-030-52780-8_11

395

This is contrary to several subfamilies of K^+ channels as well as cotransporters and exchange mechanisms that may serve both transepithelial transport and cell volume regulation. In the same cell, these functions may be maintained by different ion pathways that are separately regulated. RVD is often preceded by an increase in cytosolic free Ca^{2+} probably via influx through TRP channels and/or release from intracellular stores. Cell volume regulation is associated with specific ATP release mechanisms and involves mitogen-activated protein kinases, WNKs, and Ste20-related kinases that are modulated by osmotic stress.

Keywords Regulatory volume decrease (RVD) · Regulatory volume increase (RVI) · Aniso-osmotic and isoosmotic volume regulation · Cross talk · Absorbing epithelia · Exocrine glands · Heterocellular epithelia · Osmotic permeability · VRAC · K^+ channel subfamilies · NKCC · NHE · TRP channels · Osmo-sensing · Osmo-signaling

11.1 Introduction

Extracellular volume and electrolyte homeostasis depend on water flows and solute fluxes through epithelia. High resistance or "tight" epithelia, which have developed paracellular junctions of low ion permeability, have capacity to maintain large transepithelial gradients of solute concentrations and osmolality. The transported fluid may be hypo- or hyperosmotic to the body fluids. Both extracellular water volume and electrolyte concentrations are regulated by these epithelia. Distal tubules of the vertebrate kidney, amphibian epidermis and urinary bladder, vertebrate colon and gills of teleost fish are examples of tight epithelia. Mammalian airway epithelia and ducts of exocrine glands also belong to this class. They regulate ion concentrations and osmolality of the airway surface layer and ductal fluids, respectively. Low resistance or "leaky" epithelia of significant paracellular ion conductance have capacity for transporting large volumes of isoosmotic fluid between body compartments at transepithelial osmotic equilibrium. Kidney proximal tubules, small intestine, gallbladder, and the acinus epithelium of exocrine glands belong to this class. This chapter is concerned with the question: How do epithelia take care of their own volume and ion concentrations while serving whole-body fluid homeostasis? The pertinent issues concern: (*i*) osmotic water permeability of the epithelial cell membranes, (*ii*) cell volume response to osmotic challenge, (*iii*) cell volume as a signal of isoosmotic transport, (*iv*) molecular identity of channels and transporters, (*v*) nature of the sensing mechanisms of cell volume and/or cytoplasmic osmolality, and (*vi*) design of coordinating signaling pathways.

11.2   Concepts in Cell Volume and Electrolyte Homeostasis

11.2.1   Application of the van't Hoff Law

Animal body cells are in osmotic equilibrium with the extracellular fluid. A change of the solute concentration of the extracellular fluid, ΔC changes its osmotic pressure by $\Delta \pi$ according to the van't Hoff law:

$$\Delta \pi = \phi R T \Delta C \tag{11.1}$$

where ϕ is the osmotic coefficient, which is 0.93 for a 150 mM NaCl solution, R is the Universal gas constant (8.317 $J \cdot K^{-1} \cdot mol^{-1}$) and T is the temperature in K. With few exceptions, the water permeability of the plasma membrane is orders of magnitudes larger than the permeability of common solutes implying that reflection coefficients would be near unity. Thus, according to Eq. (11.1), cell volume $V^{(c)}$ would respond to the above change in extracellular concentration such that:

$$\Delta V^{(c)} \Delta \pi = \phi R T \sum m^{(c)} \tag{11.2}$$

where $\sum m^{(c)}$ is the total amount of intracellular osmolytes. If a fraction of $V^{(c)}$ is occupied by an osmotically inactive non-solvent volume v' the osmotically active solution of the cell occupies a volume that is equal to $V^{(c)} - v'$. Let $V_0^{(c)}$ be the volume at the osmotic equilibrium of the cell in an isoosmotic solution of osmotic pressure $\pi_0^{(c)}$, and $V^{(c)}$ be the volume at equilibrium in a hypo- or hyperosmotic solution with the osmotic pressure $\pi^{(c)}$ then:

$$\pi_0^{(c)} \left(V_0^{(c)} - v' \right) = \pi^{(c)} \left(V^{(c)} - v' \right) \tag{11.3}$$

which can be rearranged to give:

$$V^{(c)}/V_0^{(c)} = \left[\left(V_0^{(c)} - v' \right)/V_0^{(c)} \right] \left(\pi_0^{(c)}/\pi^{(c)} - 1 \right) + 1 \tag{11.4}$$

The above relationship between $V^{(c)}/V_0^{(c)}$ and $\left(\pi_0^{(c)}/\pi^{(c)} - 1 \right)$ depicts a straight line of slope:

$$\alpha = \left(V_0^{(c)} - v' \right)/V_0^{(c)} \tag{11.5}$$

Equation (11.4) applies if the cell does not eliminate or accumulate osmolytes during transition from one steady state to the other, which provides a simple way of calculating v' by Eq. (11.5) after having obtained α by curve fitting of Eq. (11.4) to experimental data. Many types of animal cells regulate their volume by eliminating osmolytes when swelled or accumulating osmolytes when shrunken. If the rate of volume regulation is slow relative to the speed of solution shifts and volume recording, the above Eqs. (11.4) and (11.5) would still be useful when applied to "instantaneous" changes. Subsequent change in $V^{(c)}$ as a function of time would then be governed by cell volume regulatory processes, which sets the stage for their experimental investigation. Epithelial studies have reported $v'/V_0^{(c)}$ ratios that range from 0.13 in rabbit kidney cortical collecting duct (Strange and Spring 1987b) to 0.41 in isolated epithelial cells of mouse choroid plexus (Hughes et al. 2010).

11.2.2 Cell Water Homeostasis Depends on Metabolic Energy

A cell of non-permeant negatively charged intracellular macroions and diffusible small anions kept electroneutral by K^+, which is in osmotic equilibrium with a surrounding NaCl solution, constitutes a Gibbs–Donnan system that will tend to swell. This is because Na^+ and Cl^- by electrodiffusion enter the cell together with an osmotically equivalent amount of water, a process that cannot proceed to equilibrium before the cell bursts. See Sten-Knudsen (2002) for a comprehensive mathematical treatment. This fatal situation is prevented by the Na^+/K^+ pump, which powers the back flux of Na^+ (and Cl^-) to the extracellular fluid fuelled by ATP hydrolysis. In the classical "pump-leak" description of the cell's osmotic equilibrium with the extracellular fluid, Cl^- was assumed to be in thermodynamic equilibrium across the plasma membrane (Leaf 1959; Tosteson and Hoffman 1960; Ussing 1960; Kay and Blaustein 2019). This applies to a number of body cells and as discussed below also to some epithelial cells. In most cells, however, Cl^- is kept above thermodynamic equilibrium by, e.g., inwardly directed cotransport with Na^+ that enters the cell along this ion's electrochemical potential gradient. Independent of the energetics of the Cl^- distribution, the intracellular Cl^- pool and the pools of other intracellular small osmolytes are the most important determinants of the volume of intracellular water. In cells with Cl^- being above thermodynamic equilibrium, the water volume associated with the Cl^- pool is under dynamic control by a balance between passive flows of Cl^- and K^+ through specific channels and oppositely directed cotransport of these two ions with Na^+ (Fig. 11.1a). Thus, an osmotically swelled cell may restore its volume by leaking K^+ and Cl^- to the extracellular fluid either through specific channels or by cotransport, termed *regulatory volume decrease* (RVD), whereas the volume of shrunken cells would be restored by uptake of Cl^- and K^+ energized by being coupled to cotransport with Na^+, which is termed *regulatory volume increase* (RVI). As indicated in Fig. 11.1a, a set of Na^+/H^+- and Cl^-/HCO_3^- exchange mechanisms may serve this latter mentioned function. The intracellular pools of other small osmolytes like amino acids and sugars may also contribute to cell volume regulation by similar dynamic balance between solute accumulation (uptake or synthesis) and solute elimination (release or degradation). This framework was developed for a small number of cell types but is now accepted

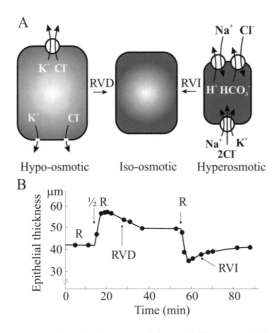

Fig. 11.1 Ion mechanisms of cell volume regulation. (**a**) The osmotically swollen cell may eliminate K^+ and Cl^- either through ion-selective channels or by coupled transport. By eliminating an osmotically equivalent amount of water cell volume returns toward the isoosmotic volume (RVD). The osmotically shrunken cell may regain volume by water uptake associated with uptake K^+ and Cl^- that are coupled to and driven by the uptake of Na^+, or mediated by a set of Na^+/H^+ and a Cl^-/HCO_3^- exchange mechanisms (RVI). The transporters' localization to epithelial membrane domain depends on cell type and is here randomly indicated for convenience. (**b**) Several years before the above concepts were introduced (see text for references), cell volume regulation was discovered in the first real-time study of volume responses of an epithelium (frogskin) to external osmotic perturbations. The preparation was exposed to 1/20 Ringer on the outside, while the inside bath was either Ringer (R) or Ringer diluted by distilled water (1/2 R). Redrawn from MacRobbie and Ussing (1961)

as a general scheme of cell volume regulation (see reviews by Lang et al. 1998b, Choe and Strange 2009, Hoffmann et al. 2009, Hoffmann and Pedersen 2011, Lang 2013).

11.2.3 Isoosmotic and Aniso-osmotic Cell Volume Regulation

The volume regulatory machinery may be activated by changes in the intracellular amount of osmolytes, termed *isoosmotic cell volume regulation* or by changes in extracellular osmolality, i.e., *aniso-osmotic cell volume regulation*. Isoosmotic cell volume changes are brought about by upsetting the balance between entry and exit of osmolytes, by synthesis/metabolism of organic osmolytes, and by the turnover of intracellular macromolecules that leads to change in the amount of their constituent monomers. In turn, this drives water into or out of the cell until osmotic equilibrium is reestablished. This activates volume regulation for bringing cell water volume back to its normal isoosmotic state. In mammals, the kidneys secure appropriate extracellular ion and water homeostasis. Therefore, body cells are more often challenged by isoosmotic than by aniso-osmotic cell volume perturbations (see Chamberlin and Strange 1989). Perturbation of cell volume by change in extracellular osmolality is a common challenge to transporting epithelia. This applies to epithelia that generate or maintain transcellular osmotic gradients and probably also to low-resistance fluid transporting epithelia that seem to depend on cell volume regulation for securing the transported fluid being isoosmotic to interstitial fluid.

11.2.4   *"Cross Talk" Between Membrane Domains of Transporting Epithelia*

Coordinated activity of solute entry and exit pathways constitutes an important homeostatic mechanism of transporting epithelia, which has been named "cross talk" (Schultz 1981, 1992, Diamond 1982, Harvey 1995). Examples are the activity of the Na^+/K^+ pump versus that of apical Na^+ channels and basolateral K^+ channels of tight epithelia (MacRobbie and Ussing 1961; Larsen 1973; Finn 1982), coupling of apical rheogenic alanine/$2Na^+$ uptake and activity of K^+ channels of the intestinal mucosa (Gunter-Smith et al. 1982), interaction between the apical Cl^- conductance and the basolateral K^+ conductance of a Cl^- secreting cell (Welsh and McCann 1985), coordinated activity of apical Na^+ channels and basolateral K^+ channels in frogskin (Harvey et al. 1988), Na^+ absorption and H^+ secretion by α-type mitochondrion-rich (MR) cells (Harvey 1992), apical Cl^- channel activation and multiple basolateral Na^+ transport pathways of secretory acinar cells (Robertson and Foskett 1994) and coupled activity of apical Cl^- channels of mitochondrion rich (MR) cells and apical Na^+ channels of principal cells of amphibian skin (Larsen 2011). Different types of signals govern these regulations as exemplified by intracellular pH (Harvey and Ehrenfeld 1988; Harvey et al. 1988), intracellular free-$[Ca^2$ $^+]$ (Grinstein and Erlij 1978; Taylor and Windhager 1979; Chase and Alawqati 1983; Welch and McCann 1985; Tinel et al. 2000), ATP (Tsuchiya et al. 1992; Welling 1995; Urbach et al. 1996), membrane-bound G proteins (Cantiello et al. 1989), sub-membrane actin filaments (Wang et al. 2005), a concomitant fall in intracellular $[Cl^-]$ and rise in intracellular $[Ca^{2+}]$ (Robertson and Foskett 1994) and membrane potential (Larsen and Harvey 1994; Larsen 2011).

Quite generally, because of the voltage dependence of sodium pump flux (Gadsby and Nakao 1989; Wu and Civan 1991; Larsen et al. 2009), perturbation of any rheogenic apical transport system (e.g., a Na^+ channel, the SGLT2 mechanism or the $3Na^+$:$1HPO_4^-$ transporter) would result in an instantaneous change in the lateral Na^+/K^+-pump fluxes. Recently, this type of cross talk mechanism was investigated quantitatively in a study of Na^+ and glucose fluxes in kidney proximal tubule (Larsen and Sørensen 2019). It was indicated that activation of glucose uptake across the luminal brush border membrane via SGLT2 would lead to an instantaneous transepithelial fluid reabsorption as well, energized by the lateral Na^+/K^+-pumps. Among several other novel findings, this computational study further indicated that rate of fluid reabsortion is given by the rate at which Na^+ is pumped into the lateral intercellular space, while osmolarity of absorbed fluid depends on the hydraulic conductance of apical (luminal) membranes.

Volume regulation and cross talk operate together in securing homeostasis in cell volume, electrolyte concentrations, and driving forces of transmembrane solute fluxes, which otherwise would be submitted to large variations depending on the functional activity of the epithelium. As discussed below, this may result in surprising responses to experimentally induced cell volume perturbations.

11.3 Osmotic Permeability of Epithelial Cell Membranes

Studies on the water content of tissue slices corrected for interstitial water by an impermeable extracellular maker proved the significance of energy metabolism and active Na^+ transport for maintaining epithelial cell volume (Macknight and Leaf 1977). There were problems in following up on these studies with sufficient time and space resolution as it turned out that the osmotic permeability (hydraulic conductance) of individual cell membranes in normal functioning epithelia is very high. Added to this, some epithelia are heterocellular, which requires visual identification of cell types of specific functions to be studied individually. In the body, polarized epithelia are often exposed to solutions of different composition on the two sides, which imposes a further challenge to in vitro studies. Epithelial cell volume regulation is therefore studied in an Ussing chamber for flat-type preparations and in a Burg microperfusion setup for tubular preparations, both of which permit unilateral perturbation of bathing solutions (MacRobbie and Ussing 1961; Burg and Knepper 1986). For measuring the osmotic permeability of individual cell membranes (P_f) by recording "instantaneous" volume responses to rapid external osmotic perturbations, unstirred layer artifacts must be eliminated which has been accomplished by several laboratories (Schafer et al. 1978; Spring and Hope 1979; Carpi-Medina et al. 1983, 1984; Welling et al. 1983b; Cotton et al. 1989). An advanced technology applies quantitative DIC microscopy by online computer-controlled fast optical sectioning, storing of video images of section frames, and offline computation of cell volume as a function of time (Marsh et al. 1985; Strange and Spring 1986; Guggino et al. 1990). Another powerful technology relies on measuring the concentration of a non-permeable cytosolic probe molecule for calculating time-dependent changes of cell water volume in response to unilateral manipulation of bath composition (Reuss 1985; Ford et al. 2005). A quite different approach, only aiming at P_f, applies stopped-flow and light scattering techniques for following volume changes of vesicles of the isolated apical or basolateral membrane, respectively (van Heeswijk

Table 11.1 Examples of osmotic permeabilities (P_f) of luminal and basolateral plasma membranes of transporting epithelia; areas corrected for apical microvilli and basolateral infoldings, respectively

Epithelium	Luminal membrane P_f ($\mu m \cdot s^{-1}$)	Basolateral membrane P_f ($\mu m \cdot s^{-1}$)	Method [Ref.]
Rana temporaria skin, control[a] + ADH[a]	≤ 1 3.8 ± 1.2	24 ± 1 23.5 ± 1.7	Optical [MacRobbie and Ussing (1961)]
Necturus gallbladder	545	347	Optical [Persson and Spring (1982)]
Necturus gallbladder	660	440	TMA$^+$ as volume marker [Cotton et al. (1989)]
Rat small intestine	60 ± 0.4	60 ± 15	Stopped flow, light scattering [van Heeswijk and van Os (1986)]
Rat proximal tubule, isolated plasma membrane vesicles	604 ± 20	557 ± 137	Stopped flow, light scattering [van Heeswijk and van Os (1986)]
Rabbit proximal tubule, S_2 segment	70–250	100–375	Optical [Gonzáles et al. (1984), Carpi-Medina and Whittembury (1988)]
Amphiuma kidney, diluting segment	1.0 ± 0.3	2.5 ± 0.3	Optical [Guggino et al. (1985)]
Rabbit cortical collecting tubule-PC, control + ADH	19.8 ± 1.9[b] 92.2 ± 8.6[b]	68.2 ± 3.5[c] 74.1 ± 5.8[c]	Optical [Strange and Spring (1987b)]
Rabbit cortical collecting tubule-IC, control + ADH	25.0 ± 2.3[b] 86.2 ± 6.0[b]	70.5 ± 4.6[c] 61.2 ± 9.6[c]	Optical [Strange and Spring (1987b)]
WT-RCCD1 cells[d]	10 ± 0.6		BCECF as volume marker [Ford et al. (2005)]
AQP2-RCCD1 cells[e]	123 ± 7		BCECF as volume marker [Ford et al. (2005)]

PC principal cells, *IC* intercalated cells
[a]Normalized for apical membrane area
[b]At a 100 mosmol/kg osmotic gradient
[c]At a 100 mosmol/kg osmotic gradient; tubule lumen filled with oil shortly before basolateral solution shift
[d]Clonal cell line of rat cortical collecting tubule, wild type
[e]RCCD1 stably transfected with AQP2

and van Os 1986). Examples of P_f values corrected for membrane area-amplification by apical microvilli or basolateral infoldings are listed in Table 11.1 chosen here for illustrating the range covered by epithelial cell membranes. Except for small intestine, P_f of leaky epithelia are significantly larger than P_f of tight epithelia. It is seen that the antidiuretic hormone (ADH) stimulates apical P_f of tight epithelia with no effect on the basolateral membrane. This finding of considerable biological significance was first reported for frogskin by MacRobbie and Ussing (1961). Much later it was reproduced and generalized by Strange and Spring (1987b) who studied cellular regulation of the osmotic membrane permeability of rabbit cortical collecting duct. The osmotic permeability is associated with the expression of aquaporin water channels (Preston et al. 1992; Agre et al. 1993a; Nielsen and Agre 1995; Nielsen et al. 2002, 2013), which is discussed in Chap. 30 of Vol. 3 of this book series. While P_f values listed in Table 11.1 are a measure of the density of water channels, the total osmotic membrane permeability may be significantly larger due to the gain in membrane area by luminal microvilli and basolateral infoldings, respectively. For example, the total osmotic permeability of the S2-segment of rabbit proximal tubule is 1260–4500 $\mu m \cdot s^{-1}$ for the luminal membrane and 1400–5000 $\mu m \cdot s^{-1}$ for the basolateral membrane (Gonzalez et al. 1982; Schafer 1990). Likewise, taking into account area-gains in intact cortical collecting tubules the ratio of the total membrane osmotic permeability of apical and basolateral membrane would be 1:7 after hormone stimulation rather than about 1:1 as indicated in Table 11.1 (Strange and Spring 1987b).

Studies by Arthur K. Solomon and his coworkers of isosmotic transport by kidney proximal tubule concluded that "the driving force for water movement arises from the efflux of NaCl from tubule to plasma" (Windhager et al. 1959). Recently, this notion was analyzed in detail by a computational method by Larsen and Sørensen (2019) showing that *at transepithelial osmotic equilibrium* the rate of transepithelial fluid transport is simply given by the rate at which Na^+ is pumped into the lateral intercellular space, which constitutes the compartment coupling the water flow and the active Na^+ flux. This was independent of the osmotic permeability of the epithelial plasma membranes. In particular, in an AQP-1($-/-$) engineered epithelium the water flow was shown to shift route from being translateral to being paracellular with no change in flow rate as long as the Na^+ pump flux is the same. On this notion, the observed reduced fluid reabsorption by proximal tubule of AQP-1($-/-$) mice (Schnermann et al. 1998) would be the result of the concomitantly observed reduced active Na^+ reabsorption rather than of the lack of AQP-1 water channels, which was verified in a similar way by the abovementioned quantitative computational method.

In the literature on epithelial water transport one is frequently confronted with a measure of what is considered to be the "osmotic permeability of the intact epithelium." This quantity is obtained by exposing the epithelial preparation to a brief unilateral transepithelial pulse of a non-permeable electroneutral molecule, $\Delta\pi$, for measuring the associated *steady-state* response of the transepithelial water flux, ΔJ_V^{epit}. From this protocol, an (apparent) osmotic permeability of the intact epithelium is calculated by:

$$P_f^{epit} = \Delta J_V^{epit} / \left(\Delta\pi \cdot \overline{V}_W \right) \tag{11.6}$$

where $\overline{V}_W$ is the molar volume of water. It has been of considerable concern that the osmotic permeability of individual cell membranes is two to three orders of magnitude larger than the above apparent osmotic permeability (Eq. 11.6), e.g., Weinstein (2013), which has never been given a satisfactory biophysical explanation. Recently, Larsen and Sørensen (2019) suggested a logical and simple solution to the problem. Applying the computer-assisted method they studied intraepithelial volume responses to a fast unilateral osmotic pulse. It was shown that the transepithelial osmotic pulse evokes time-dependent cell volume change comprising an instantaneous shrinkage and a subsequent membrane-compliance dependent return of cell volume toward the volume prior to the osmotic pulse. Therefore, at the new steady state, the cell volume is changed by a relatively small amount and so is the associated steady-state transepithelial water flux. This time-dependent response of the transepithelial water flux readily explains the fairly low "transepithelial osmotic permeability" calculated by the expression above concerning the steady-state response (Eq. 11.6). This of course is an artifact which has no biophysical relationship to an osmotic permeability of the epithelium. Because the above, P_f^{epit} has ambiguous meaning devoid of physiological significance it has to be abandoned (Larsen and Sørensen 2019).

11.4 Cell Volume Response to Osmotic Challenges in Extrarenal Epithelia

11.4.1 Amphibian Skin

Amphibian skin epithelium is heterocellular with Na^+ transporting principal cells and mitochondria-rich (MR) cells that are excluded from the functional syncytium of principal cells (Rick et al. 1984; Larsen 1988). Populations of MR cells are differentiated for passive and active Cl^- uptake, and acid/base regulation, respectively. Reviewed by Larsen (2011), the γ-type MR cell expresses functional CFTR (Sørensen and Larsen 1996) and a proton ATPase (Jensen et al. 1997) in the apical membrane similar to the newly identified "unknown" airway "ionocyte" of mammalian trachea (Montoro et al. 2018; Plasschaert et al. 2018; Travaglini and Krasnow 2018).

11.4.1.1 Principal Cells

The first real-time study of epithelial cell volume aimed at testing the transport model for frogskin (MacRobbie and Ussing 1961) indicated that ADH selectively increases apical P_f, see Table 11.1. While some of their observations were in good agreement with predicted model behavior, the study also raised problems that were not solved until about 20 years later. Thus, with sulfate as a major anion, the epithelium behaved as expected for a simple osmometer of a relatively large non-osmotic volume (Eq. 11.3). However, as shown in Fig. 11.1b, after substitution with chloride on the inside, the acute osmotic volume expansion in the hypoosmotic solution was followed by a return of the cell volume toward its initial value. When reexposed to normal osmolality, the epithelium shrank initially, as expected, but then took up fluid for returning toward the original isoosmotic volume. These unexpected delayed responses were not predicted by any ideas of the time. In addition, it was troublesome that pump inhibition by ouabain did not result in swelling of the cells as predicted for a Gibbs–Donnan system. As it turned out, because of inhibition of the apical Na^+- and the basolateral K^+ permeability, cell volume stayed the same for

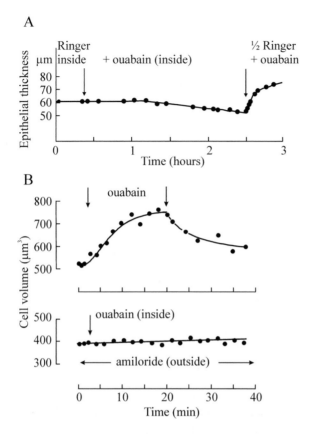

Fig. 11.2 Response of anuran epidermal cells to ouabain. (**a**) Principal cells of frogskin; the immediate response to the block of the Na^+/K^+ pump is not swelling. Rather, the volume stayed constant for about 1/2 h. This is due to inhibition of the apical and basolateral cation conductances, later denoted cross talk. Thereafter, the cell volume decreased. See text for references and further discussion. Ringer solution on the outside diluted 20 times by distilled water (see text for details) and Ringer's solution or a 50% hypoosmotic Ringer's solution (indicated by 1/2 R) on the inside as indicated. Redrawn from MacRobbie and Ussing (1961). (**b**) The mitochondria rich (MR) cell bathed with Ringer on both sides swelled with ouabain on the inside, which was very significantly reduced by blocking the apical Na^+ entrance mechanism by amiloride. Redrawn from Larsen et al. (1987)

about 30 min. As discussed in Sect. 11.2.4 above, this response was later denoted cross talk. It was even more surprising, however, that the cross talk response was often followed by cell shrinkage, see Fig. 11.2a. The ouabain induced cell shrinkage found its logical explanation once it was realized that the Cl^- pool in frogskin cells is maintained above thermodynamic equilibrium by regulated basolateral Na^+, K^+, 2 Cl^- (NKCC) cotransport in parallel with dynamic Cl^- and K^+ permeabilities (Ussing 1982, 1985; Giraldez and Ferreira 1984; Harvey and Kernan 1984; Dörge et al. 1985; Willumsen and Larsen 1986) similar to what is seen in non-epithelial cells (Kregenow 1971; Hoffmann et al. 1979; 1983; Gelfand et al. 1984; Sarkadi et al. 1984). Computer-assisted analysis showed that the frogskin model with NKCC in the serosal membrane has the capacity to reproduce the abovementioned perplex time course of epithelial cell volume following ouabain inhibition of the Na^+/K^+ pump. In particular, on the way toward Gibbs–Donnan equilibrium transient cell shrinkage results from Cl^- pool depletion associated with decreased Cl^- uptake via the serosal cotransporter caused by a passive increase in $[Na^+]_c$ (Larsen 1991). This particular volume response to ouabain is further discussed in Sect. 11.5.1 on kidney proximal tubule.

11.4.1.2 Mitochondria-Rich Cells

When exposed on the outside to Ringer's solution, apical Cl^- channels of MR cells can be activated by membrane depolarization (Larsen and Rasmussen 1982), which mediates passive Cl^- uptake together with active Na^+ uptake by principal cells (Larsen 1991). All evidence considered indicates that this is an adaptation to life on land where the skin is covered by subepidermal gland secretions from which ions are reabsorbed during evaporative water loss (Larsen 2011; Larsen and Ramløv 2013; Larsen et al. 2014). Serosal ouabain evoked relatively fast cell swelling of toad skin-MR cells, which was reversible upon wash with fresh Ringer and could be prevented by amiloride in apical bath (Fig. 11.2b). In bilateral Ringer $[Cl^-]_c$ was passively distributed at a concentration of 19.8 ± 1.7 mM. The cell swelled in parallel with voltage activation of apical Cl^- channels due to uptake of Cl^- across the apical membrane and K^+ uptake across the membrane facing the lateral intercellular space. Upon bilateral changes of external osmolality, the volume responded in agreement with Eq. 11.4 of $v'/V_0^{(c)} = 0.21$. Following logically from passive Cl^- distribution neither RVD nor RVI were observed in response to cell swelling and cell shrinkage, respectively (Larsen et al. 1987). Frogskin exposed on the outside to a Ringer's solution that is diluted 20 times by distilled water ($[Na^+] = [Cl^-] = 5.5$ mM) leads to closure of apical Cl^- channels of MR cells as reviewed by Larsen (2011). This was shown to result in a significant osmotic increase in MR cell volume and in a cytosolic osmolality calculated to be about 166 mosM (Spring and Ussing 1986). Applying this protocol, it was found that subsequent dilution of the serosal Ringer's solution by 50% with distilled water resulted in the expected acute further osmotic cell volume expansion. With the very small cellular Cl^- pool left—most surprisingly—this initiated a complete RVD, which therefore was considered to be caused by the release of unidentified nega- tively charged diffusible osmolytes together with the release of K^+ (Spring and Ussing 1986).

11.4.2   Gallbladder

Gallbladder epithelium modifies the bile secreted by the liver cells by isoosmotic reabsorption of Na^+, Cl^-, and HCO_3^-. The epithelium is formed by a monolayer of regular slender cells of an apical membrane with short 0.4–5-μm microvilli and a flat basal membrane resting on the basal lamina (Tormey and Diamond 1969). The studies by Kenneth Spring and his coworkers used *Necturus maculosus* gallbladder with epithelial cells of simple geometry (Spring and Hope 1978, 1979). Thus, Persson and Spring (1982) measured for the first time individual membrane permeabilities of a leaky epithelium (Table 11.1). The cells effectively regulate the volume in hyperosmotic (RVI) and hypoosmotic (RVD) solutions of an initial volume change to a fast change in external osmolarity in agreement with Eq. 11.4 with $v'/V_0^{(c)} = 0.18$ (see Fig. 11.3a). With both Cl^-- and K^+ concentration being well above electrochemical equilibrium at control conditions, RVD was caused by transient activation of basolateral electrogenic K^+ and Cl^- transport (Larson and Spring 1984). RVI depended on parallel operation of apical Na^+/H^+ and Cl^-/HCO_3^- exchange mechanisms (Fisher et al. 1981; Ericson and Spring 1982a, b) regardless of the sidedness of the hyperosmotic challenge (Marsh and Spring 1985). Mucosal bumetanide at low concentrations inhibited apical NaCl uptake (Larson and Spring 1983) and transcellular fluid flow (Davis and Finn 1985) indicating the role of apical NKCC2 in both NaCl uptake and cell volume regulation. The relative importance of

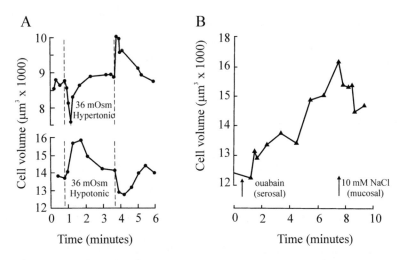

Fig. 11.3 Necturus gallbladder. (**a**) In response to the addition or removal of mannitol the cell lost or gained water. Subsequent RVI or RVD proceeded at a significant rate. The osmotic challenges shown in the graphs were applied on the mucosal side of the epithelium. RVI was observed also with a hyperosmotic mucosal bath achieved by adding NaCl or urea (not shown). (**b**) Serosal ouabain caused rapid swelling of the cell bathed with Ringer + 36 mM mannitol on both sides. At the right arrow, 90% of mucosal NaCl was replaced by an osmotically equivalent amount of mannitol, which reduced mucosal [NaCl] to 10 mM. Redrawn from Persson and Spring (1982)

the above apical ion uptake mechanisms is cAMP dependent (Garvin and Spring 1992). Serosal ouabain caused acute cell swelling (see Fig. 11.3b) at a rate depending on mucosal [NaCl], which was abolished if all NaCl was replaced by mannitol (Ericson and Spring 1982a), or in K^+ free mucosal bath (Davis and Finn 1985).

11.4.3   Small Intestine

Intestinal homeostatic mechanisms are important both because the osmolality of intestinal fluid varies with water and food intake, and because the pools of cellular osmolytes and metabolites depend on whether the intestine is in a resting or an absorbing/secreting state. Villus cells are mainly engaged in the absorption of NaCl, nutrients, and water, while crypts are mainly engaged in fluid secretion. The absorptive surface of the large columnar cells is increased by elaborated microvilli. The serosal membrane is flat but the area of lateral membranes is increased by interdigitating with neighboring cells. Absorption of nutrients is linked to Na^+ transport at the apical membrane, which also harbors Na^+/H^+ exchangers, $Cl^-/$ HCO_3^- exchangers, and CFTR Cl^- channels (Xia et al. 2014). K^+ channels, Cl^- channels, and NKCC1 are expressed at the basolateral membrane. Na^+/K^+ pumps at the lateral membranes create hyperosmotic lateral spaces that governs water uptake by osmosis. In human small intestine, this amounts to ~5.5 liter/day. Stimulation by cAMP, cGMP, Ca^{2+}, or enterotoxins inhibits the apical electroneutral exchange mechanisms and stimulates the apical CFTR channel, whereby the epithelium switches from absorption to secretion. Human small intestine secretes about 2 liters of fluid per day. A detailed discussion of intestinal ion transporters and their functions is given by Kato and Romero (2011).

11.4.3.1   Intestinal Crypt Cells

Isolated guinea pig intestinal crypts regulate cell volume in both hypoosmotic (O'Brien et al. 1991) and hyperosmotic solutions (O'Brien et al. 1993). Their technique prevented the analysis of the sidedness of the osmotic membrane response. The RVD was almost complete within 20 min and effectively inhibited by K^+ channel blockers (Ba^{2+}, quinine) and a Cl^- channel blocker (9ACA). The RVI led to complete volume recovery in about 20 min (Fig. 11.4a) and was abolished by bumetanide inhibition of the NKCC cotransporter (Fig. 11.4b). RVI is decoupled if the added osmolyte was D-mannitol (Fig. 11.4c), probably due to a lack of appropriate ion gradients for NKCC1. Secretagogues (vasoactive intestinal peptide and carbachol) initiated cell shrinkage with no spontaneous RVI, but the volume loss was reversible upon removal of the hormone (Fig. 11.4d). Interestingly, volume regulation of intestinal crypts in hypoosmotic solutions is impaired in cystic fibrosis mice (Valverde et al. 1995). This is not due to lack of a Cl^- exit channel, but by the

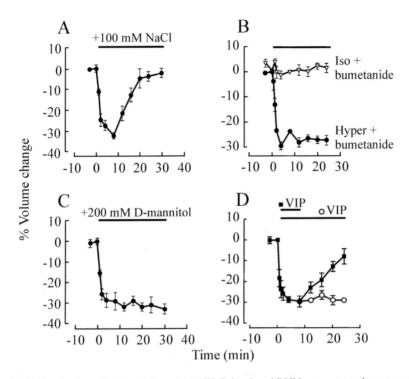

Fig. 11.4 Intact guinea pig intestinal crypts. (**a**) Well-developed RVI in response to hyperosmotic challenge achieved with 100 mM NaCl in bath. (**b**) RVI inhibited by bumetanide, hyperosmolality obtained by adding 100 mM NaCl to bath. (**c**) Cell shrinkage with 200 mM D-mannitol added to bath; RVI decoupled. (**d**) Crypts stimulated to fluid secretion by vasoactive intestinal peptide (VIP) exhibits isoosmotic shrinkage and remain shrunken until of VIP is being removed. Redrawn from O'Brien et al. (1993)

absence of a K^+ pathway, because the K^+ selective valinomycin-ion pore restored full RVD (Valverde et al. 2000). In human intestine 407 cells, RVD was associated with the release of organic osmolytes through a non-identified PKC activated pathway (Tomassen et al. 2004).

11.4.3.2 Intestinal Villus Cells

Schulz and coworkers discovered cross talk between apical nutrient uptake and activation of a basolateral K^+ channel, which in turn restores the apical membrane potential and the driving force for the cotransporter (Gunter-Smith et al. 1982; Grasset et al. 1983). The addition of D-glucose or L-alanine, but not L-glucose or D-alanine, caused immediate and fast swelling of isolated villi enterocytes followed by RVD, which was associated with loss of K^+ and Cl^- through volume-activated ion channels (MacLeod and Hamilton 1991). Like intestinal crypts, RVI of hyperosmotically shrunken cells from jejunum villi also depended on NKCC cotransport and was inhibited by bumetanide (Macleod and Hamilton 1990). Ouabain-induced swelling proceeded slowly for leveling off already at about 110% of the isoosmotic cell volume (Macleod and Hamilton 1990). The mechanism preventing further development of the Gibbs–Donnan response was not investigated.

11.4.4 Upper Airways

The upper airways from the nasal cavity to bronchi are lined by ciliated columnar cells covered by an airway surface layer (ASL) of a near isoosmotic periciliary liquid (PCL) with mucous on top. Of importance for pulmonary host defense, ciliary beating moves mucus and periciliary fluid toward the throat where it is swallowed; dysfunction may lead to severe pathological conditions like pulmonary obstruction and cystic fibrosis. ASL is renewed by flow from distal areas and by submucosal gland secretions. Recently, the inner surface of the human trachea was shown to harbor multiple cells types (Montoro et al. 2018; Plasschaert et al. 2018; Travaglini and Krasnow 2018) of which a so-called "unknown ionocyte" in the apical membrane expresses CFTR and a proton ATPase similar to the γ-type mitochondrion-rich cell of amphibian skin epithelium (Sørensen and Larsen 1996; Jensen et al. 1997). From an evolutionary perspective, this is an interesting notion because soft-skinned amphibians, which were the first air-breathing vertebrates inhabiting terrestrial environments, on land are challenged by evaporative water loss stemming from a cutaneous surface layer secreted by subcutaneous mucous glands (Larsen 2011; Larsen and Ramløv 2013; Larsen et al. 2014). Likewise, upper airways replenish evaporative water loss by secreting saline for maintaining a wet extracellular surface layer that prevents desiccation of airway cells. Recently, Wittekindt and Dietl (2018) discussed the role of AQPs in the lung with regard to fluid homeostasis of the respiratory epithelium. Cell volume regulatory mechanisms would be of importance for securing robust functioning of airway epithelia.

The surface of the small airways is relatively very large, and the height of ASL is the same in distal and proximal regions. Because the volume flows relative to the epithelial surface area are about the same, it follows that during the passage from distal to proximal airways water and ions are absorbed probably at isoosmotic proportions. This is energized by Na^+ uptake via apical ENaC and basolateral Na^+/K^+ pumps (Willumsen and Boucher 1991; Boucher 1994, 1999, 2003; Grubb et al. 1997; Tarran et al. 2001). The apical membrane also harbors cAMP-activated CFTR and Ca^{2+}-activated Cl^- channels. A basolateral NKCC1 cotransporter maintains $[Cl^-]_c$ above thermodynamic equilibrium across the basolateral membrane. In the absorbing state, uptake of Cl^- is passive probably through the paracellular pathway. Secretion is driven by basolateral NKCC1 cotransport and downhill movement of Cl^- through the apical membrane's Cl^- channels (Willumsen et al. 1989; Willumsen and Boucher 1991). Thus, PCL volume of a depth $\leq$ 7 μm is balanced by ENaC-activated fluid absorption and CFTR-stimulated fluid secretion regulated by not yet fully understood complex interactions between ATP and ADP of ASL (Lazarowski et al. 2004; Tarran et al. 2006). The volume-sensing mechanism is unknown, but previous studies indirectly indicated that osmotically induced epithelial volume changes may provide an early signal. Thus, cell shrinkage was seen only in response to mucosal hyperosmolality, and the cells remained shrunken until physiological saline again was applied on the mucosal surface as expected for an epithelial cell that serves sensing of ASL osmolality (Willumsen et al. 1994). RVD is developed in human tracheal and bronchial cells and is defective in tracheal cells from CF patients. This was caused by altered regulation of an intermediate conductance, Ca^{2+} regulated K^+ channel. The recording technique did not allow for analysis of the sidedness of the osmotic membrane response (Vazquez et al. 2001; Fernandez-Fernandez et al. 2002). A swelling-activated entry of Ca^2 via the TRPV4 channel was abolished in cells of CF patients, which is compatible with the hypothesis that TRPV4 is a mechano- and osmo-receptor (see Sect. 11.8.2), and is of significance for the maintenance of mechanosensitivity and isoosmotic equilibrium (Arniges et al. 2004), see Sect. 11.9 for a further discussion.

11.4.5   Exocrine Glands

The acinar epithelium of exocrine glands generates an isoosmotic primary secretion of a $[K^+]$ higher than that of plasma (Petersen and Gallacher 1988; Petersen 1993), which may be modified by the epithelium lining the excretory duct (Novak and Greger 1988; Novak 2011). Carbachol-induced increase in acinar $[Ca^{2+}]_c$ results in Cl^- and K^+ loss and isoosmotic cell shrinkage (Nauntofte and Dissing 1988; Takemura et al. 1991; Petersen 1992; Foskett et al. 1994; Speake et al. 1998), and active Cl^- secretion (Cook et al. 1994; Nielsen and Nielsen 1999) driven by increased activity of the NKCC1 cotransporter (Mills 1985; Foskett et al. 1994) and/or the NKE1 mechanism (Manganel and Turner 1991). In frogskin, stimulation by cAMP-linked agonists (prostaglandin E_2, isoproterenol) leads to ion secretion through coexpressed apical CFTR Cl^- channels and maxi-K^+ channels (Sørensen et al. 2001). Stimulation by Ca^{2+}-linked agonists (carbachol) activates Ca^{2+}-dependent apical TMEM16A Cl^- channels (Romanenko et al. 2010). Experiments on the perfused shark rectal gland showed that it is the transient cell shrinkage that triggers cotransport by NKCC1 (Greger et al. 1999). In single isolated parotid cells, large amplitude oscillations of volume and intracellular Na^+, K^+ and Cl^- concentrations driven by oscillations in intracellular $[Ca^{2+}]$ have been demonstrated by Foskett and his coworkers (Wong and Foskett 1991; Foskett et al. 1994). Frogskin acinar cells in nystatin-permeabilized whole-cell patch-clamp configuration did not exhibit membrane potential oscillations neither during Ca^{2+}- nor during cAMP-coupled secretion (Sørensen and Larsen 1999). This indicates that oscillations in $[K^+]_c$ (and cell volume) are not associated with continued secretion by an intact acinar epithelium. The studies discussed above, however, have shown that an isoosmotic cell volume decrease precedes the secretory response similar to the secretion-stimulated intestinal crypt epithelium (see Fig. 11.4d).

Acinar cells isolated from rat lacrimal glands exhibit aniso-osmotic volume regulation. The cells lost volume in hyperosmotic solutions, but a regulatory volume increase (RVI) was only observed under certain conditions depending on temperature and buffer system (Douglas and Brown 1996). Thus, in HEPES-buffered solutions at 37 °C, RVI was dependent on ion uptake via an NKCC mechanism. In contrast, at 20 °C and in HCO_3^- buffered solutions inhibitor studies indicated that RVI required ion uptake via Cl^-/HCO_3^- and Na^+/H^+ exchangers. RVI was also supported by uptake of a mixture of neutral amino acids via system-A Na^+-coupled amino acid cotransport. RVD was observed in hypoosmotic solutions probably controlled by channel-mediated inflow of Ca^{2+} (Speake et al. 1998). Since the above studies were performed on isolated cells, the sidedness of the above osmotic membrane responses could not be assessed. Isolated parotid cells also exhibit classical RVD (Foskett et al. 1994).

11.4.6   Teleost Gill and Opercular Epithelium

The euryhaline killifish, *Fundulus heteroclitus*, tolerates salinities ranging from freshwater (FW) to seawater (SW) and adapts quickly to changes in salinity. The Killifish opercular epithelium has a high density of chloride secreting (MR) cells and is an excellent model of an osmosensing Cl^- secreting epithelium (Foskett 1982; Foskett et al. 1983; Marshall 2011). The MR cell has an ouabain-inhibitable Na^+, K^+ ATPase and a bumetanide-sensitive NKCC1 cotransporter in the basolateral membrane with a CFTR anion channel in the apical membrane (Evans et al. 2005; Marshall and Grosell 2005). There is an increase of $\sim$70 mosmol/kg in plasma osmolality during the transition of the killifish to high salinity (Zadunaisky et al. 1995) and this increase in osmolality results in a rapid increase in salt secretion across the epithelium. This seems to involve a hyperosmotic (shrinkage) activation of the basolateral NKCC1 cotransporter and a suite of the protein kinases PKC, MLCK, p38 MAPK, JUNK, oxidation stress response kinase 1 (OSR1) and Ste20-related proline–alanine-rich kinase (SPAK) (Hoffmann et al. 2002; Marshall et al. 2005), see Fig. 11.5. A major decrease in plasma osmolality accompanies freshwater transition of the Killifish, especially in the first 12 h after transfer (Marshall et al. 2000). This results in hypoosmotically (cell swelling) controlled deactivation of the NKCC cotransporter involving Ca^{2+} and an unidentified protein phosphatase. The mitogen-activated protein kinase (MAPK) p38 has an inhibitory role in the hypoosmotic deactivation of NKCC1 (Marshall et al. 2005). Finally, the focal adhesion kinase (FAK), which co-localizes with NKCC1 at the basolateral membrane, seems to have an important role to set the levels of NKCC activity, and it is likely that chloride cells respond to hypoosmotic shock using integrin β1 as an osmosensor coupled to dephosphorylation of FAK that leads to NKCC1 deactivation (Marshall et al. 2008).

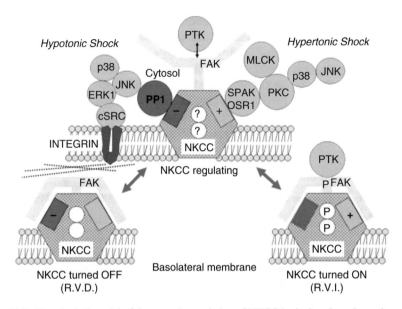

Fig. 11.5 Hypothetical model of the osmotic regulation of NKCC in the basolateral membrane of killifish opercular epithelium. *Right-hand side*. JNK, MLCK, and P38 MAPK kinases near the basolateral membrane are all activated by hyperosmotic stimuli and by NKCC-activation associated with regulatory volume increase (RVI). Kinases co-localized with NKCC are also included (FAK, OSR1, and SPAK). Because SPAK and OSR1 in other systems co-immunoprecipitate with NKCC, they are placed nearest to NKCC. The order of the other kinases in the cascade is unknown. *Left-hand side*. The inhibitory cascade terminating with PP1 shown in other systems to co-immunoprecipitate with NKCC. Known players are JNK and P38 MAPK. Shown are also integrin, ERK, and cSRC because they mediate hypoosmotic stimuli in other cell types. The model proposes three conditions: an unstable regulating phase that results in NKCC becoming phosphorylated and stimulated during RVI (right), or dephosphorylated and inactivated during RVD (left). The reason why FAK is depicted as occluding the phosphorylation sites of NKCC is because of the unusual dual effect of the PTK inhibitor genistein on the chloride current: when the chloride current is high it is decreased by genistein; when the chloride current is low it is the increased by genistein. From Hoffmann, Schettino, and Marshall, Comparative Biochemistry and Physiology A. 148, 29–43, 2007. With permission

11.4.7   Intestine of European Eel

In eel, intestinal epithelium transepithelial Cl^- absorption is via an apical NKCC2 cotransporter and a basolateral Cl^- channel. K^+ is returned to intestinal lumen through an apical K^+ channel. The model for the eel intestine is of the type described for the mammalian cTAL (see below). Downstream migratory eels drink seawater resulting in hyperosmotic exposure of the apical side of the intestinal epithelium (Lionetto et al. 2001). It has been shown that shrinkage-induced stimulation of NKCC2 and RVI plays a central role in the response to hyperosmotic challenge, which is dependent on the integrity of the F-actin microfilament system and involves PKC and MLCK (Lionetto et al. 2002). When the eel moves back to freshwater, the intestine becomes exposed to hypoosmotic fluids both on the basolateral and the apical side resulting in cell swelling and an RVD (recorded by morphometric measurement of the epithelium height) associated with swelling activation of several anion and K^+ conductive pathways in the apical and the basolateral membrane (Lionetto et al. 2008). The Ca^{2+}-activated large conductance (BK) K^+ channel in the apical membrane is a particularly important participant in this response (Lionetto et al. 2008) where it in all cases is found to be activated under hypoosmotic conditions, reviewed in Hoffmann et al. (2007).

11.5 Cell Volume Response to Osmotic Challenges in Renal Epithelia

Most segments of vertebrate nephron have evolved the ability to volume regulate when exposed to aniso-osmotic solutions. In the sections below, examples of principal mechanisms are discussed based on representative studies. Comprehensive reviews are given by Hebert (1987), Montroserafizadeh and Guggino (1990), and Lang (2013).

11.5.1 Kidney Proximal Tubules

The proximal tubules of human kidneys reabsorb ~120 liters/day of a primary filtrate of 180 liters. Ion and water turnovers are fast, requiring rapid regulation of cell volume and intracellular ion pools to maintain normal function. The tubule is heterocellular by being composed of three consecutive segments, S1–S3 (Maunsbach 1966). The cells have apical microvilli and elaborated lateral interdigitations near the bottom of the cells (Welling and Welling 1988). All three segments reabsorb solutes and water in isosmotic proportion with the water flow being translateral via AQP1 in brush border and lateral membranes (Agre et al. 1993b). The active translateral sodium reabsorption is energized by Na^+, K^+ pumps at the plasma membranes lining the intercellular space of complicated microanatomy (Welling and Welling 1988). The large literature on electrolyte and acid–base transports by the different segments was recently reviewed in Zhuo and Li (2013). Stationary and nonstationary ion and water flux interactions in proximal convoluted tubule were discussed in a comprehensive study by Larsen and Sørensen (2019), which includes novel analysis of the electrophysiology of this segment of mammalian kidney.

    When challenged by hypoosmotic peritubular solution, the cells of perfused rabbit proximal straight tubule swelled immediately but reversed to the isoosmotic control volume in the continued presence of the hypoosmotic challenge (Fig. 11.6a). When returning to the control bath of 290 mosmol/kg, the cells shrank but reversed relatively fast to the volume prior to the osmotic perturbation (Kirk et al. 1987b). In contrast, following a hyperosmotic peritubular challenge (390 mosmol/kg), the cells shrank as expected for a perfect osmometer but stayed shrunken for 15–20 min until

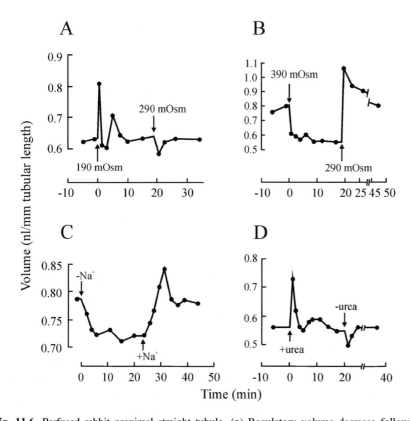

Fig. 11.6 Perfused rabbit proximal straight tubule. (**a**) Regulatory volume decrease following hypoosmotic challenge by removal of NaCl from the luminal perfusion solution and peritubular bath. (**b**) Response of the tubule to a hyperosmotic challenge by adding NaCl to the luminal perfusion solution and peritubular bath. Note the absence of RVI. (**c**) Reversible isoosmotic volume decrease obtained by exchanging luminal Na⁺ for choline⁺. Note the absence of RVI. (**d**) Regulatory volume decrease by isoosmotic substitution of 50 mM NaCl for 100 mM urea in luminal perfusion solution and peritubular bath. The apparently "heavily damped oscillations" after the second solution shifts in (**b**) and (**c**) were suggested to reflect periods where entrance and exit of osmolytes across the opposite membranes are out of phase. A similar explanation might apply to the transient volume attenuation during RVD in (**a**) and (**d**). Redrawn from Kirk et al. (1987b)

the return to the isoosmotic solution of 290 mosmol/kg (Fig. 11.6b). The reintroduction of isoosmotic bath resulted in a transient volume increase that exceeded the control volume before the slow return to the isoosmotic control value (discussed below). When luminal Na⁺ was replaced by choline, the cells shrank reflecting the expected decreased inflow of Na⁺ through the apical membrane and continued pumping of Na⁺ out of the cells across the lateral membrane (Fig. 11.6c). Like the aniso-osmotic shrinkage in Fig. 11.6b, the isoosmotic shrinkage shown in Fig. 11.6c did not elicit back-regulation of cell volume. Interestingly, the tubule cells regulate volume following isoosmotic swelling obtained by replacing a fraction of NaCl of the luminal perfusion solution and the peritubular bath with an osmotically

Fig. 11.7 Perfused rabbit proximal straight tubule. The immediate response to pump inhibition by ouabain is not cell swelling, but cell shrinkage. Furthermore, pump inhibition decouples RVD. Both of these nontrivial observations are discussed in the text. Redrawn from Kirk et al. (1987a)

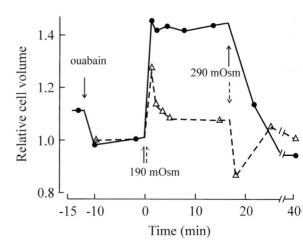

equivalent amount of urea (Fig. 11.6d). This indicates that RVD can be elicited by swelling per se in the absence of cytosolic dilution. Like in the experiment on cytosolic dilution at aniso-osmotic conditions (Fig. 11.6b), following return to the control solution the cells swelled transiently over and above the isoosmotic control volume. The authors pointed out that such a cell volume response in excess of the steady-state isoosmotic volume may reflect a period where the rate of solute entry across the apical membrane transiently is out of phase with solute exit across the basolateral membrane. The RVD is associated with loss of K^+ through channels in the peritubular membrane (Kirk et al. 1987a; Beck and Potts 1990; Schild et al. 1991).

The immediate response of rabbit proximal straight tubule to pump inhibition by ouabain is not cell swelling, but cell shrinkage (see Fig. 11.7; Kirk et al. 1987a). The authors mention that the cells remained shrunken for at least 25 min, whereafter they swelled continually during 30–60 min until structural changes took place like blebbing and vacuolation. Thus, this would indicate initial transient loss of cellular osmolytes kept above thermodynamic equilibrium by the Na^+/K^+ pump that preceded the inevitable swelling. One such candidate would be Cl^- of a cellular activity that is 1.3–2 times higher than predicted for passive distribution (Cassola et al. 1983; Ishibashi et al. 1988). The apical Cl^- uptake mechanisms comprise a set of Na^+/H^+ and Cl^-/HCO_3^- exchangers (Alpern 1987), and uptake of Cl^- in exchange with organic anions in parallel with V-type proton pump, reviewed by Weinstein (2013). Their coordinated operation would result in cellular accumulation of Cl^- of which the set of exchangers depends on Na^+/K^+ pump activity. Thus, there is an explanation to hand for the ouabain induced initial cell shrinkage that complies with the mechanism of frogskin discussed above (Fig. 11.2a). Furthermore, as shown in Fig. 11.7, pump inhibition by ouabain decouples RVD. The authors hypothesized that this is caused by inhibition of the basolateral K^+ permeability secondary to pump inhibition (cross talk) as described in previous studies on (amphibian) proximal tubule (Matsumura et al. 1984; Messner et al. 1985).

11.5.2   Cortical Collecting Tubule

Cortical collecting tubule (CCT) is heterocellular with principal and intercalated (IC) cells (Welling et al. 1983a). Principal cells reabsorb Na^+ through ENaC and secrete K^+ through ROMK- and maxi K^+ (BK) channels at the luminal membrane (Wang and Giebisch 2009). Of the two types of intercalated cells, the apical membrane of α-MR cells (Type-A ICs) is configured for acid secretion via apical H^+ pumps, K^+ reabsorption via apical H^+/K^+ exchangers, and K^+ secretion through apical BK channels. The β-MR cell (Type-B IC) is configured for base secretion with H^+ pumps at the basolateral membrane and Cl^-/HCO_3^- exchangers at the apical membrane, reviewed by Breton and Brown (2013). The osmotic permeability of CCT is controlled by ADH (Table 11.1) that stimulates the translocation of aquaporin-2 water channels from intracellular vesicles to the apical membrane (Nielsen et al. 1995). In situ, this results in reabsorption of water from tubular fluid. Thereby, this portion of the nephron participates in controlling the volume and the degree of dilution of the urine. A relatively large P_f of the lateral membrane was suggested to protect ADH-simulated cells from swelling and cytoplasmic dilution (Strange and Spring 1987a; Strange 1988).

Both principal and intercalated cells exhibit typical RVD (Figs. 11.8a and b). Inhibitor and ion substitution studies did not lead to clear answers about the release mechanisms for the efflux of osmolytes during RVD. RVI was observed only in one type of ICs, whereas principal cells and the other IC-type remained shrunken for 20–40 min following a serosal hyperosmolarity (Strange and Spring 1987b). The RCCT-28A CCT cell line harbor a cell swelling-activated ~300 pS apical Cl^- channel that is activated by membrane stretch and disruption of F-actin (Schwiebert et al. 1994). It is interesting to compare the time course of RVD in Fig. 11.8 with RVD of rabbit proximal tubule (Fig. 11.6a) and RVD of Necturus gallbladder (Fig. 11.3a); they were all studied under conditions of practically eliminated unstirred layer effects and of about similar osmolality challenges, i.e., nearly similar "instantaneous" driving forces for water. Yet RVD proceeds with a significantly slower rate in cells of smallest water permeability (see Sect. 11.3 above), indicating that expression levels of water channels, ion transporters, and ion channels are positively correlated.

Ouabain caused swelling of principal cells, but not of intercalated cells. The swelling was immediate and fast and abolished, or reduced, by 95% in Na^+-free luminal perfusion solution or application of luminal amiloride, respectively (Strange 1989). With an anion-selective microelectrode $[Cl]_c$ in rat CCT was estimated to be near equilibrium (Schlatter et al. 1990), and subsequent studies have adopted the assumption of passive Cl^- distribution in principal cells of CCT (Staruschenko 2012). Thus, ouabain-induced increase in $[Na^+]_c$ instantly would be followed by an increase in $[Cl^-]_c$ and an associated cell swelling, which should continue until the cell bursts if the cell behaves in accordance with a nonregulated Gibbs–Donnan system. Indeed, cell swelling commenced with a fast rate right after the addition of ouabain. Surprisingly, however, rather than continued swelling, the cells exhibited

Fig. 11.8 Perfused rabbit renal cortical collecting tubule. During the control period, the tubules were exposed to bath and perfusion solutions of about 240 mosmol/Kg. RVD was elicited by a change of the osmolality of peritubular bath to 190 mosmol/kg. The initial response to bath dilution indicates complete swelling in less than 1 s. For this reason, the cross-sectional area of the mid-zone of the cell was used for estimating water exchange rates rather than the time consuming whole-cell optical sectioning technique. *Panel above,* Principal cell; *Panel below,* Intercalated cell. Redrawn from Strange (1988)

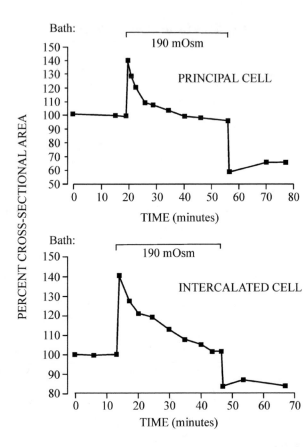

spontaneous isoosmotic RVD (see Fig. 11.9a). Thus according to the author, in CCT principal cells pump inhibition seems to reduce apical P_{Na} (cross talk) (Strange 1989). Following long-term exposure and removal of ouabain, cell volume rapidly decreased to a level that was significantly lower than that of the isoosmotic control volume, and a second ouabain exposure had little or no effect on cell volume (Fig. 11.9b). This supported the hypothesis that pump inhibition results in a rapid inhibition of P_{Na}. It was more difficult to provide an explanation for the late and complete RVD during the continued presence of peritubular ouabain (Fig. 11.9a). Kevin Strange suggested that ouabain binding would be transient upon prolonged exposure so that eventually pumps are released for resuming active Na^+ transport. According to this hypothesis, it would be the pumps of lost ouabain sensitivity that energizes the decrease of cell volume that went below the isoosmotic control volume. In a subsequent investigation of this late RVD in the presence of ouabain, Strange (1990) provided evidence for K^+ loss via an apical K^+ pathway different from the conductance mediating the resting K^+ secretion.

Fig. 11.9 Perfused rabbit renal cortical collecting tubule. (**a**) Complex response to long-term exposure of ouabain on principal cell volume. Pump inhibition induced immediate and fast cell swelling. However, this was followed by spontaneous isoosmotic cell shrinkage. After wash, at 80 min., a second addition of ouabain had practically no effect. (**b**) Effects of ouabain removal shortly after activation of isoosmotic cell shrinkage of principal cell. The cell rapidly shrank to a volume significantly below the isoosmotic volume prior to ouabain treatment. The second application of ouabain failed to initiate any significant cell swelling. Redrawn from Strange (1989)

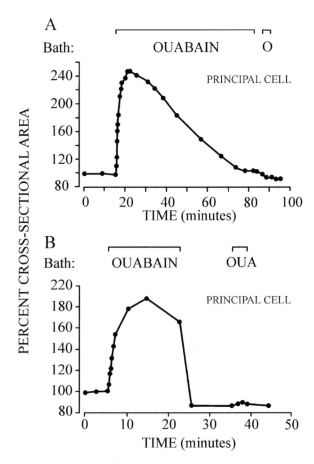

11.5.3   Medullary and Papillary Portions of Mammalian Nephron

By counter current multiplication, the loop of Henle has the capacity to generate an osmotic gradient in the interstitial space that increases linearly toward the tip of the papilla. Thus, the epithelial cells may be facing both hypoosmotic and hyperosmotic conditions depending on whether the organism is in a diuretic or antidiuretic state. Transport of NaCl across the thin descending limb (tDLH) is passive with Na^+/K^+ pumps serving maintenance of ionic gradients, which are important for RVD during hypoosmotic volume regulation. Both Cl^- and HCO_3^- participate in RVD, which is associated with K^+ loss across the peritubular membrane (Lopes et al. 1988). In the absence of ADH, cells of mouse thick ascending limb (TAL) shrank and remained shrunken as a perfect osmometer when exposed to a hyperosmotic bath. In the presence of ADH, however, TAL cells exhibited perfect RVI (Hebert 1986a). During RVI in mouse thick ascending limb, intracellular Na^+ and Cl^- were accumulated via parallel operation of Na^+/H^+ and Cl^-/HCO_3^- exchange across the basolateral membrane (Hebert 1986b). In TAL of rabbit kidney, NaCl would accumulate via a basolateral NKCC1 of increased activity in hyperosmotic media (Eveloff and Calamia 1986). In the papillary region, the cells are capable of preventing shrinkage to hyperosmotic challenge by net uptake of NaCl and/or by genome directed synthesis of intracellular organic osmolytes (Montroserafizadeh and Guggino 1990; Burg et al. 1997; Ferraris and Burg 2006; Jeon et al. 2006). Polyols (sorbitol, inositol), trimethylamines (betaine, glycerophosphocholine), and urea constitute the common organic osmolytes of inner medulla during antidiuresis that maintain cell volume and function in the very strong hyperosmotic environment (Bagnasco et al. 1986; Balaban and Burg 1987).

With an apical NKCC2 cotransporter, the early distal tubule of *Amphiuma* kidney is regarded as equivalent to mammalian TAL. Ouabain-induced cell swelling of perfused isolated tubules was prevented by inhibition of the cotransporter (Guggino et al. 1985). The cells did not show RVD in response to bath dilution. The cells just swelled as expected for a perfect osmometer of a nonsolvent volume of 18%. Independent $[Cl^-]_c$ measurements confirmed unchanged cellular Cl^- pool (Guggino et al. 1985). Cell swelling after the addition of luminal furosemide, on the other hand, was followed by a significant loss of cellular Cl^-. The authors pointed out that the unchanged Cl^- pool after aniso-osmotic bath dilution indicates cross talk between the entrance and exit pathways so that the presumed increased ion loss through the peritubular membrane was compensated by a stimulated ion uptake via the apical cotransporter.

11.6 Epithelial Cell Volume as a Signal for Regulating Isoosmotic Transport

It is a distinct feature of both low-resistance (Schatzmann et al. 1958; Curran 1960; Diamond 1964) and high-resistance epithelia (Schafer 1993; Nielsen and Larsen 2007; Gaeggeler et al. 2011) that the fluid transport is energized by Na^+/K^+ pumps with the water flow being due to osmosis (Altenberg and Reuss 2013). In all investigated epithelia the pumps are abundantly expressed at the membranes lining the lateral intercellular spaces (Stirling 1972; Mills and Ernst 1975; Ernst and Mills 1977; Mills et al. 1977; Mills and DiBona 1978; Kashgarian et al. 1985; Maunsbach and Boulpaep 1991; Pihakaski-Maunsbach et al. 2003; Grosell 2007). Thus, at transepithelial osmotic equilibrium, the lateral intercellular space constitutes an intraepithelial compartment that couples a passive flow of water with the active flux of Na^+. Since the fluid of the osmotic coupling compartment is hyperosmotic relative to the solution of departure, truly isoosmotic transport would be possible if a fraction of the pumped Na^+ is recirculated back into the coupling compartment (Ussing and Nedergaard 1993; Nedergaard et al. 1999). The physical justification for this theory is reinforced by the fact that the exit of ions from the above lateral coupling compartment is due to convection–diffusion across an interface of relatively high ion diffusion coefficient implying that the osmolality of the emerging fluid would be larger than that of the coupling compartment; for discussions and mathematical treatments, see Diamond and Bossert (1967), Larsen et al. (2000, 2003, 2009) and Larsen and Sørensen (2019). Hoffmann and Ussing (1992) discussed that changes in cell volume act as a signal for regulating the recirculation flux and inferred that cell volume regulation is an integrated mechanism of isoosmotic transport. Below, this hypothesis is discussed for an absorbing and a secreting epithelium.

In isolated toad skin epithelium stimulation of β-adrenergic receptors targets three transport pathways: (*i*) a passive Cl^- conductance of MR cells is stimulated (Willumsen et al. 1992; Larsen et al. 2003) by activation of an 8-pS Cl^- channel in the apical membrane (Sørensen and Larsen 1996); (*ii*) AQP2a isoform water channels are inserted into the apical membrane of principal cells (Bellantuono et al. 2008); and (*iii*) ENaC of principal cells is activated (Ogushi et al. 2010). As a result, the epithelium generates a transepithelial flow of near isoosmotic fluid (Nielsen and Larsen 2007). In principal cells $[Na^+]_c$ is governed by apical Na^+ uptake via ENaC, pumping by the Na^+, K^+-ATPase of Na^+ into the lateral space, and uptake ("recirculation") of Na^+ via the NKCC cotransporter at the serosal membrane (Rick et al. 1978; Dörge et al. 1985). Since the cotransport fluxes are stimulated in shrunken cells (Ussing 1982, 1985), this mechanism is a strong candidate for the cell volume-regulated recirculation pathway. Thus, if the transported fluid becomes hyperosmotic leading to cell shrinkage, or hypoosmotic leading to cell swelling, the recirculation flux would be stimulated or reduced, respectively, whereby ion fluxes and water flow return toward isoosmotic proportion (Hoffmann and Ussing 1992).

It was Ussing who introduced the seminal idea that isosmotic transport must be due to regulation at the cell level and suggested that ion recirculation would constitute the associated mechanism (reviewed in Larsen 2009). The theory was originally developed in studies of exocrine glands (Ussing and Eskesen 1989; Ussing et al. 1996) which provided the experimental evidence for reuptake of Na^+ by acini of frogskin submucosal glands during adrenaline-stimulated secretion. Subsequent whole-cell patch-clamp studies of microdissected frog glands showed that both cAMP- and Ca^{2+}-dependent Cl^- channels are present in the acinar epithelium and are activated by physiological stimulation together with both K^+ and Na^+ channels (Sørensen and Larsen 1999). Single-channel recording from apical membrane patches disclosed a CFTR Cl^- channel coexpressed with a maxi K^+ channel and a 5-pS Na^+ channel (Sørensen and Larsen 1998; Sørensen et al. 1998, 2001). The Na^+ channel of the apical membrane shown in Fig. 11.10 is the obvious candidate for the Na^+ recirculation pathway in gland acinus. In Ehrlich ascites tumor cells P_{Na} is a continuous function of cell volume in such a way that P_{Na} decreases upon cytoplasmic dilution and increases in response to cell shrinkage (Hoffmann 1978; Hoffmann et al. 2009). If this is applied to acinar cells, as suggested by Hoffmann and Ussing (1992) an increase in luminal fluid osmolality that leads to cell shrinkage would result in increased P_{Na} and thus increased Na^+ recirculation, i.e., an RVI response. In turn, this would restore isoosmotic volume

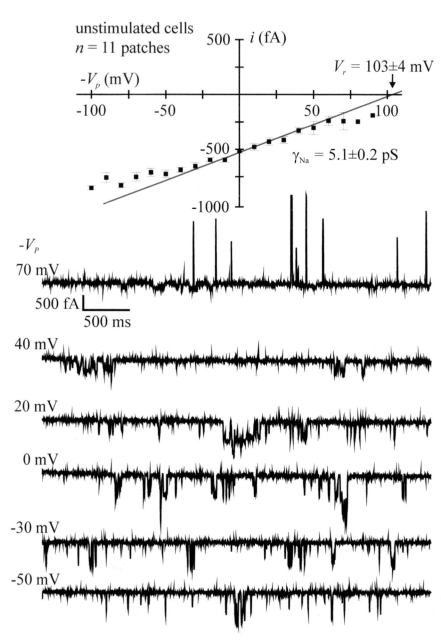

Fig. 11.10 The apical (luminal) Na$^+$ recirculation pathway of frogskin submucosal gland epithelium. Sodium channels in cell-attached patches on the apical membrane of micro-dissected frogskin mucous gland recorded with Ringer in the pipette. *Upper panel*: The *curve* was obtained by linear regression to points, $-V_p > 0$ mV giving a single-channel conductance, $\gamma_{Na} = 5.1 \pm 0.2$ pS, and a reversal potential, $-V_r = 103 \pm 4$ mV. With a membrane potential of unstimulated acinar cells of $V_m = -69.5 \pm 0.7$ mV (Sørensen and Larsen 1999), according to usual sign convention, the reversal potential would be +33 mV, thus accounting quantitatively for the reversal potential of +103 mV in a cell attached patch. *Lower panels*: Examples of single-channel recordings at indicated pipette potentials $(-V_p)$. Redrawn from Sørensen et al. (1998)

flow and cell volume. In contrast, if the luminal fluid becomes diluted, the acinar cells swell. The evoked RVD will then decrease P_{Na} whereby isoosmotic transport and cell volume are being restored. If acinar cells function as sensors of luminal osmolality, it is expected that P_f of the apical membrane is much larger than that of the basolateral membrane. Immunofluorescence labeling of amphibian AQP-x5 (homologous to mammalian AQP5) verified exclusive expression in the apical membrane of subepidermal glands. An immune-fluorescent signal of AQP-h3BL was similarly localized in the basolateral membrane (Hillyard et al. 2009). The relative osmotic permeability of the two membrane domains has not been studied.

11.7 Molecular Identity of Channels and Transporters Involved in Epithelial Cell Volume Regulation

As mentioned in the Introduction, transport of electrolytes, nutrients, and water in epithelia is a challenge to cellular electrolyte homeostasis and epithelial cell volume. In the following sections, we discuss examples of how cell volume regulation is being achieved during the variation in epithelial transport activity and to what extent cell volume changes contribute to the regulation of secretion/absorption. Thus, below we focus on ion transporters and channels that may serve both transepithelial transport and cell volume regulation. From the above description of various *absorptive epithelia*, it is seen that the mechanism of transepithelial transport involves apical nutrient uptake which would tend to swell the cells with resulting activation of basolateral K^+ channels, which in turn restore the apical membrane potential and the driving force for the cotransporter (Grasset et al. 1983). Recovery of cell volume under these conditions is achieved through activation of the basolateral swelling-activated K^+ channels like KCNK1 and two-pore K^+ channels (K_{2P}), volume-regulated anion channels (VRAC) and increased activity of the Na^+/K^+ pump followed by the exit of osmotically obliged water across the basolateral membrane (Lang et al. 1998a; Vanoye and Reuss 1999; Schultz and Dubinsky 2001; Hoffmann et al. 2009; Bachmann et al. 2011). With respect to *secretory epithelia*, the most common mechanisms for initiating fluid secretion is the opening of luminal Cl^- channels and luminal and basolateral K^+ channels leading to a cell volume decrease. Known shrinkage-activated proteins are NKCC1 cotransporters (Manganel and Turner 1991; Petersen 1992; Foskett et al. 1994), which provide ions for luminal exit and thus secretion, and potentially lead to regain of cell volume. Cell volume of many native epithelial cells recover cell volume only partially in the acute/secretory state, or they do not recover it at all until the stimulus is withdrawn (Bachmann et al. 2007). Thus, some secretory cells shrink by more than 20% during stimulation and remain shrunken until the stimulus is withdrawn (Lee and Foskett 2010).

11.7.1   Chloride Channels

A key player in RVD is the Volume-Regulated Anion Channel (VRAC), which is expressed in all vertebrate cells investigated (Hoffmann et al. 2009). The first studies were carried out with non-polarized Ehrlich ascites tumor cells (EATC) (Hoffmann 1978) and human lymphocytes (Grinstein et al. 1982). The dynamic permeability response to cell volume perturbations was measured by isotope tracer technique, as illustrated in Fig. 11.11a, indicating the characteristic cell volume dependence on Cl^- efflux in cells under isosmotic, hypoosmotic, and hyperosmotic conditions, respectively. Under isosmotic conditions, less than 5% represents the conductive (net) Cl^- flux on the background of a much larger electroneutral anion exchange (AE) flux. The increased Cl^- flux under hypoosmotic or hyperosmotic conditions is carried by VRAC and the NKCC cotransporter, respectively. Figure 11.11b shows the volume regulatory response in EATC under hypoosmotic conditions where the intrinsic volume sensitive K^+ channel was blocked by quinine. The large K^+ permeability was artificially established by the cation–ionophore gramicidin, hence making Cl^- movement rate-limiting for volume changes. With this protocol, it became evident that the Cl^- permeability increases abruptly when the cell swells (seen as an accelerated RVD response) for being deactivated again within the following 10 min. These early studies were followed up by patch-clamp experiments that revealed the outwardly rectifying nature of the swelling-activated anion current as first observed in human intestinal cells (Hazama and Okada 1990). In most cell types, the K^+ conductance is significantly larger than the Cl^- conductance. Therefore, the swelling-activated Cl^- conductance is essential for volume regulation. The VRAC current which is reversibly activated by cell swelling (see Fig. 11.12a) has characteristic biophysical fingerprints (Worrell et al. 1989; Nilius et al. 1997; Okada 1997; Nilius and Droogmans, 2003; Stutzin and Hoffmann 2006; Hoffmann et al. 2009; Okada et al. 2009): (*i*) Slow inactivation of VRAC at hyperpolarizing membrane potentials, see Fig. 11.12b and c; (*ii*) mild outward rectification in symmetrical solutions (Fig. 11.11c) and excess GHK rectification in asymmetrical configuration (Fig. 11.12c); and (*iii*) Eisenman type I anion-permeability sequence as listed in Figs. 11.11d and 11.12d showing VRAC permeabilities and anion permeabilities relative to Cl^-. Notably, compared to other halide permeabilities, the I^- permeability of VRAC is relatively large.

Single-channel recordings of VRAC have been performed in rat and mouse renal IMCD cells (Boese et al. 1996; Meyer and Korbmacher 1996) and human bronchial epithelium (Toczylowska-Maminska and Dolowy 2012). The channel exhibits conductances between 10 and 20 pS at negative potentials, and between 40 and 80 pS at positive potentials, which qualitatively conforms to the instantaneous nonlinear I–V relationship in whole-cell configuration indicated in Fig. 11.12c. Another type of directly swelling-activated channel is the maxi-anion channel (MAC; also called the Maxi-Cl) which is distinctly characterized by large unitary ohmic conductance of more than 100 pS (see comprehensive review by Okada et al. 2019).

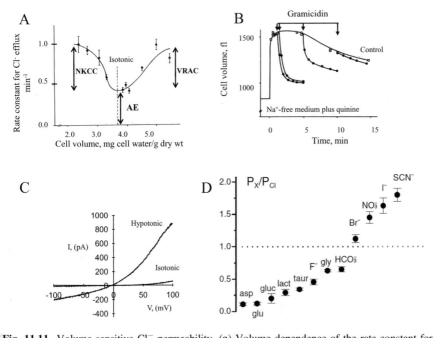

Fig. 11.11 Volume sensitive Cl⁻ permeability. (**a**) Volume dependence of the rate constant for ^{36}Cl efflux measured as unidirectional steady-state efflux in EATC. Under isosmotic conditions 95% of the Cl⁻ efflux was mediated by an electroneutral anion exchange (AE). The remaining efflux constituted the conductive Cl⁻ flux which increased significantly after cell swelling and was taken to represent VRAC activity. Under hyperosmotic conditions an increasing part of the Cl⁻ efflux constitutes cation-dependent electroneutral Na, K, 2Cl cotransporter (NKCC). Values from Hoffmann et al. (1979) and Hoffmann (1982). (**b**) Transient increases in the Cl⁻ permeability following hypoosmotic cell swelling. At time zero EATC cells were exposed to hypoosmotic, Na⁺-free medium containing quinine to block intrinsic, volume sensitive K⁺ channels. Gramicidin was added as indicated by the arrow to ensure high cation permeability, i.e., to establish conditions where the rate of RVD is dictated by the Cl⁻ permeability. Modified from Hoffmann et al. (1986). (**c**) Current/voltage (I–V) relationship obtained by whole-cell patch-clamp recordings (−140 mV to +80 mV, fast ramp-protocol) of the Cl⁻ current in EATC under isosmotic and hypoosmotic (27% decrease) conditions. Results from Pedersen et al. (1998). (**d**) VRAC permeation properties. VRAC permeabilities to organic substrates [aspartate (asp), glutamate (glu), gluconate (gluc), lactate (lact), taurine (taur), glycine (gly)] and anions are given relative to the Cl⁻ permeability calculated from the shift in E_{rev} by substitution of Cl⁻. Results from Nilius, Droogmans (2003). The figure is reproduced from Hoffmann et al. (2015) with permission from Taylor, Francis

11.7.1.1 Activation and Modulation of Chloride Channels

The description of VRAC activation is still incomplete, although many details are known such as: (*i*) The current can be activated in the absence of free intracellular Ca^{2+}, but requires binding of intracellular ATP to the channel protein whereas there is an open channel block by extracellular ATP (Voets et al. 1999; Eggermont et al. 2001; Okada et al. 2009, 2019); (*ii*) reduced intracellular ionic strength triggers activation of volume-regulated anion channels (Voets et al. 1999) probably by

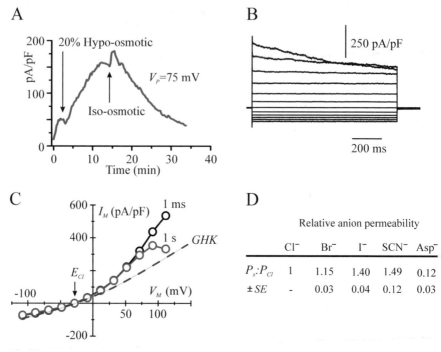

Fig. 11.12 Macroscopic phenotype of VRAC of human embryonic kidney cells studied by whole-cell patch-clamp technique with 40 mM Cl⁻ in the pipette solution. (**a**) Reversible channel activation by diluting [NaCl] of bath-PBS to 20% hypo-osmolality. V_P is the pipette potential. (**b**) VRAC currents obtained from a holding potential of -20 mV by a step protocol with steps ranging from $V_P = -100$ mV to 120 mV with 20-mV increments and a step length of 1 s. Note the slow inactivation of VRAC currents at hyperpolarizing membrane potentials. (**c**) Current–voltage relationships obtained from currents of panel-B, 1 ms (black symbols) and 1 s (red symbols), respectively, after voltage step. V_M is membrane potential, which is V_P corrected for liquid junction potential. Blue dashed line indicated by "*GHK*" is fit of the GHK-current equation to data points about the reversal potential with $P_{Cl} = 57.7 \cdot 10^{-14}$ cm³/s/pF. The deviation from the 1-ms relationship indicates the rectification of VRAC currents in excess of simple electrodiffusion. (**d**) Relative anion permeabilities of VRAC characterized by I⁻ > Br⁻ > Cl⁻ ($n = 3$). E. H. Larsen unpublished

affecting associated signaling protein molecules; (*iii*) the small G protein RhoA seems to play a role in the activation of VRAC (Nilius et al. 1999; Pedersen et al. 2002a) probably by modulation of F-actin polymerization which is known to be affected by osmotic cell swelling (Klausen et al. 2006); (*iv*) Various lipids have been shown to affect VRAC activity, and thus, arachidonic acid directly inhibits VRAC in most cell types (Lambert 1987; Lambert and Hoffmann 1994), whereas PIP3 might stimulate it (Yamamoto et al. 2008); (*v*) pharmacological evidence indicates a role for tyrosine kinases in the regulation of VRAC (see Hoffmann et al. 2009); and (*vi*) reactive oxygen species (ROS), such a H_2O_2 released upon cell swelling, stimulate VRAC (Browe and Baumgarten 2004; Shimizu et al. 2004; Varela et al. 2004). The role of phosphorylation in the regulation of VRAC is still unclear because of

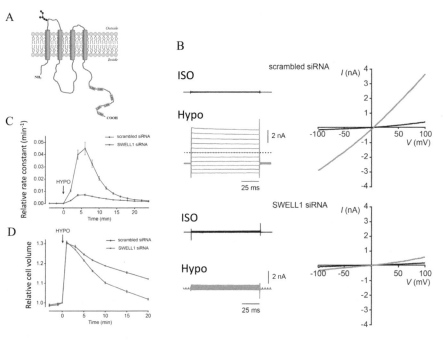

Fig. 11.13 LRRC8A/SWELL1 knockdown abolish RVD, endogenous VRAC currents, and VSOAC activity following hypoosmotic cells swelling. (**a**) LRCC8A model. LRR indicates leucine-rich repeats. (**b**) Whole-cell currents (right) monitored by voltage step protocols (left) under isosmotic conditions (black) and after 6 min in hypoosmotic solution (210 mosmol/kg Hypo) in HEK293T cells transfected with either scrambled or SWELL1 siRNA. (**c**) Rate constant for swelling-induced 3H-taurine efflux in HeLa cells transfected with either scrambled siRNA or siRNA against SWELL1. Shift in tonicity to hypoosmotic solution (210 mosmol/kg) is indicated by an arrow. (**d**) Cell volume under isosmotic conditions and following exposure to hypoosmotic solution (220 mOsm/kg; indicated by the arrow) in HeLa cells transfected with either scrambled siRNA or siRNA against SWELL1. Calculated from the shift in E_{rev} by substitution of Cl^-. Results from Nilius and Droogmans (2003). The figure is reproduced from Hoffmann et al. (2015) with permission from Taylor, Francis

contradictory findings. However, pharmacological evidence indicates a role for tyrosine kinases (Hoffmann et al. 2009).

The molecular identity of VRAC was unknown for manny years and several proteins have wrongly been suggested to be VRAC including Ca^{2+} activated channels ANO1 and ANO6 (Almaca et al. 2009). With respect to ANO1 and ANO6, it was shown that neither of these contributed to the volume sensitive current in ANO overexpressing HEK293 cells in the absence of Ca^{2+} (Juul et al. 2014). In 2014, two groups independently identified members of the plasma membrane spanning protein leucine-rich repeat containing (LRRC8) family as being essential compounds/regulators of VRAC (Fig. 11.13a), e.g., Qiu et al. (2014), Voss et al. (2014) and Stauber (2015). The LRRC8 family is composed of four transmembrane segments located at the N-terminal half of the protein as illustrated for LRRC8A in Fig. 11.13a. A leucine-rich repeat domain carrying up to 17 leucine-rich repeats (only 6 are shown in LRRC8A) is located at the C-terminal end (Abascal and Zardoya 2012). Sequence comparison of the LRRC8 family has revealed that the LRRC8 proteins share a common ancestor with pannexins, a protein family considered to form hexameric channels and known to be involved in leakage of Ca^{2+} from the ER and ATP-dependent cell death (Qu et al. 2011, Abascal and Zardoya 2012). Patapoutian and coworkers (Qiu et al. 2014) demonstrated that LRRC8A localizes to the plasma membrane and that LRRC8A current resembles VRAC (here called I_{Cl}^{swell}) with respect to ion selectivity ($I^- > Cl^-$), sensitivity to inhibitors, and a mild outwardly directed rectification. Knockdown with siRNA against LRRC8A prevented endogenous VRAC current (Fig. 11.13b), volume sensitive taurine efflux (Fig. 11.13c), and the overall ability to perform RVD (Fig. 11.13d). At the same time Jentsch and coworkers (Voss et al. 2014) demonstrated that (*i*) the LRRC8A ion selectivity was similar to that of VRAC ($I^- > NO_3^- > Cl^- >$ gluconate); (*ii*) LRRC8B-E translocation to the plasma membrane required LRRC8A; and (*iii*) disruption of LRRC8B-E in contrast to LRRC8A disruption had no effect on VRAC current. Because I_{Cl}^{swell} (VRAC) is abolished in LRRC8A$^{-/-}$ cells like I_{Cl}^{swell} in LRRC8 (B/C/D/E)$^{-/-}$ cells, they suggested that VRAC activity depends on LRRC8 heteromerization (Qiu et al. 2014). The LRRC8 family includes five members (LRRC8A to LRRC8E); it has been suggested that the functional channel is formed by six proteins from the LRRC8 family (Abascal and Zardoya 2012) and that a shift in stoichiometry between LRRC8 members affects channel activity/selectivity (Sorensen et al. 2014). It was demonstrated that the swelling-induced Cl^- conductance via VRAC is mediated by LRRC8A, LRRC8C, and LRRC8E (Voss et al. 2014), whereas swelling-induced taurine efflux via VSOAC relies on LRRC8A and LRRC8D (Planells-Cases et al. 2015). LRRC8A has been identified as an essential component for VRAC activity, as reduced expression of LRRC8A correlates with a reduced ability to regulate cell volume in human ovary (A2780), colon (HCT116), cervical (HeLa), and embryonic kidney (HEK) cells, as well as in T-lymphocytes (Qiu et al. 2014; Sorensen et al. 2014; Voss et al. 2014). Furthermore, LRRC8A expression was shown to correlate with the swelling-induced release of the organic osmolyte taurine from, e.g., HeLa (Qiu et al. 2014), HCT116 (Voss et al. 2014), human ovarian A2780 cancer cells (Sorensen et al. 2014) and primary rat astrocytes (Hyzinski-Garcia et al. 2014). Finally, LRRC8A is also required for ATP-induced release of glutamate and taurine from non-swollen rat astrocytes (Hyzinski-Garcia et al. 2014).

11.7.2 Potassium Channels

Members of nine different potassium subfamilies have been shown to be activated after cell swelling including several examples in epithelia. The cell volume sensitive K^+ channels discussed below are indicated in red in Fig. 11.14.

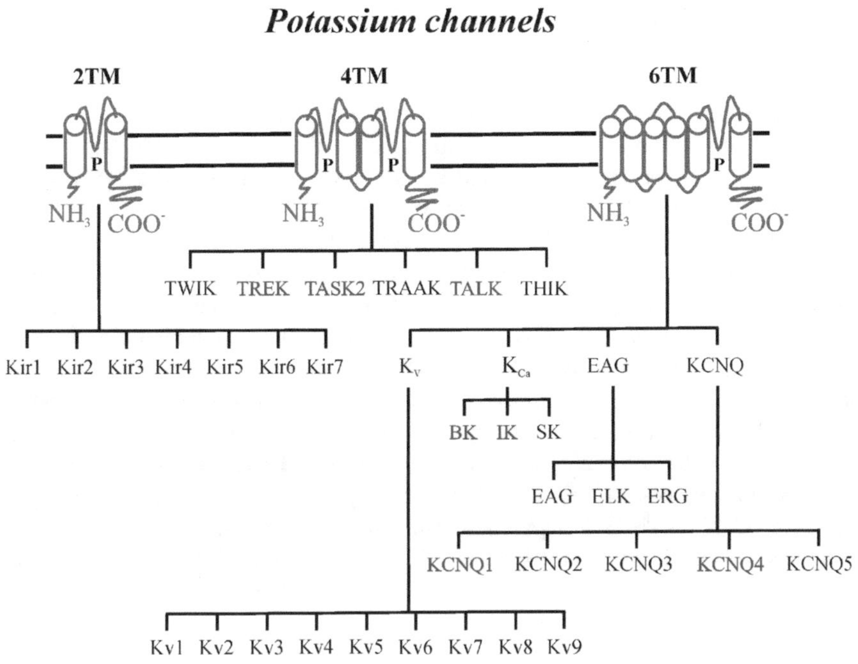

Potassium channels

Fig. 11.14 K$^+$ channel families. Channels found to be cell volume sensitive in epithelia are given in red color. Modified from Hoffmann et al. (2009)

11.7.2.1 Large Conductance or Maxi-(BK) K$^+$ Channel

The BK channel is expressed in the apical membrane of rat medullary thick ascending limb (MTAL) (Taniguchi and Guggino 1989), rat collecting tubule (Stoner and Morley 1995), rabbit kidney proximal tubule cells (Dube et al. 1990), basolateral and apical membrane of acinar cells in frogskin glands (Andersen et al. 1994; Sørensen et al. 2001), human bronchial epithelial cells (Fernandez-Fernandez et al. 2002), and eel intestinal epithelium (Lionetto et al. 2008). In most cases, it has been shown to be activated under hypoosmotic conditions. It was suggested that BK channels themselves are stretch-activated. However, examined for MTAL this was found not to be the case. Rather, stretch appears to elevate cytosolic Ca^{2+} which in turn activates Ca^{2+}-controlled BK channels (Taniguchi and Guggino 1989). Similarly, in the majority of cells, activation of BK channels after cell swelling is secondary to an increase in [Ca^{2+}]$_c$ (Calderone 2002), which in some cases is coupled to an autocrine/paracrine swelling-induced ATP release (Wehner 2006).

11.7.2.2 RVD-Mediating Intermediate Conductance (IK) Channels

IK potassium channels apparently unrelated to cell volume regulation were discovered in patch-clamp studies of chloride secreting airway cells (Welsh and McCann 1985) and in a chloride secretory colonic cell line (T84 cells) (Devor and Frizzell 1993). This issue is covered in detail in Chap. 22 of Vol. 3 of this book series. Functions of the IK channel related to cell volume regulation have been studied in renal A6 cells (Urbach et al. 1999), human embryonic kidney (HEK293) cells (Jorgensen et al. 2003), human intestinal epithelial cells (Wang et al. 2003), mouse proximal tubule cells (Barriere et al. 2003), human parotid gland cells (Begenisich et al. 2004; Thompson and Begenisich 2006), and human lens epithelial cells (Lauf et al. 2008). Swelling activation of IK channels occurs in some cells, e.g., the intestine 407 cell line, to be dependent on an increase in $[Ca^{2+}]_c$. The increase in $[Ca^{2+}]_c$ is receptor-mediated and coupled to an autocrine/paracrine swelling-induced release of ATP (Okada et al. 2001). When expressed in HEK cells, activation of hIK channels by cell swelling does however not depend on an increase in $[Ca^{2+}]_c$. A certain level of $[Ca^{2+}]_c$ is though required and the swelling-induced activation of hIK channels is dependent upon the integrity of the F-actin cytoskeleton (Jorgensen et al. 2003).

11.7.2.3 The Two-Pore Domain K⁺ (K$_{2P}$) Channel KCNK5

KCNK5 (also known as TASK-2, or K$_{2P}$5.1) is a two-pore domain K⁺ channel belonging to the TALK (TWIK-related alkaline pH-activated K⁺ channels) subgroup of the K$_{2P}$ family (Fig. 11.14), see (Cid et al. 2013). TASK channels are gated only by extracellular pH (Enyedi and Czirjak 2010) but not affected by intracellular pH, whereas TALK channels like KCNK5 are sensitive to intracellular pH changes (Niemeyer et al. 2010). KCNK5 was the first member of the TALK subfamily to be cloned, which was done from the human kidney in 1998 by Reyes and coworkers (Reyes et al. 1998). Northern blot analysis has shown KCNK5 to be present in human kidney, pancreas, liver, placenta, lung, and small intestine (Medhurst et al. 2001), whereas RT-PCR from mouse tissue showed abundant KCNK5 presence in the liver, kidney, small intestine, lung, and uterus (Reyes et al. 1998; Medhurst et al. 2001). For a review on KCNK5, see Cid et al. (2013).

KCNK5 is a volume-sensitive K⁺ channel in Ehrlich cells (Niemeyer et al. 2001a, b), in mouse kidney proximal tubule (Barriere et al. 2003), and in T lymphocytes (Bittner et al. 2010; Bobak et al. 2011). It has been demonstrated how protein tyrosine kinase activity is involved in RVD (Kirkegaard et al. 2010) and how cell swelling elicits a time-dependent tyrosine phosphorylation of the channel itself upon activation. In the same study, Kirkegaard et al. (2010) showed that inactivation of the swelling-activated channel involves protein tyrosine phosphatases. Furthermore, swelling activation of KCNK5 has in kidney cells been proposed to be caused by an extracellular alkalinization due to Cl^-/HCO_3^-

exchanger mediated Cl⁻ influx (L'Hoste et al. 2007). Leukotriene D-4 (LTD4) has been implicated in the swelling activation of KCNK5 because the addition of LTD4 resulted in the activation of a K^+ current similar to the swelling-activated K^+ current (Hougaard et al. 2000) and hypoosmotic conditions and cell swelling is followed by LTD4 release (Lambert 1987; Lambert et al. 1987). In addition, it was shown that the desensitization of the LTD4 receptor impairs RVD in EAT cells (Jørgensen et al. 1997). A more recent study indicated how KCNK5 is inhibited by heterotrimeric G protein subunits Gβγ an effect that could be eliminated by mutation of lysine residues of the C terminus (Anazco et al. 2013; Cid et al. 2013). Despite all of these findings, there are still many questions to be answered in regard to KCNK5 channel activation; thus we have yet to identify the tyrosine sites on KCNK5 that are being phosphorylated upon swelling-mediated activation (Kirkegaard et al. 2010). Furthermore, a direct connection between KCNK5 and LTD4 has still not been shown, and though the effect of G protein subunits Gβγ has been established, the answer to how exactly it works and through which signaling pathway remains to be answered. The associated signaling pathways were studied in detail in EAT cells by Niemeyer et al. (2001a, b). Response to hypoosmotic conditions can be divided into an acute and a long-term phase. The acute phase, which includes RVD, happens within minutes of cell swelling. During longtime exposure to osmolarity changes, the cell uses other mechanisms to ensure a steady intracellular environment, such as an altered gene expression and an altered protein synthesis (Cohen 2005). The effect of long-term hypoosmotic conditions (24–48 h) on KCNK5 in Ehrlich cells on the mRNA, protein, and physiological levels was investigated. The data suggest that a strong physiological impairment of KCNK5 in Ehrlich cells after long-term hypoosmotic stimulation is predominantly due to the downregulation of the KCNK5 protein synthesis (Kirkegaard et al. 2013).

An interesting potential role of KCNK5 in activated T cell physiology has recently been described. So far KCNK5 has been seen to be upregulated in T cells in multiple sclerosis patients and to be implicated in the regulatory volume decrease (RVD) in T cells (Andronic et al. 2013). Recently, we found that KCNK5 is highly upregulated in CD3/CD28 activated T cells both at mRNA- (after 24 h) and at protein level (72 and 144 h). However, despite this upregulation, the RVD response is inhibited and the cells swell. This was shown to be because the swelling-activated Cl⁻ permeability in activated T cells is strongly decreased and the RVD inhibited. Thus the upregulated KCNK5 in activated human T cells does not play a volume regulatory role, but rather plays a role in hyperpolarization of the cell membrane leading to increased Ca^{2+} influx and proliferation of T cells (Kirkegaard et al. 2016). Sustained increase in $[Ca^{2+}]_i$ is essential for the transcription and expression of interleukin 2 (IL-2) which in turn is vital in keeping the T cells activated without the requirement of further antigen stimulation. Other volume-sensitive (or "mechano-gated") members of the KCNK family are KCNK2 and KCNK4 (Maingret et al. 1999a, b, 2000).

11.7.2.4 KCNQ Channels

Finally, in several epithelial cell types, an apical KCNQ1 potassium channel was identified as an essential player in RVD as exemplified in cells from the inner ear (the KCNQ1 β-subunit KCNE1), (Wangemann et al. 1995), in murine tracheal epithelial cells (the KCNQ1/KCNE1-complex (Lock and Valverde 2000)), in human mammary epithelial cell line MCF-7 (vanTol et al. 2007), and in secretion-associated shrinkage in rabbit parietal cells (Bachmann et al. 2007). The general implication of KCNQ channels in the RVD response was indicated in studies by Grunnet et al. (2003) and Hougaard et al. (2004) who showed that the KCNQ4 channel is sensitive to changes in volume in non-epithelial cells.

11.7.3 Na^+/H^+ Exchangers

The Na^+/H^+ exchanger NHE-1 (*SLC9A1*), of the *SLC9A* family was initially described as a pH regulator but is also involved in cell volume regulation as it is activated after cell shrinkage which results in the uptake of osmotically active sodium ions (Rotin and Grinstein 1989) together with Cl^- via a Cl^-/HCO_3^- exchanger as it was first suggested by Cala (1980). NHE1 is essentially ubiquitous and localizes to the basolateral membrane of polarized epithelial cells as shown in ileus villus, in crypt epithelial cells, in rat alveolar epithelial cells as well as in multiple nephron segments (Wakabayashi et al. 1997; Orlowski and Grinstein 2011; Donowitz et al. 2013). NHE1 is both shrinkage activated (Orlowski and Grinstein 2011) and swelling inhibited (Elsing et al. 2007).

While other plasma membrane NHE isoforms, including NHE2 and NHE4, are also activated by cell shrinkage (see below), their expression is more restricted, and NHE1 is by far the most important mechanism of NHE-mediated RVI in most cells investigated (see Hoffmann et al. 2009). The activation of NHE1 by cell shrinkage involves an alkaline shift in the H^+ sensitivity of the exchanger (e.g., Wakabayashi et al. 1997; Hendus-Altenburger et al. 2014) and a change in the allosteric affinity constant of NHE1 in favor of the activated state, but does not involve changes in V_{max} (Lacroix et al. 2008). Several molecular mechanisms for the activation by cell shrinkage-induced NHE1 activity have been described and a valuable discussion can be found in, e.g., Hendus-Altenburger et al. (2014). It has been demonstrated that NHE1 activation by cell shrinkage is a consequence of shrinkage per se rather than increased osmotic strength (Krump et al. 1997). NHE1 is regulated by membrane lipids via interactions with the C-terminal NHE1 tail (Shimada-Shimizu et al. 2014; Webb et al. 2016). While this cannot fully account for its shrinkage sensitivity, NHE1 responds directly to mechanical stress applied to the membrane or to changes in lipid packing and cholesterol concentration (Fuster et al. 2004; Lacroix et al. 2008; Tekpli et al. 2008), suggesting that NHE1 mechanosensitivity is associated with an ability to sense membrane differential packing (Pang et al. 2012) and/or is due to direct interaction between NHE1 and specific lipids (Wakabayashi et al. 1997). The precise mechanisms through which cell shrinkage eventually translate into activation of NHE1 remain to be elucidated. Posttranslational mechanisms, directly or indirectly, play a role in congruence with the known roles of protein kinases and phosphatases in the modulation of shrinkage-activated NHE1 activity (Pedersen et al. 2002b). Shrinkage activation of NHE1 was originally thought not to involve altered phosphorylation of NHE1 directly (Grinstein et al. 1992), but a recent mass spec analysis of *Amphiuma tridactylum* NHE1 demonstrated that NHE1 is phosphorylated on multiple residues during osmotic cell shrinkage (Rigor et al. 2011). Posttranslational regulation of NHE1 during cell shrinkage could also involve altered interaction of NHE1 with its multiple binding partners, most of which are regulated by phosphorylation (see Hendus-Altenburger et al. 2014). For instance, osmotic activation of Janus kinase was suggested to elicit phosphorylation of calmodulin, calmodulin-NHE1 interaction, and NHE1 activation (Garnovskaya et al. 2003).

In addition to NHE1, the plasma membrane-NHEs include NHE2-NHE5. NHE2 (*SLCA2*) and NHE3 (*SLCA3*) are apical isoforms restricted to various types of epithelial cells in the stomach/gastrointestinal tract, kidney, and gallbladder (Wakabayashi et al. 1997; Donowitz et al. 2013). For NHE2 shrinkage activation, shrinkage inhibition and lack of effect of shrinkage have been reported for different cell types. There is agreement that NHE3 is inhibited by cell shrinkage and activated by cell swelling (see Hendus-Altenburger et al. 2014). NHE4 is basolateral and found mainly in the stomach, at intermediate levels in small intestine and colon, and at a low level in the kidney (Orlowski and Grinstein 2004). Finally, NHE4 is potently shrinkage activated in a manner dependent on the cytoskeleton and is important for RVI in kidney medulla (Bookstein et al. 1994), whereas NHE5 is inhibited by cell shrinkage (Attaphitaya et al. 2001).

11.7.4 Na^+-K^+-$2Cl^-$ Cotransporters (NKCC)

NKCC1 is encoded by the *SLC12A2* gene, whereas NKCC2 is encoded by *SLC12A1*. Both NKCC1 and NKCC2 form homodimers (Moore-Hoon and Turner 2000). NKCC1 and NKCC2 are both activated by cell shrinkage and by a decrease in $[Cl^-]_c$ (Markadieu and Delpire 2014). The two stimuli are often working in opposite directions as cell shrinkage caused by an increase in external osmolality is accompanied by an increase in intracellular Cl^-. Thus, a decrease in cell volume gives an activation signal and an increase in intracellular Cl^- an inhibition signal. As a result, many cell types fail to exhibit RVI when exposed directly to a hyperosmotic solution but perform RVI only when they are returned to isoosmotic conditions after a hypoosmotic pretreatment that results in Cl^- loss as exemplified in Fig. 11.1a and b (see review by Hoffmann et al. 2009).

11.7.4.1 NKCC1

NKCC1 is widely expressed in vertebrates and is a central effector in cell volume regulation as well as in epithelial cells with secretory and absorptive functions as for example, exocrine glands, airways, intestinal tract, inner medullary collecting duct, and nonmammalian rectal gland cells, where it is localized to the basolateral membrane (Hebert et al. 2004; Markadieu and Delpire 2014). In salivary glands, where NKCC1 is involved in the secretion of fluid, there is a severe impairment of salivation in NKCC1-deficient mice (Evans et al. 2000). In the small intestine, NKCC1 together with the Na^+/K^+ pump drive fluid secretion (Jakab et al. 2011), see Sect. 11.4.3. In marginal cells of the inner ear, NKCC1 has been proposed as a component of the entry pathway for K^+ that is secreted into the endolymph, thus playing a critical role in hearing (Flagella et al. 1999).

11.7.4.2 NKCC2

NKCC2 exhibits a highly restricted pattern of expression. It is expressed on the apical membrane of the epithelial cells of the thick ascending limb of Henle (TAL) where it reabsorbs a large amount of NaCl (Ares et al. 2011). NKCC2 homologs are also found in tissues from nonmammalian vertebrates, including shark kidney (Gagnon et al. 2002) and eel intestine (Lionetto and Schettino 2006).

Activation by Cell Shrinkage

Two families of protein kinases, the WNKs and the Ste20-related kinases called Ste20-related proline/alanine-rich kinase (SPAK) and oxidative stress-responsive kinase (OSR1) play major roles in the shrinkage activation (see Sect. 11.9.4 for references). The regulation of NKCC1/NKCC2 by the kinases is complex but has been excellently described in a recent review (Gagnon and Delpire 2012). Beyond SPAK/OSR1, other ser/thr protein kinases have also been suggested to be involved in shrinkage-induced NKCC1 activation, including PKC (Gagnon and Delpire 2012), MLCK (Klein and O'Neill 1995; Lionetto et al. 2002; Hoffmann and Pedersen 2007), and FAK (Hoffmann et al. 2007). In NIH3T3 fibroblasts we find that Src, FAK, and Jak2 are rapidly activated by osmotic shrinkage and that Src-dependent signaling, possibly through Jak2, contributes substantially to shrinkage activation of NKCC1 (Rasmussen et al. 2015). PKC-δ acts upstream of SPAK and OSR1 in the hyperosmotic-induced activation of NKCC1. Thus, SPAK kinase activity increased in airway epithelial cells after treatment with a PKC-δ activator and decreased after treatment with a PKC-δ inhibitor (see Gagnon and Delpire 2012). Both PKC and MLCK were also found to contribute to the hyperosmotic activation of the apical NKCC2 in intestinal epithelial cells of European eel, which was strongly dependent on the integrity of both F-actin and microtubules (Lionetto et al. 2002; Lionetto and Schettino 2006).

11.8 Putative Sensors of Cell Volume and Cell Volume Changes

Cell volume changes result in alterations in (i) macromolecular crowding, (ii) cellular ionic strength or concentrations of specific ions, or (iii) mechanical/chemical changes in the lipid bilayer. All three mechanisms play roles in the response to volume perturbations. None of them, however, have been specifically studied in epithelia and will not be dealt with here. The reader is referred to reviews by Burg (2000), Minton (2006), and Pedersen and Nilius (2007). For our purpose, we focus on the nature of the "osmosensors." The mechanism of osmosensing is still incompletely understood in vertebrate cells as compared to osmosensing in bacteria archaea, fungi, and plants (Bahn et al. 2007) where the volume sensor consists of a two-component histidine kinase system, resulting in the activation of the high osmolarity glycerol 1-mitogen activated protein (HOG1-MAP) kinase (Li et al. 1998; Bahn et al. 2006). Several membrane proteins, including integrins, growth factor receptors (GFRs), cytokine receptor tyrosine kinases (RTKs), calcium-sensing receptors (CaRs), transient receptor potential channels (TRPs), phospholipases (PLA_2 and PLC), lipid kinases (PI3K and PI5K), the cytoskeleton (e.g., actin), and small GTP-binding proteins (e.g., Rho, Rac, and Cdc42), as reviewed in Bahn et al. (2006) seem to play a role as sensors of cell volume changes. Schematically, Fig. 11.13 indicates how a membrane protein may sense changes in cell volume (Pedersen et al. 2011).

11.8.1 Integrins and Other Receptors

Several volume-sensitive K^+ channels are found to be regulated by integrin β_1, including $K_V1.3$ (Levite et al. 2000; Artym and Petty 2002; Cahalan and Chandy 2009). The basolateral membrane contains integrins, which connect the cytoskeleton to extracellular matrix proteins within the basement membrane. Given that many tissues are lined by epithelia with apical cell membranes facing the lumen, polarization allows epithelial cells to transport molecules across the surface in a directional manner. Integrins link the internal actin cytoskeleton to the extracellular matrix that may mediate mechanotransduction. Integrin heterodimers associate with cytoskeletal and signaling proteins in focal adhesions (FA), and upon integrin activation, the focal adhesion kinase (FAK) is phosphorylated, which mediates a coordinating role in the signaling events (see Pedersen et al. 2011). We find that integrin β_1 has a function in volume sensing in adherent ELA cells but not in the non-adherent suspension cells EATC-WT (Sorensen et al. 2015). Direct evidence for roles of integrins in osmo/volume sensing in epithelia is so far rather limited. Integrin alpha1beta1 ($\alpha_1\beta_1$-Integrin) is a player in the regulation of renal medullary osmolyte concentration, and hyperosmotic conditions induce mRNA and protein expression of β_1-integrin in MDCK cells (see Burg et al. 2007). The increase in osmolyte accumulation induced by hyperosmotic treatment was strongly impaired in kidney medulla of α_1 integrin-null mice (Moeckel et al. 2006). Integrins contributes to the osmosensitivity of some ion transport proteins. Examples are the TRPV4-mediated Ca^{2+} influx in epithelial cells that occurs under hypoosmotic stress, and during activation by cell shrinkage of NKCC1 (see Hoffmann et al. 2009).

The inhibition of growth factor signaling by hyperosmotic cell shrinkage was studied in kidney cells where it was found that hyperosmotic conditions block signaling by growth factor receptor protein tyrosine kinases immediately downstream of Ras (Copp et al. 2005). The role of CaRs has been studied as salinity sensors in fish (Okada et al. 2001; Nearing et al. 2002; Fiol and Kultz 2007).

11.8.2 Transient Receptor Potential (TRP) Channels

The role of TRP channels in cell volume regulation is probably to mediate an increase in $[Ca^{2+}]_c$, which stimulates RVD by activating Ca^{2+}-controlled K^+ and Cl^- channels. Thus, the increase in $[Ca^{2+}]_c$ after cell swelling, especially in epithelia, seems to be dependent on Ca^{2+} influx via TRP channels (Strotmann et al. 2000; Arniges et al. 2004; Cohen 2005; Liedtke and Kim 2005; Numata et al. 2007). All TRPV ("V" for vanilloid) channels are Ca^{2+} permeable with $P_{Ca}/P_{monovalent}$ between 1 and 10 for TRPV1–4, and $P_{Ca}/P_{monovalent} > 100$ for TRPV5 and 6 (Owsianik et al. 2006). Several TRPs are sensitive to cell volume and/or to membrane stretch (Christensen and Corey 2007; Pedersen and Nilius 2007) and TRPC1, TRPC6, TRPV4, and TRPM7 are all candidates as volume sensors, which is best documented for TRPV4 (Nilius et al. 2007). A full description of the osmotic activation of TRPV4 is still awaiting (Everaerts et al. 2010; Pochynyuk et al. 2013). "With No Lysine Kinase" (WNK4) and aquaporin 5 are involved in the osmosensitivity (Fu et al. 2006; Liu et al. 2006) and recently hypoosmotic TRPV4 activation was found to depend on phosphatidylinositol-4,5-diphosphate (PIP2) binding to the cytosolic N-terminus (Garcia-Elias et al. 2013). Finally, a series of highly conserved aromatic residues in transmembrane (TM) helices 5–6 is shown to have profound importance for TRPV4 activity in response to cell swelling (Klausen et al. 2014).

11.8.3 Phospholipases of the Phospholipase 2 (PLA2) Family

PLA2-mediated hydrolysis of phospholipids for releasing arachidonic acid initiates a swelling-induced signaling cascade that involves potent biological substances of the eicosanoid family some of which are involved in volume regulation (Hoffmann et al. 2007). Thus, the osmotic sensitivity of TRPV4 (Vriens et al. 2004) depends on the activation of phospholipase A_2 and subsequent production of arachidonic acid-metabolite $5'$-$6'$-epoxyeicosatrienoic acid (EET). In Ehrlich cells, the arachidonic acid metabolite leukotriene D4 (LTD$_4$) is released upon hypoosmotic stress that activates the TASK-2 channel (Hougaard et al. 2000). Several PLA2 isoforms may be stimulated during cell swelling. The swelling-induced arachidonic acid release and RVD thus involve a Ca^{2+}-dependent PLA2 activity in rat inner medullar collecting duct cells (Tinel et al. 2000).

11.8.4   Cytoskeleton

The cytoskeleton, specifically F-actin, is important for sensing and transduction of cell volume perturbations (Ingber 2006). Reorganization of the cortical F-actin occurs rapidly after volume changes involving a net increase and a net decrease in actin polymerization after shrinkage and swelling, respectively (Pedersen and Hoffmann 2002; Hoffmann et al. 2009). This change in cortical F-actin is however most normal in non-adherent cells. In epithelial cells, the cellular content of polymerized F-actin changes very little, and the changes in cortical F-actin goes along with a change in stress fibers (Di Ciano-Oliveira et al. 2006). Little is known about the significance of the intermediate filaments or microtubules. In shark, rectal gland hypoosmotic swelling was reported neither to affect microtubules nor the intermediate filaments (Henson et al. 1997), whereas in opossum kidney cells vimentin was reorganized into short fragments after hypoosmotic treatment (Dartsch et al. 1994).

There are several mechanisms by which the cytoskeleton might influence volume regulation: (*i*) The transport protein may interact with the cytoskeleton, (*ii*) the cytoskeleton may regulate signaling events that are controlling transport proteins, or (*iii*) dependent on the cytoskeleton, cell shrinkage may inhibit and cell swelling stimulates exocytosis (van der Wijk et al. 2003). In turn, this regulates transporters/channels through regulation of signal molecules like the swelling-induced release of ATP in intestinal 407-cells that is vesicle recycling-dependent, through the insertion of transport proteins, as for example NKCC2, in eel intestine (Lionetto Schettino 2006), or swelling activation of Cl$^-$ channels in nonpigmented ciliary epithelial cells (Vinnakota et al. 1997). The Rho family small G proteins, Rho, Rac, and Cdc42, are regulators of actin organization (Hall 1998; Lewis et al. 2002) and sensitive to cell volume changes. Therefore, they may well be involved in cytoskeletal reorganizations during RVD and RVI. Hyperosmotic cell shrinkage in kidney epithelial cells was thus shown to increase Rho activity (Di Ciano-Oliveira et al. 2003). Opposite to that, Rho was activated by cell swelling in Intestine-407 cells (Tilly et al. 1996).

11.9   Signal Transduction in Response to Cell Volume

11.9.1   Free Intracellular Ca^{2+} Concentration

An increase in cytosolic-free Ca^{2+} ($[Ca^{2+}]_c$) is involved in RVD in many epithelial cells. Thus, in culture human Intestine 407 cells a biphasic rise in $[Ca^{2+}]_c$ leading to an increase in K^+ and Cl^- conductances involves both Ca^{2+} influx and Ca^{2+} release from intracellular stores (Hazama and Okada 1990). These two sources of Ca^{2+} may act synergistically to increase the level of $[Ca^{2+}]_c$ and turn on the Ca^{2+}-activated K^+ and Cl^- channels. The importance for RVD of Ca^{2+} entry is also reported in, e.g., intestinal epithelial cells (MacLeod and Hamilton 1999), renal thick ascending loop of Henle (Montrose-Rafizadeh et al. 1991; Tinel et al. 2002), rabbit corneal epithelial cells (Wu et al. 1997), and mouse distal colon (Mignen et al. 1999). In some epithelia as rat inner medullary collecting duct (IMCD) cells (Tinel et al. 2000), this is coupled with release from intracellular stores, but this is not a general mechanism (see Hoffmann et al. 2009). The influx of Ca^{2+} is probably predominantly via TRP channels which play a crucial role in the response to mechanical and osmotic perturbations in a wide range of cell types (Pedersen Nilius 2007), see Sect. 11.8. In renal cortical collecting tubule, an association between TRPV4 and AQP2 is involved in the hypoosmotic activation of TRPV4 resulting in an influx of Ca^{2+} and regulation of cellular volume (Galizia et al. 2012). Ca^{2+}-dependent processes do not seem to be involved in the swelling-induced VRAC channel activation (see Hoffmann et al. 2009). However, as suggested by Okada's group (Akita et al. 2011), it is a possibility that the VRAC channel activation is accomplished in the immediate vicinity of individual open Ca^{2+}-permeable channels, where a very high intracellular Ca^{2+} concentration could be generated in so-called "Ca^{2+} nanodomains."

11.9.2   Role of ATP Release

ATP is released during cell swelling in several epithelial cells, which will result in the stimulation of purinergic receptors and Ca^{2+} influx (Novak 2011). In many secretory epithelia, luminally applied UTP/ATP increases intracellular Ca^{2+} and stimulates Ca^{2+}-activated Cl^- channels (CaCC) most likely via P2Y2 receptors. Thus, in mouse colonic epithelium and salivary gland the Ca^{2+}-activated Cl^- channel Anoctamin-1 together with other anoctamin proteins are activated by cell swelling through such an autocrine mechanism that involves ATP release and binding to purinergic P2Y2 receptors and increase in $[Ca^{2+}]_c$ (Almaca et al. 2009). A number of ATP release mechanisms have been proposed, including ATP release via anion channels (Sabirov and Okada 2005).

11.9.3 Mitogen-Activated Protein Kinases (MAPKs)

p38 MAPKs, c-Jun N-terminal kinases (JNK1,2), and extracellular signal-regulated kinase 1/2 (ERK1/2) are modulated by osmotic stress in many mammalian epithelial cells including the cells in medullary thick ascending limb (Roger et al. 1999) and human tracheal epithelial cells (Liedtke Cole 2002). Cell swelling is generally associated with ERK1/2 stimulation and cell shrinkage with stimulation of p38 MAPK and JNK, but there are many exceptions where opposite effects are seen (see Hoffmann et al. 2009) supporting the idea that the MAPKs may be activated by cell volume disturbances in general rather than by swelling or shrinkage as such.

There is evidence for the involvement of ERK, p38 MAPK, and JNK in the control of RVI in certain epithelia (Roger et al. 1999; Liedtke and Cole 2002) and in RVD in others, e.g., rabbit corneal epithelial cells where it is shown that swelling-induced ERK and JNK stimulation precedes Cl^- and K^+ channel activation, whereas p38 activation occurs as a consequence of RVD (Pan et al. 2007). The most well-studied role of MAPKs activation during osmotic stress is in the long-time adaptive changes at the level of gene transcription regulation predominantly via p38 MAPK-mediated effects on the osmoprotective transcription factor TonEBP (tonicity-responsive enhancer-binding protein) resulting in increased osmolyte transporter expression (Tsai et al. 2007). It is well established that MAPKs activation by osmotic stress in the yeast is via the p38 MAPK orthologue HOG1, which is activated by small G proteins and some scaffold proteins. However, the mechanism by which cell shrinkage leads to activation of the MAPKs is still incompletely understood, but there might well be parallels to the yeast pathway. Thus, the small G protein Cdc42 and/or Rac can be important upstream activators of p38 MAPK after cell shrinkage, see e.g. Uhlik et al. (2003) and Friis et al. (2005).

11.9.4 With No Lysine Kinases (WNKs) and Ste20-Related Kinases

The With-No lysine (K) kinases (WNK) got the name due to the lack of lysine in subdomain II (Huang et al. 2007). The best-described role of the WNKs is the regulation of Na^+, K^+, Cl^-, HCO_3^-, and Ca^{2+} transporters in epithelia (Vitari et al. 2005; McCormick and Ellison 2011). Among these are the cation-chloride cotransporters, the renal outer medulla K^+ channel (ROMK1), ENaC, CFTR and TRPV4 and TRPV5 (see Peng and Warnock 2007). WNKs are ser/thr protein kinases that include four isoforms in mammalian cells, WNK1–4 (Verissimo and Jordan 2001; Kahle et al. 2006; Peng and Warnock 2007). They are volume-sensitive kinases and are thus very important in secretory epithelia as they regulate NKCC1 during cell shrinkage (Park et al. 2012). As mentioned above, NKCC1 plays an essential role in salt and water secretion or reabsorption and in volume regulation.

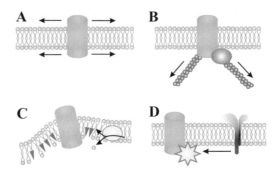

Fig. 11.15 Possible mechanisms through which a membrane protein may sense cell volume perturbations. In all panels, the violet cylinder illustrates the volume sensitive protein. (**a**) Direct stretch sensitivity. (**b**) Direct or indirect tethering to the cytoskeleton. (**c**) Changes in membrane curvature, either mechanical or through changes in membrane composition. (**d**) Via interaction with integrins activated by the cell volume perturbation. See the text for examples and details. Modified from Pedersen et al. (2011)

Moreover, WNKs play a role in the regulation of the NaCl cotransporter NCC and mutations in WNK1 and WNK4 are the basis for familial hyperkalemic hypertension (FHHt) or Gordon's syndrome, a rare form of congenital hypertension with hyperkalemia and metabolic acidosis (Farfel et al. 1978). Briefly, WNK4 reduces the number of NaCl cotransporter (NCC) proteins at the plasma membrane in the distal nephron and thus inhibits NCC and WNK1 inhibits the effect of WNK4 on the transporter (Yang et al. 2003; Cai et al. 2006; Golbang et al. 2006). The WNK mutations associated with (FHHt) results in increased WNK1 expression and decreased WNK4 expression, and the resulting increased NCC activity is essential in the pathophysiology of FHHt, see Kahle et al. (2006) and Peng and Warnock (2007).

The WNKs do not in general act directly on the target transporters but instead phosphorylate the two sterile 20 family stress kinases (Ste20-like kinases) called Ste20-related proline/alanine-rich kinase (SPAK) and oxidative stress responsive kinase (OSR1) (Vitari et al. 2005; Richardson and Alessi 2008; Delpire 2009; Delpire and Austin 2010). SPAK and OSR1 are widely expressed (Gagnon and Delpire 2012). Figure 11.15 shows that mutations in the binding pocket of SPAK results in loss of WINK4-SPAK activation of NKCC1. High expression of SPAK found in several Cl⁻ secreting epithelia like gastric mucosa, sublingual gland, and salivary gland. The expression of SPAK in the salivary gland is localized to the basolateral membrane where NKCC1 is highly expressed and SPAK and OSR1 co-localize with NKCC2 in the thick ascending limb of Henle (Gagnon and Delpire 2012). The WNKs seem to have a scaffolding role when activating SPAK/OSR1 as shown when studying the regulation of NKCC1 (Piechotta et al. 2003). Thus during cell shrinkage WNK1 and WNK4 bind, phosphorylate, and activate OSR1 and SPAK (Richardson and Alessi 2008; Delpire and Austin 2010). The effect of SPAK/OSR1 is rather specific for a given transporter. For details on the effect on

Fig. 11.16 Loss of NKCC1 function of SPAK F481A mutant and in the absence of WNK4. Expressed in *Xenopus laevis* oocytes with the K+ influx via NKCC1 monitored by 86Rb+. Bars represent means ±SE (*n* = 20–25 oocytes). Data from Delpire Laboratory, redrawn from Gagnon and Delpire (2012)

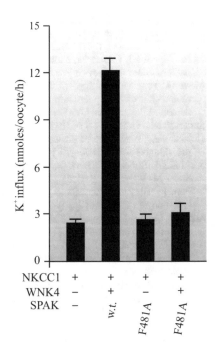

NKCC1, see Richardson and Alessi (2008), Delpire and Austin (2010), and Gagnon and Delpire (2012) (Fig. 11.16).

Acknowledgments Stine F Pedersen is acknowledged for suggestions to and critical reading of Sect. 11.7.3. Work in the authors' laboratories is supported by grant CF17–0186 from the Carlsberg Foundation, the Natural Science Foundation, the Augustinus Foundation, and Brødrene Hartmann Foundation.

References

Abascal F, Zardoya R (2012) LRRC8 proteins share a common ancestor with pannexins, and may form hexameric channels involved in cell-cell communication. Bioassays 34:551–560

Agre P, Christensen EI, Smith BL, Nielsen S (1993a) Distribution of the aquaporin CHIP in secretory and resorbtive epithelia and capillary endothelia. Proc Natl Acad Sci U S A 90:7275–7279

Agre P, Knepper MA, Christensen EI, Smith BL, Nielsen S (1993b) CHIP28 water channels are localized in constitutively water-permeable segments of the nephron. J Cell Biol 120:371–383

Akita T, Fedorovich SV, Okada Y (2011) Ca^{2+} nanodomain-mediated component of swelling-induced volume-sensitive outwardly rectifying anion current triggered by autocrine action of ATP in mouse astrocytes. Cell Physiol Biochem 28:1181–1190

Almaca J, Tian YM, Aldehni F, Ousingsawat J, Kongsuphol P, Rock JR, Harfe BD, Schreiber R, Kunzelmann K (2009) TMEM16 proteins produce volume-regulated chloride currents that are reduced in mice lacking TMEM16A. J Biol Chem 284:28571–28578

Alpern RJ (1987) Apical membrane chloride base-exchange in the rat proximal convoluted tubule. J Clin Invest 79:1026–1030

Altenberg GA, Reuss L (2013) Mechanisms of water transport across cell membranes and epithelia. In: Alpern RJ, Caplan M, Moe OW (eds) Seldin and Giebisch's the kidney physiology and pathophysiology. Elsevier, Academic Press, pp 95–120

Anazco C, Pena-Munzenmayer G, Araya C, Cid LP, Sepulveda FV, Niemeyer MI (2013) G protein modulation of K2P potassium channel TASK-2: a role of basic residues in the C terminus domain. Pflugers Arch 465:1715–1726

Andersen HK, Urbach V, Van Kerkhove E, Prosser E, Harvey BJ (1994) Maxi K^+ channels in the basolateral membrane of the exocrine frog skin gland regulated by intracellular calcium and pH. Pflugers Arch 431:52–65

Andronic J, Bobak N, Bittner S, Ehling P, Kleinschnitz C, Herrmann AM, Zimmermann H, Sauer M, Wiendl H, Budde T, Meuth SG, Sukhorukov VL (2013) Identification of two-pore domain potassium channels as potent modulators of osmotic volume regulation in human T lymphocytes. Biochim Biophys Acta 1828:699–707

Ares GR, Caceres PS, Ortiz PA (2011) Molecular regulation of NKCC2 in the thick ascending limb. Am J Cell Physiol Cell Physiol 301:F1143–F1159

Arniges M, Vazquez E, Fernandez-Fernandez JM, Valverde MA (2004) Swelling-activated Ca^{2+} entry via TRPV4 channel is defective in cystic fibrosis airway epithelia. J Biol Chem 279:54062–54068

Artym VV, Petty HR (2002) Molecular proximity of Kv1.3 voltage-gated potassium channels and beta(1)-integrins on the plasma membrane of melanoma cells: effects of cell adherence and channel blockers. J Gen Physiol 120:29–37

Attaphitaya S, Nehrke K, Melvin JE (2001) Acute inhibition of brain-specific Na^+/H^+ exchanger isoform 5 by protein kinases A and C and cell shrinkage. Am J Physiol Cell Physiol 281:C1146–C1157

Bachmann O, Heinzmann A, Mack A, Manns MP, Seidler U (2007) Mechanisms of secretion-associated shrinkage and volume recovery in cultured rabbit parietal cells. Am J Physiol Gastrointest Liver 292:G711–G717

Bachmann O, Juric M, Seidler U, Manns MP, Yu H (2011) Basolateral ion transporters involved in colonic epithelial electrolyte absorption, anion secretion and cellular homeostasis. Acta Physiol 201:33–46

Bagnasco S, Balaban RS, Fales HM, Yang YM, Burg M (1986) Predominant osmotically active organic solutes in rat and rabbit renal medullas. J Biol Chem 261:5872–5877

Bahn YS, Kojima K, Cox GM, Heitman J (2006) A unique fungal two-component system regulates stress responses, drug sensitivity, sexual development, and virulence of Cryptococcus neoformans. Mol Biol Cell 17:3122–3135

Bahn YS, Xue CY, Idnurm A, Rutherford JC, Heitman J, Cardenas ME (2007) Sensing the environment: lessons from fungi. Nat Rev Microbiol 5:57–69

Balaban RS, Burg MB (1987) Osmotically active organic solutes in the renal inner medulla. Kidney Int 31:562–564

Barriere H, Belfodil R, Rubera I, Tauc M, Lesage F, Poujeol C, Guy N, Barhanin J, Poujeol P (2003) Role of TASK2 potassium channels regarding volume regulation in primary cultures of mouse proximal tubules. J Gen Physiol 122:177–190

Beck JS, Potts DJ (1990) Cell swelling, cotransport activation and potassium conductance in isolated perfused rabbit kidney proximal tubules. J Physiol 425:369–378

Begenisich T, Nakamoto T, Ovitt CE, Nehrke K, Brugnara C, Alper SL, Melvin JE (2004) Physiological roles of the intermediate conductance, Ca^{2+}-activated potassium channel Kcnn4. J Biol Chem 279:47681–47687

Bellantuono V, Cassano G, Lippe C (2008) The adrenergic receptor substypes in frog (Rana esculenta) skin. Comp Biochem Physiol C Toxicol Pharmacol 148:160–164

Bittner S, Bobak N, Herrmann AM, Gobel K, Meuth P, Hohn KG, Stenner MP, Budde T, Wiendl H, Meuth SG (2010) Upregulation of K2P5.1 potassium channels in multiple sclerosis. Ann Neurol 68:58–69

Bobak N, Bittner S, Andronic J, Hartmann S, Muhlpfordt F, Schneider-Hohendorf T, Wolf K, Schmelter C, Gobel K, Meuth P, Zimmermann H, Doring F, Wischmeyer E, Budde T, Wiendl H, Meuth SG, Sukhorukov VL (2011) Volume regulation of murine T lymphocytes relies on voltage-dependent and two-pore domain potassium channels. Biochim Biophys Acta 1808:2036–2044

Boese SH, Kinne RK, Wehner F (1996) Single channel properties of swelling-activated anion conductance in rat inner medullary collecting duct cells. Am J Physiol Renal Physiol 271: F1224–F1233

Bookstein C, Musch MW, Depaoli A, Xie Y, Villereal M, Rao MC, Chang EB (1994) A unique sodium-hydrogen exchange isoform (Nhe-4) of the inner medulla of the rat-kidney is induced by hyperosmolarity. J Biol Chem 269:29704–29709

Boucher RC (1994) Human airway ion transport. Part I. Am J Respir Crit Care 150:271–281

Boucher RC (1999) Molecular insights into the physiology of the 'thin film' of airway surface liquid. J Physiol 516:631–638

Boucher RC (2003) Regulation of airway surface liquid volume by human airway epithelia. Pflugers Arch 445:495–498

Breton S, Brown D (2013) Regulation of luminal acidification by the V-ATPase. Physiology 28:318–329

Browe DM, Baumgarten CM (2004) Angiotensin II (AT1) receptors and NADPH oxidase regulate Cl- current elicited by beta 1 integrin stretch in rabbit ventricular myocytes. J Gen Physiol 124:273–287

Burg MB (2000) Macromolecular crowding as a cell volume sensor. Cell Physiol Biochem 10:251–256

Burg MB, Knepper MA (1986) Single tubule perfusion techniques. Kidney Int 30:166–170

Burg MB, Kwon ED, Kultz D (1997) Regulation of gene expression by hypertonicity. Annu Rev Physiol 59:437–455

Burg MB, Ferraris JD, Dmitrieva NI (2007) Cellular response to hyperosmotic stresses. Physiol Rev 87:1441–1474

Cahalan MD, Chandy KG (2009) The functional network of ion channels in T lymphocytes. Immunol Rev 231:59–87

Cai L, Friedman N, Xie XS (2006) Stochastic protein expression in individual cells at the single molecule level. Nature 440:358–362

Cala PM (1980) Volume regulation by amphiuma red-blood-cells - nature of the ion flux pathways. Fed Proc 39:379–379

Calderone V (2002) Large-conductance, Ca^{2+}-activated K^+ channels: function, pharmacology and drugs. Curr Med Chem 9:1385–1395

Cantiello HF, Patenaude CR, Ausiello DA (1989) G-protein subunit, alpha-I-3, activates a pertussis toxin-sensitive Na^+ channel from the epithelial-cell line, A6. J Biol Chem 264:20867–20870

Carpi-Medina P, Whittembury G (1988) Comparison of transcellular and transepithelial water osmotic permeabilities (P_{os}) in the isolated proximal straight tubule (PST) of the rabbit kidney. Pflugers Arch 412:66–74

Carpi-Medina P, Gonzáles E, Whittembury G (1983) Cell osmotic water permeability of isolated rabbit proximal convoluted tubules. Am J Phys 244:F554–F563

Carpi-Medina P, Lindemann B, Gonzáles E, Whittembury G (1984) The continous measurement of tubular volume changes in response to step changes in contraluminal osmolarity. Pflugers Arch 400:343–348

Cassola AC, Mollenhauer M, Frömter E (1983) The intracellular chloride activity of rat kidney proximal tubular cells. Pflugers Arch 399:259–265

Chamberlin ME, Strange K (1989) Anisosmotic cell volume regulation: a comparative view. Am J Phys 257:C159–C173

Chase HS, Alawqati Q (1983) Calcium reduces the sodium permeability of luminal membrane-vesicles from toad bladder - studies using a fast-reaction apparatus. J Gen Physiol 81:643–665

Choe K, Strange K (2009) Volume regulation and osmosensing in animal cells. In: Evans DH (ed) Osmotic and ionic regulation cells and animals. CRC Press, Boca Raton, pp 37–67

Christensen AP, Corey DP (2007) TRP channels in mechanosensation: direct or indirect activation? Nat Rev Neurosci 8:510–521

Cid LP, Roa-Rojas HA, Niemeyer MI, Gonzalez W, Araki M, Araki K, Sepulveda FV (2013) TASK-2: a K2P K^+ channel with complex regulation and diverse physiological functions. Front Physiol 4:198

Cohen DM (2005) TRPV4 and the mammalian kidney. Pflugers Arch 451:168–175

Cook DI, Van Lennep EW, Roberts ML, Young JA (1994) Secretion by the major salivary glands. In: Physiology of the gastrointestial tract, pp 1061–1117

Copp J, Wiley S, Ward MW, van der Geer P (2005) Hypertonic shock inhibits growth factor receptor signaling, induces caspase-3 activation, and causes reversible fragmentation of the mitochondrial network. Am J Physiol Cell Physiol 288:C403–C415

Cotton CU, Weinstein AM, Reuss L (1989) Osmotic water permeability of Necturus gallbladder epithelium. J Gen Physiol 93:649–679

Curran PF (1960) Na, Cl, and water transport by rat ileum in vitro. J Gen Physiol 43:1137–1148

Dartsch PC, Kolb HA, Beckmann M, Lang F (1994) Morphological alterations and cytoskeletal reorganization in opossum kidney (Ok) cells during osmotic swelling and volume regulation. Histochemistry 102:69–75

Davis CW, Finn AL (1982) Sodium transport inhibition by amiloride reduces basolateral membrane potassium conductance in tight epithelia. Science 216:525–527

Davis CW, Finn AL (1985) Effects of mucosal sodium removal on cell volume in *Necturus* gallbladder epithelium. Am J Physiol Cell Physiol 249:C304–C312

Delpire E (2009) The mammalian family of sterile 20p-like protein kinases. Pflugers Arch 458:953–967

Delpire E, Austin TM (2010) Kinase regulation of Na^+-K^+-$2Cl^-$ cotransport in primary afferent neurons. J Physiol 588:3365–3373

Devor DC, Frizzell RA (1993) Calcium-mediated agonists activate an inwardly rectified K^+ channel in colonic secretory cells. Am J Physiol Cell Physiol 265:C1271–C1280

Di Ciano-Oliveira C, Sirokmany G, Szaszi K, Arthur WT, Masszi A, Peterson M, Rotstein OD, Kapus A (2003) Hyperosmotic stress activates Rho: differential involvement in Rho kinase-dependent MLC phosphorylation and NKCC activation. Am J Physiol Cell Physiol 285:C555–C566

Di Ciano-Oliveira C, Thirone ACP, Szaszi K, Kapus A (2006) Osmotic stress and the cytoskeleton: the R(h)ole of Rho GTPases. Acta Physiol 187:257–272

Diamond JM (1964) Transport of salt and water in rabbit and Guinea pig gallbladder. J Gen Physiol 48:1–14

Diamond JM (1982) Trans-cellular cross-talk between epithelial-cell membranes. Nature 300:683–685

Diamond JM, Bossert WH (1967) Standing-gradient osmotic flow. A mechanism for coupling of water and solute transport in epithelia. J Gen Physiol 50:2061–2083

Donowitz M, Tse CM, Fuster D (2013) SLC9/NHE gene family, a plasma membrane and organellar family of Na^+/H^+ exchangers. Mol Asp Med 34:236–251

Dörge A, Rick R, Beck F-X, Thurau K (1985) Cl transport across the basolateral membrane in frog skin epithelium. Pflugers Arch 405(Suppl 1):S8–S11

Douglas IJ, Brown PD (1996) Regulatory volume increase in rat lacrimal gland acinar cells. J Membr Biol 150:209–217

Dube L, Parent L, Sauve R (1990) Hypotonic shock activates a maxi K^+ channel in primary cultured proximal tubule cells. Am J Physiol Renal Physiol 259:F348–F356

Eggermont J, Trouet D, Carton I, Nilius B (2001) Cellular function and control of volume-regulated anion channels. Cell Biochem Biophys 35:263–274

Elsing C, Gosch I, Hennings JC, Hubner CA, Herrmann T (2007) Mechanisms of hypotonic inhibition of the sodium, proton exchanger type 1 (NHE1) in a biliary epithelial cell line (Mz-Cha-1). Acta Physiol 190:199–208

Enyedi P, Czirjak G (2010) Molecular background of leak K^+ currents: two-pore domain potassium channels. Physiol Rev 90:559–605

Ericson A, Spring KR (1982a) Coupled NaCl entry into *Necturus* gallbladder epithelial-cells. Am J Physiol Cell Physiol 243:C140–C145

Ericson AC, Spring KR (1982b) Volume regulation by *Necturus* gallbladder - apical Na^+-H^+ and Cl^--HCO_3^- exchange. Am J Physiol Cell Physiol 243:C146–C150

Ernst SA, Mills JW (1977) Basolateral plasma membrane localization of ouabain-sensitive sodium transport sites in the secretory epithelium of the avian salt gland. J Cell Biol 75:74–94

Evans RL, Park K, Turner RJ, Watson GE, Nguyen HV, Dennett MR, Hand AR, Flagella M, Shull GE, Melvin JE (2000) Severe impairment of salivation in $Na^+/K^+/2Cl^-$ cotransporter (NKCC1)-deficient mice. J Biol Chem 275:26720–26726

Evans DH, Piermarini PM, Choe KP (2005) The multifunctional fish gill: dominant site of gas exchange, osmoregulation, acid-base regulation, and excretion of nitrogenous waste. Physiol Rev 85:97–177

Eveloff JL, Calamia J (1986) Effect of osmolarity on cation fluxes in medullary thick ascending limb cells. Am J Phys 250:F176–F180

Everaerts W, Nilius B, Owsianik G (2010) The vanilloid transient receptor potentialchannel TRPV4: from structure to disease. Prog Biophys Mol Biol 103:2–17

Farfel Z, Iaina A, Levi J, Gafni J (1978) Proximal renal tubular-acidosis - association with familial normaldosteronemic hyper-potassemia and hypertension. Arch Intern Med 138:1837–1840

Fernandez-Fernandez JM, Nobles M, Currid A, Vazquez E, Valverde MA (2002) Maxi K^+ channel mediates regulatory volume decrease response in a human bronchial epithelial cell line. Am J Physiol Cell Physiol 283:C1705–C1714

Ferraris JD, Burg MB (2006) Tonicity-dependent regulation of osmoprotective genes in mammalian cells. Contrib Nephrol 152:125–141

Fiol DF, Kultz D (2007) Osmotic stress sensing and signaling in fishes. FEBS J 274:5790–5798

Fisher RS, Persson BE, Spring KR (1981) Epithelial cell volume regulation - bicarbonate dependence. Science 214:1357–1359

Flagella M, Clarke LL, Miller ML, Erway LC, Giannella RA, Andringa A, Gawenis LR, Kramer J, Duffy JJ, Doetschman T, Lorenz JN, Yamoah EN, Cardell EL, Shull GE (1999) Mice lacking the basolateral Na-K-2Cl cotransporter have impaired epithelial chloride secretion and are profoundly deaf. J Biol Chem 274:26946–26955

Ford P, Rivarola V, Chara O, Blot-Chabaud M, Cluzeaud F, Farman N, Parisi M, Capurro C (2005) Volume regulation in cortical collecting duct cells: role of AQP2. Biol Cell 97:687–697

Foskett JK (1982) The chloride cell: definitive identification as the salt-secretory cell in teleosts. Science 215:164–166

Foskett JK, Bern HA, Machen TE, Connor M (1983) Chloride cells and the hormonal control of teleost fish osmoregulation. J Exp Biol 106:255–281

Foskett JK, Wong MMM, Sueaquan G, Robertson MA (1994) Isosmotic modulation of cell-volume and intracellular ion activities during stimulation of single exocrine cells. J Exp Zool 268:104–110

Friis MB, Friborg CR, Schneider L, Nielsen MB, Lambert IH, Christensen ST, Hoffmann EK (2005) Cell shrinkage as a signal to apoptosis in NIH 3T3 fibroblasts. J Physiol 567:427–443

Fu Y, Subramanya A, Rozansky D, Cohen DM (2006) WNK kinases influence TRPV4 channel function and localization. Am J Physiol Renal Physiol 290:F1305–F1314

Fuster D, Moe OW, Hilgemann DW (2004) Lipid- and mechanosensitivities of sodium/hydrogen exchangers analyzed by electrical methods. Proc Natl Acad Sci U S A 101:10482–10487

Gadsby DC, Nakao M (1989) Steady-state current-voltage relationship of the Na/K pump in Guinea-pig ventricular myocytes. J Gen Physiol 94:511–537

Gaeggeler HP, Guillod Y, Loffing-Cueni D, Loffing J, Rossier BC (2011) Vasopressin-dependent coupling between sodium transport and water flow in a mouse cortical collecting duct cell line. Kidney Int 79:843–852

Gagnon KB, Delpire E (2012) Molecular physiology of Spak and Osr1: two Ste20-related protein kinases regulating ion transport. Physiol Rev 92:1577–1617

Gagnon E, Forbush B, Flemmer AW, Caron L, Isenring P (2002) Functional and molecular characterization of the shark renal Na-K-Cl cotransporter: novel aspects. Am J Physiol Renal Physiol 283:F1046–F1055

Galizia L, Pizzoni A, Fernandez J, Rivarola V, Capurro C, Ford P (2012) Functional interaction between AQP2 and TRPV4 in renal cells. J Cell Biochem 113:580–589

Garcia-Elias A, Mrkonjic S, Pardo-Pastor C, Inada H, Hellmich UA, Rubio-Moscardo F, Plata C, Gaudet R, Vicente R, Valverde MA (2013) Phosphatidylinositol-4,5-biphosphate-dependent rearrangement of TRPV4 cytosolic tails enables channel activation by physiological stimuli. Proc Natl Acad Sci U S A 110:9553–9558

Garnovskaya MN, Mukhin YV, Vlasova TM, Raymond JR (2003) Hypertonicity activates Na^+/H^+ exchange through Janus kinase 2 and calmodulin. J Biol Chem 278:16908–16915

Garvin JL, Spring KR (1992) Regulation of apical membrane ion transport in *Necturus* gallbladder. Am J Physiol Cell Physiol 263:C187–C193

Gelfand EW, Cheung RKK, Ha K, Grinstein S (1984) Volume regulation in lymphoid leukemia-cells and assignment of cell lineage. New Engl J Med 311:939–944

Giraldez F, Ferreira KTG (1984) Intracellular chloride activity and membrane potential in stripped frog skin (*Rana temporaria*). Biochim Biophys Acta 769:625–628

Golbang AP, Cope G, Hamad A, Murthy M, Liu CH, Cuthbert AW, O'Shaughnessy KM (2006) Regulation of the expression of the Na/Cl cotransporter by WNK4 and WNK1: evidence that accelerated dynamin-dependent endocytosis is not involved. Am J Physiol Renal Physiol 291: F1369–F1376

Gonzáles E, Carpi-Medina P, Linares H, Whittembury G (1984) Osmotic water permeability of the apical membrane of proximal straight tubular (PST) cells. Pflugers Arch 402:337–339

Gonzalez E, Carpimedina P, Whittembury G (1982) Cell osmotic water permeability of isolated rabbit proximal straight tubules. Am J Physiol Renal Physiol 242:F321–F330

Grasset E, Gunter-Smith P, Schultz SG (1983) Effects of Na-coupled alanine transport on intracellular K-activities and the K-conductance of the basolateral membranes of Necturus small intestine. J Membr Biol 71:89–94

Greger R, Heitzmann D, Hug MJ, Hoffmann EK, Bleich M (1999) The Na^+, $2Cl^-$, K^+ cotransporter in the rectal gland of *Squalus acanthias* is activated by cell shrinkage. Pflüg Arch Eur J Physiol 438:165–176

Grinstein S, Erlij D (1978) Intracellular calcium and regulation of sodium-transport in frog skin. Proc R Soc Ser B-Bio 202:353–360

Grinstein S, Clarke CA, Rothstein A (1982) Increased anion permeability during volume regulation in human-lymphocytes. Philos Trans Roy Soc Ser B-Bio 299:509–518

Grinstein S, Woodside M, Sardet C, Pouyssegur J, Rotin D (1992) Activation of the Na^+/H^+ antiporter during cell-volume regulation - evidence for a phosphorylation-independent mechanism. J Biol Chem 267:23823–23828

Grosell M (2007) Intestinal transport processes in marine fish osmoregulation. In: Baldisserotto B, Mancera JM, Kapoor BG (eds) Fish osmoregulation. Science Publishers, New Hamshire, pp 332–357

Grubb BR, Schiretz FR, Boucher RC (1997) Volume transport across tracheal and bronchial airway epithelia in a tubular culture system. Am J Physiol Cell Physiol 273:C21–C29

Grunnet M, Jespersen T, MacAulay N, Jorgensen NK, Schmitt N, Pongs O, Olesen SP, Klaerke DA (2003) KCNQ1 channels sense small changes in cell volume. J Physiol 549:419–427

Guggino WB, Oberleithner H, Giebisch G (1985) Relationship between cell volume and ion transport in the early distal tubule of the Amphiuma kidney. J Gen Physiol 86:31–58

Guggino WB, Markakis D, Amzel LM (1990) Measurements of volume and shape changes in isolated tubules. Methods Enzymol 191:371–379

Gunter-Smith PJ, Grasset E, Schultz SG (1982) Sodium-coupled amino-acid and sugar transport by Necturus small intestine - an equivalent electrical circuit analysis of a rheogenic cotransport system. J Membr Biol 66:25–39

Hall A (1998) Rho GTPases and the actin cytoskeleton. Science 279:509–514

Harvey BJ (1992) Energization of sodium absorption by the H^+-ATPase pump in mitochondria-rich cells of frog skin. J Exp Biol 172:289–309

Harvey BJ (1995) Cross-talk between sodium and potassium channels in tight epithelia. Kidney Int 48:1191–1199

Harvey BJ, Ehrenfeld J (1988) Role of Na^+/H^+ exchange in the control of intracellular pH and cell membrane conductance in frog skin epithelium. J Gen Physiol 92:793–810

Harvey BJ, Kernan RP (1984) Intracellular ion activities in frog skin in relation to external sodium and effects of amiloride and/or ouabain. J Physiol 349:501–517

Harvey BJ, Thomas SR, Ehrenfeld J (1988) Intracellular pH controls cell membranes Na^+ and K^+ conductances and transport in frog skin epithelium. J Gen Physiol 92:767–791

Hazama A, Okada Y (1990) Biphasic rises in cytosolic free Ca^{2+} in association with activation of K^+ and Cl^- conductance during the regulatory volume decrease in cultured human epithelial-cells. Pflugers Arch Eur J Physiol 416:710–714

Hebert SC (1986a) Hypertonic cell-volume regulation in mouse thick limbs. 1. ADH dependency and nephron heterogeneity. Am J Physiol Cell Physiol 250:C907–C919

Hebert SC (1986b) Hypertonic cell-volume regulation in mouse thick limbs. 2. Na^+-H^+ and Cl^--HCO_3^- exchange in basolateral membranes. Am J Physiol Cell Physiol 250:C920–C931

Hebert SC (1987) Volume regulation in renal epithelial-cells. Seminars Nephr 7:48–46

Hebert SC, Mount DB, Gamba G (2004) Molecular physiology of cation-coupled Cl- cotransport: the SLC12 family. Pflugers Arch Eur J Physiol 447:580–593

Hendus-Altenburger R, Kragelund BB, Pedersen SF (2014) Structural dynamics and regulation of the mammalian SLC9A family of Na^+/H^+ exchangers. In: Bevensee MO (ed) Exchangers chapter 2: current topics in membranes, vol 73. Academic Press, Burlington, pp 69–148

Henson JH, Roesener CD, Gaetano CJ, Mendola RJ, Forrest JN, Holy J, Kleinzeller A (1997) Confocal microscopic observation of cytoskeletal reorganizations in cultured shark rectal gland cells following treatment with hypotonic shock and high external K^+. J Exp Zool 279:415–424

Hillyard SD, Møbjerg N, Tanaka S, Larsen EH (2009) Osmotic and ion regulation in amphibians. In: Evans DH (ed) Osmotic and ionic regulation cells and animals. Taylor Francis Group, Boca Raton, pp 367–441

Hoffmann EK (1978) Regulation of cell volume by selective changes in the leak permeabilities of Ehrlich ascites tumore cells. In: Jørgensen CB, Skadhauge E (eds) Proc Alfr Benzon Symp XI osmotic and volume regulation. Munksgaard, Copenhagen, pp 397–417

Hoffmann EK (1982) Anion exchange and anion-cation co-transport systems in mammalian cells. Phil Trans R Soc Lond B Biol 299:519–535

Hoffmann EK, Pedersen SF (2007) Shrinkage insensitivity of NKCC1 in myosin II-depleted cytoplasts from Ehrlich ascites tumor cells. Am J Physiol Cell Physiol 292:C1854–C1866

Hoffmann EK, Pedersen SF (2011) Cell volume homeostatic mechanisms: effectors and signalling pathways. Acta Physiol 202:465–485

Hoffmann EK, Ussing HH (1992) Membrane mechanisms in volume regulation in vertebrane cells and epithelia. In: Giebisch G, Schafer JA, Ussing H, Kristensen P (eds) Membrane transport in biology. Springer, Cham, pp 317–399

Hoffmann EK, Simonsen LO, Sjøholm C (1979) Membrane potential, chloride exchange, and chloride conductance in ehrlich mouse ascites tumour cells. J Physiol 296:61–84

Hoffmann EK, Sjøholm C, Simonsen LO (1983) Na^+, Cl^- co-transport in Ehrlich ascites tumor cells activated during volume regulation (regulatory volume increase). J Membr Biol 76:269–280

Hoffmann EK, Lambert IH, Simonsen LO (1986) Separate, Ca^{2+}-activated K^+ and Cl^- transport pathways in Ehrlich ascites tumor cells. J Memb Biol 91:227–244

Hoffmann EK, Hoffmann E, Lang F, Zadunaisky JA (2002) Control of Cl^- transport in the operculum epithelium of Fundulus heteroclitus: long- and short-term salinity adaptation. Biochim Biophys Acta 1566:129–139

Hoffmann EK, Schettino T, Marshall WS (2007) The role of volume-sensitive ion transport systems in regulation of epithelial transport. Comp Biochem Physiol A 148:29–43

Hoffmann EK, Lambert IH, Pedersen SF (2009) Physiology of cell volume regulation in vertebrates. Physiol Rev 89:193–277

Hoffmann EK, Sorensen BH, Sauter DP, Lambert IH (2015) Role of volume-regulated and calcium-activated anion channels in cell volume homeostasis, cancer and drug resistance. Channels 9:380–396

Hougaard C, Niemeyer MI, Hoffmann EK, Sepulveda FV (2000) K^+ currents activated by leukotriene D-4 or osmotic swelling in Ehrlich ascites tumour cells. Pflugers Arch Eur J Physiol 440:283–294

Hougaard C, Klaerke DA, Hoffmann EK, Olesen SP, Jorgensen NK (2004) Modulation of KCNQ4 channel activity by changes in cell volume. Biochim Biophys Acta 1660:1–6

Huang CL, Cha SK, Wang HR, Xie J, Cobb MH (2007) WNKs: protein kinases with a unique kinase domain. Exp Mol Med:565–573

Hughes ALH, Pakhomova A, Brown PD (2010) Regulatory volume increase in epithelial cells isolated from the mouse fourth ventricle choroid plexus involves Na^+-H^+ exchange but not Na^+-K^+-$2Cl^-$ cotransport. Brain Res 1323:1–10

Hyzinski-Garcia MC, Rudkouskaya A, Mongin AA (2014) LRRC8A protein is indispensable for swelling-activated and ATP-induced release of excitatory amino acids in rat astrocytes. J Physiol 592:4855–4862

Ingber DE (2006) Cellular mechanotransduction: putting all the pieces together again. FASEB J 20:811–827

Ishibashi K, Sasaki S, Yoshiyama N (1988) Intracellular chloride activity of rabbit proximal straight tubule perfused in vitro. Am J Physiol Renal Physiol 255:F49–F56

Jakab RL, Collaco AM, Ameen NA (2011) Physiological relevance of cell-specific distribution patterns of CFTR, NKCC1, NBCe1, and NHE3 along the crypt-villus axis in the intestine. Am J Physiol Gastrointest Liver Physiol 300:G82–G98

Jensen LJ, Sørensen JN, Larsen EH, Willumsen NJ (1997) Proton pump activity of mitochondria-rich cells. The interpretation of external proton-concentration gradients. J Gen Physiol 109:73–91

Jeon US, Kim J-A, Sheen MR, Kwon HM (2006) How tonicity regulates genes: story of TonEBP transcriptional activator. Acta Physiol 187:241–247

Jørgensen NK, Christensen S, Harbak H, Brown AM, Lambert IH, Hoffmann EK (1997) On the role of calcium in the regularory volume decrease (RVD) response in Ehrlich mouse ascites tumor cells. J Membr Biol 157:281–299

Jorgensen NK, Pedersen SF, Rasmussen HB, Grunnet M, Klaerke DA, Olesen SP (2003) Cell swelling activates cloned Ca^{2+}-activated K^+ channels: a role for the F-actin cytoskeleton. Biochim Biophys Acta Biomembr 1615:115–125

Juul CA, Grubb S, Poulsen KA, Kyed T, Hashem N, Lambert IH, Larsen EH, Hoffmann EK (2014) Anoctamin 6 differs from VRAC and VSOAC but is involved in apoptosis and supports volume regulation in the presence of Ca^{2+}. Pflugers Arch Eur J Physiol 466:1899–1910

Kahle KT, Rinehart J, Ring A, Gimenez I, Gamba G, Hebert SC, Lifton RP (2006) WNK protein kinases modulate cellular Cl^- flux by altering the phosphorylation state of the Na-K-Cl and K-Cl cotransporters. Physiology 21:326–335

Kashgarian M, Biemesderfer D, Caplan M, Forbush B III (1985) Monoclonal antibody to Na,K-ATPase: immunocytochemical localization along nephron segments. Kidney Int 28:899–913

Kato A, Romero MF (2011) Regulation of electroneutral NaCl absorption by the small intestine. Annu Rev Physiol 73:261–281

Kay AR, Blaustein MP (2019) Evolution of our understanding of cell volume regulation by the pump-leak mechanism. J Gen Physiol 151(4):407–416. https://doi.org/10.1085/jgp.201812274

Kirk KL, DiBona DR, Schafer JA (1987a) Regulatory volume decrease in perfused proximal nephron: evidence for a dumping of cell K^+. Am J Physiol Renal Physiol 252:F933–F942

Kirk KL, Schafer JA, DiBona DR (1987b) Cell volume regulation in rabbit proximal straight tubule perfused in vitro. Am J Physiol Renal Physiol 252:F922–F932

Kirkegaard SS, Lambert IH, Gammeltoft S, Hoffmann EK (2010) Activation of the TASK-2 channel after cell swelling is dependent on tyrosine phosphorylation. Am J Physiol Cell Physiol 299:C844–C853

Kirkegaard SS, Wulff T, Gammeltoft S, Hoffmann EK (2013) KCNK5 is functionally down-regulated upon long-term hypotonicity in Ehrlich ascites tumor cells. Cell Physiol Biochem 32:1238–1246

Kirkegaard SS, Strom PD, Gammeltoft S, Hansen AJ, Hoffmann EK (2016) The volume activated potassium channel KCNK5 is up-regulated in activated human T cells, but volume regulation is impaired. Cell Physiol Biochem 38:883–892

Klausen TK, Hougaard C, Hoffmann EK, Pedersen SF (2006) Cholesterol modulates the volume-regulated anion current in Ehrlich-Lettre ascites cells via effects on Rho and F-actin. Am J Physiol Cell Physiol 291:C757–C771

Klausen TK, Janssens A, Prenen J, Owsianik G, Hoffmann EK, Pedersen SF, Nilius B (2014) Single point mutations of aromatic residues in transmembrane helices 5 and-6 differentially affect TRPV4 activation by 4 alpha-PDD and hypotonicity: implications for the role of the pore region in regulating TRPV4 activity. Cell Calcium 55:38–47

Klein JD, O'Neill WC (1995) Volume-sensitive myosin phosphorylation in vascular endothelial cells: correlation with Na-K-2Cl cotransport. Am J Phys 269:C1524–C1531

Kregenow FM (1971) The response of duck erythrocytes to hypertonic media. Further evidence for a volume-controlling mechanism. J Gen Physiol 58:396–412

Krump E, Nikitas K, Grinstein S (1997) Induction of tyrosine phosphorylation and Na^+/H^+ exchanger activation during shrinkage of human neutrophils. J Biol Chem 272:17303–17311

L'Hoste S, Barriere H, Belfodil R, Rubera I, Duranton C, Tauc M, Poujeol C, Barhanin J, Poujeol P (2007) Extracellular pH alkalinisation by Cl^-/HCO_3^- exchanger is crucial for TASK2 activation by hypotonic shock in proximal cell lines from mouse kidney. Am J Physiol Renal Physiol 292:F628–F638

Lacroix J, Poet M, Huc L, Morello V, Djerbi N, Ragno M, Rissel M, Tekpli X, Gounon P, Lagadic-Gossmann D, Counillon L (2008) Kinetic analysis of the regulation of the Na^+/H^+ exchanger NHE-1 by osmotic shocks. Biochemist 47:13674–13685

Lambert IH (1987) Effect of arachidonic acid, fatty acids, prostaglandins, and leukotrienes on volume regulation in Ehrlich ascites tumor cells. J Membr Biol 98:207–221

Lambert IH, Hoffmann EK (1994) Cell swelling activates separate taurine and chloride channels in Ehrlich mouse ascites tumor cells. J Membr Biol 142:289–298

Lambert IH, Hoffmann EK, Christensen P (1987) Role of prostaglandins and leukotrienes in volume regulation by Ehrlich ascites tumor cells. J Membr Biol 98:247–256

Lang F (2013) Cell volume control. In: Alpern RJ, Moe OW, Caplan M (eds) Seldin and Giebisch's the kidney. Academic Press, Cambridge, MA, pp 121–141

Lang F, Busch GL, Ritter M, Volkl H, Waldegger S, Gulbins E, Haussinger D (1998a) Functional significance of cell volume regulatory mechanisms. Physiol Rev 78:247–306

Lang F, Busch GL, Volkl H (1998b) The diversity of volume regulatory mechanisms. Cell Physiol Biochem 8:1–45

Larsen EH (1973) Effect of amiloride, cyanide and ouabain on the active transport pathway in toad skin. In: Proc Alfred Benzon Symp V transport mechanisms in epithelia, pp 131–147

Larsen EH (1988) NaCl transport in amphibian skin. In: NaCl transport in epithelia. Springer, Berlin

Larsen EH (1991) Chloride transport by high-resistance heterocellular epithelia. Physiol Rev 71:235–283

Larsen EH (2009) Hans Henriksen Ussing 30 December 1911–22 December 2000. Biogr Mem Fellows Royal Soc 55:305–335

Larsen EH (2011) Reconciling the Krogh and Ussing interpretations of epithelial chloride transport - presenting a novel hypothesis for the physiological significance of the passive cellular chloride uptake. Acta Physiol 202:435–464

Larsen EH, Harvey BJ (1994) Chloride currents of single mitochondria-rich cells of toad skin epithelium. J Physiol 478:7–15

Larsen EH, Ramløv H (2013) Role of cutaneous surface fluid in frog osmoregulation. Comp Biochem Physiol A Mol Integr Physiol 165:365–370

Larsen EH, Rasmussen BE (1982) Chloride channels in toad skin. Phil Trans R Soc Lond B Biol 299:413–434

Larsen EH, Sørensen JN (2019) Stationary and non-stationary ion- and water flux interactions in kidney proximal tubule. Mathematical analysis of isosmotic transport by a minimalistic model. Rev Physiol Biochem Pharm DOI: 10.1007/112_2019_16

Larsen EH, Ussing HH, Spring KR (1987) Ion transport by mitochondria-rich cells in toad skin. J Membr Biol 99:25–40

Larsen EH, Sørensen JB, Sørensen JN (2000) A mathematical model of solute coupled water transport in toad intestine incorporating recirculation of the actively transported solute. J Gen Physiol 116:101–112

Larsen EH, Amstrup J, Willumsen NJ (2003) β-Adrenergic receptors couple to CFTR chloride channels of intercalated mitochondria-rich cells in the heterocellular toad skin epithelium. Biochim Biophys Acta 1618:140–152

Larsen EH, Willumsen NJ, Møbjerg N, Sørensen JN (2009) The lateral intercellular space as osmotic coupling compartment in isotonic transport. Acta Physiol 195:171–186

Larsen EH, Deaton LE, Onken H, O'Donnell M, Grosell M, Dantzler WH, Weihrauch D (2014) Osmoregulation and excretion. Compr Physiol 4:405–573

Larson M, Spring KR (1983) Bumetanide inhibition of NaCl transport by Necturus gallbladder. J Membr Biol 74:123–129

Larson M, Spring KR (1984) Volume regulation by Necturus gallbladder - basolateral KCl exit. J Membr Biol 81:219–232

Lauf PK, Misri S, Chimote AA, Adragna NC (2008) Apparent intermediate K conductance channel hyposmotic activation in human lens epithelial cells. Am J Physiol Cell Physiol 294:C820–C832

Lazarowski ER, Tarran R, Grubb BR, van Heusden CA, Okada S, Boucher RC (2004) Nucleotide release provides a mechanism for airway surface liquid homeostasis. J Biol Chem 279:36855–36864

Leaf A (1959) Maintenance of concentration gradients and regulation of cell volume. Ann N Y Acad Sci 72:396–404

Lee RJ, Foskett JK (2010) cAMP-activated Ca^{2+} signaling is required for CFTR-mediated serous cell fluid secretion in porcine and human airways. J Clin Invest 120:3137–3148

Levite M, Cahalon L, Peretz A, Hershkoviz R, Sobko A, Ariel A, Desai R, Attali B, Lider O (2000) Extracellular K^+ and opening of voltage-gated potassium channels activate T cell integrin function: physical and functional association between Kv1.3 channels and beta1 integrins. J Exp Med 191:1167–1176

Lewis A, Di Ciano C, Rotstein OD, Kapus A (2002) Osmotic stress activates Rac and Cdc42 in neutrophils: role in hypertonicity-induced actin polymerization. Am J Physiol Cell Physiol 282:C271–C279

Li S, Ault A, Malone CL, Raitt D, Dean S, Johnston LH, Deschenes RJ, Fassler JS (1998) The yeast histidine protein kinase, Sln1p, mediates phosphotransfer to two response regulators, Ssk1p and Skn7p. EMBO J 17:6952–6962

Liedtke CM, Cole TS (2002) Activation of NKCC1 by hyperosmotic stress in human tracheal epithelial cells involves PKC-delta and ERK. Biochem Biophys Acta Mol Cell Res 1589:77–88

Liedtke W, Kim C (2005) Functionality of the TRPV subfamily of TRP ion channels: add mechano-TRP and osmo-TRP to the lexicon! Cell Mol Life Sci 62:2985–3001

Lionetto MG, Schettino T (2006) The Na^+-K^+-$2Cl^-$ cotransporter and the osmotic stress response in a model salt transport epithelium. Acta Physiol 187:115–124

Lionetto MG, Giordano ME, Nicolardi G, Schettino T (2001) Hypertonicity stimulates cl- transport in the intestine of fresh water acclimated eel, *Anguilla anguilla*. Cell Physiol Biochem 11:41–54

Lionetto MG, Pedersen SF, Hoffmann EK, Giordano ME, Schettino T (2002) Roles of the cytoskeleton and of protein phosphorylation events in the osmotic stress response in eel intestinal epithelium. Cell Physiol Biochem 12:163–163

Lionetto MG, Rizzello A, Giordano ME, Maffia M, De Nuccio F, Nicolardi G, Hoffmann EK, Schettino T (2008) Molecular and functional expression of high conductance Ca^{2+} activated K^+ channels in the eel intestinal epithelium. Cell Physiol Biochem 21:373–384

Liu XB, Bandyopadhyay B, Nakamoto T, Singh B, Liedtke W, Melvin JE, Ambudkar I (2006) A role for AQP5 in activation of TRPV4 by hypotonicity - concerted involvement of AQP5 and TRPV4 in regulation of cell volume recovery. J Biol Chem 281:15485–15495

Lock H, Valverde MA (2000) Contribution of the IsK (MinK) potassium channel subunit to regulatory volume decrease in murine tracheal epithelial cells. J Biol Chem 275:34849–34852

Lopes AG, Amzel LM, Markakis D, Guggino WB (1988) Cell volume regulation by the thin descending limb og Henle's loop. Proc Natl Acad Sci U S A 85:2873–2877

Macknight ADC, Leaf A (1977) Regulation of cellular volume. Physiol Rev 57:510–573

Macleod RJ, Hamilton JR (1990) Regulatory volume increase in mammalian jejunal villus cells is due to bumetanide-sensitive NaKCl2 cotransport. Am J Physiol Gastrointest Liver Physiol 258: G665–G674

MacLeod RJ, Hamilton JR (1991) Volume regulation initiated by Na^+-nutrient cotransport in isolated mammalian villus enterocytes. Am J Physiol Gastrointest Liver Physiol 260:G26–G33

MacLeod RJ, Hamilton JR (1999) Ca^{2+}/calmodulin kinase II and decreases in intracellular pH are required to activate K^+ channels after substantial swelling in villus epithelial cells. J Membr Biol 172:59–66

MacRobbie EAC, Ussing HH (1961) Osmotic behaviour of the epithelial cells of frog skin. Acta Physiol Scand 53:348–365

Maingret F, Fosset M, Lesage F, Lazdunski M, Honore E (1999a) TRAAK is a mammalian neuronal mechano-gated K^+ channel. J Biol Chem 274:1381–1387

Maingret F, Patel AJ, Lesage F, Lazdunski M, Honore E (1999b) Mechano- or acid stimulation, two interactive modes of activation of the TREK-1 potassium channel. J Biol Chem 274:26691–26696

Maingret F, Patel AJ, Lesage F, Lazdunski M, Honore E (2000) Lysophospholipids open the two-pore domain mechano-gated K^+ channels TREK-1 and TRAAK. J Biol Chem 275:10128–10133

Manganel M, Turner RJ (1991) Rapid secretagogue-induced activation of Na^+/H^+ exchange in rat parotid acinar-cells - possible interrelationship between volume regulation and stimulus-secretion coupling. J Biol Chem 266:10182–10188

Markadieu N, Delpire E (2014) Physiology and pathophysiology of SLC12A1/2 transporters. Pflugers Arch Eur J Physiol 466:91–105

Marsh DJ, Spring KR (1985) Polarity of volume-regulatory increase by *Necturus* gallbladder epithelium. Am J Phys 249:C471–C475

Marsh DJ, Jensen PK, Spring KR (1985) Computer-based determination of size and shape in living cells. J Microsc 137:281–292

Marshall WS (2011) Mechanosensitive signalling in fish gill and other ion transporting epithelia. Acta Physiol 202:487–499

Marshall WS, Grosell M (2005) Ion transport, osmoregulation, and acid-base balance. In: Evans DH, Claiborne JB (eds) The physiology of fishes, 3rd edn. CRC Press, Boca Raton, pp 177–230

Marshall WS, Bryson SE, Luby T (2000) Control of epithelial Cl^- secretion by basolateral osmolality in the euryhaline teleost *Fundulus heteroclitus*. J Exp Biol 203:1897–1905

Marshall WS, Ossum CG, Hoffmann EK (2005) Hypotonic shock mediation by p38 MAPK, JNK PKC, FAK, OSR1 and SPAK in osmosensing chloride secreting cells of killifish opercula epithelium. J Exp Biol 208:1063–1077

Marshall WS, Katoh F, Main HP, Sers N, Cozzi RRF (2008) Focal adhesion kinase and β1 integrin regulation of Na⁺, K⁺, 2Cl⁻ cotransporter in osmosensing ion transporting cells of killifish *Fundulus heteroclitus*. Comp Biochem Phys A 150:288–300

Matsumura Y, Cohen B, Guggino WB, Giebisch G (1984) Regulation of the basolateral potassium conductance of the Necturus proximal tubule. J Membr Biol 79:153–161

Maunsbach AB (1966) Observations on the segmentation of the proximal tubule in the rat kidney Comparison of results from phase contrast, fluorescence and electron microscopy. J Ultrastr Res 16:239–258

Maunsbach AB, Boulpaep EL (1991) Immunoelectron microscope localization of Na,K-ATPase in transport pathways in proximal tubule epithelium. Micr Microscop Acta 22:55–56

McCormick JA, Ellison DH (2011) The WNKs: atypical protein kinases with pleiotropic actions Physiol Rev 91:177–219

Medhurst AD, Rennie G, Chapman CG, Meadows H, Duckworth MD, Kelsell RE, Gloger II Pangalos MN (2001) Distribution analysis of human two pore domain potassium channels in tissues of the central nervous system and periphery. Brain Res Mol Brain Res 86:101–114

Messner G, Wang W, Paulmichl M, Oberleithner H, Lang F (1985) Ouabain decreases apparen potassium-conductance in proximal tubules of the amphibian kidney. Pflugers Arch Eur J Physiol 404:131–137

Meyer K, Korbmacher C (1996) Cell swelling activates ATP-dependent voltage-gated chloride channels in M-1 mouse cortical collecting duct cells. J Gen Physiol 108:177–193

Mignen O, Le Gall C, Harvey BJ, Thomas S (1999) Volume regulation following hypotonic shock in isolated crypts of mouse distal colon. J Physiol 515:501–510

Mills JW (1985) Ion transport across the exocrine glands of the frog skin. Pflüg Arch Eur J Physio 405(suppl 1):S44–S49

Mills JW, DiBona DR (1978) Distribution of Na⁺-pump sites in the frog gallbladder. Nature 271:273–275

Mills JW, Ernst SA (1975) Localization of sodium pump sites in frog urinary bladder. Biochim Biophys Acta 375:268–273

Mills JW, Ernst SA, DiBona DR (1977) Localization of Na⁺-pump sites in frog skin. J Cell Bio 73:88–110

Minton AP (2006) Macromolecular crowding. Curr Biol 16:R269–R271

Moeckel GW, Zhang L, Chen X, Rossini M, Zent R, Pozzi A (2006) Role of integrin alpha1beta1 in the regulation of renal medullary osmolyte concentration. Am J Physiol Renal Physiol 290 F223–F231

Montoro DT, Haber AL, Biton M, Vinarsky V, Lin B, Birket SE, Yuan F, Chen SJ, Leung HM Villoria J, Rogel N, Burgin G, Tsankov AM, Waghray A, Slyper M, Waldman J, Nguyen L Dionne D, Rozenblatt-Rosen O, Tata PR, Mou HM, Shivaraju M, Bihler H, Mense M, Tearney GJ, Rowe SM, Engelhardt JF, Regev A, Rajagopal J (2018) A revised airway epithelial hierarchy includes CFTR-expressing ionocytes. Nature 560:319–324

Montroserafizadeh C, Guggino WB (1990) Cell-volume regulation in the nephron. Annu Rev Physiol 52:761–772

Montrose-Rafizadeh C, Guggino WB, Montrose MH (1991) Cellular differentiation regulates expression of Cl⁻ transport and cystic fibrosis transmembrane conductance regulator mRNA in human intestinal cells. J Biol Chem 266:4495–4499

Moore-Hoon ML, Turner RJ (2000) The structural unit of the secretory Na⁺-K⁺-2Cl⁻ cotransporter (NKCCl) is a homodimer. Biochemist 39:3718–3724

Nauntofte B, Dissing S (1988) Cholinergic-induced electrolyte transport in rat parotid acini. Comp Biochem Physiol 90A:739–746

Nearing J, Betka M, Quinn S, Hentschel H, Elger M, Baum M, Bai M, Chattopadyhay N, Brown EM, Hebert SC, Harris HW (2002) Polyvalent cation receptor proteins (CaRs) are salinity sensors in fish. Proc Natl Acad Sci U S A 99:9231–9236

Nedergaard S, Larsen EH, Ussing HH (1999) Sodium recirculation and isotonic transport in toad small intestine. J Membr Biol 168:241–251

Nielsen S, Agre P (1995) The aquaporin family of water channels in kidney. Kidney Int 48:1057–1068

Nielsen R, Larsen EH (2007) Beta-adrenergic activation of solute coupled water uptake by toad skin epithelium results in near-isosmotic transport. Comp Bioch Physiol Mol Integr Physiol 148A:64–71

Nielsen MS, Nielsen R (1999) Effect of carbachol and prostaglandin E2 on chloride secretion and signal transduction in the exocrine glands of frog skin (*Rana esculenta*). Pflugers Arch Eur J Physiol 438:732–740

Nielsen S, Chou C-L, Marples D, Christensen EI, Kishore BM, Knepper M (1995) Vasopressin increases water permeability of kidney collecting duct by inducing translocation of aquaporin-CD water channels to plasma membrane. Proc Natl Acad Sci U S A 92:1013–1017

Nielsen S, Frokiaer J, Marples D, Kwon TH, Agre P, Knepper MA (2002) Aquaporins in the kidney: from molecules to medicine. Physiol Rev 82:205–244

Nielsen S, Kwon T-H, Dimke H, Skott M, Frøkiær J (2013) Aquaporin water channels in mammalian kidney. In: Alpern RJ, Moe OW, Caplan M (eds) Seldin and Giebisch's the kidney. Academic Press, Cambridge, MA, pp 1405–1439

Niemeyer MI, Cid LP, Barros LF, Sepulveda FV (2001a) Modulation of the two-pore domain acid-sensitive K$^+$ channel TASK-2 (KCNK5) by changes in cell volume. J Biol Chem 276:43166–43174

Niemeyer MI, Cid LP, Sepulveda FV (2001b) K$^+$ conductance activated during regulatory volume decrease. The channels in Ehrlich cells and their possible molecular counterpart. Comp Biochem Physiol A Mol Integr Physiol 130:565–575

Niemeyer MI, Cid LP, Pena-Munzenmayer G, Sepulveda FV (2010) Separate gating mechanisms mediate the regulation of K2P potassium channel TASK-2 by intra- and extracellular pH. J Biol Chem 285:16467–16475

Nilius B, Droogmans G (2003) Amazing chloride channels: an overview. Acta Physiol Scand 177:119–147

Nilius B, Eggermont J, Voets T, Buyse G, Manolopoulos V, Droogmans G (1997) Properties of volume-regulated anion channels in mammalian cells. Prog Biophys Mol Bio 68:69–119

Nilius B, Voets T, Prenen J, Barth H, Aktories K, Kaibuchi K, Droogmans G, Eggermont J (1999) Role of Rho and Rho kinase in the activation of volume-regulated anion channels in bovine endothelial cells. J Physiol 516:67–74

Nilius B, Owsianik G, Voets T, Peters JA (2007) Transient receptor potential cation channels in disease. Physiol Rev 87:165–217

Novak I (2011) Purinergic signalling in epithelial ion transport: regulation of secretion and absorption. Acta Physiol 202:501–522

Novak I, Greger R (1988) Properties of the luminal membrane of isolated perfused rat pancreatic ducts. Effect of cyclic AMP and blockers of chloride transport. Pflugers Arch Eur J Physiol 411:546–553

Numata T, Shimizu T, Okada Y (2007) TRPM7 is a stretch- and swelling-activated cation channel involved in volume regulation in human epithelial cells. Am J Physiol Cell Physiol 292:C460–C467

O'Brien JA, Walters RJ, Valverde MA, Sepulveda FV (1993) Regulatory volume increase after hypertonicity- or vasoactive-intestinal-peptide-induced cell-volume decrease in small-intestinal crypts is dependent on Na$^+$-K$^+$-2Cl$^-$ cotransport. Pflugers Arch Eur J Physiol 423:67–73

O'Brien JA, Walters RJ, Sepulveda FV (1991) Regulatory volume decrease in small intestinal crypts is inhibited by K$^+$ and Cl$^-$ channel blockers. Biochim Biophys Acta 1070:501–504

Ogushi Y, Kitagawa D, Hasegawa T, Suzuki M, Tanaka S (2010) Correlation between aquaporin and water permeability in response to vasotocin, hydrin and β-adrenergic effectors in the ventral pelvic skin of the tree frog *Hyla japonica*. J Exp Biol 213:288–294

Okada Y (1997) Volume expansion-sensing outward-rectifier Cl- channel: fresh start to the molecular identity and volume sensor. Am J Physiol Cell Physiol 273:C755–C789

Okada Y, Maeno E, Shimizu T, Dezaki K, Wang J, Morishima S (2001) Receptor-mediated control of regulatory volume decrease (RVD) and apoptotic volume decrease (AVD). J Physiol 532:3–16

Okada Y, Sato K, Numata T (2009) Pathophysiology and puzzles of the volume-sensitive outwardly rectifying anion channel. J Physiol 587:2141–2149

Okada Y, Okada T, Sato-Numata K, Islam MR, Ando-Akatsuka Y, Numata T, Kubo M, Shimizu T, Kurbannazarova RS, Marunaka Y, Sabirov RZ (2019) Cell volume-activated and volume-correlated anion channels in mammalian cells: their biophysical, molecular, and pharmacological properties. Pharmacol Rev 71:49–88

Orlowski J, Grinstein S (2004) Diversity of the mammalian sodium/proton exchanger SLC9 gene family. Pflugers Arch Eur J Physiol 447:549–565

Orlowski J, Grinstein S (2011) Na^+/H^+ exchangers. Compr Physiol 1:2083–2100

Owsianik G, Talavera K, Voets T, Nilius B (2006) Permeation and selectivity of TRP channels. Annu Rev Physiol 68:685–717

Pan Z, Capo-Aponte JE, Zhang F, Wang Z, Pokorny KS, Reinach PS (2007) Differential dependence of regulatory volume decrease behavior in rabbit corneal epithelial cells on MAPK superfamily activation. Exp Eye Res 84:978–990

Pang V, Counillon L, Lagadic-Gossmann D, Poet M, Lacroix J, Sergent O, Khan R, Rauch C (2012) On the role of the difference in surface tensions involved in the allosteric regulation of NHE-1 induced by low to mild osmotic pressure, membrane tension and lipid asymmetry. Cell Biochem Biophys 63:47–57

Park S, Hong JH, Ohana E, Muallem S (2012) The WNK/SPAK and IRBIT/PP1 pathways in epithelial fluid and electrolyte transport. Physiology 27:291–299

Pedersen SF, Hoffmann EK (2002) Possible interrelationship between changes in F-actin and myosin II, protein phosphorylation, and cell volume regulation in Ehrlich ascites tumor cells. Exp Cell Res 277:57–73

Pedersen SF, Nilius B (2007) Transient receptor potential channels in mechanosensing and cell volume regulation. Methods Enzymol 428:183–207

Pedersen SF, Prenen J, Droogmans G, Hoffmann EK, Nilius B (1998) Separate swelling- and Ca^{2+}-activated anion currents in Ehrlich ascites tumor cells. J Membr Biol 163:97–110

Pedersen SF, Beisner KH, Hougaard C, Willumsen BM, Lambert IH, Hoffmann EK (2002a) Rho family GTP binding proteins are involved in the regulatory volume decrease process in NIH3T3 mouse fibroblasts. J Physiol 541:779–796

Pedersen SF, Varming C, Christensen ST, Hoffmann EK (2002b) Mechanisms of activation of NHE by cell shrinkage and by calyculin A in ehrlich ascites tumor cells. J Membr Biol 189:67–81

Pedersen SF, Kapus A, Hoffmann EK (2011) Osmosensory mechanisms in cellular and systemic volume regulation. J Am Soc Nephrol 22:1587–1597

Peng JB, Warnock DG (2007) WNK4-mediated regulation of renal ion transport proteins. Am J Physiol Renal Physiol 293:F961–F973

Persson B-E, Spring KR (1982) Gallbladder epithelial cell hydraulic water permeablity and volume regulation. J Gen Physiol 79:481–505

Petersen OH (1992) Stimulus-secretion coupling - cytoplasmic calcium signals and the control of ion channels in exocrine acinar-cells. J Physiol 448:1–51

Petersen OH (1993) Regulation of isotonic fluid secretion in exocrine acini. In: Ussing HH, Fischbarg J, Sten-Knudsen O, Larsen EH, Willumsen NJ (eds) Proc Alfred Benzon Symp 34 isotonic transport in leaky epithelia. Munksgaard, Copenhagen, pp 103–146

Petersen OH, Gallacher DV (1988) Electrophysiology of pancreatic and salivary acinar cells. Annu Rev Physiol 50:65–80

Piechotta K, Garbarini N, England R, Delpire E (2003) Characterization of the interaction of the stress kinase SPAK with the Na^+-K^+-$2Cl^-$ cotransporter in the nervous system - evidence for a scaffolding role of the kinase. J Biol Chem 278:52848–52856

Pihakaski-Maunsbach K, Vorum H, Locke EM, Garty H, Karlish SJ, Maunsbach AB (2003) Immunocytochemical localization of Na,K-ATPase gamma subunit and CHIF in inner medulla of rat kidney. Ann N Y Acad Sci 986:401–409

Planells-Cases R, Lutter D, Guyader C, Gerhards NM, Ullrich F, Elger DA, Kucukosmanoglu A, Xu G, Voss FK, Reincke SM, Stauber T, Blomen VA, Vis DJ, Wessels LF, Brummelkamp TR, Borst P, Rottenberg S, Jentsch TJ (2015) Subunit composition of VRAC channels determines substrate specificity and cellular resistance to Pt-based anti-cancer drugs. EMBO J 34:2993–3008

Plasschaert LW, Zilionis R, Choo-Wing R, Savova V, Knehr J, Roma G, Klein AM, Jaffe AB (2018) A single-cell atlas of the airway epithelium reveals the CFTR-rich pulmonary ionocyte. Nature 560:377–381

Pochynyuk O, Zaika O, O'Neil RG, Mamenko M (2013) Novel insights into TRPV4 function in the kidney. Pflugers Arch Eur J Physiol 465:177–186

Preston GM, Carroll TP, Guggino WB, Agre P (1992) Appearance of water channels in *Xenopus* oocytes expressing red cell CHIP28 protein. Science 256:385–387

Qiu Z, Dubin AE, Mathur J, Tu B, Reddy K, Miraglia LJ, Reinhardt J, Orth AP, Patapoutian A (2014) SWELL1, a plasma membrane protein, is an essential component of volume-regulated anion channel. Cell 157:447–458

Qu Y, Misaghi S, Newton K, Gilmour LL, Louie S, Cupp JE, Dubyak GR, Hackos D, Dixit VM (2011) Pannexin-1 is required for ATP release during apoptosis but not for inflammasome activation. J Immunol 186:6553–6561

Rasmussen LJ, Muller HS, Jorgensen B, Pedersen SF, Hoffmann EK (2015) Osmotic shrinkage elicits FAK- and Src phosphorylation and Src-dependent NKCC1 activation in NIH3T3 cells. Am J Physiol Cell Physiol 308:C101–C110

Reuss L (1985) Changes in cell volume measured with an electrophysiologic technique. Proc Natl Acad Sci U S A 82:6014–6018

Reyes R, Duprat F, Lesage F, Fink M, Salinas M, Farman N, Lazdunski M (1998) Cloning and expression of a novel pH-sensitive two pore domain K^+ channel from human kidney. J Biol Chem 273:30863–30869

Richardson C, Alessi DR (2008) The regulation of salt transport and blood pressure by the WNK-SPAK/OSR1 signalling pathway. J Cell Sci 121:3293–3304

Rick R, Dörge A, Von Arnim E, Thurau K (1978) Electron microprobe analysis of frog skin epithelium: evidence for a syncytial sodium transport compartment. J Membr Biol 39:313–331

Rick R, Roloff C, Dörge A, Beck F-X, Thurau K (1984) Intracellular electrolyte concentrations in the frog skin epithelium: effect of vasopressin and dependence on the Na concentration in the bathing media. J Membr Biol 78:129–145

Rigor RR, Damoc C, Phinney BS, Cala PM (2011) Phosphorylation and activation of the plasma membrane Na^+/H^+ exchanger (NHE1) during osmotic cell shrinkage. PLoS One 6:e29210

Robertson MA, Foskett JK (1994) Na^+ transport pathways in secretory acinar-cells - membrane cross-talk mediated by $[Cl^-]_i$. Am J Physiol Cell Physiol 267:C146–C156

Roger F, Martin PY, Rousselot M, Favre H, Feraille E (1999) Cell shrinkage triggers the activation of mitogen-activated protein kinases by hypertonicity in the rat kidney medullary thick ascending limb of the Henle's loop - requirement of p38 kinase for the regulatory volume increase response. J Biol Chem 274:4103–34110

Romanenko VG, Catalan MA, Brown DA, Putzier I, Hartzell HC, Marmorstein AD, Gonzalez-Begne M, Rock JR, Harfe BD, Melvin JE (2010) TMEM16A encodes the Ca^{2+}-activated Cl^- channel in mouse submandibular salivary gland acinar cells. J Biol Chem 285:2990–13001

Rotin D, Grinstein S (1989) Impaired cell-volume regulation in Na$^+$-H$^+$ exchange-deficient mutants. Am J Phys 257:C1158–C1165

Sabirov RS, Okada Y (2005) ATP release via anion channels. Purinerg Sign 1:311–328

Sarkadi B, Attisano L, Grinstein S, Buchwald M, Rothstein A (1984) Volume regulation of Chinese-Hamster ovary cells in anisoosmotic media. Biochim Biophys Acta 774:159–168

Schafer JA (1990) Transepithelial osmolality differences, hydraulic conductivities, and volume absorption in the proximal tubule. Annu Rev Physiol 52:709–726

Schafer JA (1993) The rat collecting duct as an isosmotic volume reabsorber. In: Ussing HH, Fischbarg J, Sten-Knudsen O, Larsen EH, Willumsen NJ (eds) Proc Alfred Benzon Symp 34 isotonic transport in leaky epithelia. Munksgaard, Copenhagen, pp 339–354

Schafer JA, Patlak CS, Troutman SL, Andreoli TE (1978) Volume absorption in the parts recta. II. Hydraulic conductivity coefficient. Am J Physiol Renal Physiol 234:F340–F348

Schatzmann HJ, Windhager EE, Solomon AK (1958) Single proximal tubules of the Necturus kidney. II. Effect of 2, 4-dinitro-phenol and ouabain on water reabsorption. Am J Phys 195:570–574

Schild L, Aronson PS, Giebisch G (1991) Basolateral transport pathways for K$^+$ and Cl$^-$ in rabbit proximal tubule: effects on cell volume. Am J Physiol Renal Physiol 260:F101–F109

Schlatter E, Greger R, Schafer JA (1990) Principal cells of cortical collecting ducts of the rat are not a route of transepithelial Cl$^-$ transport. Pflugers Archiv Eur J Physiol 417:317–323

Schnermann J, Chou C-L, Ma T, Traynor T, Knepper MA, Verkman AS (1998) Defective proximal tubular fluid reabsorption in transgenic aquaporin-1 null mice. Proc Natl Acad Sci U S A 95:9660–9664

Schultz SG (1981) Homocellular regulatory mechanisms in sodium-transporting epithelia: avoidance of extinction by "flush-through". Am J Physiol Renal Physiol 10:F579–F590

Schultz SG (1992) Membrane crosstalk in sodium-absorbing epithelial cells. In: Seldin DW, Giebisch G (eds) The kidney: physiology and pathophysiology. Raven Press, New York, pp 287–299

Schultz SG, Dubinsky WP (2001) Sodium absorption, volume control and potassium channels: a tribute to a great biologist. J Membr Biol 184:255–261

Schwiebert EM, Mills JW, Stanton BA (1994) Actin-based cytoskeleton regulates a chloride channel and cell volume in a renal cortical collecting duct cell line. J Biol Chem 269:7081–7089

Shimada-Shimizu N, Hisamitsu T, Nakamura TY, Hirayama N, Wakabayashi S (2014) Na$^+$/H$^+$ exchanger 1 is regulated via its lipid-interacting domain, which functions as a molecular switch: a pharmacological approach using indolocarbazole compounds. Mol Pharmac 85:18–28

Shimizu T, Numata T, Okada Y (2004) A role of reactive oxygen species in apoptotic activation of volume-sensitive Cl$^-$ channel. Proc Natl Acad Sci U S A 101:6770–6773

Sørensen JB, Larsen EH (1996) Heterogeneity of chloride channels in the apical membrane of isolated mitochondria-rich cells from toad skin. J Gen Physiol 108:421–433

Sørensen JB, Larsen EH (1998) Patch clamp on the luminal membrane of exocrine gland acini from frog skin (*Rana esculenta*) reveals the presence of cystic fibrosis transmembrane conductance regulator-like Cl$^-$ channels activated by cyclic AMP. J Gen Physiol 112:19–31

Sørensen JB, Larsen EH (1999) Membrane potential and conductance of frog skin gland acinar cells in resting conditions and during stimulation with agonists of macroscopic secretion. Pflugers Arch Eur J Physiol 439:101–112

Sørensen JB, Nielsen MS, Nielsen R, Larsen EH (1998) Luminal ion channels involved in isotonic secretion by Na$^+$-recirculation in exocrine gland-acini. Roy Dan Acad Sci Lett Biol Ser 49:179–191

Sørensen JB, Nielsen MS, Gudme CN, Larsen EH, Nielsen R (2001) Maxi K$^+$ channels co-localised with CFTR in the apical membrane of an exocrine gland acinus: possible involvement in secretion. Pflugers Arch Eur J Physiol 442:1–11

Sorensen BH, Thorsteinsdottir UA, Lambert IH (2014) Acquired cisplatin resistance in human ovarian A2780 cancer cells correlates with shift in taurine homeostasis and ability to volume regulate. Am J Phys 307:C1071–C1080

Sorensen BH, Rasmussen LJH, Broberg BS, Klausen TK, Sauter DPR, Lambert IH, Aspberg A, Hoffmann EK (2015) Integrin beta(1), osmosensing, and chemoresistance in mouse ehrlich carcinoma cells. Cell Physiol Biochem 36:111–132

Speake T, Douglas IJ, Brown PD (1998) The role of calcium in the volume regulation of rat lacrimal acinar cells. J Membr Biol 164:283–291

Spring KR, Hope A (1978) Size and shape of the lateral intercellular spaces in a living epithelium. Science 200:54–57

Spring KR, Hope A (1979) Fluid transport and the dimensions of cell and interspaces of living *Necturus* gallbladder. J Gen Physiol 73:287–305

Spring KR, Ussing HH (1986) The volume of mitochondria-rich cells of frog skin epithelium. J Membr Biol 92:21–26

Staruschenko A (2012) Regulation of transport in the connecting tubule and cortical collecting duct. Compr Physiol 2:1541–1584

Stauber T (2015) The volume-regulated anion channel is formed by LRRC8 heteromers - molecular identification and roles in membrane transport and physiology. Biol Chem 396:975–990

Sten-Knudsen O (2002) Biological membranes. Theory of transport, potentials and electric impulses. Cambridge University Press, Cambridge

Stirling CE (1972) Radioautographic localization of sodium pump sites in rabbit intestine. J Cell Biol 53:704–714

Stoner LC, Morley GE (1995) Effect of basolateral or apical hyposmolarity on apical maxi-K channels of everted rat collecting tubule. Am J Physiol Renal Physiol 268:F569–F580

Strange K (1988) RVD in principal and intercalated cells of rabbit cortical collecting tubule. Am J Physiol Cell Physiol 255:C612–C621

Strange K (1989) Ouabain-induced cell swelling in rabbit cortical collecting tubule: NaCl transport by principal cells. J Memb Biol 107:249–261

Strange K (1990) Volume regulation following Na+ pump inhibition in CCT principal cells: apical K+ loss. Am J Physiol Renal Physiol 258:F732–F740

Strange K, Spring K (1986) Methods for imaging renal tubule cells. Kid Intern 30:192–200

Strange K, Spring KR (1987a) Absence of significant cellular dilution during ADH-stimulated water reabsorption. Science 235:1068–1070

Strange K, Spring KR (1987b) Cell membrane water permeability of rabbit cortical collecting duct. J Membr Biol 96:27–43

Strotmann R, Harteneck C, Nunnenmacher K, Schultz G, Plant TD (2000) OTRPC4, a nonselective cation channel that confers sensitivity to extracellular osmolarity. Nat Cell Biol 2:695–702

Stutzin A, Hoffmann EK (2006) Swelling-activated ion channels: functional regulation in cell-swelling, proliferation and apoptosis. Acta Physiol 187:27–42

Takemura T, Sato F, Suzuki Y, Sato K (1991) Intracellular ion concentrations and cell volume during cholinergic stimulation of eccrine secretory coil cells. J Membr Biol 119:211–219

Taniguchi J, Guggino WB (1989) Membrane stretch - a physiological stimulator of Ca2+-activated K+ channels in thick ascending limb. Am J Physiol Renal Physiol 257:F347–F352

Tarran R, Grubb BR, Gatzy JT, Davis CW, Boucher RC (2001) The relative roles of passive surface forces and active ion transport in the modulation of airway surface liquid volume and composition. J Gen Physiol 118:223–236

Tarran R, Trout L, Donaldson SH, Boucher RC (2006) Soluble mediators, not cilia, determine airway surface liquid volume in normal and cystic fibrosis superficial airway epithelia. J Gen Physiol 127:591–604

Taylor A, Windhager EE (1979) Possible role of cytosolic calcium and Na-Ca exchange in regulation of transepithelial sodium transport. Am J Physiol Renal Physiol 236:F505–F512

Tekpli X, Huc L, Lacroix J, Rissel M, Poet M, Noel J, Dimanche-Boitrel MT, Counillon L, Lagadic-Gossmann D (2008) Regulation of Na+/H+ exchanger 1 allosteric balance by its localization in cholesterol- and caveolin-rich membrane microdomains. J Cell Physiol 216:207–220

Thompson J, Begenisich T (2006) Membrane-delimited inhibition of maxi-K channel activity by the intermediate conductance Ca^{2+}-activated K channel. J Gen Physiol 127:159–169

Tilly BC, Edixhoven MJ, Tertoolen LGJ, Morii N, Saitoh Y, Narumiya S, deJonge HR (1996) Activation of the osmo-sensitive chloride conductance involves p21(rho) and is accompanied by a transient reorganization of the F-actin cytoskeleton. Mol Biol Cell 7:1419–1427

Tinel H, Kinne-Saffran E, Kinne RKH (2000) Calcium signalling during RVD of kidney cells. Cell Physiol Biochem 10:297–302

Tinel H, Kinne-Saffran E, Kinne RKH (2002) Calcium-induced calcium release participates in cell volume regulation of rabbit TALH cells. Pflugers Arch Eur J Physiol 443:754–761

Toczylowska-Maminska R, Dolowy K (2012) Ion transporting proteins of human bronchial epithelium. J Cell Biochem 113:426–432

Tomassen SFB, Fekkes D, de Jonge HR, Tilly BC (2004) Osmotic swelling-provoked release of organic osmolytes in human intestinal epithelial cells. Am J Phys 286:C1417–C1422

Tormey JM, Diamond JM (1969) The ultrastructure route of fluid transport in rabbit gall bladder. J Gen Physiol 50:2031–2060

Tosteson DC, Hoffman JF (1960) Regulation of cell volume by active cation transport in high and low potassium sheep red cells. J Gen Physiol 44:169–194

Travaglini KJ, Krasnow MA (2018) Profile of an unknown airway cell. Nature 560:313–314

Tsai TT, Guttapalli A, Agrawal A, Albert TJ, Shapiro IM, Risbud MV (2007) MEK/ERK signaling controls osmoregulation of nucleus pulposus cells of the intervertebral disc by transactivation of TonEBP/OREBP. J Bone Miner Res 22:965–974

Tsuchiya K, Wang W, Giebisch G, Welling PA (1992) ATP is a coupling modulator of parallel Na, K-ATPase-K-channel activity in the renal proximal tubule. Proc Natl Acad Sci U S A 89:6418–6422

Uhlik MT, Abell AN, Johnson NL, Sun WY, Cuevas BD, Lobel-Rice KE, Horne EA, Dell'Acqua ML, Johnson GL (2003) Rac-MEKK3-MKK3 scaffolding for p38 MAPK activation during hyperosmotic shock. Nat Cell Biol 5:1104–1110

Urbach V, Van Kerkhove E, Maguire D, Harvey BJ (1996) Cross-talk between ATP-regulated K^+ channels and Na^+ transport via cellular metabolism in frog skin principal cells. J Physiol 491:99–109

Urbach V, Leguen I, O'Kelly I, Harvey BJ (1999) Mechanosensitive calcium entry and mobilization in renal A6 cells. J Membr Biol 168:29–37

Ussing HH (1960) Active and passive transport of the alkali metal ions. In: Ussing HH, Kruhøffer P, Thaysen JH, Thorn NA (eds) The alkali metal ions in biology. Springer, Berlin, pp 45–143

Ussing HH (1982) Volume regulation of frog skin epithelium. Acta Physiol Scand 114:363–369

Ussing HH (1985) Volume regulation and basolateral co-transport of sodium, potassium, and chloride ion in frog skin epithelium. Pflugers Arch Eur J Physiol 405(Suppl 1):S2–S7

Ussing HH, Eskesen K (1989) Mechanism of isotonic water transport in glands. Acta Physiol Scand 136:443–454

Ussing HH, Nedergaard S (1993) Recycling of electrolytes in small intestine of toad. In: Ussing HH, Fischbarg J, Sten-Knudsen O, Larsen EH, Willumsen NJ (eds) Proc Alfred Benzon symposium 34 isotonic transport in leaky epithelia. Munksgaard, Copenhagen, pp 26–36

Ussing HH, Lind F, Larsen EH (1996) Ion secretion and isotonic transport in frog skin glands. J Membr Biol 152:101–110

Valverde MA, O'Brien JA, Sepúlveda FV, Ratcliff RA, Evans MJ, Colledge WH (1995) Impaired cell volume regulation in intestinal crypt epithelia of cystic fibrosis mice. Proc Natl Acad Sci U S A 92:9038–9041

Valverde MA, Vazquez E, Munoz FJ, Nobles M, Delaney SJ, Wainwright BJ, Colledge WH, Sheppard DN (2000) Murine CFTR channel and its role in regulatory volume decrease of small intestine crypts. Cell Physiol Biochem 10:321–328

van der Wijk T, Tomassen SFB, Houtsmuller AB, de Jonge HR, Tilly BC (2003) Increased vesicle recycling in response to osmotic cell swelling - cause and consequence of hypotonicity-provoked ATP release. J Biol Chem 278:40020–40025

van Heeswijk MPE, van Os CH (1986) Osmotic water permeabilities of brush border and basolateral membrane vesicles from rat renal cortex and small intestine. J Membr Biol 92:183–193

Vanoye CG, Reuss L (1999) Stretch-activated single K^+ channels account for whole-cell currents elicited by swelling. Proc Natl Acad Sci U S A 96:6511–6516

vanTol BL, Missan S, Crack J, Moser S, Baldridge WH, Linsdell P, Cowley EA (2007) Contribution of KCNQ1 to the regulatory volume decrease in the human mammary epithelial cell line MCF-7. Am Physiol Cell Physiol 293:C1010–C1019

Varela D, Simon F, Riveros A, Jorgensen F, Stutzin A (2004) NAD(P)H oxidase-derived H_2O_2 signals chloride channel activation in cell volume regulation and cell proliferation. J Biol Chem 279:13301–13304

Vazquez E, Nobles M, Valverde MA (2001) Defective regulatory volume decrease in human cystic fibrosis tracheal cells because of altered regulation of intermediate conductance Ca^{2+}-dependent potassium channels. Proc Natl Acad Sci U S A 98:5329–5334

Verissimo F, Jordan P (2001) WNK kinases, a novel protein kinase subfamily in multi-cellular organisms. Oncogene 20:5562–5569

Vinnakota S, Qian XJ, Egal H, Sarthy V, Sarkar HK (1997) Molecular characterization and in situ localization of a mouse retinal taurine transporter. J Neurochem 69:2238–2250

Vitari AC, Deak M, Morrice NA, Alessi DR (2005) The WNK1 and WNK4 protein kinases that are mutated in Gordon's hypertension syndrome phosphorylate and activate SPAK and OSR1 protein kinases. Biochem J 391:17–24

Voets T, Droogmans G, Raskin G, Eggermont J, Nilius B (1999) Reduced intracellular ionic strength as the initial trigger for activation of endothelial volume-regulated anion channels. Proc Natl Acad Sci U S A 96:5298–5303

Voss FK, Ullrich F, Munch J, Lazarow K, Lutter D, Mah N, Andrade-Navarro MA, von Kries JP, Stauber T, Jentsch TJ (2014) Identification of LRRC8 heteromers as an essential component of the volume-regulated anion channel VRAC. Science 344:634–638

Vriens J, Watanabe H, Janssens A, Droogmans G, Voets T, Nilius B (2004) Cell swelling, heat, and chemical agonists use distinct pathways for the activation of the cation channel TRPV4. Proc Natl Acad Sci U S A 101:396–401

Wakabayashi S, Shigekawa M, Pouyssegur J (1997) Molecular physiology of vertebrate Na^+/H^+ exchangers. Physiol Rev 77:51–74

Wang W-H, Giebisch G (2009) Regulation of potassium (K) handling in the renal collecting duct. Pflugers Arch Eur J Physiol 458:157–168

Wang J, Morishima S, Okada Y (2003) IK channels are involved in the regulatory volume decrease in human epithelial cells. Am J Physiol Cell Physiol 284:C77–C84

Wang GX, Dai YP, Bongalon S, Hatton WJ, Murray K, Hume JR, Yamboliev IA (2005) Hypotonic activation of volume-sensitive outwardly rectifying anion channels (VSOACs) requires coordinated remodeling of subcortical and perinuclear actin filaments. J Membr Biol 208:15–26

Wangemann P, Liu J, Shen Z, Shipley A, Marcus DC (1995) Hypoosmotic challenge stimulates transepithelial K^+ secretion and activates apical I-Sk channel in vestibular dark cells. J Membr Biol 147:263–273

Webb BA, White KA, Grillo-Hill BK, Schonichen A, Choi C, Barber DL (2016) A histidine cluster in the cytoplasmic domain of the Na-H exchanger NHE1 confers pH-sensitive phospholipid binding and regulates transporter activity. J Biol Chem 291:24096–24104

Wehner F (2006) Cell volume-regulated cation channels. Contr Nephr 152:25–53

Weinstein AM (2013) Sodium and chloride transport: proximal nephron. In: Alpern RJ, Moe OW, Caplan MJ (eds) Seldin and Giebisch's the kidney. Elsevier, Amsterdam, pp 1081–1141

Welling PA (1995) Cross-talk and role of K_{ATP} channels in the proximal tubule. Kidney Internat 48:1017–1023

Welling LW, Welling DJ (1988) Relationship between structure and function in renal proximal tubule. J Electr Microsc Techn 9:171–185

Welling LW, Evan AP, Welling DJ, Gattone VH III (1983a) Morphometric comparison of rabbit cortical connecting tubules and collecting ducts. Kidney Intern 23:358–367

Welling LW, Welling DJ, Ochs TJ (1983b) Video measurement of basolateral membrane hydraulic conductivity in the proximal tubule. Am J Physiol Renal Physiol 245:F123–F129

Welsh MJ, McCann JD (1985) Intracellular calcium regulates basolateral potassium channels in a chloride-secreting epithelium. Proc Natl Acad Sci U S A 82:8823–8826

Willumsen NJ, Boucher RC (1991) Sodium transport and intracellular sodium activity in cultured human nasal epithelium. Am J Physiol Cell Physiol 261:C319–C331

Willumsen NJ, Larsen EH (1986) Membrane potentials and intracellular Cl^- activity of toad skin epithelium in relation to activation and deactivation of the transepithelial Cl^- conductance. J Membr Biol 94:173–190

Willumsen NJ, Davis CW, Boucher RC (1989) Intracellular Cl^- activity and cellular Cl^- pathways in cultured human airway epithelium. Am J Physiol Cell Physiol 256:C1033–C1044

Willumsen NJ, Vestergaard L, Larsen EH (1992) Cyclic AMP-and β-agonist-activated chloride conductance of a toad skin epithelium. J Physiol 449:641–653

Willumsen NJ, Davis CW, Boucher RC (1994) Selective responce of human airway epithelia to luminal but nor serosal solution hypertonicity. Possible role for proximal airway epithelia as an osmolality transducer. J Clin Invest 94:779–787

Windhager EE, Whittembury G, Oken DE, Schatzmann HJ, Solomon AK (1959) Single proximal tubules of the Necturus kidney. III. Dependence of H_2O movement on NaCl concentration. Am J Phys 197:313–318

Wittekindt OH, Dietl P (2018) Aquaporins in the lung. Pflugers Arch Eur J Physiol 471 (4):519–532. https://doi.org/10.1007/s00424-018-2232-y

Wong MMY, Foskett JK (1991) Oscillations of cytosolic sodium during calcium oscillations in exocrine acinar cells. Science 254:1014–1016

Worrell RT, Butt AG, Cliff WH, Frizzell RA (1989) Cell physiology - a volume-sensitive chloride conductance in human colonic cell-line T84. Am J Physiol Cell Physiol 256:C1111–C1119

Wu MM, Civan MM (1991) Voltage dependence of current through the Na,K-exchange pump of Rana oocytes. J Membr Biol 121:23–36

Wu X, Yang H, Iserovich P, Fischbarg J, Reinach PS (1997) Regulatory volume decrease by SV40-transformed rabbit corneal epithelial cells requires ryanodine-sensitive Ca^{2+}-induced Ca^{2+} release. J Membr Biol 158:127–136

Xia W, Yu Q, Riederer B, Singh AK, Engelhardt R, Yeruva S, Song P, Tian D-A, Soleimani M, Seidler U (2014) The distinct roles of anion transporters Slc26a3 (DRA) and Slc26a6 (PAT-1) in fluid and electrolyte absorption in the murine small intestine. Pflugers Arch Eur J Physiol 466:1541–1556

Yamamoto S, Ichishima K, Ehara T (2008) Regulation of volume-regulated outwardly rectifying anion channels by phosphatidylinositol 3,4,5-trisphosphate in mouse ventricular cells. Biomed Res-Tokyo 29:307–315

Yang CL, Angell J, Mitchell R, Ellison DH (2003) WNK kinases regulate thiazide-sensitive Na-Cl cotransport. J Clin Invest 111:1039–1045

Zadunaisky JA, Cardona S, Au L, Roberts DM, Fisher E, Lowenstein B, Cragoe EJ, Spring KR (1995) Chloride transport activation by plasma osmolarity during rapid adaptation to high salinity of Fundulus heteroclitus. J Membr Biol 143:207–217

Zhuo JL, Li XC (2013) Proximal nephron. Compr Physiol 3:1079–1123

Chapter 12
Fundamentals of Bicarbonate Secretion in Epithelia

Ivana Novak and Jeppe Praetorius

Abstract Certain epithelia secrete HCO_3^- to drive fluid secretion, to modify luminal pH and properties of secreted mucus, and to fulfill other functions of a given epithelium. Dysregulation of HCO_3^- secretion can lead to conditions such as malabsorption, acid/base disturbances, cystic fibrosis, biliary cirrhosis, peptic, and duodenal ulcers. In addition to the transport of HCO_3^- across the epithelium, epithelial cells also need to maintain intracellular pH, despite significant HCO_3^- extrusion and sometimes even despite exposure to external acid. In this chapter, we will introduce the main plasma membrane acid/base transporters and describe their role in general cellular homeostasis. The same transporters are also used in building the molecular machinery for vectorial HCO_3^- transport, i.e., bicarbonate secretion. We will highlight HCO_3^- secreting epithelia by examples from the digestive system (pancreas, salivary glands, hepatobiliary system, and duodenum), the renal collecting duct B-intercalated cell, as well as the choroid plexus epithelium of the brain. We seek an integrative approach to understand the HCO_3^- secretion processes by combining historical perspectives with molecular and genetic studies as well as studies of selected regulatory systems.

Keywords Secretory epithelia · Bicarbonate secretion · Pancreas · Salivary glands · Duodenum

I. Novak (✉)
Cell Biology and Physiology, Department of Biology, University of Copenhagen, Copenhagen, Denmark
e-mail: inovak@bio.ku.dk

J. Praetorius
Department of Biomedicine, Health, Aarhus University, Aarhus, Denmark
e-mail: jp@biomed.au.dk

© The American Physiological Society 2020
K. L. Hamilton, D. C. Devor (eds.), *Basic Epithelial Ion Transport Principles and Function*, Physiology in Health and Disease,
https://doi.org/10.1007/978-3-030-52780-8_12

12.1 Introduction

12.1.1 Overview

A number of epithelia in our body secrete significant amount of HCO_3^-, which is often accompanied by fluid secretion. One of the important early observations was made on the pancreas, which secretes pancreatic juice rich in HCO_3^- and poor in Cl^-, a relation between two anions that together with later studies on isolated pancreatic duct epithelium became important steps for understanding general cellular models for HCO_3^- secretion (Fig. 12.1a and b). The purpose of epithelial HCO_3^- secretion is manifold, as presented by examples of epithelia chosen for this chapter. For example, HCO_3^- secretion can set extracellular pH, buffer and protect cells against acids produced and secreted by cells during digestive or metabolic processes, and solubilize proteins and other macromolecules. Dysregulation of these processes can lead to serious diseases such as cystic fibrosis, biliary cirrhosis, peptic, and duodenal ulcers. In addition to transporting significant amounts of HCO_3^- from interstitium to lumen, epithelia face another major challenge—they have to defend their intracellular pH (pH_i). This fact is a challenge to scientists, as it is often difficult to study and separate the transepithelial acid/base transport as opposed to the transport across the single plasma cell membrane exerted for the purpose of pH_i regulation. In the first part of the chapter, we will introduce the main H^+/HCO_3^- transporters and describe their role in general cellular acid/base homeostasis. These "building blocks" will then be used to equip epithelial cells so that they can perform vectorial HCO_3^- transport, i.e., secretion. Other ion channels and transporters necessary for overall transepithelial HCO_3^- transport will be given in specific tissues/organs. We will focus on HCO_3^- secreting epithelia of the digestive system (pancreas, salivary glands, hepatobiliary system, and duodenum), choroid plexus epithelium of the brain, and renal collecting ducts. Combining the historical perspectives with molecular and genetic studies in this chapter, we hope to mark a more integrative approach that will help us to understand the challenges of HCO_3^- secretion.

12.1.2 Cellular Acid/Base Homeostasis

In secretory epithelial cells, as in most other cells, the intracellular pH is maintained within the range 7.1–7.4, as most cellular processes have a pH optimum within this range (Boron and Boulpaep 2017). The balance between production, consumption, and transmembrane movement of acid/base equivalents determines the intracellular pH (pH_i). The cellular buffering capacity is not regulating the steady-state pH_i, but determines the size and rate of the pH change inflicted by an acute acid or base challenge (Boron and Boulpaep 2017). The intrinsic buffering capacity is set by the cellular weak acid/base pairs such as phosphate, bicarbonate, and anionic proteins.

Fig. 12.1 (a) The classical electrolyte excretion curves showing the relationship between secretory rates and electrolyte concentrations in pancreatic juice collected from the dog pancreas stimulated with secretin. Reproduced with permission from (Bro-Rasmussen et al. 1956). Similar excretory curves were obtained for the cat pancreas (Case et al. 1969). Similar excretory curves are expected for the human pancreas. (b) The cellular model of ion transport in a pancreatic duct cell as established from electrophysiological studies of isolated perfused rat pancreatic ducts. Reproduced with permission from (Novak and Greger 1988b). (c) The relation between secretory rates and HCO_3^- concentrations in the pancreatic juice of various species. Secretion was stimulated with secretin and secretory rates were corrected for body weights. Reproduced with permission from (Novak et al. 2011)

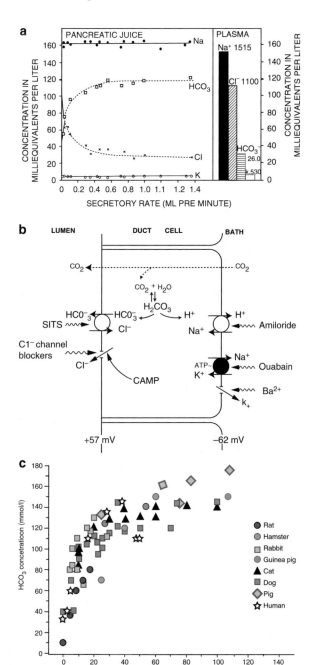

In addition to the intrinsic buffers, the open buffer system of CO_2/HCO_3^- enables very efficient buffering of pH. Virtually all cells express plasma membrane ion transporters that contribute to cellular pH homeostasis. Some of these exploit the inward gradient for Na^+ to drive acid or base transport. Other transporters are dependent on, for example, the Cl^- gradient, the HCO_3^- gradient, electrical gradient, or ATP hydrolysis to drive the transport. Also, some ion channels may contribute to acid/base transport. Most cells express acid/base transporters (and channels), depending on the function of the specific cell, and especially in HCO_3^--secreting epithelia a great variety of such transporters are found.

12.1.2.1   Sodium Hydrogen Exchangers (NHEs, *SLC9*)

Most of the NHEs mediate the electroneutral exchange of intracellular H^+ for extracellular Na^+ given typical ionic distribution and intracellular pH. Of the nine members of the NHE gene family, only NHE1 (*SLC9A1*) seems ubiquitously expressed and is therefore regarded as the central cellular acid extruder (Orlowski and Grinstein 2004). Linked to this function, NHE1 plays a role in cell volume regulation, cell migration, and cell cycle regulation in various health cells, including cancer cells (Flinck et al. 2018). NHE2 and NHE3 (*SLC9A2* and *A3*, respectively) are luminal proteins mainly found in Na^+ absorptive epithelia. Nevertheless, both can be found alongside potent HCO_3^- secretory machinery in the alkaline secretory cells of the stomach surface, duodenal villus cells, and exocrine gland ducts. In these cases, NHE2 and NHE3 could potentially favor HCO_3^- absorption rather than secretion (Praetorius et al. 2000). The last plasma membrane *SLC9* member, NHE4 (*SLC9A4*), is a basolateral alternative to NHE1 in specialized cells of the kidney, stomach, salivary glands, and liver. NHEs are inhibited by amiloride and its derivatives, as well as cariporide with the highest potency toward NHE1 (Scholz et al. 1995).

12.1.2.2   Sodium Bicarbonate Cotransporters (NBCs and NDCBEs, *SLC4*)

The electrogenic Na^+-HCO_3^- cotransporter NBCe1 (*SLC4A4*) was the first Na^+-HCO_3^- cotransporter to be identified at the molecular level (Romero et al. 1997). NBCe1 mediates electrogenic Na^+-HCO_3^- cotransport with either 1:2 or 1:3 stoichiometry depending on the tissue and is localized to the basolateral surface in epithelial cells involved in vectorial HCO_3^- transport in the kidney, intestine and pancreatic ducts (Boron and Boulpaep 1983; Schmitt et al. 1999). The second electrogenic NBC, i.e., NBCe2 (*SLC4A5*), displays similar transport properties as NBCe1, also with varying Na^+:HCO_3^- stoichiometries (Pushkin et al. 2000; Sassani et al. 2002; Virkki et al. 2002). NBCe2 expression pattern is more controversial; NBCe2 is described in epithelial tissues such as liver, testis, kidney, lung, and the choroid plexus, where it is localized to the luminal membrane (Abuladze et al. 2004; Pushkin et al. 2000; Virkki et al. 2002). The Na^+-dependent exchange of Cl^- and HCO_3^- has been found in various tissues (Boron and Knakal 1992; Liu et al. 1990; Schlue and Deitmer 1988). Two Na^+-dependent Cl^- and HCO_3^- exchangers have been described NDCBE1 (*SLC4A8*) and NCBE (*SLC4A10*) (Grichtchenko et al. 2001; Virkki et al. 2003; Wang et al. 2000). These transporters were characterized as electroneutral, DIDS-sensitive, and work with an apparent stoichiometry of $1Na^+:1Cl^-:2HCO_3^-$, where Na^+ and HCO_3^- are normally imported and Cl^- extruded from the cells. The Cl^- dependence of NCBE has been challenged by compelling experiments in a study by Parker and colleagues (Parker et al. 2008). The only epithelial expression sites described thus far is the choroid plexus and connecting tubules for NCBE and hepatobiliary system for NDCBE1 (Grichtchenko et al. 2001; Wang et al. 2000; Strazzabosco et al. 1997; Banales et al. 2006b). At these sites, the transporters may well take part in transepithelial movement of both Na^+ and HCO_3^-.

The electroneutral NBC, NBC3, or NBCn1 (Choi et al. 2000; Pushkin et al. 1999), also belongs to the *SLC4* gene family (*SLC4A7*). As the name indicates, the apparent $Na^+:HCO_3^-$ stoichiometry is 1:1. This means that it is normally importing the two ions into cells. NBCn1 is expressed in the basolateral membranes in many epithelia including HCO_3^- secretory epithelia, such as the stomach surface cells, duodenum, colon, and choroid plexus. Except for epithelial variants of NBCn1, the NBCs and NDCBEs are inhibited by stilbene derivatives such as DIDS and SITS (Aalkjaer and Cragoe Jr. 1988; Boedtkjer et al. 2006; Bouzinova et al. 2005; Odgaard et al. 2004; Praetorius et al. 2001, 2004a). The drug S0859 seems to be a general inhibitor of Na^+-HCO_3^- cotransporter (Larsen et al. 2012). For further details on NBCs see Chap. 4 of Vol. 3.

12.1.2.3 Classical Anion Exchangers (AE, *SLC4*)

AE1-3 are Na^+-independent Cl^-/HCO_3^- exchangers which are electroneutral and all belong to the *SLC4* gene family (*SLC4A1-3*). AE1, or band-3 protein, was first demonstrated in red blood cells where it exports HCO_3^- (Lux et al. 1989). After entry into the red blood cells, CO_2 is hydrated and carbonic anhydrases accelerate the subsequent formation of HCO_3^- and H^+. The H^+ is buffered by hemoglobin and HCO_3^- extrudes in exchange for Cl^- (the chloride shift) by AE1. Type-A-intercalated cells of the renal collecting duct express an epithelial variant of AE1. This basolateral plasma membrane anion exchanger may play a supportive role for the apical acid secretion by extruding HCO_3^- to the blood side (Kollert-Jons et al. 1993). Another member, AE2 is expressed basolaterally in most epithelia, except for the hepatobiliary system, and is involved in the protection of the cells against alkalization (Alper 2006). AE2 deletion in mice results in a severe phenotype with growth retardation, gastrointestinal dysplasia, biliary cirrhosis, and death before weaning (Gawenis et al. 2004; Concepcion et al. 2013). AE3 is expressed mainly in excitable tissues, such as brain and heart, but is also found in gastrointestinal enterocytes (Yannoukakos et al. 1994). Human AE3 point mutations have been associated with seizures, most likely as a consequence of the impaired neuronal pH_i regulation (Hentschke et al. 2006). The AE's like NBC's and NDCBE's are inhibited by DIDS.

12.1.2.4 Promiscuous Anion Exchangers

A separate gene family of anion exchangers with a more promiscuous anion transport profile has taken a central position in understanding transepithelial HCO_3^- movement. The *SLC26* genes give rise to 12 transporters, which are expressed in many different tissues and mediate very diverse functions, transporting anions, such as sulfate, oxalate, phosphate, chloride, bicarbonate, iodide, and formate to a variable extent (Alper and Sharma 2013; Mount and Romero 2004). Several of the gene family members encode HCO_3^- transporters: *SLC26A3,-4,-6,-7*, and *-9* (Alper and Sharma 2013). The stoichiometry and thereby electrogenic properties of the HCO_3^- transport some of these proteins is debated (for detailed review (Alper and Sharma 2013; Cordat and Reithmeier 2014)). While DRA (Down-Regulated in Adenoma, *SLC26A3*) and Pendrin (*SLC26A4*) mediate electroneutral Cl^-/HCO_3^- transport (Chernova et al. 2003; Shcheynikov et al. 2008), CFEX/PAT1 (*SLC26A6*), *SUT2* (*SLC26A7*), and *SLC26A9* has been described as both electrogenic and electroneutral Cl^-/HCO_3^- exchangers, the latter two in some reports even as ion channels, probably depending on expression system and species (Chernova et al. 2005; Petrovic et al. 2004; Kim et al. 2005; Kosiek et al. 2007). Dysfunction of the intestinally expressed DRA produces congenital chlorodiarrhoea (Hoglund et al. 1996), which is caused by reduced luminal Cl^-/HCO_3^- exchange in the intestinal tract (Melvin et al. 1999). Pendrin, *SLC26A4*, is defective in the Pendred syndrome, in which patients suffer from impaired hearing and thyroid function. The symptoms result from dysfunctional thyroid Cl^-/I^- exchange, defective Cl^-/HCO_3^- exchange in the stria vascularis of the inner ear, and mice probably also decreased renal collecting duct HCO_3^- reabsorption (Masmoudi et al. 2000; Royaux et al. 2000, 2001). PAT-1 (or CFEX) also exchanges Cl^- with HCO_3^- and seems necessary for normal pancreatic and duodenal bicarbonate secretion (Ko et al. 2002b; Shcheynikov et al. 2006). Finally, deletion of *SLC26A9* has been shown to impair luminal alkalization in the gastric mucosa (Demitrack et al. 2010) and duodenal HCO_3^- secretion as well as worsening intestinal function and survival of CFTR-deficient mice (Liu et al. 2015)

12.1.2.5 Anion Channels

One of the long-lasting challenges in the bicarbonate transport field is the question of whether Cl^-/anion channels can conduct HCO_3^- in physiological conditions. Many patch-clamp studies of the cystic fibrosis transmembrane regulator Cl^- channel (CFTR) have shown that in physiological-like conditions, the permeability ratio

$PHCO_3^-/PCl^-$ is 0.2–0.5, implying that secretion of Cl^- would dominate. Several studies suggest that CFTR could become more HCO_3^- permeable if intracellular Cl^- was reduced (Ishiguro et al. 2009; Park et al. 2010, 2012). It is proposed that some cell volume/Cl^- regulatory mechanisms could be contributing to the regulation of HCO_3^- permeability and this will be discussed in relation to the pancreas (see Sect. 12.2.2.1). For additional information about CFTR, see Chaps. 15 and 16 of Vol. 3. Bicarbonate secreting epithelia also express Ca^{2+} activated Cl^- channels, CaCC. The identity of CaCC channels has been difficult to pinpoint (see Duran et al. 2010). After suggestions of CCl-2 and bestrophins, the TMEM16/ANO family was discovered (Caputo et al. 2008; Schroeder et al. 2008; Yang et al. 2008). One of the members, TMEM16A/ANO1, is regarded as a good candidate for CaCC in epithelia and again relevance for HCO_3^- secretion has been raised. Modulation of the channel HCO_3^- permeability by Calmodulin, and not With No Lysine kinases (WNKs), has been tested and discussed (Jung et al. 2013; Yu and Chen 2015). Further information about TMEM16 can be found in Chap. 17 of Vol. 3. Recently, it was proposed that pore dilatation of CFTR, TMEM16A/ANO1, and glycine receptor increases $PHCO_3^-/PCl^-$ (Jun et al. 2016). Interestingly, Bestrophins (BEST1) have relatively high HCO_3^- permeability in HEK293 cells (Qu and Hartzell 2008; Kane Dickson et al. 2014). Bestrophin function is well documented in retinal diseases and in HCO_3^--secreting epithelia, it is less clear (see below). The volume-regulated anion channels, VRACs, are ubiquitously expressed channels composed of LRRC8 heteromers (Jentsch 2016; Jentsch et al. 2016) (see Chap. 11 of this volume). Volume regulation is important in epithelia (Pedersen et al. 2013b), though the direct role of VRAC in ion/HCO_3^- secretion is difficult to unmask (Catalan et al. 2015).

As mentioned above, SLC26A9 also behaves as a Cl^- channel, which is constitutively active and has a minimal conductance to HCO_3^-, but HCO_3^- can facilitate Cl^- transport (Loriol et al. 2008). SLC16A9 is often co-expressed with CFTR and there may be direct physical interactions with CFTR mediated by PDZ proteins (Bertrand et al. 2017). Potential role of SLC16A9 channels in cystic fibrosis and other diseases is proposed, but the detailed role of SLC26A9 as a channel, anion exchanger, or modulator of other channels/transporters is yet to be elucidated in specific tissues (Balazs and Mall 2018; Liu et al. 2018).

12.1.2.6 Vacuolar H⁺-ATPase and H⁺/K⁺-ATPase

Vacuolar H^+-ATPases, (V-ATPases), are expressed ubiquitously in the lysosomal system, but certain cells are known to express V-ATPases on the plasma membrane (Breton and Brown 2013; Cotter et al. 2015). The V-ATPases are large transmembrane protein complexes consisting of several subunits and resembles the mitochondrial ATP synthases. Only complexes with the certain specific composition of subunits can reside in the plasma membrane. The energy resulting from ATP hydrolysis is exploited to move H^+ across the membrane without counterion transport. Hence, V-ATPases are electrogenic. Epithelial cells such as renal intercalated cells and epididymis use V-ATPase for transepithelial transport (Brown et al. 2009; Pastor-Soler et al. 2008). The V-ATPase is inhibited by bafilomycin and concanamycin (Huss and Wieczorek 2009). A separate group of P-type ATPases mediate the exchange of H^+ and K^+, i.e., the H^+/K^+-ATPases. The H^+/K^+-ATPases are classified in two families: gastric and non-gastric (also called colonic); and α-subunits are coded by *ATP4A* and *ATP12A (ATP1AL1)* genes, respectively (Modyanov et al. 1991, 1995; Sachs et al. 2007; Forte and Zhu 2010; Sangan et al. 2000). The pump complex consists of two α- and two β-subunits, whereby the gastric α assembles with gastric β subunit (*ATP4B*), while non-gastric α subunits can assemble with gastric or Na^+/K^+-ATPase β subunits. The gastric H^+/K^+-ATPase is best known from the gastric corpus/fundus glands, where it mediates potent H^+ secretion for gastric acid, but is also expressed in kidney and cochlea; and the non-gastric form is expressed in colon, kidney, skin, and placenta (Pestov et al. 1998). Some HCO_3^- secreting epithelia also express these pumps (see below). Proton pump inhibitors such as omeprazole are potent inhibitors of gastric H^+/K^+-ATPases, while high concentrations of potassium-competitive acid blockers and ouabain probably inhibit the non-gastric type (Grishin and Caplan 1998; Grishin et al. 1996; Swarts et al. 2005).

12.1.2.7 Carbonic Anhydrases

HCO_3^- and H^+ are the major biologically relevant base and acid, respectively. The most potent cellular pH homeostasis and base secretion relies on a steady supply of these ion species. The hydration of CO_2 occurs spontaneously at a sufficient rate, while the uncatalyzed hydrolysis of H_2CO_3 is quite slow for biological purposes. Thus, the carbonic anhydrases, which catalyze the conversion of $CO_2 + H_2O$ to HCO_3^- and H^+, are of major importance both for pH homeostasis and bicarbonate secretion and there are 14 different CA isoenzymes: CA I–III and VII are cytosolic; IV, IX, and XII are membrane associated; V is mitochondrial and VI is secreted (Supuran 2008). The canonical and ubiquitously expressed form is CAII. This enzyme has one of the fastest turnovers in mammalian biology (Maren 1962) and is a soluble cytosolic protein. In recent years, it has become evident that other forms of carbonic anhydrases are resident in the plasma membrane, either with extracellular enzyme activity (GPI anchored) or with cytosolic sub-membrane activity. Acetazolamide has been used to block carbonic anhydrases for decades, and seem to block both cytosolic and membrane-associated enzyme forms. In any case, inhibition of HCO_3^- formation by this drug has a profound impact on epithelial bicarbonate secretion in several tissues. Interestingly, some investigators have reported the physical as well as functional interaction between carbonic anhydrases and bicarbonate transporters, such as AE2 and NBCe1 and such interactions could facilitate transport by securing the substrate to the HCO_3^- transporters (McMurtrie et al. 2004; Becker et al. 2014).

12.1.3 Vectorial Bicarbonate Transport

It is evident that bicarbonate secreting epithelia need to employ one of the abovementioned mechanisms to extrude HCO_3^- from the cytosol into the luminal/apical compartment. Epithelia with potent bicarbonate extrusion are generally equipped with anion channels, with promiscuous anion exchangers or with an electrogenic Na^+-HCO_3^- cotransporter at the site of exit. However, to avoid cellular acidification, the cells must have just as effective means of getting rid of the H^+ across the opposite plasma membrane to avoid damaging acidification and to sustain the production of new HCO_3^-. So, in the case of luminal HCO_3^- secretory epithelia, the cells must have sufficient acid extrusion mechanisms such as NHE1/4 or an H^+-ATPase. Alternatively, the luminal HCO_3^- secretion can be supported by basolateral HCO_3^- entry through any of the Na^+ driven HCO_3^--transporters.

In the following sections, selected epithelia with distinct acid/base transport characteristics will be described: pancreatic ducts, salivary glands, hepatobiliary system, duodenum, collecting duct, and choroid plexus. While the first four organs have high to very high HCO_3^- output, the choroid plexus epithelium has an intermediate HCO_3^- output, and the terminal renal collecting ducts exemplifies epithelia with little or no transepithelial HCO_3^- movement, where a subset of specialized cells mediate HCO_3^- secretion depending on the acid/base status. Thus, similarities and differences in molecular machinery for HCO_3^- transport between these tissues may help establishing hypotheses regarding the functional roles of specific acid-base transporters and ion channels.

12.2 Pancreas

12.2.1 The Prototype of a Bicarbonate Secretor Is a Complex Gland: Integrated Function and Morphology

The pancreas and other exocrine glands are composed of at least two main types of epithelia—secretory acini/endpieces and excretory ducts. Thaysen and coworkers (Bro-Rasmussen et al. 1956) proposed the two-stage hypothesis of secretion for complex exocrine glands and this can still be used as a starting point to understand their integrated function. Basically, it says that acini/endpieces secrete fluid similar to that in plasma in their electrolyte composition, and they secrete macromolecules such as enzymes. The ducts may modify this secretion, and in the pancreas, they do so by adding a secretion of their own (Fig. 12.1a). Pancreatic ducts are generally regarded as leaky epithelia expressing aquaporins and they are able to secrete a fluid that is HCO_3^--rich and alkaline (Bro-Rasmussen et al. 1956; Steward and Ishiguro 2009; Wilschanski and Novak 2013; Ishiguro et al. 1998; Fernandez-Salazar et al. 2004; Novak et al. 2011; Wang et al. 2015). In humans, the maximum HCO_3^- output in the secretin-stimulated gland is about 500 μmol/h/g pancreas tissue weight.

This output would be at least five times higher if it is assumed that it arises from pancreatic ducts contributing around 20% to pancreas mass in humans. The well-established function of pancreatic bicarbonate secretion is that it contributes to buffering of acid chyme entering duodenum; the other contributors are duodenal epithelium and bile duct epithelium (Ainsworth et al. 1992). Recently, it has been discussed whether the bicarbonate secretion has an additional function already within the pancreas (Hegyi et al. 2011; Wilschanski and Novak 2013; Novak et al. 2013). That is, there are some indications that the acinar secretion might be acidic and the function of adjoining ducts may be therefore to alkalinize this acinar secretion very early in its passage through the duct tree, and thus prevent premature activation of digestive enzymes and maintain the balance in exo-/endocytosis in acini (Freedman and Scheele 1994; Freedman et al. 2001; Behrendorff et al. 2010; Hegyi and Petersen 2013). The third possible function for bicarbonate secretion is to solubilize mucins, and although this has not been proven for the pancreas (Quinton 2008, 2010), it has been shown that the very early key symptom in cystic fibrosis is reduced HCO_3^- secretion and mucoviscidosis in the pancreas (Andersen 1938; Kopelman et al. 1985, 1988).

Pancreatic juice collected from the pancreas stimulated with the main "secretagogue" in many species, secretin, has electrolyte composition that depends on secretory rates (Fig. 12.1a). Basically, at high secretory rates, the pancreatic juice is rich in HCO_3^- and poor in Cl^- and as secretion decreases HCO_3^- falls and Cl^- increases in a mirror-like fashion. Na^+ concentrations do not change with flow rate and are plasma-like. K^+ concentrations are similar to or higher than in plasma.

Over many years, it has been regarded that some animals produce juice low in HCO_3^- concentrations (mice, rats, rabbits), while the pancreas of other species produces secretion with high HCO_3^- concentrations (man, dog, cat, pig, and guinea pig). Nevertheless, close analysis shows that HCO_3^- concentrations in pancreatic juice among different species may depend on the relative proportion of acinar to duct cells contributing to the final secretion. If corrected for this, it becomes apparent that secretion of all species, summarized in Fig. 12.1c, falls within one excretory curve, which implicates similar secretory mechanisms in all species. But why does HCO_3^- fall and Cl^- increases with the falling secretory rate (Fig. 12.1a)? These curves are often pictured but are rarely elaborated. One explanation is provided by the ad-mixture hypothesis, which states that the final secretion is a mix of fluids with different compositions (acini and ducts) (Fig. 12.2a). Another theory is the exchange theory (implying exchange of HCO_3^- for Cl^-). This is most apparent at low secretory rates, that is, when secretion from acini and proximal ducts is low, distal ducts are not overridden by incoming secreted fluid and thus they use their full capacity to simply exchange HCO_3^- for Cl^-, a process referred to recently as "HCO_3^- salvaging". This HCO_3^-/Cl^- exchange was demonstrated many years ago on the cat main pancreatic duct (see below) that was perfused with various solutions and could carry out such an exchange (Case et al. 1969). Yet another, so far theoretical possibility is that pancreatic ducts can also secrete H^+, a process most obvious in low secretory rates. These explanations might not be mutually exclusive and they implicate that the ductal tree is heterogeneous.

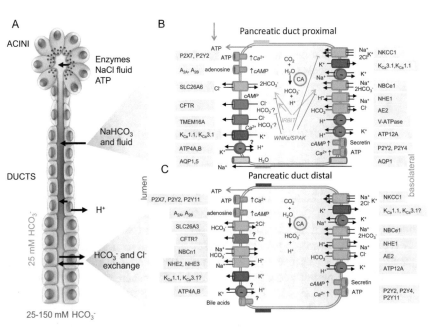

Fig. 12.2 (**a**) Schematic diagram showing simplified pancreas with acini, proximal and distal ducts, and pancreatic juice with a typical range of HCO_3^- concentrations. Inserts show two types of cells with the cellular models for a HCO_3^- secreting cell (**b**) and a cell that is exchanging HCO_3^- for Cl^- (**c**). Interstitial/plasma HCO_3^- concentration is 25 mmol/l, pancreatic juice contains 25–150 mmol/l HCO_3^- and depends on the secretory rate and stimulation (see Fig. 12.1). Molecular identities of ion transporters, channels, and receptors are discussed in the text; question marks indicate unclear identities, localization, or functions. Intracellular signaling is simplified, stimulatory (green) and inhibitory (red) pathways, and other interaction between cAMP and Ca^{2+} signaling (double-headed arrow) are discussed in Sect. 12.2.4.3. The ion transport model for pancreatic acinar cells is reviewed elsewhere (Heitzmann and Warth 2008) and is similar to Cl^- secreting salivary acini (see Sect. 12.3)

The ductal tree comprises 5–20% of the pancreas tissue mass, depending on the species, and morphologically ducts are quite different—progressing from intercalated, small intralobular, larger intralobular, inter-/extralobular and eventually joining into main ducts that might join the bile duct in some species (Kodama 1983; Ashizawa et al. 1997; Githens 1988; Bouwens and Pipeleers 1998; Gmyr et al. 2004). The cell types progress from the flat small cells with large long primary cilia to cuboidal and later columnar cells with short primary cilia. Large ducts contain several cell types, including mucus-secreting cells and single endocrine cells. The centroacinar cells are very flat cells extending from intercalated cells into the acinar lumen and their physiological function is not established. Recently, pancreatic duct glands have been described as a potential progenitor niche (Yamaguchi et al. 2015).

Given the morphological heterogeneity of the ductal tree, one could expect some functional heterogeneity. However, functional studies are limited to ducts that can be isolated or micro-dissected from animals and to culture models, mostly pancreatic ductal adenocarcinoma cell models. Nevertheless, combined with data acquired from transgenic cell/animal models, immunohistochemistry, and many other techniques, coherent models can be proposed.

12.2.2 HCO_3^- and H^+ Transporters in Pancreatic Ducts

12.2.2.1   CFTR and Cl^-/HCO_3^- Exchangers

The first studies of cellular mechanisms for pancreatic duct HCO_3^- transport showed, surprisingly, that secretin/cAMP activated Cl^- channels on the luminal membranes of isolated rat pancreatic ducts (Fig. 12.1b) (Novak and Greger 1988b; Gray et al. 1988). Almost in parallel, the cystic fibrosis transmembrane conductance regulator, CFTR, was discovered (Riordan et al. 1989; Kerem et al. 1989), and it was shown to have properties of a Cl^- channel, also in the pancreatic ducts (Tabcharani et al. 1991; Gray et al. 1993) (Fig. 12.2b). Subsequently, CFTR was immunolocalized in human and rodent pancreas to intercalated and small intralobular ducts, which also express other key proteins in HCO_3^- secretion, aquaporins and carbonic anhydrases (Hyde et al. 1997; Marino et al. 1991; Kumpulainen and Jalovaara 1981; Burghardt et al. 2003). Since pancreatic HCO_3^- and fluid secretion is a defect in cystic fibrosis, and since the underlying signatures are mutations in CFTR, the channel has been considered as the key element in the pancreatic duct secretion (Wilschanski and Novak 2013). Nevertheless, the question whether and how CFTR Cl^- channels could transport HCO_3^- has been a long-lasting challenge (see Sects. 12.1.2 and 12.2.4.3).

The first proposal for HCO_3^- exit pathway was the Cl^-/HCO_3^- exchange mechanism coupled to luminal Cl^- channels, thus allowing Cl^- recirculation and net HCO_3^- secretion, though this would only account for 60–80 mM HCO_3^- in secretion (Novak and Greger 1988a, b). Through intensive efforts in molecular and cell biology, the following Cl^-/HCO_3^- exchangers were identified. The anion exchanger *SLC26A3*, also known as DRA and *SLC26A6*, also known as PAT-1, was found expressed on the luminal membrane of large mouse and human pancreatic ducts and (Greeley et al. 2001; Lohi et al. 2000). These two exchangers have different stoichiometry showing $Cl^-:HCO_3^-$ of 2:1 for *SLC26A3* and 1:2 for *SLC26A6*. It was proposed that *SLC26A3* was expressed in more distal ducts. *SLC26A6* was more proximal on the luminal membrane of intralobular ducts, and rarely on larger ones (Ko et al. 2002b, 2004). Theoretically, the Cl^-/HCO_3^- exchange of 1:2 for *SLC26A6* would be thermodynamically more favorable for HCO_3^- secretion, while *SLC26A3* would favor HCO_3^- absorption (Fig. 12.2b and c). There is a functional coupling between *SLC26A6* and CFTR, and this involves the R domain of CFTR and sulfate transporter anti-sigma (STAS) domains of *SLC26A6* exchanger (Ko et al. 2004; Dorwart et al. 2008; Wang et al. 2006; Stewart et al. 2009). Nevertheless, studies using knockout strategy for *SLC26A* exchanger showed some interdependence between the two isoforms and varied effects on duct/pancreas secretion (Ishiguro et al. 2007; Song et al. 2012).

AE2 (*SLC4A2*), another anion exchanger from the *SLC4* family, was demonstrated, usually on the basolateral membranes, in a number of pH_i studies in pancreatic ducts (Stuenkel et al. 1988; Zhao et al. 1994; Rakonczay Jr. et al. 2006). However, immunohistochemical studies are not congruent as to which membrane the transporter is localized (Hyde et al. 1999; Roussa et al. 2001; Kulaksiz and Cetin 2002). Most likely, AE2 is more involved in pH_i regulation rather than transepithelial HCO_3^- transport.

SLC26A9 is also weakly expressed in the pancreas (Liu et al. 2015), but whether and how it contributes to pancreatic HCO_3^- and fluid secretion remains to be explored in detail. Nevertheless, interesting speculations of whether SLC26A9 as a Cl^- channel potentiates HCO_3^-/Cl^- exchange, or is itself the exchanger and/or regulates CFTR (Balazs and Mall 2018; Liu et al. 2018).

12.2.2.2 Calcium-Activated Cl^- channels

In addition to cyclic AMP regulated secretion, a number of studies show that agonists such as acetylcholine and extracellular nucleotides (see below) act via Ca^{2+} signaling to stimulate Ca^{2+}-activated Cl^- channels (CaCC) (see Chap. 17 of Vol. 3), and thus could support duct secretion (Gray et al. 1989, 1994; Pahl and Novak 1993; Hug et al. 1994; Winpenny et al. 1998; Szalmay et al. 2001; Pascua et al. 2009). In pancreatic ducts, studies on human duct cell lines show that they express TMEM16A/ANO1, which targets to the luminal membrane upon stimulation and gives rise to the secretory potential in polarized duct epithelia (Wang et al. 2013; Wang and Novak 2013). This channel could be relevant for pancreatic HCO_3^- secretion (see Sect. 12.1.2). There is also one immunohistological study on human pancreatic sections showing that the ANO1/DOG-1 antibody localizes to centroacinar and small ducts cells, and the channel is grossly over-expressed in pancreatic cancer and other gastrointestinal stromal tumors (Bergmann et al. 2011; Sauter et al. 2015). BEST1 is expressed in CFPAC-1 cells (Marsey and Winpenny 2009), but whether it plays an important role as CaCC in normal pancreatic ducts is not clear.

12.2.2.3 NBCs, NHEs, and Carbonic Anhydrases

HCO_3^- transport across the luminal membrane, whatever the mechanism is, relies on the provision of cellular HCO_3^-. One well-supported solution is the import of HCO_3^- across the basolateral membrane via a Na^+ coupled process, i.e., Na^+-HCO_3^- cotransporters, NBC. One NBC isoform was cloned from the pancreas, pNBC (NBCe1B, *SLC4A4*) and it transports 1 Na^+: 2 HCO_3^- and putative inhibitor H_2DIDS inhibits about 50% of duct secretion (Abuladze et al. 1998; Ishiguro et al. 1998; Choi et al. 1999). pNBC is found on the basolateral membranes of epithelial cells of a duct tree in the human pancreas, though in rat pancreas it was also acinar and duct labeling was occasionally on both membranes (Marino et al. 1999; Satoh et al. 2003). If pancreatic ducts were relying predominantly on this transporter, secretion would be highly dependent on the provision of HCO_3^- from interstitium/plasma rather than endogenous CO_2 production, CA activity, and H^+ extrusion mechanism. This was found to be the case for isolated cat pancreas and guinea pig ducts, though H_2DIDS inhibited about 50% of secretions (Schulz 1971; Case et al. 1970; Ishiguro et al. 1998).

Another isoform, the electroneutral NBCn1 (NBC3, *SLC4A7*), is also expressed in pancreas (Damkier et al. 2006), though one study shows that in mouse ducts it interacts with CFTR, it is inhibited by cAMP and therefore should be placed on the luminal membrane and possibly regulate HCO_3^- salvage (Park et al. 2002b) (Fig. 12.2c).

An alternative or additional solution for cellular HCO_3^- (and H^+) provision is carbonic anhydrase (CA) catalyzed hydration of CO_2, provided from metabolism and/or HCO_3^-/CO_2 buffer system. Isoforms CAII, IV, VI, IX, and XII are expressed in the human pancreas and cultured duct tumor cells (Kumpulainen and Jalovaara 1981; Nishimori et al. 1999; Nishimori and Onishi 2001). CAII and CAIV interact with H^+/HCO_3^- transporters, however, localization of the CA isoforms do not always match the predicted localization of the transporters in pancreatic ducts. For example, CAII is found intracellularly and on the luminal membrane (Alvarez et al. 2001), and it seems to interact with NHE1 and NBC3 (Li et al. 2002; Loiselle et al. 2003). CAIV is expressed in the luminal membrane of the ductal tree (centroacinar cells and in intercalated, intralobular, and interlobular ductal cells) (Fanjul et al. 2004; Mahieu et al. 1994), but it interacts with NBCe1 in expression studies (Alvarez et al. 2003). CA IX and XII are expressed on the basolateral membranes of normal and pathological samples of the pancreas (Kivela et al. 2000; Juhasz et al. 2003). Carbonic anhydrases have been somewhat neglected in pancreatic duct studies in recent years. Nevertheless, CAs are key enzymes in pancreatic duct function, as their inhibition leads to marked effects on pH_i and pancreatic secretion (Hollander and Birnbaum 1952; Case et al. 1979; Cheng et al. 1998; Steward et al. 2005; Rakonczay Jr. et al. 2006).

Intracellular H^+, generated from CA activity or metabolism, can be extruded out of the cell by a Na^+/H^+ exchanger (NHE). Such exchanger was proposed based on the observation that pancreatic duct secretion could be maintained efficiently without HCO_3^- by a number of weak lipid-soluble acids, such as acetate (Schulz et al. 1971; Case et al. 1979). NHE, sensitive to amiloride and derivatives, has been detected in many studies monitoring secretion of the whole pancreas in different species and later on isolated pancreatic ducts (Wizemann and Schulz 1973; Veel et al. 1992; Novak and Greger 1988a; Ishiguro et al. 1998; de Ondarza and Hootman 1997; Fernandez-Salazar et al. 2004; Szucs et al. 2006). Nevertheless, amiloride type inhibitors could decrease secretion by about 20–50%. One of the NHE isoforms, the ubiquitous NHE1 (*SLC9A1*) is the major pH_i regulator. In functional studies, it was revealed that NHE contributed significantly to pH_i regulation in many duct preparations including pig, guinea pig, rat and mice ducts and human duct cell lines

(Veel et al. 1992; Szucs et al. 2006; de Ondarza and Hootman 1997; Ishiguro et al. 2000; Novak and Christoffersen 2001; Lee et al. 2000; Demeter et al. 2009; Rakonczay Jr. et al. 2006; Olszewski et al. 2010). There is some molecular evidence for NHE1 expression is normal ducts and localization appears to be on the basolateral membrane (Lee et al. 2000; Roussa et al. 2001), thus function in secretion (and pH_i regulation) could be supported. In addition, the NHE2 and 3 isoforms are expressed on the luminal membrane of main ducts and are proposed to interact with CFTR via PDZ domains (Lee et al. 2000; Ahn et al. 2001; Marteau et al. 1995). These exchangers would then not support secretion, but conduct HCO_3^- salvage (or pH_i regulation) (Fig. 12.2c).

12.2.2.4 Proton Pumps

The above models do provide a number of answers, but still, we are left with the problem of how to explain high HCO_3^- concentrations and why inhibitors of NHE1, NBC, and CA are relatively ineffective in blocking secretion (Fernandez-Salazar et al. 2004; Grotmol et al. 1986). The above transporters and ion channels rely on gradients that are created by the Na^+/K^+-ATPase. Another solution to create HCO_3^- / H^+ gradients would be to extrude H^+ via the V-ATPase. In one early study V-ATPase on the basolateral membrane was proposed (Villanger et al. 1995), and V-ATPase was detected on basolateral membrane of intralobular ducts, although occasionally some cells had luminal staining (Roussa et al. 2001). A number of functional studies gave contradictory findings (Zhao et al. 1994; Ishiguro et al. 1996; de Ondarza and Hootman 1997; Cheng et al. 1998), perhaps depending on which parameters were measured. It seems that the contribution of the pump to pH_i regulation is relatively small (compared to NHE1), but inhibition of secretion or short-circuit currents with V-ATPase blockers can be significant.

Recently, other types of H^+ pumps have been detected in pancreatic ducts. Both rodent (and human) ducts express the gastric type H^+/K^+-ATPases (*ATP4A* and *ATP4B*) and non-gastric types H^+/K^+-ATPase (*ATP12A*) (Novak et al. 2011; Wang et al. 2015). Inhibition of these with proton pump inhibitors such as omeprazole and SCH28080 reduced pH_i recovery in response to acid loads, and more importantly, they reduced secretion in isolated pancreatic ducts and in the whole pancreas tested *in vivo* (Novak et al. 2011; Wang et al. 2015). The immunohistochemical study showed that the H^+/K^+-ATPases (mainly colonic type) are localized to the basolateral membrane, and thus is consistent with HCO_3^-- secretion model. However, some H^+/K^+-ATPases were also localized to the luminal membrane, especially the gastric form (Novak et al. 2011). At present, the function of these pumps in pancreatic ducts is unclear, similar to other HCO_3^--secreting epithelia such as the airway epithelia (Novak et al. 2013), but interestingly *ATP12A* is upregulated in CF airways (Shah et al. 2016; Scudieri et al. 2018). It is speculated that these luminal pumps are creating a buffer zone protecting cells against bulk secretion which is pH > 8. In addition, luminal H^+/K^+ pumps in distal ducts would by virtue of H^+ secretion have more impact on pancreatic juice composition at low flow rates and minor at high flow rates, thus explaining excretory curves for HCO_3^- (Fig. 12.1). Furthermore, the luminal H^+/K^+ pumps would recirculate K^+ extruded by the luminal K^+ channels (Hayashi et al. 2012; Novak et al. 2013; Wang et al. 2015).

12.2.2.5 K^+ Channels

In addition to HCO_3^-/H^+ transporters, it K^+ channels are important for pancreatic duct secretion. They maintain the resting potential, and during stimulation opening of K^+ channels and hyperpolarization of the membrane potential maintains the driving force for Cl^- or HCO_3^- exit (Novak and Greger 1988a, 1991). Conductance for K^+ (G_K) is both present on the basolateral and luminal membranes and equivalent-circuit analysis has shown that the luminal K^+ channels contribute with at least with 10% to the total conductance in stimulated duct (Novak and Greger 1988a, 1991). Modeling in salivary glands confirms that such a ratio of luminal to basolateral K^+ channels would optimize secretion without destroying the transepithelial potential and transport (Almassy et al. 2012; Cook and Young 1989). Another function of luminal K^+ channels could be to contribute to secreted K^+, as pancreatic juice contains 4–8 mM K^+ (Sewell and Young 1975; Caflisch et al. 1979; Seow et al. 1991; Wang et al. 2015). The molecular identities and function of only some K^+ channels are known (see Hayashi and Novak 2013). The best studied until now are the Ca^{2+}-activated K^+ channels. The $K_{Ca}1.1$ channels (maxi-K, BK, coded by *KCNMA1*) are present in pancreatic ducts (Hede et al. 2005; Venglovecz et al. 2011) (Fig. 12.2b). Earlier patch-clamp studies indicate that these channels are located basolaterally (Gray et al. 1990; Hede et al. 1999). However, recent studies indicate that these channels are expressed on the luminal membrane and activated by, e.g., low concentrations of bile acids (see below). Evidence for another K^+ channel activated by extracellular ATP via purinergic P2 receptors was provided in studies of rat and dog duct epithelia (Hug et al. 1994; Nguyen et al. 1998) and later the intermediate conductance, KCa3.1 channel (IK, SK4, coded by *KCNN4*) was documented (Hede et al. 2005; Jung et al. 2006; Hayashi et al. 2012). Immunolocalization indicates that $K_{Ca}3.1$ is expressed on both luminal and basolateral membranes (Fig. 12.2b). The $K_{Ca}3.1$ channel activator EBIO enhanced secretion potential in Capan-1 monolayer indicating that these channels are important in pancreatic duct secretion (Hayashi et al. 2012; Wang et al. 2013). Recent studies on pancreatic ducts offer molecular identities of several other K^+ channels, including KVLQT1, HERG, EAG2; Slick and Slack (Hayashi et al. 2012), and interestingly the pH sensors TASK-2 and TREK-1 (Fong et al. 2003; Sauter et al. 2016). Nevertheless, the function and regulation of these channels in pancreatic physiology need to be studied.

12.2.2.6 Aquaporins and NKCC1

Taking that pancreatic ducts are secretory, water follows paracellularly and transcellulary via aquaporins (AQP). AQP1 is expressed on centroacinar cells and luminal and basolateral membrane of intercalated ducts and AQP5 is expressed luminally and labeling decreases in larger ducts in the human pancreas and is more distal in rodent pancreas (Burghardt et al. 2003, 2006). Notably, AQPs are co-expressed with CFTR in the same cells.

Upon stimulation of secretion, there would be a significant reduction in cell volume due to solute transport followed by osmotically obliged water. Subsequently, the cell volume would need to be reinstituted and one of the most important transporters in that respect is the Na^+-K^+-$2Cl^-$ cotransporter (NKCC1, *SLC12A2*). This transporter is expressed in pancreatic ducts, however, it is not clear whether shrinkage-activation of NKCC1 occurs in pancreatic ducts, or first following withdrawal of stimuli, as is the case in salivary acinar cells (see Sect. 12.3.2). Additionally, NKCC1 could provide cellular Cl^- for Cl^--driven fluid transport. In fact, diuretics such as bumetanide can inhibit duct/pancreas secretion, but the effect depends on the species (Fernandez-Salazar et al. 2004; Grotmol et al. 1986).

12.2.3 Integrating Ion Channels and Transporters to Pancreatic Ducts

Taking the above-described channels and ion transporters and placing them into one cell model becomes rather problematic—such cell would secrete and absorb at the same time. Therefore, it should be also considered where these transporters are localized within the pancreatic duct tree. Since there are only very few functional studies on native ducts from different regions of the ductal tree, we have to resort to taking into account heterogeneity in duct morphology and immunohistochemistry, and interaction between channels/transporters in expression studies. Studies summarized in the preceding sections have indicated that small proximal ducts express CFTR, CA, SLC26A, AE2, NBC1e, and AQP1. The larger, distal ducts express SLC26A3, and possibly CFTR, NBCn1, and NHE3, as well. It is not yet clear whether AE2, K^+ channels, H^+-pumps, and NKCC1 are differentially expressed.

For simplicity, if one inserts these transporters into two cells (Fig. 12.2b and c), it becomes apparent that the first cell has a potential to secrete, while the second has the possibility to exchange HCO_3^- for Cl^-. It cannot be excluded that these two models are two different states of one cell at different times or conditions. However, a very likely scenario is that one cell represents small proximal ducts that are secreting HCO_3^--rich fluid, and the other represents large interlobular/lobar ducts that are modifying incoming fluid but not contributing with a net fluid secretion. The simplest interpretation for these data obtained for large distal ducts is that they reabsorb HCO_3^- in exchange for Cl^-—thus giving the rise to the excretory curve

(Fig. 12.1). Nevertheless, with maximal stimulation and maximal secretory rates, pancreatic secretion is HCO_3^- rich. Potentially, one would expect acidic interstitial pH, which could favor pathogenic processes in pancreatitis and pancreatic cancer (Novak et al. 2013; Pedersen et al. 2017).

12.2.4 Regulation of Pancreatic Duct Secretion

The classical HCO_3^--evoking secretagogue is secretin, though a number of other hormones and transmitters can also evoke and co-regulate HCO_3^- secretion. Even cholinergic stimulation and cholecystokinin (CCK) can evoke HCO_3^- secretion in some species, and they can potentiate the secretin effect on the volume of secretion (Hickson 1970; Holst 1993; Park et al. 1998; You et al. 1983; Evans et al. 1996; Chey and Chang 2001; Szalmay et al. 2001). Here, we will consider the novel paracrine and autocrine regulators of pancreatic ducts—those secreted by acini and ducts themselves (nucleotides) and those that are entering the duct via the retrograde route (bile acids). Subsequently, we will consider novel interaction between signaling pathways and ion transporters and how they can in an integrative way affect pancreatic duct secretion.

12.2.4.1 Purinergic Signaling

In physiological settings, the function of pancreatic acini and ducts is coordinated (Hegyi and Petersen 2013). It has become accepted that purinergic signaling contributes to integrating acinar and duct functions, in particular fine-tuning duct performance. Pancreatic acini release ATP, some of which is stored in zymogen granules and released upon hormonal and cholinergic stimulation (Sorensen and Novak 2001; Haanes and Novak 2010; Haanes et al. 2014) and small amounts of ATP can be also detected in pancreatic juice, though most is hydrolyzed by ecto-nucleotidases (Kordas et al. 2004; Yegutkin et al. 2006) (Fig. 12.2a). ATP is also most likely released from nerves, as well as from acini and duct cells in response to cell volume changes, mechanical and chemical stress. In contrast to acini, pancreatic ducts express a number of functional purinergic (P2Y2, P2Y4, P2Y11, P2X4, and P2X7) and adenosine (A2A and A2B) receptors that regulate various epithelial transporters (see Novak 2008, 2011) (Fig. 12.2b and c). For example, luminal ATP (or UTP) can increase anion and fluid secretion, and this involves the regulation of TMEM16A/ANO1 and CFTR, as well as $K_{Ca}3.1$ and Cl^-/HCO_3^- exchange (Chan et al. 1996; Hug et al. 1994; Ishiguro et al. 1999; Namkung et al. 2003; Hede et al. 2005; Jung et al. 2006; Novak et al. 2010; Hayashi et al. 2012; Wang et al. 2013). In addition, luminal ATP stimulates P2X7 receptors and potentiates cholinergically evoked ductal secretion (Novak et al. 2010). Furthermore, ATP/UTP also potentiates cAMP-evoked mucin secretion (Jung et al. 2010). Ca^{2+} signaling and P2Y2 and P2X7 receptors in particular have been considered in these actions.

Adenosine receptors via cAMP signaling regulate CFTR (Novak et al. 2008). From the basolateral side, ATP released by nerves and/or distended epithelium can also affect the secretion and some purinergic receptors are inhibitory to secretion (e.g., P2Y2 receptors inhibit $K_{Ca}1.1$ channels), while other P2 receptors, including P2Y11 receptors, may have positive effects on secretion (Hede et al. 1999, 2005; Ishiguro et al. 1999; Nguyen et al. 2001; Wang et al. 2013). A number of processes in purinergic signaling are pH sensitive, and it will be relevant to investigate those in the microenvironment of the duct epithelium (Novak et al. 2013; Kowal et al. 2015b). Due to the fact that nucleotides/side could stimulate a multitude of P2 and adenosine receptors acting via Ca^{2+} and cAMP signaling, interactions need to be considered. For further information about P2X receptors in epithelial transport, the reader is directed to Chap. 28 of Vol. 3.

12.2.4.2 Bile Acids

Systemic bile acids (primary or secondary) produced in the liver and by gut microbiota, are becoming regarded as important physiological regulators of a wide range of cells. In the exocrine pancreas, though, bile acids (BA) may exert additional effects, as they can enter pancreatic duct tree by reflux following outflow obstruction by gallstones, and apparently affect both duct and acinar cells. Biliary acute pancreatitis (AP), or gallstone obstruction-associated AP, account for a significant percentage of clinical cases of AP and animal and cellular models are important tools for understanding development of this disease (Wan et al. 2012). Many studies have been carried out on acinar cells, and it has been shown that at high concentrations (mM) of, for example, taurine-conjugated BA cause large increases in intracellular Ca^{2+}, activation of intracellular trypsinogen and necrosis (Voronina et al. 2002, 2004; Gerasimenko et al. 2006; Kim et al. 2002). These BA effects are mediated by TGR5/Gpbar1 receptor, which is expressed on the apical surface of pancreatic acini in mice (Perides et al. 2010).

Only a few studies show that BA also have effects on pancreatic ducts, but these may be important since pancreatic ducts are pivotal for the maintenance of the physiological function of the whole pancreas. BA have a bimodal effect on pancreatic ducts inducing pancreatic fluid hypersecretion in the early stages of pancreatitis and hyposecretion during the onset of the disease (Czako et al. 1997). Studies on isolated ducts and duct epithelia show that this bimodal effect may be related to concentration of BA used. At high (>mM) concentrations BA have a detrimental effect. For example, in guinea pig ducts non-conjugated BA, chenodeoxycholate acid (CDCA) at 1 mM, caused large sustained increase in Ca^{2+}, inhibited HCO_3^- transport, caused mitochondrial damage and increased permeability of duct cells, and caused mitochondrial damage (Venglovecz et al. 2008; Maleth et al. 2011). In bovine duct cell monolayer 5 mM taurodeoxycholic acid (TDCA) decreased epithelial resistance due to damage of the epithelial barrier (Alvarez et al. 1998). At lower concentrations, BAs have positive effects. For example, in epithelium derived from the dog main duct, TDCA increased luminal G_{Cl} and basolateral G_K

in Ca^{2+}-dependent manner (Okolo et al. 2002). In other studies on guinea pig ducts and CFPAC-cells, it was shown that 0.1 mM CDCA increased Ca^{2+} via PLC and IP_3 (inositol 1,4,5-trisphosphate) and stimulated HCO_3^- transport (i.e., pH_i monitoring), though NBC, NHE, AE or CFTR or other Cl^- channels seem not to be primary targets (Venglovecz et al. 2008; Ignath et al. 2009). Studies on guinea pig ducts show that a low dose CDCA activated maxi-K^+ channels on the luminal membrane and thereby could initiate the secretory machinery (Venglovecz et al. 2011). Thus, it is proposed at high concentrations BA are damaging, but at low concentrations BA would be able to promote duct secretion, and thus wash out refluxed bile. It was not yet clear whether these physiological-like BA signals are mediated via TGR5/ Gpbar1 receptors One study offers another explanation. CDAC can evoke ATP release from duct cells, which then stimulates purinergic receptors and thereby increases cellular Ca^{2+}. TGR5 receptor is not involved in this process but can play a protective role at high Ca^{2+} conditions by stimulating Na^+/Ca^{2+} exchanger (Kowal et al. 2015a).

12.2.4.3 Synergistic Intracellular Signaling: Calcium, cAMP, and Cell Volume

In pancreatic ducts, as in other biological systems, physiological regulation would involve stimulation of several types of receptors and coordination of several signaling pathways to stimulate relevant ion transporters on both luminal and basolateral membranes to achieve transcellular secretion of ions/fluid, as well balancing cell volume and pH_i changes. Utilizing synergism of signaling pathways would ensure maximum effect without running each pathway at a maximum capacity, which could be detrimental to cell survival, as exemplified by Ca^{2+}-mediated cellular toxicity (Berridge 2012). Here, we summarize the evidence for the interaction of Ca^{2+} and cAMP signaling pathways and their effect on pancreatic duct ion transport.

First of all, some agonists, such as ubiquitous nucleotides/sides signal via multiple receptors: coupled to G_q, G_s, G_i proteins (P2Y and adenosine receptors) and ligand-gated ion channels (P2X receptors) (Jacobson and Muller 2016; Burnstock 2017). Pancreatic ducts express G-protein coupled P2Y receptors, P2X receptors, and various nucleotidases, such that ATP would have multiple effects via cAMP and Ca^{2+} signaling (Novak 2008, 2011) (Fig. 12.2).

Another synergistic mechanism occurs at the ion channel level, where Ca^{2+}-sensitive K^+ channels can alter the driving force for anion secretion through cAMP/PKA regulated CFTR. Further potentially synergistic mechanism to increase secretory output is the parallel anion transport through CaCC channels and CFTR channels in pancreatic ducts. However, there is evidence that there is some interdependence between CFTR and CaCC, such that malfunctioning CFTR (CF models) down-regulates expression or function of CaCC in parallel to CFTR (Gray et al. 1994; Winpenny et al. 1995; Pascua et al. 2009; Wang et al. 2013).

The central channel in the pancreatic duct is CFTR, which is a part of signaling complex that includes scaffolds, adaptors, and many regulatory enzymes associated with cAMP/PKA signaling (Frizzell and Hanrahan 2012). Several studies show that there is cross-talk between Ca^{2+} signaling and CFTR activation. There are Ca^{2+} sensitive adenylate cyclases 1 and 8 (Namkung et al. 2010; Martin et al. 2009) and Ca^{2+}-dependent activation of tyrosine kinases (Src2/Pyk complex), both of which could eventually alter the activity of CFTR, as shown for airway and intestinal epithelia (Billet and Hanrahan 2013; Billet et al. 2013). Another effect at the CFTR level would be priming of some PKC isoforms that enhance the activity of CFTR (see Billet and Hanrahan 2013).

Furthermore, synergy between Ca^{2+} and cAMP signaling could be exerted by the third messenger IRBIT (IP_3 receptor-binding protein released with IP_3). Agonists that couple to G_s, increase cAMP and via PKA phosphorylation of the IP_3 receptor, and receptors that activate G_q increase level of IP_3. Increased affinity of IP_3R to IP_3 facilitates the release of IRBIT from the apical pools, which then translocates and coordinates epithelial fluid and HCO_3^- secretion by stimulating NBCe1B and CFTR and SLC26A6 (Yang et al. 2009, 2011). This type of synergy is well studied in pancreatic ducts using genetic modifications of SLC26A6 and IRBIT (Park et al. 2013). Using cAMP/Ca^{2+} signaling agonist pairs such as forskolin/carbachol, secretin/carbachol, forskolin/carbachol—synergy in fluid secretion and HCO_3^- flux is revealed.

Lastly, the inhibitory pathways, which are downstream of Ca^{2+} and cAMP should be considered. Cell signaling pathways involving volume- and low Cl^--sensitive With No Lysine kinases (WNKs), acting via Ste20-like kinases, SPS-related proline/alanine-rich kinase (SPAK) and oxidative stress responsive kinase (OSR1), may be key factors in secretory epithelia, since they regulate NKCC1 and other transporters (Kahle et al. 2006; McCormick and Ellison 2011). Basically, these kinases are activated by hyperosmolarity (cell shrinkage) and low intracellular Cl^-, and thus would restore cell volume. In relation to pancreas, WNK1 and 4 are expressed in lateral membranes of interlobular and main pancreatic ducts and they inhibit NKCC1 and SLC26A6 (Choate et al. 2003; Kahle et al. 2004). Other studies show that WNKs inhibit CFTR (Yang et al. 2007, 2011). For example, in mice ducts, it was shown that WNKs and SPAK reduced expression of CFTR and NBCe1—and duct secretion (Yang et al. 2011). IRBIT increases membrane surface expression of NBCe1B, CFTR, and SCL26A6 and thus overcomes antagonizing WNK/SPAK signaling, which otherwise reduces secretion (see above). It seemed somewhat surprising then that the WNK/OSR1/SPAK system stimulated by low intracellular Cl^- could change the permeability of CFTR in favor of HCO_3^-, i.e., $PHCO_3^-/PCl^-$ increased from 0.24 to 1.09, and at the same time inhibited SLC26A6 and A3 (Park et al. 2010, 2012). It is proposed that WNK signaling for distal ducts and IRBIT signaling for proximal ducts could be a part of the mechanisms underlying overall pancreatic duct function (see Lee et al. 2012; Park and Lee 2012).

The above section indicates that cell volume regulation, e.g., via WNK/OSR1/SPAK system may be important for pancreatic ducts. Similarly, autocrine and paracrine signaling via volume-sensitive ATP release must be a key regulator in short- and long-term cell volume and ion transport in epithelia, including the pancreatic duct. Although cell volume regulation it is a cornerstone in epithelial physiology and pathophysiology we know very little about this process in in pancreatic ducts (Pedersen et al. 2013b).

12.3   Salivary Glands

12.3.1   Salivary Glands: Heterogenous Structures and Functions

Saliva is a complex mixture of fluid containing amylase, lipase, glycoproteins (e.g., mucins, vitamin B12 binding haptocorrin), proline-rich proteins and proteins regulating calcium phosphate and hydroxyapatite formation (e.g., statherins, histatins, and cystatins), growth factors (e.g., EGF), antibacterial agents (immunoglobulins, lysozyme, lactoferrin), water, and electrolytes (including varied concentrations of HCO_3^-) and minerals. The major function of the salivary glands is to protect the teeth and oral-oesophageal mucosa (by modulating re-/demineralization of teeth enamel, protecting gingiva, and antibacterial actions); initiate digestive processes; enhance taste perception and provide lubrication; and provide pH buffering capacity (bicarbonate, phosphate, proteins) (Humphrey and Williamson 2001; Matsuo 2000; Pedersen et al. 2002, 2013a). In some animals, salivary glands have additional functions, e.g., in grooming and evaporative cooling keeping oral cavity moist in panting, regulating salt homeostasis (e.g., in crocodiles). Major human salivary glands supply about 90% of the whole saliva and comprise of three pairs of exocrine glands: parotid glands (P), submandibular (submaxillary) (SM) glands and sublingual glands (SL). The rest of the secretion is provided by hundreds of minor glands spread throughout the oral cavity (Pedersen et al. 2013a). The largest glands in human, parotid glands, contain serous acinar cells and secrete amylase-rich secretions. The submandibular glands contain seromucous and serous acini and produce mucous saliva. The sublingual glands are the smallest glands and contain prevalently mucous acini and produce mucin-rich viscous secretion. Secretions that originate in acini are conducted through short intercalated ducts (that may be secretory) to striated ducts, characterized by many mitochondria and folds on the basal membrane, and to excretory (extralobular) ducts leading to the main excretory duct of Stensen (P), Wharton (SM) or series of excretory ducts (SL) (Pedersen et al. 2013a). In some glands/animals, e.g. male rodents, the striated ducts of SM contain granules and are referred to as granular ducts (Schneyer et al. 1972). Salivary glands are under the control of both branches of autonomic nervous systems, as well as higher brain centers and autocrine/paracrine regulation (Schneyer et al. 1972; Garrett 1987; Pedersen et al. 2002, 2013a; Proctor and Carpenter 2014).

The major component of saliva is water (99%) and electrolytes and there are special relationships between those and secretory rates (see Fig. 12.3). Based on these relationships Thaysen and coworkers proposed that saliva was formed in two stages—in acini and ducts (Thaysen et al. 1954). Similar hypotheses were proposed

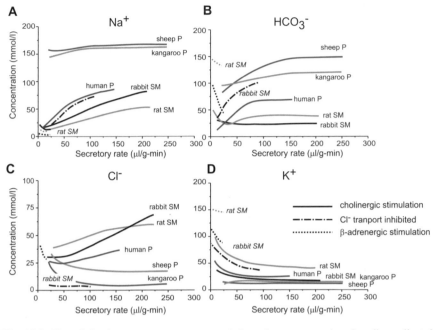

Fig. 12.3 The relation between secretory rates and electrolyte concentrations in saliva collected from salivary glands of various species. Preparations were *in vivo* glands or *ex vivo* perfused glands stimulated with cholinergic agonist, e.g., pilocarpine or carbachol, (full lines). In some experiments on perfused glands, Cl⁻ transport was inhibited, e.g., with furosemide (dot-dash line). Some glands were stimulated with β-adrenergic agonist, isoproterenol (dotted line). Secretory rates were corrected for gland weights and data were redrawn from publications on the rabbit submandibular gland (Case et al. 1980, 1984; Novak and Young 1986); the rat submandibular gland (Young and Martin 1971); the human parotid gland (Thaysen et al. 1954); the sheep parotid gland (Compton et al. 1980) and the kangaroo parotid gland (Beal 1984)

for the pancreas and other exocrine glands, though importantly, the function of ducts differs among various gland types (Bro-Rasmussen et al. 1956; Schwartz and Thaysen 1956). In the following paragraphs, the simplest scenario valid for most salivary glands will be outlined (Fig. 12.4). In the first stage, salivary acini generate primary saliva that is isotonic plasma-like fluid that is high in Na⁺ and Cl⁻ concentrations, K⁺ concentrations are slightly above the plasma (Schneyer et al. 1972; Young et al. 1980). The anion-gap is most likely HCO₃⁻ secreted at about plasma-concentrations (see Sect. 12.3.4 for exceptions). In the second stage, the ducts reabsorb Na⁺ and Cl⁻ and partially compensate electrolytes by secreting some K⁺ and HCO₃⁻. The ductal epithelium is electrically tight and water impermeable (Young et al. 1980). In this respect, the salivary ducts are fundamentally different from the pancreatic ducts, the latter being a leaky and secretory epithelium (see Sect. 12.2.1). Due to hypertonic salt transport in salivary ducts, final saliva in many gland types and species is hypotonic. Importantly though, tonicity and electrolyte patterns depend on the secretory rate (acini) and saturation of various transporters on the

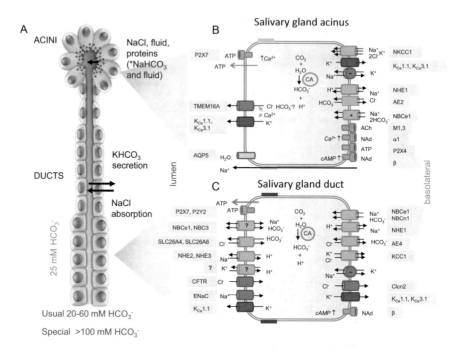

Fig. 12.4 (a) Schematic diagram showing a simplified salivary gland with acini and excretory ducts and saliva with a typical range of HCO_3^- concentrations for most cholinergically stimulated salivary glands and special salivary glands and/or special circumstance (see Fig. 12.3). (b) Inserts show the cellular models for acinar cells that can secrete Cl^- relatively independent of HCO_3^- and rely on NKCC1 and double exchange system (e.g., SM glands). Some acini rely primarily on HCO_3^- transport (e.g., parotid glands) and express NBCe1, marked with *. See also Fig. 12.3. (c) Cellular model of a duct cell that is absorbing Na^+ and Cl^- (via luminal ENaC-CFTR channels or double exchange system) and secreting K^+ and HCO_3^-

downstream ducts. Above-described processes would give rise to the simplest excretory curves (Fig. 12.3 rat, rabbit, human glands).

A number of studies were conducted to verify the two-stage theory, including micropuncture studies that involve sampling and analyzing fluid at or close to acini/ intercalated ducts and in downstream ducts, as well as studies on isolated salivary ducts. These are summarized in earlier reviews in this field (Martinez et al. 1966; Young and Schögel 1966; Young et al. 1980). In the current research on exocrine glands, many advanced techniques on cellular/genetic level are used, but it is still very valuable to take the integrative approach and return to the whole gland secretion and electrolyte patterns. Nonetheless, understanding of ion transport in salivary glands is particularly challenging as there are three different major glands (parotid, submandibular, and sublingual), there are large interspecies and even male/female variations in structure and regulation, and these may reflect very varied salivary gland functions.

In the following section, we will describe the basic ion transport mechanisms in acini and ducts of most common experimental animals (rat, mouse, rabbit) stimulated with cholinergic stimulation that evokes the largest fluid secretion rates (Figs. 12.3 and 12.4). Later (Sects. 12.3.4 and 12.3.5) we will consider other modes of stimulation, e.g., sympathetic, as well as specific glands/animals and experimental conditions that evoke saliva with unusual ion compositions. This holds in particular HCO_3^- secretion, which shows the most bewildering variety of excretion patterns (Fig. 12.3) and may originate in acini and/or ducts (Young et al. 1980; Novak 1993) (Fig. 12.3b). In humans, parotid glands can secrete up to 40–60 mM HCO_3^-, though mixed saliva from major and minor salivary glands rarely exceeds 20–25 mM HCO_3^- (Thaysen et al. 1954; Bardow et al. 2000). Human parotid glands have HCO_3^- output ranging from about 40 to 500 µmol/h/g gland weight. Rat and rabbit submandibular stimulated with cholinergic stimulus have similar HCO_3^- output ranging from about 40 to 400 µmol/h/g gland weight. In sheep and kangaroo parotid the output ranges from around 100 to 1500 µmol/h/g gland weight (see Fig. 12.3).

12.3.2   Ion Channels and Transporters in Salivary Gland Acini

Since many salivary glands commonly studied (e.g., rodent P and SM) can secrete very efficiently without exogenous HCO_3^- and/or with CA inhibitors, the ion transport models for acini are based predominantly on the transport of Cl^-. Cl^- is transported across the basolateral membrane via loop diuretic sensitive Na^+-Cl^- cotransporter, later identified as NKCC1 in several preparations including human parotid acini (Case et al. 1982, 1984; Martinez and Cassity 1983; Turner et al. 1986; Nauntofte and Poulsen 1986; Moore-Hoon and Turner 1998; Evans et al. 2000; Nakamoto et al. 2007) (Fig. 12.4b). An alternative mechanism for NaCl transport is the parallel transport via Na^+/H^+ and Cl^-/HCO_3^- exchangers, as proposed from studies of isolated glands (Novak and Young 1986; Turner and George 1988). Interestingly, the Cl^-/HCO_3^- exchanger, most likely AE2, is upregulated when NKCC1 is inhibited or genetically silenced (Evans et al. 2000). AE2 is expressed together with NHE1 on the basolateral membrane of acini (He et al. 1997; Lee et al. 1998; Park et al. 1999). In *NHE1–/–* mice, parotid acini express higher activity of AE2 and CAII, as determined by pH_i measurements, indicating increased Cl^- and HCO_3^- transport across the plasma membranes, though no data on salivary secretion are available (Gonzalez-Begne et al. 2007). There is also evidence for NBC1e expression on salivary gland acini (see Sect. 12.3.4), but in many glands transport of Cl^- is sufficient to drive full fluid secretion.

Parasympathetic stimulation produces large saliva volumes (see Sect. 12.3.5), e.g., acetylcholine acting via muscarinic receptors, increases intracellular Ca^{2+} concentrations and Ca^{2+} signaling has been well-studied mode of stimulus-secretion coupling (Petersen 2014), though more recent studies show that similar to the pancreas there is a synergism between Ca^{2+} and cAMP signaling (see Jung and Lee 2014; Ahuja et al. 2014). In salivary acini, Ca^{2+} signals are essential in regulating Cl^- efflux via the luminal channels. These CaCC properties are corresponding to recently identified Cl^- channel TMEM16A/ANO1, and submandibular glands from *TMEM16a-/-* mice produce lower amount of saliva (Romanenko et al. 2010; Catalan et al. 2015). Some TMEM16A/ANO1 is also expressed on the luminal membrane of intercalated ducts, though another CaCC candidate Besthropin-2 may be relevant to duct function (Romanenko et al. 2010). Expression of TMEM16A/ANO1 is also found on the luminal membrane of human parotid acini and intercalated ducts (Chenevert et al. 2012) (Fig. 12.4b). Studies in HEK293 cells and SM gland acini indicate that TMEM16A/ANO1 anion selectivity is dynamically modulated by Ca^{2+}/calmodulin, possibly increasing $PHCO_3^-/PCl^-$ (Jung et al. 2013). Regarding CFTR, the protein is expressed on the luminal membrane of ducts, but there are contradicting reports regarding the expression of CFTR in SM of rodent acini. Nevertheless, since in mice with $\Delta F508$ mutation in CFTR or inhibition of CFTR had no significant effect on salivary secretion rate, other Cl^- channel must have rescued secretion (He et al. 1997; Zeng et al. 1997; Catalan et al. 2010).

Sympathetic stimulation leading to β-adrenoceptor stimulation and cAMP signaling produces lower volumes of HCO_3^-- and protein-rich saliva (Case et al. 1980). One study shows that isoproterenol stimulates secretion in salivary glands of mice where TMEM16A and CFTR have been ablated, and inhibitor sensitivity profiles indicate VRAC channels may be involved (Catalan et al. 2015).

One of the most marked effects in salivary acini is the loss of intracellular K^+ upon stimulation, as observed in initial studies on in vivo glands and isolated acini (Burgen 1956; Nauntofte 1992). Also, many electrophysiological studies on plasma membrane potentials in acini reported hyperpolarizing "secretory potentials", which would be consistent with increased K^+ conductance, and Ca^{2+}-activated maxi-K^+ channels were characterized in patch-clamp studies (Imai 1965; Petersen and Poulsen 1967; Maruyama et al. 1983; Petersen and Gallacher 1988). Ca^{2+}-activated K^+ channels ($BK-K_{Ca}1.1$ and $IK-K_{Ca}3.1$) have been identified on the basolateral membrane (Wegman et al. 1992; Park et al. 2001a; Nehrke et al. 2003; Begenisich et al. 2004; Romanenko et al. 2007) (Fig. 12.4b). It is now well accepted that the basolateral K^+ channels serve for K^+ recirculation necessary for the operation of the Na^+/K^+ pump and thus secretion. However, micropuncture studies and analysis of fluid close to acini, indicated that the primary secretion has K^+ concentrations higher than the plasma, e.g., up to 10 mM K^+ (Young and Schögel 1966; Mangos et al. 1966, 1973) (see Schneyer et al. 1972), indicating that there may be some secretory mechanisms for K^+ on the luminal membrane. Hence, it was proposed that K^+ channels are also present on the luminal membrane of salivary acini and various mathematical models verified that luminal K^+ channels are necessary for creating the driving force for Cl^- exit and account for at least 10–20% of total K^+ conductance in acinar cells (Cook and Young 1989; Palk et al. 2010). Recent studies on mouse parotid acini using spatially localized manipulation of Ca^{2+} and whole cell patch clamp show that the very small area of the luminal membrane, e.g., approximately 3–8% of the overall plasma membrane (Poulsen and Bundgaard 1994), expresses high density of $K_{Ca}1.1$ and $K_{Ca}3.1$ channels (Almassy et al. 2012). These channels exhibit some interdependence/interaction (Thompson and Begenisich 2006, 2009). In submandibular acini, it seems that the apical Ca^{2+} signals stimulate CaCC, and only when signals spread to the basolateral membrane and/or the membrane is depolarized, then the basolateral K^+ channels are activated. It also seems that either $K_{Ca}1.1$ or $K_{Ca}3.1$ can support full secretion in the mouse submandibular gland; and only a double knockout of these K^+ channels reduces secretion significantly (Romanenko et al. 2007). Whether incongruence between parotid and submandibular acini is related to different Ca^{2+} signaling or patterns of K^+ channel expression is not clear yet. Nevertheless, the most significant K^+ secretion is contributed by the ducts (see Sect. 12.3.3) (Fig. 12.4c).

In normal salivary secretion water transport, occurring by transcellular and paracellular routes and the cell volume regulation is dependent on the expression of water channels, aquaporins. The most important aquaporin in salivary gland acini is AQP5, as determined in knockout studies on mice (Ma et al. 1999; Krane et al. 2001; Murakami et al. 2006; Kawedia et al. 2007). Cell volume regulation is important in many cellular functions, including epithelial transport (Pedersen et al. 2013b). In secreting epithelia, physiological stimulus leads to opening of luminal Cl^- channels and basolateral/luminal K^+ channels and osmotically driven water movement leads to shrinkage of secreting cells. Basolateral transporters and pH_i regulating mechanisms need to be activated to provide ions for luminal exit. Nevertheless, in the secretory state these mechanisms are unable to maintain the cell volume of salivary secretory cells which remain shrunken, until the stimulus is terminated, after which, cell volume recovers (Dissing et al. 1990; Foskett 1990; Nakahari et al. 1990, 1991; Lee and Foskett 2010). This seems to be the case for the Ca^{2+} signaling pathways, as cAMP-mediated signaling leads to increased cell volume in salivary acini and VRAC may be involved (Catalan et al. 2015).

12.3.3 Ion Channels and Transporters in Salivary Gland Ducts

The cornerstone in salivary duct ion transport is NaCl absorption (and $KHCO_3$ secretion), and it is apparent that Na^+ and Cl^- excretion curves are following each other (Fig. 12.3a and c). One possible mechanism for NaCl absorption is the electroneutral model—luminal Na^+/H^+ and Cl^-/HCO_3^- exchangers. The alternative model is the parallel activity of epithelial Na^+ channels (ENaC) and CFTR (Fig. 12.4c). There is evidence for both systems and it has been proposed that ducts of low-HCO_3^- secretors (mouse and rabbit SM) are dominated by Na^+ and Cl^- channels on the luminal membrane, while ducts of high HCO_3^- secretors (rat SM) are dominated by the double exchangers (Chaturapanich et al. 1997).

There is solid evidence for ENaC expression on the luminal membrane of salivary ducts (Fig. 12.4c), and ENaC is regulated by ubiquitin-protein ligase Nedd4 (Komwatana et al. 1996b; Dinudom et al. 1998, 2001; Cook et al. 2002). There are number of electrophysiological and inhibitors studies on isolated ducts and glands supporting the evidence for ENaC, e.g., low concentrations of amiloride leads to increased NaCl content in saliva (Bijman et al. 1981; Komwatana et al. 1996a). CFTR is expressed on the luminal membrane of salivary ducts and its inhibition by specific inhibitors and CFTR knockout leads to decreased NaCl absorption and increased salt excretion in saliva, as seen in murine models of CF (Dinudom et al. 1995; Zeng et al. 1997; Catalan et al. 2010). If CFTR is to transport Cl^- from lumen to the cell, it requires markedly depolarized luminal membrane potential, which has not been measured, but quantitative modeling of salivary ion transporters strongly supports this model (Patterson et al. 2012). Exit pathway for Cl^- on the basolateral membrane is not clear and proposals include a hyperpolarization-activated Cl^- channel (Clcn2), KCl cotransporter (KCC1), or Cl^-/HCO_3^- exchanger (AE4) (Romanenko et al. 2008; Roussa et al. 2002; Ko et al. 2002a). Note that the basolateral membrane needs to be more hyperpolarized than the luminal to permit Cl^- exit out of the cell toward interstitium.

The molecular basis for the alternative electroneutral NaCl transport model is more difficult to pinpoint. Salivary ducts express NHE2 and NHE3 on the luminal membrane and their function is not clear, as knockout of either NHE isoform has no effect on salivary secretion in mice (Park et al. 2001b; Lee et al. 1998). Salivary ducts also express SLC26A4 and A6 (Shcheynikov et al. 2008) and although they differ in coupling Cl^- : HCO_3^- (i.e., 1:1 versus 1:2), they could ensure Cl^- influx into duct cells and HCO_3^- efflux, and it seems that either one can explain the excretory curves in a model simulation of salivary ducts (Patterson et al. 2012).

There are several other transporters that could contribute to duct HCO_3^- secretion (or HCO_3^- absorption). On the basolateral membrane, the means of HCO_3^- import into duct cells could be NBCe1 (e.g., guinea pig SM) or NBCn1 (e.g., rat SM ducts, human SM, and P) (Li et al. 2006; Gresz et al. 2002). Alternatively, or in addition, NHE1 on the basolateral membrane together with CAII could be a part of HCO_3^-/H^+ system involved in secretion and/or pH_i regulation (Park et al. 1999). Interestingly, some NBC transporters, NBC3 and NBCe1, are also expressed on the luminal membrane of several types of salivary ducts, and their proposed functions are to absorb (salvage) HCO_3^- (Park et al. 2002b; Li et al. 2006; Gresz et al. 2002). Presumably, HCO_3^- ductal absorption would occur if salivary acini were secreting primary fluid rich in HCO_3^- (see below).

Regarding H^+ pumps, immunohistochemical studies show a presence of V-ATPases in intracellular compartments in SM granular and main ducts in rats with normal acid-base balance, but in the rat parotid gland, the pumps are close to the luminal membrane of the striated and excretory ducts (Roussa et al. 1998; Roussa and Thevenod 1998). The pump has not been studied functionally, though heterogenic distribution may indicate that the parotid glands have special acid/base challenges. The H^+/K^+ pump (not the putative passive H^+/K^+ exchanger, Fig. 12.4c) has not been seriously considered in salivary glands, except for one study where proton pump inhibitors (omeprazole and SCH28080) did not have effect on pH_i on striated ducts from the rat parotid in a given experimental condition (Paulais et al. 1994).

In many species, the collected saliva has K^+ concentrations several-fold higher than the plasma and it is inversely related to secretory rate, and often HCO_3^- excretion follows similar pattern (see exception Sect. 12.3.4). Therefore, it is not surprising that one of the original proposals to explain $KHCO_3$ secretion (Knauf et al. 1982; Paulais et al. 1994) was a K^+/H^+ exchanger working in parallel with Na^+/H^+ exchange and Cl^-/HCO_3^- exchange (Fig. 12.4c). Together these would perform a functional K^+/HCO_3^- cotransport on luminal membrane and provide Na^+ and Cl^- for the basolateral exit. Such K^+/H^+ transporter has not been cloned and several studies show that K^+ and HCO_3^- transport is not very tightly coupled (Nakamoto et al. 2008; Chaturapanich et al. 1997). Therefore, other K^+ exit pathways have been considered, such as the luminal K^+ channels. Striated and excretory SM ducts express $K_{Ca}1.1$ channels and knockout studies show that much decreased K^+ secretion in whole glands, indicating that indeed these channels are important for ductal secretion (Nakamoto et al. 2008).

Using the above transport components for salivary glands ducts, and experimental values obtained from various studies and modeling, it has been possible to reproduce excretory curves similar to Fig. 12.3 for major ions in saliva, including HCO_3^- concentrations 35–45 mM and K^+ of 20–60 mM (Patterson et al. 2012).

12.3.4   Salivary Glands Can Secrete Very High Bicarbonate and/or Potassium: Where and When?

Now that the basics of ion transport in salivary gland acini and ducts have been presented, we can consider three special circumstances in which some salivary glands secrete saliva with very high HCO_3^- concentrations, with/without accompanying K^+ (Fig. 12.3b and d), and try to resolve how this happens.

First of all, as introduced above, ducts can secrete HCO_3^- (and K^+), though without accompanying water fluxes and the lower the secretory rate of saliva is, the higher the concentrations of HCO_3^- and K^+ are (Fig. 12.3b and d hyperbolic curves). These patterns can be particularly dominant with some forms of stimulation (see Sect. 12.3.5).

Salivary glands of some animals have a large capacity to secrete HCO_3^- and concentrations are almost as high as in the pancreas. This is the case for foregut fermenters such as sheep, cattle, camels, and kangaroos, where usually parotid glands supply well-buffered saliva to stabilize pH of the fermenting digesta and saliva can contain 110–120 mM HCO_3^- (and also 20–60 mM phosphate), though K^+ is relatively low 5–15 mM (Kay 1960; Young and van Lennep 1979; Beal 1984) (Fig. 12.3). The analysis of the relationship between the salivary flow rate and electrolyte concentrations suggests that the primary secretion itself must be high in

HCO_3^- concentration and that the duct contribution is relatively small. This conclusion is supported by micropuncture studies on sheep parotid, which showed that Cl^- concentrations were about 50 mM (Compton et al. 1980). By inferences then, HCO_3^- must have been correspondingly high (due to very small volumes H^+/HCO_3^- could not be measured) and that furosemide, and inhibitor of NKCC1, had a relatively small effect on saliva secretion (Wright et al. 1986). Further studies showed that the sheep parotid acini exhibited acetylcholine stimulated a Na^+-dependent HCO_3^--influx and a Na^+-HCO_3^- cotransport was proposed (Poronnik et al. 1995; Steward et al. 1996). Subsequently, it was verified that bovine parotid acini express NBCe1B at high levels (on mRNA and protein levels) and show large electrogenic currents (Yamaguchi and Ishikawa 2005) (Fig. 12.4b). The transporter is localized on the basolateral membrane of acini and interacts with IRBIT (Yamaguchi and Ishikawa 2012).

The NBCe1 transporter is also expressed in other acini, such as rat and human parotid acini, and this is in accordance with observations that parotid glands can secrete saliva with relatively high HCO_3^- concentrations (Roussa et al. 1999; Park et al. 2002a; Bardow et al. 2000). Correspondingly, the excretory curve patterns are consistent with the acinar origin of HCO_3^- secretion (Fig. 12.3b). In submandibular glands, acinar secretion is not dependent on HCO_3^- driven transport normally, as the Cl^- driven transport is dominating (Case et al. 1982, 1984). Nevertheless, if Cl^- transport is inhibited, these glands can also produce saliva with concentrations >100 mM HCO_3^- (although secretion rate is diminished) (Fig. 12.3b rabbit SM). This indicates that, most likely, SM acini also have the machinery to transport and drive HCO_3^- transport (Novak and Young 1986). This is in fact revealed if glands are stimulated with β-adrenergic rather than muscarinic agonists (Case et al. 1980).

12.3.5 Regulation of Salivary Gland Secretion

The main regulation of saliva secretion is via the autonomic nervous system, and little if any by hormones (Proctor and Carpenter 2014). Parasympathetic and sympathetic stimulation leads to activation of muscarinic M1 and M3 receptors and α1 receptors on salivary gland cells, respectively, that via $G_{q/11}$ eventually leading to Ca^{2+} signaling. Furthermore, noradrenaline via β receptors and G_s and stimulates cAMP/PKA signaling pathway. Both parasympathetic and sympathetic stimulations result in approximately similar plasma-like primary secretion, as measured in, e.g., rat SM (Young and Martin 1971). However, parasympathetic stimulation causes 6–8 times higher fluid secretion compared to sympathetic one, the latter resulting in more concentrated protein secretion. Salivary ducts are also well innervated by autonomic nerves and they modify secretion mainly by stimulating K^+ and HCO_3^- secretion (Young and Martin 1971; Martin and Young 1971). In particular, β-adrenergic receptors stimulation can result in disproportionately greater effect on K^+ and HCO_3^- concentrations in small volumes of final saliva (Case et al. 1980) (Figs. 12.3b and d dotted lines). Furthermore, sympathetic stimulation via β-adrenergic receptors (at low concentrations) can modulate saliva composition by activation of CFTR, and thereby increasing the rate of Na^+ absorption (Dinudom et al. 1995).

Effects of acetylcholine and noradrenaline are modulated by a number of non-adrenergic non-cholinergic (NCNA) co-transmitters, including VIP, substance P, neuropeptide Y, NO, ATP, and others (Pedersen et al. 2013a; Proctor and Carpenter 2014). Under physiological conditions, it may be expected that multiple transmitters would collaborate to induced salivary fluid and protein secretion. The synergistic effects of cAMP/PKA and Ca^{2+} signaling have been demonstrated in many studies and recent data indicate that IRBIT mediates synergism between these signaling pathways (Bruce et al. 2002; Ahuja et al. 2014; Jung and Lee 2014).

Extracellular nucleotides/sides are probably NANC and also autocrine/paracrine signaling molecules that have an important function in coordinating salivary gland functions including HCO_3^- secretion. The purinergic signaling has been pioneered in salivary glands (see Novak 2011). It is most likely that ATP is a cotransmitter with acetylcholine or noradrenaline and a number of P2 receptors are expressed and functional on salivary gland acini. In particular, P2X4 and P2X7 receptors have been described in early studies and it is clear that in physiological conditions they do not form permeable pores, but rather remain cation channels/receptor that co-regulate secretion, for example, by aiding Ca^{2+} signaling and stimulating K^+ and Cl^- channels. Recent studies show that in the rat parotid acini P2X4 receptors are located on the basolateral membrane, while P2X7 receptors are enriched on the luminal membrane, and they evoke spatially distinct Ca^{2+} signals and effects on protein exocytosis (Bhattacharya et al. 2012), and fluid secretion in mouse salivary glands (Nakamoto et al. 2009; Novak et al. 2010) (Fig. 12.4). Interestingly, the P2Y receptors are expressed transiently, e.g., during gland development (P2Y1) and stress (P2Y2). Furthermore, ATP is most likely also stored in secretory granules and it is secreted into duct lumen and ATP can be detected in the whole saliva (Ishibashi et al. 2008; Novak et al. 2010). In addition, the duct epithelium also releases ATP spontaneously or in response to, for example, mechanical stimulation (Shitara et al. 2009; Ryu et al. 2010). Here it acts via P2Y2 receptors to increase Cl^- absorption by stimulating CFTR (Ishibashi et al. 2008). It is not clear how HCO_3^- transport is affected and how luminally expressed P2X7 receptors regulates salivary duct function.

12.4 Hepatobiliary System

12.4.1 Hepatobiliary System: Concerted Action of Several Types of Epithelial Cells

The hepatobiliary system shares many similarities with the pancreas. It secretes highly complex bile containing 95% water with electrolytes, bile salts and bilirubin, lipids, proteins, enzymes, peptides and amino acids, nucleotides, heavy metals, and vitamins (Boyer 2013). Most importantly, bile contains 15–55 mM HCO_3^-. The hepatobiliary system is composed of several types of epithelial cells: hepatocytes, cholangiocytes, and gallbladder epithelial cells; all of which contribute to the formation of bile in several stages. Hepatocytes, the major liver cell population (65%) secrete primary or canalicular bile, which is driven by bile salt-dependent transport and bile salt-independent transport of bicarbonate and reduced glutathione, and osmotically obliged water flux. Canalicular bile is delivered to the intrahepatic bile ducts, which are lined with cholangiocytes, forming about 3–5% of liver cell population, and these modify bile by secretion and absorption (Fig. 12.5a). The intrahepatic biliary duct system is the most important epithelium in the liver regarding the ability to secrete HCO_3^- containing fluid in response to secretin and other regulators. This secretion contributes to alkalization and fluidization of bile, which prevents protonation and absorption of weak lipophilic acids. In addition, distally it provides the buffer function in the duodenum—similar to the pancreatic duct system (Boyer 2013). In the last stage, gallbladder concentrates and stores bile, but also modifies it by secretion in some species, and lastly delivers it to the duodenum during feeding. Assuming that in a human the canalicular bile acid-independent flow and bile duct flow together are about 4 µl/min/kg body weight and final bile contains 15–55 mM HCO_3^- (Banales et al. 2006b; Boyer 2013) it can be approximated that HCO_3^- output is about 0.3–0.7 µmol/h/g liver weight. Considering the whole organ though, HCO_3^- output of 300–700 µmol/h is a significant output.

12.4.2 Canalicular Bile Salt-Independent Flow Generated by Hepatocytes

Canalicular bile fluid is secreted by hepatocytes and secretion has two components (see above). The bile-salt dependent fraction is driven by a series of organic anionic transporters, BA transporters, multidrug resistance transporters (MRP) (Esteller 2008). The bile salt-independent fraction is driven by the transport of glutathione and HCO_3^-, each contributing about 50% to this fraction (Banales et al. 2006b; Boyer 2013; Esteller 2008). Glutathione is secreted via organic anion transporter MPR2/ABCC2. Bicarbonate is secreted via a DIDS-sensitive Na^+-independent Cl^-/HCO_3^- exchanger (AE2) expressed on the luminal membrane (Fig. 12.5b) (Meier et al. 1985; Martinez-Anso et al. 1994). Apparently, CFTR is not expressed in

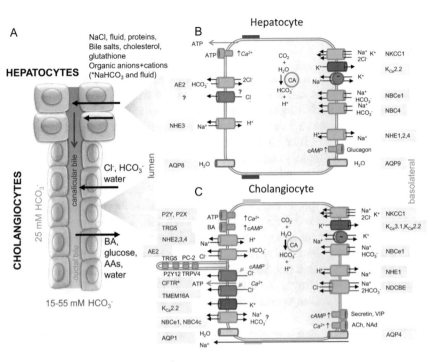

Fig. 12.5 (**a**) Schematic diagram showing a simplified hepatobiliary system composed of hepatocytes and cholangiocytes. Hepatocytes secrete canalicular bile fluid driven by bile-salt dependent fraction and bile-salt independent fraction. Cholangiocytes secrete Cl^-, HCO_3^- and fluid and absorb bile acids, glucose, amino acids, and water. (**b**) Insert shows cellular model for bile-salt independent HCO_3^-—driven fluid secretion only. (**c**) Insert shows cellular model for ion transport and regulation in cholangiocytes with a primary cilium (expressing several receptors and channels). CFTR* indicates that CFTR is functional in large but not small cholangiocytes

hepatocytes and it might be another Cl^- channel that together with AE2 coordinates or performs HCO_3^- secretion, and water follows via AQP8 (Banales et al. 2006b). HCO_3^- loaders on the basolateral membrane of hepatocytes are most likely NHE1, and also NHE2 and 4 (Moseley et al. 1986; Pizzonia et al. 1998). In addition, NHE3 is expressed on the canalicular membrane and more importantly on the apical membrane of cholangiocytes, where it most likely participates in fluid absorption (Mennone et al. 2001). Electrogenic Na^+-HCO_3^- cotransport was demonstrated in studies using plasma membrane vesicles and microlectrodes and subsequently, NBCe1/*SLC4A4* and NBCe2/NBC4/*SLC4A5* were identified (Fitz et al. 1989; Renner et al. 1989). However, the expression of NBCe1 is low while NBC4, especially NBC4c variant, is high in the liver (Pushkin et al. 2000; Abuladze et al. 2004).

12.4.3 Intrahepatic Biliary Duct System: Ion Transport in Cholangiocytes

Intrahepatic biliary duct system is a network of interconnecting bile ductules and ducts that extend from canals of Hering, partly lined with hepatocytes, to ductules (<15 μm) lined with cholangiocytes, and then converge to large bile ducts (>15 μm) of increasing diameter to interlobular, septal, segmental ducts and then hepatic and common hepatic duct (Tabibian et al. 2013). Along the duct system, cholangiocytes change in size, morphology, regulation, and function. Although they account only for 3–5% of the liver population, they are responsible for about 30–40% of bile volume, depending on species. The magnitude of the ductal contribution of basal bile flow depends on species and can vary from 10 to 30%. Primarily cholangiocytes secrete Cl^-, HCO_3^- and fluid, and they reabsorb bile acids, glucose, and amino acids. Secretin stimulates HCO_3^- secretion and this determines the pH and hydration of bile, and thereby decreases protonation of glycine-conjugated bile acids and thus their passive reabsorption (see Sect. 12.4.5), and lastly, it contributes to alkalization of duodenal contents (Boyer 2013; Beuers et al. 2012). In the following section, we will only focus on cholangiocyte ion channels and transporters involved in HCO_3^- transport.

Ductal bile secretion is a regulated process that is initiated by the transport of Cl^- across the luminal membrane into the ductal lumen and coupling to AE2, as verified in isolated bile duct unit preparations (IBDU). CFTR has been identified and it is functional in large but not small IBDU (Fitz et al. 1993; Alpini et al. 1997; Cohn et al. 1993; Dutta et al. 2011). However, some studies have challenged the role of CFTR as the primary route for Cl^- exit; rather they ascribed the protein a regulatory role in ATP release (see below). In response to mechanical stimulation (fluid flow and cell swelling) and ATP stimulation, cholangiocytes isolated from various species exhibit CaCC (Tabibian et al. 2013). The molecular identity of these channels was unknown until recent studies showed that TMEM16A/ANO1 is a good candidate (Dutta et al. 2011, 2013). It seems that CFTR is only expressed in larger bile ducts, while TMEM16A is expressed in both small and large bile ducts (Dutta et al. 2011). Interestingly, CaCC contribution to anion transport is several-fold higher than CFTR.

Similar to pancreatic ducts, Cl^- channels are functionally coupled to Cl^-/HCO_3^- exchanger and AE2 is indeed the main effector of both basal and stimulated Cl^-/HCO_3^- exchange (Banales et al. 2006a, b; Strazzabosco et al. 1997; Aranda et al. 2004; Concepcion et al. 2013). Interestingly, AE2–/– mice develop biochemical, histological, and immunological alterations resembling primary biliary cirrhosis (Concepcion et al. 2013). The luminal membrane of cholangiocytes also expresses amiloride-insensitive NHE2/SLC9A2 and amiloride-sensitive NHE3/SCL9A3 and NHE4/SLC9A4 (Banales et al. 2006b). The function of these is not clear, though NHE3 knockout studies showed that the exchanger is important in fluid absorption in resting duct epithelium and also interacts with CFTR (Mennone et al. 2001). The Cl^- secretory response is maintained by small conductance Ca^{2+}-activated K^+

channels, SK2 ($K_{Ca}2.2$, *KCNN2*) expressed on the luminal membrane (Feranchak et al. 2004).

NHE1 is expressed on the basolateral membrane of cholangiocytes and it has a multitude of functions, including regulation of pH_i and cell volume, and transepithelial HCO_3^- transport in rat and human bile ducts (Spirli et al. 1998; Banales et al. 2006b). In addition, another acid extruder, V-ATPase was identified and functional in pig cholangiocytes (Villanger et al. 1993), though it seems not important in human cholangiocytes (Strazzabosco et al. 1997). CA is also expressed in bile ducts, though compared to other HCO_3^- secretory organs CA activity is rather low (Banales et al. 2006b).

Regarding the Na^+ coupled HCO_3^- import mechanisms across the basolateral membrane, there seem to be some interspecies variations (Fig. 12.5c). Na^+-dependent Cl^-/HCO_3^- exchanger, NDCBE1/*SLC4A8*, which has been cloned in humans (Grichtchenko et al. 2001) is assumed to import HCO_3^- across the basolateral membrane of cholangiocytes (Strazzabosco et al. 1997; Banales et al. 2006b). In rat cholangiocytes, several HCO_3^- transport mechanisms are functional as revealed by pH_i measurements (Strazzabosco et al. 1997). But apparently in the rat liver only the NBC4c variant of NBCe2/NBC4/*SLC4A5* is expressed in cholangiocytes and hepatocytes, though in cholangiocytes it is expressed apically (Abuladze et al. 2004). In murine cholangiocytes, NBCe1/*SLC4A4* is expressed on the luminal membrane (Uriarte et al. 2010), and in order to secrete HCO_3^- into the lumen, it would require coupling of several HCO_3^- to one Na^+ and highly hyperpolarized membrane potential.

K^+ channels are providing the driving force secretion and at least two types of channels are expressed on the basolateral membrane—$K_{Ca}3.1$ and SK2/$K_{Ca}2.2$ (*KCNN2*) (Dutta et al. 2009; Feranchak et al. 2004). NKCC1 and Na^+/K^+-pump are also expressed on the basolateral membranes. A number of AQP are expressed in cholangiocytes, in particular, AQP4 on the basolateral membrane and AQP1 on the luminal membrane, though there are species differences (Banales et al. 2006b; Tabibian et al. 2013).

There are additional transport systems in cholangiocytes in addition to those dealing with anion transport. Cholangiocytes take up bile salts via Na^+-dependent transporter ABAT or ASBT (*SLC10A2*) and released them across the basolateral membrane through truncated isoform of the same carrier (Tabibian et al. 2013; Banales et al. 2006b). Glucose is taken up by sodium-glucose transporter SGLT1 (*SLC5A1*) and released by GLUT1 (*SLC2A1*) (Tabibian et al. 2013). Glutathione, the tripeptide that is one of the principal driving forces for canalicular bile salt-independent secretion, is degraded by γ-glutamyltranspeptidase, expressed in cholangiocyte apical membranes, and resulting glutamate, cysteine and glycine are reabsorbed by Na^+-dependent transport in cholangiocytes (Tabibian et al. 2013).

12.4.4   Gallbladder Epithelium

The function of the gallbladder is to store bile and concentrate it during interdigestive phases by salt-dependent water reabsorption. Most studies on the rabbit and *Necturus* gallbladders showed that this transport was electroneutral (Na^+/H^+ and Cl^-/HCO_3^- exchangers in parallel) and these epithelia have low resistance and are regarded as a leaky type (Petersen and Reuss 1983; Petersen et al. 1990). However, in studies on human and primate gallbladders it was discovered that these electrically silent absorptive organs can secrete, under the influence of secretin, HCO_3^--containing fluid after meals (Igimi et al. 1992; Svanvik et al. 1984). This secretion seems to depend on CFTR-like cAMP activated channels that have relatively high permeability to HCO_3^- (Meyer et al. 2005). The importance of CFTR is highlighted in pig models of cystic fibrosis, where the disruption of CFTR leads to gallbladder and bile duct abnormalities (Rogers et al. 2008). Furthermore, prairie dog gallbladder shows forskolin-induced short-circuit current that was due to CFTR (Moser et al. 2007). This preparation is often used as an experimental model of human cholelithiasis, due to its unique propensity for developing gallstones on high-cholesterol chow. The basolateral HCO_3^- transport in gallbladder epithelium is carried out by pNBC1 and the driving force for secretion is maintained by cAMP stimulated K^+ channels that are also sensitive to Ca^{2+} and pH, but their identity is not clear (Moser et al. 2007; Meyer et al. 2005).

12.4.5   Regulation of Bile Formation

The bile formation that occurs in at least three steps is also regulated at three levels. The first step is the canalicular bile fluid secretion by hepatocytes that occurs continuously and is relatively poorly regulated (Esteller 2008; Banales et al. 2006b). Interestingly, one regulator of canalicular HCO_3^- secretion is glucagon (Lenzen et al. 1997; Alvaro et al. 1995), which increases cAMP and stimulates the insertion of vesicles containing Cl^-/HCO_3^- exchanger (AE2, *SLC4A2*) and AQP8 (Gradilone et al. 2003). In the last step, gallbladder secretion is regulated by secretin and gallbladder contraction is stimulated by CCK.

The most extensive regulation of bile modification (secretion and absorption) occurs in bile ducts (Banales et al. 2006b; Tabibian et al. 2013). Firstly, there are regulatory factors modifying the basal secretion of cholangiocytes. The most important hormone is secretin that is released as a physiological response to meals; the other two are bombesin (gastrin-releasing peptide) and vasoactive intestinal peptide (also a neurotransmitter). Secondly, acetylcholine and noradrenaline potentiate secretin-stimulated HCO_3^- and fluid secretion. Thirdly, many factors such as somatostatin, gastrin, insulin, dopaminergic agonists, α2 adrenergic receptor agonists, endothelin, GABA, and cytokines (e.g., IL1, IL6, TNFα), inhibit basal and secretin-stimulated cholangiocyte secretion. Lastly, there a number of bile-borne factors (flow and osmolality, amino acids, glucose, nucleotides, bile acids), that regulate cholangiocyte function. Here, we will focus on regulation by nucleotides and bile acids, similar to the section above dealing with pancreatic ducts.

12.4.5.1 Purinergic Signaling

Both hepatocytes and cholangiocytes release adenosine nucleotides, they express a number of ecto-nucleotidases, and both cell types express a number of P2 and P1 receptors (Schlosser et al. 1996; Chari et al. 1996; Fausther and Sevigny 2011). Cholangiocytes exhibit shear/flow-sensitive and cell volume sensitive ATP is release, as well as release stimulated by forskolin and ionomycin, and it has been proposed that CFTR is regulating this ATP release (Schlosser et al. 1996; Chari et al. 1996; Fiorotto et al. 2007; Minagawa et al. 2007). Nevertheless, it seems that ATP release is larger in small upstream cholangiocytes compared to downstream ones, indicating that smaller ducts could signal to larger ones via paracrine ATP signal (Woo et al. 2008), and that other mechanism than CFTR are important in regulating ATP secretion. In fact, recent studies indicate that ATP is also released by vesicular exocytosis similar to the pancreas (Feranchak et al. 2010; Woo et al. 2010).

Both P2 and adenosine receptors regulate cholangiocytes by stimulating K^+, Cl^- and HCO_3^- secretion. P2 receptors on the basolateral membrane of cholangiocytes, similar to acetylcholine evoked Ca^{2+} signaling, seem to have minimal effect on bile duct secretion (Nathanson et al. 1996), possibly due to rapid hydrolysis of nucleotides (Dranoff et al. 2001). The most prominent effects are relayed by the P2 receptors on the luminal membrane, many of which stimulate Ca^{2+} signaling or induce Ca^{2+} influx, which leads to stimulation of CaCC, short-circuit current and ductal alkalization (Woo et al. 2010; Dranoff et al. 2001). Mechanosensitive ATP release stimulates TMEM16A/ANO1 in human cholangiocytes from small and large biliary ducts (Dutta et al. 2011, 2013). There a number of receptors expressed on cholangiocytes, e.g., mouse and rat express P2X4 and P2Y2, as well as P2Y1, P2Y4, P2Y6, P2Y12, and P2Y13 receptors (Woo et al. 2010; Dranoff et al. 2001; Doctor et al. 2005). Interestingly, the P2Y12 receptor is expressed on the primary cilium and induces cAMP signaling (Masyuk et al. 2008).

12.4.5.2 Bile Acids

Bile acids are synthesized from cholesterol in the liver and excreted into bile and subsequently into the small intestine, where they facilitate digestion and absorption of dietary fats and fat-soluble vitamins. Bile acids are also signaling hormone-like molecules that act on many tissues/cells in our body and do so by activating several different receptors and sensing proteins including nuclear farnesoid X receptor (FXR), pregnane X receptor (PXR) and G-protein coupled receptor TGR5 (Gpbar-1) (Keitel and Haussinger 2013; Pols et al. 2011). FXR and PXR are strongly expressed in hepatocytes, while TGR5 is strongly expressed in cholangiocytes.

Cholangiocytes are exposed to millimolar concentrations of bile acids and conjugated salts. Such concentrations would be toxic to cholangiocytes and several defense mechanisms are at hand. One of these is the formation of micelles of phospholipids and bile salts. Another defense mechanism proposed is the biliary HCO_3^- secretion (van Niekerk et al. 2018). It has been suggested that BA profile/composition changes radially from the midstream to the apical membrane of cholangiocytes (Keitel and Haussinger 2013). Cholangiocytes express TGR5 in the primary cilia, the apical membrane, and sub-cellular (Keitel et al. 2010; Masyuk et al. 2013). In particular, the primary cilia, which are mechano-, chemo-, and osmosensors, could sense BA and via TGR5 receptors that trigger cAMP signaling pathways that stimulate CFTR and AE2 and thus HCO_3^- secretion. Bile acids also stimulate CaCC directly (Shimokura et al. 1995), or through CFTR regulated ATP release and P2 receptor-mediated stimulation of CaCC (Fiorotto et al. 2007). The outcome of these events would be the formation of unstirred HCO_3^- layer, so-called "HCO_3^- umbrella", which together with dense glycocalyx creates a protective microenvironment at the apical membrane of cholangiocytes and prevents the protonation of glycine-conjugated bile acids dominating in human bile (Beuers et al. 2010; Keitel and Haussinger 2013; van Niekerk et al. 2018). Thereby, cholangiocytes would be protected from diffusion of these polar and pro-apototic acids. Relevance of this system is demonstrated in patients with primary biliary cirrhosis, in which AE2 expression in bile ducts is reduced, leading to defective HCO_3^- secretion and BA mediated cell injury (Keitel and Haussinger 2013).

12.5 Duodenum

The pH of the duodenal luminal contents is more or less dictated by the acid chyme arising from the stomach until the pH is neutralized by the pancreatic and hepatic outlets. In order to withstand the potentially damaging effect of low bulk pH, the duodenal epithelium creates a mucous layer. This layer is supplemented with HCO_3^- from the epithelial cells to create a protective barrier toward acid and contributes to pH neutralization of duodenal contents (Allen and Flemstrom 2005; Flemstrom and Kivilaakso 1983; Williams and Turnberg 1981; Ainsworth et al. 1990, 1992). In rabbit duodenum, HCO_3^- secretion amounts to 154 µEQ/cm/h, while other species have lower rates (Flemstrom et al. 1982). Duodenal ulcers result from an unbalance between protective factors such as mucus and HCO_3^-, and the aggressive factors gastric acid, pepsin, and the infestation with pathogenic *Helicobacter pylori*. The long-prevailing model for duodenal epithelial HCO_3^- secretion was inspired by studies of the exocrine pancreas given above. As mentioned, the functional coupling of a luminal surface Cl^- channel and a Cl^-/HCO_3^- exchanger is fuelled by intracellular conversion of CO_2 and H_2O to form HCO_3^- and H^+ catalyzed by intracellular carbonic anhydrases. The intracellular pH is normalized by basolateral NHE mediated extrusion of protons. The essence of this model is still standing, but a number of recent investigations have greatly expanded and

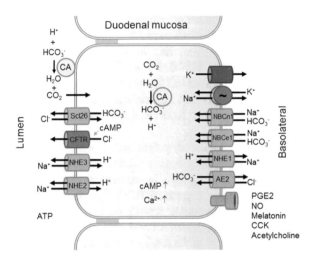

Fig. 12.6 Schematic diagram showing a current model for HCO_3^- secretion from duodenal enterocytes. In this model, secreted HCO_3^- does not arise from the blood side CO_2 and intracellular conversion to HCO_3^- and H^+. Instead, CO_2 enters from the luminal side, which has high pCO_2, and potentially counteracting the HCO_3^- secretion and challenging the cellular pH homeostasis. HCO_3^- for maintaining intracellular pH and for secretion may arise from the basolateral HCO_3^- import

refined the understanding of the molecular machinery for duodenal bicarbonate secretion (Fig. 12.6).

The involvement of interstitial CO_2, which enters the cells and CA and generates HCO_3^- for secretion, has been challenged. In the duodenum, the luminal pCO_2 is higher than anywhere else in the mammalian body (Rune and Henriksen 1969), due to the massive neutralization of gastric acid mainly by the pancreatic bicarbonate and hepatobiliary outlets. CO_2 crosses the luminal membrane to allow the enterocytes to sense and respond to luminal acid in a complex interplay with cytosolic carbonic anhydrase and a moderating effect of luminal carbonic anhydrases (Holm et al. 1998; Kaunitz and Akiba 2002, 2006a, b; Sjoblom et al. 2009). Beside the cytosolic CAII, the duodenal enterocytes express the membrane-bound luminal CAXIV and basolateral CAIX (Lonnerholm et al. 1989; Saarnio et al. 1998). As reviewed recently, it is most likely that the CA activity is centrally involved in acid sensing and signaling in the duodenal enterocytes rather than participating in HCO_3^- secretion directly (Sjoblom et al. 2009).

Isolated duodenal epithelial cells exhibit NHE activity (Ainsworth et al. 1996, 1998; Isenberg et al. 1993). The findings were first interpreted as basolateral NHE1 activity according to the general working model. However, a subsequent study of mouse duodenal enterocytes reported several distinct NHE activities and evidenced the molecular expression of NHE1 in the basolateral membrane, and NHE2 and NHE3 in the luminal membrane (Praetorius et al. 2000; Praetorius 2010). NHE1 seems expressed at low abundance as compared to NHE2 and NHE3 and is an unlikely mechanism for eliminating intracellularly produced protons from duodenal enterocytes. Interestingly, Na^+-HCO_3^- cotransport may serve an important role in the duodenum as a supply for intracellular HCO_3^- as a defense against intracellular acid formed from CO_2 entering from the lumen (Akiba et al. 2001a, b). The robust expression of the HCO_3^- loaders NBCe1 and NBCn1 in the basolateral membrane of duodenal epithelial cells spurred the hypothesis that HCO_3^- is transported from the interstitial space to the lumen via a transcellular route, where the two transporters import HCO_3^- in symport with Na^+ from the interstitium to the cell (Praetorius et al. 2001).

Acidified rabbit enterocytes display potent Na^+-dependent recovery of pH_i in the presence of the CO_2/HCO_3^- buffer system (Ainsworth et al. 1996; Isenberg et al. 1993). DIDS and H_2DIDS only slightly affected the recovery rate from the acid load. A similar DIDS-insensitive Na^+ dependent HCO_3^- import was observed in mouse (Praetorius et al. 2001) and rabbit duodenal enterocytes (Jacob et al. 2000b). Human duodenal enterocytes express both NBCn1 and NBCe1, although immunoreactivity for the latter is subtle compared to the renal and pancreatic staining (Damkier et al. 2007). Nevertheless, the DIDS-sensitive electrogenic NBCe1 has been put forward by several investigators as the main HCO_3^- import mechanism for transcellular HCO_3^- secretion, although both NBCn1 transport and the duodenal Na^+-HCO_3^- import are relatively DIDS insensitive (Jacob et al. 2000a; Praetorius et al. 2001). Importantly, based on NBCn1 knockout studies in mice, it was recently shown that NBCn1 is the major contributor to Na^+-dependent HCO_3^- transport as well as basal and acid-induced HCO_3^- secretion (Chen et al. 2012; Singh et al. 2013b).

The mechanism by which duodenal enterocytes extrude HCO_3^- across the luminal membrane has been a matter of intense debate. Many transporters and channels have been investigated, among which CFTR and the proteins DRA (*SLC26A3*), PAT1 (*SLC26A6*), and *SLC26A9* have been the most prominent examples. First, it is firmly established that basal and agonist-induced duodenal HCO_3^- secretion strongly depends on CFTR expression (Clarke and Harline 1998; Hogan et al. 1997a, b; Seidler et al. 1997). It is still not clear how much CFTR conductance contributes to HCO_3^- secretion directly. First, HCO_3^- extrusion does not depend on CFTR, at least in isolated duodenal villus cells (Praetorius et al. 2002). Second, DRA mediates electroneutral Cl^-/HCO_3^- exchanger (Alper et al. 2011), is involved in NaCl absorption throughout the leaky intestinal segments, and seems mainly expressed in the lower villus/crypt axis in the duodenum (Walker et al. 2009). The study indicates that DRA contributes to bicarbonate secretion mainly in the lower villus/crypt. PAT1 is mainly expressed in the proximal small intestine and is prevalent in the villus region where it also functions as the predominant anion exchanger (Simpson et al. 2007; Singh et al. 2013a; Walker et al. 2009). Knockout of PAT1 reduces basal duodenal HCO_3^- secretion by approximately 70%, while knockout of DRA only induced a 30% reduction (Simpson et al. 2007). However, only a minor role for PAT1 was found in acid-stimulated HCO_3^- secretion (Singh et al. 2013a). The relevance of this transport system was highlighted in a recent study. Here, *H. pylori* infestation decreased duodenal mucosal CFTR and SLC26A6 expression via increased serum transforming growth factor β (TGFβ) in both mice and a human duodenal cell line (Wen et al. 2018). The *SLC26A9* gene product is mainly expressed in the proximal gastrointestinal tract and found in the crypt region of the duodenum (Liu et al. 2015). SLC26A9 is associated with anion conductance and the corresponding knockdown mice have reduced acid-induced duodenal HCO_3^- secretion and worsened intestinal function and survival of CFTR-deficient mice (Singh et al. 2013a; Walker et al. 2009; Liu et al. 2015). An integrative model of duodenal bicarbonate secretion is shown in Fig. 12.6.

Among the many stimulators of duodenal bicarbonate secretion, luminal acid is a physiological and effective agonist (Flemstrom and Kivilaakso 1983; Isenberg et al. 1986; Allen and Flemstrom 2005). The neurohumoral control of duodenal HCO_3^- transport is mediated either directly by cholinergic innervation acting on M3 receptors and indirectly via paracrine melatonin secretion from enterochromaffin cells, both using Ca^{2+} signaling to induce stimulation (Sjoblom and Flemstrom 2003). The same study shows that CCK also enhances HCO_3^- secretion through Ca^{2+} signaling. One of the first agonists to be discovered was PGE2, which acts via cAMP as well as Ca^{2+} signaling (Aoi et al. 2004), while the stimulatory effect of enterotoxin and guanylins occurs through cGMP increases (Guba et al. 1996; Joo et al. 1998). The intracellular signals are most likely to stimulate HCO_3^- secretion via regulation of the luminal membrane transporters and channels, with CFTR being the best documented.

12.6 Renal Intercalated Cells

The overall function of the kidneys is broadly to secure homeostasis and this is achieved by maintaining water and electrolyte balance, acid/base balance, blood pressure, and rid the organism of waste products (Boron and Boulpaep 2012, 2017). The kidneys filter the blood plasma in the renal corpuscles and then modify the filtrate to conserve appropriate amounts of water, nutrients, and salt by secretion and reabsorption processed in an intricate system of renal tubules. As an integral part of the process, the kidney helps to maintain normal acid/base balance by adjusting the secretion of acid/base equivalents. Among the many epithelial cell types in the renal tubular system, a single one has the capacity of secreting HCO_3^- which is the type-B intercalated cell (IC). These cells are restricted to the late distal convoluted tubules, the connecting tubules, and cortical collecting ducts, and are functionally mirror images of the acid-secreting type-A ICs (Kwon et al. 2012). In the mentioned renal cortical segments, the intercalated cells comprise almost 50% of the tubular cells. Most of the cortical ICs are type-B cells, whereas medullary collecting ducts exclusively contain type-A ICs. Interestingly, the numbers of intercalated cell types can vary in a regulated and dynamic fashion. The number of type-A ICs is increased in response to metabolic acidosis (Schwartz et al. 1985). It seems that the change in relative numbers relies on the ability of type-B ICs to differentiate into type-A ICs, given the proper stimulus (Schwartz et al. 1985). Hensin is a protein that is secreted by the epithelial cells and incorporated in the extracellular matrix and is necessary for cell differentiation (Gao et al. 2010; Al-Awqati 2013; Schwartz et al.

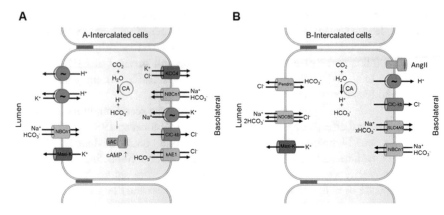

Fig. 12.7 (**a**) Schematic diagram showing the molecular machinery for acid/base transport in type-A intercalated cells from the renal tubular system. (**b**) Schematic diagram showing the molecular machinery for acid/base transport in type-B intercalated cells from the renal tubular system

2002). It stimulated differentiation of type-B ICs to type-A ICs only in its polymeric forms induced by binding of certainly activated integrins and the activity of extracellular galectin 3 (Hikita et al. 2000; Schwaderer et al. 2006; Vijayakumar et al. 2008). Figures 12.7a and b depict the two types of cells and evidence for ion transporters is elaborated below.

The bicarbonate secreting type-B IC (or β-IC) was first described in the turtle bladder (Leslie et al. 1973; Stetson and Steinmetz 1985), which like similar cells in toad skin shares the functional significance for HCO_3^- secretion, as well as transport properties with mammalian renal type-B ICs (Schwartz et al. 1985). The studies showed that these cells were characterized by the luminal extrusion of HCO_3^- in exchange for Cl^-, thereby affecting pH homeostasis in the opposite direction as the already described type-A ICs. The opposite function of the two cell types is accompanied by different polarization of V-ATPases and anion exchangers. The acid-secreting type-A ICs expresses the V-ATPase in the luminal membrane, whereas type-B ICs have basolateral V-ATPase (Brown et al. 1988a). It is now evident that the type-B ICs use the *SLC26A4* gene product pendrin to extrude bicarbonate apically into the urine. Pendrin is an anion exchanger shown to function as a Cl^-/HCO_3^- exchanger in the type-B ICs (Royaux et al. 2001). This was the first study to show that pendrin is expressed in the apical membrane of type-B ICs and is involved in HCO_3^- secretion when animals were challenged by alkalosis, but not under baseline conditions. Importantly, the renal tubules lose apical Cl^-/HCO_3^- exchange and the capability to rid the body of excess base by the urine when the pendrin gene is disrupted (Amlal et al. 2010; Royaux et al. 2001). Type-A ICs expresses a renal specific variant of the anion exchanger AE1 (*SLC4A1*) in the basolateral membrane to extrude excess HCO_3^- (Alper et al. 1989). For both types of ICs, the source for acid/base extrusion is the intracellular conversion of CO_2 and H_2O to protons and HCO_3^-. The reaction rate is critically dependent on the intracellular carbonic anhydrase, CAII, and therefore inhibited by acetazolamide or

CAII gene deletion (Breton et al. 1995; Sly et al. 1983). Pendrin function is sensitive to changes in local pH (Azroyan et al. 2011), whereas the expression level and function are reduced by metabolic acidosis and increased in alkalosis (Frische et al. 2003; Petrovic et al. 2003; Wagner et al. 2002). The key modulator of pendrin expression may not be systemic pH alone but include a concomitant change in circulating Cl^- (Hafner et al. 2008; Vallet et al. 2006; Verlander et al. 2006). At the hormone level, there is an agreement that angiotensin II stimulates pendrin either directly or via activation of the V-ATPase (Wagner et al. 2011).

Thus, the acid/base transport models for ICs are quite simple and well established, but have nevertheless been challenged somewhat by more recent findings. First, antibodies against the human NBC3, also known as NBCn1 (*SLC4A7*) colocalized the protein with the V-ATPase in vesicles and the apical membrane of rat type-A ICs and to the basolateral membrane of type-B ICs (Kwon et al. 2000; Pushkin et al. 1999). The authors detected DIDS-sensitive NBC activity in the apical surface of type–A ICs. These findings were surprising, as NBC would function as an electroneutral alternative transporter for type-A ICs to acidify the urine. In the type-B ICs the function would be to acidify the interstitium in parallel to direct H^+ transport by V-ATPase. However, antibodies against the corresponding rat protein, NBCn1, localized the protein basolaterally in rat type-A ICs (Vorum et al. 2000). This localization of NBC is counterintuitive, as a main task for that membrane is in fact base extrusion. It should be mentioned that neither antibody produced labeling of human or mouse renal intercalated cells, but was expressed elsewhere in the kidney in humans (Damkier et al. 2007). Thus, data on NBCn1 expression in the ICs are inconsistent and the lack of renal phenotype of *SLC4A7* depleted mice indicates that its function may be much less important than the H^+-ATPase function.

Second, recent discoveries have led to a change of the paradigm that intercalated cells are simply acid/base regulating cells, but participate also in Cl^- transport. Previously, Cl^- was expected to take a paracellular route depending solely on a transepithelial potential difference and Cl^- permeable tight junctions such as the aldosterone regulated claudin-4 (Le et al. 2005). Pendrin was established as a transcellular route for connecting tubule and collecting duct Cl^- reabsorption (Wall et al. 2004). Genetic deletion of pendrin was shown to eliminate collecting duct Cl^- reabsorption in mineralocorticoid- and bicarbonate-treated mice. Thus, pendrin knockout mice are protected from mineralocorticoid-induced hypertension, while overexpression of pendrin leads to hypertension in mice on high-NaCl diet (Jacques et al. 2013; Verlander et al. 2003). The significance of HCO_3^- secretion by pendrin was also suggested to directly affect the function of the principal cell Na^+ loader ENaC (Pech et al. 2010). Whereas single-gene knockout for the distal tubule NaCl reabsorption protein NCC and pendrin are surprisingly benign, mice with NCC-pendrin double knockout have a severe salt and volume wasting alkalotic phenotype, even under control conditions (Soleimani et al. 2012). Thus, according to the current model of collecting duct, Na^+ reabsorption and Cl^- reabsorption takes place in separate cells—principal cells and type-B ICs, respectively (Wall and Pech 2008). Pendrin now seems to take part in the electroneutral NaCl reabsorption mechanism in addition to its role in alkalization in metabolic alkalosis. It is not known how these roles are separated and regulated.

The *SLC4A8* gene product NDCBE1 has been associated with electroneutral and thiazide sensitive NaCl reabsorption in the cortical collecting duct (Leviel et al. 2010). It may reside and function in the luminal membrane of the type-B ICs and in the proposed model (Fig. 12.7), NDCBE1 and pendrin work in concert with two turnovers of pendrin for each NDCBE1 turnover to yield electroneutral, pH neutral NaCl reabsorption through type-B ICs. The model is based on the observations that Na^+ reabsorption in collecting ducts are not completely inhibited by amiloride, but contains a thiazide component under some experimental conditions. Also, the authors find luminal Cl^- and CO_2/HCO_3^- dependence of the thiazide sensitive Na^+ transport in isolated cortical collecting ducts from Na^+ depleted animals. Furthermore, they detect NDCBE1 protein expression and mRNA in the relevant tubular segments and demonstrate lack of such transport in tubules from *NDCBE1* knockout mice (Leviel et al. 2010). In a separate study, the *SLC4A9* gene product, usually named AE4, was suggested as the basolateral Na^+ exit pathway in these cells (Chambrey et al. 2013). It was shown, that the *AE4* gene deletion greatly reduced thiazide sensitive Na^+ and Cl^- reabsorption. Interestingly, AE4 is suggested to work as an electrogenic Na^+-HCO_3^- extruder in their model. This would, however, render the total reabsorption process electrogenic, as extrusion of Na^+ and HCO_3^- (depending on the electrochemical gradients) would usually be in a 1:3 stoichiometry and the model, only 1 Cl^- follows through ClC channels in the basolateral membrane. As all these discoveries have a profound impact of our understanding of renal electrolyte handling, it is highly important to establish the electrochemical gradients of the type-B ICs in order to validate the model, which also suggests the V-ATPase as the creating mechanism for transmembrane gradients (as opposed to the Na^+/K^+-ATPase (Chambrey et al. 2013)). It is also necessary to localize the NDCBE1 protein in the apical membrane of type-B ICs, and directly evidence that thiazides target this protein e.g., by ligand binding assays. In *Xenopus laevis* oocytes, neither NDCBE1 nor pendrin seemed to be the site of action for thiazides, and carbonic anhydrase was only discharged as target in control experiments in untreated collecting ducts (Leviel et al. 2010). Furthermore, before NDCBE1 is recognized as important in renal Na^+ and Cl^- reabsorption and in blood pressure regulation, it is necessary to report normal urine electrolytes, pH, osmolarity as well as blood pressure values at baseline and in Na^+ depleted NDCBE1 KO mice. Whereas principal cell-specific ENaC subunit knockout has a profound impact on baseline Na^+ and K^+ homeostasis, as well as urine output (Christensen et al. 2010), both pendrin KO and NDCBE1 KO phenotypes seem mild as far as they have been reported.

12.7   Choroid Plexus Epithelium

The choroid plexus epithelium (CPE) produces the majority of the intraventricular cerebrospinal fluid, CSF. Rougemont and colleagues directly demonstrated the involvement of the choroid plexus in CSF formation in 1960 (Rougemont et al. 1960), although this had already been suggested by J. Faivre and H. Cushing (1914) and Faivre (1854). Rougemont's group established that the solute contents in the nascent CSF differed from a simple plasma ultrafiltrate. Taken together with later studies, the evidence that CSF cannot be a simple ultrafiltrate includes (1) the CSF is approximately 5 mOsM hypertonic compared to plasma (Davson and Purvis 1954); (2) the [Na^+] and [HCO_3^-] are slightly higher, whereas [K^+] and [Cl^-] are lower than expected in the CSF at equilibrium (Ames III et al. 1964); and (3) there is a 5 mV lumen positive electrical potential difference across the choroid plexus epithelium. Therefore, the choroid plexus must sustain a secretory function, given that most intraventricular CSF arises from this structure. The CSF has many important functions in the central nervous system. CSF fills the internal and external fluid spaces and thereby cushions the central nervous system, keeps the intracranial pressure at controlled levels, and it stabilizes the composition of the brain extracellular fluid. Primarily, CPE secreted fluid contains $NaHCO_3$, and therefore, can neutralize acid produced by neuronal activity. The choroid plexus epithelium also secretes growth factors and nutrients into the CSF that are crucial for the brain development and function (Damkier et al. 2013). Importantly, CSF also helps to remove, i.e., absorb, K^+ produced by active neurons, as well as breakdown products of serotonin and dopamine. The composition of newly formed CSF is roughly 150 mM Na^+, 3 mM K^+, 130 mM Cl^-, and 25 mM HCO_3^- (Davson and Segal 1996; Johanson and Murphy 1990). The choroid plexus produces up to 0.4 ml/g/min fluid, which gives approximately 60 µmol HCO_3^-/h/g tissue output.

While some mechanisms of CSF secretion are well established, other aspects of secretion are not fully understood. For example, the main transport mechanisms for Na^+ into the choroid plexus epithelium from the blood side are not defined. Another central question is how HCO_3^- is extruded from the cells into the CSF. Clues into the main transport pathways are reflected by the wide profile of drugs inhibiting CSF secretion: ouabain, acetazolamide, amiloride, DIDS, bumetanide, and furosemide (Ames III et al. 1965; Davson and Segal 1970; Johanson and Murphy 1990; Melby et al. 1982; Wright 1972). The secretion rate of CSF is also influenced by the HCO_3^- concentration at the basolateral side (Mayer and Sanders-Bush 1993; Saito and Wright 1983). Thus, CO_2/HCO_3^- metabolism seems important for CSF secretion. Before describing the putative HCO_3^- extrusion mechanisms, we will give the general transport machinery that creates and sustains the gradients permitting HCO_3^- secretion. Lastly, other acid/base transporters that may assist or modify HCO_3^- secretion are considered.

12.7.1 Basic Secretory Machinery

As opposed to most other epithelia, the choroid plexus Na^+/K^+-ATPase is localized to the luminal membrane. Ventricular application of ouabain blocks the transepithelial net Na^+ flux as well as the CSF secretion (Ames III et al. 1965; Davson and Segal 1970; Wright 1978). The luminal Na^+/K^+-ATPase expression is confirmed by both immunoreactivity and ouabain binding studies (Ernst et al. 1986; Masuzawa et al. 1984; Praetorius and Nielsen 2006; Quinton et al. 1973; Siegel et al. 1984). The luminal Na^+/K^+-ATPase directly transports Na^+ into the CSF and also maintains the Na^+ and K^+ gradients that drive most other transport processes in CSF secretion (Fig. 12.8). Na^+/K^+-ATPase complexes consisting of $\alpha 1$, and either $\beta 1$ or $\beta 2$ subunits, which are expressed in the choroid plexus together with the γ-subunit,

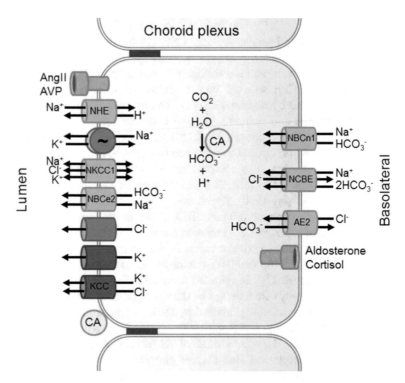

Fig. 12.8 Schematic diagram showing a simplified model of the CSF and thereby HCO_3^- secretion process by the choroid plexus epithelium. Although HCO_3^- secretion greatly depends on carbonic anhydrase activity, it also fully relies on a DIDS-sensitive basolateral supply of Na^+ and HCO_3^-. The luminal HCO_3^- extrusion is electrogenic and CSF pH increases are followed by an increase in CSF Na^+ content, indicating a molecular or functional coupling of the extrusion of both ions. Interconversion between $CO_2 + H_2O$ and $H^+ + HCO_3^-$ is catalyzed by carbonic anhydrases, i.e., luminal CAXII and intracellular CAII. Cl^- conductances are characterized as a volume sensitive anion conductance (VRAC) and an inward rectifying Cl^- conductance (Clir), whereas the K^+ channels probably include Kv1.3, Kv1.1, and Kir7.1. KCCs are also a luminal K^+ exit pathway

phospholemman (FXYD1) (Feschenko et al. 2003; Klarr et al. 1997; Zlokovic et al. 1993). These subunits were confirmed in a recent proteomic study on FACS isolated mouse choroid plexus epithelial cells (Damkier et al. 2018). This study also revealed the expression of $\alpha 2$, $\alpha 4$, and $\beta 3$ Na^+/K^+-ATPase subunits.

K$^+$ channels expressed on the luminal membrane have several functions. Firstly, they are important for K$^+$ recycling in connection with the Na^+/K^+-ATPase activity. Secondly, K$^+$ channels are the main determinant for the membrane potential. This is important for the luminal HCO_3^- secretion, which is electrogenic. Thirdly, certain K$^+$ channels (inward rectifiers) are also important for cellular uptake of K$^+$ in hyperpolarized voltages in order to provide a K$^+$ sink, e.g., removal of neuron derived K$^+$. K$^+$ channels were first identified in the choroid plexus by patch-clamping the luminal membrane of *Necturus maculosa* cells (Brown et al. 1988b; Christensen and Zeuthen 1987; Zeuthen and Wright 1981). Several K$^+$ conductances have been identified: an inward-rectifying conductance (Kir7.1 channels), and outward-rectifying conductances (Kv1.1 and Kv1.3) (Doring et al. 1998; Kotera and Brown 1994; Speake and Brown 2004). KCNQ1/KCNE2 channels may also contribute to the outward-rectifying conductance (Roepke et al. 2011). All these K$^+$ channel proteins are expressed in the luminal membrane of rat and mouse choroid plexus (Nakamura et al. 1999; Roepke et al. 2011; Speake and Brown 2004). Proteomic analysis confirmed the expression of Kir7.1 and voltage-gated potassium channel KCNE2, and detected potassium channel subfamily K$^+$ member 1 KCNK1/ TWIK-1 as well (Damkier et al. 2018).

The CPE cells express abundant NKCC1 in the luminal membrane (Keep et al. 1994; Plotkin et al. 1997). With the typical ionic distribution across the luminal membrane in CPE, NKCC1 is close to equilibrium (Keep et al. 1994). Thus, both inward and outward NKCC1 transport has been observed (Steffensen et al. 2018; Gregoriades et al. 2018). Measurements of the precise ionic gradients operating in vivo will hopefully soon settle this discrepancy. The K^+-Cl^--cotransporters (KCCs) are electroneutral and transport of the ions out the cells is driven by the outward [K$^+$] gradient across cell membranes. KCCs are like NKCC1 inhibited by furosemide (Russell 2000). KCCs were described as a furosemide-sensitive K$^+$- and Cl^--dependent transport in the luminal membrane of bullfrog choroid plexus (Zeuthen 1991). KCC1 and KCC4 (*SLC12A7*) most likely mediate the observed transport in the luminal membrane, whereas KCC3a (*SLC12A6*) is expressed in the basolateral membrane (Karadsheh et al. 2004; Pearson et al. 2001). KCC4 may, therefore, contribute to the recycling of K$^+$ across the luminal membrane helping to sustain the Na^+/K^+-ATPase activity (Fig. 12.8). Interestingly, the proteomic study mentioned above identified only NKCC1, KCC1, KCC4, and included a new candidate, CCC1/Cip1 (*SLC12A9*) in mouse choroid plexus epithelial cells (Damkier et al. 2018).

12.7.2 Luminal HCO₃⁻ Extrusion

A substantial Cl^- and HCO_3^- efflux across the luminal membrane of the choroid plexus occurs via electrogenic pathways, mainly anion channels for Cl^-. The mechanisms involved in HCO_3^- efflux are less well described, but may overlap to some extend with the Cl^- extrusion. The majority of Cl^- moves via a transcellular route, because luminal DIDS minimizes Cl^- secretion (Deng and Johanson 1989). There are several potential candidates for Cl^- channels. Inward-rectifying anion conductances (Clir) have been shown by patch-clamping in CPE cells from several species (Kajita et al. 2000; Kibble et al. 1996, 1997) and Clir channel activity is augmented by protein kinase A (Kibble et al. 1996), and inhibited by protein kinase C (Kajita et al. 2000). For the amphibian choroid plexus, Saito and Wright proposed that cAMP–regulated ion channels conducting Cl^- and HCO_3^- form the main efflux pathway for the anions in the luminal membrane (Saito and Wright 1984). Inward rectifier Cl^- (Clir) channels have high HCO_3^- permeability in mammals (Kibble et al. 1996). The molecular identity of the Clir channel remains to be determined. Apart from the intracellular chloride channels CLIC1, CLIC3-6, and CLCC1, only voltage-dependent anion-selective channel VDAC1, VDAC2, VDAC3 were detected at the molecular level (Damkier et al. 2018)

The CFTR does not seem to contribute to cAMP-regulated Cl^- conductance in the choroid plexus (Kibble et al. 1996, 1997). ClC-2 channels are also unlikely to play a significant role in the choroid plexus transport, as Cl^- conductance is unaffected in the choroid plexus epithelium from ClC-2 knockout mice (Speake et al. 2002). VRACs were also identified in the choroid plexus epithelium (Kibble et al. 1996, 1997). VRACs are most likely less important for CSF secretion, as their Cl^- conductance at normal cell volumes is low (Kibble et al. 1996; Millar and Brown 2008). LRRC8A was the only subunit of volume-regulated anion channels detected in the recent mouse proteomic study (Damkier et al. 2018).

An alternative mechanism for luminal HCO_3^- extrusion to Cl^- channels is the electrogenic cotransporter NBCe2 (or NBC4), which mediates the transport of 1 Na^+ with 2–3 HCO_3^- ions (Sassani et al. 2002; Virkki et al. 2002). In rat choroid plexus, NBCe2 is situated in the luminal membrane (Bouzinova et al. 2005). In the mouse, it mediates the export of 1 Na^+ for 3 HCO_3^- across the luminal membrane (Millar and Brown 2008). A significant increase in the CSF Na^+ content without a reciprocal change in CSF K^+ occurs in rodents exposed to 11% CO_2 at the CSF side (Nattie 1980). This finding indicates that a Na^+-dependent acid/base transporters, such as NBCe2, is involved in compensating CSF pH for the increased pCO_2. Indeed, the composition of the CSF is changed in NBCe2 gene trap knockout mice with a significant decrease in $[HCO_3^-]$ from 24 to 20 mM (Kao et al. 2011). Direct evidence for a role in CSF pH regulation was shown recently in another NBCe2 knockout mouse model. When HCl was injected into the ventricle cavity, only NBCe2 wildtype mice efficiently normalized the low CSF pH (Christensen et al. 2018). This effect was confirmed by NBCe2 siRNA injections. These findings strongly suggest that NBCe2 participates directly in the regulation of CSF pH by extruding HCO_3^-.

12.7.3 Other Acid/Base Transporters of Consequence for HCO_3^- Secretion

The V-ATPase does not seem to be expressed in the luminal plasma membrane of the choroid plexus epithelium. Among the many V-ATPase subunits only a few are associated with plasma membrane expression of the entire complex. The B1 subunit, which is involved in plasma membrane targeting, is not expressed at mRNA level in the choroid plexus, but B2 mRNA is detectable in these cells (Christensen, Damkier, and Praetorius, unpublished data). Thus, the V-ATPase complexes seem restricted to the lysosomal system in these cells. The rabbit choroid plexus expresses mRNA encoding the non-gastric H^+/K^+-ATPase (Lindvall-Axelsson et al. 1992; Pestov et al. 1998). Interestingly, luminally applied omeprazole inhibited CSF secretion in another study indicating the gastric H^+/K^+-ATPase activity gastric (Lindvall-Axelsson et al. 1992). Both V-ATPase and H^+/K^+-ATPase are expressed at the luminal membrane together with HCO_3^- transporters, thus, counteracting ongoing HCO_3^- secretion. Thus also in this epithelium, the role of H^+-pumps is not clear.

Two NHE forms are expressed in the choroid plexus epithelium: NHE1 and NHE6. The mRNA encoding the NHE1 isoform was detected in the mouse choroid plexus (Damkier et al. 2009; Kalaria et al. 1998) and the Na^+/H^+ exchange activity of isolated epithelial cells from rats and mice is inhibited by EIPA (Bouzinova et al. 2005; Damkier et al. 2009). In most epithelia, NHE1 is localized to the basolateral membrane (Orlowski and Grinstein 2004), and this was also believed to be the case for the choroid plexus epithelium, as intravenous application of amiloride was reported to inhibit the rate of CSF secretion *in vivo* (Murphy and Johanson 1989). Nevertheless, NHE1 is located in the luminal membrane of both mouse and human choroid plexus (Damkier et al. 2009; Kao et al. 2011). NHE6 mRNA was recently detected in the mouse choroid plexus epithelium and the protein was mainly localized to the luminal membrane (Damkier et al. 2018). Functionally, NHE activity in the choroid plexus seems confined to the luminal plasma membrane (Damkier et al. 2009). The luminal membrane location of the NHE1 and NHE6 in CPE predicts that the protein mediates H^+ extrusion from the epithelium to the CSF, Thus, NHE1 and/or NHE6 may serve roles in preventing increases in CSF pH, thereby counteracting efficient HCO_3^- secretion.

The *SLC4A10* gene product NCBE or NBCn2 has been described as a DIDS-sensitive, electroneutral Na^+-HCO_3^- cotransporter driven by the inward Na^+ gradient (Damkier et al. 2010; Giffard et al. 2003; Wang et al. 2000). These authors found that the Na^+ dependent HCO_3^- transport required intracellular Cl^- in mammalian cell lines, and the transporter was therefore called NCBE for Na^+-dependent Cl^-/HCO_3^- exchanger. However, Parker *et al.* showed that the human *SLC4A10* gene product did not extrude Cl^- when expressed in *Xenopus laevis* oocytes, and protein was renamed the protein NBCn2 by these scientists (Parker et al. 2008). This controversy is not resolved and in this chapter, we use the rodent name NCBE for simplicity.

The NCBE is a basolateral plasma membrane protein in the choroid plexus (Praetorius and Nielsen 2006; Praetorius et al. 2004b), and it is believed to contribute to both Na^+ and HCO_3^- uptake from the blood side. The involvement in cellular Na^+ and HCO_3^- uptake is supported by a finding that reports a 70% decrease in the Na^+-dependent HCO_3^- import into choroid plexus cells in an NCBE knockout mouse model (Jacobs et al. 2008). In the murine choroid plexus, NCBE activity can account for almost all of the DIDS-sensitive Na^+-dependent HCO_3^- import (Damkier et al. 2009).

The electroneutral Na^+-HCO_3^- cotransporter 1, NBCn1, is also expressed in the choroid plexus epithelium (Praetorius et al. 2004b). It is normally found in the basolateral membrane of mammalian choroid plexus epithelial cells, but it is localized to the luminal membrane in certain mouse strains and regionally in human choroid plexus (Damkier et al. 2009; Praetorius and Nielsen 2006). The epithelial NBCn1 form seems to be DIDS insensitive. In rat choroid plexus, a large fraction of the pH_i recovery from acid load was mediated by DIDS insensitive Na^+-HCO_3^- cotransport (Bouzinova et al. 2005). However, the NBC activity in choroid plexus from NBCn1 knockout mice is indistinguishable from that of wild type littermates (Damkier, unpublished results). Thus, NBCn1 protein is not an obvious candidate to participate in transcellular HCO_3^- transport.

AE2 mediates electroneutral uptake of Cl^- in exchange for HCO_3^- and is the only anion exchanger from the *SLC4* family in the choroid plexus. AE2 was first localized in the basolateral membrane of mouse choroid plexus epithelium (Lindsey et al. 1990) and later found at the same site in choroid plexus tissue from other mammals (Alper et al. 1994; Praetorius and Nielsen 2006). AE2 is the only known basolateral entry pathway for Cl^- and may, therefore, be very important for CSF secretion. Two lines of evidence suggest the implication of basolateral AE in CSF formation. The inward chemical gradient for Cl^- would favor AE2 mediated Cl^- uptake across the basolateral membrane in the choroid plexus. Furthermore, DIDS applied from the blood side greatly reduce Cl^- transport into the CSF (Deng and Johanson 1989; Frankel and Kazemi 1983).

12.7.4 Model for Bicarbonate Secretion by the Choroid Plexus

A significant body of evidence shows that HCO_3^- transport is an essential element of the CSF secretion and here we summarize the most important transporters (Fig. 12.8). CSF secretion is greatly decreased by carbonic anhydrase inhibition (Vogh et al. 1987; Ames III et al. 1965; Davson and Segal 1970; Welch 1963), and

Na^+/K^+-ATPase activity is diminished in the absence of CO_2/HCO_3^- in amphibian CPE (Saito and Wright 1983). In humans, the intracranial pressure is also reduced by the carbonic anhydrase inhibition in humans (Cowan and Whitelaw 1991). It appears that the majority of the ions secreted to the CSF take a transcellular route. Na^+, Cl, HCO_3^-, and Ca^{2+} are secreted into the CSF, while K^+ is reabsorbed. The transport of these ions is carefully regulated to provide a relatively constant CSF composition despite fluctuations in their plasma or brain extracellular fluid concentrations (Husted and Reed 1976; Jones and Keep 1988; Murphy et al. 1986). NCBE may be a good candidate for mediating HCO_3^- uptake from the blood side, while NBCn1 can only assist in cases where this protein is situated in the basolateral membrane. In the choroid plexus as elsewhere, cellular HCO_3^- is also generated from CO_2 and H_2O catalyzed by carbonic anhydrases, and inhibition of the carbonic anhydrases by acetazolamide reduced CSF secretion by approximately 50% (Vogh et al. 1987). The choroid plexus expresses the cytosolic carbonic anhydrase CAII as well as the membrane-associated isoforms CAXII and CAIX (Kallio et al. 2006). For the actual extrusion of HCO_3^- to the lumen, NBCe2 is identified as a key contributor for regulating CSF pH (Christensen et al. 2018). This does not rule out that the luminal anion conductances also provide HCO_3^- transport across the luminal membrane.

12.7.5 Regulation of CP Bicarbonate Secretion

The control and regulation of choroid plexus secretion of HCO_3^- and CSF is a topic of major clinical importance. Several life-threatening conditions resulting in an increased intracranial pressure, such as head trauma, tumors, stroke, and irradiation damage, can potentially be alleviated by acute reduction of CSF secretion. The relevance of the bicarbonate transporters was underscored in a recent study, where inhibition of NCBE was found efficient as a therapeutic target in post-hemorrhagic hydrocephalus in rat pups (Li et al. 2018). The CSF contains very little protein buffers and the appropriate pH level is most likely kept solely by the CO_2/HCO_3^- buffer system. Adjustments of the HCO_3^- extrusion from the choroid plexus can potentially control the deviations of CSF pH from neutral values. However, virtually nothing is known about the potential acid/base sensing.

Arginine vasopressin (AVP, or antidiuretic hormone ADH) decreases the CSF secretion and reduces the blood flow to the choroid plexus (Johanson et al. 1999). The choroid plexus expresses V1a receptors (Ostrowski et al. 1992; Phillips et al. 1988). AVP is actually produced by CPE stimulated by local angiotensin II through AT1 receptor-mediated activation. Activation of V1a receptor in the luminal membrane decreases CSF secretion by the CPE (Szmydynger-Chodobska et al. 2004). Accordingly, Ca^{2+} increases transiently in the choroid plexus epithelium upon vasopressin administration (Battle et al. 2000), without affecting cAMP levels (Crook et al. 1984). A local system of renin, angiotensinogen as well as angiotensin-converting enzyme (ACE) and AT1 receptors are expressed in the choroid plexus epithelium (Chai et al. 1987a, b; Gehlert et al. 1986; Imboden et al.

1987; Inagami et al. 1980). However, the significance of this local RAS system in the choroid plexus is unclear apart from increasing AVP production. The mineralocorticoid and glucocorticoid receptors are expressed in the choroid plexus (de Kloet et al. 2000; Weber et al. 2003). Plasma glucocorticoid levels usually greatly exceed aldosterone levels, and in aldosterone sensitive cells, the specificity of MR receptor activation is assured by intracellular conversion of cortisol to inactive metabolites by enzyme 11β-hydroxysteroid dehydrogenase type 2 (11βHSD2). Importantly, the choroid plexus expresses 11βHSD type 1 instead of 11βHSD2 (Sinclair et al. 2007) and it is therefore unlikely that aldosterone exerts specific MR mediated actions on the choroid plexus epithelium.

Regarding intracellular signaling, it is unknown how Ca^{2+} inhibits CSF secretion by the choroid plexus. There are substantial numbers of expressed transport mechanisms that are activated by cAMP and/or PKA. However, cAMP and/or PKA induction has only been shown for isoproterenol, prostaglandin, serotonin, and histamine (Crook et al. 1984). Finally, it is noted that the choroid plexus also expresses receptors for atrial natriuretic factor, endothelin-1, serotonin, bradykinin, and insulin, all of which could affect HCO_3^- secretion (Chodobski and Szmydynger-Chodobska 2001).

12.8   Conclusions and Perspectives

The importance of HCO_3^- secretion is clearly demonstrated by the large selection of ion transporters and channels involved in the process, and the fact that mutations or dysregulation of these can lead to a large spectrum of diseases. In this chapter, we have covered a wide selection of, but not all, epithelia that secrete HCO_3^-. In leaky epithelia, HCO_3^- secretion is accompanied by fluid secretion, while in tight epithelia (such as salivary gland ducts and CCDs), HCO_3^- is usually exchanged for another anion. One of the analyses made in this chapter is that there is a great spectrum of secreted HCO_3^- concentrations, secretion rates and therefore HCO_3^- outputs, which range from 1 to 1000 μmol/h/g tissue or organ. These values could be more revealing if corrected for the mass of cells actually performing HCO_3^- secretion in the given organ. Nevertheless, it should be pointed out that HCO_3^- secretion is usually only a part of a complex function any given epithelium performs, and it is regulated by specific neurotransmitters/hormones and locally generated agents or agents from neighboring organs or tissues. Therefore, it is not unexpected to see that there are many molecularly different transporters and constellations for bicarbonate secreting epithelia presented in this chapter. Nevertheless, we seek to deduce whether there are any common mechanisms for HCO_3^- secretion. It appears that the basolateral membranes of epithelia express one or more members of the NBC transporter family, and/or NHEs and intracellular CA generating H^+ and HCO_3^- from CO_2. The HCO_3^- exit on the luminal membrane is a little more controversial. It seems that Cl^- channels in the combination of HCO_3^-/Cl^- exchangers (*SLC26* family), especially the electrogenic type of exchanger, are favored and suffice for many epithelia.

Also, permeability of Cl^- channels to HCO_3^- has been proposed. In a few epithelia, electrogenic NBC has been proposed, but functional evidence for general epithelial HCO_3^- secretion is required. Nevertheless, transport mechanism producing fluid with more than 100 mM HCO_3^- concentrations are still elusive and we are missing information about the electrochemical gradients (membrane voltages, intracellular anion concentrations), CO_2 permeability, and the question still remains whether passive anion transporters are sufficient to explain HCO_3^- exit across the luminal membrane. Key components in the secretory models are also various Cl^- and K^+ channels, that can be stimulated by agonists to open and provide accompanying ions and appropriate membrane voltages.

The most potent HCO_3^- secretors, such as pancreas, some salivary glands, and duodenum could potentially affect body acid/base balance. Therefore, it is not surprising that these epithelia are very finely regulated, perform only periodically when stimulated, and at low secretion rates, HCO_3^- is reabsorbed or salvaged. This requires HCO_3^- reabsorbing mechanism, usually localized at the distal part of the complex structures of glands or gastrointestinal tract. These mechanisms could include certain isoforms of NBC transporters, or NHEs, and H^+ transporters.

Since the 1990s research has led to molecular identification of many transport proteins involved in bicarbonate secretion. Future challenges are to pinpoint the most important pH and CO_2 sensors, elucidate the effect of local regulating agents, and clarify which acid/base transporters are involved in pH_i regulation as opposed to pH_e regulation carried out by HCO_3^- secretion. Most importantly, studies of bicarbonate secretion and regulation on more integrated organ and whole-body settings could provide some important answers.

Acknowledgments Research projects founding basis for this chapter were supported by The Danish Council for Independent Research I Natural Sciences, The Danish Council for Independent Research I Medical Sciences, The Lundbeck Foundation, The Novo Nordisk Foundation, and The Carlsberg Foundation.

References

Aalkjaer C, Cragoe EJ Jr (1988) Intracellular pH regulation in resting and contracting segments of rat mesenteric resistance vessels. J Physiol 402:391–410

Abuladze N, Lee I, Newman D, Hwang J, Boorer K, Pushkin A, Kurtz I (1998) Molecular cloning, chromosomal localization, tissue distribution, and functional expression of the human pancreatic sodium bicarbonate cotransporter. J Biol Chem 273:17689–17695

Abuladze N, Pushkin A, Tatishchev S, Newman D, Sassani P, Kurtz I (2004) Expression and localization of rat NBC4c in liver and renal uroepithelium. Am J Physiol Cell Physiol 287:781–789

Ahn W, Kim KH, Lee JA, Kim JY, Choi JY, Moe OW, Milgram SL, Muallem S, Lee MG (2001) Regulatory interaction between the cystic fibrosis transmembrane conductance regulator and HCO_3^- salvage mechanism in model systems and the mouse pancreatic duct. J Biol Chem 276:17236–17243

Ahuja M, Jha A, Maleth J, Park S, Muallem S (2014) cAMP and Ca signaling in secretory epithelia: crosstalk and synergism. Cell Calcium 55:385–393

Ainsworth MA, Kjeldsen J, Olsen O, Christensen P, Schaffalitzky de Muckadell OB (1990) Duodenal disappearance rate of acid during inhibition of mucosal bicarbonate secretion. Digestion 47:121–129

Ainsworth MA, Svendsen P, Ladegaard L, Cantor P, Olsen O, Schaffalitzky de Muckadell OB (1992) Relative importance of pancreatic, hepatic, and mucosal bicarbonate in duodenal neutralization of acid in anaesthetized pigs. Scand J Gastroenterol 27:343–349

Ainsworth MA, Amelsberg M, Hogan DL, Isenberg JI (1996) Acid-base transport in isolated rabbit duodenal villus and crypt cells. Scand J Gastroenterol 31:1069–1077

Ainsworth MA, Hogan DL, Rapier RC, Amelsberg M, Dreilinger AD, Isenberg JI (1998) Acid/base transporters in human duodenal enterocytes. Scand J Gastroenterol 33:1039–1046

Akiba Y, Furukawa O, Guth PH, Engel E, Nastaskin I, Kaunitz JD (2001a) Acute adaptive cellular base uptake in rat duodenal epithelium. Am J Physiol Gastrointest Liver Physiol 280:1083–1092

Akiba Y, Furukawa O, Guth PH, Engel E, Nastaskin I, Sassani P, Dukkipatis R, Pushkin A, Kurtz I, Kaunitz JD (2001b) Cellular bicarbonate protects rat duodenal mucosa from acid-induced injury. J Clin Invest 108:1807–1816

Al-Awqati Q (2013) Cell biology of the intercalated cell in the kidney. FEBS Lett 587:1911–1914

Allen A, Flemstrom G (2005) Gastroduodenal mucus bicarbonate barrier: protection against acid and pepsin. Am J Physiol Cell Physiol 288:1–19

Almassy J, Won JH, Begenisich TB, Yule DI (2012) Apical Ca^{2+}-activated potassium channels in mouse parotid acinar cells. J Gen Physiol 139:121–133

Alper SL (2006) Molecular physiology of SLC4 anion exchangers. Exp Physiol 91:153–161

Alper SL, Sharma AK (2013) The SLC26 gene family of anion transporters and channels. Mol Aspects Med 34:494–515

Alper SL, Natale J, Gluck S, Lodish HF, Brown D (1989) Subtypes of intercalated cells in rat kidney collecting duct defined by antibodies against erythroid band 3 and renal vacuolar H^+-ATPase. Proc Natl Acad Sci U S A 86:5429–5433

Alper SL, Stuart-Tilley A, Simmons CF, Brown D, Drenckhahn D (1994) The fodrin-ankyrin cytoskeleton of choroid plexus preferentially colocalizes with apical Na^+K^+-ATPase rather than with basolateral anion exchanger AE2. J Clin Invest 93:1430–1438

Alper SL, Stewart AK, Vandorpe DH, Clark JS, Horack RZ, Simpson JE, Walker NM, Clarke LL (2011) Native and recombinant Slc26a3 (downregulated in adenoma, Dra) do not exhibit properties of 2Cl-/1HCO3- exchange. Am J Physiol Cell Physiol 300:276–286

Alpini G, Glaser S, Robertson W, Rodgers RE, Phinizy JL, Lasater J, LeSage GD (1997) Large but not small intrahepatic bile ducts are involved in secretin-regulated ductal bile secretion. Am J Physiol 272:1064–1074

Alvarez C, Fasano A, Bass BL (1998) Acute effects of bile acids on the pancreatic duct epithelium in vitro. J Surg Res 74:43–46

Alvarez L, Fanjul M, Carter N, Hollande E (2001) Carbonic anhydrase II associated with plasma membrane in a human pancreatic duct cell line (CAPAN-1). J Histochem Cytochem 49:1045–1053

Alvarez BV, Loiselle FB, Supuran CT, Schwartz GJ, Casey JR (2003) Direct extracellular interaction between carbonic anhydrase IV and the human NBC1 sodium/bicarbonate co-transporter. Biochemistry 42:12321–12329

Alvaro D, Della GP, Bini A, Gigliozzi A, Furfaro S, La RT, Piat C, Capocaccia L (1995) Effect of glucagon on intracellular pH regulation in isolated rat hepatocyte couplets. J Clin Invest 96:665–675

Ames A III, Sakanoue M, Endo S (1964) Na, K, Ca, Mg, and Cl concentrations in choroid plexus fluid and cisternal fluid compared with plasma ultrafiltrate. J Neurophysiol 27:672–681

Ames A III, Higashi K, Nesbett FB (1965) Effects of Pco2 acetazolamide and ouabain on volume and composition of choroid-plexus fluid. J Physiol 181:516–524

Amlal H, Petrovic S, Xu J, Wang Z, Sun X, Barone S, Soleimani M (2010) Deletion of the anion exchanger Slc26a4 (pendrin) decreases apical Cl^-/HCO_3^- exchanger activity and impairs bicarbonate secretion in kidney collecting duct. Am J Physiol Cell Physiol 299:33–41

Andersen DH (1938) Cystic fibrosis of the pancreas and its relation to celiac disease: a clinical and pathological study. Am J Dis Child 56:344–399

Aoi M, Aihara E, Nakashima M, Takeuchi K (2004) Participation of prostaglandin E receptor EP4 subtype in duodenal bicarbonate secretion in rats. Am J Physiol Gastrointest Liver Physiol 287:96–103

Aranda V, Martinez I, Melero S, Lecanda J, Banales JM, Prieto J, Medina JF (2004) Shared apical sorting of anion exchanger isoforms AE2a, AE2b1, and AE2b2 in primary hepatocytes. Biochem Biophys Res Commun 319:1040–1046

Ashizawa N, Endoh H, Hidaka K, Watanabe M, Fukumoto S (1997) Three-dimensional structure of the rat pancreatic duct in normal and inflammated pancreas. Microsc Res Tech 37:543–556

Azroyan A, Laghmani K, Crambert G, Mordasini D, Doucet A, Edwards A (2011) Regulation of pendrin by pH: dependence on glycosylation. Biochem J 434:61–72

Balazs A, Mall MA (2018) Role of the SLC26A9 chloride channel as disease modifier and potential therapeutic target in cystic fibrosis. Front Pharmacol 9:1112

Banales JM, Arenas F, Rodriguez-Ortigosa CM, Saez E, Uriarte I, Doctor RB, Prieto J, Medina JF (2006a) Bicarbonate-rich choleresis induced by secretin in normal rat is taurocholate-dependent and involves AE2 anion exchanger. Hepatology 43:266–275

Banales JM, Prieto J, Medina JF (2006b) Cholangiocyte anion exchange and biliary bicarbonate excretion. World J Gastroenterol 12:3496–3511

Bardow A, Madsen J, Nauntofte B (2000) The bicarbonate concentration in human saliva does not exceed the plasma level under normal physiological conditions. Clin Oral Investig 4:245–253

Battle T, Preisser L, Marteau V, Meduri G, Lambert M, Nitschke R, Brown PD, Corman B (2000) Vasopressin V1a receptor signaling in a rat choroid plexus cell line. Biochem Biophys Res Commun 275:322–327

Beal AM (1984) Electrolyte composition of parotid saliva from sodium-replete red kangaroos (Macropus rufus). J Exp Biol 111:225–237

Becker HM, Klier M, Deitmer JW (2014) Carbonic anhydrases and their interplay with acid/base-coupled membrane transporters. Subcell Biochem 75:105–134

Begenisich T, Nakamoto T, Ovitt CE, Nehrke K, Brugnara C, Alper SL, Melvin JE (2004) Physiological roles of the intermediate conductance, Ca^{2+}-activated potassium channel Kcnn4. J Biol Chem 279:47681–47687

Behrendorff N, Floetenmeyer M, Schwiening C, Thorn P (2010) Protons released during pancreatic acinar cell secretion acidify the lumen and contribute to pancreatitis in mice. Gastroenterology 139:1711–1720

Bergmann F, Andrulis M, Hartwig W, Penzel R, Gaida MM, Herpel E, Schirmacher P, Mechtersheimer G (2011) Discovered on gastrointestinal stromal tumor 1 (DOG1) is expressed in pancreatic centroacinar cells and in solid-pseudopapillary neoplasms--novel evidence for a histogenetic relationship. Hum Pathol 42:817–823

Berridge MJ (2012) Calcium signalling remodelling and disease. Biochem Soc Trans 40:297–309

Bertrand CA, Mitra S, Mishra SK, Wang X, Zhao Y, Pilewski JM, Madden DR, Frizzell RA (2017) The CFTR trafficking mutation F508del inhibits the constitutive activity of SLC26A9. Am J Physiol Lung Cell Mol Physiol 312:912–925

Beuers U, Hohenester S, de Buy Wenniger LJ, Kremer AE, Jansen PL, Elferink RP (2010) The biliary HCO_3^- umbrella: a unifying hypothesis on pathogenetic and therapeutic aspects of fibrosing cholangiopathies. Hepatology 52:1489–1496

Beuers U, Maroni L, Elferink RO (2012) The biliary HCO_3^- umbrella: experimental evidence revisited. Curr Opin Gastroenterol 28:253–257

Bhattacharya S, Verrill DS, Carbone KM, Brown S, Yule DI, Giovannucci DR (2012) Distinct contributions by ionotropic purinoceptor subtypes to ATP-evoked calcium signals in mouse parotid acinar cells. J Physiol 590:2721–2737

Bijman J, Cook DI, van Os CH (1981) Effect of amiloride on electrolyte transport parameters of the main duct of the rabbit mandibular gland. Pflügers Arch 398:96–102

Billet A, Hanrahan JW (2013) The secret life of CFTR as a calcium-activated chloride channel. J Physiol 591(21):5273–5278

Billet A, Luo Y, Balghi H, Hanrahan JW (2013) Role of tyrosine phosphorylation in the muscarinic activation of the cystic fibrosis transmembrane conductance regulator (CFTR). J Biol Chem 288:21815–21823

Boedtkjer E, Praetorius J, Aalkjaer C (2006) NBCn1 (slc4a7) mediates the Na+-dependent bicarbonate transport important for regulation of intracellular pH in mouse vascular smooth muscle cells. Circ Res 98:515–523

Boron WF, Boulpaep EL (1983) Intracellular pH regulation in the renal proximal tubule of the salamander. Basolateral HCO_3^- transport. J Gen Physiol 81:53–94

Boron WF, Boulpaep EL (2012) Medical physiology. Elsevier, Philadelphia

Boron WF, Boulpaep EL (2017) Medical physiology. Elsevier, Philadelphia

Boron WF, Knakal RC (1992) Na(+)-dependent Cl-HCO3 exchange in the squid axon: dependence on extracellular pH. J Gen Physiol 99:817–837

Bouwens L, Pipeleers DG (1998) Extra-insular beta cells associated with ductules are frequent in adult human pancreas. Diabetologia 41:629–633

Bouzinova EV, Praetorius J, Virkki LV, Nielsen S, Boron WF, Aalkjaer C (2005) Na+-dependent HCO3-uptake into the rat choroid plexus epithelium is partially DIDS sensitive. Am J Physiol Cell Physiol 289:1448–1456

Boyer JL (2013) Bile formation and secretion. Compr Physiol 3:1035–1078

Breton S, Brown D (2013) Regulation of luminal acidification by the V-ATPase. Physiology (Bethesda) 28:318–329

Breton S, Alper SL, Gluck SL, Sly WS, Barker JE, Brown D (1995) Depletion of intercalated cells from collecting ducts of carbonic anhydrase II-deficient (CAR2 null) mice. Am J Physiol Renal Physiol 269:761–774

Bro-Rasmussen F, Killmann SA, Thaysen JH (1956) The composition of pancreatic juice as compared to sweat, parotid saliva and tears. Acta Physiol Scand 37:97–113

Brown D, Hirsch S, Gluck S (1988a) An H^+-ATPase in opposite plasma membrane domains in kidney epithelial cell subpopulations. Nature 331:622–624

Brown PD, Loo DD, Wright EM (1988b) Ca^{2+}-activated K^+ channels in the apical membrane of Necturus choroid plexus. J Membr Biol 105:207–219

Brown D, Paunescu TG, Breton S, Marshansky V (2009) Regulation of the V-ATPase in kidney epithelial cells: dual role in acid-base homeostasis and vesicle trafficking. J Exp Biol 212:1762–1772

Bruce JI, Shuttleworth TJ, Giovannucci DR, Yule DI (2002) Phosphorylation of inositol 1,4,5-trisphosphate receptors in parotid acinar cells. A mechanism for the synergistic effects of cAMP on Ca^{2+} signaling. J Biol Chem 277:1340–1348

Burgen ASV (1956) The secretion of potassium in saliva. J Physiol (Lond) 132:20–39

Burghardt B, Elkaer ML, Kwon TH, Racz GZ, Varga G, Steward MC, Nielsen S (2003) Distribution of aquaporin water channels AQP1 and AQP5 in the ductal system of the human pancreas. Gut 52:1008–1016

Burghardt B, Nielsen S, Steward MC (2006) The role of aquaporin water channels in fluid secretion by the exocrine pancreas. J Membr Biol 210:143–153

Burnstock G (2017) Purinergic signalling: therapeutic developments. Front Pharmacol 8:661

Caflisch CR, Solomon S, Galey WR (1979) Exocrine ductal pCO_2 in the rabbit pancreas. Pflugers Arch 380:121–125

Caputo A, Caci E, Ferrera L, Pedemonte N, Barsanti C, Sondo E, Pfeffer U, Ravazzolo R, Zegarra-Moran O, Galietta LJ (2008) TMEM16A, a membrane protein associated with calcium-dependent chloride channel activity. Science 322:590–594

Case RM, Harper AA, Scratcherd T (1969) The secretion of electrolytes and enzymes by the pancreas of the anaesthetized cat. J Physiol (Lond) 201:335–348

Case RM, Scratcherd T, Wynne RDA (1970) The origin and secretion of pancreatic juice bicarbonate. J Physiol (Lond) 210:1–15

Case RM, Hotz J, Hutson D, Scratcherd T, Wynne RDA (1979) Electrolyte secretion by the isolated cat pancreas during replacement of extracellular bicarbonate by organic anions and chloride by inorganic anions. J Physiol (Lond) 286:563–576

Case RM, Conigrave AD, Novak I, Young JA (1980) Electrolyte and protein secretion by the perfused rabbit mandibular gland stimulated with acetylcholine or catecholamines. J Physiol (Lond) 300:467–487

Case RM, Conigrave AD, Favaloro EJ, Novak I, Thompson CH, Young JA (1982) The role of buffer anions and protons in secretion by the rabbit mandibular salivary gland. J Physiol (Lond) 322:273–286

Case RM, Hunter M, Novak I, Young JA (1984) The anionic basis of fluid secretion by the rabbit mandibular gland. J Physiol (Lond) 349:619–630

Catalan MA, Nakamoto T, Gonzalez-Begne M, Camden JM, Wall SM, Clarke LL, Melvin JE (2010) Cftr and ENaC ion channels mediate NaCl absorption in the mouse submandibular gland. J Physiol 588:713–724

Catalan MA, Kondo Y, Pena-Munzenmayer G, Jaramillo Y, Liu F, Choi S, Crandall E, Borok Z, Flodby P, Shull GE, Melvin JE (2015) A fluid secretion pathway unmasked by acinar-specific Tmem16A gene ablation in the adult mouse salivary gland. Proc Natl Acad Sci U S A 112:2263–2268

Chai SY, McKinley MJ, Mendelsohn FA (1987a) Distribution of angiotensin converting enzyme in sheep hypothalamus and medulla oblongata visualized by in vitro autoradiography. Clin Exp Hypertens A 9:449–460

Chai SY, Mendelsohn FA, Paxinos G (1987b) Angiotensin converting enzyme in rat brain visualized by quantitative in vitro autoradiography. Neuroscience 20:615–627

Chambrey R, Kurth I, Peti-Peterdi J, Houillier P, Purkerson JM, Leviel F, Hentschke M, Zdebik AA, Schwartz GJ, Hubner CA, Eladari D (2013) Renal intercalated cells are rather energized by a proton than a sodium pump. Proc Natl Acad Sci U S A 110:7928–7933

Chan HC, Cheung WT, Leung PY, Wu LJ, Chew SB, Ko WH, Wong PY (1996) Purinergic regulation of anion secretion by cystic fibrosis pancreatic duct cells. Am J Physiol 271:469–477

Chari RS, Schutz SM, Haebig JE, Shimokura GH, Cotton PB, Fitz JG, Meyers WC (1996) Adenosine nucleotides in bile. Am J Physiol 270:246–252

Chaturapanich G, Ishibashi H, Dinudom A, Young JA, Cook DI (1997) H^+ transporters in the main excretory duct of the mouse mandibular salivary gland. J Physiol 503(Pt 3):583–598

Chen M, Praetorius J, Zheng W, Xiao F, Riederer B, Singh AK, Stieger N, Wang J, Shull GE, Aalkjaer C, Seidler U (2012) The electroneutral Na(+):HCO(3)(-) cotransporter NBCn1 is a major pHi regulator in murine duodenum. J Physiol 590:3317–3333

Chenevert J, Duvvuri U, Chiosea S, Dacic S, Cieply K, Kim J, Shiwarski D, Seethala RR (2012) DOG1: a novel marker of salivary acinar and intercalated duct differentiation. Mod Pathol 25:919–929

Cheng HS, Leung PY, Cheng Chew SB, Leung PS, Lam SY, Wong WS, Wang ZD, Chan HC (1998) Concurrent and independent HCO_3^- and Cl^- secretion in a human pancreatic duct cell line (CAPAN-1). J Membr Biol 164:155–167

Chernova MN, Jiang L, Shmukler BE, Schweinfest CW, Blanco P, Freedman SD, Stewart AK, Alper SL (2003) Acute regulation of the SLC26A3 congenital chloride diarrhoea anion exchanger (DRA) expressed in Xenopus oocytes. J Physiol 549:3–19

Chernova MN, Jiang L, Friedman DJ, Darman RB, Lohi H, Kere J, Vandorpe DH, Alper SL (2005) Functional comparison of mouse slc26a6 anion exchanger with human SLC26A6 polypeptide variants: differences in anion selectivity, regulation, and electrogenicity. J Biol Chem 280:8564–8580

Chey WY, Chang T (2001) Neural hormonal regulation of exocrine pancreatic secretion. Pancreatology 1:320–335

Choate KA, Kahle KT, Wilson FH, Nelson-Williams C, Lifton RP (2003) WNK1, a kinase mutated in inherited hypertension with hyperkalemia, localizes to diverse Cl^- - transporting epithelia. Proc Natl Acad Sci U S A 100:663–668

Chodobski A, Szmydynger-Chodobska J (2001) Choroid plexus: target for polypeptides and site of their synthesis. Microsc Res Tech 52:65–82

Choi I, Romero MF, Khandoudi N, Bril A, Boron WF (1999) Cloning and characterization of a human electrogenic Na^+-HCO_3^- cotransporter isoform (hhNBC). Am J Physiol 276:576–584

Choi I, Aalkjaer C, Boulpaep EL, Boron WF (2000) An electroneutral sodium/bicarbonate cotransporter NBCn1 and associated sodium channel. Nature 405:571–575

Christensen O, Zeuthen T (1987) Maxi K^+ channels in leaky epithelia are regulated by intracellular Ca^{2+}, pH and membrane potential. Pflügers Arch 408:249–259

Christensen BM, Perrier R, Wang Q, Zuber AM, Maillard M, Mordasini D, Malsure S, Ronzaud C, Stehle JC, Rossier BC, Hummler E (2010) Sodium and potassium balance depends on alphaENaC expression in connecting tubule. J Am Soc Nephrol 21:1942–1951

Christensen HL, Barbuskaite D, Rojek A, Malte H, Christensen IB, Fuchtbauer AC, Fuchtbauer EM, Wang T, Praetorius J, Damkier HH (2018) The choroid plexus sodium-bicarbonate cotransporter NBCe2 regulates mouse cerebrospinal fluid pH. J Physiol 596:4709–4728

Clarke LL, Harline MC (1998) Dual role of CFTR in cAMP-stimulated HCO_3^- secretion across murine duodenum. Am J Physiol 274:718–726

Cohn JA, Strong TV, Picciotto MR, Nairn AC, Collins FS, Fitz JG (1993) Localization of the cystic fibrosis transmembrane conductance regulator in human bile duct epithelial cells. Gastroenterology 105:1857–1864

Compton JS, Nelson J, Wright RD, Young JA (1980) A micropuncture investigation of electrolyte transport in the parotid glands of sodium-replete and sodium-deplete sheep. J Physiol (Lond) 309:429–446

Concepcion AR, Lopez M, Ardura-Fabregat A, Medina JF (2013) Role of AE2 for pH regulation in biliary epithelial cells. Front Physiol 4:413

Cook DI, Young JA (1989) Effect of K^+ channels in the apical plasma membrane on epithelial secretion based on secondary active Cl^- transport. J Membr Biol 110:139–146

Cook DI, Dinudom A, Komwatana P, Kumar S, Young JA (2002) Patch-clamp studies on epithelial sodium channels in salivary duct cells. Cell Biochem Biophys 36:105–113

Cordat E, Reithmeier RA (2014) Structure, function, and trafficking of SLC4 and SLC26 anion transporters. Curr Top Membr 73:1–67

Cotter K, Stransky L, McGuire C, Forgac M (2015) Recent insights into the structure, regulation, and function of the V-ATPases. Trends Biochem Sci 40:611–622

Cowan F, Whitelaw A (1991) Acute effects of acetazolamide on cerebral blood flow velocity and pCO2 in the newborn infant. Acta Paediatr Scand 80:22–27

Crook RB, Farber MB, Prusiner SB (1984) Hormones and neurotransmitters control cyclic AMP metabolism in choroid plexus epithelial cells. J Neurochem 42:340–350

Cushing H (1914) Studies on the cerebrospinal fluid. J Med Res 26:1–19

Czako L, Yamamoto M, Otsuki M (1997) Exocrine pancreatic function in rats after acute pancreatitis. Pancreas 15:83–90

Damkier HH, Nielsen S, Praetorius J (2006) An anti-NH2-terminal antibody localizes NBCn1 to heart endothelia and skeletal and vascular smooth muscle cells. Am J Physiol Heart Circ Physiol 290:172–180

Damkier HH, Nielsen S, Praetorius J (2007) Molecular expression of SLC4-derived Na+-dependent anion transporters in selected human tissues. Am J Physiol Regul Integr Comp Physiol 293:2136–2146

Damkier HH, Prasad V, Hubner CA, Praetorius J (2009) Nhe1 is a luminal Na^+/H^+ exchanger in mouse choroid plexus and is targeted to the basolateral membrane in Ncbe/Nbcn2-null mice. Am J Physiol Cell Physiol 296:1291–1300

Damkier HH, Aalkjaer C, Praetorius J (2010) Na^+-dependent HCO_3^- import by the slc4a10 gene product involves Cl^- export. J Biol Chem 285:26998–27007

Damkier HH, Brown PD, Praetorius J (2013) Cerebrospinal fluid secretion by the choroid plexus. Physiol Rev 93:1847–1892

Damkier HH, Christensen HL, Christensen IB, Wu Q, Fenton RA, Praetorius J (2018) The murine choroid plexus epithelium expresses the 2Cl⁻/H⁺ exchanger ClC-7 and Na⁺/H⁺ exchanger NHE6 in the luminal membrane domain. Am J Physiol Cell Physiol 314:439–448

Davson H, Purvis C (1954) Cryoscopic apparatus suitable for studies on aqueous humour and cerebro-spinal fluid. J Physiol 124:12–3P

Davson H, Segal MB (1970) The effects of some inhibitors and accelerators of sodium transport on the turnover of 22Na in the cerebrospinal fluid and the brain. J Physiol 209:131–153

Davson H, Segal MB (1996) Physiology of the CSF and blood-brain barriers. CRC Press, Boca Raton

de Kloet ER, Van Acker SA, Sibug RM, Oitzl MS, Meijer OC, Rahmouni K, de Jong JW (2000) Brain mineralocorticoid receptors and centrally regulated functions. Kidney Int 57:1329–1336

de Ondarza J, Hootman SR (1997) Confocal microscopic analysis of intracellular pH regulation in isolated guinea pig pancreatic ducts. Am J Physiol 272:124–134

Demeter I, Hegyesi O, Nagy AK, Case MR, Steward MC, Varga G, Burghardt B (2009) Bicarbonate transport by the human pancreatic ductal cell line HPAF. Pancreas 38:913–920

Demitrack ES, Soleimani M, Montrose MH (2010) Damage to the gastric epithelium activates cellular bicarbonate secretion via SLC26A9 Cl⁻/HCO₃⁻. Am J Physiol Gastrointest Liver Physiol 299:255–264

Deng QS, Johanson CE (1989) Stilbenes inhibit exchange of chloride between blood, choroid plexus and cerebrospinal fluid. Brain Res 501:183–187

Dinudom A, Komwatana P, Young JA, Cook DI (1995) A forskolin-activated Cl⁻ current in mouse mandibular duct cells. Am J Physiol 268:806–812

Dinudom A, Harvey KF, Komwatana P, Young JA, Kumar S, Cook DI (1998) Nedd4 mediates control of an epithelial Na⁺ channel in salivary duct cells by cytosolic Na⁺. Proc Natl Acad Sci U S A 95:7169–7173

Dinudom A, Harvey KF, Komwatana P, Jolliffe CN, Young JA, Kumar S, Cook DI (2001) Roles of the C termini of alpha-, beta-, and gamma- subunits of epithelial Na⁺ channels (ENaC) in regulating ENaC and mediating its inhibition by cytosolic Na⁺. J Biol Chem 276:13744–13749

Dissing S, Hansen HJ, Undén M, Nauntofte B (1990) Inhibitory effects of amitriptyline on the stimulation-induced Ca²⁺ increase in parotid acini. Eur J Pharmacol 177:43–54

Doctor RB, Matzakos T, McWilliams R, Johnson S, Feranchak AP, Fitz JG (2005) Purinergic regulation of cholangiocyte secretion: identification of a novel role for P2X receptors. Am J Physiol Gastrointest Liver Physiol 288:779–786

Doring F, Derst C, Wischmeyer E, Karschin C, Schneggenburger R, Daut J, Karschin A (1998) The epithelial inward rectifier channel Kir7.1 displays unusual K⁺ permeation properties. J Neurosci 18:8625–8636

Dorwart MR, Shcheynikov N, Yang D, Muallem S (2008) The solute carrier 26 family of proteins in epithelial ion transport. Physiology (Bethesda) 23:104–114

Dranoff JA, Masyuk AI, Kruglov EA, LaRusso NF, Nathanson MH (2001) Polarized expression and function of P2Y ATP receptors in rat bile duct epithelia. Am J Physiol Gastrointest Liver Physiol 281:1059–1067

Duran C, Thompson CH, Xiao Q, Hartzell HC (2010) Chloride channels: often enigmatic, rarely predictable. Annu Rev Physiol 72:95–121

Dutta AK, Khimji AK, Sathe M, Kresge C, Parameswara V, Esser V, Rockey DC, Feranchak AP (2009) Identification and functional characterization of the intermediate-conductance Ca²⁺-activated K⁺ channel (IK-1) in biliary epithelium. Am J Physiol Gastrointest Liver Physiol 297:1009–1018

Dutta AK, Khimji AK, Kresge C, Bugde A, Dougherty M, Esser V, Ueno Y, Glaser SS, Alpini G, Rockey DC, Feranchak AP (2011) Identification and functional characterization of TMEM16A, a Ca2+-activated Cl- channel activated by extracellular nucleotides, in biliary epithelium. J Biol Chem 286:766–776

Dutta AK, Woo K, Khimji AK, Kresge C, Feranchak AP (2013) Mechanosensitive Cl⁻ secretion in biliary epithelium mediated through TMEM16A. Am J Physiol Gastrointest Liver Physiol 304:87–98

Ernst SA, Palacios JR, Siegel GJ (1986) Immunocytochemical localization of Na+,K+-ATPase catalytic polypeptide in mouse choroid plexus. J Histochem Cytochem 34:189–195

Esteller A (2008) Physiology of bile secretion. World J Gastroenterol 14:5641–5649

Evans RL, Ashton N, Elliott AC, Green R, Argent BE (1996) Interaction between secretin and acetylcholine in the regulation of fluid secretion by isolated rat pancreatic ducts. J Physiol (Lond) 461:265–273

Evans RL, Park K, Turner RJ, Watson GE, Nguyen HV, Dennett MR, Hand AR, Flagella M, Shull GE, Melvin JE (2000) Severe impairment of salivation in Na⁺/K⁺/2Cl⁻ cotransporter (NKCC1)-deficient mice. J Biol Chem 275:26720–26726

Faivre J (1854) Structure du conarium et des plexus chorode chez lhommes et des animaux. Gaz Med Paris 9:555–556

Fanjul M, Alvarez L, Salvador C, Gmyr V, Kerr-Conte J, Pattou F, Carter N, Hollande E (2004) Evidence for a membrane carbonic anhydrase IV anchored by its C-terminal peptide in normal human pancreatic ductal cells. Histochem Cell Biol 121:91–99

Fausther M, Sevigny J (2011) Extracellular nucleosides and nucleotides regulate liver functions via a complex system of membrane proteins. C R Biol 334:100–117

Feranchak AP, Doctor RB, Troetsch M, Brookman K, Johnson SM, Fitz JG (2004) Calcium-dependent regulation of secretion in biliary epithelial cells: the role of apamin-sensitive SK channels. Gastroenterology 127:903–913

Feranchak AP, Lewis MA, Kresge C, Sathe M, Bugde A, Luby-Phelps K, Antich PP, Fitz JG (2010) Initiation of purinergic signaling by exocytosis of ATP-containing vesicles in liver epithelium. J Biol Chem 285:8138–8147

Fernandez-Salazar MP, Pascua P, Calvo JJ, Lopez MA, Case RM, Steward MC, San Roman JI (2004) Basolateral anion transport mechanisms underlying fluid secretion by mouse, rat and guinea-pig pancreatic ducts. J Physiol (Lond) 556:415–428

Feschenko MS, Donnet C, Wetzel RK, Asinovski NK, Jones LR, Sweadner KJ (2003) Phospholemman, a single-span membrane protein, is an accessory protein of Na,K-ATPase in cerebellum and choroid plexus. J Neurosci 23:2161–2169

Fiorotto R, Spirli C, Fabris L, Cadamuro M, Okolicsanyi L, Strazzabosco M (2007) Ursodeoxycholic acid stimulates cholangiocyte fluid secretion in mice via CFTR-dependent ATP secretion. Gastroenterology 133:1603–1613

Fitz JG, Persico M, Scharschmidt BF (1989) Electrophysiological evidence for Na+-coupled bicarbonate transport in cultured rat hepatocytes. Am J Physiol 256:491–500

Fitz JG, Basavappa S, McGill J, Melhus O, Cohn JA (1993) Regulation of membrane chloride currents in rat bile duct epithelial cells. J Clin Invest 91:319–328

Flemstrom G, Kivilaakso E (1983) Demonstration of a pH gradient at the luminal surface of rat duodenum in vivo and its dependence on mucosal alkaline secretion. Gastroenterology 84:787–794

Flemstrom G, Garner A, Nylander O, Hurst BC, Heylings JR (1982) Surface epithelial HCO₃⁻ transport by mammalian duodenum in vivo. Am J Physiol 243:348–358

Flinck M, Kramer SH, Pedersen SF (2018) Roles of pH in control of cell proliferation. Acta Physiol (Oxf) 223:e13068

Fong P, Argent BE, Guggino WB, Gray MA (2003) Characterization of vectorial chloride transport pathways in the human pancreatic duct adenocarcinoma cell line, HPAF. Am J Physiol Cell Physiol 285:433–445

Forte JG, Zhu L (2010) Apical recycling of the gastric parietal cell H,K-ATPase. Annu Rev Physiol 72:273–296

Foskett JK (1990) [Ca²⁺]ᵢ modulation of Cl⁻ content controls cell volume in single salivary acinar cells during fluid secretion. Am J Physiol 259:998–1004

Frankel H, Kazemi H (1983) Regulation of CSF composition--blocking chloride-bicarbonate exchange. J Appl Physiol Respir Environ Exerc Physiol 55:177–182

Freedman SD, Scheele GA (1994) Acid-base interactions during exocrine pancreatic secretion. Primary role for ductal bicarbonate in acinar lumen function. Ann N Y Acad Sci 713:199–206

Freedman SD, Kern HF, Scheele GA (2001) Pancreatic acinar cell dysfunction in CFTR-/- mice is associated with impairments in luminal pH and endocytosis. Gastroenterology 121:950–957

Frische S, Kwon TH, Frokiaer J, Madsen KM, Nielsen S (2003) Regulated expression of pendrin in rat kidney in response to chronic NH4Cl or NaHCO3 loading. Am J Physiol Renal Physiol 284:584–593

Frizzell RA, Hanrahan JW (2012) Physiology of epithelial chloride and fluid secretion. Cold Spring Harb Perspect Med 2:a009563

Gao X, Eladari D, Leviel F, Tew BY, Miro-Julia C, Cheema FH, Miller L, Nelson R, Paunescu TG, McKee M, Brown D, Al-Awqati Q (2010) Deletion of hensin/DMBT1 blocks conversion of beta- to alpha-intercalated cells and induces distal renal tubular acidosis. Proc Natl Acad Sci U S A 107:21872–21877

Garrett JR (1987) The proper role of nerves in salivary secretion: a review. J Dent Res 66:387–397

Gawenis LR, Ledoussal C, Judd LM, Prasad V, Alper SL, Stuart-Tilley A, Woo AL, Grisham C, Sanford LP, Doetschman T, Miller ML, Shull GE (2004) Mice with a targeted disruption of the AE2 Cl-/HCO3- exchanger are achlorhydric. J Biol Chem 279:30531–30539

Gehlert DR, Speth RC, Wamsley JK (1986) Distribution of [125I]angiotensin II binding sites in the rat brain: a quantitative autoradiographic study. Neuroscience 18:837–856

Gerasimenko JV, Flowerdew SE, Voronina SG, Sukhomlin TK, Tepikin AV, Petersen OH, Gerasimenko OV (2006) Bile acids induce Ca2+ release from both the endoplasmic reticulum and acidic intracellular calcium stores through activation of inositol trisphosphate receptors and ryanodine receptors. J Biol Chem 281:40154–40163

Giffard RG, Lee YS, Ouyang YB, Murphy SL, Monyer H (2003) Two variants of the rat brain sodium-driven chloride bicarbonate exchanger (NCBE): developmental expression and addition of a PDZ motif. Eur J Neurosci 18:2935–2945

Githens S (1988) The pancreas duct cell: proliferative capabilities, specific characteristics, metaplasia, isolation, and culture. J Pediatr Gastroenterol Nutr 7:486–506

Gmyr V, Belaich S, Muharram G, Lukowiak B, Vandewalle B, Pattou F, Kerr-Conte J (2004) Rapid purification of human ductal cells from human pancreatic fractions with surface antibody CA19-9. Biochem Biophys Res Commun 320:27–33

Gonzalez-Begne M, Nakamoto T, Nguyen HV, Stewart AK, Alper SL, Melvin JE (2007) Enhanced formation of a HCO_3^- transport metabolon in exocrine cells of Nhe1-/- mice. J Biol Chem 282:35125–35132

Gradilone SA, Garcia F, Huebert RC, Tietz PS, Larocca MC, Kierbel A, Carreras FI, LaRusso NF, Marinelli RA (2003) Glucagon induces the plasma membrane insertion of functional aquaporin-8 water channels in isolated rat hepatocytes. Hepatology 37:1435–1441

Gray MA, Greenwell JR, Argent BE (1988) Secretin-regulated chloride channel on the apical plasma membrane of pancreatic duct cells. J Membr Biol 105:131–142

Gray MA, Harris A, Coleman L, Greenwell JR, Argent BE (1989) Two types of chloride channel on duct cells cultured from human fetal pancreas. Am J Physiol 257:240–251

Gray MA, Greenwell JR, Garton AJ, Argent BE (1990) Regulation of maxi-K^+ channels on pancreatic duct cells by cyclic AMP-dependent phosphorylation. J Membr Biol 115:203–215

Gray MA, Plant S, Argent BE (1993) cAMP-regulated whole cell chloride currents in pancreatic duct cells. Am J Physiol Cell Physiol 264:591–602

Gray MA, Winpenny JP, Porteous DJ, Dorin JR, Argent BE (1994) CFTR and calcium-activated chloride currents in pancreatic duct cells of a transgenic CF mouse. Am J Physiol 266:213–221

Greeley T, Shumaker H, Wang Z, Schweinfest CW, Soleimani M (2001) Downregulated in adenoma and putative anion transporter are regulated by CFTR in cultured pancreatic duct cells. Am J Physiol Gastrointest Liver Physiol 281:G1301–G1308

Gregoriades JMC, Madaris A, Alvarez FJ, Alvarez-Leefmans FJ (2018) Genetic and pharmacologic inactivation of apical NKCC1 in choroid plexus epithelial cells reveals the physiological function of the cotransporter. Am J Physiol Cell Physiol 316(4):525–544

Gresz V, Kwon TH, Vorum H, Zelles T, Kurtz I, Steward MC, Aalkjaer C, Nielsen S (2002) Immunolocalization of electroneutral Na^+-HCO_3^- cotransporters in human and rat salivary glands. Am J Physiol Gastrointest Liver Physiol 283:473–480

Grichtchenko II, Choi I, Zhong X, Bray-Ward P, Russell JM, Boron WF (2001) Cloning, characterization, and chromosomal mapping of a human electroneutral Na^+-driven Cl-HCO_3 exchanger. J Biol Chem 276:8358–8363

Grishin AV, Caplan MJ (1998) ATP1AL1, a member of the non-gastric H,K-ATPase family, functions as a sodium pump. J Biol Chem 273:27772–27778

Grishin AV, Bevensee MO, Modyanov NN, Rajendran V, Boron WF, Caplan MJ (1996) Functional expression of the cDNA encoded by the human ATP1AL1 gene. Am J Physiol 271:539–551

Grotmol T, Buanes T, Bros O, Raeder MG (1986) Lack of effect of amiloride, furosemide, bumetanide and triamterene on pancreatic $NaHCO_3$ secretion in pigs. Acta Physiol Scand 126:593–600

Guba M, Kuhn M, Forssmann WG, Classen M, Gregor M, Seidler U (1996) Guanylin strongly stimulates rat duodenal HCO_3^- secretion: proposed mechanism and comparison with other secretagogues. Gastroenterology 111:1558–1568

Haanes KA, Novak I (2010) ATP storage and uptake by isolated pancreatic zymogen granules. Biochem J 429:303–311

Haanes KA, Kowal JM, Arpino G, Lange SC, Moriyama Y, Pedersen PA, Novak I (2014) Role of vesicular nucleotide transporter VNUT (SLC17A9) in release of ATP from AR42J cells and mouse pancreatic acinar cells. Purinergic Signal 10:431–440

Hafner P, Grimaldi R, Capuano P, Capasso G, Wagner CA (2008) Pendrin in the mouse kidney is primarily regulated by Cl^- excretion but also by systemic metabolic acidosis. Am J Physiol Cell Physiol 295:1658–1667

Hayashi M, Novak I (2013) Molecular basis of potassium channels in pancreatic duct epithelial cells. Channels (Austin) 7:432–441

Hayashi M, Wang J, Hede SE, Novak I (2012) An intermediate-conductance Ca^{2+}-activated K^+ channel is important for secretion in pancreatic duct cells. Am J Physiol Cell Physiol 303:151–159

He X, Tse C-M, Danowitz M, Alper SL, Gabriel SE, Baum BJ (1997) Polarized distribution of key membrane transport proteins in the rat submandibular gland. Pflügers Arch 433:260–268

Hede SE, Amstrup J, Christoffersen BC, Novak I (1999) Purinoceptors evoke different electrophysiological responses in pancreatic ducts. P2Y inhibits K^+ conductance, and P2X stimulates cation conductance. J Biol Chem 274:31784–31791

Hede SE, Amstrup J, Klaerke DA, Novak I (2005) P2Y2 and P2Y4 receptors regulate pancreatic Ca^{2+}-activated K^+ channels differently. Pflugers Arch 450:429–436

Hegyi P, Petersen OH (2013) The exocrine pancreas: the acinar-ductal tango in physiology and pathophysiology. Rev Physiol Biochem Pharmacol 165:1–30

Hegyi P, Maleth J, Venglovecz V, Rakonczay Z Jr (2011) Pancreatic ductal bicarbonate secretion: challenge of the acinar acid load. Front Physiol 2:36

Heitzmann D, Warth R (2008) Physiology and pathophysiology of potassium channels in gastrointestinal epithelia. Physiol Rev 88:1119–1182

Hentschke M, Wiemann M, Hentschke S, Kurth I, Hermans-Borgmeyer I, Seidenbecher T, Jentsch TJ, Gal A, Hubner CA (2006) Mice with a targeted disruption of the Cl^-/HCO_3^- exchanger AE3 display a reduced seizure threshold. Mol Cell Biol 26:183–191

Hickson JC (1970) The secretion of pancreatic juice in response to stimulation of the vagus nerves in the pig. J Physiol 206:275–297

Hikita C, Vijayakumar S, Takito J, Erdjument-Bromage H, Tempst P, Al-Awqati Q (2000) Induction of terminal differentiation in epithelial cells requires polymerization of hensin by galectin 3. J Cell Biol 151:1235–1246

Hogan DL, Crombie DL, Isenberg JI, Svendsen P, Schaffalitzky de Muckadell OB, Ainsworth MA (1997a) Acid-stimulated duodenal bicarbonate secretion involves a CFTR-mediated transport pathway in mice. Gastroenterology 113:533–541

Hogan DL, Crombie DL, Isenberg JI, Svendsen P, Schaffalitzky de Muckadell OB, Ainsworth MA (1997b) CFTR mediates cAMP- and Ca2+-activated duodenal epithelial HCO3- secretion. Am J Physiol 272:872–878

Hoglund P, Haila S, Socha J, Tomaszewski L, Saarialho-Kere U, Karjalainen-Lindsberg ML, Airola K, Holmberg C, Albert de la CA, Kere J (1996) Mutations of the down-regulated in adenoma (DRA) gene cause congenital chloride diarrhoea. Nat Genet 14:316–319

Hollander F, Birnbaum D (1952) The role of carbonic anhydrase in pancreatic secretion. Trans N Y Acad Sci 15:56–58

Holm M, Johansson B, Pettersson A, Fandriks L (1998) Carbon dioxide mediates duodenal mucosal alkaline secretion in response to luminal acidity in the anesthetized rat. Gastroenterology 115:680–685

Holst JJ (1993) Neural regulation of pancreatic exocrine function. In: VLW G, EP DM, Gardner JD, Lebenthal E, Reber HA, Scheele GA (eds) The pancreas: biology, pathobilogy, and disease. Raven Press, New York, pp 381–402

Hug M, Pahl C, Novak I (1994) Effect of ATP, carbachol and other agonists on intracellular calcium activity and membrane voltage of pancreatic ducts. Pflügers Arch 426:412–418

Humphrey SP, Williamson RT (2001) A review of saliva: normal composition, flow, and function. J Prosthet Dent 85:162–169

Huss M, Wieczorek H (2009) Inhibitors of V-ATPases: old and new players. J Exp Biol 212:341–346

Husted RF, Reed DJ (1976) Regulation of cerebrospinal fluid potassium by the cat choroid plexus. J Physiol 259:213–221

Hyde K, Reid CJ, Tebbutt SJ, Weide L, Hollingsworth MA, Harris A (1997) The cystic fibrosis transmembrane conductance regulator as a marker of human pancreatic duct development. Gastroenterology 113:914–919

Hyde K, Harrison D, Hollingsworth MA, Harris A (1999) Chloride-bicarbonate exchangers in the human fetal pancreas. Biochem Biophys Res Commun 263:315–321

Igimi H, Yamamoto F, Lee SP (1992) Gallbladder mucosal function: studies in absorption and secretion in humans and in dog gallbladder epithelium. Am J Physiol 263:69–74

Ignath I, Hegyi P, Venglovecz V, Szekely CA, Carr G, Hasegawa M, Inoue M, Takacs T, Argent BE, Gray MA, Rakonczay Z Jr (2009) CFTR expression but not Cl⁻ transport is involved in the stimulatory effect of bile acids on apical Cl⁻/HCO₃⁻ excahnge activity in human pancreatic duct cells. Pancreas 38:921–929

Imai Y (1965) Study of the secretion mechanism of the submaxillary gland of dog. Part 2. Effects of exchanging ions in the perfusate on salivary secretion and secretory potentials, with special reference to the ionic distribution in the gland tissue. Jpn J Physiol 27:313–324

Imboden H, Harding JW, Hilgenfeldt U, Celio MR, Felix D (1987) Localization of angiotensinogen in multiple cell types of rat brain. Brain Res 410:74–77

Inagami T, Celio MR, Clemens DL, Lau D, Takii Y, Kasselberg AG, Hirose S (1980) Renin in rat and mouse brain: immunohistochemical identification and localization. Clin Sci (Lond) 59 (Suppl 6):49s–51s

Isenberg JI, Hogan DL, Koss MA, Selling JA (1986) Human duodenal mucosal bicarbonate secretion. Evidence for basal secretion and stimulation by hydrochloric acid and a synthetic prostaglandin E1 analogue. Gastroenterology 91:370–378

Isenberg JI, Ljungstrom M, Safsten B, Flemstrom G (1993) Proximal duodenal enterocyte transport: evidence for Na⁺-H⁺ and Cl⁻-HCO₃⁻ exchange and NaHCO3 cotransport. Am J Physiol 265:677–685

Ishibashi K, Okamura K, Yamazaki J (2008) Involvement of apical P2Y2 receptor-regulated CFTR activity in muscarinic stimulation of Cl$^-$ reabsorption in rat submandibular gland. Am J Physiol Regul Integr Comp Physiol 294:1729–1736

Ishiguro H, Steward MC, Wilson RW, Case RM (1996) Bicarbonate secretion in interlobular ducts from guinea-pig pancreas. J Physiol (Lond) 495(1):179–191

Ishiguro H, Naruse S, Steward MC, Kitagawa M, Ko SB, Hayakawa T, Case RM (1998) Fluid secretion in interlobular ducts isolated from guinea-pig pancreas. J Physiol (Lond) 511:407–422

Ishiguro H, Naruse S, Kitagawa M, Hayakawa T, Case RM, Steward MC (1999) Luminal ATP stimulates fluid and HCO_3^- secretion in guinea-pig pancreatic duct. J Physiol (Lond) 519:551–558

Ishiguro H, Naruse S, Kitagawa M, Suzuki A, Yamamoto A, Hayakawa T, Case RM, Steward MC (2000) CO2 permeability and bicarbonate transport in microperfused interlobular ducts isolated from guinea-pig pancreas. J Physiol 528(Pt 2):305–315

Ishiguro H, Namkung W, Yamamoto A, Wang Z, Worrell RT, Xu J, Lee MG, Soleimani M (2007) Effect of Slc26a6 deletion on apical Cl$^-$/HCO_3^- exchanger activity and cAMP-stimulated bicarbonate secretion in pancreatic duct. Am J Physiol Gastrointest Liver Physiol 292:447–455

Ishiguro H, Steward MC, Naruse S, Ko SB, Goto H, Case RM, Kondo T, Yamamoto A (2009) CFTR functions as a bicarbonate channel in pancreatic duct cells. J Gen Physiol 133:315–326

Jacob P, Christiani S, Rossmann H, Lamprecht G, Vieillard-Baron D, Muller R, Gregor M, Seidler U (2000a) Role of Na(+)HCO(3)(-) cotransporter NBC1, Na(+)/H(+) exchanger NHE1, and carbonic anhydrase in rabbit duodenal bicarbonate secretion. Gastroenterology 119:406–419

Jacob P, Christiani S, Rossmann H, Lamprecht G, Vieillard-Baron D, Muller R, Gregor M, Seidler U (2000b) Role of Na$^+$-HCO_3^- cotransporter NBC1, Na$^+$/H$^+$ exchanger NHE1, and carbonic anhydrase in rabbit duodenal bicarbonate secretion. Gastroenterology 119:406–419

Jacobs S, Ruusuvuori E, Sipila ST, Haapanen A, Damkier HH, Kurth I, Hentschke M, Schweizer M, Rudhard Y, Laatikainen LM, Tyynela J, Praetorius J, Voipio J, Hubner CA (2008) Mice with targeted Slc4a10 gene disruption have small brain ventricles and show reduced neuronal excitability. Proc Natl Acad Sci U S A 105:311–316

Jacobson KA, Muller CE (2016) Medicinal chemistry of adenosine, P2Y and P2X receptors. Neuropharmacology 104:31–49

Jacques T, Picard N, Miller RL, Riemondy KA, Houillier P, Sohet F, Ramakrishnan SK, Busst CJ, Jayat M, Corniere N, Hassan H, Aronson PS, Hennings JC, Hubner CA, Nelson RD, Chambrey R, Eladari D (2013) Overexpression of pendrin in intercalated cells produces chloride-sensitive hypertension. J Am Soc Nephrol 24:1104–1113

Jentsch TJ (2016) VRACs and other ion channels and transporters in the regulation of cell volume and beyond. Nat Rev Mol Cell Biol 17:293–307

Jentsch TJ, Lutter D, Planells-Cases R, Ullrich F, Voss FK (2016) VRAC: molecular identification as LRRC8 heteromers with differential functions. Pflugers Arch 468:385–393

Johanson CE, Murphy VA (1990) Acetazolamide and insulin alter choroid plexus epithelial cell [Na+], pH, and volume. Am J Physiol 258:1538–1546

Johanson CE, Preston JE, Chodobski A, Stopa EG, Szmydynger-Chodobska J, McMillan PN (1999) AVP V1 receptor-mediated decrease in Cl$^-$ efflux and increase in dark cell number in choroid plexus epithelium. Am J Physiol Cell Physiol 276:82–90

Jones HC, Keep RF (1988) Brain fluid calcium concentration and response to acute hypercalcaemia during development in the rat. J Physiol 402:579–593

Joo NS, London RM, Kim HD, Forte LR, Clarke LL (1998) Regulation of intestinal Cl$^-$ and HCO_3^- secretion by uroguanylin. Am J Physiol 274:633–644

Juhasz M, Chen J, Lendeckel U, Kellner U, Kasper HU, Tulassay Z, Pastorekova S, Malfertheiner P, Ebert MP (2003) Expression of carbonic anhydrase IX in human pancreatic cancer. Aliment Pharmacol Ther 18:837–846

Jun I, Cheng MH, Sim E, Jung J, Suh BL, Kim Y, Son H, Park K, Kim CH, Yoon JH, Whitcomb DC, Bahar I, Lee MG (2016) Pore dilatation increases the bicarbonate permeability of CFTR, ANO1 and glycine receptor anion channels. J Physiol 594:2929–2955

Jung J, Lee MG (2014) Role of calcium signaling in epithelial bicarbonate secretion. Cell Calcium 55:376–384

Jung SR, Kim K, Hille B, Nguyen TD, Koh DS (2006) Pattern of Ca^{2+} increase determines the type of secretory mechanism activated in dog pancreatic duct epithelial cells. J Physiol 576:163–178

Jung SR, Hille B, Nguyen TD, Koh DS (2010) Cyclic AMP potentiates Ca^{2+}-dependent exocytosis in pancreatic duct epithelial cells. J Gen Physiol 135:527–543

Jung J, Nam JH, Park HW, Oh U, Yoon JH, Lee MG (2013) Dynamic modulation of ANO1/ TMEM16A. Proc Natl Acad Sci U S A 110:360–365

Kahle KT, Gimenez I, Hassan H, Wilson FH, Wong RD, Forbush B, Aronson PS, Lifton RP (2004) WNK4 regulates apical and basolateral Cl^- flux in extrarenal epithelia. Proc Natl Acad Sci U S A 101:2064–2069

Kahle KT, Rinehart J, Ring A, Gimenez I, Gamba G, Hebert SC, Lifton RP (2006) WNK protein kinases modulate cellular Cl- flux by altering the phosphorylation state of the Na-K-Cl and K-Cl cotransporters. Physiology (Bethesda) 21:326–335

Kajita H, Whitwell C, Brown PD (2000) Properties of the inward-rectifying Cl^- channel in rat choroid plexus: regulation by intracellular messengers and inhibition by divalent cations. Pflugers Arch 440:933–940

Kalaria RN, Premkumar DR, Lin CW, Kroon SN, Bae JY, Sayre LM, LaManna JC (1998) Identification and expression of the Na^+/H^+ exchanger in mammalian cerebrovascular and choroidal tissues: characterization by amiloride-sensitive [3H]MIA binding and RT-PCR analysis. Brain Res Mol Brain Res 58:178–187

Kallio H, Pastorekova S, Pastorek J, Waheed A, Sly WS, Mannisto S, Heikinheimo M, Parkkila S (2006) Expression of carbonic anhydrases IX and XII during mouse embryonic development. BMC Dev Biol 6:22

Kane Dickson V, Pedi L, Long SB (2014) Structure and insights into the function of a Ca^{2+}-activated Cl^- channel. Nature 516:213–218

Kao L, Kurtz LM, Shao X, Papadopoulos MC, Liu L, Bok D, Nusinowitz S, Chen B, Stella SL, Andre M, Weinreb J, Luong SS, Piri N, Kwong JM, Newman D, Kurtz I (2011) Severe neurologic impairment in mice with targeted disruption of the electrogenic sodium bicarbonate cotransporter NBCe2 (Slc4a5 gene). J Biol Chem 286:32563–32574

Karadsheh MF, Byun N, Mount DB, Delpire E (2004) Localization of the KCC4 potassium-chloride cotransporter in the nervous system. Neuroscience 123:381–391

Kaunitz JD, Akiba Y (2002) Luminal acid elicits a protective duodenal mucosal response. Keio J Med 51:29–35

Kaunitz JD, Akiba Y (2006a) Duodenal carbonic anhydrase: mucosal protection, luminal chemosensing, and gastric acid disposal. Keio J Med 55:96–106

Kaunitz JD, Akiba Y (2006b) Review article: duodenal bicarbonate - mucosal protection, luminal chemosensing and acid-base balance. Aliment Pharmacol Ther 24(Suppl 4):169–176

Kawedia JD, Nieman ML, Boivin GP, Melvin JE, Kikuchi K, Hand AR, Lorenz JN, Menon AG (2007) Interaction between transcellular and paracellular water transport pathways through Aquaporin 5 and the tight junction complex. Proc Natl Acad Sci U S A 104:3621–3626

Kay RNB (1960) The rate of flow and composition of various salivary secretions in sheep and calves. J Physiol (Lond) 150:515–537

Keep RF, Xiang J, Betz AL (1994) Potassium cotransport at the rat choroid plexus. Am J Physiol 267:1616–1622

Keitel V, Haussinger D (2013) TGR5 in cholangiocytes. Curr Opin Gastroenterol 29:299–304

Keitel V, Ullmer C, Haussinger D (2010) The membrane-bound bile acid receptor TGR5 (Gpbar-1) is localized in the primary cilium of cholangiocytes. Biol Chem 391:785–789

Kerem B, Rommens JM, Buchanan JA, Markiewicz D, Cox TK, Chakravarti A, Buchwald M, Tsui LC (1989) Identification of the cystic fibrosis gene: genetic analysis. Science 245:1073–1080

Kibble JD, Trezise AE, Brown PD (1996) Properties of the cAMP-activated Cl- current in choroid plexus epithelial cells isolated from the rat. J Physiol 496(Pt 1):69–80

Kibble JD, Garner C, Colledge WH, Brown S, Kajita H, Evans M, Brown PD (1997) Whole cell Cl$^-$ conductances in mouse choroid plexus epithelial cells do not require CFTR expression. Am J Physiol Cell Physiol 272:1899–1907

Kim JY, Kim KH, Lee JA, Namkung W, Sun AQ, Ananthanarayanan M, Suchy FJ, Shin DM, Muallem S, Lee MG (2002) Transporter-mediated bile acid uptake causes Ca^{2+}-dependent cell death in rat pancreatic acinar cells. Gastroenterology 122:1941–1953

Kim KH, Shcheynikov N, Wang Y, Muallem S (2005) SLC26A7 is a Cl- channel regulated by intracellular pH. J Biol Chem 280:6463–6470

Kivela AJ, Parkkila S, Saarnio J, Karttunen TJ, Kivela J, Parkkila AK, Pastorekova S, Pastorek J, Waheed A, Sly WS, Rajaniemi H (2000) Expression of transmembrane carbonic anhydrase isoenzymes IX and XII in normal human pancreas and pancreatic tumours. Histochem Cell Biol 114:197–204

Klarr SA, Ulanski LJ, Stummer W, Xiang J, Betz AL, Keep RF (1997) The effects of hypo- and hyperkalemia on choroid plexus potassium transport. Brain Res 758:39–44

Knauf H, Lubcke R, Kreutz W, Sachs G (1982) Interrelationships of ion transport in rat submaxillary duct epithelium. Am J Physiol 242:132–139

Ko SB, Luo X, Hager H, Rojek A, Choi JY, Licht C, Suzuki M, Muallem S, Nielsen S, Ishibashi K (2002a) AE4 is a DIDS-sensitive Cl$^-$/HCO$_3$$^-$ exchanger in the basolateral membrane of the renal CCD and the SMG duct. Am J Physiol Cell Physiol 283:1206–1218

Ko SB, Shcheynikov N, Choi JY, Luo X, Ishibashi K, Thomas PJ, Kim JY, Kim KH, Lee MG, Naruse S, Muallem S (2002b) A molecular mechanism for aberrant CFTR-dependent HCO$_3$$^-$ transport in cystic fibrosis. EMBO J 21:5662–5672

Ko SB, Zeng W, Dorwart MR, Luo X, Kim KH, Millen L, Goto H, Naruse S, Soyombo A, Thomas PJ, Muallem S (2004) Gating of CFTR by the STAS domain of SLC26 transporters. Nat Cell Biol 6:343–350

Kodama T (1983) A light and electron microscopic study on the pancreatic ductal system. Acta Pathol Jpn 33:297–321

Kollert-Jons A, Wagner S, Hubner S, Appelhans H, Drenckhahn D (1993) Anion exchanger 1 in human kidney and oncocytoma differs from erythroid AE1 in its NH2 terminus. Am J Physiol 265:813–821

Komwatana P, Dinudom A, Young JA, Cook DI (1996a) Control of the amiloride-sensitive Na$^+$ current in salivary duct cells by extracellular sodium. J Membr Biol 150:133–141

Komwatana P, Dinudom A, Young JA, Cook DI (1996b) Cytosolic Na$^+$ controls and epithelial Na$^+$ channel via the Go guanine nucleotide-binding regulatory protein. Proc Natl Acad Sci U S A 93:8107–8111

Kopelman H, Durie P, Gaskin K, Weizman Z, Forstner G (1985) Pancreatic fluid secretion and protein hyperconcentration in cystic fibrosis. N Engl J Med 312:329–334

Kopelman H, Corey M, Gaskin K, Durie P, Weizman Z, Forstner G (1988) Impaired chloride secretion, as well as bicarbonate secretion, underlies the fluid secretory defect in the cystic fibrosis pancreas. Gastroenterology 95:349–355

Kordas KS, Sperlagh B, Tihanyi T, Topa L, Steward MC, Varga G, Kittel A (2004) ATP and ATPase secretion by exocrine pancreas in rat, Guinea pig, and human. Pancreas 29:53–60

Kosiek O, Busque SM, Foller M, Shcheynikov N, Kirchhoff P, Bleich M, Muallem S, Geibel JP (2007) SLC26A7 can function as a chloride-loading mechanism in parietal cells. Pflugers Arch 454:989–998

Kotera T, Brown PD (1994) Evidence for two types of potassium current in rat choroid plexus epithelial cells. Pflugers Arch 427:317–324

Kowal JM, Haanes KA, Christensen NM, Novak I (2015a) Bile acid effects are mediated by ATP release and purinergic signalling in exocrine pancreatic cells. Cell Commun Signal 13:28

Kowal JM, Yegutkin GG, Novak I (2015b) ATP release, generation and hydrolysis in exocrine pancreatic duct cells. Purinergic Signal 11:533–550

Krane CM, Melvin JE, Nguyen HV, Richardson L, Towne JE, Doetschman T, Menon AG (2001) Salivary acinar cells from aquaporin 5-deficient mice have decreased membrane water permeability and altered cell volume regulation. J Biol Chem 276:23413–23420

Kulaksiz H, Cetin Y (2002) The electrolyte/fluid secretion stimulatory peptides guanylin and uroguanylin and their common functional coupling proteins in the rat pancreas: a correlative study of expression and cell-specific localization. Pancreas 25:170–175

Kumpulainen T, Jalovaara P (1981) Immunohistochemical localization of carbonic anhydrase isoenzymes in the human pancreas. Gastroenterology 80:796–799

Kwon TH, Pushkin A, Abuladze N, Nielsen S, Kurtz I (2000) Immunoelectron microscopic localization of NBC3 sodium-bicarbonate cotransporter in rat kidney. Am J Physiol Renal Physiol 278:327–336

Kwon TH, Fenton RA, Praetorius J, Nielsen S (2012) Section I: normal renal function: molecular, cellular, structural, and physiologic principles; chapter 2: anatomy and topography. In: Taal MW, Chertow GM, Marsden PA, Skoreckiand K, Yu ASL (eds) Brenner & Rector's the kidney. Saunders, New York

Larsen AM, Krogsgaard-Larsen N, Lauritzen G, Olesen CW, Honore HS, Boedtkjer E, Pedersen SF, Bunch L (2012) Gram-scale solution-phase synthesis of selective sodium bicarbonate co-transport inhibitor S0859: in vitro efficacy studies in breast cancer cells. ChemMedChem 7:1808–1814

Le MC, Boulkroun S, Gonzalez-Nunez D, Dublineau I, Cluzeaud F, Fay M, Blot-Chabaud M, Farman N (2005) Aldosterone and tight junctions: modulation of claudin-4 phosphorylation in renal collecting duct cells. Am J Physiol Cell Physiol 289:1513–1521

Lee RJ, Foskett JK (2010) Mechanisms of Ca^{2+}-stimulated fluid secretion by porcine bronchial submucosal gland serous acinar cells. Am J Physiol Lung Cell Mol Physiol 298:210–231

Lee MG, Schultheis PJ, Yan M, Shull GE, Bookstein C, Chang E, Tse M, Donowitz M, Park K, Muallem S (1998) Membrane-limited expression and regulation of Na^+-H^+ exchanger isoforms by P2 receptors in the rat submandibular gland duct. J Physiol 513:341–357

Lee MG, Ahn W, Choi JY, Luo X, Seo JT, Schultheis PJ, Shull GE, Kim KH, Muallem S (2000) Na^+-dependent transporters mediate HCO_3^- salvage across the luminal membrane of the main pancreatic duct. J Clin Invest 105:1651–1658

Lee MG, Ohana E, Park HW, Yang D, Muallem S (2012) Molecular mechanism of pancreatic and salivary gland fluid and HCO_3^- secretion. Physiol Rev 92:39–74

Lenzen R, Elster J, Behrend C, Hampel KE, Bechstein WO, Neuhaus P (1997) Bile acid-independent bile flow is differently regulated by glucagon and secretin in humans after orthotopic liver transplantation. Hepatology 26:1272–1281

Leslie BR, Schwartz JH, Steinmetz PR (1973) Coupling between Cl- absorption and HCO3- secretion in turtle urinary bladder. Am J Physiol 225:610–617

Leviel F, Hubner CA, Houillier P, Morla L, El MS, Brideau G, Hassan H, Parker MD, Kurth I, Kougioumtzes A, Sinning A, Pech V, Riemondy KA, Miller RL, Hummler E, Shull GE, Aronson PS, Doucet A, Wall SM, Chambrey R, Eladari D (2010) The Na^+-dependent chloride-bicarbonate exchanger SLC4A8 mediates an electroneutral Na^+ reabsorption process in the renal cortical collecting ducts of mice. J Clin Invest 120:1627–1635

Li X, Alvarez B, Casey JR, Reithmeier RA, Fliegel L (2002) Carbonic anhydrase II binds to and enhances activity of the Na^+/H^+ exchanger. J Biol Chem 277:36085–36091

Li J, Koo NY, Cho IH, Kwon TH, Choi SY, Lee SJ, Oh SB, Kim JS, Park K (2006) Expression of the Na^+-HCO_3^- cotransporter and its role in phi regulation in guinea pig salivary glands. Am J Physiol Gastrointest Liver Physiol 291:1031–1040

Li Q, Ding Y, Krafft P, Wan W, Yan F, Wu G, Zhang Y, Zhan Q, Zhang JH (2018) Targeting germinal matrix hemorrhage-induced overexpression of sodium-coupled bicarbonate exchanger reduces posthemorrhagic hydrocephalus formation in neonatal rats. J Am Heart Assoc 7: e007192

Lindsey AE, Schneider K, Simmons DM, Baron R, Lee BS, Kopito RR (1990) Functional expression and subcellular localization of an anion exchanger cloned from choroid plexus. Proc Natl Acad Sci U S A 87:5278–5282

Lindvall-Axelsson M, Nilsson C, Owman C, Winbladh B (1992) Inhibition of cerebrospinal fluid formation by omeprazole. Exp Neurol 115:394–399

Liu S, Piwnica-Worms D, Lieberman M (1990) Intracellular pH regulation in cultured embryonic chick heart cells. Na^+-dependent Cl^-/HCO_3^- exchange. J Gen Physiol 96:1247–1269

Liu X, Li T, Riederer B, Lenzen H, Ludolph K, Yeruva S, Tuo B, Soleimani M, Seidler U (2015) Loss of Slc26a9 anion transporter alters intestinal electrolyte and HCO_3^- transport and reduces survival in CFTR-deficient mice. Pflugers Arch 467:1261–1275

Liu X, Li T, Tuo B (2018) Physiological and pathophysiological relevance of the anion transporter Slc26a9 in multiple organs. Front Physiol 9:1197

Lohi H, Kujala M, Kerkela E, Saarialho-Kere U, Kestila M, Kere J (2000) Mapping of five new putative anion transporter genes in human and characterization of SLC26A6, a candidate gene for pancreatic anion exchanger. Genomics 70:102–112

Loiselle FB, Morgan PE, Alvarez BV (2003) Regulation of the human NBC3 Na^+/HCO_3^- co-transporter by carbonic anhydrase II and protein kinase A. Am J Physiol Cell Physiol 286:1423–1433

Lonnerholm G, Knutson L, Wistrand PJ, Flemstrom G (1989) Carbonic anhydrase in the normal rat stomach and duodenum and after treatment with omeprazole and ranitidine. Acta Physiol Scand 136:253–262

Loriol C, Dulong S, Avella M, Gabillat N, Boulukos K, Borgese F, Ehrenfeld J (2008) Characterization of SLC26A9, facilitation of Cl(-) transport by bicarbonate. Cell Physiol Biochem 22:15–30

Lux SE, John KM, Kopito RR, Lodish HF (1989) Cloning and characterization of band 3, the human erythrocyte anion-exchange protein (AE1). Proc Natl Acad Sci U S A 86:9089–9093

Ma T, Song Y, Gillespie A, Carlson EJ, Epstein CJ, Verkman AS (1999) Defective secretion of saliva in transgenic mice lacking aquaporin-5 water channels. J Biol Chem 274:20071–20074

Mahieu I, Becq F, Wolfensberger T, Gola M, Carter N, Hollande E (1994) The expression of carbonic anhydrases II and IV in the human pancreatic cancer cell line (Capan 1) is associated with bicarbonate ion channels. Biol Cell 81:131–141

Maleth J, Venglovecz V, Razga Z, Tiszlavicz L, Rakonczay Z Jr, Hegyi P (2011) Non-conjugated chenodeoxycholate induces severe mitochondrial damage and inhibits bicarbonate transport in pancreatic duct cells. Gut 60:136–138

Mangos JA, Braun G, Hamann KF (1966) Micropuncture study of sodium and potassium excretion in the rat parotid saliva. Pflügers Arch 291:99–106

Mangos JA, McSherry NR, Nousia-Arvanitakis S, Irwin K (1973) Secretion and transductal fluxes of ions in exocrine glands of the mouse. Am J Physiol 225:18–24

Maren TH (1962) The binding of inhibitors to carbonic anhydrase in vivo: drugs as markers for enzyme. Biochem Pharmacol 9:39–48

Marino CR, Matovcik LM, Gorelick FS, Cohn JA (1991) Localization of the cystic fibrosis transmembrane conductance regulator in pancreas. J Clin Invest 88:712–716

Marino CR, Jeanes V, Boron WF, Schmitt BM (1999) Expression and distribution of the Na^+-HCO_3^- cotransporter in human pancreas. Am J Physiol 277:487–494

Marsey LL, Winpenny JP (2009) Bestrophin expression and function in the human pancreatic duct cell line, CFPAC-1. J Physiol 587:2211–2224

Marteau C, Silviani V, Ducroc R, Crotte C, Gerolami A (1995) Evidence for apical Na^+/H^+ exchanger in bovine main pancreatic duct. Dig Dis Sci 40:2336–2340

Martin CJ, Young JA (1971) A microperfusion investigation of the effects of a sympathomimetic and a parasympathomimetic drug on water and electrolyte fluxes in the main duct of the rat submaxillary gland. Pflügers Arch 327:303–323

Martin AC, Willoughby D, Ciruela A, Ayling LJ, Pagano M, Wachten S, Tengholm A, Cooper DM (2009) Capacitative Ca2+ entry via Orai1 and stromal interacting molecule 1 (STIM1) regulates adenylyl cyclase type 8. Mol Pharmacol 75:830–842

Martinez JR, Cassity N (1983) Effects of transport inhibitors on secretion by perfused rat submandibular gland. Am J Physiol 245:711–716

Martinez JR, Holzgreve H, Frick A (1966) Micropuncture study of submaxillary glands of adult rats. Pflügers Arch 290:124–133

Martinez-Anso E, Castillo JE, Diez J, Medina JF, Prieto J (1994) Immunohistochemical detection of chloride/bicarbonate anion exchangers in human liver. Hepatology 19:1400–1406

Maruyama Y, Gallacher DV, Petersen OH (1983) Voltage and Ca^{2+}-activated K^+ channel in basolateral acinar cell membranes of mammalian salivary glands. Nature 302:827–829

Masmoudi S, Charfedine I, Hmani M, Grati M, Ghorbel AM, Elgaied-Boulila A, Drira M, Hardelin JP, Ayadi H (2000) Pendred syndrome: phenotypic variability in two families carrying the same PDS missense mutation. Am J Med Genet 90:38–44

Masuzawa T, Ohta T, Kawamura M, Nakahara N, Sato F (1984) Immunohistochemical localization of Na+, K+-ATPase in the choroid plexus. Brain Res 302:357–362

Masyuk AI, Gradilone SA, Banales JM, Huang BQ, Masyuk TV, Lee SO, Splinter PL, Stroope AJ, LaRusso NF (2008) Cholangiocyte primary cilia are chemosensory organelles that detect biliary nucleotides via P2Y12 purinergic receptors. Am J Physiol Gastrointest Liver Physiol 295:725–734

Masyuk AI, Huang BQ, Radtke BN, Gajdos GB, Splinter PL, Masyuk TV, Gradilone SA, LaRusso NF (2013) Ciliary subcellular localization of TGR5 determines the cholangiocyte functional response to bile acid signaling. Am J Physiol Gastrointest Liver Physiol 304:1013–1024

Matsuo R (2000) Role of saliva in the maintenance of taste sensitivity. Crit Rev Oral Biol Med 11:216–229

Mayer SE, Sanders-Bush E (1993) Sodium-dependent antiporters in choroid plexus epithelial cultures from rabbit. J Neurochem 60:1308–1316

McCormick JA, Ellison DH (2011) The WNKs: atypical protein kinases with pleiotropic actions. Physiol Rev 91:177–219

McMurtrie HL, Cleary HJ, Alvarez BV, Loiselle FB, Sterling D, Morgan PE, Johnson DE, Casey JR (2004) The bicarbonate transport metabolon. J Enzyme Inhib Med Chem 19:231–236

Meier PJ, Knickelbein R, Moseley RH, Dobbins JW, Boyer JL (1985) Evidence for carrier-mediated chloride/bicarbonate exchange in canalicular rat liver plasma membrane vesicles. J Clin Invest 75:1256–1263

Melby JM, Miner LC, Reed DJ (1982) Effect of acetazolamide and furosemide on the production and composition of cerebrospinal fluid from the cat choroid plexus. Can J Physiol Pharmacol 60:405–409

Melvin JE, Park K, Richardson L, Schultheis PJ, Shull GE (1999) Mouse down-regulated in adenoma (DRA) is an intestinal Cl^-/HCO_3^- exchanger and is up-regulated in colon of mice lacking the NHE3 Na+/H+ exchanger. J Biol Chem 274:22855–22861

Mennone A, Biemesderfer D, Negoianu D, Yang CL, Abbiati T, Schultheis PJ, Shull GE, Aronson PS, Boyer JL (2001) Role of sodium/hydrogen exchanger isoform NHE3 in fluid secretion and absorption in mouse and rat cholangiocytes. Am J Physiol Gastrointest Liver Physiol 280:247–254

Meyer G, Guizzardi F, Rodighiero S, Manfredi R, Saino S, Sironi C, Garavaglia ML, Bazzini C, Botta G, Portincasa P, Calamita G, Paulmichl M (2005) Ion transport across the gallbladder epithelium. Curr Drug Targets Immune Endocr Metabol Disord 5:143–151

Millar ID, Brown PD (2008) NBCe2 exhibits a 3 HCO_3^-:1 Na^+ stoichiometry in mouse choroid plexus epithelial cells. Biochem Biophys Res Commun 373:550–554

Minagawa N, Nagata J, Shibao K, Masyuk AI, Gomes DA, Rodrigues MA, Lesage G, Akiba Y, Kaunitz JD, Ehrlich BE, LaRusso NF, Nathanson MH (2007) Cyclic AMP regulates bicarbonate secretion in cholangiocytes through release of ATP into bile. Gastroenterology 133:1592–1602

Modyanov NN, Petrukhin KE, Sverdlov VE, Grishin AV, Orlova MY, Kostina MB, Makarevich OI, Broude NE, Monastyrskaya GS, Sverdlov ED (1991) The family of human Na,K-ATPase genes. ATP1AL1 gene is transcriptionally competent and probably encodes the related ion transport ATPase. FEBS Lett 278:91–94

Modyanov NN, Mathews PM, Grishin AV, Beguin P, Beggah AT, Rossier BC, Horisberger JD, Geering K (1995) Human ATP1AL1 gene encodes a ouabain-sensitive H-K-ATPase. Am J Physiol 269:992–997

Moore-Hoon ML, Turner RJ (1998) Molecular and topological characterization of the rat parotid Na$^+$-K$^+$-2Cl$^-$ cotransporter1. Biochim Biophys Acta 1373:261–269

Moseley RH, Meier PJ, Aronson PS, Boyer JL (1986) Na-H exchange in rat liver basolateral but not canalicular plasma membrane vesicles. Am J Physiol 250:35–43

Moser AJ, Gangopadhyay A, Bradbury NA, Peters KW, Frizzell RA, Bridges RJ (2007) Electrogenic bicarbonate secretion by prairie dog gallbladder. Am J Physiol Gastrointest Liver Physiol 292:1683–1694

Mount DB, Romero MF (2004) The SLC26 gene family of multifunctional anion exchangers. Pflugers Arch 447:710–721

Murakami M, Murdiastuti K, Hosoi K, Hill AE (2006) AQP and the control of fluid transport in a salivary gland. J Membr Biol 210:91–103

Murphy VA, Johanson CE (1989) Alteration of sodium transport by the choroid plexus with amiloride. Biochim Biophys Acta 979:187–192

Murphy VA, Smith QR, Rapoport SI (1986) Homeostasis of brain and cerebrospinal fluid calcium concentrations during chronic hypo- and hypercalcemia. J Neurochem 47:1735–1741

Nakahari T, Murakami M, Yoshida H, Miyamoto M, Sohma Y, Imai Y (1990) Decrease in rat submandibular acinar cell volume during ACh stimulation. Am J Physiol 258:878–886

Nakahari T, Murakami M, Sasaki Y, Kataoka T, Imai Y, Shiba Y, Kanno Y (1991) Dose effects of acetylcholine on the cell volume of rat mandibular salivary acini. Jpn J Physiol 41:153–168

Nakamoto T, Srivastava A, Romanenko VG, Ovitt CE, Perez-Cornejo P, Arreola J, Begenisich T, Melvin JE (2007) Functional and molecular characterization of the fluid secretion mechanism in human parotid acinar cells. Am J Physiol Regul Integr Comp Physiol 292:2380–2390

Nakamoto T, Romanenko VG, Takahashi A, Begenisich T, Melvin JE (2008) Apical maxi-K (KCa1.1) channels mediate K$^+$ secretion by the mouse submandibular exocrine gland. Am J Physiol Cell Physiol 294:810–819

Nakamoto T, Brown DA, Catalan MA, Gonzalez-Begne M, Romanenko VG, Melvin JE (2009) Purinergic P2X7 receptors mediate ATP-induced saliva secretion by the mouse submandibular gland. J Biol Chem 284:4815–4822

Nakamura N, Suzuki Y, Sakuta H, Ookata K, Kawahara K, Hirose S (1999) Inwardly rectifying K$^+$ channel Kir7.1 is highly expressed in thyroid follicular cells, intestinal epithelial cells and choroid plexus epithelial cells: implication for a functional coupling with Na$^+$,K$^+$-ATPase. Biochem J 342(Pt 2):329–336

Namkung W, Lee JA, Ahn W, Han W, Kwon SW, Ahn DS, Kim KH, Lee MG (2003) Ca^{2+} activates cystic fibrosis transmembrane conductance regulator- and Cl$^-$-dependent HCO$_3$ transport in pancreatic duct cells. J Biol Chem 278:200–207

Namkung W, Finkbeiner WE, Verkman AS (2010) CFTR-adenylyl cyclase I association responsible for UTP activation of CFTR in well-differentiated primary human bronchial cell cultures. Mol Biol Cell 21:2639–2648

Nathanson MH, Burgstahler AD, Mennone A, Boyer JL (1996) Characterization of cytosolic Ca2+ signaling in rat bile duct epithelia. Am J Physiol 271:86–96

Nattie EE (1980) Brain and cerebrospinal fluid ionic composition and ventilation in acute hypercapnia. Respir Physiol 40:309–322

Nauntofte B (1992) Regulation of electrolyte and fluid secretion in salivary acinar cells. Am J Physiol Gastrointest Liver Physiol 263:823–837

Nauntofte B, Poulsen JH (1986) Effects of Ca^{2+} and furosemide on Cl$^-$ transport on O$_2$ uptake in rat parotid acini. Am J Physiol 251:175–185

Nehrke K, Quinn CC, Begenisich T (2003) Molecular identification of Ca2+-activated K+ channels in parotid acinar cells. Am J Physiol Cell Physiol 284:535–546

Nguyen TD, Moody MW, Savard CE, Lee SP (1998) Secretory effects of ATP on nontransformed dog pancreatic duct epithelial cells. Am J Physiol 275:104–113

Nguyen TD, Meichle S, Kim US, Wong T, Moody MW (2001) P2Y$_{11}$, a purinergic receptor acting via cAMP, mediates secretion by pancreatic duct epithelial cells. Am J Physiol Gastrointest Liver Physiol 280:795–804

Nishimori I, Onishi S (2001) Carbonic anhydrase isozymes in the human pancreas. Dig Liver Dis 33:68–74

Nishimori I, FujikawaAdachi K, Onishi S, Hollingsworth MA (1999) Carbonic anhydrase in human pancreas: hypotheses for the pathophysiological roles of CA isozymes. Ann N Y Acad Sci 880:5–16

Novak I (1993) Cellular mechanism of salivary gland secretion. In: Gilles R (ed) Advances in comparative and environmental physiology, vol 15. Springer, Berlin, pp 1–43

Novak I (2008) Purinergic receptors in the endocrine and exocrine pancreas. Purinergic Signal 4:237–253

Novak I (2011) Purinergic signalling in epithelial ion transport: regulation of secretion and absorption. Acta Physiol (Oxf) 202:501–522

Novak I, Christoffersen BC (2001) Secretin stimulates HCO$_3$$^-$ and acetate efflux but not Na$^+$/HCO$_3$$^-$ uptake in rat pancreatic ducts. Pflügers Arch 441:761–771

Novak I, Greger R (1988a) Electrophysiological study of transport systems in isolated perfused pancreatic ducts: properties of the basolateral membrane. Pflügers Arch 411:58–68

Novak I, Greger R (1988b) Properties of the luminal membrane of isolated perfused rat pancreatic ducts: effect of cyclic AMP and blockers of chloride transport. Pflügers Arch 411:546–553

Novak I, Greger R (1991) Effect of bicarbonate on potassium conductance of isolated perfused rat pancreatic ducts. Pflügers Arch 419:76–83

Novak I, Young JA (1986) Two independent anion transport systems in rabbit mandibular salivary glands. Pflugers Arch 407:649–656

Novak I, Hede SE, Hansen MR (2008) Adenosine receptors in rat and human pancreatic ducts stimulate chloride transport. Pflugers Arch 456:437–447

Novak I, Jans IM, Wohlfahrt L (2010) Effect of P2X7 receptor knockout on exocrine secretion of pancreas, salivary glands and lacrimal glands. J Physiol (Lond) 588(18):3615–3627

Novak I, Wang J, Henriksen KL, Haanes KA, Krabbe S, Nitschke R, Hede SE (2011) Pancreatic bicarbonate secretion involves two proton pumps. J Biol Chem 286:280–289

Novak I, Haanes KA, Wang J (2013) Acid-base transport in pancreas-new challenges. Front Physiol 4:380. https://doi.org/10.3389/fphys.2013.00380

Odgaard E, Jakobsen JK, Frische S, Praetorius J, Nielsen S, Aalkjaer C, Leipziger J (2004) Basolateral Na+-dependent HCO$_3$$^-$ transporter NBCn1-mediated HCO$_3$$^-$ influx in rat medullary thick ascending limb. J Physiol 555:205–218

Okolo C, Wong T, Moody MW, Nguyen TD (2002) Effects of bile acids on dog pancreatic duct epithelial cell secretion and monolayer resistance. Am J Physiol Gastrointest Liver Physiol 283:1042–1050

Olszewski U, Hlozek M, Hamilton G (2010) Activation of Na$^+$/H$^+$ exchanger 1 by neurotensin signaling in pancreatic cancer cell lines. Biochem Biophys Res Commun 393:414–419

Orlowski J, Grinstein S (2004) Diversity of the mammalian sodium/proton exchanger SLC9 gene family. Pflugers Arch 447:549–565

Ostrowski NL, Lolait SJ, Bradley DJ, O'Carroll AM, Brownstein MJ, Young WS III (1992) Distribution of V1a and V2 vasopressin receptor messenger ribonucleic acids in rat liver, kidney, pituitary and brain. Endocrinology 131:533–535

Pahl C, Novak I (1993) Effect of vasoactive intestinal peptide, carbachol and other agonists on cell membrane voltage of pancreatic duct cells. Pflügers Arch 424:315–320

Palk L, Sneyd J, Shuttleworth TJ, Yule DI, Crampin EJ (2010) A dynamic model of saliva secretion. J Theor Biol 266:625–640

Park HW, Lee MG (2012) Transepithelial bicarbonate secretion: lessons from the pancreas. Cold Spring Harb Perspect Med 2:155–170

Park HS, Lee YL, Kwon HY, Chey WY, Park HJ (1998) Significant cholinergic role in secretin-stimulated exocrine secretion in isolated rat pancreas. Am J Physiol 274:413–418

Park K, Olschowka JA, Richardson LA, Bookstein C, Chang EB, Melvin JE (1999) Expression of multiple Na^+/H^+ exchanger isoforms in rat parotid acinar and ductal cells. Am J Physiol 276:470–478

Park K, Case RM, Brown PD (2001a) Identification and regulation of K+ and Cl- channels in human parotid acinar cells. Arch Oral Biol 46:801–810

Park K, Evans RL, Watson GE, Nehrke K, Richardson L, Bell SM, Schultheis PJ, Hand AR, Shull GE, Melvin JE (2001b) Defective fluid secretion and NaCl absorption in the parotid glands of Na+/H+ exchanger-deficient mice. J Biol Chem 276:27042–27050

Park K, Hurley PT, Roussa E, Cooper GJ, Smith CP, Thevenod F, Steward MC, Case RM (2002a) Expression of a sodium bicarbonate cotransporter in human parotid salivary glands. Arch Oral Biol 47:1–9

Park M, Ko SB, Choi JY, Muallem G, Thomas PJ, Pushkin A, Lee MS, Kim JY, Lee MG, Muallem S, Kurtz I (2002b) The cystic fibrosis transmembrane conductance regulator interacts with and regulates the activity of the HCO_3^- salvage transporter human Na^+-HCO_3^- cotransport isoform 3. J Biol Chem 277:50503–50509

Park HW, Nam JH, Kim JY, Namkung W, Yoon JS, Lee JS, Kim KS, Venglovecz V, Gray MA, Kim KH, Lee MG (2010) Dynamic regulation of CFTR bicarbonate permeability by $[Cl^-]_i$ and its role in pancreatic bicarbonate secretion. Gastroenterology 139:620–631

Park S, Hong JH, Ohana E, Muallem S (2012) The WNK/SPAK and IRBIT/PP1 pathways in epithelial fluid and electrolyte transport. Physiology (Bethesda) 27:291–299

Park S, Shcheynikov N, Hong JH, Zheng C, Suh SH, Kawaai K, Ando H, Mizutani A, Abe T, Kiyonari H, Seki G, Yule D, Mikoshiba K, Muallem S (2013) Irbit mediates synergy between Ca^{2+} and cAMP signaling pathways during epithelial transport in mice. Gastroenterology 145:232–241

Parker MD, Musa-Aziz R, Rojas JD, Choi I, Daly CM, Boron WF (2008) Characterization of human SLC4A10 as an electroneutral Na/HCO_3 cotransporter (NBCn2) with Cl^- self-exchange activity. J Biol Chem 283:12777–12788

Pascua P, Garcia M, Fernandez-Salazar MP, Hernandez-Lorenzo MP, Calvo JJ, Colledge WH, Case RM, Steward MC, San Roman JI (2009) Ducts isolated from the pancreas of CFTR-null mice secrete fluid. Pflugers Arch 459:203–214

Pastor-Soler NM, Hallows KR, Smolak C, Gong F, Brown D, Breton S (2008) Alkaline pH- and cAMP-induced V-ATPase membrane accumulation is mediated by protein kinase A in epididymal clear cells. Am J Physiol Cell Physiol 294:488–494

Patterson K, Catalan MA, Melvin JE, Yule DI, Crampin EJ, Sneyd J (2012) A quantitative analysis of electrolyte exchange in the salivary duct. Am J Physiol Gastrointest Liver Physiol 303:1153–1163

Paulais M, Cragoe EJ Jr, Turner RJ (1994) Ion transport mechanisms in rat parotid intralobular striated ducts. Am J Physiol 266:1594–1602

Pearson MM, Lu J, Mount DB, Delpire E (2001) Localization of the K^+-Cl^- cotransporter, KCC3, in the central and peripheral nervous systems: expression in the choroid plexus, large neurons and white matter tracts. Neuroscience 103:481–491

Pech V, Pham TD, Hong S, Weinstein AM, Spencer KB, Duke BJ, Walp E, Kim YH, Sutliff RL, Bao HF, Eaton DC, Wall SM (2010) Pendrin modulates ENaC function by changing luminal HCO_3^-. J Am Soc Nephrol 21:1928–1941

Pedersen AM, Bardow A, Jensen SB, Nauntofte B (2002) Saliva and gastrointestinal functions of taste, mastication, swallowing and digestion. Oral Dis 8:117–129

Pedersen AML, Sørensen CE, Dynesen AW, Jensen SB (2013a) Salivary glands structure and function and regulation of saliva secretion in health and disease. In: Braxton L, Quinn L (eds)

Salivary glands: anatomy, functions in digestion and rrole in disease. Nova Science, New York, pp 1–43

Pedersen SF, Hoffmann EK, Novak I (2013b) Cell volume regulation in epithelial physiology and cancer. Front Physiol 4:233

Pedersen SF, Novak I, Alves F, Schwab A, Pardo LA (2017) Alternating pH landscapes shape epithelial cancer initiation and progression: Focus on pancreatic cancer. Bioessays 39:1600253

Perides G, Laukkarinen JM, Vassileva G, Steer ML (2010) Biliary acute pancreatitis in mice is mediated by the G-protein-coupled cell surface bile acid receptor Gpbar1. Gastroenterology 138:715–725

Pestov NB, Romanova LG, Korneenko TV, Egorov MV, Kostina MB, Sverdlov VE, Askari A, Shakhparonov MI, Modyanov NN (1998) Ouabain-sensitive H,K-ATPase: tissue-specific expression of the mammalian genes encoding the catalytic alpha subunit. FEBS Lett 440:320–324

Petersen OH (2014) Calcium signalling and secretory epithelia. Cell Calcium 55:282–289

Petersen OH, Gallacher DV (1988) Electrophysiology of pancreatic and salivary acinar cells. Annu Rev Physiol 50:65–80

Petersen OH, Poulsen JH (1967) The effects of varying the extracellular potassium concentration on the secretory rate and on resting and secretory potentials in the perfused cat submandibular gland. Acta Physiol Scand 70:293–298

Petersen KU, Reuss L (1983) Cyclic AMP-induced chloride permeability in the apical membrane of Necturus gallbladder epithelium. J Gen Physiol 81:705–729

Petersen KU, Wehner F, Winterhager JM (1990) Transcellular bicarbonate transport in rabbit gallbladder epithelium: mechanisms and effects of cyclic AMP. Pflugers Arch 416:312–321

Petrovic S, Wang Z, Ma L, Soleimani M (2003) Regulation of the apical Cl^-/HCO_3^- exchanger pendrin in rat cortical collecting duct in metabolic acidosis. Am J Physiol Renal Physiol 284:103–112

Petrovic S, Barone S, Xu J, Conforti L, Ma L, Kujala M, Kere J, Soleimani M (2004) SLC26A7: a basolateral Cl^-/HCO_3^- exchanger specific to intercalated cells of the outer medullary collecting duct. Am J Physiol Renal Physiol 286:161–169

Phillips PA, Abrahams JM, Kelly J, Paxinos G, Grzonka Z, Mendelsohn FA, Johnston CI (1988) Localization of vasopressin binding sites in rat brain by in vitro autoradiography using a radioiodinated V1 receptor antagonist. Neuroscience 27:749–761

Pizzonia JH, Biemesderfer D, Abu-Alfa AK, Wu MS, Exner M, Isenring P, Igarashi P, Aronson PS (1998) Immunochemical characterization of Na+/H+ exchanger isoform NHE4. Am J Physiol 275:510–517

Plotkin MD, Kaplan MR, Peterson LN, Gullans SR, Hebert SC, Delpire E (1997) Expression of the $Na^+-K^+-2Cl^-$ cotransporter BSC2 in the nervous system. Am J Physiol Cell Physiol 272:173–183

Pols TW, Noriega LG, Nomura M, Auwerx J, Schoonjans K (2011) The bile acid membrane receptor TGR5: a valuable metabolic target. Dig Dis 29:37–44

Poronnik P, Schumann SY, Cook DI (1995) HCO_3^--dependent ACh-activated Na^+ influx in sheep parotid secretory endpieces. Pflugers Arch 429:852–858

Poulsen JH, Bundgaard M (1994) Quantitative estimation of the area of luminal and basolateral membranes of rat parotid acinar cells: some physiological applications. Pflugers Arch 429:240–244

Praetorius J (2010) Na^+-coupled bicarbonate transporters in duodenum, collecting ducts and choroid plexus. J Nephrol 23(Suppl 16):35–42

Praetorius J, Nielsen S (2006) Distribution of sodium transporters and aquaporin-1 in the human choroid plexus. Am J Physiol Cell Physiol 291:59–67

Praetorius J, Andreasen D, Jensen BL, Ainsworth MA, Friis UG, Johansen T (2000) NHE1, NHE2, and NHE3 contribute to regulation of intracellular pH in murine duodenal epithelial cells. Am J Physiol Gastrointest Liver Physiol 278:197–206

Praetorius J, Hager H, Nielsen S, Aalkjaer C, Friis UG, Ainsworth MA, Johansen T (2001) Molecular and functional evidence for electrogenic and electroneutral Na^+-HCO_3^- cotransporters in murine duodenum. Am J Physiol Gastrointest Liver Physiol 280:332–343

Praetorius J, Friis UG, Ainsworth MA, Schaffalitzky de Muckadell OB, Johansen T (2002) The cystic fibrosis transmembrane conductance regulator is not a base transporter in isolated duodenal epithelial cells. Acta Physiol Scand 174:327–336

Praetorius J, Kim YH, Bouzinova EV, Frische S, Rojek A, Aalkjaer C, Nielsen S (2004a) NBCn1 is a basolateral Na^+-HCO_3^- cotransporter in rat kidney inner medullary collecting ducts. Am J Physiol Renal Physiol 286:903–912

Praetorius J, Nejsum LN, Nielsen S (2004b) A SLC4A10 gene product maps selectively to the basolateral plasma membrane of choroid plexus epithelial cells. Am J Physiol Cell Physiol 286:601–610

Proctor GB, Carpenter GH (2014) Salivary secretion: mechanism and neural regulation. Monogr Oral Sci 24:14–29

Pushkin A, Abuladze N, Lee I, Newman D, Hwang J, Kurtz I (1999) Cloning, tissue distribution, genomic organization, and functional characterization of NBC3, a new member of the sodium bicarbonate cotransporter family. J Biol Chem 274:16569–16575

Pushkin A, Abuladze N, Newman D, Lee I, Xu G, Kurtz I (2000) Cloning, characterization and chromosomal assignment of NBC4, a new member of the sodium bicarbonate cotransporter family. Biochim Biophys Acta 1493:215–218

Qu Z, Hartzell HC (2008) Bestrophin Cl^- channels are highly permeable to HCO_3^-. Am J Physiol Cell Physiol 294:1371–1377

Quinton PM (2008) Cystic fibrosis: impaired bicarbonate secretion and mucoviscidosis. Lancet 372:415–417

Quinton PM (2010) Role of epithelial HCO_3^- transport in mucin secretion: lessons from cystic fibrosis. Am J Physiol Cell Physiol 299:1222–1233

Quinton PM, Wright EM, Tormey JM (1973) Localization of sodium pumps in the choroid plexus epithelium. J Cell Biol 58:724–730

Rakonczay Z Jr, Fearn A, Hegyi P, Boros I, Gray MA, Argent BE (2006) Characterization of H^+ and HCO_3^- transporters in CFPAC-1 human pancreatic duct cells. World J Gastroenterol 12:885–895

Renner EL, Lake JR, Scharschmidt BF, Zimmerli B, Meier PJ (1989) Rat hepatocytes exhibit basolateral Na^+/HCO_3^- contransport. J Clin Invest 83:1225–1235

Riordan JR, Rommens JM, Kerem B, Alon N, Rozmahel R, Grzelczak Z, Zielenski J, Lok S, Plavsic N, Chou JL (1989) Identification of the cystic fibrosis gene: cloning and characterization of complementary DNA. Science 245:1066–1073

Roepke TK, Kanda VA, Purtell K, King EC, Lerner DJ, Abbott GW (2011) KCNE2 forms potassium channels with KCNA3 and KCNQ1 in the choroid plexus epithelium. FASEB J 25:4264–4273

Rogers CS, Stoltz DA, Meyerholz DK, Ostedgaard LS, Rokhlina T, Taft PJ, Rogan MP, Pezzulo AA, Karp PH, Itani OA, Kabel AC, Wohlford-Lenane CL, Davis GJ, Hanfland RA, Smith TL, Samuel M, Wax D, Murphy CN, Rieke A, Whitworth K, Uc A, Starner TD, Brogden KA, Shilyansky J, McCray PB Jr, Zabner J, Prather RS, Welsh MJ (2008) Disruption of the CFTR gene produces a model of cystic fibrosis in newborn pigs. Science 321:1837–1841

Romanenko VG, Nakamoto T, Srivastava A, Begenisich T, Melvin JE (2007) Regulation of membrane potential and fluid secretion by Ca^{2+}-activated K^+ channels in mouse submandibular glands. J Physiol 581:801–817

Romanenko VG, Nakamoto T, Catalan MA, Gonzalez-Begne M, Schwartz GJ, Jaramillo Y, Sepulveda FV, Figueroa CD, Melvin JE (2008) Clcn2 encodes the hyperpolarization-activated chloride channel in the ducts of mouse salivary glands. Am J Physiol Gastrointest Liver Physiol 295:1058–1067

Romanenko VG, Catalan MA, Brown DA, Putzier I, Hartzell HC, Marmorstein AD, Gonzalez-Begne M, Rock JR, Harfe BD, Melvin JE (2010) Tmem16A encodes the Ca^{2+}-activated Cl^- channel in mouse submandibular salivary gland acinar cells. J Biol Chem 285:12990–13001

Romero MF, Hediger MA, Boulpaep EL, Boron WF (1997) Expression cloning and characterization of a renal electrogenic Na+/HCO3- cotransporter. Nature 387:409–413

Rougemont JD, Ames A III, Nesbett FB, Hofmann HF (1960) Fluid formed by choroid plexus; a technique for its collection and a comparison of its electrolyte composition with serum and cisternal fluids. J Neurophysiol 23:485–495

Roussa E, Thevenod F (1998) Distribution of V-ATPase in rat salivary glands. Eur J Morphol 36:147–152

Roussa E, Thevenod F, Sabolic I, Herak-Kramberger CM, Nastainczyk W, Bock R, Schulz I (1998) Immunolocalization of vacuolar-type H^+-ATPase in rat submandibular gland and adaptive changes induced by acid-base disturbances. J Histochem Cytochem 46:91–100

Roussa E, Romero MF, Schmitt BM, Boron WF, Alper SL, Thevenod F (1999) Immunolocalization of anion exchanger AE2 and Na^+-HCO_3- cotransporter in rat parotid and submandibular glands. Am J Physiol 277:1288–1296

Roussa E, Alper SL, Thevenod F (2001) Immunolocalization of anion exchanger AE2, Na^+/H^+ exchangers NHE1 and NHE4, and vacuolar type H^+-ATPase in rat pancreas. J Histochem Cytochem 49:463–474

Roussa E, Shmukler BE, Wilhelm S, Casula S, Stuart-Tilley AK, Thevenod F, Alper SL (2002) Immunolocalization of potassium-chloride cotransporter polypeptides in rat exocrine glands. Histochem Cell Biol 117:335–344

Royaux IE, Suzuki K, Mori A, Katoh R, Everett LA, Kohn LD, Green ED (2000) Pendrin, the protein encoded by the pendred syndrome gene (PDS), is an apical porter of iodide in the thyroid and is regulated by thyroglobulin in FRTL-5 cells. Endocrinology 141:839–845

Royaux IE, Wall SM, Karniski LP, Everett LA, Suzuki K, Knepper MA, Green ED (2001) Pendrin, encoded by the pendred syndrome gene, resides in the apical region of renal intercalated cells and mediates bicarbonate secretion. Proc Natl Acad Sci U S A 98:4221–4226

Rune SJ, Henriksen FW (1969) Carbon dioxide tensions in the proximal part of the canine gastrointestinal tract. Gastroenterology 56:758–762

Russell JM (2000) Sodium-potassium-chloride cotransport. Physiol Rev 80:211–276

Ryu SY, Peixoto PM, Won JH, Yule DI, Kinnally KW (2010) Extracellular ATP and P2Y2 receptors mediate intercellular Ca^{2+} waves induced by mechanical stimulation in submandibular gland cells: role of mitochondrial regulation of store operated Ca^{2+} entry. Cell Calcium 47:65–76

Saarnio J, Parkkila S, Parkkila AK, Waheed A, Casey MC, Zhou XY, Pastorekova S, Pastorek J, Karttunen T, Haukipuro K, Kairaluoma MI, Sly WS (1998) Immunohistochemistry of carbonic anhydrase isozyme IX (MN/CA IX) in human gut reveals polarized expression in the epithelial cells with the highest proliferative capacity. J Histochem Cytochem 46:497–504

Sachs G, Shin JM, Vagin O, Lambrecht N, Yakubov I, Munson K (2007) The gastric H,K ATPase as a drug target: past, present, and future. J Clin Gastroenterol 41(Suppl 2):226–242

Saito Y, Wright EM (1983) Bicarbonate transport across the frog choroid plexus and its control by cyclic nucleotides. J Physiol 336:635–648

Saito Y, Wright EM (1984) Regulation of bicarbonate transport across the brush border membrane of the bull-frog choroid plexus. J Physiol 350:327–342

Sangan P, Thevananther S, Sangan S, Rajendran VM, Binder HJ (2000) Colonic H-K-ATPase alpha- and beta-subunits express ouabain-insensitive H-K-ATPase. Am J Physiol Cell Physiol 278:182–189

Sassani P, Pushkin A, Gross E, Gomer A, Abuladze N, Dukkipati R, Carpenito G, Kurtz I (2002) Functional characterization of NBC4: a new electrogenic sodium-bicarbonate cotransporter. Am J Physiol Cell Physiol 282:408–416

Satoh H, Moriyama N, Hara C, Yamada H, Horita S, Kunimi M, Tsukamoto K, Iso O, Inatomi J, Kawakami H, Kudo A, Endou H, Igarashi T, Goto A, Fujita T, Seki G (2003) Localization of

Na⁺-HCO₃⁻ cotransporter (NBC-1) variants in rat and human pancreas. Am J Physiol Cell Physiol 284:729–737

Sauter DR, Novak I, Pedersen SF, Larsen EH, Hoffmann EK (2015) ANO1 (TMEM16A) in pancreatic ductal adenocarcinoma (PDAC). Pflugers Arch 467:1495–1508

Sauter DR, Sorensen CE, Rapedius M, Bruggemann A, Novak I (2016) pH-sensitive K+ channel TREK-1 is a novel target in pancreatic cancer. Biochim Biophys Acta 1862:1994–2003

Schlosser SF, Burgstahler AD, Nathason MH (1996) Isolated rat hepatocytes can signal to other hepatocytes and bile duct cells by release of nucleotides. Proc Natl Acad Sci USA 93:9948–9953

Schlue WR, Deitmer JW (1988) Ionic mechanisms of intracellular pH regulation in the nervous system. Ciba Found Symp 139:69

Schmitt BM, Biemesderfer D, Romero MF, Boulpaep EL, Boron WF (1999) Immunolocalization of the electrogenic Na⁺-HCO₃⁻ cotransporter in mammalian and amphibian kidney. Am J Physiol Renal Physiol 276:27–38

Schneyer LH, Young JA, Schneyer CA (1972) Salivary secretion of electrolytes. Physiol Rev 52:720–777

Scholz W, Albus U, Counillon L, Gogelein H, Lang HJ, Linz W, Weichert A, Scholkens BA (1995) Protective effects of HOE642, a selective sodium-hydrogen exchange subtype 1 inhibitor, on cardiac ischaemia and reperfusion. Cardiovasc Res 29:260–268

Schroeder BC, Cheng T, Jan YN, Jan LY (2008) Expression cloning of TMEM16A as a calcium-activated chloride channel subunit. Cell 134:1019–1029

Schulz I (1971) Influence of bicarbonate-CO₂ - and glycodiazine buffer on the secretion of the isolated cat's pancreas. Pflugers Arch 329:283–306

Schulz I, Ströver F, Ullrich KJ (1971) Lipid soluble weak organic acid buffers as "substrate" for pancreatic secretion. Pflügers Arch 323:121–140

Schwaderer AL, Vijayakumar S, Al-Awqati Q, Schwartz GJ (2006) Galectin-3 expression is induced in renal beta-intercalated cells during metabolic acidosis. Am J Physiol Renal Physiol 290:148–158

Schwartz IL, Thaysen JH (1956) Excretion of sodium and potassium in human sweat. J Clin Invest 35:114–120

Schwartz GJ, Barasch J, Al-Awqati Q (1985) Plasticity of functional epithelial polarity. Nature 318:368–371

Schwartz GJ, Tsuruoka S, Vijayakumar S, Petrovic S, Mian A, Al-Awqati Q (2002) Acid incubation reverses the polarity of intercalated cell transporters, an effect mediated by hensin. J Clin Invest 109:89–99

Scudieri P, Musante I, Caci E, Venturini A, Morelli P, Walter C, Tosi D, Palleschi A, Martin-Vasallo P, Sermet-Gaudelus I, Planelles G, Crambert G, Galietta LJ (2018) Increased expression of ATP12A proton pump in cystic fibrosis airways. JCI Insight 3(20):123616

Seidler U, Blumenstein I, Kretz A, Viellard-Baron D, Rossmann H, Colledge WH, Evans M, Ratcliff R, Gregor M (1997) A functional CFTR protein is required for mouse intestinal cAMP-, cGMP- and Ca²⁺-dependent HCO₃⁻ secretion. J Physiol 505(Pt 2):411–423

Seow KTFP, Case RM, Young JA (1991) Pancreatic secretion by the anaesthetized rabbit in response to secretin, cholecystokinin, and carbachol. Pancreas 6:385–391

Sewell WA, Young JA (1975) Secretion of electrolytes by the pancreas of the anaesthetized rat. J Physiol (Lond) 252:379–396

Shah VS, Meyerholz DK, Tang XX, Reznikov L, Abou AM, Ernst SE, Karp PH, Wohlford-Lenane CL, Heilmann KP, Leidinger MR, Allen PD, Zabner J, McCray PB Jr, Ostedgaard LS, Stoltz DA, Randak CO, Welsh MJ (2016) Airway acidification initiates host defense abnormalities in cystic fibrosis mice. Science 351:503–507

Shcheynikov N, Wang Y, Park M, Ko SB, Dorwart M, Naruse S, Thomas PJ, Muallem S (2006) Coupling modes and stoichiometry of Cl⁻/HCO₃⁻ exchange by slc26a3 and slc26a6. J Gen Physiol 127:511–524

Shcheynikov N, Yang D, Wang Y, Zeng W, Karniski LP, So I, Wall SM, Muallem S (2008) The Slc26a4 transporter functions as an electroneutral $Cl^-/I^-/HCO_3^-$ exchanger: role of Slc26a4 and Slc26a6 in I^- and HCO_3^- secretion and in regulation of CFTR in the parotid duct. J Physiol 586:3813–3824

Shimokura GH, McGill JM, Schlenker T, Fitz JG (1995) Ursodeoxycholate increases cytosolic calcium concentration and activates Cl- currents in a biliary cell line. Gastroenterology 109:965–972

Shitara A, Tanimura A, Sato A, Tojyo Y (2009) Spontaneous oscillations in intracellular Ca^{2+} concentration via purinergic receptors elicit transient cell swelling in rat parotid ducts. Am J Physiol Gastrointest Liver Physiol 297:1198–1205

Siegel GJ, Holm C, Schreiber JH, Desmond T, Ernst SA (1984) Purification of mouse brain (Na^++K^+)-ATPase catalytic unit, characterization of antiserum, and immunocytochemical localization in cerebellum, choroid plexus, and kidney. J Histochem Cytochem 32:1309–1318

Simpson JE, Schweinfest CW, Shull GE, Gawenis LR, Walker NM, Boyle KT, Soleimani M, Clarke LL (2007) PAT-1 (Slc26a6) is the predominant apical membrane Cl^-/HCO_3^- exchanger in the upper villous epithelium of the murine duodenum. Am J Physiol Gastrointest Liver Physiol 292:1079–1088

Sinclair AJ, Onyimba CU, Khosla P, Vijapurapu N, Tomlinson JW, Burdon MA, Stewart PM, Murray PI, Walker EA, Rauz S (2007) Corticosteroids, 11beta-hydroxysteroid dehydrogenase isozymes and the rabbit choroid plexus. J Neuroendocrinol 19:614–620

Singh AK, Liu Y, Riederer B, Engelhardt R, Thakur BK, Soleimani M, Seidler U (2013a) Molecular transport machinery involved in orchestrating luminal acid-induced duodenal bicarbonate secretion in vivo. J Physiol 591:5377–5391

Singh AK, Xia W, Riederer B, Juric M, Li J, Zheng W, Cinar A, Xiao F, Bachmann O, Song P, Praetorius J, Aalkjaer C, Seidler U (2013b) Essential role of the electroneutral Na^+-HCO_3^- cotransporter NBCn1 in murine duodenal acid-base balance and colonic mucus layer build-up in vivo. J Physiol 591:2189–2204

Sjoblom M, Flemstrom G (2003) Melatonin in the duodenal lumen is a potent stimulant of mucosal bicarbonate secretion. J Pineal Res 34:288–293

Sjoblom M, Singh AK, Zheng W, Wang J, Tuo BG, Krabbenhoft A, Riederer B, Gros G, Seidler U (2009) Duodenal acidity "sensing" but not epithelial. Proc Natl Acad Sci U S A 106:13094–13099

Sly WS, Hewett-Emmett D, Whyte MP, Yu YS, Tashian RE (1983) Carbonic anhydrase II deficiency identified as the primary defect in the autosomal recessive syndrome of osteopetrosis with renal tubular acidosis and cerebral calcification. Proc Natl Acad Sci U S A 80:2752–2756

Soleimani M, Barone S, Xu J, Shull GE, Siddiqui F, Zahedi K, Amlal H (2012) Double knockout of pendrin and Na-Cl cotransporter (NCC) causes severe salt wasting, volume depletion, and renal failure. Proc Natl Acad Sci U S A 109:13368–13373

Song Y, Yamamoto A, Steward MC, Ko SB, Stewart AK, Soleimani M, Liu BC, Kondo T, Jin CX, Ishiguro H (2012) Deletion of Slc26a6 alters the stoichiometry of apical Cl^-/HCO_3^- exchange in mouse pancreatic duct. Am J Physiol Cell Physiol 303:815–824

Sorensen CE, Novak I (2001) Visualization of ATP release in pancreatic acini in response to cholinergic stimulus. Use of fluorescent probes and confocal microscopy. J Biol Chem 276:329525–329532

Speake T, Brown PD (2004) Ion channels in epithelial cells of the choroid plexus isolated from the lateral ventricle of rat brain. Brain Res 1005:60–66

Speake T, Kajita H, Smith CP, Brown PD (2002) Inward-rectifying anion channels are expressed in the epithelial cells of choroid plexus isolated from ClC-2 'knock-out' mice. J Physiol 539:385–390

Spirli C, Granato A, Zsembery K, Anglani F, Okolicsanyi L, LaRusso NF, Crepaldi G, Strazzabosco M (1998) Functional polarity of Na^+/H^+ and Cl^-/HCO_3^- excahngers in a rat choliangiocyte cell line. Am J Physiol 275:1236–1245

Steffensen AB, Oernbo EK, Stoica A, Gerkau NJ, Barbuskaite D, Tritsaris K, Rose CR, MacAulay N (2018) Cotransporter-mediated water transport underlying cerebrospinal fluid formation. Nat Commun 9:2167

Stetson DL, Steinmetz PR (1985) Alpha and beta types of carbonic anhydrase-rich cells in turtle bladder. Am J Physiol 249:553–565

Steward MC, Ishiguro H (2009) Molecular and cellular regulation of pancreatic duct cell function. Curr Opin Gastroenterol 25:447–453

Steward MC, Poronnik P, Cook DI (1996) Bicarbonate transport in sheep parotid secretory cells. J Physiol 494(Pt 3):819–830

Steward MC, Ishiguro H, Case RM (2005) Mechanisms of bicarbonate secretion in the pancreatic duct. Annu Rev Physiol 67:377–409

Stewart AK, Yamamoto A, Nakakuki M, Kondo T, Alper SL, Ishiguro H (2009) Functional coupling of apical Cl^-/HCO_3^- exchange with CFTR in stimulated HCO_3^- secretion by guinea pig interlobular pancreatic duct. Am J Physiol Gastrointest Liver Physiol 296:1307–1317

Strazzabosco M, Joplin R, Zsembery A, Wallace L, Spirli C, Fabris L, Granato A, Rossanese A, Poci C, Neuberger JM, Okolicsanyi L, Crepaldi G (1997) Na^+-dependent and -independent Cl^-/HCO_3^- exchange mediate cellular HCO_3^- transport in cultured human intrahepatic bile duct cells. Hepatology 25:976–985

Stuenkel EL, Machen TE, Williams JA (1988) pH regulatory mechanism in rat pancreatic ductal cells. Am J Physiol 254:925–930

Supuran CT (2008) Carbonic anhydrases--an overview. Curr Pharm Des 14:603–614

Svanvik J, Allen B, Pellegrini C, Bernhoft R, Way L (1984) Variations in concentrating function of the gallbladder in the conscious monkey. Gastroenterology 86:919–925

Swarts HG, Koenderink JB, Willems PH, De Pont JJ (2005) The non-gastric H,K-ATPase is oligomycin-sensitive and can function as an H^+,NH_4^+-ATPase. J Biol Chem 280:33115–33122

Szalmay G, Varga G, Kajiyama F, Yang XS, Lang TF, Case RM, Steward MC (2001) Bicarbonate and fluid secretion evoked by cholecystokinin, bombesin and acetylcholine in isolated guinea-pig pancreatic ducts. J Physiol (Lond) 535:795–807

Szmydynger-Chodobska J, Chung I, Kozniewska E, Tran B, Harrington FJ, Duncan JA, Chodobski A (2004) Increased expression of vasopressin v1a receptors after traumatic brain injury. J Neurotrauma 21:1090–1102

Szucs A, Demeter I, Burghardt B, Ovari G, Case RM, Steward MC, Varga G (2006) Vectorial bicarbonate transport by Capan-1 cells: a model for human pancreatic ductal secretion. Cell Physiol Biochem 18:253–264

Tabcharani JA, Chang X-B, Riordan JR, Hanrahan JW (1991) Phosphorylation-regulated Cl^- channel in CHO cells stably expressing the cystic fibrosis gene. Nature 352:628–631

Tabibian JH, Masyuk AI, Masyuk TV, O'Hara SP, LaRusso NF (2013) Physiology of cholangiocytes. Compr Physiol 3:541–565

Thaysen JH, Thorn NA, Schwartz IL (1954) Excretion of sodium, potassium, chloride, and carbon dioxide in human parotid saliva. Am J Physiol 178:155–159

Thompson J, Begenisich T (2006) Membrane-delimited inhibition of maxi-K channel activity by the intermediate conductance Ca^{2+}-activated K channel. J Gen Physiol 127:159–169

Thompson J, Begenisich T (2009) Mechanistic details of BK channel inhibition by the intermediate conductance, Ca^{2+}-activated K channel. Channels (Austin) 3:194–204

Turner RJ, George JN (1988) Cl^--HCO_3^- exchange is present with Na^+-K^+-Cl^- cotransport in rabbit parotid acinar basolateral membranes. Am J Physiol 254:391–396

Turner RJ, George JN, Baum BJ (1986) Evidence for a $Na^+/K^+/Cl^-$ cotransport system in basolateral membrane vesicles from the rabbit parotid. J Membr Biol 94:143–152

Uriarte I, Banales JM, Saez E, Arenas F, Oude Elferink RP, Prieto J, Medina JF (2010) Bicarbonate secretion of mouse cholangiocytes involves Na^+-HCO_3^- cotransport in addition to Na^+-independent Cl^-/HCO_3^- exchange. Hepatology 51:891–902

Vallet M, Picard N, Loffing-Cueni D, Fysekidis M, Bloch-Faure M, Deschenes G, Breton S, Meneton P, Loffing J, Aronson PS, Chambrey R, Eladari D (2006) Pendrin regulation in mouse kidney primarily is chloride-dependent. J Am Soc Nephrol 17:2153–2163

van Niekerk J, Kersten R, Beuers U (2018) Role of bile acids and the biliary HCO_3^- umbrella in the pathogenesis of primary biliary cholangitis. Clin Liver Dis 22:457–479

Veel T, Villanger O, Holthe MR, Cragoe EJ, Raeder MG (1992) Na^+-H^+ exchange is not important for pancreatic HCO_3^- secretion in the pig. Acta Physiol Scand 144:239–246

Venglovecz V, Rakonczay Z Jr, Ozsvari B, Takacs T, Lonovics J, Varro A, Gray MA, Argent BE, Hegyi P (2008) Effects of bile acids on pancreatic ductal bicarbonate secretion in guinea pig. Gut 57:1102–1112

Venglovecz V, Hegyi P, Rakonczay Z Jr, Tiszlavicz L, Nardi A, Grunnet M, Gray MA (2011) Pathophysiological relevance of apical large-conductance Ca^{2+}-activated potassium channels in pancreatic duct epithelial cells. Gut 60:361–369

Verlander JW, Hassell KA, Royaux IE, Glapion DM, Wang ME, Everett LA, Green ED, Wall SM (2003) Deoxycorticosterone upregulates PDS (Slc26a4) in mouse kidney: role of pendrin in mineralocorticoid-induced hypertension. Hypertension 42:356–362

Verlander JW, Kim YH, Shin W, Pham TD, Hassell KA, Beierwaltes WH, Green ED, Everett L, Matthews SW, Wall SM (2006) Dietary Cl^- restriction upregulates pendrin expression within the apical plasma membrane of type B intercalated cells. Am J Physiol Renal Physiol 291:833–839

Vijayakumar S, Erdjument-Bromage H, Tempst P, Al-Awqati Q (2008) Role of integrins in the assembly and function of hensin in intercalated cells. J Am Soc Nephrol 19:1079–1091

Villanger O, Veel T, Raeder MG (1993) Secretin causes H^+ secretion from intrahepatic bile ductules by vacuolar-type H^+-ATPase. Am J Physiol 265:719–724

Villanger O, Veel T, Ræder MG (1995) Secretin causes H^+/HCO_3^- secretion from pig pancreatic ductules by vacuolar-type H^+-adenosine triphosphatase. Gastroenterology 108:850–859

Virkki LV, Wilson DA, Vaughan-Jones RD, Boron WF (2002) Functional characterization of human NBC4 as an electrogenic Na^+-HCO_3^- cotransporter (NBCe2). Am J Physiol Cell Physiol 282:1278–1289

Virkki LV, Choi I, Davis BA, Boron WF (2003) Cloning of a Na^+-driven Cl/HCO_3 exchanger from squid giant fiber lobe. Am J Physiol Cell Physiol 285:771–780

Vogh BP, Godman DR, Maren TH (1987) Effect of AlCl3 and other acids on cerebrospinal fluid production: a correction. J Pharmacol Exp Ther 243:35–39

Voronina S, Longbottom R, Sutton R, Petersen OH, Tepikin A (2002) Bile acids induce calcium signals in mouse pancreatic acinar cells: implications for bile-induced pancreatic pathology. J Physiol 540:49–55

Voronina SG, Barrow SL, Gerasimenko OV, Petersen OH, Tepikin AV (2004) Effects of secreta-gogues and bile acids on mitochondrial membrane potential of pancreatic acinar cells: compar-ison of different modes of evaluating DeltaPsim. J Biol Chem 279:27327–27338

Vorum H, Kwon TH, Fulton C, Simonsen B, Choi I, Boron W, Maunsbach AB, Nielsen S, Aalkjaer C (2000) Immunolocalization of electroneutral Na-HCO_3^- cotransporter in rat kidney. Am J Physiol Renal Physiol 279:901–909

Wagner CA, Finberg KE, Stehberger PA, Lifton RP, Giebisch GH, Aronson PS, Geibel JP (2002) Regulation of the expression of the Cl-/anion exchanger pendrin in mouse kidney by acid-base status. Kidney Int 62:2109–2117

Wagner CA, Mohebbi N, Capasso G, Geibel JP (2011) The anion exchanger pendrin (SLC26A4) and renal acid-base homeostasis. Cell Physiol Biochem 28:497–504

Walker NM, Simpson JE, Brazill JM, Gill RK, Dudeja PK, Schweinfest CW, Clarke LL (2009) Role of down-regulated in adenoma anion exchanger in HCO_3^- secretion across murine duodenum. Gastroenterology 136:893–901

Wall SM, Pech V (2008) The interaction of pendrin and the epithelial sodium channel in blood pressure regulation. Curr Opin Nephrol Hypertens 17:18–24

Wall SM, Kim YH, Stanley L, Glapion DM, Everett LA, Green ED, Verlander JW (2004) NaCl restriction upregulates renal Slc26a4 through subcellular redistribution: role in Cl⁻ conservation. Hypertension 44:982–987

Wan MH, Huang W, Latawiec D, Jiang K, Booth DM, Elliott V, Mukherjee R, Xia Q (2012) Review of experimental animal models of biliary acute pancreatitis and recent advances in basic research. HPB (Oxford) 14:73–81

Wang J, Novak I (2013) Ion transport in human pancreatic duct epithelium, Capan-1 cells, is regulated by secretin, VIP, acetylcholine, and purinergic receptors. Pancreas 42:452–460

Wang CZ, Yano H, Nagashima K, Seino S (2000) The Na⁺-driven Cl⁻/HCO₃⁻ exchanger. Cloning, tissue distribution, and functional characterization. J Biol Chem 275:35486–35490

Wang Y, Soyombo AA, Shcheynikov N, Zeng W, Dorwart M, Marino CR, Thomas PJ, Muallem S (2006) Slc26a6 regulates CFTR activity in vivo to determine pancreatic duct HCO3- secretion: relevance to cystic fibrosis. EMBO J 25:5049–5057

Wang J, Haanes KA, Novak I (2013) Purinergic regulation of CFTR and Ca²⁺-activated Cl⁻ channels and K⁺ channels in human pancreatic duct epithelium. Am J Physiol Cell Physiol 304:673–684

Wang J, Barbuskaite D, Tozzi M, Giannuzzo A, Sorensen CE, Novak I (2015) Proton pump inhibitors inhibit pancreatic secretion: role of gastric and non-gastric H⁺/K⁺-ATPases. PLoS One 10:e0126432

Weber KT, Sun Y, Wodi LA, Munir A, Jahangir E, Ahokas RA, Gerling IC, Postlethwaite AE, Warrington KJ (2003) Toward a broader understanding of aldosterone in congestive heart failure. J Renin Angiotensin Aldosterone Syst 4:155–163

Wegman EA, Ishikawa T, Young JA, Cook DI (1992) Cation channels in basolateral membranes of sheep parotid secretory cells. Am J Physiol Gastrointest Liver Physiol 263:786–794

Welch K (1963) Secretion of cerebrospinal fluid by the choroid plexus of the rabbit. Am J Physiol 205:617–624

Wen G, Deng S, Song W, Jin H, Xu J, Liu X, Xie R, Song P, Tuo B (2018) Helicobacter pylori infection downregulates duodenal CFTR and SLC26A6 expressions through TGFbeta signaling pathway. BMC Microbiol 18:87

Williams SE, Turnberg LA (1981) Demonstration of a pH gradient across mucus adherent to rabbit gastric mucosa: evidence for a 'mucus-bicarbonate' barrier. Gut 22:94–96

Wilschanski M, Novak I (2013) The cystic fibrosis of exocrine pancreas. Cold Spring Harb Perspect Med 3:a009746. https://doi.org/10.1101/cshperspect.a009746

Winpenny JP, Verdon B, McAlroy HL, Colledge WH, Ratcliff R, Evans MJ, Gray MA, Argent BE (1995) Calcium-activated chloride conductance is not increased in pancreatic duct cells of CF mice. Pflügers Arch 430:26–33

Winpenny JP, Harris A, Hollingsworth MA, Argent BE, Gray MA (1998) Calcium-activated chloride conductance in a pancreatic adenocarcinoma cell line of ductal origin (HPAF) and in freshly isolated human pancreatic duct cells. Pflugers Arch 435:796–803

Wizemann V, Schulz I (1973) Influence of amphotericin, amiloride, ionophores, and 2,4- dinitrophenol on the secretion of the isolated cats pancreas. Pflügers Arch 339:317–338

Woo K, Dutta AK, Patel V, Kresge C, Feranchak AP (2008) Fluid flow induces mechanosensitive ATP release, calcium signalling and Cl- transport in biliary epithelial cells through a PKCzeta-dependent pathway. J Physiol 586:2779–2798

Woo K, Sathe M, Kresge C, Esser V, Ueno Y, Venter J, Glaser SS, Alpini G, Feranchak AP (2010) Adenosine triphosphate release and purinergic (P2) receptor-mediated secretion in small and large mouse cholangiocytes. Hepatology 52:1819–1828

Wright EM (1972) Mechanisms of ion transport across the choroid plexus. J Physiol 226:545–571

Wright EM (1978) Transport processes in the formation of the cerebrospinal fluid. Rev Physiol Biochem Pharmacol 83:3–34

Wright RD, Blair-West JR, Nelson JF (1986) Effects of ouabain, amiloride, monenesin, and other agents on ovine parotid secretion. Am J Physiol 250:503–510

Yamaguchi S, Ishikawa T (2005) Electrophysiological characterization of native Na$^+$-HCO$_3^-$ cotransporter current in bovine parotid acinar cells. J Physiol 568:181–197

Yamaguchi S, Ishikawa T (2012) IRBIT reduces the apparent affinity for intracellular Mg^{2+} in inhibition of the electrogenic Na$^+$-HCO$_3^-$ cotransporter NBCe1-B. Biochem Biophys Res Commun 424:433–438

Yamaguchi J, Liss AS, Sontheimer A, Mino-Kenudson M, Castillo CF, Warshaw AL, Thayer SP (2015) Pancreatic duct glands (PDGs) are a progenitor compartment responsible for pancreatic ductal epithelial repair. Stem Cell Res 15:190–202

Yang CL, Liu X, Paliege A, Zhu X, Bachmann S, Dawson DC, Ellison DH (2007) WNK1 and WNK4 modulate CFTR activity. Biochem Biophys Res Commun 353:535–540

Yang YD, Cho H, Koo JY, Tak MH, Cho Y, Shim WS, Park SP, Lee J, Lee B, Kim BM, Raouf R, Shin YK, Oh U (2008) TMEM16A confers receptor-activated calcium-dependent chloride conductance. Nature 455:1210–1215

Yang D, Shcheynikov N, Zeng W, Ohana E, So I, Ando H, Mizutani A, Mikoshiba K, Muallem S (2009) IRBIT coordinates epithelial fluid and HCO$_3^-$ secretion by stimulating the transporters pNBC1 and CFTR in the murine pancreatic duct. J Clin Invest 119:193–202

Yang D, Li Q, So I, Huang CL, Ando H, Mizutani A, Seki G, Mikoshiba K, Thomas PJ, Muallem S (2011) IRBIT governs epithelial secretion in mice by antagonizing the WNK/SPAK kinase pathway. J Clin Invest 121:956–965

Yannoukakos D, Stuart-Tilley A, Fernandez HA, Fey P, Duyk G, Alper SL (1994) Molecular cloning, expression, and chromosomal localization of two isoforms of the AE3 anion exchanger from human heart. Circ Res 75:603–614

Yegutkin GG, Samburski SS, Jalkalen S, Novak I (2006) ATP-consuming and ATP-generating enzymes secreted by pancreas. J Biol Chem 281:29441–29447

You CH, Rominger JM, Chey WY (1983) Potentiation effect of cholecystokinin-octapeptide on pancreatic bicarbonate secretion stimulated by a physiologic dose of secretin in humans. Gastroenterology 85:40–45

Young JA, Martin CJ (1971) The effect of a sympatho- and a parasympathomimetic drug on the electrolyte concentrations of primary and final saliva of the rat submaxillary gland. Pflügers Arch 327:285–302

Young JA, Schögel E (1966) Micropuncture investigation of sodium and potassium excretion in rat submaxillary saliva. Pflügers Arch 291:85–98

Young JA, van Lennep EW (1979) Transport in salivary and salt glands. In: Giebisch G, Tosteson DC, Ussing HH (eds) Membrane transport in biology, vol IVB. Springer, Berlin, pp 563–692

Young JA, Case RM, Conigrave AD, Novak I (1980) Transport of bicarbonate and other anions in salivary secretion. Ann NY Acad Sci 341:172–190

Yu Y, Chen TY (2015) Purified human brain calmodulin does not alter the bicarbonate permeability of the ANO1/TMEM16A channel. J Gen Physiol 145:79–81

Zeng W, Lee MG, Yan M, Diaz J, Benjamin I, Marino CR, Kopito R, Freedman S, Cotton C, Muallem S, Thomas P (1997) Immuno and functional characterization of CFTR in submandibular and pancreatic acinar and duct cells. Am J Physiol 273:442–455

Zeuthen T (1991) Secondary active transport of water across ventricular cell membrane of choroid plexus epithelium of Necturus maculosus. J Physiol 444:153–173

Zeuthen T, Wright EM (1981) Epithelial potassium transport: tracer and electrophysiological studies in choroid plexus. J Membr Biol 60:105–128

Zhao H, Star RA, Muallem S (1994) Membrane localization of H$^+$ and HCO$_3^-$ transporters in the rat pancreatic ducts. J Gen Physiol 104:57–85

Zlokovic BV, Mackic JB, Wang L, McComb JG, McDonough A (1993) Differential expression of Na,K-ATPase alpha and beta subunit isoforms at the blood-brain barrier and the choroid plexus. J Biol Chem 268:8019–8025

Chapter 13
MicroRNA Regulation of Channels and Transporters

Kelly M. Weixel and Michael B. Butterworth

Abstract The mechanisms regulating ion and water homeostasis are well established, with most of the protein components and signaling cascades identified and studied. To ensure a near-constant internal environment the coordinated actions of key hormones to regulate channels and transporters are required. The signaling cascades maintain electrolyte and water balance in the body. Emerging evidence indicates that, like the protein constituents of signaling pathways, noncoding RNAs (ncRNAs) may also be required to achieve an integrated hormonal response. These ncRNAs contribute to both the heterogeneity of signaling outcomes and facilitate the fine-tuning of hormonal responses in target tissues. The best studied of the ncRNAs are the small microRNA species. MicroRNA expression is regulated by hormonal stimulation, and they modulate protein expression of many transporters and proteins critical for hormone signaling cascades, including all components of the renin-angiotensin-aldosterone-signaling system (RAAS). In this chapter, the discovery and synthesis of microRNAs will be introduced followed by a discussion of ion channel regulation by microRNAs. MicroRNA dysregulation in disease is highlighted. The reciprocal roles of hormone-microRNA regulation in maintaining electrolyte homeostasis and the possible feedback regulation by microRNAs in signaling cascades will be addressed.

Keywords MicroRNAs · RAAS · Vasopressin · Renin · ncRNA

K. M. Weixel
Department of Biology, Washington and Jefferson College, Washington, PA, USA
e-mail: kwexel@washjeff.edu

M. B. Butterworth (✉)
Department of Cell Biology, School of Medicine, University of Pittsburgh, Pittsburgh, PA, USA
e-mail: michael7@pitt.edu

© The American Physiological Society 2020   543
K. L. Hamilton, D. C. Devor (eds.), *Basic Epithelial Ion Transport Principles and Function*, Physiology in Health and Disease,
https://doi.org/10.1007/978-3-030-52780-8_13

13.1 Introduction

13.1.1 Background and History of miRNAs

Organisms hold increasingly larger and more complex genomes the higher up the evolutionary tree they are located. The compact genomes in prokaryotes are composed predominantly of protein-coding regions (>90%) with a small fraction of untranscribed DNA (Table 13.1 and (Lynch and Conery 2003)). This pattern remains true in simple eukaryotes like yeast, however, a proportionately larger fraction of noncoding DNA contains both untranscribed and transcribed regions that are not translated into proteins.

When the human genome was sequenced and analyzed it was found that most of the genome consisted of untranscribed DNA. Over 95% of the human genome is composed of DNA that does not code for proteins, with transcribed and non-translated regions far exceeding the protein-coding regions (Gardiner 1995; Zuckerkandl 1997). Initially this "junk" DNA was ignored, but subsequent studies identified transcribed but noncoding RNA (ncRNA) that served numerous regulatory roles even though no protein was produced. Several species of small ncRNA have been identified in eukaryotes (Eddy 2001). These ncRNAs are classified by their length (20–30 nucleotides) as well as their association with the Argonaute family proteins (Ha and Kim 2014). The shared characteristic of these small ncRNAs is that they selectively silence genetic transcripts. Small ncRNAs have been classified into three groups: microRNAs (miRNAs), small interfering RNA (siRNA), and PIWI-interacting RNA (piRNA) (Ha and Kim 2014; Ishizu et al. 2012).

Of these, miRNAs constitute the major class and are considered one of the principal posttranscriptional regulators of gene expression (Ha and Kim 2014). This characteristic has made them an attractive target for drug discovery and potential biomarkers for disease. MiRNAs are typically 18–25 nucleotides in length

Table 13.1 List of increasingly complex organisms with their approximate genome size and estimated percentage of the genome that codes for proteins

Organism	Approximate genome size (Mb)	Estimated % of genome that is protein-coding	Approximate gene number
Escherichia coli	5	90	4000
Saccharomyces cerevisiae	10	70	6000
Caenorhabditis elegans	100	25	19,000
Arabidopsis thaliana	125	28	25,000
Drosophila melanogaster	200	12	14,000
Gallus gallus	1000	3	20,000
Mus Musculus	2500	5	30,000
Homo sapiens	3000	1.5	25,000

and have been shown to bind predominantly to the 3'untranslated regions (UTRs) of messenger RNAs (mRNAs) to induce translational suppression or mRNA degradation (Brennecke et al. 2003; Reinhart et al. 2000; Zeng and Cullen 2003) However, many exceptions to this canonical pathway are known (Chua et al. 2009; Flynt et al. 2010; Ha and Kim 2014; Kim et al. 2009; Miyoshi et al. 2010; Yang and Lai 2010). The function of miRNAs was first noted with their discovery in *Caenorhabditis elegans* in 1993 (Lee et al. 1993). In these studies, miRNAs were shown to be critical in the posttranslational control of developmental gene expression (Wightman et al. 1993). Depending on the organism, it is estimated that miRNAs comprise anywhere from 1–5% of the genome, and are capable of targeting most protein-coding transcripts (Bentwich et al. 2005; Lewis et al. 2005). Therefore, it is likely that miRNAs are involved in most cellular processes including development, growth, and signaling across animal species. In humans, it is estimated that >60% of the protein-coding genes contain a conserved miRNA binding site (Friedman et al. 2009; Lujambio and Lowe 2012). This, combined with the number of non-conserved binding sites, makes it likely that most protein-coding genes exhibit some form of miRNA regulation. It is not surprising, then, that the biogenesis of miRNAs is tightly regulated. Deviations from the spatial and temporal control relieve regulated protein suppression and result in aberrant signaling in chronic diseases like cancer (Lujambio and Lowe 2012). While understanding miRNA dysregulation is important for disease states, constitutive miRNA regulation taking place in every cell facilitates many of the homeostatic feedback loops that maintain essential physiological processes and are the major focus of this chapter.

13.1.2 MiRNA Biogenesis

Like protein-coding genes, miRNAs are encoded in the genome and transcribed as longer precursor molecules in the nucleus ending with a mature miRNA in the cytoplasm. RNA polymerase II (and in some cases RNA polymerase III), transcribes miRNA genes as the immature long primary miRNAs (pri-miRNAs) (Lee et al. 2002). While there is a diverse strategy to transcribe miRNAs, a common feature of most miRNAs is one or more stem-loop structures, each with a 7-methylguanosine cap and poly A tail (Chhabra et al. 2010; Janga and Vallabhaneni 2011). The sequences that produce this stem-loop structure are encoded throughout the genome, some within protein-coding genes or as stand-alone sequences. MiRNAs can be transcribed by a unique promoter or by one contained within genes to share the promoters of the encoded gene (Breving and Esquela-Kerscher 2010; Hinske et al. 2010; Lee et al. 2008; Liu et al. 2007; Marson et al. 2008). This allows for independent regulation of miRNA expression as well as modulation linked to the gene in which the miRNA is embedded. Pri-miRNAs are initially trimmed by a complex in the nucleus composed of the RNase III, Drosha and Pasha (double-stranded RNA-binding protein) to precursor miRNAs (pre-miRNAs), which maintain the stem-loop structure but are smaller at about 70 nucleotides (Ha and Kim

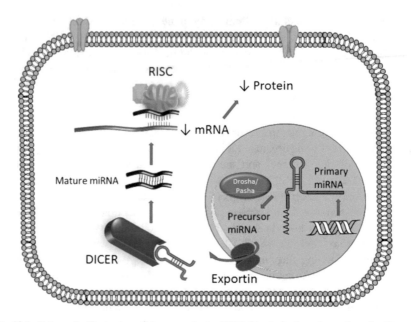

Fig. 13.1 Schematic illustration of the canonical miRNA biosynthetic pathway in cells. Steps in miRNA synthesis from transcription in the nucleus, export from the nucleus, and processing in the cytoplasm are described in the text. Assembly of the mature miRNA into the RISC results in mRNA destruction and a decrease in target protein expression

2014; Han et al. 2013; Kim 2005; Lee et al. 2003; Lee and Kim 2007). Pre-miRNAs are exported from the nucleus to the cytoplasm by the RanGTP-dependent nuclear transport receptor Exportin 5. Once in the cytoplasm, the pre-miRNA is subject to further processing by Dicer, another RNase, which trims the elongated tail and clips the loop to produce a mature double-stranded miRNA 22–23 nucleotides long (Kim 2004; Lund et al. 2004; Sontheimer 2005). The mature duplex can now be loaded onto an Argonaute family protein (AGO2), generating an effector complex known as the RNA-induced silencing complex (RISC) (Carmell et al. 2002). One strand of the miRNA is degraded, while the other strand remains bound to AGO2, which functions as an endonuclease. With the miRNA as a guide, RISC binds to target sequences and induces mRNA degradation or translational repression (Meister et al. 2004; Pillai et al. 2004). The biosynthetic steps in miRNA production are schematically illustrated in Fig. 13.1.

This is the most well-established pathway of miRNA synthesis, but there are exceptions (Cheloufi et al. 2010; Okamura et al. 2007; Ruby et al. 2007). Indeed, at nearly every step described above, one can observe alternative pathways. For example, non-canonical biogenesis has been identified for a specific type of miRNAs that are encoded in full introns, known as mirtrons (Okamura et al. 2007; Ruby et al. 2007). These bypass the micro-processing and instead can be directly excised by spliceosome action to generate pre-miRNA. Dicer and Drosha independent processing has also been observed. MiRNAs can be produced from a variety of sources including siRNAs, small nucleolar RNAs and short hairpin RNAs (shRNAs) and do not require further processing via Drosha (Breving and Esquela-Kerscher 2010; Suzuki and Miyazono 2011; Wahid et al. 2010).

13.2 General miRNA Function

The rate-limiting step of mRNA translation is initiation, and therefore, provides a powerful target for translational control. MiRNAs reduce protein expression by either repressing translation or cleaving mRNA. Most studies indicate that the RISC is directed to bind to the mRNA strand by the miRNAs which bind to the 3′ untranslated region (UTR) of their target via base-pairing to induce degradation through mRNA deadenylation and decapping, or repress translation by impeding access of the translational machinery to the target mRNA (Hutvagner and Zamore 2002; Tomari and Zamore 2005a, b). Most miRNAs investigated bind to multiple, partially complementary sites in the 3' UTR. MiRNAs have been shown to bind to other regions of mRNA, including the 5'UTR, coding sequence, as well as regions within the promoter. Binding to the 5' UTR and coding regions have been associated with gene silencing, but binding to the promoter regions appear to induce transcription (Felekkis et al. 2010; Valinezhad Orang et al. 2014). While the major role of miRNAs is to downregulate protein expression, understanding the functional significance of the different modes of interaction will be critical in establishing their interplay with other posttranscriptional modifiers.

The extent of the base-pairing of the miRNA to the mRNA target plays a role in the mechanism of inhibition by the miRNA. If there is a perfect or near-perfect base-pair match, then AGO-mediated endonucleolytic cleavage will be initiated with the cleavage of a single phosphodiester bond in the target mRNA (Elbashir et al. 2001). Partial sequence complementarity induces a mechanism of miRNA translation inhibition, often without strongly affecting mRNA levels (Wightman et al. 1993; Zeng et al. 2003). While all human AGO proteins are capable of binding both miRNAs and siRNAs, it is only the AGO2-containing complexes that will induce mRNA cleavage events (Meister et al. 2004).

13.3 Regulation of miRNAs

Like protein-coding genes, genes encoding miRNAs are controlled by transcription factors and are subject to the positive and negative cues driving tissue-specific and development-specific expression. For example, p53 has been shown to enhance the expression of families of miRNAs, including the miR-34 and miR-107 families which are involved in apoptosis as well as cell cycle arrest (He et al. 2007; Okada et al. 2014; Yamakuchi et al. 2010). Furthermore, miRNA expression is regulated by chemical modifications such as methylation of the promotor. Most promoter regions identified for miRNAs have CpG islands and gene promoter methylation status has been observed in numerous miRNAs, resulting in changes in their gene transcription (Behm-Ansmant et al. 2006; Diederichs and Haber 2007). Hypermethylation has been observed in miR-132, miR-34b/c, and miR33b, which have been associated with disease progression and poor prognosis in cancers (Behm-Ansmant et al. 2006; Han et al. 2007). Another important mechanism in determining miRNA expression patterns is posttranscriptional regulation. The expression of miRNAs can be downregulated with changes in the enzymes important for their biogenesis including Dicer and Drosha. Both Dicer and Drosha work in complexes with RNA-binding partners (RBP), and the levels and activities of RBPs can serve to regulate the levels of miRNAs (Diederichs and Haber 2007; Kawahara 2014). Similarly, changes in the AGO family of proteins have been shown to cause significant changes in miRNA expression. The stability/longevity of miRNA may depend upon the developmental stage and cell type studied. For example, some studies demonstrate that members of the ElaV family of RNA-binding proteins may bind to miRNAs and affect their half-life (Meisner and Filipowicz 2011). Overexpression of AGO proteins has also been shown to increase miRNA stability by decreasing their degradation. This is an area of recent scrutiny and will be critical in understanding the temporal effects of miRNAs in the pathways in which they exert control.

13.4   Role of miRNAs in Channel Physiology

MiRNAs play a profound role in development in eukaryotic systems due to their ability to influence cell growth and differentiation. Alterations in the patterns and distribution of miRNA in cells are also implicated in disease. Deficiencies or excesses in miRNAs have been associated with a wide range of pathologies including cardiovascular and autoimmune diseases, and reproductive cancers, reviewed here and elsewhere (Li and Rana 2014; Soifer et al. 2007) There is increasing evidence that miRNAs fine tune the expression and activity of ion channels/transporters in numerous tissues. In neuronal systems miRNAs have been shown to target ion channel genes to influence neuroadaptation to alcohol (Bekdash and Harrison 2015; Eddy 2001), modulate circadian rhythms (Chen et al. 2013) and alter sensory fiber activity in pain systems (Li et al. 2019; Su et al. 2017). In other excitable tissues like cardiac muscle, miRNAs have been shown to target genes for ion channels involved in electrical conduction and repolarization, affecting the arrhythmogenic potential of the heart (Wang 2010). Yang et al. (2007) showed that miR-1 silenced the expression of the KCNJ2, which encodes the main K^+ channel subunit Kir2.1 and is responsible for the maintenance of the cardiac resting potential, as well as the GJA1 gene which encodes connexin 43, the main gap junction channel responsible for intercellular conduction. MiR-1 expression is enhanced in individuals with coronary artery disease, and its overexpression exacerbates arrhythmias in normal as well as infarcted rat models (Yang et al. 2007). The effects of aldosterone and its mineralocorticoid receptor (MR) on cardiomyocyte excitability are mediated at least in part by the regulation of multiple ion channels (see for instance (Benitah and Vassort 1999; Mesquita et al. 2018; Ouvrard-Pascaud et al. 2005; Sabourin et al. 2016)). The MR blocker spironolactone reduces ventricular arrhythmias by decreasing the expression of the hyperpolarization-activated cyclic nucleotide-gated channel (HCN) (Song et al. 2011). Spironolactone was found to increase miR-1 expression in rat left ventricle in a model of myocardial infarction, contributing to repression of HCN protein expression, which may be related to decreased ventricular arrhythmia (Yu et al. 2015).

In absorptive and secretory tissues, miRNAs have been shown to play a role in disease progression. Recent studies have explored the role of miRNAs in steady-state homeostatic functions. In the kidney, miRNAs including miR-192, miR-194, miR-204, miR-215, and miR-216 are abundant and play a vital part in regulating renal development and maintenance of transport mechanisms. The profile of miRNA expression differs between the regions of the kidney. For example, Tian et al. (2008) showed that miR-192 is highly expressed in the cortex compared to the medulla, where it works to modulate sodium transport. Mladinov et al. (2013) demonstrated that the target of miR-192 is the beta subunit of the Na/K ATPase. Downregulation of miR-192 in mice led to upregulation of Atp1b1 protein in the kidney cortex within 48 h. Furthermore, miR-192 expression could be suppressed by a low Na^+ diet, resulting in an increase in Atp1b1 expression, establishing a role for miR-192 in maintaining the specific transport processes associated with the proximal convoluted tubule and medullary thick ascending limb (TAL) (Mladinov et al. 2013). Further analysis of the mechanism of action of miR-192 demonstrated that it targets the 5′, not the 3′UTR (Mladinov et al. 2013).

Acute kidney injury (AKI) is characterized by a sudden decrease in kidney function, reduced urine output, and increased serum creatinine levels and can lead to chronic kidney disease (CKD) (Lameire et al. 2013). Excessive aldosterone plays a role in the development of acute kidney injury (AKI) and MR antagonists have been proposed as therapeutic tools to limit kidney damage associated with this condition (Barrera-Chimal et al. 2016; Ramirez et al. 2009). MiRNAs have been implicated in AKI (Xue et al. 2016). Kidney injury molecule-1 (Kim-1) is a biomarker of AKI that associates with a tubular injury during ischemia and proteinuria. Xiao et al. (2016) found that MR blocker spironolactone partially attenuates increased Kim-1 expression in an ischemia-reperfusion (I/R) mouse model and in NRK-52E renal cells treated with antimycin-A to induce fibrosis or with a hypoxia/reoxygenation (H/R) protocol. Kim-1 is a target of miR-203 and was downregulated by I/R in vivo or by antimycin-A or H/R treatment in cultured cells. Aldosterone-induced apoptosis of NRK-52E cells was associated with rno-miR-203 promoter hypermethylation, which represses its expression, triggering increased levels of Kim-1 (Xiao et al. 2016).

MicroRNAs can also exert their effect on transport function by modulating factors that influence trafficking pathways regulating channel membrane insertion and retrieval. Lin et al. showed that miR-194 expression in the cortical collecting duct was enhanced by high dietary K^+ intake (HK) which increased ROMK channel currents (Lin et al. 2014). This miRNA does not directly interact with ROMK transcripts, instead, miR-194 targets the 3'UTR of intersectin1 (ITSN1), a protein involved in With-No-Lysine-Kinase (WNK)-induced endocytosis. The authors concluded that the stimulatory effect of HK on ROMK currents is due to the inhibition of ITSN1/WNK-dependent endocytosis of the channel as opposed to a direct modulation of ROMK transcripts (Lin et al. 2014). An analogous mechanism was uncovered in response to elevated aldosterone stimulation in cortical collecting duct (CCD) epithelial cells of the distal nephron. In this case, expression of miRNAs-23,-24 & -27 was increased in response to low Na^+ diets in mice. MiRNA-27 was shown to target intersectin 2 (ITSN2), but the result on the regulation of the epithelial sodium channel (ENaC) was the same, namely a decrease in ENaC endocytosis and increase in the reabsorption of Na^+ (Liu et al. 2017).

The regulation of transporters and ion channels expressed in multiple epithelial tissues is cell type, time, and developmentally specific. This can make it difficult to tease out the role of miRNAs in regulating their activity. An example of this is CFTR, whose transcription can be directly regulated by multiple miRNAs (Gillen et al. 2011) and whose tissue expression can dramatically change during development and after birth. Pulmonary CFTR expression is high prior to birth but falls dramatically after birth (Marcorelles et al. 2007; Viart et al. 2015). Adult cell lung cell lines can be negatively regulated by miR-101, while fetal lines fail to be influenced by it. It is likely that multiple miRNAs work together to suppress transcription. Megiorini et al. (2011) showed that miR-101 in combination with miR-494 act synergistically on CFTR-reporter inhibition in human embryonic kidney cells, a phenomenon confirmed for other miRNA combinations including miR-509-3p and miR-494 (Megiorni et al. 2011; Oglesby et al. 2013; Ramachandran et al. 2013). The role of miRNAs in CFTR expression is complicated by the observation of variability of genotype-phenotype correlation in patients with the same mutation in the CFTR gene. Indeed, polymorphisms have been identified in the miR-99b/let-7e/miR-125a cluster, perhaps modulating the expression of miRNAs and impacting patient phenotypes (Endale Ahanda et al. 2015). Furthermore, Amato et al. (2013) describe polymorphisms within the 3'-UTR region of CFTR responsible for the binding of miR-509-3p that results in decreased CFTR expression.

While many studies have focused on the direct interaction of miRNAs with CFTR, its surface expression and activity have shown to be enhanced by a network of proteins influenced by the miRNA-138 (Ramachandran et al. 2012). Here, Ramachandran et al. (2013) showed that miR-138 interacts with SIN3A, a transcriptional regulatory protein, relieving a transcriptional repressor resulting in increased CFTR mRNA and CFTR-mediated Cl^- permeability in epithelial cells. They also observed that in cells overexpressing CFTR treated with miR-138 cell surface abundance of CFTR increased independent of mRNA abundance, suggesting that miR-138 and SIN3A can regulate processes that influence posttranslational maturation of the channel (Ramachandran et al. 2013). They could also rescue CFTR-F508del-mediated Cl^- transport in CF primary airway by overexpressing miR-138, indicating miRNAs have the impressive capacity of coordinating a repertoire of targets to facilitate changes in biosynthesis and trafficking responses. Given the miR-138 mediated rescue of CFTR activity, miRNA modulation may provide therapeutic interventions in CF, a topic reviewed elsewhere (Bardin et al. 2018).

Because changes in membrane permeability are dependent on an orchestrated response of many different channels and transporters, it is not surprising that miRNAs that influence one channel may coordinate the effects of related transporters. Gillen described at least 12 miRNAs that targeted CFTR expression in the Caco-2 cell line, three of which, hsa-miR-384, hsa-miR-494, and hsa-miR-1246, that were capable of repressing a reporter carrying the 3' UTR of the gene that encodes the cotransporter NKCCl (Na^+-K^+-2Cl- cotransport protein) (Gillen et al. 2011).

13.5 MiRNAs as Components of Feedback Regulation

Much of the early investigations into miRNA function in humans implicated their roles in disease, and from this was evident that miRNAs are closely involved in signaling pathways. However, miRNAs are engaged in homeostatic feedback loops to serve as a regulatory rheostat under normal physiological states too. In the context of hormonal signaling pathways, a role for miRNAs in altering the functional activity of transporters in response to signaling cascades is beginning to emerge.

13.6 MiRNAs in RAAS Signaling

The renin-angiotensin-aldosterone-signaling system (RAAS) is a well-established hormonal cascade that leads to the release of aldosterone in response to decreased plasma sodium, increased circulating potassium, or decreased effective circulating volume (Ibrahim et al. 1997). RAAS constitutes a critical homeostatic feedback loop that coordinates the activity of renal and vascular tissues to regulate blood volume, maintaining blood pressure within narrow limits (Campese and Park 2006; Hsueh and Wyne 2011; Ibrahim et al. 1997). In the kidney, aldosterone modulates the expression of proteins responsible for increasing Na^+ and water reabsorption in the distal nephron, leading to volume expansion. In this cascade, aldosterone action is largely driven via the serum and glucocorticoid-induced kinase (SGK1) and glucocorticoid-induced leucine zipper (GILZ) proteins. SGK1 alters the regulation of several proteins that coordinate an increase in Na^+ transport. This includes the E3 ubiquitin ligase Nedd4l which is phosphorylated by SGK1, to reduce the ubiquitination status of the epithelial sodium channel (ENaC) and results in a prolonged half-life of ENaC at the apical membrane and augmented Na^+ transport (Bhalla et al. 2006; Debonneville et al. 2001; Muller et al. 2003; Rotin 2000; Rotin and Staub 2011; Soundararajan et al. 2005; Staub and Verrey 2005; Wiemuth et al. 2007; Zhou et al. 2007). The RAAS cascade (Fig. 13.2) is composed of three major signaling compounds: renin, angiotensin II, and aldosterone and is responsible for more sustained control over arterial pressure than beta-agonism. MiRNAs play a role

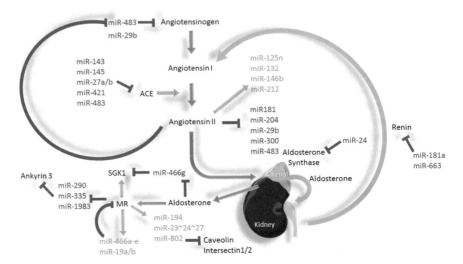

Fig. 13.2 Schematic representation of the RAAS signaling pathway with the miRs involved in regulating RAAS components listed (see text for details). Red arrows and text represent inhibition/ repression, green arrows and text represent stimulation. Several miRs feedback to alter upstream signaling pathways in the RAAS cascade

in regulating protein expression at each stage of the RAAS cascade, exerting influence over a homeostatic control system that largely governs salt and water homeostasis.

13.6.1 Renin

Aldosterone release is the result of a coordinated production and release of renin from the juxtaglomerular (JG) cells of the kidney. JG cells are specialized smooth muscle cells in the afferent and efferent arterioles, they secrete renin in response to a decrease in renal perfusion. JG cells can also be stimulated by macula densa cells, located in the distal convoluted tubule, which monitors Na^+ in the tubular fluid (Friis et al. 2013). There is evidence that renin production can be directly downregulated by increased expression of miR-181a and miR-633 (Marques et al. 2011). Increases in these miRNAs depress renin expression by binding to the 3'UTR of renin (encoded by REN). Marques et al. (2011) observed that these miRNAs are found in samples from normal and hypertensive subjects, but are significantly reduced in hypertensive kidneys. In a separate study, Morris (2015) showed that transfection of miR-181a mimics caused a reduction in renal renin, further indicating this miRNA as a modulator of renin expression (Morris 2015). As these miRNAs downregulate renin, the decreased expression of miR-181a and miR-633 and subsequent increase in renin mRNA has been offered as one avenue to explain elevated blood pressure in hypertensive patients (Marques et al. 2011). Increases in renin production could result in elevated aldosterone, culminating in hypernatremia, and blood volume expansion.

13.6.2 Angiotensin

The release of renin into the bloodstream leads to the conversion of angiotensinogen to ANG I, (see Fig. 13.1). Levels of circulating angiotensinogen have been shown to be influenced by miR-483 in a mechanism that involves feedback via ANG II. Specifically, ANGII downregulates miR-483, which then alters the expression of several RAAS components, including angiotensinogen (Kemp et al. 2014). MiR-483 targets the 3'-UTR of ANG-converting enzymes 1 and 2 as well as ANG II receptor 2 (Kemp et al. 2014). The results of miR-483 repression of these RAAS targets are downregulation of the RAAS pathway and consequently, a reduction in aldosterone expression (Kemp et al. 2014). While aldosterone levels were not measured in this study, the fact that miR-483 could be altered by ANG II and that this miRNA could exert control over RAAS proteins is a new layer of control within a homeostatic feedback loop controlling Na^+ levels. Additional control of angiotensin has been observed with miRs-421 and -143 which target angiotensin-converting enzyme 2 (Gu et al. 2014; Lambert et al. 2014), whereas ANG II decreased the expression of miR-29b in the renal cortex of spontaneously hypertensive rats (Pan et al. 2014). In studies with adult rat cardiac fibroblasts, ANG II upregulates miR-132, -125n-3p, and miR-146b but decreases the expression of miR-300-5 pm, -204-3p and -181b (Jiang et al. 2013).

As noted above with CFTR, the contribution of single-nucleotide polymorphisms in the target UTRs could also provide another means to alter endogenous feedback by miRNAs. Nossent et al. (2011) identified SNPs in canonical miRNA binding sites in UTRs of several RAAS genes. Presumably, SNPs would disrupt miRNA binding and result in a de-repression of the targets and upregulation of those proteins. SNPs associated with increased blood pressure were linked to mutations in ANG II, ANG II receptors 1 and 2, renin as well as the MR (Nossent et al. 2011). It is possible that SNPs that affect miRNA can provide links between genetic variation and pathophysiology and disease progression, and account for the heterogeneity of etiologies associated with hypertension onset in humans.

13.6.3 Aldosterone

Aldosterone production in the adrenal cortex is governed by aldosterone synthase (AS) and mainly stimulated by ANG II or increased plasma K^+. AS is a steroidogenic enzyme encoded by the gene *CYP11B2*, which is a target for miR-24. Interestingly, miR-24 also targets the related 11B-hydrolase (CYP11B1) which is responsible for cortisol production (Robertson et al. 2013). Overexpression of miR-24 reduced the expression of both enzymes, and depletion of miR-24 increased mRNA levels of both enzymes and led to increased aldosterone production in human adrenal cells (Robertson et al. 2013). In a separate study, miR-24 has been shown to be upregulated by aldosterone signaling via its MR (Lin et al. 2009), indicating a novel role for miR-24 in a feedback loop, regulating the expression of AS when aldosterone levels are elevated.

The release of aldosterone is the final element of the RAAS pathway that regulates Na$^+$ and blood volume. Aldosterone binds to its MR (encoded by *NR3C2*) in its target cells and is translocated to the nucleus where it binds to appropriate response elements to initiate transcription of aldosterone responsive genes. This is a well-studied paradigm of hormonal signaling, where the specificity of aldosterone signaling is protected by the action of 11β-hydroxysteroid dehydrogenase (11β-HSD2) which converts cortisol to cortisone to regulate access to the MR (Funder et al. 1988; Odermatt and Kratschmar 2012). However, it is evident that miRNAs exert regulation over MR, altering expression of the steroid receptor and 11β-HSD2. The miRs-124 and -135a bind to and downregulate the expression of an MR 3'-UTR luciferase reporter construct, while miR-20a was shown to be enriched in the cortical collecting duct where it reduced the expression of 11β-HSD2 in rat models (Rezaei et al. 2014). MiRNAs are involved in modulating the production of aldosterone at every step in the RAAS cascade. What is also evident is that the miRNAs act as effectors of aldosterone signaling. This is true for aldosterone and other hormone signaling pathways like vasopressin (each to be discussed in turn).

13.7 Aldosterone Regulated miRNAs

The major function of aldosterone in renal tissue is the regulation of tubular Na$^+$ and K$^+$ transport. In principal cells, aldosterone modulates the expression of many proteins including ENaC, leading to an increase in Na$^+$ reabsorption. Aldosterone stimulation of a mouse cortical collecting duct cell line (mCCDc11) both upregulates and downregulates a set of miRNAs that recapitulate regulation in vivo in CCD cells isolated from mice placed on low Na$^+$ diets. Edinger et al. (2014) showed that the downregulated miRNAs, including miR-335-3p, miR-290-5p, and miR-1983, targeted ankyrin 3 (Ank3), a membrane regulatory protein. CCD cells stimulated with aldosterone showed a corresponding increase in Ank3 expression, which was also observed when miRs-335-3p, -290-5p, and -1983 were inhibited in the absence of aldosterone stimulation. This increase in Ank3 expression correlated with an insertion of ENaC into the apical surface of mCCD cells, increased ENaC abundance, and enhanced Na$^+$ reabsorption (Klemens et al. 2017). Aldosterone also induces the expression of a miRNA cluster in the distal nephron, mouse mmu-miR23 ~ 24 ~ 27 which targets the 3'-UTR of intersectin-2 (ITSN2) to enhance ENaC-mediated Na$^+$ transport (see above). In another CCD cell line, (mpkCCDc14), aldosterone acutely alters the expression of miRNAs and their targets. The expression of miR-466g, which targets the 3'-UTR of SGK1, was significantly reduced following a 1-h stimulation with aldosterone (Luo et al. 2014). As mentioned previously, SGK1 upregulation decreases ENaC endocytosis, increasing the apical expression of the channel, and ENaC-mediated Na^+ transport. By inhibiting miR-466g, SGK1 activity is increased which would enhance the effects of aldosterone on ENaC. Extended treatment with aldosterone, above 3 days, appears to provide a different panel of miRNA regulation. In the same mpkCCDc14 cells, long-term aldosterone stimulation resulted in miR-34c-5p downregulation with a corresponding increase in Ca^{2+}/calmodulin-dependent kinase type II β-chain. The authors postulated that this upregulated kinase activity could be linked to aldosterone-induced fibrosis as an increase in other fibrotic markers were observed (Park et al. 2018).

There are a small number of studies that have examined the aldosterone regulation of K^+ via miRNA. Expression of miR-192 as a regulator of K^+ secretion was observed in mice under conditions that increase aldosterone secretion: increased K^+ load, salt depletion, and aldosterone infusions which downregulated miR-192 (Elvira-Matelot et al. 2010). The target of miR-192 was the long form of WNK1 (L-WNK1), a well-known modulator of Na^+ and K^+ transport processes (Subramanya et al. 2006; Wade et al. 2006; Yang et al. 2005). MiRNAs regulate multiple mechanisms to alter ion permeabilities, likely acting to integrate time-dependent and spatial signaling cues and assist in homeostatic control of ion transport regulation.

13.8 MiRNAs in Vasopressin Signaling

Like aldosterone, the antidiuretic hormone, arginine vasopressin (AVP) works to coordinate multiple systems in a homeostatic feedback loop. The mechanism of AVP action to increase water reabsorption is well-described; it primarily functions in the cortical collecting duct principal cells via binding to its Vasopressin type 2 receptor (V2R), inducing a cyclic AMP-dependent signaling cascade that results in the translocation of aquaporin-2 (AQP2) water channels into the apical membrane. This results in an increased water permeability of the cortical collecting duct, water exits the filtrate and enters the cell via AQP2 and exits via the basolateral AQP3 or AQP4, returning water into the circulation, restoring osmolarity or increasing blood volume, depending on the volume status (Knepper 1997; Nielsen et al. 1993, 1995, 2000). AVP has a second action on vascular smooth muscle where it binds to V1Rs and triggers an IP-3 mediated release of intracellular calcium resulting in contraction, thereby increasing peripheral resistance and hence increasing blood pressure (Holmes et al. 2004; Nemenoff 1998). Together with an increase in blood volume, this mechanism elevates blood pressure, crucial in maintaining tissue perfusion in hypovolemic states.

Recently, miRNAs were shown to play a role in AVP production and modulation of its signaling cascade. AVP is a peptide hormone synthesized by neurons in the hypothalamic paraventricular nucleus (PVN) and supraoptic nucleus (SON) and secreted into the systemic circulation from the posterior pituitary, reviewed elsewhere (Leng and Russell 2019). Osmoreceptors on these neurons are sensitive to blood osmolarity and modest elevations result in the secretion of AVP. AVP secretion is also triggered by hypovolemia, as baroceptors detect decreased arterial blood volume and signal the direct release of AVP via the vagus nerve (Thrasher 1994). While osmolarity and volume are the key drivers for AVP release, AVP-dependent water homeostasis can be modulated by other endocrine signals including ANG II and prostaglandins as well as other factors including hypoglycemia, pain, and nausea. Recent studies by Bijkerk et al. (2018) show that the hypothalamic AVP synthesis is attenuated by a process involving miR-132 repression of the methyl-CpG-binding protein-2 (MECP2). These studies demonstrated that silencing miR-132 in mice causes acute diuresis and increased plasma osmolarity. Total and apical membrane AQP2 expression was reduced in these mice compared to controls, due to the reduction in AVP mRNA levels. In experiments where antagomir-132 was delivered by central and intracerebroventricular injection, hypothalamic AVP mRNA levels were significantly decreased. They further demonstrated that the target for miR-132 is MECP2, which had previously been shown to inhibit AVP gene expression in PVN neurons (Murgatroyd and Spengler 2014). MiR-132 expression in hypothalamic cells is salt-dependent, such that hyperosmolarity increased miR-132 expression and decreases MECP2 expression, relieving the repression of the *Avp* gene. This is the first example of posttranscriptional regulation of AVP by miRs and offers an intriguing opportunity to explore miR-mediated control over congenital water balance disorders, particularly since miR-132 is highly conserved among mammals (Cheng et al. 2007).

Recent studies indicate that in addition to the well-described control over AQP2 targeting, AVP can induce specific miRNA expression in the inner medullary collecting duct that target AQP2 transcripts. Kim et al. (2015) identified four AVP-induced miRNAs, miR-32, -137, -216a, and -216b that target the 3'-UTR of rat AQP2 mRNA. Mir-32 and miR-137 target regions were also identified in mouse AQP2 (Kim et al. 2015). They showed that AVP-induced AQP2 expression and translocation to the apical membrane was reduced in mpkCCDc14 cells when transfected with mimics of the miR-32 or miR-137 (Kim et al. 2015). This suggests that AVP-responsive miRNAs provide an additional level of control of apical water channels in the distal nephron, it will be interesting to pursue the precise level feedback that this attenuation provides.

13.9   Conclusions

We provide examples of signaling cascades that regulate and are impacted by miRNAs. The involvement of miRNAs is undoubtedly recurrent in most of the homeostatic feedback pathways represented in higher organisms. The more complex the signaling cascade, the greater chance there is for it to become dysregulated over time. This could be due to, for example, mutations, infections, trauma, or environmental extremes which cause the homeostatic balance to be disturbed past the intrinsic limits of the system. As an organisms' survival depends on adapting to an ever-changing environment, more controls and safeguards to respond to the external changes will aid in its success. One of these safeguard roles appears to be filled by ncRNAs. MiRNAs are ideally suited to this task, as they often shadow mRNA and protein regulation, and guard against extreme swings in protein expression or signaling. An example of this buffering role can be seen in the alteration of miRNA expression by aldosterone or vasopressin that feedback into the same pathway to downregulate cellular responses to extended hormone release. Conversely, miRNAs can facilitate the action of these hormones by augmenting the response to receptor-initiated cascades. Therefore, the known action of aldosterone to increase SGK1 expression and activity is further enhanced when miRNAs know to downregulate SGK1 mRNA are rapidly repressed by aldosterone signaling. In addition to this, the expression of positive regulators of Na^+ transport is increased and negative regulators decreased by longer aldosterone stimulation.

Just as the hormonal signaling cascades for ion homeostasis were elucidated by understanding the protein mediators of ion transport, an underlying ncRNA signaling network has started to emerge that works in parallel with the protein components of these cascades. This "silent" network may be as important as changes to protein expression in maintaining electrolyte and water balance. It is likely that a combination of coding and noncoding RNAs are co-regulated to effectively manage the homeostatic networks essential for life.

Acknowledgments MBB is supported by a National Institutes of Health grant (NIDDK DK102843).

References

Amato F, Seia M, Giordano S, Elce A, Zarrilli F, Castaldo G, Tomaiuolo R (2013) Gene mutation in microRNA target sites of CFTR gene: a novel pathogenetic mechanism in cystic fibrosis? PLoS One 8:e60448

Bardin P, Sonneville F, Corvol H, Tabary O (2018) Emerging microRNA therapeutic approaches for cystic fibrosis. Front Pharmacol 9:1113

Barrera-Chimal J, Bobadilla NA, Jaisser F (2016) Mineralocorticoid receptor antagonism: a promising therapeutic approach to treat ischemic AKI. Nephron 134:10–13

Behm-Ansmant I, Rehwinkel J, Izaurralde E (2006) MicroRNAs silence gene expression by repressing protein expression and/or by promoting mRNA decay. Cold Spring Harb Symp Quant Biol 71:523–530

Bekdash RA, Harrison NL (2015) Downregulation of Gabra4 expression during alcohol withdrawal is mediated by specific microRNAs in cultured mouse cortical neurons. Brain Behav 5:e00355

Benitah JP, Vassort G (1999) Aldosterone upregulates Ca(2+) current in adult rat cardiomyocytes. Circ Res 85:1139–1145

Bentwich I, Avniel A, Karov Y, Aharonov R, Gilad S, Barad O, Barzilai A, Einat P, Einav U, Meiri E et al (2005) Identification of hundreds of conserved and nonconserved human microRNAs. Nat Genet 37:766–770

Bhalla V, Soundararajan R, Pao AC, Li H, Pearce D (2006) Disinhibitory pathways for control of sodium transport: regulation of ENaC by SGK1 and GILZ. Am J Physiol Renal Physiol 291:714–721

Bijkerk R, Trimpert C, van Solingen C, de Bruin RG, Florijn BW, Kooijman S, van den Berg R, van der Veer EP, Bredewold EOW, Rensen PCN et al (2018) MicroRNA-132 controls water homeostasis through regulating MECP2-mediated vasopressin synthesis. Am J Physiol Renal Physiol 315:F1129–F1138

Brennecke J, Hipfner DR, Stark A, Russell RB, Cohen SM (2003) Bantam encodes a developmentally regulated microRNA that controls cell proliferation and regulates the proapoptotic gene hid in Drosophila. Cell 113:25–36

Breving K, Esquela-Kerscher A (2010) The complexities of microRNA regulation: mirandering around the rules. Int J Biochem Cell Biol 42:1316–1329

Campese VM, Park J (2006) The kidney and hypertension: over 70 years of research. J Nephrol 19:691–698

Carmell MA, Xuan Z, Zhang MQ, Hannon GJ (2002) The Argonaute family: tentacles that reach into RNAi, developmental control, stem cell maintenance, and tumorigenesis. Genes Dev 16:2733–2742

Cheloufi S, Dos Santos CO, Chong MM, Hannon GJ (2010) A dicer-independent miRNA biogenesis pathway that requires Ago catalysis. Nature 465:584–589

Chen R, D'Alessandro M, Lee C (2013) miRNAs are required for generating a time delay critical for the circadian oscillator. Curr Biol 23:1959–1968

Cheng HY, Papp JW, Varlamova O, Dziema H, Russell B, Curfman JP, Nakazawa T, Shimizu K, Okamura H, Impey S et al (2007) microRNA modulation of circadian-clock period and entrainment. Neuron 54:813–829

Chhabra R, Dubey R, Saini N (2010) Cooperative and individualistic functions of the microRNAs in the miR-23a~27a~24-2 cluster and its implication in human diseases. Mol Cancer 9:232

Chua JH, Armugam A, Jeyaseelan K (2009) MicroRNAs: biogenesis, function and applications. Curr Opin Mol Ther 11:189–199

Debonneville C, Flores SY, Kamynina E, Plant PJ, Tauxe C, Thomas MA, Munster C, Chraibi A, Pratt JH, Horisberger JD et al (2001) Phosphorylation of Nedd4-2 by Sgk1 regulates epithelial Na(+) channel cell surface expression. EMBO J 20:7052–7059

Diederichs S, Haber DA (2007) Dual role for argonautes in microRNA processing and posttranscriptional regulation of microRNA expression. Cell 131:1097–1108

Eddy SR (2001) Non-coding RNA genes and the modern RNA world. Nat Rev Genet 2:919–929

Edinger RS, Coronnello C, Bodnar AJ, Labarca M, Bhalla V, LaFramboise WA, Benos PV, Ho J, Johnson JP, Butterworth MB (2014) Aldosterone regulates microRNAs in the cortical collecting duct to alter sodium transport. J Am Soc Nephrol 25:2445–2457

Elbashir SM, Martinez J, Patkaniowska A, Lendeckel W, Tuschl T (2001) Functional anatomy of siRNAs for mediating efficient RNAi in Drosophila melanogaster embryo lysate. EMBO J 20:6877–6888

Elvira-Matelot E, Zhou XO, Farman N, Beaurain G, Henrion-Caude A, Hadchouel J, Jeunemaitre X (2010) Regulation of WNK1 expression by miR-192 and aldosterone. J Am Soc Nephrol 21:1724–1731

Endale Ahanda ML, Bienvenu T, Sermet-Gaudelus I, Mazzolini L, Edelman A, Zoorob R, Davezac N (2015) The hsa-miR-125a/hsa-let-7e/hsa-miR-99b cluster is potentially implicated in cystic fibrosis pathogenesis. J Cyst Fibros 14:571–579

Felekkis K, Touvana E, Stefanou C, Deltas C (2010) microRNAs: a newly described class of encoded molecules that play a role in health and disease. Hippokratia 14:236–240

Flynt AS, Greimann JC, Chung WJ, Lima CD, Lai EC (2010) MicroRNA biogenesis via splicing and exosome-mediated trimming in Drosophila. Mol Cell 38:900–907

Friedman RC, Farh KK, Burge CB, Bartel DP (2009) Most mammalian mRNAs are conserved targets of microRNAs. Genome Res 19:92–105

Friis UG, Madsen K, Stubbe J, Hansen PB, Svenningsen P, Bie P, Skott O, Jensen BL (2013) Regulation of renin secretion by renal juxtaglomerular cells. Pflugers Arch 465:25–37

Funder JW, Pearce PT, Smith R, Smith AI (1988) Mineralocorticoid action: target tissue specificity is enzyme, not receptor, mediated. Science 242:583–585

Gardiner K (1995) Human genome organization. Curr Opin Genet Dev 5:315–322

Gillen AE, Gosalia N, Leir SH, Harris A (2011) MicroRNA regulation of expression of the cystic fibrosis transmembrane conductance regulator gene. Biochem J 438:25–32

Gu Q, Wang B, Zhang XF, Ma YP, Liu JD, Wang XZ (2014) Contribution of renin-angiotensin system to exercise-induced attenuation of aortic remodeling and improvement of endothelial function in spontaneously hypertensive rats. Cardiovasc Pathol 23:298–305

Ha M, Kim VN (2014) Regulation of microRNA biogenesis. Nat Rev Mol Cell Biol 15:509–524

Han L, Witmer PD, Casey E, Valle D, Sukumar S (2007) DNA methylation regulates microRNA expression. Cancer Biol Ther 6:1284–1288

Han Y, Staab-Weijnitz CA, Xiong G, Maser E (2013) Identification of microRNAs as a potential novel regulatory mechanism in HSD11B1 expression. J Steroid Biochem Mol Biol 133:129–139

He L, He X, Lim LP, de Stanchina E, Xuan Z, Liang Y, Xue W, Zender L, Magnus J, Ridzon D et al (2007) A microRNA component of the p53 tumour suppressor network. Nature 447:1130–1134

Hinske LC, Galante PA, Kuo WP, Ohno-Machado L (2010) A potential role for intragenic miRNAs on their hosts' interactome. BMC Genomics 11:533

Holmes CL, Landry DW, Granton JT (2004) Science review: vasopressin and the cardiovascular system part 2 - clinical physiology. Crit Care 8:15–23

Hsueh WA, Wyne K (2011) Renin-Angiotensin-aldosterone system in diabetes and hypertension. J Clin Hypertens (Greenwich) 13:224–237

Hutvagner G, Zamore PD (2002) A microRNA in a multiple-turnover RNAi enzyme complex. Science 297:2056–2060

Ibrahim HN, Rosenberg ME, Hostetter TH (1997) Role of the renin-angiotensin-aldosterone system in the progression of renal disease: a critical review. Semin Nephrol 17:431–440

Ishizu H, Siomi H, Siomi MC (2012) Biology of PIWI-interacting RNAs: new insights into biogenesis and function inside and outside of germlines. Genes Dev 26:2361–2373

Janga SC, Vallabhaneni S (2011) MicroRNAs as post-transcriptional machines and their interplay with cellular networks. Adv Exp Med Biol 722:59–74

Jiang X, Ning Q, Wang J (2013) Angiotensin II induced differentially expressed microRNAs in adult rat cardiac fibroblasts. J Physiol Sci 63:31–38

Kawahara Y (2014) Human diseases caused by germline and somatic abnormalities in microRNA and microRNA-related genes. Congenit Anom (Kyoto) 54:12–21

Kemp JR, Unal H, Desnoyer R, Yue H, Bhatnagar A, Karnik SS (2014) Angiotensin II-regulated microRNA 483-3p directly targets multiple components of the renin-angiotensin system. J Mol Cell Cardiol 75:25–39

Kim VN (2004) MicroRNA precursors in motion: exportin-5 mediates their nuclear export. Trends Cell Biol 14:156–159

Kim VN (2005) MicroRNA biogenesis: coordinated cropping and dicing. Nat Rev Mol Cell Biol 6:376–385

Kim VN, Han J, Siomi MC (2009) Biogenesis of small RNAs in animals. Nat Rev Mol Cell Biol 10:126–139

Kim JE, Jung HJ, Lee YJ, Kwon TH (2015) Vasopressin-regulated miRNAs and AQP2-targeting miRNAs in kidney collecting duct cells. Am J Physiol Renal Physiol 308:749–764

Klemens CA, Edinger RS, Kightlinger L, Liu X, Butterworth MB (2017) Ankyrin G expression regulates apical delivery of the epithelial sodium channel (ENaC). J Biol Chem 292:375–385

Knepper MA (1997) Molecular physiology of urinary concentrating mechanism: regulation of aquaporin water channels by vasopressin. Am J Phys 272:3–12

Lambert DW, Lambert LA, Clarke NE, Hooper NM, Porter KE, Turner AJ (2014) Angiotensin-converting enzyme 2 is subject to post-transcriptional regulation by miR-421. Clin Sci 127:243–249

Lameire NH, Bagga A, Cruz D, De Maeseneer J, Endre Z, Kellum JA, Liu KD, Mehta RL, Pannu N, Van Biesen W et al (2013) Acute kidney injury: an increasing global concern. Lancet 382:170–179

Lee Y, Kim VN (2007) In vitro and in vivo assays for the activity of Drosha complex. Methods Enzymol 427:89–106

Lee RC, Feinbaum RL, Ambros V (1993) The C. elegans heterochronic gene lin-4 encodes small RNAs with antisense complementarity to lin-14. Cell 75:843–854

Lee Y, Jeon K, Lee JT, Kim S, Kim VN (2002) MicroRNA maturation: stepwise processing and subcellular localization. EMBO J 21:4663–4670

Lee Y, Ahn C, Han J, Choi H, Kim J, Yim J, Lee J, Provost P, Radmark O, Kim S et al (2003) The nuclear RNase III Drosha initiates microRNA processing. Nature 425:415–419

Lee JY, Kim S, Hwang DW, Jeong JM, Chung JK, Lee MC, Lee DS (2008) Development of a dual-luciferase reporter system for in vivo visualization of microRNA biogenesis and posttranscriptional regulation. J Nucl Med 49:285–294

Leng G, Russell JA (2019) The osmoresponsiveness of oxytocin and vasopressin neurones: mechanisms, allostasis and evolution. J Neuroendocrinol 31:e12662

Lewis BP, Burge CB, Bartel DP (2005) Conserved seed pairing, often flanked by adenosines, indicates that thousands of human genes are microRNA targets. Cell 120:15–20

Li Z, Rana TM (2014) Therapeutic targeting of microRNAs: current status and future challenges. Nat Rev Drug Discov 13:622–638

Li L, Shao J, Wang J, Liu Y, Zhang Y, Zhang M, Zhang J, Ren X, Su S, Li Y et al (2019) MiR-30b-5p attenuates oxaliplatin-induced peripheral neuropathic pain through the voltage-gated sodium channel Nav1.6 in rats. Neuropharmacology 153:111–120

Lin Z, Murtaza I, Wang K, Jiao J, Gao J, Li PF (2009) miR-23a functions downstream of NFATc3 to regulate cardiac hypertrophy. Proc Natl Acad Sci U S A 106:12103–12108

Lin DH, Yue P, Zhang C, Wang WH (2014) MicroRNA-194 (miR-194) regulates ROMK channel activity by targeting intersectin 1. Am J Physiol Renal Physiol 306:53–60

Liu N, Williams AH, Kim Y, McAnally J, Bezprozvannaya S, Sutherland LB, Richardson JA, Bassel-Duby R, Olson EN (2007) An intragenic MEF2-dependent enhancer directs muscle-specific expression of microRNAs 1 and 133. Proc Natl Acad Sci U S A 104:20844–20849

Liu X, Edinger RS, Klemens CA, Phua YL, Bodnar AJ, LaFramboise WA, Ho J, Butterworth MB (2017) A microRNA cluster miR-23-24-27 is upregulated by aldosterone in the distal kidney nephron where it alters sodium transport. J Cell Physiol 232:1306–1317

Lujambio A, Lowe SW (2012) The microcosmos of cancer. Nature 482:347–355

Lund E, Guttinger S, Calado A, Dahlberg JE, Kutay U (2004) Nuclear export of microRNA precursors. Science 303:95–98

Luo Y, Liu Y, Liu M, Wei J, Zhang Y, Hou J, Huang W, Wang T, Li X, He Y et al (2014) Sfmbt2 10th intron-hosted miR-466(a/e)-3p are important epigenetic regulators of Nfat5 signaling, osmoregulation and urine concentration in mice. Biochim Biophys Acta 1839:97–106

Lynch M, Conery JS (2003) The origins of genome complexity. Science 302:1401–1404

Marcorelles P, Montier T, Gillet D, Lagarde N, Ferec C (2007) Evolution of CFTR protein distribution in lung tissue from normal and CF human fetuses. Pediatr Pulmonol 42:1032–1040

Marques FZ, Campain AE, Tomaszewski M, Zukowska-Szczechowska E, Yang YH, Charchar FJ, Morris BJ (2011) Gene expression profiling reveals renin mRNA overexpression in human hypertensive kidneys and a role for microRNAs. Hypertension 58:1093–1098

Marson A, Levine SS, Cole MF, Frampton GM, Brambrink T, Johnstone S, Guenther MG, Johnston WK, Wernig M, Newman J et al (2008) Connecting microRNA genes to the core transcriptional regulatory circuitry of embryonic stem cells. Cell 134:521–533

Megiorni F, Cialfi S, Dominici C, Quattrucci S, Pizzuti A (2011) Synergistic post-transcriptional regulation of the cystic fibrosis transmembrane conductance regulator (CFTR) by miR-101 and miR-494 specific binding. PLoS One 6:e26601

Meisner NC, Filipowicz W (2011) Properties of the regulatory RNA-binding protein HuR and its role in controlling miRNA repression. Adv Exp Med Biol 700:106–123

Meister G, Landthaler M, Patkaniowska A, Dorsett Y, Teng G, Tuschl T (2004) Human Argonaute2 mediates RNA cleavage targeted by miRNAs and siRNAs. Mol Cell 15:185–197

Mesquita TR, Auguste G, Falcon D, Ruiz-Hurtado G, Salazar-Enciso R, Sabourin J, Lefebvre F, Viengchareun S, Kobeissy H, Lechene P et al (2018) Specific activation of the alternative cardiac promoter of Cacna1c by the mineralocorticoid receptor. Circ Res 122:e49–e61

Miyoshi K, Miyoshi T, Siomi H (2010) Many ways to generate microRNA-like small RNAs: non-canonical pathways for microRNA production. Mol Gen Genomics 284:95–103

Mladinov D, Liu Y, Mattson DL, Liang M (2013) MicroRNAs contribute to the maintenance of cell-type-specific physiological characteristics: miR-192 targets Na+/K+-ATPase beta1. Nucleic Acids Res 41:1273–1283

Morris BJ (2015) Renin, genes, microRNAs, and renal mechanisms involved in hypertension. Hypertension 65:956–962

Muller OG, Parnova RG, Centeno G, Rossier BC, Firsov D, Horisberger JD (2003) Mineralocorticoid effects in the kidney: correlation between alphaENaC, GILZ, and Sgk-1 mRNA expression and urinary excretion of Na+ and K+. J Am Soc Nephrol 14:1107–1115

Murgatroyd C, Spengler D (2014) Polycomb binding precedes early-life stress responsive DNA methylation at the Avp enhancer. PLoS One 9:e90277

Nemenoff RA (1998) Vasopressin signaling pathways in vascular smooth muscle. Front Biosci 3: d194–d207

Nielsen S, DiGiovanni SR, Christensen EI, Knepper MA, Harris HW (1993) Cellular and subcellular immunolocalization of vasopressin-regulated water channel in rat kidney. Proc Natl Acad Sci U S A 90:11663–11667

Nielsen S, Chou CL, Marples D, Christensen EI, Kishore BK, Knepper MA (1995) Vasopressin increases water permeability of kidney collecting duct by inducing translocation of aquaporin-CD water channels to plasma membrane. Proc Natl Acad Sci U S A 92:1013–1017

Nielsen S, Kwon TH, Frokiaer J, Knepper MA (2000) Key roles of renal aquaporins in water balance and water-balance disorders. News Physiol Sci 15:136–143

Nossent AY, Hansen JL, Doggen C, Quax PH, Sheikh SP, Rosendaal FR (2011) SNPs in microRNA binding sites in 3'-UTRs of RAAS genes influence arterial blood pressure and risk of myocardial infarction. Am J Hypertens 24:999–1006

Odermatt A, Kratschmar DV (2012) Tissue-specific modulation of mineralocorticoid receptor function by 11beta-hydroxysteroid dehydrogenases: an overview. Mol Cell Endocrinol 350:168–186

Oglesby IK, Chotirmall SH, McElvaney NG, Greene CM (2013) Regulation of cystic fibrosis transmembrane conductance regulator by microRNA-145, −223, and −494 is altered in DeltaF508 cystic fibrosis airway epithelium. J Immunol 190:3354–3362

Okada N, Lin CP, Ribeiro MC, Biton A, Lai G, He X, Bu P, Vogel H, Jablons DM, Keller AC et al (2014) A positive feedback between p53 and miR-34 miRNAs mediates tumor suppression. Genes Dev 28:438–450

Okamura K, Hagen JW, Duan H, Tyler DM, Lai EC (2007) The mirtron pathway generates microRNA-class regulatory RNAs in Drosophila. Cell 130:89–100

Ouvrard-Pascaud A, Sainte-Marie Y, Benitah JP, Perrier R, Soukaseum C, Nguyen Dinh Cat A, Royer A, Le Quang K, Charpentier F, Demolombe S et al (2005) Conditional mineralocorticoid receptor expression in the heart leads to life-threatening arrhythmias. Circulation 111:3025–3033

Pan J, Zhang J, Zhang X, Zhou X, Lu S, Huang X, Shao J, Lou G, Yang D, Geng YJ (2014) Role of microRNA-29b in angiotensin II-induced epithelial-mesenchymal transition in renal tubular epithelial cells. Int J Mol Med 34:1381–1387

Park EJ, Jung HJ, Choi HJ, Cho JI, Park HJ, Kwon TH (2018) miR-34c-5p and CaMKII are involved in aldosterone-induced fibrosis in kidney collecting duct cells. Am J Physiol Renal Physiol 314:F329–F342

Pillai RS, Artus CG, Filipowicz W (2004) Tethering of human Ago proteins to mRNA mimics the miRNA-mediated repression of protein synthesis. RNA 10:1518–1525

Ramachandran S, Karp PH, Jiang P, Ostedgaard LS, Walz AE, Fisher JT, Keshavjee S, Lennox KA, Jacobi AM, Rose SD et al (2012) A microRNA network regulates expression and biosynthesis of wild-type and DeltaF508 mutant cystic fibrosis transmembrane conductance regulator. Proc Natl Acad Sci U S A 109:13362–13367

Ramachandran S, Karp PH, Osterhaus SR, Jiang P, Wohlford-Lenane C, Lennox KA, Jacobi AM, Praekh K, Rose SD, Behlke MA et al (2013) Post-transcriptional regulation of cystic fibrosis transmembrane conductance regulator expression and function by microRNAs. Am J Respir Cell Mol Biol 49:544–551

Ramirez V, Trujillo J, Valdes R, Uribe N, Cruz C, Gamba G, Bobadilla NA (2009) Adrenalectomy prevents renal ischemia-reperfusion injury. Am J Physiol Renal Physiol 297:932–942

Reinhart BJ, Slack FJ, Basson M, Pasquinelli AE, Bettinger JC, Rougvie AE, Horvitz HR, Ruvkun G (2000) The 21-nucleotide let-7 RNA regulates developmental timing in Caenorhabditis elegans. Nature 403:901–906

Rezaei M, Andrieu T, Neuenschwander S, Bruggmann R, Mordasini D, Frey FJ, Vogt B, Frey BM (2014) Regulation of 11beta-hydroxysteroid dehydrogenase type 2 by microRNA. Hypertension 64:860–866

Robertson S, MacKenzie SM, Alvarez-Madrazo S, Diver LA, Lin J, Stewart PM, Fraser R, Connell JM, Davies E (2013) MicroRNA-24 is a novel regulator of aldosterone and cortisol production in the human adrenal cortex. Hypertension 62:572–578

Rotin D (2000) Regulation of the epithelial sodium channel (ENaC) by accessory proteins. Curr Opin Nephrol Hypertens 9:529–534

Rotin D, Staub O (2011) Role of the ubiquitin system in regulating ion transport. Pflugers Arch Eur J Physiol 461:1–21

Ruby JG, Jan CH, Bartel DP (2007) Intronic microRNA precursors that bypass Drosha processing. Nature 448:83–86

Sabourin J, Bartoli F, Antigny F, Gomez AM, Benitah JP (2016) Transient receptor potential canonical (TRPC)/Orai1-dependent store-operated Ca2+ channels: new targets of aldosterone in cardiomyocytes. J Biol Chem 291:13394–13409

Soifer HS, Rossi JJ, Saetrom P (2007) MicroRNAs in disease and potential therapeutic applications. Mol Ther 15:2070–2079

Song T, Yang J, Yao Y, Li H, Chen Y, Zhang J, Huang C (2011) Spironolactone diminishes spontaneous ventricular premature beats by reducing HCN4 protein expression in rats with myocardial infarction. Mol Med Rep 4:569–573

Sontheimer EJ (2005) Assembly and function of RNA silencing complexes. Nat Rev Mol Cell Biol 6:127–138

Soundararajan R, Zhang TT, Wang J, Vandewalle A, Pearce D (2005) A novel role for glucocorticoid-induced leucine zipper protein in epithelial sodium channel-mediated sodium transport. J Biol Chem 280:39970–39981

Staub O, Verrey F (2005) Impact of Nedd4 proteins and serum and glucocorticoid-induced kinases on epithelial Na+ transport in the distal nephron. J Am Soc Nephrol 16:3167–3174

Su S, Shao J, Zhao Q, Ren X, Cai W, Li L, Bai Q, Chen X, Xu B, Wang J et al (2017) MiR-30b attenuates neuropathic pain by regulating voltage-gated sodium channel Nav1.3 in rats. Front Mol Neurosci 10:126

Subramanya AR, Yang CL, McCormick JA, Ellison DH (2006) WNK kinases regulate sodium chloride and potassium transport by the aldosterone-sensitive distal nephron. Kidney Int 70:630–634

Suzuki HI, Miyazono K (2011) Emerging complexity of microRNA generation cascades. J Biochem 149:15–25

Thrasher TN (1994) Baroreceptor regulation of vasopressin and renin secretion: low-pressure versus high-pressure receptors. Front Neuroendocrinol 15:157–196

Tian Z, Greene AS, Pietrusz JL, Matus IR, Liang M (2008) MicroRNA-target pairs in the rat kidney identified by microRNA microarray, proteomic, and bioinformatic analysis. Genome Res 18:404–411

Tomari Y, Zamore PD (2005a) MicroRNA biogenesis: drosha can't cut it without a partner. Curr Biol 15:R61–R64

Tomari Y, Zamore PD (2005b) Perspective: machines for RNAi. Genes Dev 19:517–529

Valinezhad Orang A, Safaralizadeh R, Kazemzadeh-Bavili M (2014) Mechanisms of miRNA-mediated gene regulation from common downregulation to mRNA-specific upregulation. Int J Genomics 2014:970607

Viart V, Bergougnoux A, Bonini J, Varilh J, Chiron R, Tabary O, Molinari N, Claustres M, Taulan-Cadars M (2015) Transcription factors and miRNAs that regulate fetal to adult CFTR expression change are new targets for cystic fibrosis. Eur Respir J 45:116–128

Wade JB, Fang L, Liu J, Li D, Yang CL, Subramanya AR, Maouyo D, Mason A, Ellison DH, Welling PA (2006) WNK1 kinase isoform switch regulates renal potassium excretion. Proc Natl Acad Sci U S A 103:8558–8563

Wahid F, Shehzad A, Khan T, Kim YY (2010) MicroRNAs: synthesis, mechanism, function, and recent clinical trials. Biochim Biophys Acta 1803:1231–1243

Wang Z (2010) The role of microRNA in cardiac excitability. J Cardiovasc Pharmacol 56:460–470

Wiemuth D, Ke Y, Rohlfs M, McDonald FJ (2007) Epithelial sodium channel (ENaC) is multi-ubiquitinated at the cell surface. Biochem J 405:147–155

Wightman B, Ha I, Ruvkun G (1993) Posttranscriptional regulation of the heterochronic gene lin-14 by lin-4 mediates temporal pattern formation in C. elegans. Cell 75:855–862

Xiao X, Tang R, Zhou X, Peng L, Yu P (2016) Aldosterone induces NRK-52E cell apoptosis in acute kidney injury via rno-miR-203 hypermethylation and Kim-1 upregulation. Exp Ther Med 12:915–924

Xue Y, Teng YQ, Zhou JD, Rui YJ (2016) Prognostic value of long noncoding RNA MALAT1 in various carcinomas: evidence from nine studies. Tumour Biol 37:1211–1215

Yamakuchi M, Lotterman CD, Bao C, Hruban RH, Karim B, Mendell JT, Huso D, Lowenstein CJ (2010) P53-induced microRNA-107 inhibits HIF-1 and tumor angiogenesis. Proc Natl Acad Sci U S A 107:6334–6339

Yang JS, Lai EC (2010) Dicer-independent, Ago2-mediated microRNA biogenesis in vertebrates. Cell Cycle 9:4455–4460

Yang CL, Zhu X, Wang Z, Subramanya AR, Ellison DH (2005) Mechanisms of WNK1 and WNK4 interaction in the regulation of thiazide-sensitive NaCl cotransport. J Clin Invest 115:1379–1387

Yang B, Lin H, Xiao J, Lu Y, Luo X, Li B, Zhang Y, Xu C, Bai Y, Wang H et al (2007) The muscle-specific microRNA miR-1 regulates cardiac arrhythmogenic potential by targeting GJA1 and KCNJ2. Nat Med 13:486–491

Yu HD, Xia S, Zha CQ, Deng SB, Du JL, She Q (2015) Spironolactone regulates HCN protein expression through micro-RNA-1 in rats with myocardial infarction. J Cardiovasc Pharmacol 65:587–592

Zeng Y, Cullen BR (2003) Sequence requirements for micro RNA processing and function in human cells. RNA 9:112–123

Zeng Y, Yi R, Cullen BR (2003) MicroRNAs and small interfering RNAs can inhibit mRNA expression by similar mechanisms. Proc Natl Acad Sci U S A 100:9779–9784

Zhou R, Patel SV, Snyder PM (2007) Nedd4-2 catalyzes ubiquitination and degradation of cell surface ENaC. J Biol Chem 282:20207–20212

Zuckerkandl E (1997) Junk DNA and sectorial gene repression. Gene 205:323–343

Printed in the United States
by Baker & Taylor Publisher Services